Animal Welfare

Series Editor

Clive Phillips
School of Veterinary Science
University of Queensland
Gatton, Queensland
Australia

More information about this series at http://www.springer.com/series/5675

Andy Butterworth
Editor

Marine Mammal Welfare

Human Induced Change in the Marine Environment and its Impacts on Marine Mammal Welfare

Springer

Editor
Andy Butterworth
Clinical Veterinary Science
University of Bristol
Langford
Bristol
United Kingdom

ISSN 1572-7408
Animal Welfare
ISBN 978-3-319-83632-4 ISBN 978-3-319-46994-2 (eBook)
DOI 10.1007/978-3-319-46994-2

Softcover reprint of the hardcover 1st edition 2017

Printed on acid-free paper

This Springer imprint is published by Springer Nature
The registered company is Springer International Publishing AG
The registered company address is: Gewerbestrasse 11, 6330 Cham, Switzerland

Animal Welfare Series Preface

Animal welfare is attracting increasing interest worldwide, especially in developed countries where the knowledge and resources are available to, at least potentially, provide better management systems for farm animals, as well as companion, zoo, laboratory and performance animals. The key requirements for adequate food, water, a suitable environment, companionship and health are important for animals kept for all of these purposes.

There has been increased attention given to animal welfare in the West in recent years. This derives largely from the fact that the relentless pursuit of financial reward and efficiency, to satisfy market demands, has led to the development of intensive animal management systems that challenge the conscience of many consumers in this part of the world, particularly in the farm and laboratory animal sectors. Livestock are the world's biggest land users, and the farmed animal population is increasing rapidly to meet the needs of an expanding human population. This results in a tendency to allocate fewer resources to each animal and to value individual animals less, for example in the case of farmed poultry where flocks of over twenty thousand birds are not uncommon. In these circumstances, the importance of each individual's welfare is diminished.

In developing countries, human survival is still a daily uncertainty, so that provision for animal welfare has to be balanced against human welfare. Animal welfare is usually a priority only if it supports the output of the animal, be it food, work, clothing, sport or companionship. However, in many situations the welfare of animals is synonymous with the welfare of the humans that look after them, because happy, healthy animals will be able to assist humans best in their struggle for survival. In principle the welfare needs of both humans and animals can be provided for, in both developing and developed countries, if resources are properly husbanded. In reality, the inequitable division of the world's riches creates physical and psychological poverty for humans and animals alike in many parts of the world.

Increased attention to welfare issues is just as evident for zoo, companion, laboratory, sport and wild animals. Of growing importance is the ethical management of breeding programmes, since genetic manipulation is now technically advanced, but there is less public tolerance of the breeding of extreme animals if it comes at the expense of animal welfare. The quest for producing novel genotypes has fascinated breeders for centuries. Dog and cat breeders have produced a variety of deformities that have adverse effects on their welfare, but nowadays the breeders are just as active in the laboratory, where the mouse is genetically manipulated with equally profound effects.

The intimate connection between animals and humans that was once so essential for good animal welfare is rare nowadays, having been superseded by technologically efficient production systems where animals on farms and in laboratories are tended by increasingly few humans in the drive to enhance labour efficiency. With today's busy lifestyles, companion animals too may suffer from reduced contact with humans, although their value in providing companionship, particularly for certain groups such as the elderly, is beginning to be recognised. Animal consumers also rarely have any contact with the animals that are kept for their benefit.

In this estranged, efficient world, people struggle to find the moral imperatives to determine the level of welfare that they should afford to animals within their charge. A few people, and in particular many companion animal owners, strive for what they believe to be the highest levels of welfare provision, while others, deliberately or through ignorance, keep animals in impoverished conditions in which their health and well-being can be extremely poor. Today's multiple moral codes for animal care and use are derived from a broad range of cultural influences, including media reports of animal abuse, guidelines on ethical consumption and campaigning and lobbying groups.

This series has been designed to contribute towards a culture of respect for animals and their welfare by producing learned treatises about the provision for the welfare of the animal species that are managed and cared for by humans. The early species-focused books were not detailed management blueprints; rather they described and considered the major welfare concerns, often with reference to the behaviour of the wild progenitors of the managed animals. Welfare was specifically focused on animals' needs, concentrating on nutrition, behaviour, reproduction and the physical and social environment. Economic effects of animal welfare provision were also considered where relevant, as were key areas where further research is required.

In this volume the book series again departs from a single species focus to address the welfare of marine mammals. Editor Andy Butterworth has drawn from his research in this field to gather a large group of authors that consider the topic from a variety of angles. Given that this is an emerging science, which hitherto had attracted little attention, it is evident from the contributed chapters that there are serious welfare issues that should be a focus of immediate attention. Prominent among these are the following concerns: impacts of climate change and associated habitat destruction on the welfare of polar bears in particular, the growing volume of marine debris that damages or kills mammals when they get wrapped in or eat it, the harmful effects of ocean noise, causing cetaceans to become stranded, and the killing of marine mammals, including seals and otters, for their fur. Much of the science is so new that reliable methods of assessing welfare are only just being developed, and some are outlined in this book for the first time. The role and husbandry of marine mammals in captivity in zoos and aquaria is considered, with prominence given to the role of polar bears in raising awareness of the plight of this species in coping with climate change. The book will undoubtedly become a standard reference work in this emerging area of animal welfare science, and it is hoped that it will stimulate a new determination to address the risks to welfare that are the focus of the book.

St. Lucia, QLD, Australia Clive Phillips

Contents

About the Authors

Rebecca Aldworth is the executive director of Humane Society International in Canada. For 17 years, she has been a firsthand observer of Canada's commercial seal hunt, escorting more than 100 scientists, parliamentarians, and journalists to the ice floes to witness the killing. She has testified extensively before international government committees in support of prohibitions on seal product trade and has published multiple articles and reports on the welfare, economic, and environmental aspects of commercial sealing. She is a recipient of the 2004 Jean Taymans award for animal welfare and in 2006 was named one of nine Eco Heroes by Alternet. In 2011, she was named Activist of the Year in the Canadian Empathy Awards.

Steven C. Amstrup is chief scientist for Polar Bears International. He also is an adjunct professor at the University of Wyoming in Laramie. Prior to joining PBI, he led polar bear ecology research in Alaska for 30 years, at the U.S. Geological Survey (USGS). He is a past chairman of the IUCN Polar Bear Specialist Group and has been an active member of the group for 35 years. Dr. Amstrup has authored or coauthored more than 150 peer-reviewed articles on animal movements, distribution, and population dynamics and is the senior editor of a text on population estimation methods. In 2007, he led a USGS research team in production of nine reports that convinced the U.S. Secretary of Interior that polar bears should be declared a threatened species. In the December 2010 issue of *Nature*, he and his coauthors showed there is a linear and inverse relationship between global mean temperature and sea ice extent, and that preserving polar bears requires a halt to anthropogenic temperature increase. In 2012, Amstrup was selected as recipient of the Indianapolis Prize, and a Bambi Award for his efforts in animal conservation.

Eduardo Moraes Arraut is currently a postdoctoral fellow at the Remote Sensing Division, National Institute for Space Research, Brazil. During the preparation of this book, he was an overseas researcher of the Wildlife Conservation Research Unit, Zoology Department, Oxford University. Eduardo has been, or is, researching the ecology or conservation of humpback whales and Amazonian manatees in Brazil, common buzzards in the UK, and African lions in Zimbabwe. His research involves both extensive fieldwork, often in remote places, and the development of animal space use analytical techniques using statistics and agent-based modeling. He is Joint-Coordinator for South America of the Thematic Group for Sustainable Use and Management of Ecosystems (SUME), Commission on Ecosystem Management (CEM), International Union for Conservation of Nature (IUCN), and member of the European Sustainable Use Specialist Group (ESUSG). He is also involved in interdisciplinary collaborations with researchers in the field of education and has worked as the Executive Secretary of the two major climate change research programs in Brazil, where he managed finances, contributed to public outreach activities, and led the preparation of multidisciplinary scientific reports.

Todd Atwood is a Research Wildlife Biologist and leader of the U.S. Geological Survey's Polar Bear Research Program. His research is focused on understanding the impacts of a changing Arctic marine ecosystem on polar bear population ecology, health, and resilience. Todd is a member of the IUCN—Polar Bear Specialist Group and is co-chair of the Science and Traditional Ecological Knowledge Working Group of the U.S. Fish and Wildlife Service Polar Bear Recovery Team.

Yara Bernaldo de Quirós is a postdoctoral research associate at the Institute of Animal Health and Food Safety, University of Las Palmas de Gran Canaria, Canary Islands, Spain. Yara has also worked as a postdoctoral investigator in Texas A&M and Woods Hole Oceanographic Institution, USA. Her research focuses on comparative physiopathology, especially diving physiology. She has participated several times as a guest participant in the scientific committee of the International Whaling Commission. She is a reviewer of several journals in the field and has over 20 publications to date.

David Blyde is the veterinarian at Sea World, Gold Coast, Australia. David graduated from the University of Sydney in 1982 with a B.V.Sc. (second class honors). He was admitted to the Australian and New Zealand College of Veterinary Scientists by examination in the Medicine of Zoo Animals in 1992 and is president of the zoo and wildlife chapter. He is a member of various bodies, including the Australian Veterinary Association, the Wildlife Disease Association, the American Association of Zoo Veterinarians, and the Australian Institute of Management. His current interests include artificial reproduction, reintroduction biology, and diseases of marine mammals. He has delivered many lectures on various veterinary topics around the world and publishes in veterinary books and journals. He is an adjunct associate professor at the University of Queensland.

Robert Bonde is a Research Biologist with the U.S. Geological Survey and holds an adjunct faculty position with the graduate school of the University of Florida, College of Veterinary Medicine. Bob has been studying manatees for 38 years and specializes in their natural history, biology, genetics, and conservation. He has served on graduate committees for 29 students focusing on topics related to genetics, endocrinology, osteology, virology, behavior, physiology, contaminants, health monitoring, and blood chemistry. In his position he has over 90 academic publications and has written a book along with Dr. Roger Reep entitled *The Florida Manatee: Biology and Conservation.*

Philippa Brakes is a researcher on social learning and culture in whales and other nonhumans and is a wildlife advocate working for the NGO Whale and Dolphin Conservation (WDC). Through her research at the University of Exeter, she explores the implications of culture and other forms of social complexity for the conservation management of marine mammals. Her interest is the evolution from traditional genetic and geographic boundaries towards conservation policy, which also incorporates behavioral ecology and recognizes the importance of sociality and consistent individual difference (or personality). Philippa currently serves as a Senior Biologist with WDC.

Stewart Breck is a carnivore ecologist for the USDA-Wildlife Services-National Wildlife Research Center and focuses on studying carnivore ecology and developing methods for minimizing human-carnivore conflict. Stewart attained degrees in ecology and wildlife biology from Colorado State University (B.S. and Ph.D.) and the University of Nevada Reno (M.S.). His current interests include studying the ecology of coyotes, wolves, black bears, and polar bears in urban, agricultural, and natural systems, with a primary focus on developing nonlethal techniques for predator management. He works extensively with graduate students and is an adjunct professor at Colorado State University.

Danielle M. Brown is currently a Ph.D. student working in Dr. Chris Parsons' lab in the Environmental Science and Policy department at George Mason University. Her research focuses on the ecology of humpback whales in the New York metropolitan area and their interactions with human activities. In addition to being a graduate student, Danielle works with Gotham Whale as an Associate Researcher and has previously worked as a Field Stranding Technician with the Marine Mammal Stranding Center.

Andy Butterworth is Reader in Animal Science and Policy in the Clinical Veterinary School, University of Bristol, UK. Andy is a Vet and teaches and carries out research in the areas of animal disease and production, animal welfare and legislation, behavioral biology, and animal welfare assessment in both farm and wild animals. He is a member of the European Food Standards Agency Scientific Panel on Animal Health and Welfare, and chairs the Ethics, Economics, Education and Regulation Committee of the Farm Animal Welfare committee in the UK. He lectures widely and publishes in books, and the academic and trade press.

Judy Che-Castaldo is a research scientist at the Alexander Center for Applied Population Biology at Lincoln Park Zoo in Chicago, Illinois, USA. She is an ecologist broadly interested in applying theories and tools from population biology to conserve threatened species. She has conducted a systematic evaluation of recovery plans for species listed under the U.S. Endangered Species Act and developed methods for analyzing population viability for data-poor species. Her other research projects include modeling population dynamics and conducting cross-species comparative analyses to improve the management of captive animal populations.

Isabella Clegg is completing a Ph.D. in dolphin behaviour and welfare, as part of a collaboration between Université Paris 13 and Parc Astérix. Her master's degree in marine mammal science (University of Miami) and bachelor's degree in animal behaviour and welfare (University of Bristol) allowed her to combine these areas of expertise for this unique Ph.D. project, partially supported by the Universities' Federation for Animal Welfare (UFAW). She is on the student committee for the European Association for Aquatic Mammals (EAAM) and the Parc Astérix Animal Welfare Committee. She has participated in numerous wild and captive marine mammal research projects and is dedicated to enhancing captive dolphin welfare and developing associated methods for use in the wild.

Lucy Keith Diagne is the founder and director of the African Aquatic Conservation Fund, a nonprofit organization dedicated to research and conservation of African manatees and turtles. Lucy leads the African Manatee Project, which encompasses a collaborative network for manatee fieldwork and conservation with members in 19 African countries. To date she has trained over 90 African biologists in manatee field techniques and conservation planning. Her own research focuses on African manatee population genetics, feeding ecology, and threat assessment. She was awarded the Manatee Conservation Award by the U.S. Fish and Wildlife Service in 2003, and she is a member of International Union for the Conservation of Nature's Sirenian Specialist Group, a co-Chair of the African Manatee Regional Subgroup, and a member of the Convention of Migratory Species Scientific Council—Aquatic Mammals Working Group. She is based in Senegal, West Africa.

Sarah Dolman has a fisheries science master's degree from Aberdeen University and a bachelor's degree in electrical and electronic engineering from Bath University. Sarah has worked at Whale and Dolphin Conservation (WDC) since 1993, where she leads their international program to reduce cetacean bycatch. She is a member of the European Cetacean Society Scientific Advisory Committee, UK Wild Animal Welfare Committee (WAWC), and the Marine Animal Rescue Coalition (MARC). Her field research focuses on behavioral studies in wild cetaceans, particularly Risso's dolphins. Her policy work focuses on marine noise pollution, European environmental legislation, and more recently on the conservation and welfare issues associated with cetacean bycatch.

David Douglas is a research wildlife biologist with the U.S. Geological Survey Alaska Science Center. He studies how year-to-year variations in the timing of sea ice and snow melt affect the availability and quality of habitats for marine animals like polar bears and pinnipeds, and for terrestrial animals like caribou and geese. Arctic animals are well adapted to variable weather conditions, but global warming is rapidly changing the Arctic and challenging their resiliency. David has participated in numerous multidisciplinary studies where he brings expertise in remote sensing and spatial modeling aimed at quantifying relationships between landscape dynamics and patterns of habitat use and population dynamics.

Colleen Duncan is a faculty veterinarian at Colorado State University in the department of Microbiology, Immunology and Pathology. She is a Diplomate of both the American College of Veterinary Pathology (anatomic) and the American College of Veterinary Preventive Medicine. In her role as a diagnostic pathologist and epidemiologist, she collaborates with many agencies to investigate a variety of animal health and disease issues. Her current work is largely focused on fur seals and polar bears in Alaska.

Nicole Duplaix is a French-American zoologist, ecologist, teacher, and photographer who specializes in studying giant otters. She discovered her passion for otters while volunteering at the Bronx Zoo, where the zoo's curator introduced her to these endangered sea and river mammals. She earned her master's degree at the University of Paris in 1965 and 1966. She later studied giant otters in South America. In one year of this study, she managed to identify 249 different animals. She also helped Suriname officials learn how to preserve river otters, though they still face endangerment today. She returned to the University of Paris to complete her Ph.D., and her 1980 dissertation concerned a colony of river otters on Suriname's Kapoeri Creek. She also worked as a fund-raiser and speaker for the World Wildlife Fund. Most recently, she has settled in Oregon and is a senior lecturer at Oregon State University.

George M. Durner is a research zoologist with the U.S. Geological Survey, Alaska Science Center in Anchorage, Alaska, USA. George's research focuses on polar bear sea ice and maternal denning habitat relationships, and the behavioral response of polar bears to habitat change in a warming Arctic. He is a member of the Wildlife Society, the Arctic Institute of North America, and the International Union for the Conservation of Nature (IUCN), Polar Bear Specialist Group (PBSG). He has been an author on over 30 peer-reviewed publications on polar bears, presented at numerous professional conferences, and regularly serves as a reviewer for journals on polar bear-related research articles.

Holly Edwards is an Assistant Research Scientist for the Florida Fish and Wildlife Conservation Commission's marine mammal section where she leads the manatee aerial survey program. Holly has developed and conducted research projects on cetacean and sirenian species, in both the USA and Latin America, focusing on population assessment, distribution and movements, habitat use, survival, health assessment, risk, and climate change. These projects have involved conducting aerial, boat, and land based surveys of both manatees and dolphins (*Sotalia fluviatilis*); photo identification; and satellite/GPS telemetry tracking of manatees. Her more recent research involves using manatee telemetry and manatee diving data to estimate the risk of boat-manatee collisions, assessing the impacts of climate change on Florida manatees, and estimating manatee abundance from aerial surveys.

Antonio Fernández is Professor of veterinary histology and pathology at the veterinary school and Institute of Animal Health and Food Safety of the University of Las Palmas Gran Canaria, Canary Islands, Spain. He is a European Veterinary Pathology diplomat (ECVP) and a European Zoological Medicine (Wild Life Population) diplomat (ECZM) and a postdoctoral researcher in the Alexander von Humboldt Institute of Veterinary Pathology, Hannover, Germany, and a visiting Professor in the Veterinary Pathology Department, Cornell University, USA. He is director of M.Sc. and Ph.D. Programs in Animal Health and Food Safety and won the Canary Islands Award in Science and Innovation 2008 and the Academic Excellence Award, University of Las Palmas de Gran Canaria, 2008.

Sheryl Fink is IFAW's Director of Wildlife Campaigns in Canada. She has been active in campaigns to educate the public on the threats facing Canada's polar bears, the live capture of marine mammals for captivity, the exotic pet trade, and commercial exploitation of animals and ecological sustainability. Sheryl is a frequent speaker on marine mammal issues at conferences, drawing upon her experience with IFAW and her background as a biologist having worked in the field with harp, hooded, and grey seals. Sheryl has authored many of IFAW's submissions to the Canadian Department of Fisheries and Oceans, commenting on proposed changes to the Marine Mammal Regulations which negatively affect the seals and has also authored documents submitted by IFAW to inform the European Commission in their assessment of the potential impact on the EU of a ban on products derived from seals and in support of the EU ban before the World Trade Organization. Sheryl holds a Bachelor of Science (B.Sc., Hons) in wildlife biology from the University of Guelph, Canada.

Jaylene Flint is the Sea Turtle Research Biologist at the Florida Aquarium and a research scholar in the School of Veterinary Science at the University of Queensland. Jaylene has worked in several government, educational, and not-for-profit facilities, in both the United States and Australia, examining the effects of natural and human impacts on the survivorship of key sentinel species. She uses this data to create predictive models that assist in the management of species of concern and habitats. Jaylene is well published in these areas, with over a dozen technical reports, public outreach documents, journal articles, and a book chapter.

Mark Flint is in the Research Faculty of the School of Forest Resources and Conservation at the University of Florida, as well as serving as the Director of Research at the Florida Aquarium's Center for Conservation in the United States; Director of the Veterinary-Marine Animal Research, Teaching and Investigation unit; and an Adjunct Senior Lecturer in the School of Veterinary Science at the University of Queensland in Australia. Mark's research focuses on identifying prevalent and emerging stressors in marine and freshwater animal species as they relate to natural and anthropogenic pressures, health, welfare, environmental management, and sustainable resource use. He is widely published on these topics, with over 70 publications in continuing education, journals, and as book chapters.

Flavio Lehner is a postdoctoral researcher at the National Center for Atmospheric Research in Boulder, USA. Flavio is researching global and regional climate variability and change on centennial to decadal time scales with a focus on quantifying uncertainties in our understanding of past, current, and future climate change. Combining climate model simulations with observational evidence, he has published scientific papers on atmospheric circulation, ocean-sea ice-atmosphere interactions, carbon cycle response to climate change, and human exposure to changes in record-breaking summer temperatures.

Sarah Long is the Director of the Association of Zoos and Aquariums' (AZA's) Population Management Center (PMC) at Lincoln Park Zoo in Chicago, Illinois, USA. Sarah has been with the PMC since its establishment in 2000, conducting genetic and demographic analyses and developing breeding and transfer plans for cooperatively managed zoo populations. In addition, she teaches zoo professionals and contributes to the development of population management software (e.g., PMCTrack, PopLink, ZooRisk) and research on small population management methodology. Sarah is also a member of AZA's Small Population Management Advisory Group (SPMAG), the World Zoo & Aquarium Association's Committee for Population Management, and serves as an advisor to AZA's Wildlife Conservation and Management Committee (WCMC), which oversees AZA cooperatively managed animal programs and the policies and issues relevant to these programs. Sarah received a B.S. in biology from DePaul University and an M.S. in biology from Northern Arizona University, where she conducted field research on communication of Gunnison's prairie dogs (*Cynomys gunnisoni*).

Rob Lott is policy manager and Orca program lead for Whale and Dolphin Conservation (WDC), an international charity which campaigns to see a world where every whale and dolphin is safe and free. Based in the UK, Rob works on WDC's anti-captivity program, addressing the issue of the ongoing live capture of cetaceans in Russia and Japan, as well as improving the welfare of captive whales and dolphins through the exploration of retirement sanctuary options. He is a keen writer and photographer and has published internationally in magazines and newspapers.

Miriam Marmontel is an oceanographer who has been working with freshwater aquatic mammals since she got her Ph.D. in wildlife conservation from the University of Florida in Gainesville. She is a researcher with the Mamirauá Institute for Sustainable Development, where she leads a research group on Amazonian aquatic mammals. Working in two large sustainable development reserves in the western Brazilian Amazon as well as in coastal Amazon, she conducts research on manatees, dolphins, and otters, focusing on anatomy, movements and habitat use, population estimation and demography. Miriam is the editor-in-chief of the *Latin American Journal of Aquatic Mammals*. She is a member of the IUCN Specialist Group and co-chair of the South American Manatee Regional Subgroup.

Helene Marsh is Distinguished Professor of Environmental Science and Dean of Graduate Research at James Cook University, Australia. Helene is a conservation biologist with some 30 years' experience in research into species conservation, management, and policy with particular reference to coastal tropical marine megafauna of conservation concern. The policy outcomes of her research include significant contributions to the science base of dugong conservation in Australia and internationally. Helene has received international awards for her research and conservation from the Pew Charitable Trust, the Society of Conservation Biology, and the American Society of Mammalogists. She is Chair of the Australian Threatened Species Scientific Committee, Co-chair of the IUCN Sirenia Specialist Group, and a former President of the Society of Marine Mammalogy. She is on the editorial boards of *Conservation Biology*, *Endangered Species Research*, and *Oecologia*. Her publications include two books and some 200 papers. Helene has supervised 75 research higher degree candidates to completion and numerous postdoctoral fellows.

Randi Meyerson is the Assistant Director of Animal Programs at the Toledo Zoo. She holds a D.V.M., and M.S. in zoology. She has over 20 years' experience in animal husbandry and vet care of zoo and native wildlife and has special interest in carnivores, elephants, and Tasmanian devils. She has been the Chair of the Association of Zoos and Aquariums' (AZA) Polar Bear Species Survival Program (SSP) since its inception in 2003, as well as the program studbook keeper. Currently she is the Vice Chair of the AZA Bear Taxon Advisory Group (TAG) and a member of the Sloth Bear SSP management committee. Prior responsibilities have included being the veterinary advisor for the Mexican Wolf SSP and Bear TAG Chair. She also is an advisor for the NGO Polar Bears International.

Michael Moore has a veterinary degree from the University of Cambridge in the UK, and a Ph.D. from the Woods Hole Oceanographic Institution (WHOI) and Massachusetts Institute of Technology in the USA. He has been based at WHOI in Woods Hole, Massachusetts, since 1986. He is Director of the WHOI Marine Mammal Center and provides veterinary support to the Marine Mammal Rescue and Research Division of the International Fund for Animal Welfare, supporting their work with stranded marine mammals on Cape Cod. His research encompasses the forensic analysis of marine mammal mortalities, especially in regard to the accurate diagnosis of perceived human impacts and the prevalence of zoonotic agents, the interaction of natural and man-made impacts on fish and marine mammal stocks, development of systems to enhance medical intervention with large whales, and the pathophysiology of marine mammal diving.

Don Moore is a zoo-based endangered species conservation specialist and animal behaviorist, who has been leading and training the next generation of animal care professionals and conservation biologists, studying animal behavior in zoos and nature, and creating conservation management plans for wild animals in zoos and in nature for over 30 years. He has a special interest in the conservation and well-being of large mammals including bears, wolves, deer, and Asian elephants. Dr. Moore currently works as Director of the Oregon Zoo and as a Smithsonian senior science advisor on strategic initiatives, species sustainability, and public education. He previously led Smithsonian's National Zoo's Animal Care Sciences' team of veterinarians, curators, keepers, nutritionists, and animal behavior professionals, providing excellence in animal care and well-being for the zoo's priceless living collection. Prior to working at Smithsonian, Dr. Moore worked as a curator, animal behavior programs leader, and director in Wildlife Conservation Society's constellation of zoos and global conservation programs.

Mike Ogle has been a ranger with New Zealand Department of Conservation (DOC) for 20 years. For the past 13 years he has been based in Golden Bay monitoring and managing the varied fauna of this area. In the Golden Bay area over this period there have been dozens of marine mammal strandings, totalling over 1000 individuals. Mike attended many of these strandings, and in more recent years he has undertaken the role of operations manager at mass strandings. He is the current national lead contact within DOC for marine mammal mass strandings and scientific sampling. Other marine mammal work has seen him surveying and remote biopsying of hector's dolphins and blue whales around the top of New Zealand's South Island.

Helen Owen is a Senior Lecturer in Veterinary Pathology at the School of Veterinary Science, University of Queensland. Helen's diagnostic and research interests include pathology of marine wildlife and pathology associated with parasitic infections.

Chris Parsons has been involved in whale and dolphin research for 25 years and has conducted projects in every continent. He is currently involved in interdisciplinary research projects on coastal dolphin populations in the Caribbean and the effectiveness of marine conservation policy. Dr. Parsons started teaching at George Mason University in 2003 and is currently the coordinator for the university's environmental science undergraduate program. Dr. Parsons has been a member of the scientific committee of the International Whaling Commission (IWC) since 1999 and is currently the co-convenor of the IWC subcommittee on Environmental Concerns. He is a Fellow of the Royal Geographical Society, the Royal Society of Biology, and Marine Biological Association. He has been involved in organizing four of the International Marine Conservation Congresses (the largest conference for marine conservation scientists) and two of the International Congresses for Conservation Biology. He served two terms as the Marine Section President and three terms as a Governor of the Society for Conservation Biology, and is currently on the Board of Directors of the Society for Marine Mammalogy and the American Cetacean Society. In addition, Dr. Parsons has published over 150 scientific journal articles and book chapters and has written a textbook on marine mammal biology and conservation.

Kelly Patyk is a Veterinary Epidemiologist with the U.S. Department of Agriculture's Center for Epidemiology and Animal Health, located in Fort Collins, Colorado. Kelly serves as an analyst engaged in studying animal health and related agricultural issues to support decision making and preparedness activities. Her research interests include population health, disease transmission at the wildlife-livestock interface, survey research, and veterinary public health. She is a U.S. representative on the Quadrilateral EpiTeam and holds an affiliate faculty position in the Department of Microbiology, Immunology, and Pathology at Colorado State University.

Lily Peacock is a student of the Alaska School of Medical Education and the University of Washington School of Medicine in Seattle. She earned her undergraduate degree in biology from the University of California, Berkeley, with a minor in Russian, before she embarked on her human medical doctoral studies. For her doctoral dissertation, Lily spent 6 years in Southeast Alaska studying black bear population ecology on Kuiu Island and polar bears in the Inuit village of Igloolik (Place of Igloos), a small island off the northern tip of Baffin Island in the Canadian territory of Nunavut. She led the U.S. Geological Survey Alaska Science Center polar bear research program, which conducts long-term research on polar bears to inform local, state, national, and international policy makers regarding conservation of the species and its habitat.

Kelly J. Pearce is a Ph.D. candidate in the Marine-Estuarine-Environmental Sciences Program at the University of Maryland Center for Environmental Science. Kelly's primary research interests include the integration of social and ecological sciences, including the field of human dimensions of wildlife. She is particularly interested in exploring the relationships of people to the natural world and developing innovative ways to engage people in conservation actions. Kelly is dedicated to promoting carnivore conservation and has helped develop field education programs for students of all ages. She is a member of the International Union for Conservation of Nature (IUCN) Otter Specialist Group and IUCN Nearctic Otter Education Group.

Jan Reed-Smith served on the Board of Directors of the American Association of Zoo Keepers, has 17 years of experience working in zoos, and 23 years working on otter husbandry issues in zoos and aquariums. She served as the Chair of the International Union for Conservation of Nature/Species Survival Commission Otter Specialist Group's OZ (Otters in Zoos, Aquariums, Rehabilitation, and Wildlife Sanctuaries) Task Force for 15 years and has published several husbandry manuals and articles on otter care and welfare. She also has conducted field research on spotted-necked otters in Africa. Currently she is working on otter conservation and welfare issues in Africa.

Diana Reiss is a cognitive psychologist, a marine mammal scientist, and a professor in the Department of Psychology at Hunter College and in the Animal Behavior and Comparative Psychology Doctoral program at the Graduate Center, City University of New York. Her research focuses on dolphin cognition and communication, comparative animal cognition, and the evolution of intelligence. Dr. Reiss and her colleagues demonstrated the capacity for mirror self-awareness in bottlenose dolphins and in Asian elephants. Her professional activities also include the rescue and rehabilitation of stranded marine mammals; from 2002 to 2007, she served on the Animal Welfare Committee of the Association for Zoos and Aquariums. Dr. Reiss's work has been featured in hundreds of articles in international and national journals, science magazines, television segments and features, and newspaper articles.

Simona Sacchini is a postdoctoral associate researcher at the Institute of Animal Health and Food Safety, University of Las Palmas de Gran Canaria, Canary Islands, Spain. Her research area is on neuroanatomy and neuropathology, in cetaceans and other species. She has a European Ph.D. (University of Las Palmas de Gran Canaria, 2015) and a veterinary degree (University of Milan, Italy, 2007). She is a member of the Society for Neuroscience (SfN) and spent 8 months at the University of Bologna (Italy) studying neuroanatomical investigations on the cetacean brain (2011–2012). She is the first author of three scientific communications from three International Scientific Congresses. She was a Resident Veterinarian at the Clinical Hospital for Small Animals at the University of Las Palmas de Gran Canaria (2007–2008).

Sue Sayer began photo identifying grey seals in 2000, giving her an unparalleled insight into their ecology, having spent thousands of hours recording them from land and sea. In 2004, Sue set up the UK Cornwall Seal Group Research Trust (CSGRT), a well-respected, independent, evidence-based conservation charity with an active network of highly motivated citizen scientists. She is self-employed and her first book *Seal Secrets* was published in 2012 (republished 2013). She routinely writes on seals, lectures all over Cornwall and Devon, and has presented at conferences in Europe and the USA.

Tom Serfass is a Professor of wildlife ecology in the Department of Biology and Natural Resources at Frostburg State University. He has worked with river otter conservation for over 25 years, having conceived and supervised the Pennsylvania river otter reintroduction project. Tom also coordinated research for the New York Otter Project and served as a consultant for the reintroduction of river otters in New Mexico. Most recently Tom has been conducting research on spotted-necked otters at Rubondo Island National Park, Tanzania, and river otters in Yellowstone National Park. Tom is the North American Coordinator for the International Union for Conservation of Nature (IUCN) Otter Specialist Group.

Eva Sierra is an associated researcher at the Institute for Animal Health and Food Safety at the University of Las Palmas de Gran Canaria. Her areas of expertise include veterinary pathology (particularly whales and dolphins) and pathogenesis of animal infectious diseases. She currently leads the molecular laboratory at her institute. Her research interests also include cetaceans' muscular pathology. She is a reviewer and author in academic press, with over 45 publications in the cetacean field.

Kristin Simac is a biologist with the U.S. Geological Survey with particular interests in polar bear research, who provides logistical support for Polar Bear Research Group and coordinates sample collections for the Alaska Marine Mammal Tissue and Archival Program. Kristin has a bachelor's degree in biology from Willamette University, Oregon, a master's degree from Alaska Pacific University in environmental science, and has been involved in over 25 publications in the area of polar bear science.

Mark Peter Simmonds is a Visiting Fellow at the Veterinary School, University of Bristol, UK, and the Senior Marine Scientist of the Humane Society International. He has authored over 200 scientific papers, reports, articles and reviews, and a number of books. His scientific work has focused on the environmental threats facing marine mammals in the twenty-first century. He has chaired the UK's Marine Animal Rescue Coalition since 1994 and has been a member of the Scientific Committee of the International Whaling Commission for more than 20 years. In the 2013 Queen's Birthday Honours, he was awarded an OBE in recognition of his work on marine mammal conservation and environmental sciences.

Sarah A. Sonsthagen is a Research Geneticist. She has a Ph.D. in biological sciences from the University of Alaska Fairbanks, Fairbanks. Sarah's interests are in the evolutionary and population genetics of vertebrates, with over 40 publications in this area, and she has been a Research Geneticist with the U.S. Geological Survey Alaska Science Center since 2013.

Courtney S. Vail holds an undergraduate degree in wildlife biology/natural resources management and a master's degree in psychology. With a background in biology, law, and environmental policy, Courtney has served as the director of programs for several conservation and animal welfare organizations, including Whale and Dolphin Conservation, where she served as Campaigns and Programs Manager, providing leadership on a variety of domestic and international policy initiatives relating to whale and dolphin protection and marine affairs. Courtney currently provides consulting services focused on balanced, progressive and incremental solutions in the animal welfare, environmental, and humanitarian sectors. She has also held positions within several government agencies, including the U.S. Fish and Wildlife Service and the Environmental Protection Agency, and has worked from within and outside of the regulatory frameworks that guide national and international conservation policy.

Michael Walsh is a graduate of the University of Missouri College of Veterinary Medicine and completed an internship in small animal medicine and surgery at the

University of Pennsylvania with a strong interest in wildlife medicine and conservation. After completing a residency in wildlife and zoo medicine at the University of Florida, he has worked with marine species medicine and conservation over 30 years with the last 9 at the University of Florida Aquatic Animal Health Program as a clinical associate professor. In addition to therapeutic developments for marine species, he is interested in One Health for wild and managed species, conservation and ecosystem health, the academic education of veterinary students in marine species, and public education in the areas of health, conservation, and welfare of marine species. He received the Manatee Conservation Award in 2007 and the Southeast Regional Manatee Conservation award in 2008 for work in rehabilitation medicine. He has authored, and co-authored, a number of book chapters and articles on manatee health and medicine.

Lindy Weilgart has specialized in underwater noise pollution and its effects on cetaceans since 1994. She has studied whales since 1982, primarily sperm whales, and her M.Sc. (Memorial Univ. of Newfoundland), Ph.D. (Dalhousie), and postdoctoral studies (Cornell) were all in the field of whale acoustic communication in the wild. Lindy has served as invited expert on several panels, workshops, and committees concerned with underwater noise impacts (e.g., Department of Fisheries and Oceans Canada, U.S. Marine Mammal Commission, International Whaling Commission). She has met with members of NATO, the European Parliament, the European Commission DG Fisheries, the Convention on Biodiversity, and the United Nations to discuss ocean noise issues, given many lectures on this topic and others, and published numerous peer-reviewed papers. She is currently an Adjunct in the Department of Biology, Dalhousie University, Canada, and was previously employed as Scientific Advisor by the private Okeanos Foundation. She also serves as Scientific Advisor for the International Ocean Noise Coalition and was a Scientific Expert on the German government's Antarctic Commission.

James M. Wilder worked for 17 years in Alaska and has worked with bears since 1999. He received his M.Sc. from the University of Idaho after leading the first black and brown bear research conducted in Alaska's Wrangell—St. Elias National Park. Subsequently, he worked for the National Park Service throughout Alaska researching and managing brown and black bears. He has worked on polar bear management and research since 2003, chaired the international Range States' Human-Polar Bear Conflict Working Group from 2009 to 2015, is a member of the International Union for Conservation of Nature Polar Bear Specialist Group, and currently works on the Shoshone National Forest in Wyoming. He remains active on the international Human-Polar Bear Conflict Working Group.

Cathy Williamson leads the End Captivity program of work at WDC, Whale and Dolphin Conservation. WDC is the leading global charity dedicated to the conservation and protection of whales and dolphins, including those captured and confined by the global zoo and aquaria industries, with offices in the UK, Germany, USA, Argentina, and Australia. Cathy represents WDC's End Captivity program in fora such as Convention on International Trade in Endangered Species of Wild Fauna and Flora (CITES) and the Convention on Migratory Species, producing comprehensive information resources, aimed at a wide range of stakeholders, including detailed reports, meeting briefings, scientific papers, articles, press releases, and internet-based materials. Cathy works with a range of stakeholders from members of the public, grassroots activists around the world, civil servants, scientists, and representatives of the tourism industry to bring about an end to captures, trade, and confinement of whales and dolphins in captivity. Through a project managed by Cathy, WDC is working in partnership with the global attractions company, Merlin Entertainments, to explore and develop sanctuaries for captive bottlenose dolphins and belugas.

Lesley Wright is the Co-Deputy Chair of the International Union for Conservation of Nature International Union for Conservation of Nature (Species Survival Commission Otter Specialist Group (OSG)). Lesley was formerly OSG species coordinator for Asian Small-Clawed Otters (*Aonyx cinereus*), she is a member of OSG's Otters in Zoos taskforce, and liaison with the International Otter Survival Fund on welfare issues. She is chair of the Otter Welfare Advisory Group, and trap/release coordinator for the UK Wild Otter Trust's Otter Predation Advisory Committee.

Geoff York is Senior Director of Conservation at Polar Bears International. Geoff has 20 years of arctic field experience, most recently as the Arctic Species and Polar Bear Lead for the World Wildlife Fund's (WWF) Global Arctic Program. While at WWF, Geoff immersed himself in international policy issues and was fortunate to work on field projects in Canada, Norway, Russia, and Alaska. Prior to that, he worked as a biologist and program manager for the U.S. Geological Survey's Polar Bear Project, the leading polar bear research team in the U.S., headquartered in Anchorage, Alaska. Since joining Polar Bears International, Geoff has continued his interest in field-based work across the Arctic. He is a member of the Polar Bear Specialist Group of the International Union for the Conservation of Nature, the U.S. Polar Bear Recovery Team, chairs the Polar Bear Range States Conflict Working Group, and sits on the advisory board for the International Polar Bear Conservation Center in Winnipeg, Manitoba. He has a M.S. in biology from the University of Alaska Fairbanks and a B.A. in English from the University of Notre Dame, the perfect combination for communicating science. Following 14 consecutive seasons of capture based work on the sea ice north of Alaska, Geoff has dedicated his career to the conservation of polar bears and their arctic home.

Chapter 1
Introduction

Andy Butterworth

Abstract Do humans really have any responsibility to wild marine mammals? Marine mammals in zoos certainly come under the heading of being under human control, but, do humans really have any responsibility to the welfare of wild marine mammals? The answer to this is, I suspect, 'it depends'. The marine mammals reflect mammalian adaptations to a fully aquatic (cetaceans, sirenia), mostly aquatic (seals, sea lions) or semiaquatic (otters, polar bear) life. This is a spectrum of dependency on water - a stranded whale will be in deep distress and likely to die after half a day on a beach and out of water; a polar bear may not touch deep water for weeks or months, but, on the contrary, it can swim in deep oceanic seas for up to 12 days without touching solid ground. The chapters in this book reflect the variation in marine mammal adaptation and their responses to human pressures. The chapters also reflect the difficulties in discussing wild animal protection, the links between conservation and animal welfare, hunting, pollution, by-catch and captivity all within the same book cover. There is a profound illogicality to some marine mammal issues - for example - in one part of the world, hundreds or even thousands of whales and dolphins are being killed for meat or for use in the entertainment industry in marine parks. In another part of the world, or even in the same country, and even on the same coastline, stranded whales or dolphins are attracting crowds of people with the good intention to rescue, refloat and rehabilitate these animals. Somewhere in this confusing mix of exploitation and protection, conservation and consumption, there remains the capacity for humans to identify animal suffering and, where it seems expedient or politically or socially appropriate, to act.

A. Butterworth
Department of Clinical Veterinary Science, University of Bristol Veterinary School, Langford, N Somerset, Bristol, UK
e-mail: andy.butterworth@bris.ac.uk

A. Butterworth (ed.), *Marine Mammal Welfare*, Animal Welfare 17,
DOI 10.1007/978-3-319-46994-2_1

1.1 Wildly Irresponsible or Responsible for the Wild?

Do humans really have any responsibility to wild marine mammals? Or is 'wild irresponsibility' a feature of how mankind views wild animals, 'not in our direct control', therefore not really our responsibility? Marine mammals in zoos certainly come under the heading, 'under man's control', but, do humans really have any responsibility to the welfare of wild marine mammals? The answer to this is *it depends*. It depends on how direct the linkage is between human impacts and the welfare of marine mammals.

Hunting of seals, or whaling, or trapping of otters or boat propeller injuries to manatees - these are very direct impacts on these animals - and how the hunting, trapping, netting and culling are done, and the methods used, are impacts which can be seen to directly affect the welfare of marine mammals when they come into close contact with humans.

Less directly, and perhaps less clearly, the effects of human generated noise in the ocean, boat traffic, coastal development and interactions with fisheries and the effects of warm water outflows from power stations or algal blooms linked with a nutrient run off from agriculture are very much human impacts, but they are probably not 'intentional'. To catch fish, to travel, to produce electrical power and to farm; the impacts on the marine mammals are secondary, not really intended and to a degree unanticipated.

The least direct, or, at least, less immediately obvious, links between human activity and marine mammal welfare are through effects including climate change, pollution in the oceans, marine debris, military sonar, mining and oil exploration, and also through the pressure of human populations on food and fish resources and human impacts on coastal land, on river drainage to the sea and on coastal plant and animal communities on which some marine mammals depend.

So, in answer to the question: can, or should, or do, humans have a responsibility for the welfare of marine mammals? 'It depends on':

(a) The clarity of the association between man, marine mammals and welfare impacts
(b) The actions and reactions that mankind may be able to take; immediate, local, short-term or longer-term, wide ranging and potentially global in implications

The chapters in this book reflect this variation in the types and 'directness' of the linkages between people's actions and marine mammal welfare. As well as discussing the 'issues', the chapters also reflect the enormous differences in approachs that are, or will, or could, be required to tackle marine mammal welfare issues including 'marine debris (Chaps. 3, 13)', 'marine noise (Chap. 7)', 'climate change (Chaps. 2, 8, 15, 19, 22, 23, 24, 25)', 'pollution (Chaps. 3, 18)', 'the welfare impacts of captivity (Chaps. 11, 16, 20, 27, 31)', 'boat strike (Chaps. 4, 17)', 'environmental change (Chaps. 10, 14, 19, 24)', 'hunting (Chaps. 5, 6, 15, 26, 30)' and 'by-catch (Chap. 4)'.

1.2 The Marine Mammals

The marine mammals reflect mammalian adaptations to a fully aquatic (cetaceans, sirenia), mostly aquatic (seals, sea lions) or semiaquatic (otters, polar bear) life. This is a spectrum of dependency on water - a stranded whale will be in deep distress and likely to die after half a day stranded on a beach and out of water; a polar bear may not touch deep water for weeks or months, or it can swim continuously in deep oceanic seas for up to 12 days without touching solid ground. Some marine mammals will be wholly and entirely influenced by the sea (or rivers, estuaries, marshland), and so the welfare impacts which most strongly affect these animals will mostly reflect human impacts on the marine world: fishing, marine debris, by-catch and water pollution. The animals which have a partial dependency on the rivers and oceans will be affected by these same human-linked conditions, pollution, debris, coastal development and boat strikes, and may also be affected by coastal development and loss of coastal habitat. The terrestrially capable marine mammals (seals, otters, polar bear) will be influenced by human influences both on the water, and on land and ice. The chapters in this book reflect this diversity, and also reflect the fact that it is not only water but ice, vegetation and coastal land changes which influence marine mammal welfare.

Marine mammals can be divided into:

Cetacea (whales and dolphins, approximately in diminishing order of body size)

Rorquals—*Balaenopteridae*—9 species
Grey whale—*Eschrichtiidae*—1 species
Right and bowhead whales—*Balaenidae*—4 species
Pygmy right whale—*Cetotheriidae*—1 species
Sperm whale—*Physeteridae*—1 species
Pygmy and dwarf sperm whales—*Kogiidae*—2 species
Narwhal and beluga—*Monodontidae*—2 species
Beaked whales—*Ziphiidae*—21 species
Oceanic dolphins—*Delphinidae*—38 species
Porpoises—*Phocoenidae*—7 species

Sirenia (sea cows)

Manatee—*Trichechidae*—3 species
Dugong—*Dugongidae*—1 species

Otters—Mustelidae—The 13 extant species are divided into semiaquatic (11 species) and marine (2 species)

Polar bear—*Ursidae*—1 species

Pinnipedia (sea lions, walruses, seals)

Eared seals and sea lions—*Otariidae*—15 species
Walrus—*Odobenidae*—1 species
True seals—*Phocidae*—18 species

1.3 Welfare, Conservation and the Messy Logic of Human Effects on the Welfare of Marine Mammals

Some readers will probably ask 'what is the difference between conservation and animal welfare?' with the understandable thinking that if animals are conserved, then their welfare is probably a secondary concern, but that if they are not conserved, then welfare becomes irrelevant. Conservation concerns itself with species, and the potential for extinction if a species does not survive.

The word 'welfare' is variably understood in different parts of the world—many languages have their own word for 'welfare' as used in the context of animal welfare or well-being: in Spanish, *benestar*, state of health, prosperity; German, *wohlbefinden*, well-being, wellness, physical comfort; and French, *bien-être*, well-being, a sense of well-being. Animal welfare focuses on the individual animal. Marine mammals are sentient animals, which have a complex experiential world and mental needs and natures; are aware of their own surroundings; have an emotional dimension; are aware of what is happening to them; have the ability to learn from experience; are aware of bodily sensations—pain, hunger, heat, cold etc.; are aware of their relationships with other animals; have the ability to choose between different animals, objects and situations; and have the capacity to suffer.

Historically there has been an understandable focus on negative welfare. However, positive experiences and states are now recognised to be (at least) as important as negative states in their contribution to overall well-being. Animal welfare concerns itself, and tries in some situations, to measure (welfare science) the 'quality' of an animal's life. Welfare science and ethical debate can, and does, address animal death, as well as animal life, as the 'quality' of 'animal death' affects the 'quantity' of animal life and the overall quality of an animal's life. The animals welfare during a marine mammal's life and at the time of its death and the impacts on the quality of life, for example, being entangled for the remainder of your life in a long buoy rope, being injured by a boat impact, or your reproductive fitness and health being affected by PCBs (polychlorinated biphenyls), are discussed in the chapters of this book.

The welfare of marine mammals is starting to enter the political arena at a high level. At its 65th meeting in 2014, the International Whaling Commission (IWC) agreed to direct a programme of work to address human activities which can adversely affect cetacean welfare, including the welfare concerns that arise when large whales become entangled in fishing gear or marine debris, and to work on the methods used to euthanise stranded whales and the effectiveness of those methods. Along with some of the other authors in this book, I attended the first IWC workshop (May 2016), which had a sole focus of considering non-whaling welfare issues (See Chaps. 4, 5, 9).

In 2014, Canada and Norway appealed to the World Trade Organisation (WTO) to overturn a European Union (EU) decision to ban trade in seal products. The trade in seal products was banned by the EU to protect 'public morals', and the science they cited indicated that some shot seals took a considerable period of time to die, and some injured animals were 'unchecked' for periods of several minutes before being finally killed by clubbing. The appeal to the WTO from Canada and Norway did not actually challenge the 'poor welfare outcomes' of the seals reported by the EU; instead, the appeal concentrated on trade issues and claimed unfair restrictions. The WTO decided against the appeal, and so trade in seal products derived from commercial sealing remains restricted in the EU, based on consideration of welfare as part of public moral concerns (see Chap. 15).

Where humankind has an influence, then it seems logical that consideration should be given for ways to provide marine mammals the potential to experience a life which avoids, as far as is pragmatic, suffering which derives from the hands of humans. The term 'good animal welfare' probably denotes a state in which there is little or no 'unnecessary suffering', and 'good welfare' is not just the absence of cruelty or 'unnecessary suffering'; it is more complex than that; it includes the physical and the mental state of the animal, whether the animal can express a range of 'normal' behaviours and whether the animal can fulfil its essential nature or 'telos'.

> 'I suggest that an animal is in a poor state of welfare only when physiological systems are disturbed to the point that survival or reproduction is impaired'.
>
> McGlone

> 'Welfare defines the state of an animal as regards its attempts to cope with its environment'.
>
> Fraser and Broom

> '... neither health nor lack of stress nor fitness is necessary and/or sufficient to conclude that an animal had good welfare. Welfare is dependent upon what animals feel'.
>
> Duncan

> 'Not only will welfare mean control of pain and suffering, it will also entail nurturing and fulfilment of the animals' nature, which I call telos'. Rollin

For animals kept in captivity, the influence of the captive environment is likely to be central to the animals' experience of life and in this case (as discussed in Chaps. 11, 12, 16, 20, 21, 27, 31) the way that the marine mammals are housed, cared for, fed, treated when sick, and provided with space, companionship and an environment which provides stimulus. These become important factors in welfare considerations that extend beyond those which may be considered for wild marine mammals.

For many people, a description of an animal having 'good welfare' might include the animal being 'well' (i.e. not unwell) and also that the animal had the potential for 'well-being'—or at least is not subject to high levels of distress or high frequencies of interference. With regard to a state of 'good welfare', disease or physiological or anatomical damage, injury and trauma would provide potential welfare challenges. The term 'cost of coping' has been used in relation to welfare, implying that emotional distress, pain or increased levels of physiological or disease-related challenge would have a 'cost' to the animal and that if this cost was great, or in some cases excessive, then the animal would be less likely to 'cope'. Prolonged failure to cope would probably result in suffering, and so the link between welfare and coping, and the cost of a welfare challenges in terms of the ability for the animal to cope, and for the animal to continue to express a range of expected or anticipated behaviours would result in challenges resulting in depletion of behavioural resilience in animals which were severely challenged by a welfare insult. We have a developing understanding of the complexity of some marine mammal 'societies' and the importance of certain associations for individual and group welfare and this is explored in Chap. 10. Welfare science is now a well-developed discipline (see Chaps. 12, 16), with its own language, agreed way of looking at things and, to a degree, its own paradigm, and I would like to follow the lead of the RSPCA in the UK, by suggesting that practical welfare assessment methods for marine mammals would, or could, follow these principles:

(a) Welfare assessment methods for marine mammals should/could be based on extrapolation from evidence, experience and knowledge from other species.
(b) Interpretation of welfare states in marine mammals would permit comparison with what is considered current 'good practice' for the treatment of other (non-marine mammal) animals.
(c) The way we interpret and try to understand marine mammal welfare issues would allow application of 'reasonable/justifiable anthropomorphism'.
(d) The interpretation of what we see as welfare issues in marine mammals would make use of 'common sense', i.e. making decisions which seem to show 'good sense' (as opposed to being 'nonsense') when viewed by a general body of reasonably informed humankind.
(e) When there are ethical considerations and decisions to be made, a structured, agreed framework based approach to consideration of these ethical issues should be adopted—to allow cool discussion of sometimes emotionally charged issues.
(f) Wherever possible, the application of the 'precautionary principle' ('informed prudence') could/should be adopted so that, when this is possible, the well-being and welfare of the animals is given weight and importance.

1.4 A Changing Wild World

The wild is less wild than it used to be across many parts of the globe; human influence, powered by oil and gas, electricity, the aeroplane, the car, the gun, air and water pollution, can be felt across the entire surface of the planet now—through the creeping tentacles of human population growth. The United Nations (2015) estimate that the global human population will reach 10.1 billion in 2100. Alongside this population growth, increasingly, the world's people live in cities; Osaka, Karachi, Jakarta, Mumbai, Shanghai, Manila, Seoul, Beijing, Mexico City, São Paulo, New York, Lagos, Los Angeles and Cairo each now have close to or more than 20 million people. Delhi and Tokyo are forecast to reach 40 million people within the next decade. Humans and their cities need food and fuel and often spread across coastal land. Human waste is linked with climate change, ocean pollution, air pollution and marine debris. Even if population growth slows, humankind and its mark on the planet and its animals are already deeply scored into the surface of the earth and will be for a long, geologically long, time.

1.5 Summary

The chapters in this book reflect the difficulties in discussing wild animal protection, the links between conservation and animal welfare, hunting, pollution, by-catch and captivity all within the same book cover. There is a profound illogicality to some marine mammal issues—for example—in one part of the world, hundreds or even thousands of whales and dolphins are being killed for meat or for use in the entertainment industry in marine parks. In another part of the world, or even in the same country, and even on the same coastline, stranded whales or dolphins are attracting crowds of people with the good intention to rescue, refloat and rehabilitate these animals. The lack of logic flows into the contrast between animals protected in the wild in reserves and parks, whilst across international borders, these same animals, if they migrate across international boundaries, may be hunted, trapped or even considered as a pest species to be culled. 'You can please some of the people all of the time, you can please all of the people some of the time, but you can't please all of the people all of the time' (John Lydgate, 1370—c. 1451), and the chapters of this book are not likely to all be received positively by all readers. Somewhere in this confusing mix of exploitation and protection, conservation and consumption, there remains the capacity for humans to identify animal suffering and where it seems expedient or politically or socially appropriate to act.

Reference

United Nations (2015) UN Department of Economic and Social Affairs, Population Division. World population prospects: the 2015 revision, volume I: comprehensive tables. ST/ESA/SER.A/379

Chapter 2
Anthropogenic Ocean Change: The Consummate Threat to Marine Mammal Welfare

Steven C. Amstrup and Flavio Lehner

Abstract Global warming is the consummate conservation and animal welfare challenge of our time. It defies traditional conservation management models and requires we broaden traditional cause and effect time horizons. Continually rising concentrations of CO_2 and other greenhouse gases (GHGs) prolong retention of the sun's energy before it escapes back into space—assuring that global temperatures must rise. Oceans have absorbed ~30% of anthropogenically emitted CO_2 and over 90% of the heat trapped by the world's enhanced greenhouse effect. Sea surface temperature and global ocean heat content have been rising accordingly. Along with rising temperatures, pH, oxygen saturation, salinity, and other aspects of ocean chemistry also are changing. Cumulative interactions among all of these symptoms of anthropogenic ocean change are and will continue to impact ocean biota. In this chapter, we summarize observed and projected anthropogenically driven ocean changes that have been and will continue to compromise marine mammal welfare.

2.1 Introduction

The action required to address global warming stands the traditional model of conservation on its head. In traditional approaches to conservation, we can build a fence, establish a preserve, or hire game wardens, and at the end of the day feel like we have protected the welfare of a particular species. But we cannot build a fence to protect melting sea ice from rising temperatures. Nor can game wardens halt ocean

---

S.C. Amstrup (✉)
Polar Bears International, 810 N. Wallace, Suite E, Bozeman, MT, USA

Department of Zoology and Physiology, University of Wyoming, 1000 E University Ave., Laramie, WY, USA
e-mail: samstrup@pbears.org

F. Lehner
Research Applications Laboratory, National Center for Atmospheric Research, 3450 Mitchell Ln, Boulder, CO, USA

A. Butterworth (ed.), *Marine Mammal Welfare*, Animal Welfare 17,
DOI 10.1007/978-3-319-46994-2_2

acidification. These are the ultimate threats to marine mammal welfare, and only united societal action can combat these challenges. Yet, many in the scientific and public domains continue to be preoccupied with traditional threats, and our focus largely remains on near-term uncertainties rather than the longer-term certainties. Consequently, the understanding needed to inspire timely action often has been lacking.

Current global warming is caused by human interference with earth's energy balance (IPCC 2013). The shortwave radiation coming to earth from the sun ultimately must be balanced by the outgoing long-wave radiation emitted, from the earth and its atmosphere, back into space (Lutgens and Tarbuck 2004, Chap. 2). Without this, the 1.22×10^{17} J of energy earth absorbs from the sun each second would "raise earth's temperature to nearly 800,000 K after a billion years" (Pierrehumbert 2011, p. 33). In other words, if the sun's heat accumulated here and was not reradiated into space, earth long ago would have been reduced to a ball of molten rock or hot gas. Various "climate forcings" can perturb the balance of incoming and outgoing energy (Hansen and Sato 2004). The shading effects of aerosols released into the atmosphere by volcanoes, for example, can provide a temporary negative climate forcing—cooling the earth by reflecting the sun's energy back into space before it gets to the surface. Rising concentrations of CO_2 and other GHGs, on the other hand, provide a positive climate forcing, trapping ever-increasing amounts of the sun's energy and prolonging retention of that heat before it escapes back into space. Simply put, the laws of physics require the world to warm as long as atmospheric GHG concentrations rise (Pierrehumbert 2011).

Naturally occurring events, like volcanic eruptions or El Niño, result in short- and medium-term variation in climate and weather—with some periods cooler than average and some warmer. The important point is that when GHG levels in the atmosphere are stable, the average, or "baseline," around which natural climate fluctuates, can be represented as a level or horizontal line (Fig. 2.1). That level long-term average allows us to use our experiences of the past to plan future actions—like when and where to plant crops and when and how much we will be able to harvest. In no small part, agriculture owes its rise and success over the last several thousand years to a favorable and stable climate (Rockström et al. 2009). When GHG concentrations are steadily increasing, as they are now, the natural variation we always have experienced continues, but it occurs over a higher and rising baseline (Fig. 2.1). Natural climate fluctuations, that surround the rising average, create uncertainty in knowing exactly when particular events will occur. We cannot, for example, confidently predict the first year summer sea ice will disappear from the Arctic, nor when surface temperatures in the Mediterranean Sea will have risen 2 °C. But, without stopping the increase in GHG concentration, it is certain we ultimately will exceed both thresholds. Unlike most predictions, which become less accurate the farther out we project, predicting that various global warming-related thresholds will be exceeded becomes more certain the farther into the future we look. Society's challenge in dealing with anthropogenic climate change is to maintain focus on the ultimate certainty—that without stopping GHG rise, we will exceed all of the thresholds we care about (Steinacher et al. 2013). The only reason we might wish to focus on the near-term uncertainties is if we don't really care about the world we are leaving behind us or for the welfare of future generations.

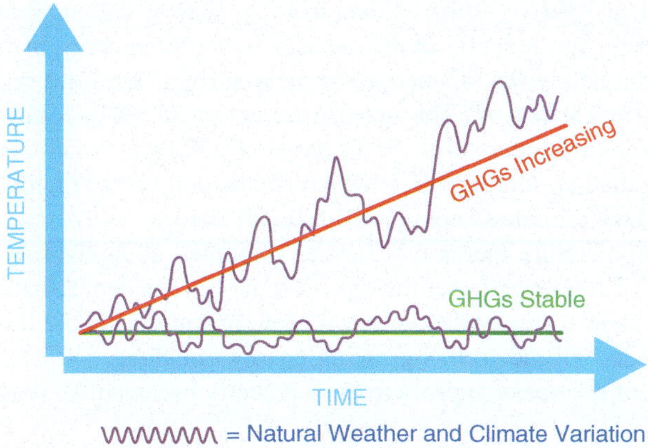

Fig. 2.1 Conceptual diagram of the impact of ever-rising atmospheric GHG concentrations. With GHG levels stable, the average of natural climate fluctuations is a level line. But with chronically rising GHG levels, average temperatures must rise. Despite the chaotic and unpredictable nature of natural variations, it always will be warmer than it would have been without the higher GHG concentrations

2.2 Anthropogenic Ocean Change

Oceans play a critical role in the earth's response to rising atmospheric concentrations of GHGs. About 30% of the CO_2 released by human activity has gone into the oceans, and, 93% of the extra solar energy captured by our enhanced greenhouse effect, has been trapped as heat by the oceans of the world (Hoegh-Guldberg et al. 2014). Climate change can be thought of as the cast of symptoms caused by the anthropogenic changes to the composition of Earth's atmosphere. In the oceans, these symptoms are many and complex. Anthropogenic ocean change includes rising temperatures, altered circulation patterns and temperature stratification, and numerous changes in ocean chemistry. In turn, all of these ocean changes influence the welfare of marine mammals and the other species upon which they depend.

Subsequent chapters will cover details of how ocean changes affect the welfare of individual species or groups of marine mammals. Here, we focus on four well-documented ocean changes the cumulative effects of which will have increasing influence on the welfare of marine mammals and other ocean biota. We review the observational record of changes already documented. We project future oceanic conditions with which marine mammals will be forced to contend, and compare them to present conditions. Finally, we provide examples of ways in which ocean changes may affect marine mammal welfare.

2.3 Temperature Effects

Sea surface temperatures (SSTs), which have been increasing at least since the middle of the twentieth century (Hoegh-Guldberg et al. 2014), may provide the most readily observed oceanic impact of the anthropogenically enhanced greenhouse effect. Riser et al. (2016) compared data from the HMS Challenger expedition with data collected by the recently deployed and widespread array of Argo profilers. They reported a nearly 0.6 °C increase in near-surface ocean average temperature during this 135-year period. The upward trajectory of SSTs follows the positive trend in surface air temperatures. SSTs, however, show less seasonal and interannual variation than air temperatures, which can respond more rapidly to short-term fluctuations in the climate system (Wijffels et al. 2016).

Ongoing temperature increase is further smoothed at depths below the ocean's surface (Fig. 2.2). Averaged over the top 2000 m of ocean depth, data from the Argo profiler array show a steady rise in ocean heat content during the last 10 years and provide a measure of the earth's growing energy imbalance (Wijffels et al. 2016). The warming of the ocean appears to have recently hastened. Wijffels et al. (2016)

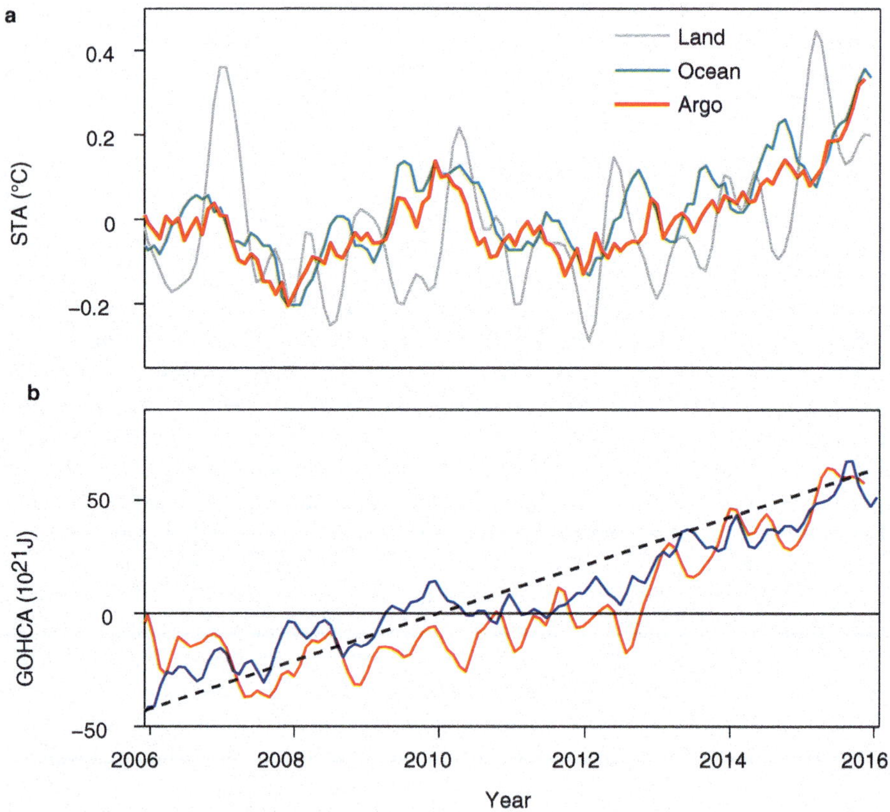

Fig. 2.2 Ocean warming rates excerpted from Fig. 1 in Wijffels et al. 2016. (**a**) Globally averaged surface temperature anomaly (STA, °C), from 5 m Argo temperature (*red*), NOAA (National Oceanic and Atmospheric Administration) global ocean (*turquoise*), and a 6-month running mean of NOAA (NOAA 2015) global land averages (*grey*). (**b**) Global ocean 0–2000 m heat content anomaly. Line plots in **b** are two interpolation methods and a robust linear fit

calculated the rate of near-surface warming at approximately 0.2 °C per decade. Gleckler et al. (2016) estimated the total ocean heat uptake since the 1870s has been 33 (±14) × 10^{22} J and calculated that approximately half of that increase occurred after 1997.

Ocean temperatures are expected to continue to rise, reflecting the guaranteed heating of the earth from increases in anthropogenically emitted GHGs. Figure 2.3 illustrates projected centennial sea surface warming for two different futures.

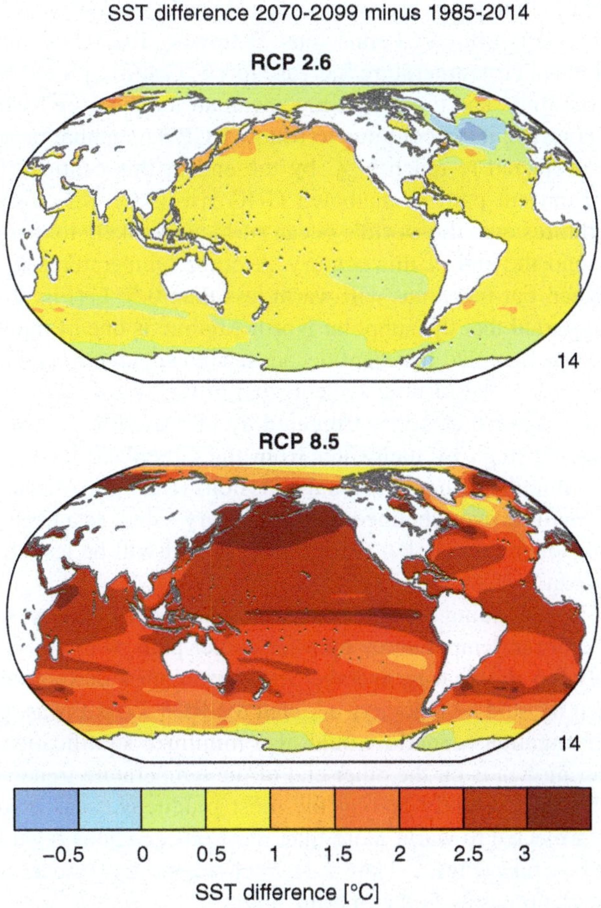

Fig. 2.3 Sea surface temperature (SST) maps showing the projected warming of the world's ocean between now and the end of this century under two greenhouse gas emissions scenarios or representative concentration pathways (RCPs). RCP 2.6 represents aggressive mitigation of emissions leading to <2 °C atmospheric warming through this century. RCP 8.5 represents a continuing path of unabated emissions and ~5 °C atmospheric warming. Climate model data used in Figs. 2.3–2.7 are taken from the CMIP5 archive (Taylor et al. 2012, BAMS). The number of different models used to form each multi-model mean map is given in the bottom right corner of each map

These are based on so-called Representative Concentration Pathways (RCPs), which provide possible future trajectories of GHG emissions and concentrations. Each RCP is tied into a coherent socioeconomic story line (Wayne 2013). Based on these GHG pathways, climate models then calculate to what extent GHG concentrations perturb earth's energy balance and what implications this forcing has for temperature and other climate variables. RCPs therefore provide a range of standardized inputs into modeling efforts around the world, focused on studying the effects of a range of future emissions choices dependent upon societal decisions. Here, we focus on two pathways, RCP 8.5, which assumes continuation of the GHG emissions path that world societies have been following, and RCP 2.6, which assumes dramatically mitigated emissions. Following RCP 2.6 would likely limit annual global mean air temperature increase to <2 °C above preindustrial levels by the end of the century, and it would allow mean air temperature to decline slightly thereafter. In contrast, if we continue to follow RCP 8.5, annual mean temperature increases are projected to reach 5 °C by the end of this century (Wayne 2013). Following our current path of unabated GHG rise (RCP 8.5) also would mean annual temperatures over the world's ocean surface are likely to increase by 2.4 °C between now and the end of this century. Average temperatures in some regions will be far higher, but few areas will warm less than 0.5 °C (Fig. 2.3). Rahmstorf et al. (2015) reported that the subpolar North Atlantic is one of the very few areas of the world to have cooled in recent decades. This apparently is due to a slowing of the Atlantic Ocean overturning currents that brings warm surface water into the region. This slowdown is possibly triggered by increased buoyancy due to warming and freshening (e.g., by meltwater from the Greenland ice sheet) of surface waters. If we maintain business-as-usual emissions (RCP 8.5), ocean cooling in the North Atlantic will be overwhelmed by this century's end, as a result of the warming of the atmosphere above, but warming there still will be less than most other areas of the ocean.

In contrast to continuing along the RCP 8.5 emissions pathway, following RCP 2.6 would take ocean temperatures on a far cooler path, with end-of-century SST increasing only 0.6 °C. That is, average SST warming on our current path will be four times what it could be if society adopted the RCP 2.6 pathway. Committing to the RCP 2.6 mitigation scenario would also minimize stratification of the upper ocean. The Bering Sea is on the other end of the temperature trend spectrum from the North Atlantic. Currently among the most productive seas of the world, the Bering Sea is projected to warm more than most other regions regardless of which emissions path we take (Figs. 2.3 and 2.4). Such warming is sure to impact the welfare of marine mammals and other marine biota.

Perhaps as important as the rise in annual average temperatures is the range of extremes that will be experienced. In some areas, like the Arabian Sea where SSTs historically have fluctuated little, seasonal and interannual variation is expected to continue to be small. In other geographic regions, seasonal or single year natural fluctuations will result in temporary periods during which high temperatures are well above the long-term mean trends. In the Bering Sea, for example, under RCP

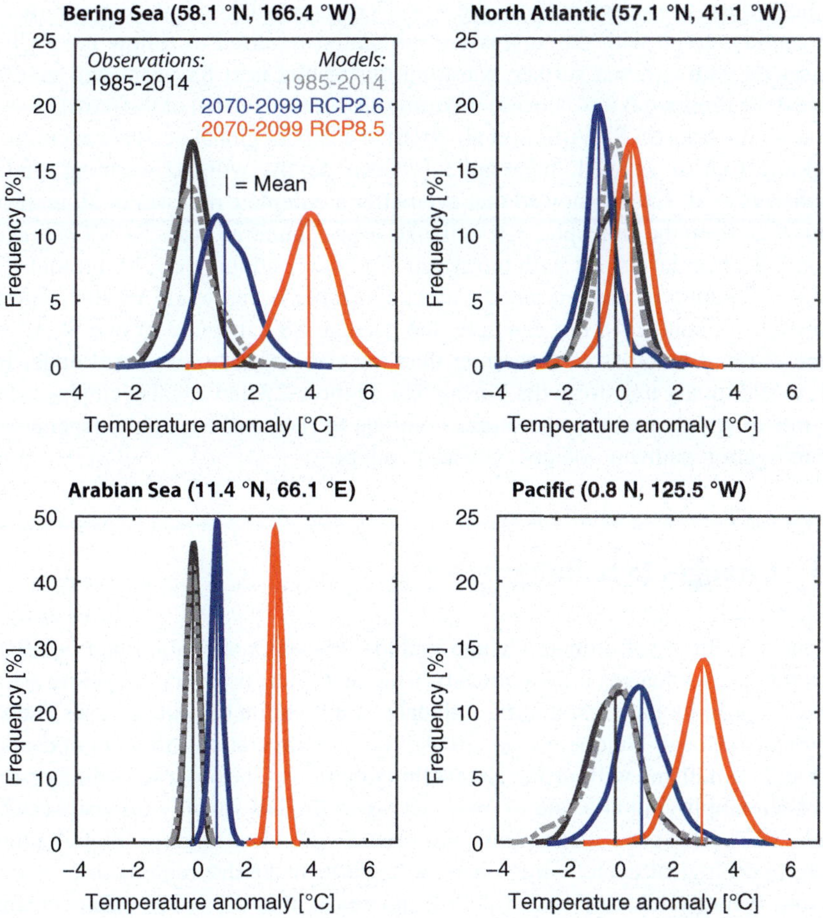

Fig. 2.4 Distribution of observed and projected sea surface temperature anomalies for four disparate oceanic regions. Frequency of occurrence (%) expressed as kernel-smoothed histograms of the distribution of sea surface temperature from each 30-year period. Note the different *y*-axis scale for the Arabian Sea, a tropical basin with little interannual variability in temperature. The climate models, emissions scenarios, and time periods used are the same as in Fig. 2.3

8.5, future annual mean temperatures are projected to be 4 °C warmer than at present, while an individual year might be 6 °C warmer than present norms (Fig. 2.4). Recent observations illustrate how seasonal and annual temperature extremes, on top of already warmer average conditions, could have major impacts on biota. In early 2016, unusually high, in historic terms, surface water temperatures caused coral bleaching across 93% of the Australian Great Barrier Reef. Investigators concluded that reaching those high water temperatures would have been nearly impossible without the chronic warming associated with rising GHG concentrations (King et al. 2016).

Regardless of societal actions, the world's oceans will continue to warm through the century. Even with the aggressive mitigation required to follow the RCP 2.6 emissions pathway, sea surface warming during the next 85 years will match or exceed the reported 0.6 °C surface warming (Riser et al. 2016) of the past 135 years. Because it tracks SST (Wijffels et al. 2016), we expect global ocean heat content to continue to rise as well, meaning all ocean depths will be warming. Indeed, Mathesius et al. (2015) showed that even after a complete removal of all anthropogenic CO_2 from the atmosphere, it will take several hundred years for SSTs to return to preindustrial levels. As with terrestrial regions, continuing on our present GHG emissions path will mean a largely unrecognizable ocean world by this century's end (Figs. 2.3 and 2.4). For example, the average SST increase of over 4 °C, with some years as much as 6 °C warmer than the current mean, projected under RCP 8.5, would totally transform the Bering Sea. Figures 2.3 and 2.4 also emphasize that the rate at which ocean temperatures continue to rise, will be highly dependent on the mitigation pathway society chooses to adopt.

2.4 Changes in Salinity

Secondary effects of climate change include altered precipitation and circulation patterns. These changes have a direct bearing on ocean chemistry, including salinity patterns. Salinity patterns, in turn, influence stratification of water masses and vary regionally (Hoegh-Guldberg et al. 2014). Rising temperatures affect both evaporation and rainfall and will further alter salinity patterns. The observed salinity pattern has been amplifying at a rate of 16% $°C^{-1}$ over the last roughly 50 years (Durack et al. 2012). This pattern of amplification reveals the clear fingerprint of an intensifying hydrological cycle, which in turn, tends to make dry regions drier and wet regions wetter (Held and Soden 2006). In oceans, this has led to enhancement of historic salinity patterns, with evaporation-dominated midlatitudes becoming more saline, while relatively fresh surface waters in rainfall-dominated tropical regions and polar regions have become fresher (Durack et al. 2012). We can anticipate even more dramatic exaggeration of ocean salinity patterns in the future. Much of the surface area of the Atlantic Ocean and large swaths in the southern Pacific are expected to become much more saline by the end of the century, while most of the Pacific Ocean and high-latitude areas will freshen (Fig. 2.5). The contrast between profound freshening in the North Atlantic and Arctic Ocean and salinity increases in most of the rest of the Atlantic will undoubtedly have major ramifications for biota. Also, because models historically have underestimated the observed rate of salinity changes (Durack et al. 2012), the scale and regional contrasts of future salinity patterns may be far greater than shown in Fig. 2.5. If society adopts an emissions pathway similar to RCP 2.6, globally averaged freshening of surface waters would only be 1/3 of what it would be under RCP 8.5.

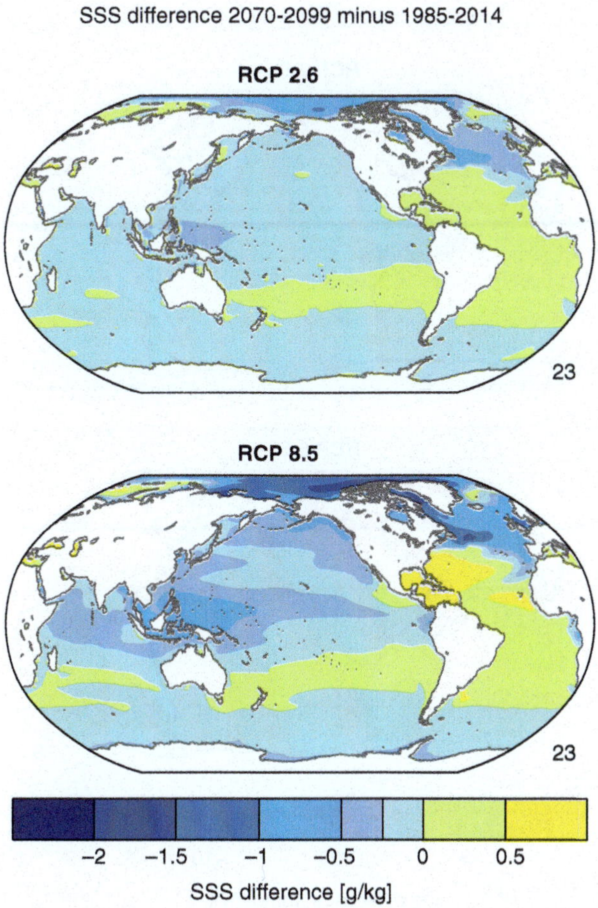

Fig. 2.5 As Fig. 2.3, but for sea surface salinity (SSS). Most of the ocean surface is projected to freshen as a consequence of an intensifying hydrological cycle and changes in ocean currents. Increased freshwater runoff from the melting of large ice sheets is typically not yet included in these climate models and would lead to additional discharge of freshwater (also contributing to sea level rise)

2.5 Changes in Oxygen Concentrations

The mean dissolved oxygen concentration in global oceans currently is ~162 μmol kg^{-1} (~162 mmol m^{-3}). This concentration varies widely among oceanic regions, with some Antarctic waters supersaturated at over 500 μmol kg^{-1}, while some coastal sediments and deep layers in the Black Sea and Cariaco Basin are essentially depleted of oxygen (Pörtner et al. 2014). Warmer water holds less oxygen and oceans globally are projected to see dramatic declines in oxygen content

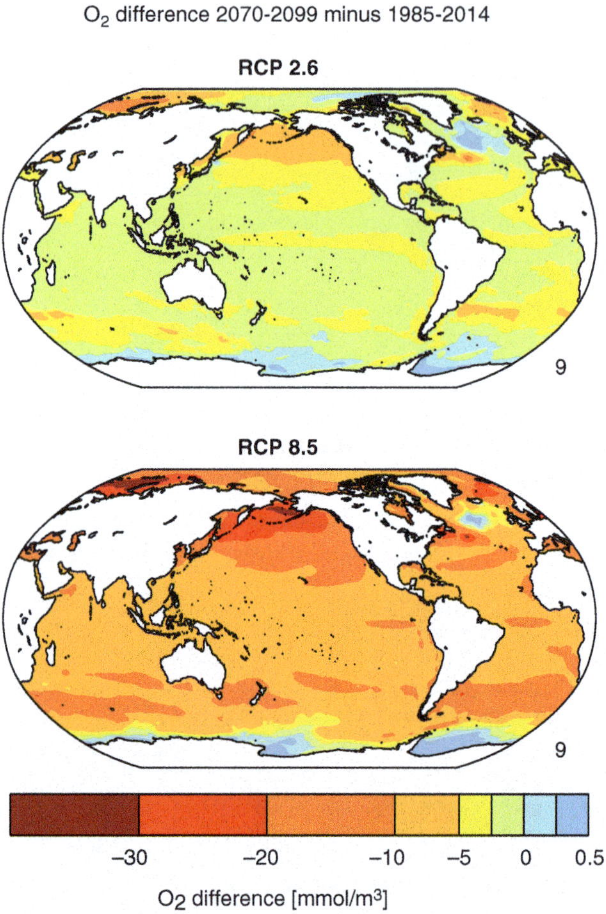

Fig. 2.6 As Fig. 2.3, but for oxygen (O_2) concentration at the ocean surface. As climate change leads to a warmer and more stratified ocean, O_2 concentrations will decline almost everywhere. O_2 reductions are also expected below the surface, leading to an expansion of so-called oxygen minimum zones, where O_2 concentrations are too low for most aerobic biota

regardless of our emissions path (Fig. 2.6). As global ocean oxygenation declines, regional contrasts also will be enhanced. The highly productive waters of the North Pacific, Bering, and Barents seas will see strong declines in oxygen available in the water column, while the subpolar North Atlantic and Antarctic seas will continue to become oxygen enriched.

The mean surface ocean oxygen concentration, between now and the end of the century, is projected to decline by 3.7% if we continue along the RCP 8.5 emissions path. On top of general declines in oxygen content with rising temperatures, increased stratification and other factors are expected to expand near shore and pelagic hypoxic zones. Oxygen concentrations below 60 μM kg^{-1} are lethal to >50% of benthos. These hypoxic areas currently include ~5% of global ocean volume (Deutsch et al. 2011). As surface layers warm, and in some areas become less saline, stratification can enhance hypoxic conditions (Deutsch et al. 2011). Assuming we continue to follow RCP 8.5, Bopp et al. (2013) projected up to 30% increase in suboxic (low oxygen carrying) waters by 2100 as a result of various global warming influences on ocean water structure and chemistry. Following RCP 2.6 through the century, however, mean global oxygen concentrations would decline by only 1% from present values, with far less drastic regional gradients (Fig. 2.6).

2.6 Changes in Ocean Acidity

Atmospheric concentrations of carbon dioxide (CO_2) have been rising, and ~30% of anthropogenically derived CO_2 has been absorbed by the ocean (Hoegh-Guldberg et al. 2014). Rising concentrations of CO_2 increase carbonic acid concentrations and acidify the ocean. Globally, the surface ocean pH declined 0.1 points (from 8.25 to 8.14) between 1751 and 2004 (Jacobson 2005). The current ocean pH ranges from 7.8 to 8.4 (Pörtner et al. 2014) and has been decreasing at a rate of −0.0013 to −0.0024 pH units per year (Pörtner et al. 2014). The observed ocean acidification (OA) rate varies greatly on a regional basis and, for example, is 50% greater in the northern Atlantic than the subtropical Atlantic (Olafsson et al. 2009). Reduced salinities due to freshwater from ice melt or precipitation can exacerbate OA by reducing availability of buffers occurring in more saline waters (Jacobs and Giulivi 2010; Vélez-Belchí et al. 2010). OA also is more severe in cold regions, which have a higher sea-air flux rate for CO_2, and because cold waters have a lower buffer capacity than warmer waters, although, both factors can vary greatly on a seasonal basis (Olafsson et al. 2009). If we continue to follow our current emissions pathway (RCP 8.5), mean surface ocean pH is projected to decline by ~0.28, from the present global mean of 8.08 to 7.80. This 3.5% decline in less than a century dwarfs the rate of pH decline Jacobson (2005) reported for the previous 250 years. Observed regional gradients also will be further enhanced (Fig. 2.7), with ramifications for large segments of ocean biota.

Of all the changing ocean chemistries, the CO_2 impact may be most significant. Higher aqueous CO_2 concentrations result in decreased carbonate ion concentrations and make it more difficult for marine organisms to form biogenic calcium carbonate ($CaCO_3$). Increasing solubility of the forms of calcium carbonate (calcite, magnesium calcite, and aragonite) that are critical components of marine organisms' shells and skeletons has important ramifications for ocean biota (Orr et al. 2005). Numerous studies have attempted to evaluate impact of various levels of OA on oceanic biota. Olafsson et al. (2009) concluded, as a result of ongoing OA, aragonite solubility has increased. Large areas of benthos that historically lived in an environment where aragonite was supersaturated, are becoming undersaturated. Controlled measurements of the impact of that undersaturation, however, are still needed. Hoegh-Guldberg et al. (2014) and Pörtner et al. (2014) provide extensive examples of the complications

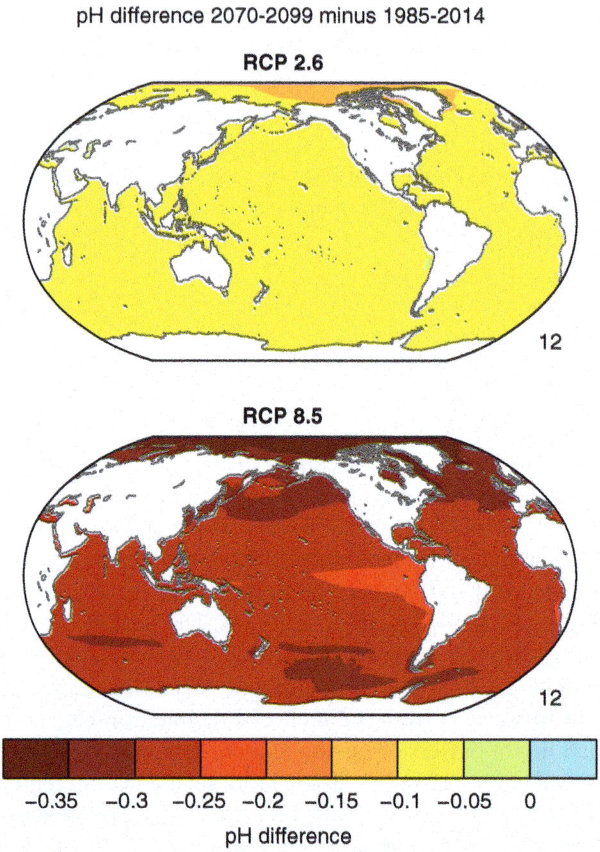

Fig. 2.7 As Fig. 2.3, but for pH at the ocean surface. The lower the pH, the more acidic the water. The projected decrease of pH in all basins is a direct consequence of the increased uptake of CO_2 by the ocean. Indeed, this is one of the most robust features of climate change, as it is a well-understood chemical process. Note that the seemingly small change under emissions scenario RCP 2.6 in fact already constitutes a significant stress for some calcifying ocean organisms

involved in such assessments, reporting a mixture of effects. Some species appeared unaffected by observed or experimental levels of OA, some showed negative impacts, and some appeared to benefit. In the long run, however, if OA continues, negative impacts on marine calcifiers are expected to dominate. Ridgwell and Schmidt (2010) calculated that we currently are on an OA path unmatched in the last 65 million years and that impacts on marine calcifiers are likely to be severe. Our projections (Fig. 2.7) make it clear that overall pH declines will be exacerbated at higher latitudes, and Orr et al. (2005) projected that detrimental effects, in those cooler, less buffered, regions, could be noticeable within decades. As with the other variables we have examined, the benefits of GHG mitigation are clear. Following RCP 2.6 would result in only 0.06

(0.76%) decline in global ocean pH, with much less amplification of regional gradients (Fig. 2.7). The potential for dramatically lowered pH to have major negative impacts on ocean productivity, and the ability of society to avoid the most profound declines, provide strong incentive to mitigate GHG rise.

2.7 Cumulative Effects

As with pH, biological ramifications of change in temperature, oxygen concentration, and salinity are individually varied and complicated and are yet to be fully understood. A full understanding of all of the complications, however, is not necessary to hypothesize a future negative trend for ocean biota. On our current emissions path, oceans will be profoundly different places than they are today. The cumulative effects of end-of-century pH decreases, along with anticipated reductions in oxygenation, changes in salinity, and warmer temperatures, are most likely to be negative with regard to forms of ocean life that we value and to which we have become accustomed (Hoegh-Guldberg et al. 2014; Pörtner et al. 2014). Bopp et al. (2013) summarized these ongoing ocean changes and projected they would result in an ~8% decline in ocean net primary productivity by 2100, if we continue to follow current emissions pathways. However, changes in all of the parameters we examined would be far less severe if society were to adopt significant and sustained emissions reductions. Bopp et al. (2013) estimated that following the RCP 2.6 path would mean only a ~2% decline in net primary productivity by the end of the century.

2.8 Impacts on Marine Mammals

Because subsequent chapters will describe examples of observed and expected impacts on individual marine mammal species, or groups of species, we provide only a few examples here. It is clear, however, that the anthropogenic ocean changes described in previous sections can affect marine mammal welfare in multiple ways. Most impacts on marine mammals will likely reflect biological productivity and food availability as mediated by changing temperatures, ocean structure, and productivity. We cannot, however, rule out more direct impacts of rising temperatures, especially in regions where the greatest water temperature rise is projected. The observed distribution of marine mammal pursuit predators may be a harbinger of direct effects of rising temperature. Because fish are ectotherms and can swim faster at warmer temperatures, ocean warming could increase energetic costs of underwater pursuits for marine mammals. Cairns et al. (2008) concluded that marine mammal (and bird) predators are limited, by the metabolic expense of pursuit, to waters cooler than ~20 °C. Similarly,

McIntyre et al. (2011) noted that southern elephant seals (*Mirounga leonina*) consistently dove deeper and stayed down longer in areas where waters were warmer. Early impacts of warming on marine mammals may therefore include range contractions to higher latitudes, as well as altered and presumably less efficient vertical stratification of foraging efforts.

Polar bears (*Ursus maritimus*) are largely restricted to catching their seal prey from the surface of the sea ice (Amstrup 2003), and there is a linear relationship between sea ice extent and global mean temperature (Amstrup et al. 2010). Declining availability of sea ice has been linked to reduced body condition, survival, and population size (Rode et al. 2010; Regehr et al. 2007). Although some species may respond positively to changes in Arctic marine productivity as sea ice cover is reduced (Crawford et al. 2015), polar bears will not have access to that productivity without the sea ice platform. And, given ongoing ocean changes, any improvements in productivity are likely to be only temporary. Also, polar bears are not likely to compensate for lost sea ice access by taking advantage of terrestrial food sources (Rode et al. 2015). Their dependence on the surface of the ice for catching prey, therefore, translates into a direct relationship between rising temperatures and polar bear food availability, regardless of potential changes in marine productivity.

Marine mammals that are tied to specific haul-out sites or rookeries could encounter higher foraging costs if altered prey distributions require longer foraging trips (Péron et al. 2012; Hazen et al. 2013). On the other hand, if prey distributions become constrained by thermal stratification or hypoxic zones, marine mammal foraging may, temporarily be enhanced by localized concentrations of prey (Hazen et al. 2009). Sea level rise (from thermal expansion and freshwater ice melt—direct consequences of rising temperatures), combined with altered prey availability, will, in the long run, negatively affect most species with high fidelity to specific locales (e.g., haul-out sites and rookeries).

Warming of ice-covered waters will alter species distributions, which could make alternate prey available but also could increase competition and even introduce new predation risks for high-latitude species adapted to ice-covered seas. Moore and Huntington (2009) hypothesized that subarctic cetaceans will move north as sea ice extent declines and open water seasons lengthen. As a result of recent declines in the spatial and temporal extent of sea ice in Hudson Bay and Hudson Strait, killer whale (*Orcinus orca*) sightings in Hudson Bay are on the increase (Oosthoek 2012), exposing resident marine mammals (and their prey) to a new predation risk. Polar bears in the Davis Strait region of Eastern Canada maintained high numbers into the early 2000s despite declining sea ice availability. There, polar bears appear to have offset some of their dependence on ringed seals (*Phoca hispida*), which themselves depend on relatively solid ice cover, with harp seals (*Pagophilus groenlandicus*) which prefer a more broken ice edge habitat. This may reflect a shift of harp seal distribution in response to northerly movement of the sea ice conditions they require for whelping (Peacock et al. 2013). Also it is likely to be a temporary condition with the harp seals following remaining ice as it continues a northerly retreat.

2.9 Conclusions

Here, we have examined ongoing ocean changes that will profoundly affect the future welfare of marine mammals and the marine environments that support them. We increasingly are aware of how climate change already has altered environments on land and at sea, with cascading impacts on welfare of the animals those environments support. Different climate models, as well as different simulations from the same model, provide a number of possible futures for any given GHG path society may take. It is important to recognize that although models project a wide range of possible future paths, we will get to realize only one path in real life. If we are lucky, our actual realization may be similar to models on the low end of the projected severity scale. All model outcomes, however, that do not include a halt to the increase in atmospheric GHG concentrations, predict a future ocean system that will be continually changing in unfavorable ways. As long as GHG concentrations increase, we will not see stability return to ocean temperatures, sea ice extent, oxygen concentrations, or pH, and marine mammals as well as other ocean biota will continually struggle to keep pace with an environment changing faster than it has in millennia. In other words, there will be no sustainable future—ocean biota and human lives depending on it, always, will be shooting at a moving target.

Although we can point to specific examples of marine mammal response to individual climate drivers, the ultimate threat anthropogenic ocean change poses for marine mammal welfare will be changes in their supporting food web caused by the cumulative effects of changing ocean temperature and chemistry. As a result of rising temperatures, lowered pH, and reduced oxygen concentration, Bopp et al. (2013) projected strong declines in global ocean net primary productivity through this century. Because these negative trends will persist over multi-centennial time frames (Mathesius et al. 2015), the impact of anthropogenic ocean change must be recognized as the consummate challenge to future welfare of all marine mammals and the ocean habitats supporting them.

The good news is that the most significant contributor to future uncertainty is in our hands. We cannot control the natural variation in the climate system (Fig. 2.1). We can, however, control the slope of the rising baseline. We can choose to keep our climate in “runaway” mode, we can choose a more gradual slope, or ideally we can choose a path (like RCP 2.6) that stops the rise in GHG emissions and bends our current upward slope to a new level baseline. Following RCP 2.6 rather than RCP 8.5 would mean less than one quarter of the global ocean SST increase toward which we are now heading. It also would result in one third of the change in ocean salinity, one quarter of the pH decline, and less than one third of the decline in ocean oxygenation. In other words, we could avoid the worst of oceanic changes that future global warming has to offer. Perhaps most importantly, following RCP 2.6 does not just reduce near-term impacts. On multi-centennial time scales, temperatures, pH, and oxygen saturation will stabilize on the RCP 2.6 emissions scenario (Mathesius et al. (2015); (Fig. 2.8)). But the urgency of action cannot be overstated. Procrastination now will assure catastrophe later. If society waits to address

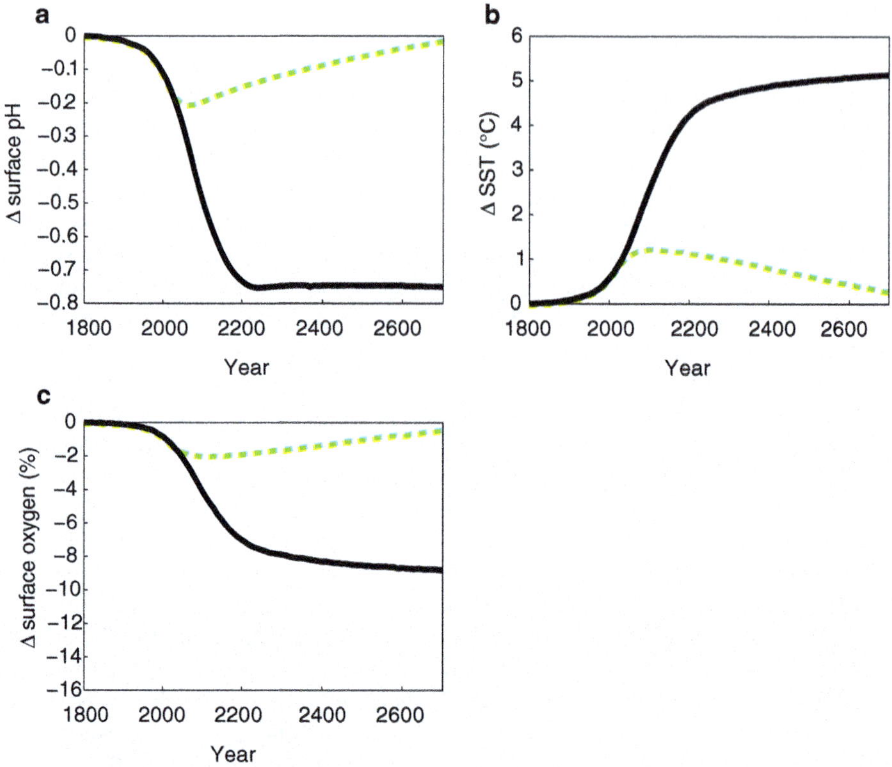

Fig. 2.8 Anomalies of globally averaged ocean variables (excerpted from Fig. 2 in Mathesius et al. 2015). Trajectories for RCP 8.5 (*black*) and RCP 2.6 (*green*), showing globally averaged anomalies of: (**a**) surface pH, (**b**) sea surface temperature (SST), and (**c**) surface dissolved oxygen. All anomalies were calculated with respect to year 1800

emissions challenges until the summer sea ice disappears, or other thresholds critical to marine mammals are exceeded, it is unlikely policy makers will have time or resources to think about, or prioritize, the welfare of marine mammals. By then, food and water shortages, refugee crises, and other human welfare challenges may trump all conservation concerns. Our current path clearly is not in the best interest of marine mammal welfare. We can assure a better future for marine mammals and the rest of us, but, time is of the essence!

References

Amstrup SC (2003) Chapter 27: Polar bear, *Ursus maritimus*. In: Feldhamer GA, Thompson BC, Chapman JA (eds) Wild mammals of North America: biology, management, and conservation. John Hopkins University Press, Baltimore, pp 587–610

Amstrup SC, DeWeaver E, Douglas DC et al (2010) Greenhouse gas mitigation can reduce sea-ice loss and increase polar bear persistence. Nature 468:955–958

Bopp L, Resplandy L, Orr JC et al (2013) Multiple stressors of ocean ecosystems in the 21st century: projections with CMIP5 models. Biogeosciences 10:6225–6245

Cairns DK, Gaston AJ, Huettmann F (2008) Endothermy, ectothermy and the global structure of marine vertebrate communities. Mar Ecol Prog Ser 356:239–250

Crawford JA, Quakenbush LT, Citta JJ (2015) A comparison of ringed and bearded seal diet, condition and productivity between historical (1975–1984) and recent (2003–2012) periods in the Alaskan Bering and Chukchi seas. Prog Oceanogr 136:133–150

Deutsch C, Brix H, Ito T, Frenzel H, Thompson L (2011) Climate-forced variability of ocean hypoxia. Science 333:336–339. doi:10.1126/science.1202422

Durack PJ, Wijffels SE, Matear RJ (2012) Ocean salinities reveal strong global water cycle intensification during 1950–2000. Science 336:455–458

Gleckler PJ, Durack PJ, Stouffer RJ, Johnson GC, Forest CE (2016) Industrial-era global ocean heat uptake doubles in recent decades. Nat Clim Chang 6:394–398

Hansen J, Sato M (2004) Greenhouse gas growth rates. Proc Natl Acad Sci U S A 101(46): 16109–16114

Hazen EL, Craig JK, Good CP, Crowder LB (2009) Vertical distribution of fish biomass in hypoxic waters on the Gulf of Mexico shelf. Mar Ecol Prog Ser 375:195–207

Hazen EL, Jorgensen S, Rykaczewski RR et al (2013) Predicted habitat shifts of Pacific top predators in a changing climate. Nat Clim Chang 3:234–238

Held IM, Soden BJ (2006) Robust responses of the hydrological cycle to global warming. J Clim 19:5686

Hoegh-Guldberg O, Cai R, Poloczanska ES et al (2014) The Ocean. In: Climate change 2014: impacts, adaptation, and vulnerability. Part B: regional aspects. Contribution of working group II to the fifth assessment report of the Intergovernmental Panel on Climate Change. Cambridge University Press, Cambridge, New York, pp 1655–1731

IPCC (2013) Summary for policymakers. In: Climate change 2013: the physical science basis. Contribution of working group I to the fifth assessment report of the Intergovernmental Panel on Climate Change. Cambridge University Press, Cambridge, New York

Jacobs SS, Giulivi CF (2010) Large multidecadal salinity trends near the Pacific-Antarctic continental margin. J Clim 23(17):4508–4524

Jacobson MZ (2005) Studying ocean acidification with conservative, stable numerical schemes for nonequilibrium air-ocean exchange and ocean equilibrium chemistry. J Geophys Res 110:D07302

King A, Karoly D, Black M et al (2016) Great barrier reef bleaching would be almost impossible without climate change. https://theconversation.com/great-barrier-reef-bleaching-would-be-almost-impossible-without-climate-change-58408. Accessed 7 April 2017

Lutgens FK, Tarbuck EJ (2004) The atmosphere. Pearson Prentice Hall, Upper Saddle River, NJ, 508 pp

Mathesius S, Hofmann M, Caldeira K, Shellnhuber HJ (2015) Long-term response of oceans to CO_2 removal from the atmosphere. Nat Clim Chang 5:1107–1114

McIntyre T, Ansorge IJ, Bornemann H et al (2011) Elephant seal dive behaviour is influenced by ocean temperature: implications for climate change impacts on an ocean predator. Mar Ecol Prog Ser 441:257–272

Moore SE, Huntington HP (2009) Arctic marine mammals and climate change: Impacts and resilience. Ecological Applications, 18(2) Supplement, 2008, pp. S157–S165

NOAA (National Oceanic and Atmospheric Administration) (2015) Global analysis - November 2015. http://wwwncdcnoaagov/sotc/global/201511. Accessed 7 April 2017

Olafsson J, Olafsdottir SR, Benoit-Cattin A, Danielsen M, Arnarson TS, Takahashi T (2009) Rate of Iceland Sea acidification from time series measurements. Biogeosciences 6;2661–2668. www.biogeosciences.net/6/2661/2009/

Oosthoek S (2012) Invasion of the bowhead snatchers. New Sci 213(2858):37–39

Orr JC et al (2005) Anthropogenic ocean acidification over the twenty-first century and its impact on calcifying organisms. Nature 437(7059):681–686

Peacock E, Taylor MK, Laake J, Stirling I (2013) Population ecology of polar bears in Davis Strait, Canada and Greenland. J Wildl Manage 77(3):463–476

Péron C, Weimerskirch H, Bost CA (2012) Projected poleward shift of king penguins' (*Aptenodytes patagonicus*) foraging range at the Crozet Islands, southern Indian Ocean. Proc Biol Sci 279(1738):2515–2523

Pierrehumbert RT (2011) Infrared radiation and planetary temperature. Phys Today 64(1):33–38

Pörtner HO, Karl DM, Boyd PW et al (2014) Ocean systems. In: Climate change 2014: impacts, adaptation, and vulnerability. Part A: global and sectoral aspects. Contribution of working group II to the fifth assessment report of the Intergovernmental Panel on Climate Change. Cambridge University Press, Cambridge, New York, pp 411–484

Rahmstorf S, Box JE, Feulner G et al (2015) Exceptional twentieth-century slowdown in Atlantic Ocean overturning circulation. Nat Clim Chang 5:475–480

Regehr EV, Lunn NJ, Amstrup SC, Stirling I (2007) Effects of earlier sea ice breakup on survival and population size of polar bears in western Hudson Bay. J Wildl Manag 71(8):2673–2683

Ridgwell A, Schmidt DN (2010) Past constraints on the vulnerability of marine calcifiers to massive carbon dioxide release. Nat Geosci 3:196–200

Riser SC, Freeland HJ, Roemmich D et al (2016) Nat Clim Chang 6:145–153

Rockström J et al (2009) A safe operating space for humanity. Nature 461:472–475

Rode KD, Amstrup SC, Regehr EV (2010) Reduced body size and cub recruitment in polar bears associated with sea ice decline. Ecol Appl 20:768–782

Rode KD, Robbins CT, Nelson L, Amstrup SC (2015) Can polar bears use terrestrial foods to offset lost ice-based hunting opportunities? Front Ecol Eviron 13(3):138–145

Steinacher M, Joos F, Stocker TF (2013) Allowable carbon emissions lowered by multiple climate targets. Nature 499:197–201

Taylor KE, Stouffer RJ, Meehl GA (2012) An overview of CMIP5 and the experiment design. Bull Am Meteorol Soc 93:485–498, doi:10.1175/BAMS-D-11-00094.1

Vélez-Belchí P, Hernández-Guerra A, Fraile-Nuez E, Benítez-Barrios V (2010) Changes in temperature and salinity tendencies of the upper subtropical North Atlantic ocean at 24.5°N. J Phy Oceanogr 40(11):2546–2555

Wayne GP (2013) The beginners guide to representative concentration pathways. https://skepticalscience.com/docs/RCP_Guide.pdf. Accessed 7 April 2017

Wijffels S, Roemmich D, Monselesan D et al (2016) Ocean temperatures chronicle the ongoing warming of earth. Nat Clim Chang 6:116–118

Chapter 3
Of Poisons and Plastics: An Overview of the Latest Pollution Issues Affecting Marine Mammals

Mark Peter Simmonds

> *The most alarming of all man's assaults on the environment is the contamination of air, earth, rivers and sea with dangerous and even lethal materials.*
>
> Rachel Carson (1962)

Abstract Persistent organic pollutants were recognised decades ago as significant threats to wildlife including marine mammals. Efforts to control certain pesticides and polychlorinated biphenyls (PCBs) and associated successful declines in environmental loadings followed. However, it has very recently become apparent that PCBs continue to pose a significant threat. This is especially the case for certain cetacean populations in Europe which now seem to be heading towards extinction because of PCBs-induced reproductive failure. The effects of such pollution on marine mammal health are a significant welfare concern, and urgent efforts to stem any further movement of PCBs into the oceans are now required. In addition, marine debris is a fast-growing threat to marine wildlife, bringing with it severe welfare concerns for some marine mammal populations. This is only set to get worse as more discarded plastics enter the oceans; again urgent action is advocated.

3.1 Enter the POPs

We live in a time when there are profound fears for the long-term survival of the human race. Given what we now expect from climate change, this concern is certainly well founded. However, it is only one *horseman* of what might be termed our

---

M.P. Simmonds
Senior Marine Scientist, Humane Society International, c/o 5 Underwood Steet, London N1 7LY, UK

Visiting Fellow, University of Bristol, School of Veterinary Sciences, Langford House, Langford, Bristol, BS40 5DU, UK
e-mail: mark.simmonds@sciencegyre.co.uk

A. Butterworth (ed.), *Marine Mammal Welfare*, Animal Welfare 17,
DOI 10.1007/978-3-319-46994-2_3

modern *apocalypse*.[1] The ongoing, precipitous and accelerating loss of biodiversity is another profound threat both to animal life and potentially ultimately to humans too, and pollution appears to be a primary driver for this.

Pollution comes in many forms, even including energy in the form of marine noise, but is more usually recognised in the shape of oil spills, sewage, radioactive substances and agricultural and industrial wastes. Oil spills in particular are highly visible environmental assaults and capture the public eye—and it is probably worth at least noting here the latest evidence that shows that there can be long-term effects of oil spills on marine mammals (e.g. Colegrove et al. 2016). However, the focus of this chapter is the invisible and insidious threat posed by certain xenobiotics (substances that do not naturally occur), starting with the persistent organic pollutants (POPs). In the late 1980s, these were the focus of my work when I was part of the Greenpeace International Science Unit, based at Queen Mary College in London. Much of our work concerned investigating and highlighting the threats from the POPs, and, at that time, Greenpeace and other organisations ran campaigns focused on controlling this group of toxic compounds which included many pesticides, such as dichlorodiphenyltrichloroethane (DDT), and the industrial compounds known as the PCBs (polychlorinated biphenyls).

DDT first came into use in 1939. It was cheap and widely used in Europe after World War II to eradicate fleas (which carried the disease typhus) and in Asia against the mosquito vectors of malaria (Parsons et al. 2013). By the 1960s, DDT could be found throughout the marine environment and even in the bodies of Antarctic wildlife. DDT was the threat that inspired Rachel Carson to write *Silent Spring* (Carson 1962), a ground-breaking book helped to bring about a ban on DDT in her native USA and encouraged an eventual ban in most countries by the 1970s. Her main theme was that DDT was not targeted in its impacts and severely affected many nontarget species.

The PCBs have a similar history, although they were never used as pesticides. Instead they were used in many other processes, including in electrical transformers and capacitors, as flame retardants and as additives in various construction and packaging materials (Parsons et al. 2013). Bans came into place from the late 1970s but, by then, much of the world production (some 1.5 million tonnes) had already made its way into the oceans. DDT and the PCBs are that part of the POP 'family' known as organochlorines, organic compounds with chlorine atoms attached. The principle problems that they present are that they are immunosuppressive—causing animals and people to be more susceptible to disease—and that they can also act as hormone mimics, leading to reproductive abnormalities (Reijnders 1996; Parsons et al. 2013; Jepson and Law 2016). They are also fat soluble and so tend to accumulate in the ample fatty tissues of marine mammals—mainly in their blubber. They may be released from this energy store during times when the animals are not feeding (e.g. on prolonged migrations) or when they are lactating.

[1] The Four Horsemen of the Apocalypse are described in the last book of the New Testament of the Bible, 'The Book of Revelation', and are often seen as harbingers of the Last Judgment and the end of the world.

3.2 The Ongoing Threat of Persistent Organic Pollutants

The levels of contaminants in the tissues of marine mammals have been documented over many decades along with associated concerns about their potential long-term impacts on populations (e.g. Holden 1978; Reijnders 1988; Tanabe and Tatsukawa 1991; Simmonds 1992; Reijnders and Simmonds 2003). Associations have been drawn between tissue concentrations and a range of health concerns, for example, for Baltic grey seals (*Halichoerus grypus*), Californian sea lions (*Zalophus californianus*), Wadden Sea harbour seals (*Phoca vitulina*), bottlenose dolphins (*Tursiops truncatus*) in the USA, striped dolphins (*Stenella coeruleoalba*) in the Mediterranean and beluga whales (*Delphinapterus leucas*) in the St. Lawrence River (Reijnders 1996). Some of these associations concern reproductive health, including implantation failure, foetal death (abortion) and abnormalities in the reproductive tracts of seals and increased first-born calf mortality in bottlenose dolphins. Additionally, severe reproductive dysfunction through the development of cancer and hermaphroditism was reported in St. Lawrence Estuary beluga whales.

It was also very well established that POPs biomagnified up food chains—being present in increased quantities at each trophic level—and that marine mammals feeding in the more polluted waters and at the apex of food chains were the most vulnerable. Toothed whales, being higher in the food chain than baleen whales, are thus typically more at risk, with those that feed on other marine mammals, such as certain populations of orcas (*Orcinus orca*) (Fig. 3.1), likely to carry the heaviest

Fig. 3.1 Polychlorinated biphenyl (PCB) concentrations have been found to have remained at elevated levels in orcas or killer whales (*Orcinus orca*) potentially threatening their survival. *Image credit: Rob Lott*

pollution burdens. POPs levels in tissues also relate to age and sex (Reijnders 1996). Levels in most cetaceans increase with age, until around sexual maturity when females mobilise their blubber to produce very fat-rich milk to feed their calves and, in doing so, also mobilise the pollutants previously held there. Hence calves can be heavily contaminated—and the rapid transfer of pollutants in their mothers' milk may also directly compromise their health.

After most organochlorine pollutants were banned in developed countries in the 1970s and 1980s, levels in the wider environment fell, and some wildlife populations recovered (Jepson et al. 2016). For example, populations and reproductive indices of grey seal, otter and white-tailed sea eagle were recovered in Sweden during the 1980s as tissue PCB and DDT concentrations fell substantially (Jepson and Law 2016). Eventually, as residues levels fell, many 'toxics' campaigners moved on—and environmental organisations moved their focus—believing that their job was done. However—and now we come to the sting in the tale of the PCBs story—some recent studies show that the problem is far from over.

Jepson et al. (2016) looked at PCB concentrations in UK harbour porpoises (*Phocoena phocoena*) (with samples collected from 1990 to 2012) and striped dolphins from the western Mediterranean Sea (specimens collected 1990–2009). In the porpoises, the PCBs declined slowly from 1990 to 1998 and then remained relatively stable from 1998 to 2012. In the Mediterranean striped dolphins, there was a marked decline from an initial peak in 1990, but then levels stabilised from 2003 to 2008. Most significantly, for many animals, blubber PCB concentrations consistently exceeded established mammalian toxicity thresholds (Fig. 3.2), the assumption being that by exceeding established toxicity limits, there is a strong likelihood that these animals are still being negatively impacted.

The work of Jepson et al. (2016) also established some global contamination hot spots: concentrations in the tissues of bottlenose dolphins and orcas from the northeast Atlantic and in bottlenose dolphins and striped dolphins from the Mediterranean were found to be among the highest recorded in any cetacean globally and markedly exceeded all known PCB toxicity thresholds for marine mammals. The researchers concluded that, in some instances, the animals have such high mean blubber PCB concentrations that they are likely to experience population declines and that these PCB levels could also suppress population recovery. In fact, some small or declining populations of bottlenose dolphins and orcas in the NE Atlantic are already known to have low recruitment (low introduction of new animals through birth), and this is consistent with PCB-induced reproductive toxicity. Whatever other protections are afforded to these 'at risk' groups of marine mammals, if they are unable to properly reproduce and replace their numbers, they will, with time, become extinct.

The same researchers also identified pathological findings that were consistent with increased susceptibility to disease (Jepson et al. 2016). This included macroparasitic and bacterial pneumonias, high lung (upper airway) and gastric macroparasite burdens and generalised bacterial infections (septicaemias). These findings appeared to indicate diminishing health status in these populations, and this prospect raises grave welfare concerns.

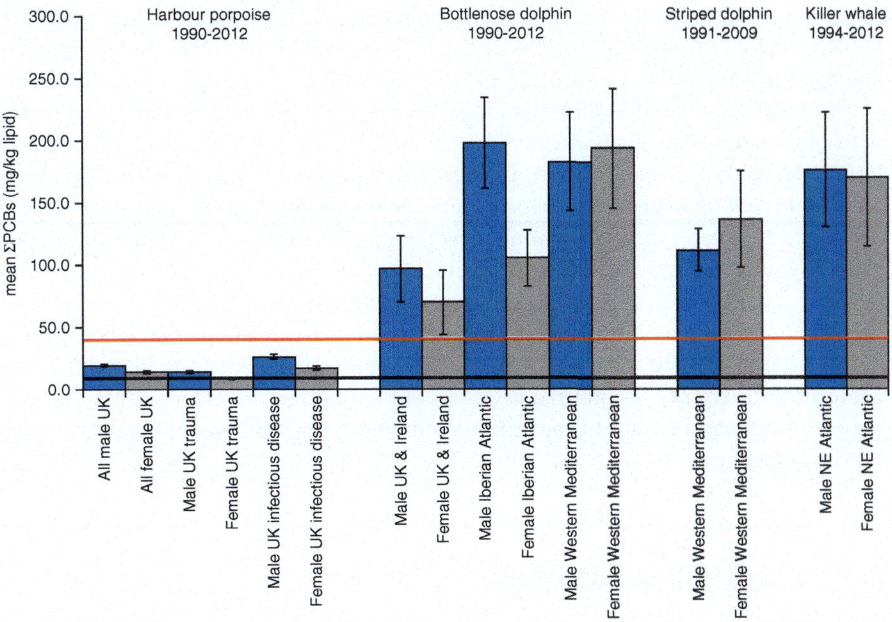

Fig. 3.2 Mean $\sum$PCBs concentrations in male and female cetaceans (four species, all ages). From: Jepson PD et al. (2016). The *blue bars* are males and the *grey bars* are females. The lower line is the equivalent $\sum$PCBs concentrations threshold (9.0 mg/kg lipid) for onset of physiological effects in experimental marine mammal studies. The upper line is the equivalent $\sum$PCBs concentrations threshold (41.0 mg/kg lipid) for the highest PCB toxicity threshold published for marine mammals based on marked reproductive impairment in ringed seals in the Baltic Sea. For full details of the animals sampled, please see original paper

The chronic and widespread nature of PCBs has also been underlined in another recent study by Murphy et al. (2015), in which these authors conducted full necropsies, and determined PCB loadings in 329 female harbour porpoises which had stranded in the UK, and were collected from 1990 to 2012. Almost 20% of the sexually mature females showed direct evidence of reproductive failure, and another 16.5% had infections or tumours of the reproductive tract that could contribute to reproductive failure. Overall, measured levels of PCBs in tissues were found to be a significant predictor of mature female reproductive status, adjusting for the effects of confounding variables. Adult female harbour porpoises that died of trauma (e.g. by-catch, caught inadvertently in fishery operations) in UK waters had pregnancy rates up to 50% lower than those normally found in harbour porpoises in more 'pristine' (i.e. less PCB-polluted) regions like Iceland and Greenland.

Looking back to the late 1980s, when Greenpeace and others ran their powerful 'toxics' (antipollution) campaigns, it is now clear that these campaigns were only partially successful. There is an ongoing threat posed by PCBs, and their conservation significance is now reasonably clear and certainly causes enough to redouble efforts to determine why these toxins are being maintained in the environment, to determine where the sources are and to work to urgently stop ongoing inputs of PCBs to the environment—for example, PCBs are still being released from poorly managed waste landfill.

But what are the welfare implications? In my mind this is reasonably clear: animals of course suffer when affected by disease, and thus, the effects of chronic pollution are likely to cause welfare impacts through disease. Do animals also suffer when deprived of young or as a result of impeded ability to reproduce—bearing in mind that this can occur at a number of points, including loss of well-developed unborn young through stillbirth? Does the absence of young animals in a pod of orcas or a school of dolphins alter the pod dynamic, the social structure, or deprive these animals of something that would otherwise positively affect their lives? Surely this is likely to be the case for these highly sophisticated social mammals. These conjectural dilemmas are not readily susceptible to scientific examination, but it is only a small leap of faith to make the assumption that the loss and lack of young causes suffering.

3.3 A New Threat Emerges

The list of different kinds of pollution given in the opening paragraph of this chapter omits one important and clearly growing category: marine debris or marine litter. This has been defined as *an environmental, economic, human health and aesthetic problem that poses a complex and multidimensional challenge with significant implications for the marine environment and human activities all over the world* (UNEP 2009). Marine debris is mainly (some 60–80%) made up of plastics (Derraik 2002). Modern human culture has developed a dependency on plastics. They pervade every aspect of our lives—we wear them, encase our technology in them, decorate our homes with them, wrap our food in them and then at some point, often after a short use, discard them. Subsequently they are increasingly found everywhere in our environment. They are arguably not essential to us (previous generations managed perfectly well before the 'plastic revolution'), and we use them mainly for our convenience and comfort.

Not surprisingly, the production of plastic resin (the basic building material of plastic items) has increased 620% since 1975, and the largest market sector is packaging material that is inherently designed for disposal (Jambeck et al. 2015). Each year, at least 8 million tonnes of plastic find their way into the ocean—which is equivalent to dumping the contents of one garbage truck into the ocean every minute (World Economic Forum 2016). If no action is taken, this is expected to increase to two truckloads per minute by 2030 and four per minute by 2050. If nothing changes, the ocean is expected to contain 1 tonne of plastic for every 3 tonnes of fish by 2025, and by 2050, more plastics than fish (by weight) (World Economic Forum 2016). Estimates of the global load of plastic on the open ocean surface are in the order of tens of thousands of tonnes, 100-fold lower than expected based on conservative estimates of plastic released into the ocean from terrestrial sources

(Cózar et al. 2014). It seems that plastics are being removed from the surface through fragmentation and then transfer into food webs and via other, as yet largely undefined, processes. Deep-sea sediments have since been identified as a likely sink for microplastics, and it seems that Arctic Sea ice has also been 'freeze storing plastic', and as the polar ice fields retreat, these plastics are now being released back into the water (Baulch and Simmonds 2015).

The effects of pieces of plastic—especially pieces which include loops—on many animals are both graphic and profoundly affecting to the entangled animals, and images of ensnared turtles and birds have become a tragic and common sight on TV and social media. Allsopp et al. (2007) recorded cetaceans, pinnipeds, turtles and seabirds as all suffering from entanglement and pointed out that pinnipeds were particularly affected (Fig. 3.3). This is born out in the USA where most reports of entanglement in marine debris involved pinnipeds, particularly northern fur seals (*Callorhinus ursinus*) and Hawaiian monk seals (*Neomonachus schauinslandi*), as well as sea turtles (NOAA 2014). However, inconsistencies in defining and distinguishing marine debris from actively deployed fishing gear do present some problems in assessing the origin of the entangling materials.

Many pinniped populations can be seen to be affected by entanglement, with seals with embedded plastic bands or netting observable on many of the sites where seals haul out. For example, at Bird Island, South Georgia, 1033 Antarctic fur seals (*Arctocephalus gazella*) were observed entangled in marine debris between 1989 and 2013 (Waluda and Staniland 2013). Most entanglements involved plastic packaging bands (43%), synthetic line (25%) or fishing net (17%). Juvenile male seals were the most commonly entangled (44%). Pinniped entanglements mainly present as loops of non-biodegradable material encircling the animals' necks, and veterinary

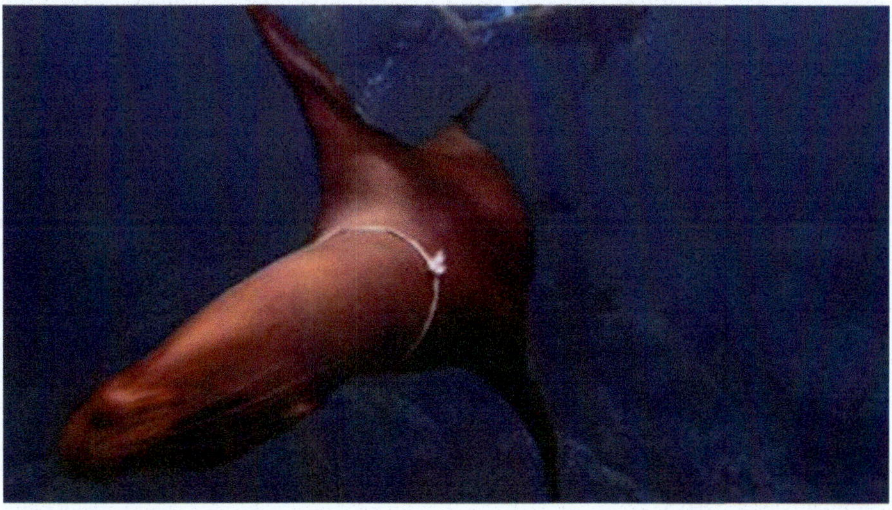

Fig. 3.3 Galapagos fur seal (*Arctocephalus galapagoensis*) with encircling rope entanglement. *Image credit: Juan Pablo Muñoz*

experts agree that these 'neck collars' can create severe welfare concerns as the animals grow (Barnett pers. comm.).

Ingestion of debris is a less well-documented cause of marine mammal morbidity and mortality, although it is well recognised as a health problem in marine birds and turtles, where ingestion of even small quantities of marine debris can have large effects (Jacobsen et al. 2010). Whilst the most obvious potential effect of ingestion is interference with alimentary processes (e.g. physically blocking or perforating the gut), another effect could be that ingested plastics could facilitate the transfer of pollutants, including PCBs, and chemicals associated directly with the plastics, such as plasticisers, into the animals' bodies (Teuten et al. 2009).

A related topic of growing concern is 'microplastics': plastic debris pieces in the size range of 0.3–5 mm (NOAA/UNEP 2011). Two categories are recognised: primary, which are either intentionally produced for direct use (such as exfoliants in personal cleaning products and particles used to blast old coatings from the outside of boats or as precursors to other products, such as pre-production plastic pellets), and secondary, formed from the breakdown of larger plastic materials. This microplastic debris now litters the global environment, and there are growing concerns about its potential impacts, including impacts on filter-feeding marine invertebrates (and hence transference onto higher levels in the food chain) and, more generally, the potential for highly mobile suspended plastic particles to transfer contaminants to wildlife.

The scientific literature relating to interactions between cetaceans and marine debris is increasing, although the issue is relatively difficult to study as many cetacean bodies are never recovered or examined, and this is likely to be especially the case for the deep-diving species which live far offshore (Baulch and Perry 2014). Another complicating factor, as mentioned, is determining whether animals have become entangled in lost or active fishing gear, and this is an important matter to determine as it will affect remedial actions. Many large whale entanglements result from encounters with fishing gear that is in use, and this calls for mitigation to focus on the fisheries concerned, something that should not be confused with entanglement in debris which would require a different response.

Nonetheless, it is growingly apparent that even the mightiest animals can be brought down by our plastic wastes. The first account of ingestion seemingly causing mortality in sperm whales comes from Jacobsen et al. (2010); in 2008, two male sperm whales were stranded along the northern California coast with large amounts of fishing net scraps, rope and other plastic debris in their stomachs. One animal had a ruptured stomach, the other was emaciated, and gastric impaction was suspected as the cause of both deaths. There were an incredible 134 different types of nets in these two animals, all made of floating material, varying in size from 10 cm^2 to about 16 m^2. Jacobsen et al. (2010) concluded that the variability in size and age of the pieces suggested the material was ingested from the surface as debris. A similar case of sperm whale death resulting from debris ingestion has been reported from the Mediterranean, and, in total, four such deaths are known worldwide (de Stephanis et al. 2013).

Whilst having some debris in the alimentary canal is not the same as knowing for sure that the material has had a deleterious effect, ingestion of these alien and potentially harmful materials is now widespread. It has been documented in more than half of all cetacean species, with occurrence as high as 31% in some populations (Baulch and Perry 2014). More generally, in the space of what seems like just a few years, marine debris has moved from something typically viewed as unappealing on the seashore to being viewed as a serious threat to wildlife. Many international initiatives are now underway to address this issue. In 2003, UNEP established a 'Global Initiative on Marine Litter' to provide an *international platform for the establishment of partnerships, co-operation and co-ordination of activities for the control and sustainable management of marine litter*. More recently, in March 2011, UNEP and the United States' National Oceanic and Atmospheric Administration (NOAA) organised the Fifth International Marine Debris Conference in Honolulu, Hawaii (NOAA/UNEP 2011). This meeting adopted the 'Honolulu Commitment', which outlined 12 key actions for the reduction of marine debris and invited international organisations, governments at national and subnational levels, industry, non-governmental organisations, citizens and other stakeholders, to commit to contribute to its development and successful implementation. Such initiatives are very welcome, but we are going to have to work very hard indeed to overcome the colossal avalanche of rubbish still heading to our seas.

At least some of the welfare concerns caused by marine debris are all too obvious. The embedded loops of material that affect a marine mammal that is entangled with fishing gear (see Chaps. 4 and 13) can cause restricted movement and eventually lead to severe wounding, amputations and death, a process that can take months if not years. Similarly the embedded noose around the neck of a growing seal is a severe welfare issue. Blockages or perforations in the gut are less visible, but would it be unreasonable to suggest that the animals suffering from such things would sometimes be in agony? As well as direct impacts on physical health, ingestion of marine debris may also affect an animal's ability to carry out normal feeding (e.g. by suppressing appetite) and other behaviours.

3.4 Conclusions

I have focused here on PCBs and marine debris because they seem especially important at this time. Their ongoing conservation and welfare implications are starting to come sharply into focus, or in the case of the PCBs, perhaps returning into focus. Looking back across the decades, there are some things that we could not have known in the 1980s, such as the remarkable persistence of PCBs despite effective production bans. We also had little knowledge back then of microplastics. An interesting comparison can be made between these two issues—PCBs and marine debris. The threat from PCBs was well established quite long ago. What is new is that it did not go away even after remedial action was taken. By contrast, the full implications of plastics in marine systems are still emerging; the fate of the smallest pieces and the role of microdebris in transfer of contaminants have only recently started to be studied and may prove to be a very significant threat indeed. Despite all the unknowns—and this includes many key aspects of the marine debris issue—there appears to be no excuse for inaction on PCBs or marine debris now. These issues need new champions to come forward to inspire and lead work to address them, including fomenting a change in our careless, wasteful societies that continue to use our oceans as a global garbage tip.

Acknowledgements Thanks to Paul Jepson, Rob Deaville and Laetitia Nunny for comments on sections of earlier drafts and to Andrew Butterworth for encouragement and the opportunity to contribute to this volume. All mistakes are my own, as are the views presented which do not necessarily reflect those of any organisation I am (or have been) affiliated with.

References

Allsopp M et al (2007) Plastic debris in the World's Oceans. Greenpeace International, Amsterdam, Netherlands, 44 pages

Baulch S, Perry C (2014) Evaluating the impacts of marine debris on cetaceans. Mar Pollut Bull 80:210–221

Baulch S, Simmonds, MP (2015) An update on research into marine debris and cetaceans. IWC/SC/66a/E/5. Paper submitted to the Scientific Committee of the International Whaling Commission. 8 pages

Carson R (1962) Silent spring. Houghton Mifflin, Boston

Colegrove KM, Venn-Watson S, Litz J et al (2016) Fetal distress and *in utero* pneumonia in perinatal dolphins during the Northern Gulf of Mexico unusual mortality event. Dis Aquat Org 119:1–16

Cózar A et al (2014) Plastic debris in the open ocean. Proc Nat Acad Sci U S A 111(28): 10239–10244

Derraik JG (2002) The pollution of the marine environment by plastic debris: a review. Mar Pollut Bull 44:842–852

Holden AV (1978) Pollution and seals – a review. Mammal Rev 8:53–66

Jacobsen JK, Massey L, Gulland F (2010) Fatal ingestion of floating net debris by two sperm whales (*Physeter macrocephalus*). Mar Pollut Bull 60(5):765–767

Jambeck J, Geyer R, Wilcox C et al (2015) Plastic waste inputs from land into the ocean. Science 437(6223):768–771

Jepson PD, Law RJ (2016) Persistent pollutants, persistent threats. Science 352(6292):1388–1389

Jepson PD et al (2016) PCB pollution continues to impact populations of orcas and other dolphins in European waters. Sci Rep 6:18573. doi:10.1038/srep18573

Murphy S, Barber JL, Learmonth JA et al (2015) Reproductive failure in UK harbour porpoises *Phocoena phocoena*: legacy of pollutant exposure? PLoS One 10(7). doi:10.1371/journal.pone.0131085

NOAA/UNEP (2011) Fifth international marine debris conference: waves of change – global lessons to improve local actions. Conference report: http://www.5imdc.org/. Accessed 19 Sept 16

NOAA Marine Debris Program (2014) Report on the entanglement of marine species in marine debris with an emphasis on species in the United States. Silver Spring, MD, 28 pp

Parsons ECM et al (2013) An introduction to marine mammal biology. Jones and Bartlett Learning, Burlington, MA

Reijnders PJH (1988) Ecotoxicological perspectives in marine mammalogy: research principles and goals for a conservation policy. Mar Mamm Sci 4:91–102

Reijnders PJH (1996) Organohalogen and heavy metal contamination in cetaceans: observed effects, potential impacts and future prospects. In: Simmonds MP, Hutchinson JD (eds) The conservation of whales and dolphins – science and practice. John Wiley and Sons Ltd., Chichester, pp 205–217

Reijnders PJH, Simmonds MP (2003) Global temporal trends of organochlorines and heavy metals in pinnipeds. In: Vos JG, Bossart GD, Fournier M, O'Shea TJ (eds) Toxicology of marine mammals. Taylor and Francis, London, pp 491–506

Simmonds MP (1992) Cetacean Mass mortalities and their potential relationship with pollution. In: Symposium on whales: biology, threats and conservation. Royal Belgium Academy of Sciences, Brussels, 5–7 June 1991, pp 217–245

de Stephanis R et al (2013) As main meal for sperm whales: plastic debris. Mar Pollut Bull 69:206–214. doi:10.1016/j.marpolbul.2013.01.033

Tanabe S, Tatsukawa R (1991) Persistent organochlorines in marine mammals. In: Jones KC (ed) Organic contaminants in the environment. Environmental pathways and effects. Elsevier, London, pp 275–289

Teuten E, Saquing JM, Knappe DRU et al (2009) Transport and release of chemicals from plastics to the environment and to wildlife. Philos Trans R Soc Lond B Biol Sci 364(1526): 2027–2045

UNEP (2009) Marine litter: a global challenge. UNEP, Nairobi. 2009

Waluda CM, Staniland IJ (2013) Entanglement of Antarctic fur seals at Bird Island, South Georgia. Mar Pollut Bull 74:244–252

World Economic Forum (2016) The new plastics economy – rethinking the future of plastics. World Economic Forum, Ellen MacArthur Foundation, McKinsey & Company. http://www.ellenmacarthurfoundation.org/publications. Accessed 19 Sept 16

Part I
Cetaceans

Chapter 4
Welfare Implications of Cetacean Bycatch and Entanglements

Sarah J. Dolman and Michael J. Moore

Abstract Each year, many cetaceans die from accidental capture in fishing gear. Despite intense study in some species, we know little about levels of bycatch and entanglement in most of the world's fisheries. Existing laws focus on maintenance of populations rather than welfare. Yet bycatch has wide-reaching welfare consequences, affecting quality of life for the many cetaceans that become injured and stressed or suffer the loss of conspecifics. For each that dies, we can expect many more to survive and suffer from such interactions. Our understanding of the welfare implications of cetacean bycatch has increased, but remains poor. As sentient, highly intelligent beings, cetaceans are considered by many to be in the highest category of animals on a scale of sensibility to pain and suffering, in the same category as primates and carnivores. Yet there has been little change in fishery management to reflect this increasing welfare knowledge and, in general, inadequate effort to reduce the numbers of cetaceans caught in gear. The assessment and awareness of welfare implications of bycaught cetaceans is several decades behind farm animal welfare. Pathological data indicate that the majority of bycaught cetaceans asphyxiate. Those that escape or are released from fishing gear can suffer a variety of injuries, high levels of stress, behavioural alterations and physiological and energetic costs that can lead to reduced long-term survival. These, along with wider social implications for conspecifics, are considered, as are ways to understand and reduce bycatch and entanglements.

S.J. Dolman (✉)
Whale and Dolphin Conservation, Brookfield House, 38 St Paul Street, Chippenham, Wiltshire SN15 1LJ, UK
e-mail: sarah.dolman@whales.org

M.J. Moore, Vet MB, PhD
Biology Department, Marine Mammal Center, Woods Hole Oceanographic Institution, Woods Hole, MA 02543, USA
e-mail: mmoore@whoi.edu

A. Butterworth (ed.), *Marine Mammal Welfare*, Animal Welfare 17,
DOI 10.1007/978-3-319-46994-2_4

4.1 Introduction

Bycatch, including entanglement, is the unintentional capture of nontarget species in fishing gear. Each year, hundreds of thousands of marine mammals, including whales, dolphins and porpoises, are believed to die from accidental capture; in waters of the USA alone, the total potential biological removal for all stocks of marine mammals is estimated at almost 60,000 annually (Read et al. 2006). At least 75% of odontocete species and 64% of mysticetes have been recorded as gill net bycatch over the past 20 plus years (Reeves et al. 2013).

Typically, the focus of bycatch is that of understanding conservation and population level impacts. Odontocetes with long, diverse lineages and few or no extant sister taxa include the obligate freshwater dolphins of the South Asian subcontinent (*Platanista*) and South America (*Inia*), the franciscana and the finless porpoises (*Neophocaena*) and are all threatened in all or parts of their range by gill net bycatch (Reeves et al. 2013). The Yangtze River dolphin or baiji (*Lipotes vexillifer*), known only from the middle-lower Yangtze River system and neighbouring Qiantang River in eastern China, is likely to be extinct, probably due to unsustainable bycatch in local fisheries (Turvey et al. 2007). Gill nets are a primary cause of decline in vaquita (*Phocoena sinus*), a small endangered porpoise endemic to Mexico's northern Gulf of California and whose numbers are less than 200 individuals (Rojas-Bracho and Reeves 2013) and may be as low as 30 (CIRVA-8, 2016). Whilst bycatch in set and drift gill nets remains a principal concern, incidental mortality in trawl nets (Fertl and Leatherwood 1997), purse seines, beach seines and longline gear is also worrisome (Reeves et al. 2004). New Zealand's Hector's dolphin (*Cephalorhynchus hectori*) populations have been fragmented, and a subspecies, Maui's dolphin (*Cephalorhynchus hectori Maui*) in the North Island, is critically endangered due to bycatch in gill nets and trawl fisheries (Slooten 2007). In one of the few examples where dedicated and long-term observations have been undertaken, population level impacts have been associated with the deliberate setting of purse seine nets around dolphins in tuna fisheries (Wade et al. 2007). This is discussed in more detail below.

In large whales, fixed fishing pot gear is a very significant cause of morbidity and mortality in addition to nets. Entanglement in static fishing gear is the leading cause of detected mortalities of large whales in the Northwest Atlantic (van der Hoop et al. 2013a). Previously hunted North Atlantic right whales (*Eubalaena glacialis*) are not recovering as fast as Southern right whales (*Eubalaena australis*) (Best et al. 2001) due to ship strikes and entanglements (Knowlton and Kraus 2001). Between 1980 and 2004, at least 83% of 493 individual North Atlantic right whales were entangled at least once. The number of entanglements ranged from 1 to 6 per individual (Knowlton et al. 2012). In SE Alaska at least 52% of humpback whales (*Megaptera novaeangliae*) have been entangled at least once, with an average of 8% of the population getting new entanglements every year (Neilson et al. 2009). Of individual Gulf of Maine humpback whales, from 2009 to 2010, 16.9 ± 6.45% ($n = 130$) exhibited new scarring and 13.5 ± 3.8% ($n = 319$) exhibited unhealed injuries likely obtained within the prior year (Robbins 2012). Smaller species appear less likely to survive any entanglement, and Leaper (Leaper et al. 2006) and Lien (1994) estimated that 70% of minke whales (*Balaenoptera acutorostrata*) that become entangled die, compared to

16% of humpback whales. In Scotland, half of minke whales stranded and post-mortemed from 1990 to 2010 showed signs of entanglement (Northridge et al. 2010).

Despite some species being the focus of intense study, we know little about current levels of bycatch in most of the world's fisheries, although experience suggests bycatch is likely to be widespread, if not universal (Read et al. 2006). Further, examples of population recovery following effective management changes including altered fishery methods are very rare (Reeves et al. 2013).

The primary enforcement of laws for cetaceans has been to support minimum species loss, or maintenance of populations, with little or no focus on the welfare aspects of incidental take in fisheries (Moore and van der Hoop 2012). Yet bycatch has wide-reaching welfare consequences, affecting quality of life (Fraser et al. 1997; Moore and van der Hoop 2012; Moore 2013) for the many whales, dolphins and porpoises that become injured and stressed or suffer the loss of conspecifics. As sentient, highly intelligent and cognitive beings, cetaceans are considered by many to be in the highest category of animals on a scale of sensibility to pain and suffering, in the same category as primates and carnivores (Porter 1992).

With many thousands of cetaceans estimated as dying each year in fishing gear (See tables in Read et al. 2006), we can expect many more to survive and suffer from such interactions. Our understanding of the welfare implications of cetacean bycatch has increased, but remains poor. There has been little change in fishery management to reflect this increasing welfare knowledge and, in general, inadequate effort to reduce the numbers of cetaceans caught in gear generally (Dolman et al. 2016). There remains no quantitative assessment and comparison of the scale of mortality and welfare implications of bycaught cetaceans. In this regard, the welfare considerations of bycaught cetaceans are decades behind farm animal welfare and slaughter (Soulsbury et al. 2008).

Increasingly, consumers want assurance about the welfare standards associated with the fish they buy. Whilst this is typically regarding the fish they choose to eat, the tuna-dolphin issue demonstrates a strong public concern about the welfare of cetaceans and other marine species accidentally caught in fishing gear.

4.2 Fishing Types

A summary of gear types is provided by the UN Food and Agriculture Organisation (FAO).[1] Mobile gears include dredgers, trawls (including twin, otter) and purse seine nets. Static nets include drift nets, gill nets (including mono-, multifilament), tangle nets, and coastal antishark nets that are anchored to the seabed, as well as longlines, creel or potting lines, mussel farm lines, aquaculture cages and fish aggregating devices. Recreational fishing gears include hooks and lines, and fishing also involves stupefying devices to stun fish, such as the use of chemicals, explosives and electrofishing. Ghost fishing occurs when abandoned, lost or discarded gear of any description continues to catch and kill organisms (Gilman 2015) and is discussed in Chap. 5.

[1] http://www.fao.org/fishery/topic/1617/en.

About four million vessels were estimated to make up the global fishing fleet in 2002.[2] Approximately two-thirds were less than 10 m in length (65% of which are not motorised), and the remaining third were 10–15 m in length. In contrast, approximately 1% of the global fleet were more than 24 m in length.

4.3 Causes and the Extent of Bycatch and Entanglement Impacts

Our understanding of marine mammal bycatch is hindered by the almost complete lack of reporting on a global scale (Read et al. 2006), and the majority of vessels are not monitored for bycatch by national or regional fishery commissions. In addition, where reporting occurs, bycatch and entanglement mortality and welfare implications are likely greatly underestimated, due to under-reporting by fishermen, with individually caught animals falling out of a net before it is brought on board, and a low probability of discovery of the bycaught animals or of recovery at sea (Cole et al. 2006; Williams et al. 2011; Kindt-Larsen et al. 2012; Bjørge et al. 2013). Even in the USA, where bycatch is monitored under the Marine Mammal Protection Act, the extent of bycatch cannot be accurately estimated in three out of four fisheries due to monitoring levels being below that which is recommended by federal fishery managers (Keledjian et al. 2014). As a result, many questions about the gear types and extent of cetacean bycatch and resulting mortality and welfare impacts remain.

At the broadest scale, spatial overlap between fishing gear and cetaceans may be related to prey distributions and associated diel (daily, diurnal) and seasonal patterns of both fishery and cetacean activity. A number of factors that influence bycatch and entanglement relate to the fishing activity itself, such as the location of fishing activity, level of effort, fishing method (active fishing methods may produce more noise) and gear used. Bycatch may occur when deploying gear, whilst gear are actively used for fishing and during hauling of gear back onto the vessel.

In some fisheries, perhaps most prominently in the tuna purse seine fisheries in the Eastern Tropical Pacific Ocean (ETP), the behaviour of the fishermen can influence levels of bycatch. Where schools of yellowfin tuna (*Thunnus albacares*) are caught by locating, chasing and encircling pods of associated dolphins, such behaviour increases the chance of capture of dolphins in the net (e.g. Cramer et al. 2008).

Other factors that influence bycatch and entanglement relate to the species being incidentally caught. Cetaceans have complex navigation systems that may enable them to detect gear under some circumstances (Kastelein et al. 2000). Detection abilities (both acoustic and visual) and distances for detection of gear by the animals vary depending upon the species, as well as on their behaviour.

However, the behaviour and the resulting method of bycatch or entanglement of individuals are largely unknown. Nielsen et al. (2012) found that porpoises do not usually actively approach gill nets, concluding that bycatch seems to be caused by individual animals accidentally being caught, likely due to attention shifts or to

[2] http://www.fao.org/fishery/topic/1616/en.

auditory masking (compromising capacity to 'hear') reducing their ability to detect the nets using echolocation. Sleep may influence odontocete entanglement in static nets, when echolocation is reduced (Goley 1999), as might scavenging, experience, curiosity, carelessness and distractions (whether predator escape or play behaviour between the cetaceans). Odontocetes feeding on hooked fish (depredation) may become entangled in the branch or main fishing lines or become hooked as they attempt to consume captured fish (Read 2008; Nitta and Henderson 1993). This behaviour leads to entanglement (Baird and Gorgone 2005) and, occasionally, to mortality. Ingestion of fishing gear, when it involves hooks becoming embedded in the throat, the 'goosebeak' formed by the epiglottis and corniculate cartilages in the throat, or the oesophagus, were found to be eventually fatal, over a period of several weeks, in all cases examined, as was line wrapped around the goosebeak (Wells et al. 2008).

Behaviours exhibited by a number of species that interacted with a bycatch reduction device in a trawl net included the animal becoming caught in the mesh by fins, head or tail; the tail being caught or stuck in the exclusion grid; the animal remaining in the net after a stressful interaction with the grid or mesh; the animal continuing to move and remaining in the net motionless after stressful interaction with grid or mesh; and, finally, the animal being assumed dead (Jaiteh et al. 2014). Baleen whales may not detect static gear or may even be attracted to areas where gear is set, due to increased productivity or because prey species are attracted to the sets (Lien 1994). The mouth is the most common attachment site for whales in static lines, followed by the peduncle and flippers (Cassoff et al. 2011), suggesting that entanglement occurs during foraging. Physiological and behavioural differences exist between species and taxonomic groups, and this variation may cause differences in the severity of certain injuries for different species (Andersen et al. 2008). Behavioural processes that vary with age, sex and reproductive status can apply to bycatch (Learmonth et al. 2014 and references therein). For example, in the albacore tuna (*Thunnus alalunga*) gill net fishery in the northeast Atlantic, young male common (*Delphinus delphis*) and striped dolphins (*Stenella coeruleoalba*) dominate the catch. Adult females constitute the second largest portion of the bycatch for both species (Brown et al. 2015). Injuries consistent with fishery interactions in Hawaiian false killer whales (*Pseudorca crassidens*) were biased towards females, suggesting a disproportionate impact of fishery-related mortality on population dynamics (Baird et al. 2014). Male humpback whales were more likely than females to exhibit entanglement-related scars, and yearlings were at the highest risk, although whales continued to become entangled when adults, in the Gulf of Maine (Robbins and Mattila 2000).

4.4 The Nature of Death, Injury and Physiological Effects

There are clear differences in the types and degree of injuries received by bycaught cetaceans (Jepson et al. 2013), and these vary with species and with age (Soulsbury et al. 2008). These differences highlight how a number of injury measures are required to diagnose bycatch.

Pathological data indicate that the majority of bycaught cetaceans asphyxiate in the nets (Soulsbury et al. 2008). Other bycaught cetaceans can suffer a variety of injuries and high levels of stress during incidental capture. Documented effects, for those that escape or are released from fishing gear, include behavioural alterations, physiological and energetic costs and associated reductions in feeding, growth or reproduction (i.e. fitness) (Wilson et al. 2014), leading to reduced long-term survival. These, along with wider social implications for conspecifics, are summarised here.

4.4.1 Asphyxiation

After being caught or entangled underwater, cetaceans may not be able to rise to the surface to breathe and may die trapped or anchored in gear. If the cetacean struggles frantically to free itself, then this effort will require an increased oxygen supply to muscles, whereas a whale that does not struggle may show the accentuated bradycardia seen in [forced] prolonged submergence (Leaper et al. 2006). Some individuals of dolphin species trapped in nets appear to go into a catatonic state even when the sea surface is accessible (Oliveros and Maldonado 2002).

Asphyxiation causes gross and histological changes to the heart and lungs (Jepson et al. 2000), and gross physical indicators are provided by Soulsbury et al. (2008). The pathological and histological changes observed in bycaught cetaceans indicate that asphyxia is the main cause of mortality. In large whales, there is often substantial bruising and oedema underlying the areas compressed by the entangling gear, reflective of a functional circulation for some time after the initial entanglement but before death (Moore and van der Hoop 2012). Criteria for the diagnosis of asphyxiation have been described under the diagnosis of peracute underwater entrapment (Moore et al. 2013).

Physiological data suggest that the cardiac changes observed in bycaught cetaceans are caused by massive releases of catecholamines in response to stress (Cowan and Curry 2002). The stress associated with capture, premortem injuries and asphyxiation are likely to be high (Soulsbury et al. 2008). It has been estimated that the time from entanglement to unconsciousness and to subsequent death in a minke whale can be considerably greater than the species' average dive times of 2–5 min (Leaper et al. 2006). Katona et al. (1993) report a single observation of a minke whale in the North Atlantic surviving submerged for 17 min as it was being freed from a fish weir. Whilst there are no quantitative data on the duration of suffering before death occurs, undoubtedly the duration can significantly exceed animal welfare standards set in other arenas, such as for the slaughter of farm animals and the catching of wild mammals in killing traps (Soulsbury et al. 2008).

4.4.2 Physical Injuries and Stress

Bycaught cetaceans suffer a variety of external and internal injuries, ranging from skin abrasions, amputations, broken teeth, broken bones, punctured or collapsed lungs and haemorrhaging (Figs. 4.1 and 4.2). The degree and severity of injuries varies with the method of fishing and species (Soulsbury et al. 2008).

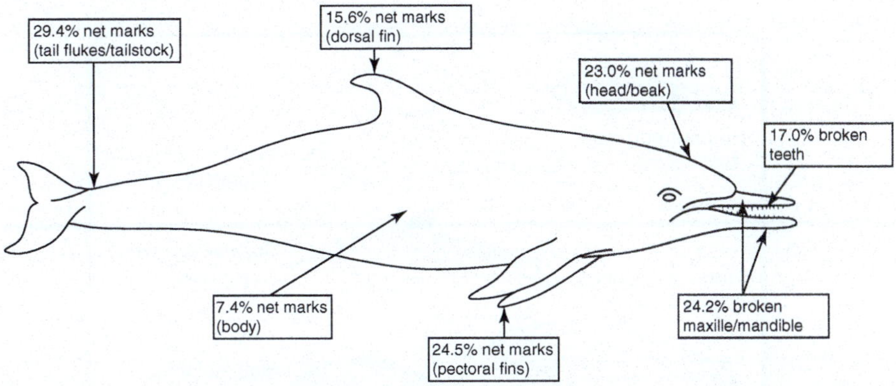

Fig. 4.1 External injuries associated with bycatch recorded from UK postmortem data from 1999 to 2005. Reproduced from Soulsbury et al. 2008

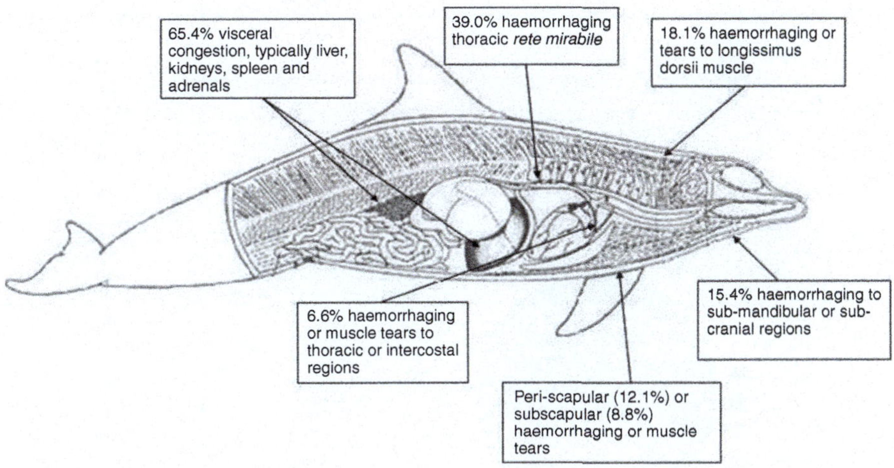

Fig. 4.2 Internal injuries associated with bycatch recorded from UK postmortem data from 1999 to 2005. Reproduced from Soulsbury et al. 2008

Injuries occur during interaction with the gear itself, through interactions with bycatch reduction devices or when the animals are hauled on board the fishing vessel, where traumatic lesions such as skull fractures have been documented (Kirkwood et al. 1997). Odontocetes can also be at risk from retaliatory measures taken by fishermen as a result of real or perceived economic losses (Read 2008). Fishermen are known to shoot at a variety of marine mammal species (Moore et al. 2013) engaging in depredation and to use other destructive means, such as small explosives, to deter such behaviour (Baird 2009).

The pathological consequences of entanglement or bycatch injuries are either anatomical or physiological (Figs. 4.3, 4.4, 4.5, 4.6, 4.7, 4.8, 4.9, 4.10, 4.11, 4.12, 4.13, 4.14). The anatomical location of an injury may lead to peracute to acute death (e.g. head trauma) or chronic debilitation (e.g. fracture of mandible or starvation). Physiological

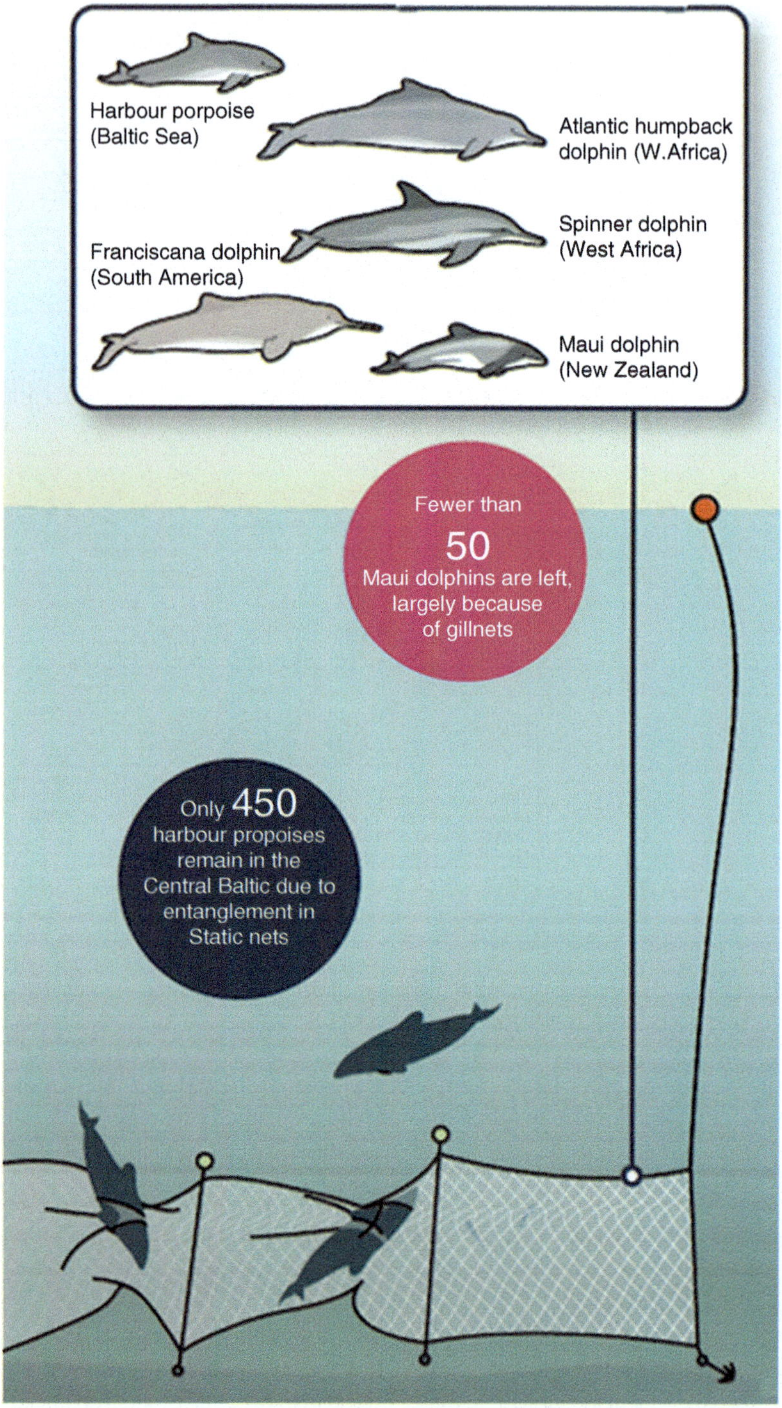

Fig. 4.3 Gill nets are set on the ocean bottom and are a huge threat to dolphins and porpoises. *Image credit: Richard Palmer Graphics, www.richardpalmergraphics.com*

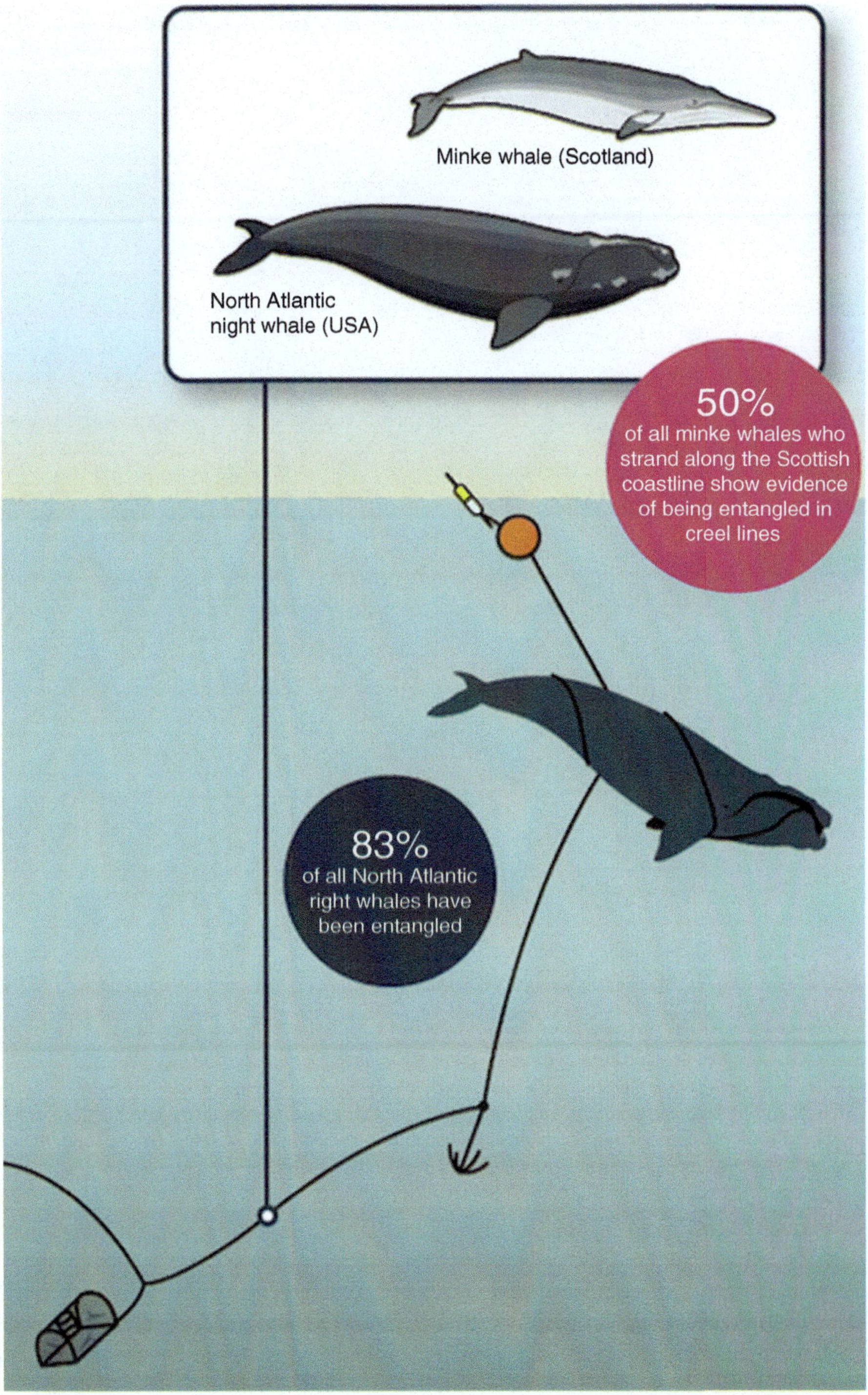

Fig. 4.4 Whales get caught in the lines either between creel pots or from the pots to buoys on the sea surface. *Image credit: Richard Palmer Graphics, www.richardpalmergraphics.com*

Fig. 4.5 Chronic entanglement with rope, most likely from pot gear, of the flipper of a North Atlantic right whale calf (*Eubalaena glacialis*). As the calf grew, the rope grew into the flesh of the flipper, ultimately constricting the radius bone such that the bone was notched by the rope. *Image credit: New England Aquarium. Permit No. 932-1905-01-MA-009536-1*

consequences of injury include shock, pain or blood loss leading to an inflammatory cascade, activation of the sympathetic nervous system, hormone release (epinephrine or norepinephrine) and vascular changes with the potential end results of hypothermia, coagulation defects, organ failure and death. However, these may not be readily determinable in an animal after initially surviving a traumatic event, and in animals that die, tissue autolysis or loss may prevent a complete assessment (Andersen et al. 2008).

Fig. 4.6 North Atlantic right whale (*Eubalaena glacialis*) with massive dissection of dorsal blubber coat, resulting from chronic constriction, subsequent to entanglement of both flippers in gill net, connecting dorsally. As the animal swam, the gear tightened and constricted. The entanglement lasted at least 4 months before the animal died. *Image credit: New England Aquarium. Permit No. 932-1905-01-MA-009536-1*

Fig. 4.7 Ventral view of a North Atlantic right whale (*Eubalaena glacialis*), showing damage caused by rope entanglement around both flippers. The animal was significantly decomposed having lost most of its skin postmortem. The animal died 6 months after it was last sighted alive, without an entanglement. *Image credit: Virginia Aquarium. Permit No. 932-1905-01-MA-009536-1*

Fig. 4.8 Rope entangled in the baleen of a North Atlantic right whale (*Eubalaena glacialis).* This animal had a complex entanglement involving the mouth, blowhole and left flipper. The flipper had been constricted such that the rope had embedded in the ulnar and radius bones and elicited a massive proliferation of new bone. The entanglement lasted between 6 and 16 months. *Image credit: Virginia Aquarium. Permit No. 932-1905-01-MA-009536-1*

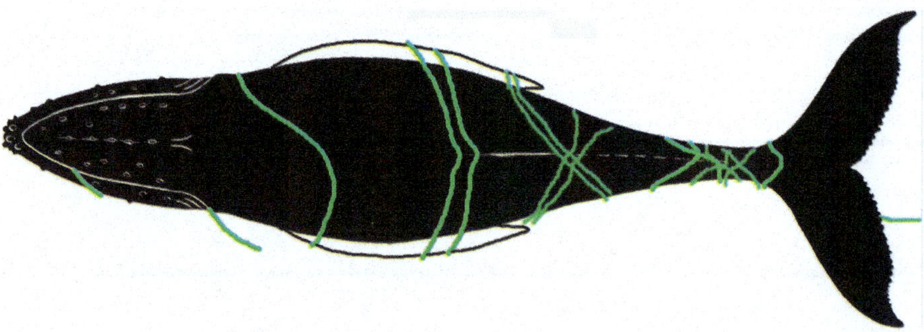

Fig. 4.9 Sketch of entanglement derived from examination of lesions on a dead humpback whale calf that had been earlier observed anchored in lobster gear. *Image credit: Scott Landry, Center for Coastal Studies*

Fig. 4.10 Humpback whale (*Megaptera novaeangliae*) entangled in lobster gear line. *Image credit: International Whaling Commission*

Fig. 4.11 Humpback whale (*Megaptera novaeangliae*) entangled in net and rope. *Image credit: International Whaling Commission*

Fig. 4.12 Immature humpback whale (*Megaptera novaeangliae*) entangled in lobster gear line being released using a long boom equipped with specially designed cutting attachments. *Image credit: International Whaling Commission*

Fig. 4.13 Entangled humpback whale (*Megaptera novaeangliae*) in Faxafloi Bay, near Reykjavik, Iceland, released by British Divers Marine Life Rescue in August 2015. *Image credit: Andy Butterworth*

Fig. 4.14 Deep incised lesion resulting from chronic entanglement with rope around the rostrum of an immature minke whale (*Balaenoptera acutorostrata*) from the North Sea. *Image credit: Andrew Brownlow, Scottish Marine Animal Stranding Scheme (SMASS)*

Postmortem of 182 cetaceans stranded in the UK (comprising 97 harbour porpoises (*Phocoena phocoena*), 80 common dolphins, 3 striped dolphins, 1 Risso's dolphin (*Griseus grampus*) and 1 minke whale) from 1999 to 2005 found evidence of complex entanglements involving multiple parts of the body. External injuries included amputations (from entanglement or being cut free), broken maxillae, mandibles and/or teeth and internal injuries consisting of organ congestion, muscle tears and haemorrhaging (either from the gear or from the cetacean struggling) (Soulsbury et al. 2008). Longer-term impacts can result after escape or release from the fishing gear. The stress and injuries for individuals that escape may cause prolonged suffering and/or subsequent mortality (Wilson et al. 2014; Soulsbury et al. 2008). Major dorsal fin disfigurements have been documented, with injuries at the base of the leading edge of the fin or a missing fin, with unknown welfare consequences (Baird and Gorgone 2005).

4.5 Case Study: North Atlantic Right Whale

Following hunting that decimated the original population, North Atlantic right whales (NARW) were listed as endangered in US waters in 1970. The population is estimated to contain less than 500 individuals (NOAA 2015). Entanglement in static pot gear lines is a primary cause of death, along with ship strikes.

Examination of large baleen whale entanglement mortalities has shown a variety of chronic impacts for persistent terminal entanglements. Juveniles and adults have a lower probability of survival following entanglement (Robbins et al. 2015). Apparent survival of entangled adults is 23% lower than other adult females and 26% lower than other adult males. The post-entanglement survival of entangled juveniles was comparable to entangled adults and 25% lower than conspecifics. This is the first estimate of survival reduction relative to unaffected animals and sets a baseline against which to evaluate the success of future mitigation efforts (Robbins et al. 2015). Larger whales breaking free of, and subsequently carrying, fixed trap and net gear are subject to a very slow and likely extremely debilitating demise, averaging 6 months in the case of the NARW, but there are cases that persist for multiple years (Moore and van der Hoop 2012). Disentanglement (gear removal) improves the survival outcome of NARW, as disentangled whales can achieve a subsequent survival rate that approaches that of unaffected animals (Robbins et al. 2015).

Protracted causes of death include impaired foraging during entanglement, resulting in emaciation through reduced mobility and foraging ability, and energy budget depletion leading to starvation after many months or years; systemic infection arising from open, unresolved entanglement wounds; and haemorrhage or debilitation due to severe gear-related damage to tissues. Serious gear-induced injury can include laceration of large vessels, occlusion of the nares, embedding of line in growing bone and massive periosteal proliferation of new bone in an attempt to wall off constricting, encircling lines (Cassoff et al. 2011). Gear-induced wounds can lead to death by impairing critical biological functions, becoming a source of haemorrhage, or providing a portal of entry for pathogens (Cassoff et al. 2011).

It has been suggested that disruption of the oral seal, holding the lower mandible closed, could have a significant impact on propulsion efficiency and energy expenditure (Lambertsen et al. 2005). Van der Hoop et al. (2013a) identified significant alteration to swimming patterns and significant drag, resulting in energy depletion in a chronically entangled NARW. Added drag of towing gear could substantially affect the energy budget of an entangled whale (Moore and van der Hoop 2012), where entanglement could increase drag and propulsive power by 1.47-fold (van der Hoop et al. 2015) and seemingly small entanglements (short pieces of line, small floats) can still impart significant drag (van der Hoop et al. 2015). Additional drag from the entangling gear has been shown to have energetic costs that can be equivalent to the cost of migration or reproduction (van der Hoop et al. 2015). Thus chronically entangled whales are perhaps best seen as 'dead whales swimming' in that females that are significantly and chronically entangled are unlikely to get pregnant. Laceration and consequent infection can be another cause of death in chronic entanglement of large whales, with secondary bronchopneumonia (Cassoff et al. 2011).

Entanglement in fishing gear is sufficiently stressful to cause both a behavioural and physiological stress response in baleen whales (Cassoff et al. 2011). Faecal glucocorticoid studies have shown markedly elevated stress hormone levels in a severely entangled NARW (Hunt et al. 2006); the relationships between entanglement stress and metabolic rate are complex. Long-term stress from being chronically wrapped in gear may explain why examined whales were unable to fight off the initial insult of infected gear lacerations, most likely leading to their demise (Cassoff et al. 2011).

Most baleen whales are able to release themselves from fishing line or net or are disentangled by humans, as evidenced by scarring patterns on many individuals (Knowlton et al. 2005; Mathewson 2012). However, visual health assessment of NARW using photographs demonstrated that stress responses existed that may have impacted health and fecundity even after the gear is no longer attached (Pettis et al. 2004). Thus the lethal entanglement events are the extreme and relatively infrequent end of a more widespread, persistent and repetitive problem, and the cost of nonlethal entanglement in terms of energy, stress and pain may significantly reduce fecundity in NARW (Moore and van der Hoop 2012). Ultimately entanglements can lead to eventual lethal trauma through a drawn-out cumulative loss of body condition and constriction of body parts, with or without secondary infection, and the level of pain and stress is presumably extreme (Moore and van der Hoop 2012). Lethal entanglements of baleen whales are, arguably, one of the worst forms of human-caused mortality in any wild animal (Cassoff et al. 2011).

4.5.1 Decompression Sickness

The stress of capture at depth appears to trigger a failure of normal gas solubility management physiology. Small cetaceans and seals bycaught in gill nets have been shown to undergo massive gas embolisation by the time they are hauled to the surface dead (Moore et al. 2009). Diagnosis of this gas as being from decompression of supersaturated tissues, rather than decomposition, was made by careful analysis of the gas components of the bubbles (Bernaldo de Quirós et al. 2013). Whilst moribund live gas-embolised marine mammals have not been described, turtles have been shown to undergo spastic morbidity when hauled to the surface in a gill net, with concomitant embolisation (García-Párraga et al. 2014). These clinical signs have been reversed with repressurisation and then controlled depressurisation. Garcia et al.'s study was the first to definitively diagnose decompression sickness, by reversal of clinical signs with repressurisation, in a breath holding, non-human diving vertebrate. This study has major potential implications for the proper treatment of moribund, but not dead vertebrates that are hauled to the surface in gill nets and other fishing gears.

4.5.2 Social Implications

Bycatch has implications for conspecifics that we are only beginning to understand. Due to the highly social nature of many odontocetes, survival and reproductive success can depend on social cohesion and organisation, and the effects of social disruption caused by bycatch mortalities may go beyond the dynamics of individual removals and impede population recovery (CMS 2015; Wade et al. 2012). Wade et al. (2012) suggest that their social and behavioural traits may contribute to a lack of resilience in odontocetes.

Observations of a bottlenose dolphin calf temporarily entangled in monofilament line showed immediate alterations in the behaviour of the mother and calf, as well as conspecifics (Mann et al. 1995). As well as causing distress to surviving family or group members, the loss of key individuals will lead to the loss of important social knowledge and will be likely to cause disruption or breakdown of social groups and networks (Soulsbury et al. 2008; Williams and Lusseau 2006). The removal of older individuals and their knowledge will have serious consequences for populations of socially advanced mammals such as cetaceans (McComb et al. 2001).

4.6 Case Study: Eastern Tropical Pacific Tuna-Dolphin Fishery

Since the 1960s, dolphins of several species continue to be used to locate, chase, herd and encircle tuna using speed boats (Fig. 4.15). The scale of death is unprecedented, with more than six million dolphins killed in the Eastern Tropical Pacific (ETP) (Perrin et al. 2002) and populations of northeastern offshore Pantropical spotted dolphins (*S. attenuata attenuata*) and eastern spinner dolphins (*S. longirostris orientalis*) reduced by an estimated 40% and 20%, respectively (Wade et al. 2007), due to these

Fig. 4.15 Dolphins in purse seine net. *Image credit: NOAA bycatch webpage*

early purse-seining techniques. Further, as many dolphins are released, a surviving dolphin may be captured and released a number of times during their lifetime.

Despite reduced kills to fewer than 1000 per year, the populations of dolphins are not showing signs of recovery (Gerrodette and Forcada 2005; Wade et al. 2002), and the rate of calf production has been declining since the 1980s (Cramer et al. 2008). Hypotheses to explain the lack of recovery (Gerrodette and Forcada 2005) have included underreporting of kills by observers; cryptic effects of the fishery not detectable by observers, such as stress-induced abortion; or the separation of mothers and calves (Noren and Edwards 2007).

An associated study on bottlenose dolphins 'carrying' a calf in echelon position found that the calf may be unable to achieve speeds which are required to be sustained by the whole group to evade fishermen (Noren 2013). To maintain proximity with the group, mothers may become separated from their calves. Permanently separated dependent calves may then represent unobserved mortality events, no doubt due to a stressful and painful demise resulting from starvation following orphaning. This may partially explain the non-recovery of depleted ETP dolphin populations (Noren 2013) where, in the case of mothers dying, a calf or dependent juvenile must be assumed to become a secondary victim (Noren and Edwards 2007). It is plausible that the chase and encircling of the dolphins has hindered or prevented recovery in these populations, animal groups which show complex social structure (Wade et al. 2012). 'Dolphin-friendly' product labelling on tuna caught in purse seine fisheries does not consider welfare implications to individuals or populations.

4.7 Conclusions: Ways to Understand and Reduce Bycatch and Entanglements

Bycatch is not an intentional practice, and many fishermen are involved in strategies to reduce the accidental capture of whales, dolphins and porpoises. The approach required will often be fishery specific, and all solutions are dependent on positive relationships and involvement with fishermen. Participation of fishermen in the management process is necessary (Bisack and Das 2015), bycatch reduction approaches can be implemented successfully from the bottom-up (Teh et al. 2015), and individual level incentive-based management measures are likely to be highly effective.

Bycatch is not inevitable. There are ways to minimise unintended mortality and welfare impacts by avoiding areas where endangered or vulnerable species or small populations are known to be present and ongoing enforcement of monitoring and mitigation to enable quantitative metrics and to improve compliance. Despite existing legislation to document, manage and reduce bycatch in some parts of the world, such as the US Marine Mammal Protection Act (MMPA) and the EU Habitats Directive (92/43/EEC), measures to do so remain mostly inadequate. For instance, even for a highly endangered whale such as the North Atlantic right whale, in perhaps the best regulated waters of the world, the mortality allowed by the US MMPA and Endangered Species Act (ESA) (zero cases per year) is exceeded every year (NOAA 2015). This periodically engenders further gear modification and spatial and temporal planning (van der Hoop et al. 2013b), but other factors such as increasing rope strength seem to more than undo whatever gains such mitigation measures might have been achieved (Knowlton et al. 2015). Existing legislation includes no provisions for the protection of cetaceans from incidental capture on welfare grounds (Soulsbury et al. 2008). Explicit policy decisions and rigorous implementation are urgently needed to bridge the gap between our poor biological and impact knowledge and what is happening at sea (Dolman et al. 2016). To this end, political will and explicit consideration of the sublethal costs of bycatch and entanglement in decision making are necessary.

Mitigation options include modifying the gear, either to make it more visible (e.g. using acoustic devices) or reduce the likelihood of entanglement once a cetacean makes contact with the gear (Soulsbury et al. 2008). There is a great need for effective mitigation measures to address bycatch of marine mammals in gill net fisheries (Read 2008). The use of active acoustic devices (such as pingers) has been demonstrated to successfully modify the behaviour of some dolphins, porpoises and small whales to reduce the frequency of their interactions with gill net fisheries (Dawson et al. 2013). Pingers on drift nets eliminated beaked whale bycatch in Californian drift gill net fishery (Carretta et al. 2008), where the species previously caught included Cuvier's beaked whales (*Ziphius cavirostris*), Hubb's beaked whales (*Mesoplodon carlhubbsi*), Stejneger's beaked whale (*M. stejnegeri*), Baird's beaked whale (*Berardius bairdii*) and unidentified *Mesoplodon* and ziphiid species. Pingers may have welfare implications themselves, causing auditory damage if loud enough (Lepper et al. 2014). The wide-scale use of acoustic devices on static fishing gear presents operational issues as well as those associated with introduced noise impacts.

There remains an urgent need for better entanglement avoidance and individual entanglement mitigation for large baleen whales. Entanglements increased in the northwest Atlantic since measures such as weak links on buoy lines and sinking ground lines between fishing traps and pots were introduced, indicating that different or additional mitigation measures are required (Pace et al. 2014). Adoption of ropes with lower breaking strengths (of 1700 lbs. or less) could reduce the number of life-threatening entanglements for large whales by at least 72% and still be strong enough to withstand the routine forces involved in many fishing operations (Knowlton et al. 2015).

Effective bycatch mitigation will require coordinated actions by the range of stakeholders to develop a combination of technological gear fixes, changes in fishing practices, modification of fishing effort and international agreements that, together, can monitor and mitigate bycatch (Lewison et al. 2004). To effectively deal with all these mortality and welfare issues, a clear strategy is required to identify the neccessary steps required by all fisheries to reduce bycatch towards zero (Dolman et al. 2016) should be implemented, and this should include welfare-specific legislation for marine species, as already exists terrestrially. Cetacean population surveillance, including effort-based collection of at-sea and stranding data, adequate monitoring of the fishery itself, reporting of lost gear, bycatch as well as mitigation measures and monitoring of effectiveness of mitigation measures are all required. Accurate estimates of cumulative incidental catch levels are typically not available due to insufficient sampling in the appropriate fisheries or areas. A high level of observer coverage allows more precise estimation of bycatch levels and would provide a conspicuous enforcement presence that could deter violations of regulations (McDonald et al. 2016).

Investigations into the sublethal effects of bycatch-related injury and stress on fitness are required, including improved understanding of the injuries suffered, the length of time to asphyxiation and the social implications of individuals dying. Better information about the nature and magnitude of the problem needs wider publicity and better public awareness. Market-based mechanisms should include retailers and suppliers working with fisheries to improve practices and governance. As a component of this, certification schemes should include the mortality and welfare considerations of bycatch in their assessments of fisheries and clear labelling of the resulting fish products. Perhaps a major effort to educate seafood consumers as to the chronic and widespread welfare concerns that marine mammal bycatch and entanglements represent would achieve their mitigation through consumer pressure, where governmental efforts on the basis of species and population conservation have largely failed.

Acknowledgements Thank you to Andy Butterworth for providing helpful comments on a draft manuscript and to Carl Soulsbury for allowing reproduction of figures.

References

Andersen MS, Forney KA, Cole TVN, Eagle T, Angliss R, Long K, Barre L, Van Atta L, Borggaard D, Rowles T, Norberg B, Whaley J, Engleby L (2008) Differentiating serious and non-serious injury of marine mammals. In: Report of the serious injury technical workshop, 10–13 September 2007, Seattle, Washington. United States Department of Commerce, NOAA Technical Memorandum NMFS-OPR-39, 94 p

Baird RW (2009) A review of false killer whales in Hawaiian waters: biology, status, and risk factors. A report prepared for the U.S. Marine Mammal Commission. http://www.mmc.gov/reports/workshop/pdf/killerwhale_review_mmc09.pdf. Accessed 21 Jan 2016

Baird RW, Gorgone AM (2005) False killer whale dorsal fin disfigurement as a possible indicator of longline fishery interactions in Hawaiian waters. Pac Sci 59:593–601

Baird RW, Mahaffy SD, Gorgone AM, Cullins T, McSweeney DJ, Oleson EM, Bradford AL, Barlow J, Webster DL (2014) False killer whales and fisheries interactions in Hawaiian waters: evidence for sex bias and variation among populations and social groups. Mar Mamm Sci 31:579–590

Bernaldo de Quirós Y, Seewald JS, Sylva SP, Greer B, Niemeyer M, Bogomolni AL, Moore MJ (2013) Compositional discrimination of decompression and decomposition gas bubbles in bycaught seals and dolphins. PLoS One 8:e83994

Best PB, Bannister JL, Brownell RL, Donovan GP (2001) Right whales: worldwide status. J Cetacean Res Manage (special issue 2). https://iwc.int/journalspec02. Accessed 12 Feb 2016

Bisack KD, Das C (2015) Understanding non-compliance with protected species regulations in the Northeast USA gillnet fishery. Front Mar Sci 2:91. doi:10.3389/fmars.2015.00091

Bjørge A, Skern-Mauritzen M, Rossman MC (2013) Estimated bycatch of harbour porpoise (*Phocoena phocoena*) in two coastal gillnet fisheries in Norway, 2006–2008. Mitigation and implications for conservation. Biol Conserv 161:164–173

Brown SL, Reid D, Rogan E (2015) Spatial and temporal assessment of potential risk to cetaceans from static fishing gears. Mar Pol 51:267–280

Carretta JV, Barlow J, Enriquez L (2008) Acoustic pingers eliminate beaked whale bycatch in a gill net fishery. Mar Mamm Sci 24:956–961

Cassoff RM, Moore KM, McLellan WA, Barco SG, Rotstein DS, Moore MJ (2011) Lethal entanglement in baleen whales. Dis Aquat Org 96:175–185

CIRVA-8 (2016) Comite Internacional Para la Recuperacion de la Vaquita / International Committee for the Recovery of the Vaquita, Scientific Report of the Eighth Meeting. Southwest Fisheries Science Centre, La Jolla, CA, USA. CMS (2015) CMS Resolution 11.23: Conservation implications of cetacean culture. http://www.ascobans.org/sites/default/files/document/AC22_Inf_4.7.b_CMSres.11.23_CetaceanCulture.pdf

Cole T, Hartley D, Garron M (2006) Mortality and serious injury determinations for baleen whale stocks along the eastern seaboard of the United States, 2000–2004. Northeast Fisheries Science Center Reference Document 06–04, Gloucester, MA

Cowan DF, Curry BE (2002) Histopathological assessment of dolphins necropsied onboard vessels in the eastern tropical Pacific tuna fishery. Administrative Report LJ- 02-24C, Southwest Fisheries Science Center, La Jolla, CA

Cramer KL, Perryman WL, Gerrodette T (2008) Declines in reproductive output in two dolphin populations depleted by the yellowfin tuna purse-seine fishery. Mar Ecol Prog Ser 369: 273–285

Dawson SM, Northridge S, Waples D, Read AJ (2013) To ping or not to ping: the use of active acoustic devices in mitigating interactions between small cetaceans and gillnet fisheries. Endang Species Res 19:201–221

Dolman SJ, Baulch S, Read F, Ritter F, Evans P (2016) Towards an EU action plan on cetacean bycatch. Mar Pol 72:67–75

Fertl D, Leatherwood S (1997) Cetacean interactions with trawls: a preliminary review. J Northw Atl Fish Sci 22:219–248

Fraser D, Weary DM, Pajor EA, Milligan BM (1997) A scientific concept of animal welfare that reflects ethical concerns. Anim Welf 6:187–205

García-Párraga D, Crespo-Picazo J, de Quirós Y, Cervera V, Martí-Bonmati L, Díaz-Delgado J, Arbelo M, Moore M, Jepson P, Fernández A (2014) Decompression sickness ('the bends') in sea turtles. Dis Aquat Org 111:191–205

Gerrodette T, Forcada J (2005) Non-recovery of two spotted and spinner dolphin populations in the eastern tropical Pacific Ocean. Mar Ecol Prog Ser 291:1–21

Gilman E (2015) Status of international monitoring and management of abandoned, lost and discarded fishing gear and ghost fishing. Mar Pol 60:225–239

Goley PD (1999) Behavioral aspects of sleep in Pacific white-sided dolphins (*Lagenorhynchus obliquidens*, Gill 1865). Mar Mamm Sci 15:1054–1064

Hunt KE, Rolland RM, Kraus SD, Wasser SK (2006) Analysis of fecal glucocorticoids in the North Atlantic right whale (*Eubalaena glacialis*). Gen Comp Endocrinol 148:260–272

Jaiteh VF, Allen SJ, Meeuwig JJ, Loneragan NR (2014) Combining in-trawl video with observer coverage improves understanding of protected and vulnerable species by-catch in trawl fisheries. Mar Fresh Res. http://dx.doi.org/10.1071/MF13130

Jepson PD, Baker JR, Kuiken T, Simpson VR, Kennedy S, Bennett PM (2000) Pulmonary pathology of harbour porpoises (*Phocoena phocoena*) stranded in England and Wales between 1990 and 1996. Vet Rec 146:721–728

Jepson PD, Barbieri M, Barco SG, Bernaldo de Quiros Y, Bogomolni A, Danil K, Rowles T (2013) Peracute underwater entrapment of pinnipeds and cetaceans. Dis Aquat Org 103:229–264

Kastelein RA, Au WWL, de Haan D (2000) Detection distances of bottom-set gillnets by harbour porpoises (*Phocoena phocoena*) and bottlenose dolphins (*Tursiops truncatus*). Mar Environ Res 49:359–375

Katona SK, Rough V, Richardson DT (1993) A field guide to whales, porpoises and seals from Cape Cod to Newfoundland. Smithsonian Institution Press, Washington

Keledjian A, Brogan G, Lowell B, Warrenchuck J, Enticknap B, Shester G, Hirshfield M, Cano-Stocco D (2014) Wasted catch: unsolved problems in U.S. fisheries. An Oceana report

Kindt-Larsen L, Dalskov J, Stage B, Larsen F (2012) Observing incidental harbour porpoise *Phocoena phocoena* bycatch by remote electronic monitoring. Endang Species Res 19:75–83

Kirkwood JK, Bennett PM, Jepson PD, Kuiken T, Simpson VR, Baker JR (1997) Entanglement in fishing gear and other causes of death in cetaceans stranded on the coasts of England and Wales. Vet Rec 141:94–98

Knowlton AR, Kraus SD (2001) Mortality and serious injury of northern right whales (*Eubalaena glacialis*) in the western North Atlantic Ocean. J Cetacean Res Manage 2:193–208

Knowlton AR, Robbins J, Landry S, McKenna HA, Kraus SD, Werner TB (2015) Effects of fishing rope strength on the severity of large whale entanglements. Conserv Biol 30(2):318–328. doi:10.1111/cobi.12590

Knowlton AR, Hamilton PK, Marx MK, Pettis HM, Kraus SD (2012) Monitoring North Atlantic right whale *Eubalaena glacialis* entanglement rates: a 30 yr retrospective. Mar Ecol Prog Ser 466:293

Knowlton AR, Marx MK, Pettis H, Hamilton PK (2005) Scarification analysis of North Atlantic right whales (*Eubalaena glacialis*): monitoring rates of entanglement interaction. Tech. Rep. 43EANF030107, National Marine Fisheries Service, England Aquarium, Central Wharf, Boston, MA, USA

Lambertsen RH, Rasmussen KJ, Lancaster WC, Hintz RJ (2005) Functional morphology of the mouth of the bowhead whale and its implications for conservation. J Mammal 86:342–352

Leaper R, Papastavrou V, Sadler L (2006) Consideration of factors affecting time to death for whales following entanglement in fishing gear. Paper IWC/58/WKM&AWI 14 presented to International Whaling Commission, 6 pages

Learmonth JA, Murphy S, Luque PL, Read RJ, Patterson AP, Brownlow A, Ross HM, Barley JP, Santos MB, Pierce GJ (2014) Life history of harbor porpoises (*Phocoena phocoena*) in Scottish (UK) waters. Mar Mamm Sci 30:1427–1455

Lepper PA, Gordon J, Booth C, Theobald P, Robinson SP, Northridge S, Wang L (2014) Establishing the sensitivity of cetaceans and seals to acoustic deterrent devices in Scotland. Scottish Natural Heritage Commissioned Report No. 517

Lewison RL, Crowder LB, Read AJ, Freeman SA (2004) Understanding impacts of fisheries bycatch on marine megafauna. TREE 19:598–604. doi:10.1016/j.tree.2004.09.004

Lien J (1994) Entrapment of large cetaceans in passive inshore fishing gear in Newfoundland and Labrador (1979–1990). Rep Int Whal Comm 15:149–157

Mann J, Smolker RA, Smuts BB (1995) Responses to calf entanglement in free-ranging bottlenose dolphins. Mar Mamm Sci 11:100–106

Mathewson A (2012) Non-lethal entanglement of minke whales (*Balaenoptera acutorostrata*) in fishing gear in the Hebrides. Bachelors thesis submitted to University of St Andrews, Fife, Scotland

McComb K, Moss C, Durant SM, Baker L, Sayialel S (2001) Matriarchs as repositories of social knowledge in African elephants. Science 292:491–494

McDonald SL, Lewison RL, Read AJ (2016) Evaluating the efficacy of environmental legislation: a case study from the US marine mammal Take Reduction Planning process. Global Ecol Conserv 5:1–11

Moore M (2013) Welfare of whales bycaught in fishing gear or struck by vessels. Anim Welf 22:117–121

Moore MJ, van der Hoop JM (2012) The painful side of trap and fixed net fisheries: chronic entanglement of large whales. J Mar Biol 2012(4). doi:10.1155/2012/230653

Moore MJ, van der Hoop J, Barco SG, Costidis AM, Gulland FM, Jepson JD, Moore KT, Raverty S, McLellan WA (2013) Criteria and case definitions for serious injury and death of pinnipeds and cetaceans caused by anthropogenic trauma. Dis Aquat Org 103:229–264

Moore MJ, Bogomolni AL, Dennison SE, Early G, Garner MM, Hayward BA, Lentell BJ, Rotstein DS (2009) Gas bubbles in seals, dolphins and porpoises entangled and drowned at depth in gill nets. Vet Pathol 46:536–547

Neilson JL, Straley JM, Gabriele CM, Hills S (2009) Nonlethal entanglement of humpback whales (*Megaptera novaeangliae*) in fishing gear in northern Southeast Alaska. J Biogeograph 36:452–464

Nielsen TP, Wahlberg M, Heikkilä S, Jensen M, Sabinsky P, Dabelsteen T (2012) Swimming patterns of wild harbour porpoises *Phocoena phocoena* show detection and avoidance of gillnets at very long ranges. Mar Ecol Prog Ser 453:241–248

Nitta E, Henderson JR (1993) A review of interactions between Hawaii's fisheries and protected species. Mar Fish Rev 55:83–92

NOAA (2015) North Atlantic right whale (*Eubalaena glacialis*): Western Atlantic stock. http://nefsc.noaa.gov/publications/tm/tm231/7_rightwhale_F2014July.pdf. Accessed 11 Feb 2016

Noren SR (2013) Altered swimming gait and performance of dolphin mothers: implications for interactions with tuna purse-seine fisheries. Mar Ecol Prog Ser 482:255–263

Noren SR, Edwards EF (2007) Physiological and behavioural development in Delphinid calves: implications for calf separation and mortality due to tuna purse-seine sets. Mar Mamm Sci 23:15–29

Northridge S, Cargill A, Coram A, Mandleberg L, Calderan S, Reid R (2010) Entanglement of minke whales in Scottish waters; an investigation into occurrence, causes and mitigation. SMRU report to Scottish Government

Oliveros ES, Maldonado FG (2002) Coping behaviors of spotted dolphins during fishing sets. Report for SW Fisheries Science Center, NMFS Administrative Report LJ-02-36C

Pace RM, Cole TVN, Henry AG (2014) Incremental fishing gear modifications fail to significantly reduce large whale serious injury rates. Endang Species Res 26:115–126. doi:10.3354/esr00635

Perrin WF, Wursig B, Thewissen JGH (eds) (2002) Encyclopedia of marine mammals. Academic Press, San Diego, CA, pp 1269–1273

Pettis HM, Rolland RM, Hamilton PK, Brault S, Knowlton AR, Kraus SD (2004) Visual health assessment of North Atlantic right whales (*Eubalaena glacialis*) using photographs. Can J Zool 82:8–19

Porter DG (1992) Ethical scores for animal experiments. Nature 356:101–102

Read AJ (2008) The looming crisis: interactions between marine mammals and fisheries. J Mammal 89:541–548

Read AJ, Drinker P, Northridge S (2006) Bycatch of marine mammals in U.S. and global fisheries. Conserv Biol 20:164. doi:10.1111/j.1523-1739.2006.00338.x

Reeves RR, McClellan K, Werner TB (2013) Marine mammal bycatch in gillnet and other entangling net fisheries, 1990 to 2011. Endang Species Res 20:71–97

Reeves RR, Berggren P, Cresop EA, Gales N, Northridge SP, Notarbartolo di Sciara G, Perrin WF, Read AJ, Rogan E, Smith BD, Van Waerebeek K (2004) Global priorities for reduction of cetacean bycatch. A report to WWF. Available from authors

Robbins J (2012) Scar-Based inference into Gulf of Maine humpback whale entanglement: 2010. Report to Northeast Fisheries Science Center, National Marine Fisheries Service, U.S. http://www.nefsc.noaa.gov/psb/docs/HUWHScarring(Robbins2012).pdf. Accessed 12 Feb 2016

Robbins J, Mattila D (2000) Gulf of Maine humpback whale entanglement scar monitoring results 1997–1999. Report to National Marine Fisheries Service

Robbins J, Knowlton AR, Landry S (2015) Apparent survival of North Atlantic right whales after entanglement in fishing gear. Biol Conserv 191:421–427

Rojas-Bracho L, Reeves RR (2013) Vaquitas and gillnets: Mexico's ultimate cetacean conservation challenge. Endang Species Res 21:77–87

Slooten E (2007) Conservation management in the face of uncertainty: effectiveness of four options for managing Hector's dolphin bycatch. Endang Species Res 3:169–179

Soulsbury CD, Iossa G, Harris S (2008) The animal welfare implications of cetacean deaths in fisheries. A University of Bristol report to the Whale and Dolphin Conservation Society (WDC)

Teh LSL, Teh LCL, Hines E, Junchompoo C, Lewison RL (2015) Contextualising the coupled socio-ecological conditions of marine megafauna bycatch. Ocean Coastal Manage 116:449–465

Turvey ST, Pitman RL, Taylor BL, Barlow J, Akamatsu T, Barrett LA, Zhao X, Reeves RR, Stewart BS, Wang K, Wei Z, Zhang X, Pusser LT, Richlen M, Brandon JR, Wang D (2007) First human-caused extinction of a cetacean species? Biol Lett 3:537–541

van der Hoop JM, Moore MJ, Fahlman A, Bocconcelli A, George C, Jackson K, Miller C, Morin D, Pitchford T, Rowles T, Smith J, Zoodsma B (2013a) Behavioral impacts of disentanglement of a right whale under sedation and the energetic cost of entanglement. Mar Mamm Sci 30(1). doi:10.1111/mms.12042

van der Hoop JM, Moore MJ, Barco SG, Cole TVN, Daoust P-Y, Henry AG, McAlpine DF, McLellan WA, Wimmer T, Solow AR (2013b) Assessment of management to mitigate anthropogenic effects on large whales. Conserv Biol 27:121–133

van der Hoop JM, Corkeron P, Kenney J, Landry S, Morin D, Smith J, Moore MJ (2015) Drag from fishing gear entangling North Atlantic right whales. Mar Mamm Sci 32(2):619–642

Wade PR, Reeves RR, Mesnick SL (2012) Social and behavioural factors in cetacean responses to overexploitation: are odontocetes less "resilient" than mysticetes? J Mar Biol. doi:10.1155/2012/567276

Wade PR, Reilly SB Gerrodette T (2002) Assessment of the population dynamics of the northeastern offshore spotted and the eastern spinner dolphin populations through 2002. NOAA report LJ-02-13, 53 pages

Wade PR, Watters GM, Gerrodette T, Reilly SB (2007) Depletion of spotted and spinner dolphins in the eastern tropical Pacific: modeling hypotheses for their lack of recovery. Mar Ecol Prog Ser 343:1–14

Wells RS, Allen JB, Hofmann S, Bassos-Hull K, Fauquier DA, Barros NB, DeLynn RE, Sutton G, Socha V, Scott MD (2008) Consequences of injuries on survival and reproduction of common bottlenose dolphins (*Tursiops truncatus*) along the west coast of Florida. Mar Mamm Sci 24:774–794

Williams R, Lusseau D (2006) A killer whale social network is vulnerable to targeted removals. Biol Lett 2:497–500

Williams R, Gero S, Bejder L, Calambokidis J, Kraus SD, Lusseau D, Read AJ, Robbins J (2011) Underestimating the damage: interpreting cetacean carcass recoveries in the context of the *Deepwater Horizon*/BP incident. Conserv Lett 4:228–233

Wilson SM, Raby GD, Burnett NJ, Hinch SG, Cooke SJ (2014) Looking beyond the mortality of bycatch: sublethal effects of incidental capture on marine animals. Biol Conserv 171:61–72

Chapter 5
From Hunting to Watching: Human Interactions with Cetaceans

E.C.M. Parsons and Danielle Monaghan-Brown

Abstract Whales and dolphins have been hunted since antiquity in many parts of the world. Although whaling started as a commercial enterprise in Europe in the eleventh century, it was not until the seventeenth century that the industry expanded rapidly. Early whaling targeted right and bowhead whales and then shifted to sperm whales. Technological developments in the nineteenth century allowed other species to be exploited, and at the end of the century, the possibilities of whaling in the Antarctic were realised. Nearly three million whales were killed in the twentieth century, with several populations reduced to 5% of their historical size, or less. The International Whaling Commission was established in 1946 and manages whaling, and it introduced a whaling moratorium in 1982. Despite this moratorium whales are still caught by Norwegian, Icelandic and Japanese whalers. Indigenous communities are given quotas to hunt whales, although not without some controversy. Small cetaceans are also hunted, by indigenous communities, and there are sizeable hunts in several locations such as Peru, the Faroe Islands and Japan. Human interactions with cetaceans also include indirect interactions via pollutants. One type of interaction for which there is growing concern is whale and dolphin watching. Despite many locations having guidelines or regulations to minimise the impacts of this type of tourism, compliance is often low. However, if properly managed, whale watching could be an economically valuable, non-consumptive, use of living whales as a resource by humans.

E.C.M. Parsons (✉) • D. Monaghan-Brown
Department of Environmental Science & Policy, George Mason University, Fairfax, VA, USA
e-mail: ecm-parsons@earthlink.net

A. Butterworth (ed.), *Marine Mammal Welfare*, Animal Welfare 17,
DOI 10.1007/978-3-319-46994-2_5

5.1 The History of Whaling

Bones of dolphins at an archaeological site in the Sea of Japan suggest that hunting of cetaceans began in this region before 2000 BC (Hiraguchi 1992; Aikens et al. 2009). In addition, harpoons from Pacific coast sites and the East China Sea, as well as cetacean-related relics of whaling that have been found in Hokkaido, Japan, show that whaling has been established in this region for well over a thousand years (Yamaura 1998). Commercial whaling where whale products became a trade commodity (rather than a local subsistence activity) did not develop until the seventeenth and eighteenth century in Japan, and even then, it was on a small scale and limited to small localised regions (Takahashi 1992; Osumi 2003). Large-scale commercial whaling and widespread consumption of whale meat did not begin in Japan until after the Second World War, and it did so at the urging and behest of the US occupying forces, particularly US General Douglas MacArthur (Scott 1999), as whale meat was considered to a means to ensure food security for a post-war Japanese population.

Alaska also has a long history of whaling: the Inuit have hunted cetaceans for over 1000 years (Whiteridge 1999). Due to this long association, whaling has an important role in the culture of native Alaskans, and various ceremonies and traditions are involved with the hunt (Freeman 2005). For example, the skull of a hunted whale should be ceremonially returned to the sea to ensure the immortality and reincarnation of the whale and, thus, future hunting success.

In 'Dark Age' (early mediaeval) Britain, Anglo-Saxons frequently used whale materials and hunted coastal species (Gardiner 1997; Gardiner et al. 1998), particularly the Atlantic grey whale (*Eschrichtius robustus, Atlantic subspecies, extinct*). Unfortunately, this particular (sub)species was so accessible to hunters that it had become extinct in the Late Middle Ages (c. 1500s; although persisting in the western Atlantic until the seventeenth century; Mead and Mitchell 1984; Bryant 1995). During the same early mediaeval period, Nordic settlers commonly used whale products from hunted and stranded whales. The use of whale (and other marine mammal) parts is particularly prevalent in archaeological materials from the Western and Northern Isles of Scotland (Szabo 1997). The Norse were so active, a group of whale hunters that the *Speculum Regale* (the King's Mirror), a Nordic saga from the mid-thirteenth century, listed 20 species of whale around the coast of Iceland (Mead and Mitchell 1984).

By the eleventh century, the Basque people from the coasts of Spain and France had begun whaling, and by the end of the twelfth century, they had expanded this into a substantive commercial enterprise (Ellis 1991). They primarily hunted North Atlantic right whales (*Eubalaena glacialis*) in the Bay of Biscay. The right whale was so named because it was the 'right' whale to hunt—it gave a high yield of oil and floated when dead and, thus, could be towed alongside the whaling vessel after it had been killed. The Basques used every part of the whales they hunted, including their bright orange faeces (thanks to a high level of carotenoid pigments in their copepod prey), which in fact was used as fashionable dye for clothes (Parsons et al. 2012). When whale catches declined, the Basques turned their attention to the northwestern Atlantic and by the 1530s were hunting North Atlantic right whales and bowhead whales (*Balaena mysticetus*) off the Atlantic coast of Canada (Barkham 1984).

In the 1610s, the UK sent whaling ships to Spitzbergen, Norway (Purchas 1625), and these were followed soon after by Dutch vessels. The focus species for these hunts were also bowhead whales. On the east coast of the USA, although indigenous populations had probably hunted cetaceans to a small extent, it was European colonists that had an interest in whales as a resource. First, stranded whales were utilised for their oil and baleen, and then in the late 1640s–1950s, colonist began to conduct boat-based operations from shore (Dolin 2008). North Atlantic right whales were the major target species initially. Sperm whaling allegedly began when Captain Christopher Hussey was blown offshore and caught a sperm whale instead of the typical right whales (Hawes 1924). Sperm whaling started to expand and by 1755 a factory specifically to process sperm whale oil was founded in New Bedford, Massachusetts (Ommanney 1971). A rapidly expanding 'Yankee' whaling fleet was curtailed by the American War of Independence (an oil-filled whaling vessel could make a valuable prize for a British naval officer), and by 1812 most of the whaling fleet had been destroyed (Ommanney 1971). However, subsequently sperm whaling activity began to expand again until its height in the 1850s, when the advent of kerosene rapidly began to make whale oil redundant (Ommanney 1971; Dolin 2008).

Sperm whales were sought because the waxy spermaceti organ in the foreheads gave a large yield of high-quality oil—this oil was used for a variety of purposes including lubricating the latest mechanical inventions such as ships' chronometers (which allowed longitudes to be calculated; Sobel 2005) and for soot-free oil lamps. The latter were particularly sought after in the eighteenth and early nineteenth centuries due to the fashion for wallpaper (often imported from Europe at great expense)—sperm whale oil lamps did not leave dirty sooty streaks on the wallpaper.

5.2 The Development of Modern Whaling

The Norwegian Svend Foyn (1809–1894) is often referred to as the 'father' of modern whaling (Ellis 1991). An ex-seal hunter, Foyn noted in the 1860s the abundance of large whales he saw on sealing trips, saying 'God had let the whales inhabit [these waters] for the benefit and blessing of mankind and, consequently, I consider it my vocation to promote these fisheries' (Ellis 1991).

Previously when whales were captured, handheld harpoons attached to ropes were cast from small rowboats. The whales, once secured, were dispatched via stabbing with a lance. As one could imagine, this was hardly a humane method of killing, and it also put the crew of the flimsy wooden whaling boats in much peril. However, in the 1850s the technology of whaling began to develop rapidly (Schmitt et al. 1980; Tonnessen and Johnsen 1982; Ellis 1991):

- 1952—the first explosive harpoon (the bomb lance) was developed.
- 1857—saw the first UK whaling boat powered by a steam engine.
- 1859—the first purpose-built steam-powered whaling ship was manufactured (although catching of whales still done from rowing boats).
- 1861—American Thomas Welcome Roys develops his 'whaling rocket'.
- 1863—Sven Foyn built the first steam-driven whaling schooner (a whale-catching boat).
- 1865—the development of a more accurate exploding harpoon.
- 1868—the cannon-fired explosive harpoon was invented.

The design of the grenade-tipped explosive harpoon has changed little since 1868, and harpoons used on whaling vessels in the present day closely resemble their Victorian predecessor. Quicker whaling vessels combined with more powerful weaponry meant that previously unharvested whales, such as the 25 m, fast-swimming fin whale (*Balaenoptera physalus*), could now be hunted.

5.3 Antarctic Whaling

Antarctic whaling began after a Scottish expedition ventured to the Falkland Islands in 1892–1893 (Watson 2004). The organisers of the expedition had speculated that as there had been an abundance of whales in the Arctic, perhaps the southern polar regions similarly had an abundance of cetaceans. Although this expedition did not actually catch any whales, the discovery of large numbers of these animals around the Subantarctic Islands and the Antarctic Peninsula—by this expedition and a following Norwegian expedition led by Carl Anton Larsen—highlighted the possibilities of Antarctic whaling (Fig. 5.1). Despite an 1895 warning by the illustrious Royal Geographic Society that 'it would be futile to start whaling in the Antarctic; a great number of the large whales had thin blubber and short baleens and it was not worth the whaler's time catching them', evidence of abundant whale populations around Antarctica inspired the establishment of shore-based whaling stations and the start of extensive commercial whaling in the Southern Ocean (Backovic 2013). The Island of South Georgia, in particular, was a major hub in the Antarctic whaling industry at this time, with the first station being established there, at Grytviken, by Larsen.

In 1925, the launch of the *Lancing*, the first whaling factory ship, led to a dramatic change in Antarctic whaling activities. These oversized vessels allowed for slaughtered whales to be processed at sea, and whaling fleets no longer had to waste

Fig. 5.1 Deserted whaling station, Deception Island, Antarctica. *Image credit: Chris Parsons*

time by returning to shore-based factories and processing plants (Clapham and Baker 2002). This greatly increased the whalers' hunting range and the whales' accessibility. For the first time, hunting the substantial numbers of animals inhabiting Antarctic waters—beyond the Antarctic Peninsula and Southern Ocean islands—became an economic viability.

A development in 1929 led to a strange new market for whale oil. The chemical process of hydrogenation had been discovered in the mid-nineteenth century, and this allowed fishy-tasting, liquid whale oil to be turned into solid margarine. However, refinement of this process resulted in a palatable whale-oil margarine that melted at body temperature, namely 'on the tongue' (Tonnessen and Johnsen 1982). The demand increased for margarine, and thus whale oil, especially during post-war years. By 1960, 17% of the fat used in margarine production came from whale oil. Strengthened by advances in vessel technology and weaponry, at its peak, commercial whaling was dangerously productive. Between 1904 and the end of the Second World War, 1.1 million whales were killed globally (Rocha et al. 2014). By the 1980s, the industry had led to the harvesting of an enormous number of whales of a variety of species, including the slaughter of approximately 350,000 blue whales, 700,000 fin whales, 1,000,000 sperm whales and 250,000 humpback whales

Fig. 5.2 Humpback whales (*Megaptera novaeangliae*) in Antarctica. *Image credit: Chris Parsons*

(*Megaptera novaeangliae*) (Fig. 5.2), and hundreds of thousands of other species such as right whales, bowhead whales, sei whales (*Balaenoptera borealis*) and northern (*B. acutorostrata*) and Antarctic minke whales (*B. bonaerensis*) (Whitehead 2002; Clapham and Baker 2002). Such vast numbers of whales were taken that the estimated worldwide blue whale population was reduced to perhaps less than 1% of its original pre-whaling abundance—at one time there may have been only 3000 blue whales left in the world—and several populations were similarly reduced to less than 5% of their historical levels (Laws 1977; Regenstein 1985; Rocha et al. 2014). From the establishment of the first South Georgia whaling station in 1904 to 2000, 2.9 million whales were killed (Rocha et al. 2014).

5.4 The International Whaling Commission

As early as 1931, whalers began to notice that some whale species were declining in numbers (Clapham and Baker 2002). In a bid to manage whale stocks, several whaling countries joined forces to enact first of all the *Convention for the Regulation of Whaling* and then the more significant *International Convention for the Regulation of Whaling*. This latter treaty led, in 1946, to the formation of the International Whaling Commission (IWC), an international forum that is still the main authority for the control of whaling and the management of whale stocks (Maffei 1997). As whale populations succumbed to exploitation, the IWC began to introduce whaling bans on each species one by one:

- 1931—Bowhead whales (*Balaena mysticetus*)
- 1935—Southern (*Eubalaena australis*) and northern right whales (*E. glacialis*)

- 1946—Grey whales (*Eschrichtius robustus*)
- 1966—Humpback whales (*Megaptera novaeangliae*)
- 1966—Blue whales (*Balaenoptera musculus*)
- 1979—Sei whales (*Balaenoptera borealis*) [except in Iceland]
- 1984—Sperm whales (*Physeter macrocephalus*)

Despite whaling bans, a large numbers of whales were taken illegally, for example, Soviet whalers conducting illegal catches decimated the North Pacific right whale (now recognised as *Eubalaena japonica*), despite right whale species being protected by the IWC (Ivashchenko and Clapham 2012). Over 30 years of illegal whaling, Soviet vessels took nearly 179,000 whales that were not reported to the International Whaling Commission (Ivashchenko and Clapham 2014). Illegal catches were also conducted by Japanese whaling vessels in the North Pacific, until the IWC introduced an international observer system in 1972 (Ivashchenko and Clapham 2015).

Whale stocks had become so depleted, and public opinion was so strongly opposed to the continuation of commercial whaling, that in 1982 the IWC voted for a global commercial whaling moratorium (effectively a zero quota for whaling)—this ban eventually came into effect in 1986 (Maffei 1997). Despite the restriction, over 40,000 whales have been killed in hunts since the moratorium was passed (IWC 2016a, b).

Why are whales still being hunted when a moratorium has been put in place? One reason is that countries that do not agree with the moratorium are not required to comply with its guidelines (Clapham and Baker 2002). When the moratorium was enacted, Norway lodged a 'reservation' to the moratorium. Norway is, therefore, not bound by the ban and currently takes an average of 650 northern minke whales per year (Glover et al. 2012), catching a total of 736 in 2014 (IWC 2016b). Iceland likewise hunted whales under a 'reservation' taking a further 24 minke whales and 137 fin whales (IWC 2016b) (Fig. 5.3). Their actions are technically legal, albeit in the face of criticism and condemnation from the majority of IWC member nations. Japan signed the whaling moratorium; however, it takes advantage of a loophole that allows for whales to be killed for scientific research. As a result, in 2014 Japan took 81 northern minke whales, 90 sei whales and 25 Bryde's whales (*Balaenoptera edeni*) in the North Pacific and 252 Antarctic minke whales in the Southern Ocean in the 2013/2014 austral (Southern hemisphere) summer (IWC 2016a).

The integrity of research whaling and its relevance to the scientific community has been questioned by the IWC's own scientists (Clapham et al. 2003, 2006). Although some blubber and stomach content samples are taken, the meat from the whale is then processed and sold in markets. While the Japanese government may state that whaling is an important economic activity for coastal communities and that scientific whaling provides important scientific data (Hirata 2005) (Figs. 5.4 and 5.5), the real reasons why the Japanese government continues to conduct whaling is arguably less about whales as a food resource but more about the politics of marine resource extraction and national pride (Parsons 2015).

Fig. 5.3 Fin whale (*Balaenoptera physalus*) hunted at sea and brought ashore for flensing (cutting up the carcass) in Iceland, 2013. *Image credits: WDC, Whale and Dolphin Conservation*

Fig. 5.4 Japanese grenade-tipped harpoon fired from the whaling vessel *Yushin Maru*. *Image credit: Jeremy Sutton-Hibbert*

Fig. 5.5 Minke whale (*Balaenoptera acutorostrata*) showing movement and, thus, showing that it is still alive at the time of this photograph, being hauled up by the harpoon line to the whaling vessel *Yushin Maru*. *Image credit: Jeremy Sutton-Hibbert*

Fig. 5.6 Whale bones derived from whaling activity at Hannah Point whaling station, Antarctica. *Image credit: Chris Parsons*

In 1994, the IWC designated the Southern Ocean Sanctuary, which includes the waters surrounding Antarctica (Fig. 5.6), as a 'whale sanctuary'. Within this sanctuary area, commercial whaling is prohibited regardless of the existence or non-existence of the moratorium (Zacharias et al. 2006). Despite this, Japan continued to hunt whales in the Antarctic for 'scientific purposes' which they consider to be exempted from the sanctuary's ban. However, in 2014, the UN's International Court of Justice (in a case brought by Australian and New Zealand governments) ruled that Japan's Antarctic 'scientific whaling programme' was not scientific research, but effectively commercial whaling (Clapham 2015; Parsons 2015). It therefore violated the Southern Ocean Sanctuary and the International Court of Justice ordered the Japanese government to cease their whaling programmes. Japan initially said that it would abide by the court's ruling, but has subsequently resumed whaling around Antarctica (Clapham 2015; Parsons 2015).

5.5 The Revised Management System

The whaling moratorium was enacted as a precaution to allow whale stocks to recover and prevent commercial whaling, theoretically until a sustainable quota system could be put in place. This quota system is called the Revised Management System (RMS) (Clapham and Baker 2002). In order for the RMS to be successful, there needs to be accurate information on the number of animals in each whale stock and accurate information on how many whales are/have been harvested.

Some antiwhaling countries and environmental NGOs want the IWC to accept the RMS and allow commercial whaling under the new quota system. Their hope is that the RMS will produce quotas smaller than the number of animals currently being taken by Norway and Japan. However, this does assume that the quotas allotted are sustainable and that the quotas are strictly adhered to, amongst other assumptions. Therefore, many animal welfare and environmental NGOs are opposed to any resumption of commercial whaling, whether under the RMS or not.

Scientists are still arguing about the accuracy of sighting surveys and numbers of animals. For example, recent circumpolar whale surveys estimated only 40% of the number of minke whales documented in the previous survey (Branch and Butterworth 2001). Why was there such a massive decrease in whale numbers? No one knows. Were previous surveys incorrect and the recent surveys more accurate? Is the decline real, perhaps the result of climate change impacting Antarctic minke whales and/or their prey? Under IUCN red list criteria, an actual, *observed* or *implied* 50% decrease of whales within a 10-year period would make the species 'endangered' (IUCN 2001); therefore, theoretically, Antarctic minke whales could be considered to be endangered, yet such a listing has not been made, to date.

5.6 Killing Methods

An additional controversy is the method used to kill whales, which has largely unrefined since the development of the cannon-fired harpoon by Svend Foyn (noted above). Detonation of a grenade at the tip of the harpoon is supposed to cause a percussive shock that renders the target whale immediately unconscious. Concerns over suffering of hunted whales led the IWC to form a working group in 1982 on "Whale Killing Methods and Associated Welfare Issues", to discuss whether or not whale-killing methods were humane. The IWC's definition of humane killing was agreed as "death brought about without pain, stress or distress perceptible to the animal". To monitor this whalers are, for example, requested to record times to death (or TTD), or instantaneous death rates (IDR), of whales after being harpooned. However, in 1993 only 54% of northern minke whales hunted by Norwegians were rendered immediately immobile (and presumed stunned/unconscious) (Øen 2001). By 2000 this had risen to approximately 80% (Øen 2001), but it still meant that 20% of whales were not immediately rendered insensible, and thus animals could have been suffering significantly (Knowles and Butterworth 2006). In the 2002/2003 Antarctic hunt, approximately 60% of whales killed were recorded as not dying 'instantaneously' (Ishikawa 2003). In Greenlandic aboriginal whaling hunts (described below), instances of northern minke whales taking 5 h to die were reported in 2002 (Anon 2003), and time to death of up to 12 h have been reported for fin whales.

5.7 Aboriginal Whaling

Some Indigenous communities who demonstrate a *traditional*, *nutritional* and *cultural* need, such as the previously mentioned Inuit of Alaska, are permitted to hunt a specific quota of whales by the IWC, in order to maintain their historical traditions and lifestyles. This currently occurs in the Russian Federation, Greenland (a protectorate of Denmark), Bequia (St. Vincent and the Grenadines) and in the USA (see Table 5.1).

Although many have sympathy towards some forms of aboriginal whaling, it can become a contentious issue. For example, in Barrow, Alaska, the Inupiat and Yup'ik have hunted bowhead whales for over 2000 years (Krupnik and Bogoslovskaya 1999). As a result, they are allocated an aboriginal whaling quota of approximately 60 whales. This hunt has been controversial because bowhead whales are considered to be endangered under US law. Moreover, the hunt does not involve traditional methods: exploding harpoons, spotter planes and motorised/speed boats are all used during the hunt (Reeves 2002). However, these nontraditional methods ultimately mean the hunt is more humane (with less time taken for the whale to die than would occur using traditional methods) and the hunt would also likely be safer for human participants. After the meat is used by the local community, small amounts of whale bone and baleen can be taken and carved into crafts, which in turn could be sold to the general public.

Table 5.1 Whale species taken under aboriginal whaling, as prescribed by the IWC

		2011	2012	2013	2014
Greenland (Denmark)	Fin whale	5	5	9	12
	Humpback whale	8	10	8	7
	Northern minke whale	189	152	181	157
St. Vincent and the Grenadines	Humpback whale	2	2	4	2
Russia	Grey whale	128	143	127	124
	Bowhead whale			1	
USA	Bowhead whale	51	69	57	53

A somewhat more controversial situation occurred with the Makah from Washington State, USA (Jenkins and Romanzo 1998). In 1996, the Makah appealed for an aboriginal whaling quota of grey whales (*Eschrichtius robustus*) which initially was rejected. In 1997, however, they issued a joint proposal to the IWC with the Chukotka people of Russia (IWC 1998). It should be noted that this was controversial as the Makah have not actually hunted whales since 1926 (Reeves 2002), and many argued that as a result, there was no subsistence nutrition need for the Makah to hunt the whales. Moreover, the hunt is supposed to be traditional but today the Makah use nontraditional methods: the hunters use speed boats and assault rifles (Kemmerer 2004).

Japan and Norway also tried to gain a quota of whales for 'small-type community-based coastal whaling'. They suggested that this type of whaling was similar to aboriginal whaling. They also suggested that the 'village' communities that would conduct this whaling were both small and impoverished (Hirata 2005). One of these whaling 'villages' is Shimonoseki—with a population of over a quarter of a million people, roughly half that of Washington, DC—as well as being a major port (Hidekazu 2013). Moreover, the type of whaling vessels operating from this 'village' would be large, deep-draft, Antarctic-style whale catcher ships, hardly comparable to the small vessels used by several aboriginal whaling groups.

5.8 Hunting Small Cetaceans

'Small cetacean' is a term that includes all toothed whales (except sperm whales) as well as dolphins and porpoises. It has been argued by some member nations that hunts for small cetaceans are not covered by the IWC, and the IWC should restrict itself to the management of baleen and sperm whales only, although other nations argue that small cetacean management is within the competence of the IWC. Several countries conduct hunts of small cetaceans, but because of the dispute at the IWC over competency, there is currently no international forum or organisation that is governing or managing these hunts.

At present, Russia, Canada, Greenland and the USA all conduct hunts of beluga whales (*Delphinapterus leucas*) and narwhals (*Monodon monoceros*) (Table 5.2). The products of these hunts are blubber and meat which are taken for human and sled dog consumption. Moreover, the tusks of the narwhals are used for tent poles, sled runners and lance shafts and sold as curios to tourists (Reeves 1992).

Table 5.2 Approximate numbers of small cetaceans taken in whaling hunts

Country	Species	Number hunter per year
Russia	Beluga whale	2000
Canada	Beluga whale	2000
	Narwhal	1000
West Greenland	Beluga whale	200
	Narwhal	1000
Alaska	Beluga whale	200

5.9 The Faroe Island Hunt

The Faroe Islands are situated north and west of the Shetland Isles and are an independent protectorate of Denmark. Hunts of long-finned pilot whales (*Globicephala melas*) have been conducted in the Faroes since 1584 and are ongoing. The methods used in these hunts have changed little since the mediaeval times: pilot whales are driven into bays by boats; hooked gaffs are driven into their blowholes and used to drag the animals to shore; the heads of the live whales are then sawn off with machete-like knives (Gibson-Lonsdale 1990; Fielding 2010). This has led to considerable concern about whether the hunt is inhumane. From 1709 until the present day, over 250,000 pilot whales have been killed in this hunt, with an average of 1200/year being taken over the last 10 years (Wade et al. 2012).

Claims are made that, despite concerns about welfare aspects of the hunt, it is an important part of Faroese culture and an essential component of the islander's nutrition. However, the Faroe islanders have an extremely high standard of living (aided by revenue from oil exploration and extraction in nearby waters), and moreover whale meat carries health warnings due to high levels of toxic mercury, cadmium and organochlorines (PCBs and pesticides) (Simmonds et al. 1994; Weihe et al. 1996).

Currently the largest commercial hunts for small cetaceans occur in Peru and Japan.

5.10 Peru

As recently as a decade ago there were active hunts for dolphins and porpoises originating from 60 ports in Peru and taking approximately 20,000 animals annually. The casualties of this hunt were mostly dusky dolphins (*Lagenorhynchus obscurus*) and Burmeister's porpoises (*Phocoena spinipinnis*) which were taken for human consumption and to be cut up for fishing bait (Alfaro-Shigueto et al. 2008). In 1990, these catches were banned, but due to lack of enforcement, the ban had little effect. In 1996, due to international pressure and media attention, enforcement of the ban improved, although catches of small cetaceans still continue (Majluf et al. 2002).

5.11 Japan

Likely the most prolific hunter of small cetaceans, Japan currently hunts for 16 different species. These include 50 Baird's beaked whales (*Berardius bairdii*) per year, hunted with explosive harpoons (Butterworth et al. 2013); Pacific white-sided dolphins (*Lagenorhynchus obliquidens*); and approximately 12,000 Dall's porpoises (*Phocoenoides dalli*) per year, which are killed with small handheld harpoons (Kasuya 2007). In addition, Japanese drive fisheries (when animals are driven into bays and inlets by fast boats and slaughtered on the shore) take striped dolphins (*Stenella coeruleoalba*), Risso's dolphins (*Grampus griseus*), bottlenose dolphins (*Tursiops truncatus*), pantropical spotted dolphins (*Stenella attenuata*), short-finned pilot whales (*Globicephala macrorhynchus*), false killer whales (*Pseudorca crassidens*) and killer whales (*Orcinus orca*) (Brownell et al. 2008). Animals killed in drive fisheries and commercial hunts are mainly rendered down to produce fertilisers and animal and fish feed—with a small fraction being sold for human consumption and with some animals taken for dolphinariums and aquariums (Shoemaker 2005; Kasuya 2007).

5.12 Cetacean Culls

Commercial whaling and hunts are not the only cause of death for cetaceans at the hands of humans. Despite lack of suitable evidence, marine mammals are often used as a scapegoat for declining fisheries, and as such cetaceans have often been culled for this reason (Plaganyi and Butterworth 2002). For example, culls of cetaceans have been conducted in the Black Sea since the 1800s. It was estimated that between 1931 and 1941, 50,000 small cetaceans were being culled each year (Birkun 2002). Other specific examples of cetacean culls include beluga whales bombed in Quebec, Canada, between 1920 and 1930 (Brennin et al. 2007) and killer whales being machine gunned and depth charged in Iceland in 1956 (Sigurjonsson 1984); and as recently as the 1980s, killer whales were culled with guns and explosives in Alaska (Ford et al. 2000).

5.13 Pollution

One of the most insidious and widespread threats to cetacean populations is marine pollution. Pollution comes in a variety of forms including industrial waste, agricultural chemicals, sewage, radioactive discharges, litter, oil and noise. Cetaceans occupy a high trophic level and as such are particularly susceptible to contaminants. Small concentrations of these contaminants can accumulate and become magnified higher up the food chain.

One class of pollutant that is of particular concern to cetaceans are organochlorines (e.g. PCBs and pesticides such as DDT). Other pollutants of concern include mercury, tributyltin (TBT) and polyaromatic hydrocarbons (PAHs). Many of these pollutants are immunosuppressive, damaging the immune system making cetaceans more susceptible to disease (Jepson et al. 2005; De Guise et al. 1995; Fossi and Marsili 2003). Organochlorines, in particular, are hormone mimics and can cause infertility, foetal abnormalities, mental retardation and growth abnormalities (Reijnders 2003). PAHs are carcinogenic and can cause DNA damage and malignant tumour growth (Misaki et al. 2015). DDT and mercury can be directly toxic, causing terminal neurological damage (Clarkson 1987; Irwin 2005). Many of the pollutants mentioned above are lipid soluble and accumulate in the blubber layer of cetaceans. While in the blubber layer, the pollutants are effectively inert; however, they can be mobilised in periods of low food availability, stress or disease or pregnancy and lactation. During the latter, lipid-soluble pollutants can be passed to the offspring both in the womb and, in particular, via lactation (Borrell 1993). Therefore, young cetaceans have been found with extremely high-contaminant burdens. These elevated levels have been linked to increases in calf mortality. To put the seriousness of the situation in context, cetaceans in the St. Lawrence Estuary (Canada) and Hong Kong have been found that are so contaminated with organochlorines that their tissues technically could be classified as toxic waste (Waldichuk 1989; Parsons 2004).

5.14 Tourism

A growing threat to cetaceans is tourism. Commercial whale watching[1] first began in 1955 with a Californian fisherman offering US$1 trips to see grey whales (*Eschrichtius robustus*) (Hoyt and Parsons 2014). It has subsequently grown into an international industry worth over US$2.1 billion (Hoyt 2009; O'Connor et al. 2009). The industry could arguably be worth significantly more if pro-whaling nations turned to cetacean tourism as a nonlethal alternative to utilising whales by hunting them (Parsons et al. 2003; Parsons and Rawles 2003; Parsons and Draheim 2009). Although whale watching can have positive educational and economic impacts and potentially assist in the conservation of cetaceans, there are many negative effects (Parsons 2012). The majority of whale watching is boat based (Figs. 5.7 and 5.8), which has a direct impact on cetacean behaviour, often causing a reduction in biologically important behaviours such as feeding or resting (Parsons 2012). As the whale-watching industry has expanded, vessels have become larger and faster, and a plethora of companies can lead to overcrowding around animals in many areas (Hoyt and Parsons 2014). In addition to disturbing the behaviour of whales, whale-watching vessels can have more direct impacts, such as boat collisions with animals leading to injuries and/or death (Laist et al. 2001). Noise produced by whale-watching vessels may also smother or 'mask' important communication calls (Jensen et al. 2008). Other behavioural changes seen as a result of whale watching include effects on their directional movements, surfacing rates or feeding activities (Parsons 2012). Repeated interruptions of natural behaviours over time, and chronic exposure to boat noise, could lead to elevated stress levels in cetaceans, which in time could impact animals' health (Orams 2004; Wright et al. 2007).

To reduce the impacts of whale-watching vessels on cetaceans, many locations have guidelines for boat operators. Some of the guidelines are legal, although the majority are only voluntary guidelines (Garrod and Fennell 2004). Even if areas have guidelines or regulations, however, it does not necessarily mean that they are enforced or com-

[1] The term whale watching generally refers to any type of viewing activity by tourists on wild cetaceans, including dolphins and porpoises.

Fig. 5.7 Watching northern bottlenose whales in Broadford Bay, Scotland. *Image credit: Chris Parsons*

Fig. 5.8 A whale-watching boat violating the US Marine Mammal Protection Act by approaching a blue whale too closely off the coast of San Diego. *Image credit: Chris Parsons*

plied with. Low levels of compliance with whale-watching guidelines have been reported from many locations. This is despite government agencies having responsibility for management of enforcement of these guidelines, or occurs even, and despite the presence of international concern and scrutiny (Scarpaci et al. 2003, 2004; Lusseau 2004; Kessler and Harcourt 2013; Sitar et al. 2016). Whale-watching researchers Brian Garrod and David Fennell (2004) warned that whale watching 'is growing very fast—several times faster than tourism more generally—and is doing so in a patchy regulatory environment—and that there must be concerns over its sustainability'.

It should be emphasised that in many areas whale-watching operators do act responsibly and are supportive of whale-watching guidelines. For example, in western Scotland, the majority of whale-watching tour operators accept and adopt codes of conduct, although they do express a preference for locally, or operator-produced, guidelines (Parsons and Woods-Ballard 2003), i.e. guidelines that are produced 'bottom up' by involving local stakeholders, rather than 'top down' ones enacted by distant government authorities.

Although whale watching can have substantive impacts on cetaceans, it can be argued that whale watching gives an important (arguably non-consumptive) economic value to living whales, as opposed to the value of a dead whale caught during commercial whaling. Moreover, if properly managed, whale watching can have

Fig. 5.9 A presentation on a whale-watching boat—the whale watchers are finding out about the planktonic prey of right whales. *Image credit: Chris Parsons*

minimal impacts on the target species and can have positive impacts on the pubic via education (Fig. 5.9) and interpretation of whales and their associated ecosystems.

5.15 Habitat Degradation

One issue that is of concern is the high rate of loss and degradation of cetacean habitat caused by human activities. Many cetacean species have very precise habitat requirements, and if they are forced from or lose this habitat, it can have serious consequences for the feeding behaviours, health and viability of populations (Bearzi et al. 2004). Habitat loss and degradation is caused by many factors including land reclamation, dams and barrages, dredging, siltation, boat traffic, oil exploration (seismic surveys) and other noise pollution (e.g. military sonar). The issues of noise and habitat degradation are dealt with elsewhere in this volume (see Chap. 7).

5.16 Concluding Remarks

Commercial whaling very nearly caused the complete extinction of several of species of cetacean, most notably the blue whale. Despite the IWC moratorium, whaling continues, and even worse, great numbers of small cetaceans are hunted without any regard or regulation. Despite the popularity of cetaceans and the degree of public sympathy of their plight, many populations are on the edge of extinction: for example, fisheries by-catch, pollution and habitat loss have driven populations of Yangtze river dolphins (*Lipotes vexillifer*) to extinction (Turvey et al. 2007). If humans can drive species with such a high media profile as whales and dolphins to this fate, one wonders how other, less charismatic species can hope to be conserved.

References

Aikens CM, Zhushchikhovskaya IS, Rhee SN (2009) Environment, ecology, and interaction in Japan, Korea, and the Russian Far East: the millennial history of a Japan Sea Oikumene. Asian Perspect 48:207–248

Alfaro-Shigueto J, Mangel JC, Van Waerebeek K (2008) Small cetacean captures and CPUE estimates in artisanal fisheries operating from a port in northern Peru, 2005–2007. Scientific Committee document SC/60/SM19, International Whaling Commission, June 2008, Santiago, Chile

Anon (2003) Times to death in Greenlandic minke and fin whale hunt in 2002. Paper presented at the 55th Meeting of the International Whaling Commission. IWC/55/WK12Rev

Backovic L (2013) Rust and ruin: the downfall of Antarctic whaling. Spiegel Online International. http://www.spiegel.de/international/zeitgeist/history-of-the-grytviken-whaling-station-near-antarctica-a-926711.html. Accessed 1 May 2016

Barkham SH (1984) The Basque whaling establishments in Labrador 1536–1632: a summary. Arctic 37:515–519

Bearzi G, Holcer D, Notarbartolo-di-Sciara G (2004) The role of historical dolphin takes and habitat degradation in shaping the present status of northern Adriatic cetaceans. Aquat Conserv Mar Freshwat Ecosyst 14:363–379

Birkun A (2002) Cetacean direct killing and live capture in the Black Sea. In: Notarbartolo di Sciara G (ed) Cetaceans of the Mediterranean and Black Seas: state of knowledge and conservation strategies. Report to the ACCOBAMS Secretariat, Monaco, February 2002, Section 6, 10 p

Borrell A (1993) PCB and DDTs in blubber of cetaceans from the Northeastern North Atlantic. Mar Pollut Bull 26:146–151

Branch TA, Butterworth DS (2001) Southern hemisphere minke whales: standardised abundance estimates from the 1978/79 to 1997/98 IDCR-SOWER surveys. J Cetacean Res Manag 3:143–174

Brennin R, Alonso E, Areitio P, Barret-Lennard L, Bill K, Binder T, Brownell R, Castellote M, Frost KJ, Geraci J, Goodwin W, Kotokak M (2007) Conference working paper for the first International Workshop on Beluga Whale Research, Husbandry and Management

Brownell RL, Nowacek DP, Ralls K (2008) Hunting cetaceans with sound: a worldwide review. J Cetacean Res Manag 10:81–88

Bryant PJ (1995) Dating remains of gray whales from the Eastern North Atlantic. J Mammal 76:857–861

Butterworth A, Brakes P, Vail CC, Reiss D (2013) A veterinary and behavioral analysis of dolphin killing methods currently used in the "drive hunt" in Taiji, Japan. J Appl Anim Welf Sci 16:184–204

Clapham PJ (2015) Japan's whaling following the International Court of Justice ruling: brave new world or business as usual? Mar Policy 51:238–241

Clapham PJ, Baker CS (2002) Whaling, modern. In: Perrin WF, Wursig B, Thewissen JGM (eds) Encyclopedia of Marine Mammals. Academic Press, New York, pp 1328–1332

Clapham PJ, Berggren P, Childerhouse S, Friday NA, Kasuya T, Kell L, Kock KH, Manzanilla-Naim S, Notabartola Di Sciara G, Perrin WF, Read AJ, Reeves RR, Rogan E, Rojas-Bracho L, Smith TD, Stachowitsch M, Taylor BL, Thiele D, Wade PR, Brownell RL (2003) Whaling as science. Bioscience 53:210–212

Clapham PJ, Childerhouse S, Gales N, Rojas L, Tillman M, Brownell B (2006) The whaling issue: conservation, confusion and casuistry. Mar Policy 31:314–319

Clarkson TW (1987) Metal toxicity in the central nervous system. Environ Health Perspect 75:59–64

De Guise S, Martineau D, Beland P, Fournier M (1995) Possible mechanisms of action of environmental contaminants on St. Lawrence beluga whales (*Delphinapterus leucas*). Environ Health Perspect 103(Suppl. 4):73–77

Dolin EJ (2008) Leviathan: the history of whaling in America. W. W. Norton and Company, New York

Ellis R (1991) Men and whales. Lyons Press, Guilford, CT

Fielding R (2010) Environmental change as a threat to the pilot whale hunt in Faroe Islands. Polar Res 19(3):430–438

Ford JK, Ellis GM, Balcomb KC (2000) Killer whales: the natural history and genealogy of *Orcinus orca* in British Columbia and Washington. UBC Press, Vancouver

Fossi MC, Marsili L (2003) Effects of endocrine disruptors in aquatic mammals. Pure Appl Chem 75(11–12):2235–2247

Freeman MMR (2005) "Just one more time before I die": securing the relationship between Inuit and whales in the Arctic Regions. Senri Ethnological Stud 67:59–76

Gardiner MF (1997) The exploitation of sea mammals in mediaeval mammals; their bones and their social context. Archaeol J 154:173–195

Gardiner M, Stewart JR, Bell G (1998) Anglo-Saxon whale exploitation: some evidence from Dengemarsh, Lydd, Kent. Medieval Archaeol 42:96–101

Garrod B, Fennell DA (2004) An analysis of whalewatching codes of conduct. Ann Tour Res 31:334–352

Gibson-Lonsdale JJ (1990) Pilot whaling in the Faroe Islands–its history and present significance. Mammal Rev 20:44–52

Glover KA, Haug T, Oien N, Walloe L, Lindblom L, Seliussen BB, Skaug HJ (2012) The Norwegian minke whale DNA register: a data base monitoring commercial harvest and trade of whale products. Fish Fish 13:313–332

Hawes CB (1924) Whaling. Heinemann, London

Hidekazu I (2013) Market area analysis of ports in Japan. In: The IAME2013 Annual Conference. International Association of Maritime Economists, Marseille, pp 1–21

Hiraguchi T (1992) Catching dolphins at the Mawaki site, central Japan, and its contribution to Jomon society. In: Aikens CM, Rhee SN (eds) Pacific Northeast Asia in prehistory: hunter-fisher-gatherers, farmers, and sociopolitical elites. Washington State University Press, Pullman, pp 35–45

Hirata K (2005) Why Japan supports whaling. J Int Wildl Law Policy 8:129–149

Hoyt E (2009) Afterword by Erich Hoyt. In: O'Connor S, Campbell R, Knowles T, Cortez H (eds) Watching worldwide: tourism numbers, expenditures and economic benefits. IFAW, Yarmouth Port, MA, pp 283–287

Hoyt E, Parsons ECM (2014) The whale watching industry: historical development. In: Higham J, Bjeder L, Williams R (eds) Whale-watching, sustainable tourism and ecological management. Cambridge University Press, Cambridge, pp 57–70

Ishikawa H (2003) Report on whale killing methods in the 2002/2003 JARPA and improvement of the time to death in the Japanese whale research programs (JARPA and JARPN). Paper presented at the 55th meeting of the International Whaling Commission, IWC/55/WK25

Irwin L (2005) Marine toxins: adverse health effects and biomonitoring with resident coastal dolphins. Aquat Mamm 31(2):195–225

IUCN (2001) IUCN red list categories and criteria. Version 3.1. IUCN, Gland, Switzerland

Ivashchenko YV, Clapham PJ (2012) Soviet catches of bowhead (*Balaena mysticetus*) and right whales (*Eubalaena japonica*) in the North Pacific and Okhotsk Sea. Endang Species Res 18:201–217

Ivashchenko YV, Clapham PJ (2014) Too much is never enough: the cautionary tale of Soviet whaling. Mar Fish Rev 76:1–21

Ivashchenko YV, Clapham PJ (2015) What's the catch? Validity of whaling data for Japanese catches of sperm whales in the North Pacific. R Soc Open Sci 2:150177

IWC (International Whaling Commission) (1998) Chairman's report of the forty-ninth annual meeting. Rep Int Whale Comm 48:17–51

IWC (International Whaling Commission) (2016a) Catches taken: special permit. https://iwcint/table_permit. Accessed 1 May 2016

IWC (International whaling Commission) (2016b) Catches taken: under objection or under reservation. https://iwcint/table_objection. Accessed 1 May 2016

Jensen FH, Wahlberg M, Bejder L, Madsen PT (2008) Noise levels and masking potential of small whale-watching and research vessels around two delphinid species. Bioacoustics 17:166–168

Jepson PD, Bennett PM, Deaville R, Allchin CR, Baker JR, Law RJ (2005) Relationships between polychlorinated biphenyls and health status in harbor porpoises (*Phocoena phocoena*) stranded in the United Kingdom. Environ Toxicol Chem 24(1):238–248

Jenkins L, Romanzo C (1998) Makah whaling: aboriginal subsistence or a stepping stone to undermine the commercial whaling moratorium? Colo J Int Environ Law Policy 9:71–144

Kasuya T (2007) Japanese whaling and other cetacean fisheries. Environ Sci Pollut Res Int 14(1):39–48

Kemmerer L (2004) Killing traditions: consistency in applied moral philosophy. Ethics Place Environ 7(3):151–171

Kessler M, Harcourt R (2013) Whale watching regulation compliance trends and the implications for management off Sydney, Australia. Mar Policy 42:14–19

Knowles TG, Butterworth A (2006) Immediate immobilisation of a minke whale using a grenade harpoon requires striking a restricted target area. Anim Welf 15:55–57

Krupnik I, Bogoslovskaya L (1999) Old records, new stories: ecosystem variability and subsistence hunting in the Bering Strait area. Arct Res U S 13:15–24

Laist DW, Knowlton AR, Mead JG, Collet AS, Podesta M (2001) Collisions between ships and whales. Mar Mamm Sci 17:35–75

Laws RM (1977) Seals and whales of the Southern Ocean. Philos Trans R Soc Lond B Biol Sci 279:81–96

Lusseau D (2004) The state of the scenic cruise industry in doubtful sound in relation to a key natural resource: bottlenose dolphins. In: Hall M, Boyd S (eds) Nature-based tourism in peripheral areas: development or disaster? Channel View Publications, Clevedon, pp 246–260
Maffei MC (1997) The International convention for the regulation of whaling. Int J Mar Coastal Law 12(3):287–305
Majluf P, Babcock EA, Riveros JC, Arias Schreiber M, Alderete W (2002) Catch and by-catch of seabirds and marine mammals in the small-scale fishery of Punta San Juan, Peru. Conserv Biol 16(5):1333–1343
Mead JG, Mitchell ED (1984) The Atlantic gray whale. In: Jones ML, Swartz SL, Leatherwood S (eds) The gray whale: *Eschrichtius robustus*. Academic Press, New York, pp 33–53
Misaki K, Takamura-Enya T, Ogawa H, Takamori K, Yanagida M (2015) Tumour-promoting activity of polycyclic aromatic hydrocarbons and their oxygenated or nitrated derivatives. Mutagenesis 31(2):205–213
O'Connor S, Campbell R, Cortez H, Knowles T (2009) Whale watching worldwide: tourism numbers, expenditures and expanding economic benefits. International Fund for Animal Welfare. Yarmouth, MA
Øen EO (2001) Norwegian minke whaling 2000. Paper presented at the 53rd Meeting of the International Whaling Commission, IWC/53/WKM and AWI6
Ommanney FD (1971) Lost leviathan: whales and whaling. Hutchinson, London
Orams M (2004) Why dolphins may get ulcers: considering the impacts of cetacean-based tourism in New Zealand. Tour Mar Environ 1(1):17–28
Osumi S (2003) Kujira to Nihonjin (Whales and the Japanese). Iwanami Shoten, Tokyo
Parsons ECM (2004) The potential impacts of pollution on humpback dolphins - with a case study on the Hong Kong population. Aquat Mamm 30:18–37
Parsons ECM (2012) Negative impacts of whale watching. J Mar Biotechnol 2012:807294
Parsons ECM (2015) Levels of marine human-wildlife conflict: a whaling case study. In: Draheim MM, Madden F, McCarthy JB, Parsons ECM (eds) Human-wildlife conflict: complexity in the marine environment. Oxford University Press, Oxford, pp 79–96
Parsons ECM, Draheim M (2009) A reason not to support whaling: a case study from the Dominican Republic. Curr Issue Tour 12(4):397–403
Parsons ECM, Rawles C (2003) The resumption of whaling by Iceland and the potential negative impact in the Icelandic whale-watching market. Curr Issue Tour 6:444–448
Parsons ECM, Woods-Ballard A (2003) Acceptance of voluntary whalewatching codes of conduct in West Scotland: the effectiveness of governmental versus industry-led guidelines. Curr Issue Tour 6:172–182
Parsons ECM, Bauer A, McCafferty D, Simmonds MP, Wright AJ (2012) An introduction to marine mammal biology and conservation. Jones and Bartlett Publishing, London
Parsons ECM, Warburton CA, Woods-Ballard A, Hughes A, Johnston P (2003) The value of conserving whales: the impacts of cetacean-related tourism on the economy of rural West Scotland. Aquat Conserv 13:397–415
Plaganyi EE, Butterworth DS (2002) Competition with fisheries. In: Perrin WF, Wursig B, Thewissen JGM (eds) Encyclopedia of marine mammals. Academic Press, New York, pp 269–275
Purchas S (1625) Hakluytus posthumus or Purchas his pilgrimes: contayning a history of the world in sea voyages and lande travells by Englishmen and others, vol 13 and 14. Reprinted 1905 by Maclehose J, Glasgow
Reeves RR (1992) Recent developments in the commerce in narwhal ivory from the Canadian Arctic. Arct Alp Res 24(2):179–187
Reeves RR (2002) The origins and character of 'aboriginal subsistence' whaling: a global review. Mammal Rev 32(2):71–106
Regenstein L (1985) Animal rights, endangered species, and human survival. In: Singer P (ed) Defense of animals. Basil Blackwell, New York, pp 118–132

Reijnders PJH (2003) Reproductive and developmental effects of environmental organochlorines on marine mammals. In: Vos JG, Bossart GD, Fournier M, O'Shea TJ (eds) Toxicology of marine mammals. Taylor and Francis, London, pp 55–66

Rocha RC, Clapham PJ, Ivashchenko YV (2014) Emptying the oceans: a summary of industrial whaling catches in the 20th century. Mar Fish Rev 76(4):37–48

Scarpaci C, Nugegoda D, Corkeron PJ (2003) Compliance with regulations by "swim-with-dolphins" operations in Port Philip Bay, Victoria, Australia. Environ Manag 31:342–347

Scarpaci C, Nugegoda D, Corkeron PJ (2004) No detectable improvement in compliance to regulations by "swim-with-dolphin" operators in Port Philip Bay, Victoria, Australia. Tour Mar Environ 1:41–48

Schmitt FP, de Jong C, Winter FH (1980) Thomas Welcome Roys, America's pioneer of modern whaling. University of Virginia Press, Charlottesville, VA

Scott S (1999) Australian diplomacy opposing Japanese Antarctic Whaling 1945–1951: the role of legal argument. Aust J Int Aff 52(2):179–192

Shoemaker N (2005) Whale meat in American history. Environ Hist 10(2):269–294

Sigurjonsson J (1984) Killer whale census off Iceland during October 1982. Rep Int Whale Comm 34:611–614

Simmonds MP, Johnston PA, French MC, Reeve R, Hutchinson JD (1994) Organochlorines and mercury in pilot whale blubber consumed by Faroe Islanders. Sci Total Environ 149(1–2):97–111

Sitar A, May-Collado LJ, Wright AJ, Peters-Burton E, Rockwood L, Parsons ECM (2016) Boat operators in Bocas del Toro, Panama display low levels of compliance with national whale-watching regulations. Mar Policy 68:221–228

Sobel D (2005) Longitude. Harper Perennial, New York

Szabo VE (1997) The use of whales in early medieval Britain. In: Lewis CP (ed) The Haskins society journal: studies in medieval history, vol 9. Boydell and Brewer, Rochester, New York, pp 137–157

Takahashi J (1992) Kujira no Nihon Bunkashi: Hogei Bunka no Kooro o Tadoru (The Japanese cultural history of whales: tracing the path of whaling culture). Tankosha, Kyoto

Tonnessen J, Johnsen A (1982) The history of modern whaling. University of California Press, Berkerley

Turvey ST, Pitman RL, Taylor BL, Barlow J, Akamatsu T, Barrett LA, Zhao X, Reeves RR, Stewart BS, Wang K, Wei Z, Zhang X, Pusser LT, Richlen M, Brandon JR, Wang D (2007) First human-caused extinction of a cetacean species? Biol Lett 3:537–540

Wade PR, Reeves RR, Mesnick SL (2012) Social and behavioral factors in cetacean responses to overexploitation: are odontocetes less "resilient" than mysticetes? J Mar Biotechnol 2012:567276

Waldichuk M (1989) St. Lawrence belugas. Mar Pollut Bull 20(10):531–533

Watson N (2004) The Dundee whalers 1750–1914. Tuckwell Press, Edinburgh

Weihe P, Grandjean P, Debes F, White R (1996) Health implications for Faroe Islanders of heavy metals and PCBs from pilot whales. Sci Total Environ 186:141–148

Whitehead H (2002) Estimates of the current global population size and historical trajectory for sperm whales. Mar Ecol Prog Ser 242:295–304

Whiteridge P (1999) The prehistory of Inuit and Yupik whale use. Rev Arqueología Am 16:99–154

Wright AJ, Soto NA, Baldwin AL, Bateson M, Beale CM, Clark C, Deak T, Edwards EF, Fernandez A, Godinho A, Hatch LT, Kakuschke A, Lusseau D, Martineau D, Romero LM, Weilgart LS, Wintle BA, Notarbartolo di Sciara G, Martin V (2007) Do marine mammals experience stress related to anthropogenic noise? J Comp Psychol 20:274–316

Yamaura K (1998) The sea mammal hunting cultures of the Okhotsk Sea with special reference to Hokkaido prehistory. Arct Anthropol 35:321–334

Zacharias MA, Gerber LH, Hyrenbach KD (2006) Review of the Southern Ocean Sanctuary: marine protected areas in the context of the International Whaling Commission Sanctuary Programme. J Cetacean Res Manag 8(1):1–2

Chapter 6
Welfare Issues Associated with Small Toothed Whale Hunts: An Example, the 'Drive Hunt' in Taiji, Japan

Andy Butterworth, Diana Reiss, Philippa Brakes, and Courtney Vail

Abstract In this chapter, we discuss in detail an example of a small toothed whale hunt, with the aim of illustrating the methods used and the welfare questions that can arise in these cetacean hunts. Annually in Japanese waters, small cetaceans are killed in drive hunts with quotas set by the government of Japan. The Taiji Fishing Cooperative in Japan has published the details of a new killing method utilized in these specific hunts that involves cutting (transecting) the spinal cord. Reports claim that this method reduces the time to death. The method involves the repeated insertion of a metal rod followed by the plugging of the wound to prevent blood loss into the water. This method does not appear to lead to an immediate death. The method employed causes damage to the vertebral blood vessels and the vascular rete from insertion of the rod and leads to significant haemorrhage, but this damage alone would not produce a rapid death in a large mammal of this type. The method induces paraplegia (paralysis of the body) and death through trauma and gradual blood loss. We discuss in this chapter how this killing method compares to the recognized requirement for 'immediate insensibility' adopted in killing procedures utilized or considered acceptable in slaughter of farmed animals.

A. Butterworth (✉)
Clinical Veterinary Science, University of Bristol Veterinary School, Langford, N. Somerset, Bristol, UK
e-mail: andy.butterworth@bris.ac.uk

D. Reiss
Department of Psychology, Hunter College, City University of New York, New York, NY, USA
e-mail: dlr28@columbia.edu

P. Brakes
Whale and Dolphin Conservation, 38 St Paul St, Chippenham, Wiltshire, UK

University of Exeter, Penryn Campus, Cornwall, UK
e-mail: p.brakes@exeter.ac.uk

C. Vail
Whale and Dolphin Conservation, 7 Nelson St, Plymouth, MA 02360, USA
e-mail: Courtney@lightkeepersfoundation.com

A. Butterworth (ed.), *Marine Mammal Welfare*, Animal Welfare 17,
DOI 10.1007/978-3-319-46994-2_6

6.1 Introduction

It is estimated that each year within Japanese waters up to 22,000 small whales, dolphins and porpoises (known collectively as 'small cetaceans') are killed in hunts that involve a range of techniques. Most of these small cetaceans are killed in a directed hunt for Dall's porpoises (*Phocoenoides dalli*), but others are taken in a particular category of hunt known as 'drive hunts' or the drive fishery (Kasuya 2007; National Research Institute of Far Seas Fisheries 2009). Drive hunts also take place in other parts of the world, for example, the *grindadráp* (*Grind*) hunt of the long-finned pilot whales (*Globicephala melas*) in the Faroe Islands (Fig. 6.1). The main species taken in the Japanese drive hunts include common bottlenose dolphins (*Tursiops truncatus*), striped dolphins (*Stenella coeruleoalba*), Risso's dolphins (*Grampus griseus*) or short-finned pilot whales (*Globicephala macrorhynchus*). These animals are herded at sea—using small fishing vessels, with underwater noise (this is referred to as the Oikomi method) (Brownell et al. 2008; Ohsumi 1972)—and driven into harbours or shallow coves which have been netted off. Here they are sometimes held for days and then killed (Fig. 6.2).

The Government of Japan sets yearly quotas that allow for up to 2000 dolphins and small whales to be killed in the drive hunts (Kasuya 2007). These hunts are conducted for several reasons: as a means of 'pest control' resulting from the perception that dolphins compete with local fisheries (Brownell et al. 2008; Morisette et al. 2012), to obtain meat for local human consumption, and to procure live

Fig. 6.1 Slaughtered animals on the beach after the *grindadráp* (Grind) hunt of long-finned pilot whales (*Globicephala melas*) in the Faroe Islands. *Image credit*: Hans Peter Roth

Fig. 6.2 Japanese drive hunt. Dolphins, herded at sea using small fishing vessels, with underwater noise, are driven into harbours or shallow coves which have been netted off where they are sometimes held for days, and then either selected for captivity or killed. *Image credits*: Hans Peter Roth

Fig. 6.2 (continued)

dolphins for marine parks and swim-with-the-dolphin programmes in Japan and internationally. After the captured dolphins are rounded up and some selected for marine parks, the remaining individuals are then killed, or occasionally, released (Vail and Risch 2006).

In 2000, fishermen began using a new killing method which Iwasaki and Kai (2010) report as an improved and more humane method of killing. Until the introduction of this new method, the primary tools used for killing were knives and spears, targeted at various parts of the dolphin or whale body. According to data published on the website of the Taiji Fishing Cooperative (Iwasaki and Kai 2010), this revised killing method—which is intended to sever the spinal cord at the junction between the occiput and first cervical vertebra—was tested from December 2000 to February 2001. When the hunt was carried out in 2008, the technique was applied comprehensively to the killing of striped dolphins, and from December 2009 a wooden wedge was driven into the wound to control bleeding and to prevent blood from 'polluting' the water (Iwasaki and Kai 2010).

The drive hunts have drawn a great deal of professional and public interest and concern internationally, particularly in relation to the killing methods used (Hemmi 2011; Reiss 2010). Similar killing methods developed over the past 10 years have also been utilized within the Faroe Island pilot whale drive hunts (Faroese Ministry of Foreign Affairs, Uttanrikisradid 2011) known as the *grindadrap* (or grinds), and these developments in hunting and killing methods have informed the methods utilized in the Japanese dolphin drive hunts (Iwasaki and Kai 2010). The following are excerpts from an English translation of Iwasaki and Kai's original *Improved Method of Killing Dolphins in the Drive Fishery in Taiji, Wakayama Prefecture* (Iwasaki and Kai 2010):

> Purpose: In the 'drive hunt' (Oikomi) in Taiji, dolphins were killed using a spear-type instrument (the conventional method, see below) and were harvested for food. However, in the Faroe Islands, methods to cut around the blood vessel plexus and cervical spine have been developed [the spinal transaction method: (Olsen 1999)]. This method results in a shorter harvest time, and is thought to improve worker safety. We report the results of the implementation of this method.
>
> From December 2000 to February 2001, the spinal cord cutting method was applied to nine Risso's dolphins, four striped dolphins, and two spotted dolphins and one pilot whale. Harvest times were recorded, using the conventional method of killing for a striped dolphin as a control. The criterion for the time of death was the termination of movement and breathing as observed by the worker (fisherman). In December 2008, the technique was applied comprehensively to the killing of striped dolphins. In December 2009, control of bleeding was attempted by driving a wooden wedge into the wound.

In their results section, Iwasaki and Kai (2010) stated that 'The spinal transection method reduced the time to death'. Iwasaki and Kai (2010) also presented images showing the use of the rod and 'the control of bleeding by using the wedge'. They describe the appropriate cutting location as follows:

> Taking the width of a fist to be approximately 10cm, and based on photographs of the events, the appropriate cutting guide was considered to be behind the blowhole by one fist width for striped and spotted dolphins (10cm), one and a half fist widths for Risso's dolphins (15cm), and 2 fist widths for the larger pilot whale (20cm).

Iwasaki and Kai (2010) also report that other methods are employed in the killing:

> Placing a vinyl sheet over the rocks facilitated the transport of the striped dolphins to the killing area and also the full application of the spinal transaction technique. In addition, by driving a wedge into the cut, bleeding was controlled. Exsanguination occurred 10 to 30 minutes later at the time of gutting, and this did not affect the quality of the meat (for consumption).

Iwasaki and Kai (2010) concluded:

> Harvest time was shortened, improving worker safety. Bleeding was controlled by the wedge, and this opens up the possibility of commercial utilization of the blood and prevents pollution of the sea with blood. The individual who developed the spinal cord transection technique has pointed out that prevention of bleeding and internal retention of blood using the wedge risks prolongation of the time to death. An additional review to compare time to death with the Faroe Islands is required.

Based on this minimal data, Iwasaki and Kai (2010) claimed that the new method was more humane. This claim was based on a shorter time to death (TTD) recorded in four species where the spinal transection technique was utilized, compared to only one instance where the conventional spear method of killing was used on a striped dolphin. TTD is defined by Iwasaki and Kai (2010) as 'the termination of movement and breathing'.

6.2 Analysis of Video of the Small Toothed Whale Hunt

In analysis of the methods used to kill toothed whales in the Taiji hunt, Butterworth et al. (2013) analysed videotape footage of a striped dolphin (*Stenella coeruleoalba*) hunt conducted in Taiji, Japan, in January 2011. The hunting events visible in the video footage were documented, tabulated and timed using the time base available on the video material. These authors compared their observations and analysis to the data and assessment reported in Iwasaki and Kai (2010).

The results of the behavioural analysis of the video documentation of the killing method presented by Butterworth et al. are reproduced in Table 6.1. Still images derived from the video material were used to overlay outlines of cetacean anatomical structures in relation to the use of the rod and wooden plug (Figs. 6.3 and 6.4). Using external landmarks (rostrum, mouth, eye, blow hole dorsal and pectoral fins), it was possible for these authors to locate with some accuracy the path and track of the insertion of the rod (Fig. 6.3). The rod appeared to enter the skin in the midline of the animal and about 10 cm caudal (behind) to the blowhole. The ease with which the rod penetrates the tissues on the first 'push' suggested that it passed only through soft tissues at this time. The soft tissues in this location—immediately caudal to the skull—would be the skin, blubber, musculature of the dorsum and the suspension of the skull, some of these tissues being associated with the cervical vertebrae and with the attachments of the very large and powerful (swimming) muscles of the dorsal

Table 6.1 Video analysis of the timing of events during a dolphin drive hunt using the new killing device and procedure

Video timecode (s)	Duration (s)	Event	Comment
—	Prolonged (video does not capture start and end of this event)	Dolphins are secured by their tail fluke and dragged by boat	These animals are unable to swim effectively and so are being repeatedly pushed under the water by the action of dragging and by pressure of other animals tied up with them. The inability to control the timing of breathing (and enforced submersion) is causing profound distress and restricted escape movements in these animals. Some will be experiencing aspects of 'forced asphyxiation' due to their inability to reach the surface to breathe
02:37	Start	Dolphin 1—first forceful insertion of metal rod	The rod pushes into tissues rapidly. It appears unlikely that this first 'push' penetrates the bone. Severing the spinal cord at the first attempt (as claimed in the description of the method) is not achieved at this first insertion
02:40–02:44	3–7	Animal moves strongly and operative redirects and re-forces the rod at multiple angles repeatedly pushing it into the animal	The animal responds strongly to the first insertion of the rod, and the operatives have to hold the animal whilst the operative with the rod redirects the rod and repeatedly pushes it into the animal
02:44–02:48	7–11	The rod appears to hit hard (bony) obstruction, and the operative pushes the rod at different angles but does not achieve deep insertion of the rod	At this point, it appears likely that the rod makes first contact with the vertebral bones of the cervical (neck) vertebrae. The rod clearly requires very significant force to push further into the tissues at this time. At the end of a period of pushing, it is possible that the cervical vertebrae have now been damaged sufficiently to allow the spinal cord to also be damaged by the rod
02:50	13	Insertion of the wooden peg	The rod is withdrawn and a wooden peg inserted. This is intended to 'reduce pollution of the sea' with blood. If 'rapid bleed out' (as is required in animals slaughtered and killed in a slaughterhouse) is part of the killing process, then blocking the bleed out passage may slow down bleed out and prolong the time to death

(continued)

Table 6.1 (continued)

Video timecode (s)	Duration (s)	Event	Comment
03:17	40	Animal with wooden peg in puncture site visible	The animal is stationary at this time, but the wooden peg is clearly visible
03:48	71	Small vertical head movements	The animal starts to make regular rhythmic vertical head movements
04:10	93	Animal stationary	The animal stops moving
04:30	113	Slow rotational movements of the body seen	The animal now makes slow regular rotational movements
04:33	116	Vertical head movements	The animal makes regular rhythmic vertical head movements
04:39	122	Vertical head tremor	The head movements become rapid and repetitive
05:07	150	Major body movements start	The entire body now makes large-scale regular repetitive movements
05:24	167	Major body movements continue with thrashing fluke causing splashing	The repetitive movements now include the whole body and the tail fluke, and this thrashing throws up considerable spray. Because this spray is interfering with the operative (who is now using the rod on another animal)—another operative puts a rope around the thrashing animal's tail fluke. Both operatives are not showing attention to the movements of the animal other than to remove it from the 'work area'
05:25	168	Operative secures thrashing fluke and drags animal away from other operative	The powerfully moving animal is dragged out of the 'work area'—but its tail fluke movement brings it back towards the operative who is using the rod on another animal[a]
05:29	172	Vigorous thrashing of the flukes	
06:02	205	Animal motionless	The animal now becomes motionless
06:36	239	Mouth visible and making small regular and co-ordinated opening and closing movements	Regular small movements of the mouth are visible[a]
06:51	254 (4 min 14 s)	Opening and closing movements of mouth continue—end of available video material	Regular small movements of the mouth are visible[a]

Table with permission originally from Butterworth et al. (2013)
[a]If the stated criteria for establishing time to death (termination of movement and breathing) are applied, then this animal has not yet achieved death

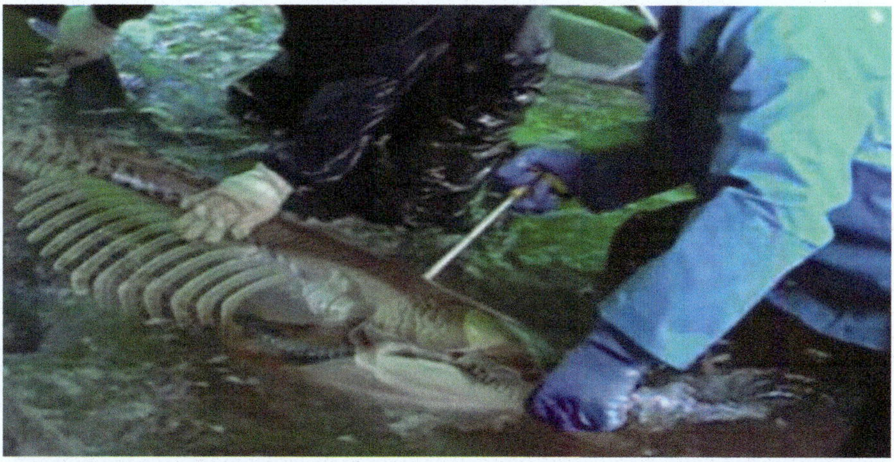

Fig. 6.3 Dolphin skeletal and soft tissue and point of insertion of the metal rod. This image shows the overlay of skeletal and soft tissues on a striped dolphin (*Stenella coeruleoalba*). This overlay shows the relationship between the skeletal and soft tissues compared with the external anatomical features (eye, mouth, blow hole, dorsal fin and pectoral fin) and with the course and positioning of the metal rod. Image with permission originally from Butterworth et al. (2013)

Fig. 6.4 The use of the wooden plug in the killing process. This image shows the use of the wooden plug inserted in the wound after the metal rod is removed. This is done to prevent the blood from escaping the body. This technique will actually most probably prolong time to death. Image with permission originally from Butterworth et al. (2013)

region including the *longissimus* and *multifidus* muscles (Rommel and Lowenstein 2001). The authors report that in his region, immediately caudal to the skull are located portions of the *rete mirabile*—a specific adaptation of the vascular system of marine mammals which appears to function to buffer pressure (and perhaps pH and oxygenation levels) in blood circulation to the brain (Lin et al. 1998; Melnikov 1986; Nagel et al. 1968). Damage to the vertebral blood vessels and the vascular rete would probably lead to significant haemorrhage, but this alone would not be capable of producing a rapid death in a large mammal (American Veterinary Medical Association 2013; Anil et al. 1995; Daly et al. 1988). In the case of the use of the rod, after the operative has used the rod to cause tissue damage, a wooden peg was seen to be inserted into the hole created by the rod (see Fig. 6.4). The bony structures in the area which are likely to be penetrated by the rod during this procedure would be the spinous neural dorsal (upward pointing) processes of the cervical vertebra and the bony bodies of the first and second cervical vertebrae (C1, C2). Cetaceans have well-developed neural processes on their vertebrae as attachments for the powerful epaxial muscles that form part of the swimming musculature. The cervical vertebrae join the skull with a bony junction at the occipital bone via the occipital condyle (the joint with the vertebrae), and in this area the spinal nerves and spinal cord emerge from the skull and enter the spinal canal. The spinal cord is well protected within the bony bodies of the cervical vertebrae and runs in a bony tunnel with the dorsal and lateral processes of the vertebrae protecting it on the upper (dorsal) side and the vertebral body protecting it on the lower (ventral) surface. To penetrate the spinal canal, the rod would have to accurately enter the space between vertebrae (which provide overlapping bony protections) or to damage the cervical vertebral bone sufficiently to allow spinal cord severance. Either of these processes, if carried out with a rod after passage through muscle and other tissues, is unlikely to be applied with a high degree of precision. It appears from the video analysis reported by Butterworth et al. (2013) that the approach was seen to be to push the rod hard and repeatedly into the tissues and that eventually this would result in very significant damage and trauma and lead (eventually, but not immediately) to the death of the animal.

6.3 Item by Item Welfare Analysis of This Small Toothed Whale Hunt

The results of the analysis of the killing methods utilized in the Taiji dolphin drive hunt were reported by Butterworth et al. (2013) to be in sharp contrast and contradictory to the descriptions and conclusions presented in Iwasaki and Kai (2010). The following points are raised by Butterworth et al. (2013) to indicate their significant concerns with this killing method:

1. After being driven into a restricted area and confined, the animals are sometimes tethered to boats by their tail flukes and pulled to the killing area (Fig. 6.1, last image). The video shows animals that are unable to swim effectively and

that are being repeatedly pushed under the water by the action of dragging and by pressure of other animals tied up alongside. The inability to control the timing of breathing (due to forced submersion) may cause distress, and escape movements are evident in these animals. It is likely that some dolphins will be experiencing aspects of 'forced asphyxiation' due to their inability to control whether they are at the surface or forced underwater. Dolphins do have the capacity to breath-hold during planned diving activity and have specific physiological adaptations (storage of oxygen in blood and muscle, bradycardia (heart slowing) and redistribution of oxygenated blood within organs to conserve the use of oxygen) (Williams et al. 1999). However, with repeated forced shallow immersion (each submersion of unknown duration and not in the control of the animal), it is unlikely that the dolphins would initiate (or be able to initiate) true deep diving responses, and so it is surmised that treatment of this type (dragging and forced submersion) is likely to be very aversive. The video material available does not allow calculation of the duration of submersion, but it is clear from the behavioural responses that the animals resist this procedure and that some are already unconscious with their heads submerged or already dead (assumed drowned or suffocated by the process). This type of treatment would not be tolerated or accepted for commercially farmed animals being prepared for slaughter in the USA or Europe.

2. The dolphins are positioned in close proximity to each other during the killing process, and struggling and whistling (which is audible on the video material despite its remote filming origins) occur throughout the process. Dolphins are highly social mammals (Connor 2007) that show advanced cognition including self-awareness as demonstrated by their capacity for mirror self-recognition (Reiss and Marino 2001). They undergo a prolonged process involving not only the herding offshore but confinement, holding and eventual corralling to the shoreline, followed by killing in close proximity to conspecifics and other members of their social and family groups. The entire process can last many hours. The American Veterinary Medical Association recommendations state 'Euthanasia should be carried out in a manner that avoids animal distress. In some cases, vocalization and release of pheromones occur during induction of unconsciousness. For that reason, other animals should not be present when euthanasia is performed' (American Veterinary Medical Association 2013). 'The regulations and guidelines governing the humane treatment and slaughter of animals in the USA and the UK 'prohibit the killing of an animal in the presence of other animals' (Humane Slaughter Act 2003; The Welfare of Animals (Slaughter or Killing) Regulations 1995). From a scientific, humane and ethical perspective, the treatment of dolphins in the drive hunts sharply contradicts current animal welfare standards employed in most modern and technologically advanced societies.
3. The use of 'termination of movement' (Iwasaki and Kai 2010) as the determinant time of death in an animal with a transected spinal cord is not a credible measure of death for a mammal. Immobility (termination of movement) will be the natural final consequence of severance of the spinal cord; however, in any

mammal (including humans), severing the spinal cord does not immediately lead to death, and this is apparent in the continued life of many human and animal patients following spinal injury. Evaluation of death when livestock are slaughtered is based on the cessation of central neurological function and respiratory activity or that the animal has been effectively exsanguinated (bled out) (American Veterinary Medical Association 2013; Commission of the European Communities COM 2006; FAO Animal Production and Health 2004; Humane Slaughter Association 2001).

4. Termination of breathing (Iwasaki and Kai 2010) is not (in the short term and certainly in the periods described in the translation above) an appropriate indicator of death in marine mammals, which have enormous capacity for breath-holding (Joulia et al. 2009), with dives of up to 40 min recorded in some of the dolphin species (Miller et al. 2006; Noren and Williams 2000; Snyder 1983; Kooyman et al. 1999). The striped dolphin does not usually breath-hold for periods of longer than 15 min, and Iwasaki and Kai (2010) claim that death can be assessed after breathing has stopped for as short a period as 5 (Risso's dolphin), 8 (spotted dolphin) or 25 (pilot whale) seconds. These periods (times of up to 25 s) are well within the 'breath-holding' capacity of many mammals and a very short breath-hold for a marine mammal.
5. The sample size for the 'control' animal (one striped dolphin) described in the paper proposing the method (Iwasaki and Kai 2010) is unlikely to be sufficient to draw any meaningful conclusions, particularly in light of the availability of a large number of animals to study for these authors.
6. The method describes the times taken for an animal to die (as defined using termination of movement and breathing) to be as short a period as 5 (Risso's dolphin), 8 (spotted dolphin) or 25 (pilot whale) seconds—with average times of 13.7 (Risso's dolphin), 9 (spotted dolphin) or 25 (pilot whale) seconds. The data derived from the analysis of a striped dolphin killed using the rod (Table 6.1) indicates that the animal was still moving after 254 s (4 min 14 s). The disparity between the published results (Iwasaki and Kai 2010) and those from this observation based assessment is considerable and calls into question the confidence that can be attributed to the data provided in the Iwasaki and Kai (2010) report.
7. Damage to the vertebral blood vessel and the vascular rete from insertion of the rod will lead to significant haemorrhage, but this alone would not produce a rapid death in a large mammal. After the operative has used the rod to cause tissue damage, a wooden peg is inserted into the hole created by the rod (Iwasaki and Kai 2010). It is likely that this would impede bleeding and so it is also possible that this process prolongs the time for the animal to die (Katsura et al. 1994). This risk is acknowledged by Iwasaki and Kai (2010) who state—'The person who developed the spinal cord transection technique has pointed out that prevention of bleeding and internal retention of blood using the wedge risks prolongation of the time to death'. This calls into question the contention that this new killing method results in reduced TTD.
8. Analysis of the video evidence suggests that the operator must make repeated 'pushes' of the rod into the tissues close to the back of the skull. The video

shows the animal making vigorous movements during the insertion of the rod and subsequently making powerful muscular movements at times after the rod has been withdrawn. This evidence strongly suggests that the method is immediately invasive and distressing and does not bring about immediate insensibility, as the brain itself remains unaffected. Complete and rapid (immediate) cord transection could result in destruction of sensory (pain) pathways, but what is observed in the animals studied is neither immediate nor appears to induce effective and assured cord transection, and so there can be no assurance that pain elimination is achieved. After a period of violent insertion of a rod into sensitive tissues, the animal becomes paraplegic (paralysis of the body) and dies through trauma and gradual blood loss. This method of killing does not conform to the recognized requirement for 'immediate insensibility' and would not be tolerated or permitted in any regulated slaughterhouse process in the developed world (American Veterinary Medical Association 2013; Commission of the European Communities COM 2006; Food and Agriculture Organization Animal Production and Health 2004; Humane Slaughter Association 2001).

9. Rapid exsanguination is usually required after stunning for either humane slaughter or euthanasia. The method described in this paper is not designed primarily for bleed out—in fact, the use of the wooden plug will, to a degree, reduce the capacity for bleeding from damaged blood vessels. This method appears to be primarily focused on causing gross neural tissue damage to the spinal cord and potentially the brainstem. This will cause, initially, immobilization and eventually death due to lack of co-ordination of respiratory and motor function. The method described does not conform to any recognized mechanism for bringing about death in accepted humane slaughter or euthanasia practice in large mammals.
10. The results presented in this paper provide strong evidence that the claims regarding the improved killing method described in Iwasaki and Kai (2010) are not substantiated. Also, this killing method cannot be considered humane as it does not fulfil the recognized requirement for immediacy and in fact may result in a prolonged aversive application of a violent and traumatic physical process followed by slow death by spinal paralysis and blood loss. This method would not be recognized or approved as a humane or acceptable method of killing for mammals in any setting.

6.4 Discussion of the Video Analysis Findings

Because the hunts are extremely controversial and hidden beneath tarpaulins that are pulled over the shoreline of the killing cove, independent video footage documenting the killing method can only be obtained through remote surveillance from public spaces. New tarpaulins and other visual obstacles had been constructed during the 2011 hunting season, further limiting access to viewing points around the killing cove. The video independently documenting the killing method used for this analysis was procured from an investigative journalist representing Atlantic Blue, a German organization. The authors were provided with two clear video accounts of the killing method being utilized in December 2009 and January 2011. Because the video footage from January 2011 was of higher quality and represents the most 'current' methods in use, it was utilized for this analysis. The absolute paucity of this kind of material makes multiple analyses impossible, and so this analysis focuses on one good quality video example where the entire process from instigation to apparent end point is visible in a continuous frame without obstruction. The authors are not familiar with any other wildlife hunts that are specifically shielded from view in this manner.

As Iwasaki and Kai (2010) reference the development and testing of this method since 2000, the authors of the Butterworth et al. (2013) paper state that they are confident that the video sample is representative of current methods being utilized in the dolphin 'drive hunts' in Taiji. Additionally, from the available video material, the paper describes how it is apparent that the same process is applied to many animals (not all observed throughout the whole process in the video material), and this analysis is representative of the approach being used on many animals. The range of social attitudes towards the killing of wild species around the world raises a number of important ethical questions. These authors go on to consider that suffering is 'undesirable, and that humans should do all that is practical to ensure that suffering is minimized at the time of death for domesticated animals which humans farm, use or consume'. They then go on to consider that 'it appears logical and consistent to also acknowledge that suffering should also be avoided for wild mammalian species' (Commission of the European Communities COM 2006; Mellor and Littin 2004). The challenges presented in achieving the same standards for killing wild animals as exist for domesticated animals have, unfortunately, led to a systematic dilution or reduction in the standards permitted for the killing of wild species.

There are precedents for applying scientific knowledge and concern for animal welfare to policy decisions regarding commercial fishing and hunting practices. In the mid-1980s, increased scientific and public concern in the USA about the welfare of dolphins caught as by-catch during tuna purse seine fishing operations led to US senate subcommittee hearings and the ultimate decision to ban the use of purse seine procedures in the eastern tropical Pacific. Studies were conducted as part of a larger research programme mandated under the 1997 International Dolphin Conservation Program Act (IDCPA) that investigated whether the eastern tropical Pacific tuna fishery was having a significant adverse impact on these dolphin stocks, known collectively as the Chase Encirclement Stress Studies (CHESS). Stress-response protein profiles and various other health parameters in offshore spotted and spinner dolphins revealed acute stress response in chased and captured dolphins, including heart lesions and other tissue damages (Forney et al. 2002). Legislative policy changes are reflected in the Marine Mammal Protection Act (MMPA) and entitled the Dolphin Protection Consumer Information Act and International Dolphin Conservation Program Act (IDCPA), recognizing the desire of congress, the public and corporate interests to incorporate dolphin protection and welfare into practice through regulations addressing the tuna fishery and product labelling standards (US Marine Mammal Protection Act 1972). Policy changes included a ban on the use of purse seine fishing in the eastern tropical Pacific and protected dolphins from being encircled by fishing boats, trapped in the purse seine nets and crushed in the fishing gear. Policy changes occurred at the corporate level in the tuna industry and offered the consumer the right to know about the fishing practices used in this commercial fishery.

Another precedent for such policy changes occurred in the UK in the case of the well-established cultural practice of hunting red deer (*Cervus elaphus*) with hounds. Increased scientific and public concern for the welfare of red deer during the prolonged hunts prompted a study to be commissioned by the National Trust to assess the physiological effects of the hunts on the deer (Bateson and Bradshaw 1997). The physiological state of hunted vs. non-hunted but humanely killed red deer was compared and the results showed '(i) depletion of carbohydrate resources for powering muscles, (ii) disruption of muscle tissue, and (iii) elevated secretion of endorphin. High concentrations of cortisol, typically associated with extreme physiological and psychological stress, were found. Damage to red blood cells occurred early in the hunts'. The authors concluded that 'red deer are not well-adapted by their evolutionary or individual history to cope with the level of activity imposed on them when hunted with hounds'. These scientific findings led to the banning of this type of hunting practice in the UK (The Hunting Act 2004).

Animals used for commercial purposes have been afforded the status of sentient beings under the Treaty of Amsterdam, amending the Treaty of the European Union (The Treaty of Amsterdam 1997). Therefore, there exists a moral and legislative obligation to exercise a high standard of care for animals under the control of humans. It would seem appropriate that those animals that fall under human control during systematic hunts at the time of their death be treated following the accepted international principles described by the Treaty. As humans determine when and where these animals die, there is an ethical obligation, as well as a practical opportunity, to control the method of death to minimize pain or suffering (Mellor and Littin 2004). Based on available scales for pain, including both the National Institutes of Health and British Pain Society numeric scales, this method would register as extremely aversive—at the highest level of gross trauma, pain and distress (National Institute of Health Pain Consortium 2007; The British Pain Society 2017).

Within Japan, domesticated animals are afforded protection under the Act on Welfare and Management of Animals, where guidelines to minimize pain and suffering are outlined for species such as horses, cattle, sheep, pigs, dogs and other animals under human care (Japan Ministry of the Environment 2007). Dolphins and whales are not protected by this law, nor are they afforded protection under the Wildlife Protection and Hunting Law which manages the keeping and custody of wild mammals in Japan and outlines procedures for the protection, management and hunting of wild mammals in Japan through the oversight of the Ministry of Environment. Instead, dolphins and whales fall under the jurisdiction of the Fisheries Agency under the Department of Agriculture, which affords them little protection. This is in marked contrast to the protection for dolphins and whales in legislation in other parts of the world such as New Zealand and the USA. The US Marine Mammal

Protection Act (MMPA) of 1972 affords full protection from the 'taking' or deliberate killing of marine mammals, except under certain conditions for scientific research, enhancement for survival or recovery and public display (MMPA 1972). In New Zealand, intentional or deliberate killing of marine mammals, notably within commercial fisheries, is prohibited, and similar provisions are provided by the Marine Mammal Protection Act (New Zealand Legislation Marine Mammal Protection Act 1978).

In contrast, and looking to other whale and dolphin hunts in Japan for comparison, Japan conducts 'special permit whaling' hunts for five species of large whales in the North Pacific and minke and fin whales in the Southern Ocean. These hunts occur in open water, at sea, and the killing methods are applied from a vessel. The proximity between the whale or dolphin and the hunter during drive hunts contrasts significantly with open sea whaling. During drive hunts, killing occurs when the hunter and the animal are next to each other on the stable ground of the shore. In contrast, whaling occurs at a distance, with the whale swimming in a moving sea and the hunter aims at the target from a moving platform. The killing methods also differ significantly, due in part to the difference in size of the animals (large baleen whales, rather than dolphins or smaller toothed whales).

The primary killing method used during Japanese whaling is a penthrite grenade harpoon that is aimed at the thorax. The objective is to cause sufficient blast-induced neurotrauma to render the whale 'instantaneously' insensible or dead (Knudsen and Øen 2003). Data show that for the Japanese hunt for minke whales (the smallest species killed during Japanese special permit whaling) in the Southern Ocean during the 2003–2005 seasons, 44% of harpooned minke whales (N = 880) were reported to have died 'instantaneously' (Ishikawa 2005). In some cases where whales do not die 'instantaneously', a secondary killing method is applied. Depending on the species, this may either be another grenade harpoon or a rifle. The rifle is aimed at the head, whilst the whale is still attached to the harpoon line at the front of the vessel. Since the meat procured from these activities is sold for commercial purposes, it is legitimate to compare both special permit hunting and drive hunts with the standards required for other commercial meat production, such as those provided by the World Organization for Animal Health (OIE) recommendations for the slaughter of animals for food (OIE Terrestrial Animal Health Code 2011). Whilst the OIE recommendations are focussed on the slaughter in slaughterhouses of various domesticated terrestrial species, it is not unreasonable to extend the principles such that mammals slaughtered outside slaughterhouses should be managed in such a manner that their restraint and slaughter should avoid causing undue stress.

What is particularly unusual about these drive hunts is the proximity of the hunter to the animal that they are killing, which provides an opportunity for a swift death with potentially less margin for error than hunting at sea. For example, euthanasia of injured or moribund dolphins stranded on the beach is usually conducted by a veterinarian or a trained individual with a rifle at very close range. Best practice for cetaceans in extremis has been developed in order to administer the swiftest and most humane death. However, the authors do not recommend the use of rifles for killing cetaceans captured during these hunts, for a number of reasons. Firstly, whilst rifles are a recommended euthanasia procedure for stranded cetaceans in some stranding protocols, the RSPCA guidelines for veterinarians attending stranded cetaceans (RSPCA 1997) do not recognize rifle shooting as the preferred method. Instead, these guidelines only recommend the use of rifles for toothed cetaceans up to 4 m in length if euthanasia drugs are unavailable. Secondly, there are many differences between an individual 'mercy killing' associated with euthanasia of a stranded cetacean and the frequent and consecutive commercial killing of dolphins on the shore. The use of rifles as a humane euthanasia method for stranded cetaceans is only recommended on the basis that the operator, usually a veterinarian, using the rifle is well trained in such procedures, and that the outcome is documented. Such caveats to the use of rifles could theoretically be applied to the use of rifles during a drive hunt, but it is in the authors' view that it is highly unlikely that even with a highly skilled operator administering the shot, there would be a humane outcome for each dolphin.

Unlike a stranded dolphin that is shot because it cannot be refloated, dolphins caught in drive hunts are not moribund, but instead are usually conscious, panicked and moving, thus increasing the likelihood of error in bullet placement to the brain. In addition, during the dolphin drive hunts, the footage shows that some of the animals are secured by their tailstock. This is a particularly aversive practice due to the risk of the dolphins drowning as a result of forcing the head and blowhole under the water. In this respect, there exist no useful comparisons with other terrestrial mammal drives or hunts. In addition, since a primary sense in these highly social mammals is hearing, the impact of hearing other cetaceans—and specifically members of their social group—being killed has the potential to further compound the negative effects of this hunting method.

The process of spinal transection carried out in a fully conscious large animal is likely to be profoundly distressing, traumatic and painful and to create unnecessary suffering and distress. The AVMA Guidelines on Euthanasia (AVMA 2007) suggest that cervical dislocation can be considered a potential method for euthanasia of rabbits weighing no more than 1 kg and in other small mammals of less than 200 g. The dolphins observed in this study weigh in the region of 200 kg and would not be considered suitable candidates for cervical dislocation under any laboratory or zoo veterinary guidelines. Additionally, the use of the puntilla (a knife designed to sever the spinal cord) is not permitted in slaughter processes in developed countries (Tidswell et al. 1987).

Pain is most often attributed to a physical condition, whereas discussions of suffering require consideration of the psychological and emotional capacity of the animals being slaughtered. Japan's own slaughter guidelines for livestock, which do not apply to the drive hunts and other whale and dolphin killing methods used around Japan's coastline, require the inducement of loss of consciousness and 'methods that are scientifically proven to minimize, as much as possible, any agony to the animal' (Japan Ministry of the Environment 2007). These guidelines also define 'agony' as pain and suffering due to the excitement of the central nervous system by stimulating pain, fear, anxiety or depression, all arguably elements of suffering in higher vertebrates. The systematic mistreatment of dolphins and whales, allowed and sanctioned by a highly developed country such as Japan, is in striking contrast to EU, the USA and even existing Japanese legislation which aims to protect the welfare and ensure the humane treatment of farm, domestic and laboratory animals.

6.5 Conclusions

In conclusion, despite profound differences in their body form, dolphins, like our closest relatives the great apes, are sentient, highly social mammals that exhibit complex cognitive abilities (Herman 2006), possess self-awareness as demonstrated by their ability for mirror self-recognition (Reiss and Marino 2001) and demonstrate epimeletic (helping and caregiving) behaviours (Connor and Norris 1982). Japanese scientists have been international leaders in great ape research, and their scientific knowledge has been used to provide the rationale to increase protection of the great apes. In 2006, Japan placed an unofficial ban on invasive chimpanzee research.

Our scientific knowledge of dolphins could and should result in similar protections against the suffering and distress resulting from this current method utilized in drive hunts. Existing scientific knowledge and understanding of cetacean anatomy, physiology, social behaviour and cognition should inform local and global animal welfare policies on the treatment of these species. There thus appears no logical reason to accept a killing method that is clearly not carried out in accordance with fundamental and globally adopted principles on the commercial utilization, care and treatment of animals.

Acknowledgements The major part of this text is derived from a modified and updated version of Butterworth et al. (2013) with permission from the Journal of Applied Animal Welfare Science. The authors would like to acknowledge the assistance of the individuals who collected and provided video footage, still photographs and translations that enabled this analysis. Specifically, we thank AtlanticBlue.de for providing the video footage from drive hunts in Taiji, Japan, and Chisa Hidaka, M.D. for providing the translation of the original document (Iwasaki and Kai 2010) upon which this analysis is based.

References

American Veterinary Medical Association (2013) Guidelines on Euthanasia, American Veterinary Medical Association 1931 N. Meacham Road Schaumburg, IL 60173 ISBN 978-1-882691-21-0

Anil MH, McKinstry JL, Gregory NG, Wotton SB, Symonds H (1995) Welfare of calves—2, increase in vertebral artery blood flow following exsanguination by neck sticking and evaluation of chest sticking as an alternative slaughter method. Meat Sci 41(2):113–123

Bateson P, Bradshaw EL (1997) Physiological effects of hunting red deer (*Cervus elaphus*). Proc Biol Sci 264:1707–1714

Brownell RL, Nowacek DP, Ralls K (2008) Hunting cetaceans with sound: a worldwide review. J Cetacean Res Manage 10(1):81–88
Butterworth A, Brakes P, Vail CS, Reiss D (2013) A veterinary and behavioral analysis of dolphin killing methods currently used in the "drive hunt" in Taiji, Japan. J Appl Anim Welf Sci 16(2):184–204
Commission of the European Communities COM (2006) Commission working document on a Community Action Plan on the Protection and Welfare of Animals 2006–2010 SEC 2006 65
Connor RC (2007) Complex alliance relationships in bottlenose dolphins and a consideration of selective environments for extreme brain size evolution in mammals. Philos Trans R Soc Lond B Biol Sci 362:587–602
Connor RC, Norris KS (1982) Are dolphins reciprocal altruists? Am Nat 119:358–374
Daly CC, Kalweit E, Ellendorf F (1988) Cortical function in cattle during slaughter: conventional captive bolt stunning followed by exsanguination compared with shechita slaughter. Vet Rec 122:325–329
Faroese Ministry of Foreign Affairs, Uttanrikisradid (2011) Whaling in the Faroe Islands
Food and Agriculture Organization Animal Production and Health (2004) Good practices for the Meat Industry Manual (FAO Publication 92-5-105146-1). FAO, Rome
Forney KA, St. Aubin DJ, Chivers SJ (2002) Chase encirclement stress studies on dolphins involved in eastern tropical Pacific Ocean purse-seine operations during 2001. Administrative report LJ-02-32. Southwest Fisheries Science Center, NOAA, La Jolla
Hemmi S (2011) The dolphin drive hunt: appropriate management? Observations from the emergency extension of the hunting season of the dolphin drive hunt in Taiji. Elsa Nature Conservancy, Japan
Herman LM (2006) Intelligence and rational behaviour in the bottlenosed dolphin. In: Hurley S, Nudds M (eds) Rational animals? Oxford University Press, Oxford, pp 439–468
Humane Slaughter Act (2003) 7 U.S.C. § 1901 et seq., 9 C.F.R § 313.1-90
Humane Slaughter Association (2001) Guidance notes No. 2 captive bolt stunning of livestock 3rd edition. Wheathampstead, Humane Slaughter Association
Hunting Act 2004 Chapter 37, 1–11 (2004). http://www.legislation.gov.uk/ukpga/2004/37/pdfs/ukpga_20040037_en.pdf. Accessed 6 July 2016
Ishikawa H (2005) Improvement of the time to death in the Japanese Whale Research Program in the Antarctic Sea (JARPA) and Northwestern Pacific Ocean. JARPN and JARPN II, IWC/57/WKM&AWI 11
Iwasaki T, Kai Y (2010) Improved method of killing dolphins in the drive fishery in Taiji, Wakayama Prefecture. Report of the Fishery Research Institute of Far Seas Fisheries and Taiji Fishery Cooperative. http://www.cypress.ne.jp/jf-taiji/geiruihosatu.pdf. Accessed 6 July 2016
Japan Ministry of the Environment (2007) Guidance on slaughter of animals (Notice No. 05, November 12, 2007 revision). Japan Ministry of the Environment, Japan
Joulia F, Lemaitre F, Fontanari P, Mille PML, Barthelemy P (2009) Circulatory effects of apnoea in elite breath-hold divers. Acta Physiol 197(1):75–82
Kasuya T (2007) Japanese whaling and other cetacean fisheries. Environ Sci Pollut Res Int 14(1):39–48
Katsura K, Kristian T, Siesjo BK (1994) Energy metabolism, ion homeostasis and cell damage in the brain. Biochem Soc Trans 22:991–996
Knudsen SK, Øen EO (2003) Blast induced neurotrauma in whales. Neurosci Res 46:377–386
Kooyman GL, Ponganis PJ, Howard RS (1999) Diving animals. In: Lundgren CEG, Miller JN (eds) The lung at depth. Marcel Dekker, Inc, New York, pp 587–614
Lin FY, Lin RP, Chou LS (1998) Structure of the thoracic retia mirabilia of three cetacean species. Acta Zool Taiwan 9(2):111–118
Mellor DJ, Littin KE (2004) Using science to support ethical decisions promoting humane livestock slaughter and vertebrate pest control. Anim Welf 13:127–132
Melnikov VV (1986) Arterial system of the head and truncal anterior part of the whale *Balaenoptera edeni*. Arkhiv Anatomii Gistologi I Embriologi 90(5):57–63

Miller NJ, Daniels CB, Schurch S, Schoel WM, Orgeig S (2006) The surface activity of pulmonary surfactant from diving mammals. Respir Physiol Neurobiol 150:220–232

Morisette L, Christensen V, Pauly D (2012) Marine mammal impacts in exploited ecosystems: Would large scale culling benefit fisheries? PLoS One 7(9):e43966. doi:10.1371/journal.pone.0043966

Nagel EL, Morgane PJ, McFarlan WL, Galliano RE (1968) Rete mirabile of dolphin—its pressure damping effect on cerebral circulation. Science 161(3844):898–902

National Institute of Health Pain Consortium (2007) Pain intensity scales. http://painconsortium.nih.gov/pain_scales/index.htm. Accessed 6 July 2016

National Research Institute of Far Seas Fisheries (2009) Japanese progress reports on small cetacean research (JapProgrep SM/2009)

New Zealand Marine Mammals Protection Act (1978) Public Act 1979 No. 80

Noren SR, Williams TM (2000) Body size and skeletal muscle myoglobin of cetaceans: adaptations for maximizing dive duration. Comp Biochem Physiol A Mol Integr Physiol 126:181–191

Ohsumi S (1972) Catch of marine mammals, mainly of small cetaceans by local fisheries along the coast of Japan. Bull Far Seas Fish Res Lab 7:137–166

OIE World Organization for Animal Health (2011) OIE terrestrial animal health code (2011). http://www.oie.int/index.php?id=169&L=0&htmfile=chapitre_1.7.5.htm. Accessed 6 July 2016

Olsen J (1999) Killing methods and equipment in the Faroese pilot whale hunt (NAMMCO/99/WS-2: 1–14). North Atlantic Marine Mammal Commission, Tromso

Reiss D (2010) Self-awareness and dolphins. In: Bird SJ (Chair) Intelligence of dolphins: ethical and policy implications. Symposium conducted at the meeting of AAAS annual meeting, San Diego, CA

Reiss D, Marino L (2001) Mirror self-recognition in the bottlenose dolphin: a case of cognitive convergence. Proc Natl Acad Sci U S A 98(10):5937–5942

Rommel SA, Lowenstein LJ (2001) Gross and microscopic anatomy. In: Dieruaf LA, Gulland FMD (eds) CRC handbook of marine mammal medicine. CRC Press LLC, Boca Raton, FL, pp 129–164. Chapter 9

RSPCA Royal Society for the Prevention of Cruelty to Animals (1997) Stranded cetaceans: guidelines for veterinary surgeons. Royal Society for the Prevention of Cruelty to Animals, Horsham

Snyder GK (1983) Respiratory adaptations in diving mammals. Respir Physiol 54:269–294

The British Pain Society (2017) Pain scales. https://www.britishpainsociety.org/static/uploads/resources/files/pain_scales_eng.pdf Accessed 6 April 2017

The Welfare of Animals (Slaughter or Killing) Regulations (1995) (UK) Council Directive 93/119/EC

Tidswell SJ, Blackmore DK, Newhook JC (1987) Slaughter methods: electroencephalographs (EEG) studies on spinal cord section, decapitation and gross trauma of the brain in lambs. N Z Vet J 35:4

Treaty of Amsterdam Amending the Treaty on European Union (1997) http://www.eurotreaties.com/amsterdamtreaty.pdf. Accessed 6 July 2016

U.S. Marine Mammal Protection Act (1972) 16 U.S.C. § 1385, p 1411–1417

Vail CS, Risch D (2006) Driven by demand: dolphin drive hunts in Japan and the involvement of the aquarium industry. Wiltshire, Whale and Dolphin Conservation Society (WDCS)

Williams TM, Haun JE, Friedl WA (1999) The diving physiology of bottlenose dolphins (*Tursiops truncatus*) 1. Balancing the demands of exercise for energy conservation at depth. J Exp Biol 202:2739–2274

Chapter 7
Din of the Deep: Noise in the Ocean and Its Impacts on Cetaceans

Lindy Weilgart

Abstract Cetaceans rely on sound and hearing for all of their vital functions, such as food finding, avoiding hazards, mating, group coordination, navigation, and orientation. Human-made noise is a substantial stressor, though the full extent of the impact remains unknown. Low-frequency anthropogenic noise has risen two orders of magnitude since the industrial age. The main noise sources are commercial shipping, seismic airgun exploration for oil and gas, and naval and mapping sonars, and the potential area of impact can extend over hundreds of thousands or even millions of square kilometres. Documented noise impacts include fatal strandings, hearing damage, longer-term avoidance of the noisy area, higher energetic costs, stress responses, changes in vocalisations which can disrupt reproductive and foraging behaviour, direct interference in foraging and migration, masking or obscuring important sounds, and effects on prey. Both acute effects, such as shorter-range fatal strandings and hearing impairment, and chronic impacts sometimes occurring over the horizon, such as stress, habitat degradation, and the loss of communication space through masking, deserve attention and concern. Studying comparable populations in real-world noisy vs. quiet areas may provide us with the best knowledge, but reducing noise levels through spatial and temporal mitigation and technological solutions should be the immediate priority. These include shipping noise reduction technologies, quieter technological alternatives to seismic airguns, avoiding areas and seasons rich in marine life when siting noisy activities such as naval sonar exercises, and establishing acoustic refuges. Governmental regulatory agencies can encourage and expedite quieter technological developments. Limiting human-made underwater noise is critical to marine mammal welfare.

L. Weilgart
Department of Biology, Dalhousie University, P.O. Box 15000, 1355 Oxford St., Halifax, B3H 4R2, Nova Scotia, Canada
e-mail: linda.weilgart@dal.ca

A. Butterworth (ed.), *Marine Mammal Welfare*, Animal Welfare 17,
DOI 10.1007/978-3-319-46994-2_7

7.1 Introduction

In the pitch-black waters of the deep ocean, a sperm whale sends out a series of biosonar clicks to scan her surroundings, listening for the faint echoes of the ocean bottom to help her to follow the 400 m depth contour which is her favourite hunting ground. She continues clicking to search for prey. The returning echoes give her information on the type of prey, its size, and its composition and density. As the sperm whale hones in on a medium-sized squid, she speeds up her clicks making them sound like a creaking door, to match the narrowing distance between her and her prey, before catching and eating it. Surfacing, she calls to her calf in a series of patterned, castanet-sounding clicks called "codas" which represent communication. Her calf responds with a coda-like sound—he's still learning how to "speak". At the same time, she is listening to the clicks of her group, mostly female family members and their calves, to coordinate deep feeding dives with them. She also has a listening ear out for mature males and their distinctive "clangs", for mating opportunities, and is on the alert for faint killer whale squeals, in case she needs to protect her calf. As she sidles up to the group, she scans her sister with her biosonar, noticing through the tell-tale echoes that she is pregnant and has a fairly full stomach. All of this information, she discovers through sound.

Marine mammals, but especially cetaceans, rely on sound for all of their vital functions such as food finding, predator or hazard avoidance, mating, group coordination, navigation, orientation, and overall sensing of their environment. In effect, their hearing is like human sight; they "see" with their ears; since underwater, vision is only useful over tens of metres. The ocean is filled with natural sounds like the roar of the surf, snapping shrimp, and the rumbling of small earthquakes. Marine mammals take advantage of this "soundscape" to orient themselves. In fact, almost all marine animals, including fish, squid, and other invertebrates, use sound or vibration, so the food (prey) of the mainly carnivorous marine mammals is also acoustically dependent. This is not a coincidence. There is a good reason why almost all marine animals use sound so extensively; sound travels very fast and very far underwater compared with air—almost five times as fast. Some low-frequency (low-pitched) sounds can travel for thousands of kilometres through the ocean. Theoretically, blue and fin whales could communicate across an entire ocean.

Now superimpose on this natural soundscape, the noise humans generate, in the form of rumbling, churning shipping noise from propellers and engines, intense shots heard every 10 s from seismic airgun surveys used to find oil and gas deposits under the ocean floor, or piercing naval sonar pings used to detect enemy submarines. The oceans may not be naturally quiet, but it is the difference between walking along the beach hearing the crashing of the surf compared to the din of traffic along a busy street, with cars honking, trucks lumbering by, and ambulance sirens wailing. Similarly, natural sound could be perceived quite differently by marine mammals than human-made sound. Just because we assume marine mammals are adapted to sounds they have heard over aeons of evolutionary time does not mean they are similarly adapted to the additional noise we have thrown at them in just the last 200 years or so since the dawn of the industrial age.

Ocean noise has become an animal welfare issue, both, for instance, in terms of potentially harming individuals' hearing, in effect "blinding" them to predators, temporarily or permanently, and through degradation of habitat important for feeding or rearing their offspring. As marine mammal scientist, Barb Taylor, indicated *"if we are going to overfish their food, entangle them in our fishing gear, and add toxins to their environment, the least we can do is keep the ocean quiet enough for them to find the few remaining fish"* (personal communication).

7.2 Sources of Man-Made Noise

Hildebrand (2009) identified the main sources of human-generated noise to be commercial shipping, seismic airgun exploration, naval and mapping sonars, and small vessels. Other human activities that add noise into the marine environment are pile driving and other construction; oil drilling; naval explosions; oceanographic experiments; acoustic deterrent and harassment devices meant to, for instance, chase seals away from fish farms; ice-breaking; and underwater communication, such as between submarines. Overall, there has been at least a 20 decibel (dB) increase in low-frequency noise since the industrial age or, in some areas, about a 3 dB increase/decade (Hildebrand 2009), which since this is a logarithmic scale is a doubling in intensity every decade, over the last several decades. Most of this increase in noise is from commercial shipping, though seismic airgun exploration also contributes, as it is occurring throughout the world's oceans (Fig. 7.1).

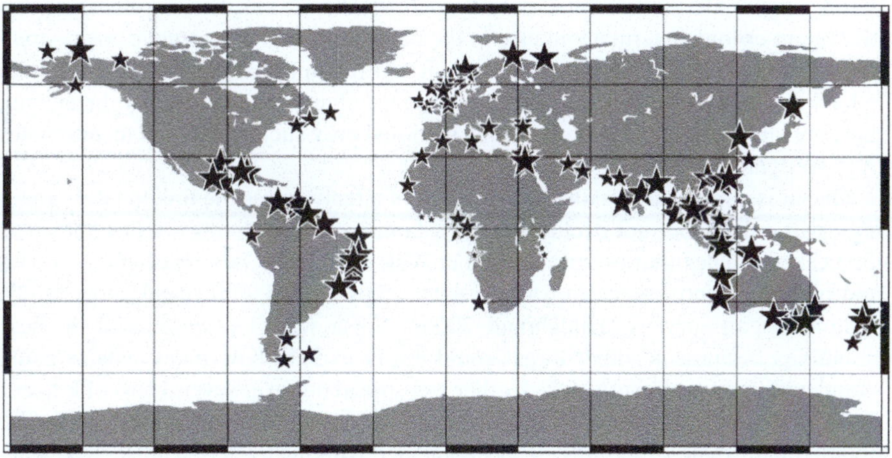

Fig. 7.1 Areas of offshore oil exploration from 1994 to 2005. Size of star denotes the relative level of activity. Data from the World Geophysical News. Reproduced with permission from Hildebrand (2009)

7.3 Area of Impact

The potential area of impact by even one noise source can be huge. The US Navy's Low-Frequency Active (LFA) Sonar, used to detect enemy submarines, could impact cetaceans at levels of over 120 dB (levels previously shown to deflect migrations) across an area of some 3.9 million square kilometres (km^2) (Johnson 2003)—about half the size of Australia—and is likely audible to them over much greater areas. Noise from a single seismic survey can blanket a region of almost 300,000 km^2, about the size of Norway, raising noise levels two orders of magnitude (20 dB), continuously for months (IWC 2005, 2007). An analysis of 10 years of recordings from the ocean floor in middle of the Atlantic found that seismic airgun noise could be heard at distances of 4000 km and were present 80–95% of the days per month for more than 12 consecutive months in some locations (Nieukirk et al. 2004, 2012). When several seismic surveys were recorded at the same time, whale sounds were masked (drowned out), and seismic airgun noise became the dominant part of background noise levels.

7.4 Documented Impacts

There is no doubt that human-made noise has an impact on marine life. Where there is still some, legitimate debate is on the extent of the impact. Impacts can vary depending on factors such as species, age, sex, context, and type of noise, making generalisations difficult. While not an exhaustive list, some examples of the types of impact from anthropogenic underwater noise that have been documented for marine mammals include fatal strandings or deaths at sea, hearing damage, displacement or longer-term avoidance of the noisy area, higher energetic costs, stress responses, changes in vocalisations which can disrupt reproductive and foraging behaviour, cessation or reduction of foraging, changes in migratory behaviour, masking, and indirect effects on marine mammal prey such as fish and invertebrates (e.g. Weilgart 2007).

One noise impact that gained a great deal of attention was the link to fatal strandings, particularly but not exclusively, of a family of deep-diving whales known as the beaked whales that occurred together with naval exercises involving powerful mid-frequency antisubmarine warfare sonar (Jepson et al. 2003) and, more rarely, seismic airgun surveys (Hildebrand 2005). Noise was first implicated in these strandings because no other threat could easily explain how many whales could strand within several hours of each other, yet spread out over several tens of kilometres of coastline. Only noise travels this fast, this far. Also, the locations and timing of individual whale strandings closely coincided with the track of a noise-producing vessel. Eventually, the necropsy findings of "severe, diffuse congestion and hemorrhage, especially around the acoustic jaw fat, ears, brain, and kidneys" and "gas bubble-associated lesions and fat embolism in the vessels…of vital organs" were determined to be consistent with, if not necessarily diagnostic of, decompression sickness or diver's "bends" (Fernández et al. 2005). Such decompression sickness may arise in whales if they change their dive pattern, affected by the loud noise, perhaps because they panicked. At least in one stranding event that was well studied, the 2002 Canary Island stranding, the time between estimated sonar exposure and death was quick—around 4 h—because vital organs like the brain or heart were affected (Fernández et al. 2005). This and other Canary Island strandings showed that animals were severely injured before stranding (Fig. 7.2), while other similarly injured carcasses were found floating at sea, implying that it was not the stranding that killed them, but the mere exposure to noise and their response to it. Thus, the number of whales killed could be greatly underestimated if just the stranded animals were counted. Whale carcasses are notoriously difficult to find at sea (or even on shore, if the location is remote), as they usually sink fairly quickly, are eaten by sharks, or are carried off by currents. Williams et al. (2011) noted that an average of only 2% of cetacean carcasses are recovered and that the "…true death toll could be 50 times the number of carcasses recovered…". Moreover, it is important to note that it took 40 years to discover the clear link between mid-frequency naval sonars and beaked whale strandings, underscoring how easy it is to miss noise impacts, even for such relatively obvious events as strandings.

Fig. 7.2 Cuvier's beaked whale haemorrhaging from the eye during the 2002 Canary Island stranding event which co-occurred with naval exercises involving mid-frequency active sonar. *Image credit*: Vidal Martín, SECAC

Hearing loss is difficult to document for marine mammals in the wild, especially cetaceans. Stranded dolphins or those severely entangled in fishing gear could, however, be tested. About 57% of such bottlenose dolphins and 36% of such rough-toothed dolphins showed either significant or profound hearing loss, implying that hearing damage could have contributed to their stranding or entanglement (Mann et al. 2010). Permanent hearing impairment can occur unexpectedly, even under careful experimental conditions. A captive harbour seal was cautiously and gradually exposed to an underwater sound stimulus to test temporary threshold shifts (TTS), which represent temporary, supposedly recoverable hearing loss in the seal, yet instead, the seal was left with a permanent threshold shift (PTS) or permanent hearing loss, without warning (Kastak et al. 2008). The start and increase of the amount of TTS did not follow the expected pattern, but rather shifted suddenly from no measureable effect to a profound hearing impairment. There are clearly many unknowns in predicting hearing loss in marine mammals, a fact that should be taken into account if populations are to be protected.

To ensure a marine mammal population's long-term welfare and health, we should be most concerned about noise impacts that affect population measures and vital rates, like birth rates, death rates, and growth rates. Such impacts are, however, very difficult to detect for marine mammals, especially cetaceans, as they spend so much time underwater. Long-term studies of known individuals are usually required to be able to determine population impacts. One such 15-year study (Claridge 2013) compared a population of Blainville's beaked whales at a US Navy range (AUTEC) in the Bahamas where mid-frequency sonar exercises were conducted, with another relatively "sonar-free" population 170 km away. It had already been established that Blainville's beaked whales move away from the range during multi-ship sonar exercises at AUTEC, only returning days later when the exercises had finished (McCarthy et al. 2011; Tyack et al. 2011). Claridge (2013) found that there were fewer calves per female at AUTEC compared with the off-range population—a population effect of sonar use at a navy range. The frequent avoidance of the naval exercises and the attendant disruption of feeding and energetic costs of travelling, together with stress, may account for the fewer births and/or calf survival at AUTEC. The beaked whale stranding that occurred in the Bahamas in 2000 as a result of the brief passage of five naval vessels using sonar also changed the Blainville's population demographics (Claridge 2013).

It is important to understand that, while some marine mammal species or individuals leave an area that becomes noisy through human activities, others remain, perhaps because the area is vital for reproduction or feeding. There are potential costs associated both with the avoidance of noise and with remaining in a noisy area. Leaving, as has been illustrated in the example above, may incur energetic costs through more travel, interfere with time spent feeding, and result in fewer surviving calves. Remaining in the area may expose populations to more stress, compromised feeding, or hearing impairment and can also cause serious impacts. The mere fact that animals stay in a region of noise should not be interpreted as the absence of impacts on them.

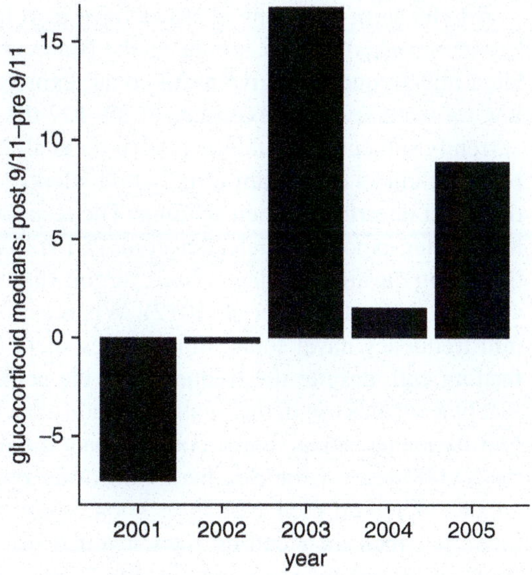

Fig. 7.3 Yearly difference in median faecal glucocorticoid (GC) levels (2001–2005) post 9/11–before 9/11. Significantly lower faecal GC levels after September 11 were only seen in 2001 and were associated with decreased underwater low-frequency noise resulting from a reduction in large vessel traffic as a result of temporary reduction around the events of 9/11 (Modified and reproduced with permission from the Royal Society from Rolland et al. 2012)

Stress can have a profound effect on an animal's ability to reproduce, to ward off infection and disease, and can cause premature ageing, and noise can function as such a stressor (Wright et al. 2007). Rolland et al. (2012) used the unusual global events of September 11, 2001, to correlate the resulting reduced ship traffic and attendant lower underwater low-frequency noise levels, with lower stress hormone levels in right whale faeces. Since right whale sounds overlap with the low-frequency noise produced by shipping, their acoustic communication space can be reduced through masking by 84% by the passage of only two ships over 13 h (Clark et al. 2009). There were six fewer large ships after September 11, 2001, than before (August 25 and 29 vs. September 12 and 13), resulting in a 6 dB decrease in overall background noise in the area. Such quantifiable physiological effects in the form of stress hormones (Fig. 7.3) could lead to biologically significant impacts on individuals and populations, especially in this highly endangered whale species, if these are indeed a measure of chronic stress from underwater noise (Rolland et al. 2012).

Long-term avoidance of presumably underwater noise and boat disturbance has been documented for harbour porpoises, where individuals seemed to move out of an area where a wind farm was being constructed (Teilmann and Carstensen 2012). This study implied that the harbour porpoise population had still not returned after 11 years, though it was gradually moving back, with porpoise biosonar activity, an indicator of their presence, increasing from 11 to 29% of baseline levels before construction. Grey whales moved away from one of their breeding lagoons for over 5 years because of industrial noise, returning only several years after the activities stopped (Jones et al. 1994). Killer whales dramatically changed locations to avoid loud acoustic harassment devices, staying away for about 6 years, returning only once the devices were discontinued in the area (Morton and Symonds 2002).

Shorter-term reactions to noise have also been shown, some over large areas. Cuvier's beaked whales, similar to the Blainville's mentioned above, avoided simulated mid-frequency active naval sonar, exhibiting strong reactions to surprisingly low received levels of exposure of 89–127 dB re 1 μPa, well below those that are currently subject to regulation (DeRuiter et al. 2013). Whales stopped emitting their biosonar clicks and swam rapidly and silently away while diving for longer durations and interrupting their feeding. The response remained pronounced for several hours after exposure (DeRuiter et al. 2013). Fin whales moved away and changed their song during a 10-day seismic airgun survey, staying away weeks after the survey ended (Castellote et al. 2012). When exposed to even relatively low levels of mid-frequency naval sonar, blue whales spent half as much time making a type of feeding call, despite the frequency of the sonar being well above the frequencies that blue whales use in their calls (Melcón et al. 2012). This reaction to even a single mid-frequency sonar source could impact the feeding of blue whales over much of the 53,000 km^2 of the Southern California Bight (Melcón et al. 2012). Bowhead whales increased their calling as soon as seismic airgun shots were detectable to them, but then their calling rates levelled off when received levels of the airgun shots increased to a certain threshold (Blackwell et al. 2015). At progressively higher received levels of airgun noise, calling rates started decreasing until finally bowheads fell silent (Blackwell et al. 2015). This meant that bowhead calling is suppressed within a radius of some 50–100 km from a seismic ship or an area of about 8000–31,000 km^2 (for reference, 31,000 km^2 is about the size of Belgium). Within around 10–40 km of a seismic ship or 300–5000 km^2, bowhead calling would be almost entirely silenced (Blackwell et al. 2015). The function of bowhead calling is unknown, but it certainly serves some important purpose, whether it be to maintain contact within the group, for mating, and/or for navigation.

Other Arctic species such as belugas and narwhals show some of the most sensitive reactions to noise documented, often responding to noise at the level they first detect it (Heide-Jørgensen et al. 2013). Hard-to-predict effects can also occur, where narwhals appeared to become entrapped in ice, often fatally, because they delayed or interrupted their annual offshore migration to avoid seismic surveys in the area (Heide-Jørgensen et al. 2013). Around 1200 narwhals died in three separate ice entrapments that were highly atypical both in timing and in area and occurred around the time of seismic surveys (Heide-Jørgensen et al. 2013). Another effect that can cause unexpectedly severe consequences is sensitisation, or a progressively greater reaction to the same or similar stimulus, each time it is encountered. An intense underwater noise, for instance, that caused a startle reflex in captive grey seals meant that these animals became sensitised to this sound and showed long-term avoidance of the tank where they first heard the sound (Götz and Janik 2011). The seals showed clear signs of rapid flight responses and fear conditioning to the point where they even avoided food that was close to the sound source. In the wild, this might mean that noise that startles a marine mammal could cause them to avoid the area where they heard the sound, over the long term, perhaps even permanently, which could have severe effects on their lifespan and ability to reproduce (Götz and Janik 2011).

Marine mammals can also be indirectly impacted through noise effects on their prey as reviewed in, e.g. Slabbekoorn et al. (2010). As mentioned, fish and invertebrates are sensitive to sound, and a wide range of documented noise impacts exist, ranging from reduced catch rates (Engås et al. 1996; Hassel et al. 2004; Skalski et al. 1992), stress (Santulli et al. 1999), higher metabolic costs (Buscaino et al. 2010), decreased foraging efficiency due to distraction (Purser and Radford 2011), reduced foraging success (Voellmy et al. 2014), impaired schooling behaviour (Sarà et al. 2007) and orientation (Holles et al. 2013), hearing damage (McCauley et al. 2003), impaired development and body malformations (Aguilar de Soto et al. 2013), massive acoustic trauma (André et al. 2011), and fatal strandings of giant squid (Guerra et al. 2004).

7.5 Acute vs. Chronic Effects

As noted above, the acute response of fatal strandings of beaked whales due to exercises involving naval mid-frequency sonar rightly garnered much attention and concern. However, at least as important are the more chronic impacts of noise that tend to occur over much larger areas. Often attention over acute hearing impairment and short-range effects has come at the expense of concern over masking, stress, and habitat degradation from chronic noise at long ranges (Simmonds et al. 2014). Loss of communication space through masking (Clark et al. 2009) is highly likely to have impacts on the welfare of populations, especially where individuals are spread out and must find each other, via long-range mating calls, to mate. Even marine mammals that do not use biosonar probably use sound for feeding to listen for the faint sounds their prey makes. Masking doesn't just mean that a sound of interest is completely obscured. It could also be that important characteristics of the sound are lost or garbled. Chronic noise even at lower levels, experienced over greater areas, can also potentially cause hearing loss.

7.6 Solutions

Different noise sources require different solutions. In the case of shipping, the noise produced is unintended and, at least to some extent, likely reduces fuel efficiency. As such, technological fixes are possible and are the focus of current studies. The International Maritime Organisation is beginning to address shipping noise, and voluntary guidelines are being developed (IMO 2013). Noise has up until now rarely been considered in ship and engine design. To address noise from seismic airgun surveys, technological alternatives to airguns are being developed that are likely less impactful on most marine life. Progress is slow, but a sound source known as Marine Vibroseis could reduce both short-range and long-range sound levels and impacts from airguns, especially for marine mammals that are sensitive to mid- and high frequencies (Weilgart 2012, 2013). Much of the sound energy airguns emit is wasted, as geophysicists only use the very low frequencies. Marine Vibroseis could eliminate broadcasting these useless higher frequencies, thus eliminating or reducing potential impacts on the species that use them. Naval sonar exercises should be sited in areas that are the equivalent of ocean deserts, poor in marine life, to reduce impacts. For marine mammals that migrate, seasonal mitigation may be useful, where noise activities are timed to avoid overlap with marine mammal presence. Vital breeding and feeding areas and seasons should be avoided wherever possible. The few remaining areas that are currently still relatively quiet could be protected as acoustic refuges.

Most importantly, we will be unable to protect marine life from noise if we require that full, incontrovertible, biological evidence of impact is obtained before we act. Many marine mammal species have long lifespans and are very slow reproducers, so they cannot genetically adapt rapidly to change. If there are population impacts, they are likely to be only discovered by the time it is too late to safeguard populations. I have noted above in various places how difficult some impacts from noise are to detect, especially in cetaceans. Noise thus presents a case where precautionary management is likely to be appropriate and may be necessary. While noise impact studies should continue to be carried out, management of noise cannot wait until our knowledge is complete. There are many variables that influence animals' responses to noise, and the ocean is not a controlled laboratory where confounding oceanographic factors like prey availability can be excluded. Rather, our emphasis is better placed on reducing noise levels through spatial and temporal mitigation and technological solutions (Simmonds et al. 2014).

Technological developments can be encouraged and expedited through actions by governmental regulatory agencies. So far, Germany is the only country that has enacted noise thresholds that are not allowed to be exceeded. These noise limits have spurred the development of quieter technologies, particularly for pile driving, used to construct offshore windfarms (Koschinski and Lüdemann 2013).

In addition to the animal welfare concerns outlined above which result from the impacts of noise, animal welfare issues also arise in the carrying out of Controlled Exposure Experiments (CEEs) where animals are artificially exposed to playbacks of noise to determine their responses under more controlled conditions. Unfortunately, while the levels of the sound source under experimental conditions can be controlled, the levels of sound experienced by the free living animals in the wild cannot. Thus, CEEs should really be called Controlled Source Experiments. Moreover, there are many more animals that are accidentally exposed than can be studied, which raises ethical concerns. While it is useful to be able to manipulate the sound source both spatially (on a vessel) and through varying its sound intensity level, if the simulated sound source does not adequately mimic the true noise source, whether airgun, naval sonar, or ship, and the context of the noisy activity, the results may not be considered fully representative. The more we study the impacts of noise on marine life, the more we discover how important context is, both in terms of the behaviour of the animals and the acoustic behaviour of the noise source. Ideally, for reactions to real (not simulated), e.g. naval sonar, actual sonar exercises, complete with all the fast-moving vessels involved, should be employed to test responses, though this can be difficult or even impractical. Studying comparable populations in real-world noisy vs. quiet areas may yield results which really allow us to understand the impact of anthropogenic ocean noise.

7.7 Conclusions

Many questions and uncertainties remain regarding the impacts of noise on marine mammals, yet we know enough to realise that noise is a substantial stressor and a problem that needs to be addressed. Striving to provide a natural environment that limits human-made noise appears critical to marine mammal welfare. Failing to do so will compromise marine mammals to some degree, either through their prey, their degraded habitat, or directly through noise interfering with their behaviour, physiology, cognition, or psychology. We cannot continue to "blindfold" marine mammals while expecting them to carry out their vital life functions without stress.

Acknowledgements I would like to sincerely thank Sigrid Lüber of OceanCare for her financial support to write this chapter.

References

Aguilar de Soto N, Delorme N, Atkins J, Howard S, Williams J, Johnson M (2013) Anthropogenic noise causes body malformations and delays development in marine larvae. Sci Rep 3:2831. doi:10.1038/srep02831

André M, Solé M, Lenoir M, Durfort M, Quero C, Mas A, Lombarte A, van der Schaar M, López-Bejar M, Morell M, Zaugg S, Houégnigan L (2011) Low-frequency sounds induce acoustic trauma in cephalopods. Front Ecol Environ. doi:10.1890/100124

Blackwell SB, Nations CS, McDonald TL, Thode AM, Mathias D, Kim KH, Greene CR Jr, Macrander AM (2015) Effects of airgun sounds on bowhead whale calling rates: evidence for two behavioral thresholds. PLoS One 10(6):e0125720. doi:10.1371/journal.pone.0125720

Buscaino G, Filiciotto F, Buffa G, Bellante A, Di Stefano V, Assenza A, Fazio F, Caola G, Mazzola S (2010) Impact of an acoustic stimulus on the motility and blood parameters of European sea bass (*Dicentrarchus labrax* L.) and gilthead sea bream (*Sparus aurata* L.) Mar Environ Res 69:136–142

Castellote M, Clark CW, Lammers MO (2012) Acoustic and behavioural changes by fin whales (*Balaenoptera physalus*) in response to shipping and airgun noise. Biol Conserv 147:115–122. doi:10.1016/j.jnc.2011.06.005

Claridge DE (2013) Population ecology of Blainville's beaked whales (*Mesoplodon densirostris*). Dissertation, University of St. Andrew's, Fife, Scotland

Clark CW, Ellison TE, Southall BL, Hatch L, Van Parijs SM, Frankel A, Ponirakis D (2009) Acoustic masking in marine ecosystems: intuitions, analysis, and implication. Mar Ecol Prog Ser 395:201–222. doi:10.3354/meps08402

DeRuiter SL, Southall BL, Calambokidis J, Zimmer WMX, Sadykova D, Falcone EA, Friedlaender AS, Joseph JE, Moretti D, Schorr GS, Thomas L, Tyack PL (2013) First direct measurements of behavioural responses by Cuvier's beaked whales to mid-frequency active sonar. Biol Lett 9:20130223. doi:10.1098/rsbl.2013.0223

Engås A, Løkkeborg S, Ona E, Soldal AV (1996) Effects of seismic shooting on local abundance and catch rates of cod (*Gadus morhua*) and haddock (*Melanogrammus aeglefinus*). Can J Fish Aquat Sci 53:2238–2249. doi:10.1139/cjfas-53-10-2238

Fernández A, Edwards JF, Rodríguez F, Espinosa de los Monteros A, Herráez P, Castro P, Jaber JR, Martín V, Arbelo M (2005) "Gas and fat embolic syndrome" involving a mass stranding of beaked whales (family Ziphiidae) exposed to anthropogenic sonar signals. Vet Pathol 42:446–457. doi:10.1354/vp.42-4-446

Götz T, Janik VM (2011) Repeated elicitation of the acoustic startle reflex leads to sensitisation in subsequent avoidance behaviour and induces fear conditioning. BMC Neurosci 12:30. doi:10.1186/1471-2202-12-30

Guerra A, González AF, Rocha F (2004) A review of records of giant squid in the north-eastern Atlantic and severe injuries in *Architeuthis dux* stranded after acoustic exploration. ICES CM 2004/CC: 29

Hassel A, Knutsen T, Dalen J, Skaar K, Løkkeborg S, Misund OA, Østensen Ø, Fonn M, Haugland EK (2004) Influence of seismic shooting on the lesser sandeel (*Ammodytes marinus*). ICES J Mar Sci 61:1165–1173. doi:10.1016/j.icesjms.2004.07.008

Heide-Jørgensen MP, Hansen RG, Westdal K, Reeves RR, Mosbech A (2013) Narwhals and seismic exploration: is seismic noise increasing the risk of ice entrapments? Biol Conserv 158:50–54

Hildebrand JA (2005) Impacts of anthropogenic sound. In: Reynolds JE III, Perrin WF, Reeves RR, Montgomery S, Ragen TJ (eds) Marine mammal research: conservation beyond crisis. Johns Hopkins University, Baltimore, pp 101–124

Hildebrand JA (2009) Anthropogenic and natural sources of ambient noise in the ocean. Mar Ecol Prog Ser 395(5):5–20. doi:10.3354/meps08353

Holles S, Simpson SD, Radford AN, Berten L, Lecchini D (2013) Boat noise disrupts orientation behaviour in a coral reef fish. Mar Ecol Prog Ser 485:295–300. doi:10.3354/meps10346

International Maritime Organization (IMO) (2013) Provisions for reduction of noise from commercial shipping and its adverse impacts on marine life. IMO Subcommittee on Ship Design and Equipment. DE 57/WP.8. www.cbd.int/doc/meetings/mar/mcbem-2014-01/other/mcbem-2014-01-submission-imo-03-en.pdf

International Whaling Commission (IWC) (2005) Report of the scientific committee. Annex K. Report of the Standing Working Group on environmental concerns. J Cetacean Res Manag 7(Suppl):267–305

IWC (International Whaling Commission) (2007) Report of the scientific committee. Annex K. Report of the Standing Working Group on environmental concerns. J Cetacean Res Manag 9(Suppl):227–296

Jepson PD, Arbelo M, Deaville R, Patterson IAP, Castro P, Baker JR, Degollada E, Ross HM, Herráez P, Pocknell AM, Rodríguez F, Howie FE, Espinosa A, Reid RJ, Jaber JR, Martín V, Cunningham AA, Fernández A (2003) Gas-bubble lesions in stranded cetaceans. Nature (London) 425:575–576. doi:10.1038/425575a

Johnson JS (2003) SURTASS LFA environmental compliance experience. Paper presented at ECOUS (Environmental Consequences of Underwater Sound), San Antonio, Texas, 12–16 May 2003

Jones ML, Swartz SL, Dahlheim ME (1994) Census of gray whale abundance in San Ignacio Lagoon: a follow-up study in response to low whale counts recorded during an acoustic playback study of noise effects on gray whales. Rep. No. NTIS PB94195062 to the US Marine Mammal Commission, Washington, DC

Kastak D, Mulsow J, Ghoul A, Reichmuth C (2008) Noise-induced permanent threshold shift in a harbor seal. J Acoust Soc Am 123(5):2986

Koschinski S, Lüdemann K (2013) Development of noise mitigation measures in offshore wind farm construction. Federal Agency for Nature Conservation (Bundesamt für Naturschutz, BfN), Hamburg, Germany

Mann D, Hill-Cook M, Manire C, Greenhow D, Montie E, Powell J, Wells R, Bauer G, Cunningham-Smith P, Lingenfelser R, DiGiovanni R Jr, Stone A, Brodsky M, Stevens R, Kieffer G, Hoetjes

P (2010) Hearing loss in stranded odontocete dolphins and whales. PLoS One 5(11):e13824. doi:10.1371/journal.pone.0013824

McCarthy E, Moretti D, Thomas L, DiMarzio N, Morrissey R, Jarvis S, Ward J, Izzi A, Dilley A (2011) Changes in spatial and temporal distribution and vocal behaviour of Blainville's beaked whales (*Mesoplodon densirostris*) during multi-ship exercises with mid-frequency sonar. Mar Mamm Sci 27:E206–E226. doi:10.1111/j.1748-7692.2010.00457.x

McCauley RD, Fewtrell J, Popper AN (2003) High intensity anthropogenic sound damages fish ears. J Acoust Soc Am 113:638–642

Melcón ML, Cummins AJ, Kerosky SM, Roche LK, Wiggins SM, Hildebrand JA (2012) Blue whales respond to anthropogenic noise. PLoS One 7(2):e32681. doi:10.1371/journal.pone.0032681

Morton AB, Symonds HK (2002) Displacement of *Orcinus orca* (L.) by high amplitude sound in British Columbia. ICES J Mar Sci 59:71–80. doi:10.1006/jmsc.2001.1136

Nieukirk SL, Stafford KM, Mellinger DK, Dziak RP, Fox CG (2004) Low-frequency whale and seismic airgun sounds recorded in the mid-Atlantic Ocean. J Acoust Soc Am 115:1832–1843

Nieukirk SL, Mellinger DK, Moore SE, Klinck K, Dziak RP, Goslin J (2012) Sounds from airguns and fin whales recorded in the mid-Atlantic Ocean, 1999–2009. J Acoust Soc Am 131(2):1102–1112

Purser J, Radford AN (2011) Acoustic noise induces attention shifts and reduces foraging performance in three-spined sticklebacks (*Gasterosteus aculeatus*). PLoS One 6(2):e17478. doi:10.1371/journal.pone.0017478

Rolland RM, Parks SE, Hunt KE, Castellote M, Corkeron PJ, Nowacek DP, Wasser SK, Kraus SD (2012) Evidence that ship noise increases stress in right whales. Proc Biol Sci 279:2363–2368

Sarà G, Dean JM, D'Amato D, Buscaino G, Oliveri A, Genovese S, Ferro S, Buffa G, Lo Martire M, Mazzola S (2007) Effect of boat noise on the behaviour of bluefin tuna *Thunnus thynnus* in the Mediterranean Sea. Mar Ecol Prog Ser 331:243–253

Simmonds MP, Dolman SJ, Jasny M, Parsons ECM, Weilgart L, Wright AJ, Leaper R (2014) Marine noise pollution—increasing recognition but need for more practical action. J Ocean Technol 9(1):70–90

Santulli A, Modica A, Messina C, Ceffa L, Curatolo A, Rivas G, Fabi G, D'amelio V (1999) Biochemical responses of European sea bass (*Dicentrarchus labrax* L.) to the stress induced by off shore experimental seismic prospecting. Mar Pollut Bull 38:1105–1114. doi:10.1016/S0025-326X(99)00136-8

Skalski JR, Pearson WH, Malme CI (1992) Effects of sounds from a geophysical survey device on catch-per-unit-effort in a hook-and-line fishery for rockfish (*Sebastes* spp.) Can J Fish Aquat Sci 49:1357–1365

Slabbekoorn H, Bouton N, van Opzeeland I, Coers A, ten Cate C, Popper AN (2010) A noisy spring: the impact of globally rising underwater sound levels on fish. Trends Ecol Evol 25(7):419–427

Teilmann J, Carstensen J (2012) Negative long term effects on harbour porpoises from a large scale offshore wind farm in the Baltic—evidence of slow recovery. Environ Res Lett 7:1–10. doi:10.1088/1748-9326/7/4/045101

Tyack PL, Zimmer WMX, Moretti D, Southall BL, Claridge DE, Durban JW, Clark CW, D'Amico A, DiMarzio N, Jarvis S, McCarthy E, Morrissey R, Ward J, Boyd IL (2011) Beaked whales respond to simulated and actual navy sonar. PLoS One 6(3):e17009 doi:10.1371/journal.pone.0017009

Voellmy IK, Purser J, Flynn D, Kennedy P, Simpson SD, Radford AN (2014) Acoustic noise reduces foraging success via different mechanisms in two sympatric fish species. Anim Behav 89:191–198

Weilgart LS (2007) The impacts of anthropogenic ocean noise on cetaceans and implications for management. Can J Zool 85:1091–1116

Weilgart L (2012) Are there technological alternatives to air guns for oil and gas exploration to reduce potential noise impacts on cetaceans? In: Popper AN, Hawkins A (eds) The effects of noise on aquatic life, Advances in experimental medicine and biology, vol 730. Springer, New York, pp 605–607

Weilgart L (2013) Alternative quieter technologies to seismic airguns for collecting geophysical data. In: von Nordheim H, Maschner K, Wollny-Goerke K (eds) Proceedings of the 3rd international conference on progress in marine conservation in Europe 2012, Stralsund, 18–22 June 2012. BfN-Skripten 339:127–134. http://www.bfn.de/0502_skripten.html

Williams R, Gero S, Bejder L, Calambokidis J, Kraus SD, Lusseau D, Read AJ, Robbins J (2011) Underestimating the damage: interpreting cetacean carcass recoveries in the context of the Deepwater Horizon/BP incident. Conserv Lett. doi:10.1111/j.1755-263X.2011.00168.x

Wright AJ, Aguilar Soto N, Baldwin AL, Bateson M, Beale CM, Clark C, Deak T, Edwards EF, Fernández A, Godinho A, Hatch LT, Kakuschke A, Lusseau D, Martineau D, Romero LM, Weilgart LS, Wintle BA, Notarbartolo-di-Sciara G, Martín V (2007) Do marine mammals experience stress related to anthropogenic noise? Int J Comp Psychol 20(2):274–316

Chapter 8
Evaluating the Welfare Implications of Climate Change for Cetaceans

Mark Peter Simmonds

'An old man was walking along the beach and the sand was littered with thousands of stranded starfish left behind by the retreating tide. The man took care not to step on any of the beautiful creatures. He knew the starfish would die if left on the hot dry sand and he considered picking some up and putting them back in the water. However, he reasoned that he could not possibly help them all, so he just continued walking carefully along. A little while later, the old man saw a small child further along the beach who was frantically throwing one starfish after another back into the sea…'.

Adapted from 'The Star Thrower' (1978) by Loren Eiseley.

Abstract Consideration of the implications of climate change for wild animal welfare is still relatively novel. The cetaceans are a very diverse group of mammals occupying a range of habitats across the world's oceans. Whilst this makes generalisations difficult, there is a growing body of scientific literature which anticipates and reports impacts. These include prey loss and associated prey stress, changes in cetacean foraging locations and other distribution shifts (including movement into higher latitudes), the use of extra energy to try to maintain body temperature and the loss of habitat for ice-dependent species. Climate change-driven changes in human behaviour, such as the introduction of new activities into increasingly ice-free polar waters, also offer challenges to marine mammals. All these impacts are predominantly considered in the literature from a conservation perspective. However, habitat destruction, pollution and the spread of disease and noise have already been cast as causes for animal welfare concern, and it is argued that climate change will further exacerbate these and other issues in many instances. Assessing the full welfare implications of climate change calls for innovative and careful application of welfare science and will be challenging, but a promising start has been made.

M.P. Simmonds
Senior Marine Scientist, Humane Society International, c/o 5 Underwood Steet, London N1 7LY, UK
e-mail: mark.simmonds@sciencegyre.co.uk

Visiting Fellow University of Bristol, School of Veterinary Sciences, Langford House, Langford, Bristol BS40 5DU, UK

A. Butterworth (ed.), *Marine Mammal Welfare*, Animal Welfare 17,
DOI 10.1007/978-3-319-46994-2_8

8.1 Introduction

Chapters 2, 14, 15, 19, 22–25 and 28 in this book are among the first attempts to consider the welfare implications of climate change for wild marine animals. Indeed, despite accelerating human encroachments into their habitats and their lives, the welfare of wild animals has itself been given little consideration. Perhaps this is because they appear to be beyond our jurisdiction, living 'wild and free'. Exceptions include matters relating to the hunting of wild animals—where it is clearly our actions that have welfare implications—and this includes whale hunting. Consideration has gone into this aspect of hunting, including work by the International Whaling Commission (Brakes et al. 2004), and the effects of ship strikes, entanglement in fishing gear and poorly conducted whale watching have also emerged as welfare concerns for wild cetaceans (Anon 2011). Meanwhile, and not surprisingly, our predominant interests, when it comes to climate change, focus around the implications for our own species, and maybe, secondarily, if we are really thinking carefully about our interlinked futures, the functionality of the ecosystems that support us. Nonetheless, it is increasingly accepted around the world that we have a responsibility to the other living beings that our actions, deliberately or otherwise, impact, and this includes taking care of the welfare needs of animals affected by human actions (Jordan 2005; Anon 2011).

Here I will seek to make the case that climate change should be viewed as a welfare issue for wild cetaceans. I will do this by considering key aspects of cetacean biology, the state of the relevant science concerning climate change and linkages to welfare issues and, finally, by comparing these topics with the established approach to animal welfare issues in non-wild species.

8.2 A Myriad of Species Within a Vast Patchwork of Habitats

There are some 90 species in the mammalian order Cetacea, and, remarkably for such large mammals, more are still being discovered. Each species has its own distinctive food and habitat requirements, and this makes generalisations difficult. However, it has long been understood that the broad-scale distributions of marine mammals worldwide are ocean temperature and food source (often linked with ocean temperature) related (e.g. Gaskin 1982; Kaschner et al. 2011). For example, some species live exclusively in the cold Arctic and some in the warm tropics. Others, famously, undertake long migrations and move from warmer water breeding grounds to their feeding grounds in the Arctic or Antarctic where they arrive to exploit the great blooms of plankton and other prey that occur there in spring. On a more local scale, the distribution of cetaceans may also reflect oceanographic features such as upwellings and fronts (again where there is high productivity and abundant prey). Depth is clearly also an important habitat feature that defines distribution, and the beaked whales (family Ziphiidae) and sperm whales (*Physeter macrocephalus*) feed at great depths (i.e. sometimes over 1000 m), whereas some other species are adapted to life in shallower waters, including inshore waters or large rivers.

An idea of how cetacean marine habitats are distributed might be achieved on a journey heading westward out to sea from the northwest corner of Scotland (Murray and Simmonds 1998). A keen observer might first witness bottlenose dolphins (*Tursiops truncatus*) close inshore; then, further out, minke whales (*Balaenoptera acutorostrata*); and, out around the offshore islands, perhaps Risso's dolphins (*Grampus griseus*). The edge of the continental shelf, some 200 miles out, then marks the transition between 'shelf species' (those whose habitat is on the continental shelf), including the white-sided dolphin (*Lagenorhynchus acutus*) and the cetaceans of the deeper sea. Sperm whales (all males in these northern waters) may be found here and also fin (*Balaenoptera physalus*) and sei (*Balaenoptera borealis)* whales, perhaps even a passing blue whale (*Balaenoptera musculus*). What a hypothetical transect like this shows is that whilst these are mainly cold-temperate water species (the sperm whale and bottlenose dolphin might be argued as exceptions), they also all have differing habitats. Surface observations miss the fact that different species are feeding at different depths. A Risso's dolphin may look superficially rather similar to a bottlenose—they are a similar size and mainly grey—but each has different feeding preferences. Risso's dolphins prefer cephalopods (cuttlefish and squid) and are deeper divers, hence often being found near deeper waters; whereas bottlenoses are more catholic in their diet but are mainly fish eaters.

It is generally difficult for us as a terrestrial species to easily comprehend the wide range of habitats in the seas and oceans and appreciate that the cetacean species and populations (and in some instances even cultural units) occupy them each according to their needs, preferences and physiological tolerances. This is a fundamental problem in addressing issues for cetaceans and other marine wildlife, because if the public and policy makers see the marine environment as a mainly homogenous environment where individuals (or even whole populations) can just move away from unpleasant, damaging or dangerous stimuli, it becomes impossible to make compelling cases to address such problems. However, in terms of how human activity is impacting cetaceans, it is important to appreciate this and, most fundamentally of all, to understand that a cetacean encountering something unpleasant—perhaps some over enthusiastic and noisy whale watchers or a stressful temperature change—may not be able to simply swim away to another sea area that will fully meet its needs.

8.3 The Underpinning Science

If the seas, oceans and even some of the larger river systems form a patchwork of cetacean habitats defined by physical (including temperature), chemical (including salinity) and biological (including suitable prey availability) features, what happens when these conditions change? Twenty years ago almost nothing had been published about the potential for marine mammals to be affected by climate change.

This is truly a rapidly emerging issue now as several hundred scientific publications can be identified which address this topic, and evidence is growing of climate change impacts (Nunny and Simmonds 2016; Simmonds 2016). The kinds of impacts that marine mammals may experience are summarised in Table 8.1.

Table 8.1 Summary of some observed and predicted effects of climate change on marine mammals, largely based on Schumann et al. (2013)

Climate phenomenon	Response from marine mammals	Possible implications
Changes in ocean and air temperature	• Changes in foraging locations	– Potential for novel competition
	• Distribution shifts, including range expansion of tropical/temperate species and range contraction of cold water species	– Loss of habitat of shelf species and their possible extinction
	• Extra energy expenditure to try to maintain body temperature	– Longer migrations
		– Regime shift
		– Exposure to novel pathogens and pollutants
		– Less energy available for reproduction
Reduction in sea ice	• Ice-dependent species lose habitat and move/decline	– Breeding impacted
	• New species enter higher latitudes	– Potential for mismatch between prey availability and critical life history stages
		– Potential for novel competition and exposure to novel pathogens
Increase in frequency and severity of extreme weather events, including flooding and increased roughness	• Reduction in coastal water quality causes prey decline and or distribution change in marine mammals	– Breeding impacted
	• Increased incidence of rough conditions cause more strandings	– Exposure to novel pathogens and pollutants
		– Animals lost from population
Changes in ocean currents, winds and circulation affecting upwelling and productivity	• Where productivity increases, more abundant prey may benefit (some) marine mammal populations, and, where it declines, marine mammals may move or decline	– Local population increase
		– Local populations decrease
Rising sea level causing coastal inundation including flooding of coastal refuse tips and similar	• Changes in breeding bays/estuaries/ inshore zones affecting breeding	– Breeding depressed
		– Increased pollution and pathogen exposure

Generally, the primary mechanisms by which climate change is expected to impact cetacean populations are by changing the physical and chemical nature of their environment and, in particular, by affecting their prey in terms of its quantity, quality and location (Simmonds 2016). On a simple level, if there is a decline in local productivity and prey availability, predators will go hungry. This does not mean that they necessarily starve, but their health may suffer and they may not have adequate energy reserves to breed successfully.

Cetaceans will certainly have some ability to adapt. Many species, including most of the dolphin species, range across large areas using their highly developed cognitive skills, echolocation ability and 'team work' to help find and efficiently exploit patches of prey. These species may be more adaptable to change than those whose abilities to respond are in some way more constrained. These might include those great whales whose energetic and migration biology hinges on the anticipated (and 'planned for') finding of certain things in certain places at specific times of the year. For example, a whale arriving at its polar feeding grounds in spring but not finding the 'bloom' of food that it needs could be severely incapacitated, and evidence has started to accrue that this is not just a hypothetical threat (as postulated by Simmonds and Eliott 2009) but already a real and current phenomenon. In particular, it has recently been revealed that two highly migratory big whale species have been arriving at their feeding grounds in the Gulf of Saint Lawrence in Canada increasingly early (Ramp et al. 2015). In the period 1984–2010, humpback (*Megaptera novaeangliae*) and fin whales have been arriving more than 1 day earlier in each subsequent year. This seems to be related to earlier ice breakup and rising sea surface temperatures. This certainly shows some adaptability on the part of the whales, but the scientists reporting this change also warn that this adaptability may be exceeded as conditions continue to change.

A similar issue may arise for the 'shelf species' that have evolved to exploit the waters of the continental shelf if the regime there changes so much (e.g. temperature rises and/or fish prey move elsewhere) that it becomes inhospitable to them. Can the 'shelf cetaceans' simply move to a new area? For example, such animals being 'pushed' off the continental shelf around the UK may not be able to find suitable habitat if they move northwards (as suggested by MacLeod et al. 2005, 2008). Similarly, for those animals whose habitat is in deep water trenches, like the beaked whales, what happens if local conditions change in terms of temperature or some other key factor such as loss of prey (Simmonds 2016)? Will these animals be able to disperse to similar habitats elsewhere?

It is perhaps easier to see the risk for a cetacean population that is actually physically constrained from moving away from adverse change. For example, the remaining river dolphin species are mainly confined to sections of specific tropical river systems, and their movements are increasingly limited by major waterway modifications, especially dams. Their capacities to alter their distributions in response to unfavourable changes are seemingly far more limited than the dolphins of the open oceans. Similarly, species living in enclosed sea areas, such as the three cetacean species found in the Black Sea (short-beaked common dolphin (*Delphinus delphis*), Black Sea bottlenose dolphin (*Tursiops truncatus ponticus*) and harbour porpoise (*Phocoena phocoena*)), may find it impossible to make mitigating range changes as conditions alter there.

Most people are now becoming acquainted with the vulnerability of the polar bear to climate change and especially the loss of its sea ice habitat (eloquently described here in Chaps. 2, 23–25 and 28). Less well appreciated is the relationship between several cetacean species and the same retreating sea ice. Narwhals (*Monodon monoceros*) in particular are viewed as especially vulnerable (Laidre et al. 2008). From the available information it is possible to diagrammatically summarise at least some of the linkages between climate change-driven effects and impacts on marine mammal populations (Fig. 3.1).

8.4 The Human Dimension

Figure 8.1 introduces the concept of impacts on marine mammals that are mediated by changes in human behaviour in response to climate change. For example, we may move our centres of population or our major marine activities, and these actions

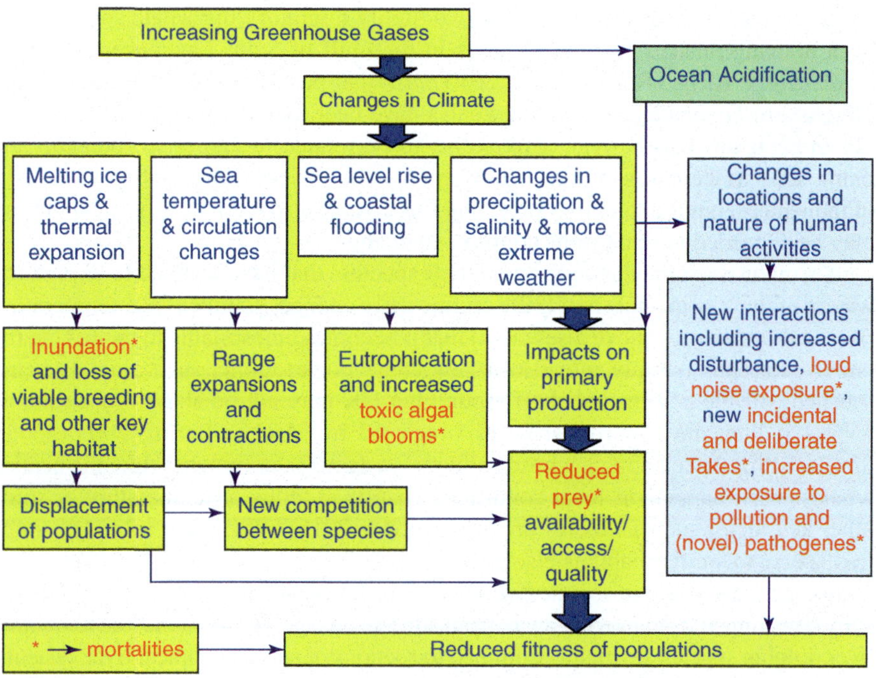

Fig. 8.1 Some of the potential linkages between climate change and effects on marine mammal populations (after Simmonds 2016). From top to bottom, physical changes (*white boxes inside yellow box*) are linked to habitat effects and then impacts on species in the *yellow boxes* below. The *red text* and *asterisks* indicate where mortalities may be precipitated. The *blue boxes* show 'tertiary effects'—where changes in human behaviour caused by climate change might be expected to impact marine mammals. The *thicker blue arrows* denote the stronger linkages. *Image credit*: Mark Simmonds

may impinge on marine mammals in new ways. As described by Alter et al. (2010), impacts on marine mammals may include:

- Increased disturbance from shipping and other marine activities—leading to stress and interference with behaviour, including communication
- Increased exposure to loud noises—leading to acoustic masking[1] or even physical harm (hearing damage and embolisms in tissues)
- Increased interference with normal behaviour
- Increased take (e.g. as by-catch when fisheries move into new areas)—causing deaths in fishing gear which may be prolonged for breath-holding mammals—and also increased incidents of chronic entanglement with severe welfare implications (see Chap. 4)
- Increased exposure to pollution with chronic health concerns (see Chaps. 3 and 32) and potentially including increased oil spills associated with vessel movements in new areas
- Increased exposure to pathogens (possibly in combination with the immunosuppressive effects of certain pollutants) potentially causing disease events

The potential for human-mediated factors to impact marine wildlife is perhaps best illustrated by the developing situation in the Arctic. Here the fast retreat of sea ice is allowing human activities to expand in the region, including increasing ship traffic, fossil fuel exploration and extraction and fishing (Reeves et al. 2014). The general prediction for cetacean species to move towards the poles in response to climate change (Whitehead et al. 2008; Kaschner et al. 2011) may favour some species—at least initially—for example, fin and humpback whales might be able to inhabit and exploit the open waters at latitudes where they did not previously occur. Meanwhile those species associated with the ice edge—bowheads (*Balaena mysticetus*), belugas (*Delphinapterus leucas*) and narwhals—may not only see their habitats shrink but may also come into contact with species that they have not previously met, leading to unknown consequences.

8.5 The Case for Climate Change as a Welfare Concern for Cetaceans

The concern that I am attempting to present here is not about climate change-driven extinction, which sadly seems likely to be the case for some wildlife populations, nor whether individuals are being killed. The issue is whether or not suffering will increase.

As noted above, reduced prey availability may occur, and this could cause 'food stress', leading to poor nutrition and potential starvation, likely invoking reproductive inhibition along the way. Poor nutrition and starvation are clearly welfare

[1] 'Masking' refers to the situation where one sound interferes with another making it difficult to comprehend key information. For example, the noise from shipping may mean that whales cannot hear the calls of conspecifics.

concerns, and, arguably, reproductive problems could be seen as such too. High temperature and nutrient loading could lead to harmful algal blooms that may poison the animals (see Chap. 19), and certainly suffering can be involved in such events as well as mortalities. It is also possible that animals that cannot move away from changing condition may be exposed to temperatures outside of their preference and tolerance. This could be distressing in the short term and may also have effects on reproductive performance and on feeding, prey sources and foraging behaviours in the longer term. The combined effects of climate change on cetaceans from changes to their habitats and prey, in combination with changing human activities, are likely to cause (and in fact already are probably already causing) increases in the incidence of disease, increased entanglement in marine debris and lost fishery material (abandoned, lost or otherwise discarded fishing gear—ALDFG) (see Chaps. 13 and 18) and wounding and deaths as by-catch in nets (see Chap. 4). These are very clearly welfare concerns, and it is thus anticipated that climate change will act to exacerbate welfare concerns for marine wildlife.

Can the study of animal welfare science help us to quantify possible impacts of climate change? For example, are these concerns minor or severe, acute or chronic (and how do we judge this and relative to what baseline); are climate changes affecting a large number of animals or only a few; and, ultimately, how concerned should we be?

Writing in 2005, Bill Jordan specifically recognised that non-intentional consequences of human actions on wild animals could be considered as legitimate animal welfare concerns. He identified habitat destruction, pollution (including specifically PCBs) and the spread of disease and noise (Jordan 2005) as causes for animal welfare concern, all matters that climate change is likely to exacerbate for cetaceans. More recently, in May 2016, experts gathered at an IWC workshop to consider non-whaling welfare issues (IWC 2016). In particular, they considered the potential application of the 'Five Domains Model' (Mellor and Reid 1994) for wild cetaceans. This provides a framework to enable assessment and grading of the severity of different impacts on welfare (Mellor 2015). Figure 8.2 shows a model based on this framework which includes a range of factors which could be integrated to provide an assessment of the welfare impact of commonly identified welfare issues for whales and dolphins.

The Five Domains Model was originally developed for livestock, and it is still early days for extrapolation to wild cetaceans, but at the IWC workshop (IWC 2016), it was apparent that the use of a structured framework to analyse and to support discussion of welfare issues in cetaceans had real merits in terms of allowing rational and clear discussion of these issues across a range of stakeholders and involved organisations. There will certainly be challenges in use of assessment/discussion methods in this way, including difficulties in trying to interpret whether these wild animals are likely to experience these welfare impacts in the way that humans might predict or anticipate. However, there is good and growing evidence that cetaceans are animals with highly developed cognitive capacities and that they have emotions (Simmonds 2006), including the ability to grieve (Reggente et al. 2016), and this structured approach to analysis of these questions is promising and should be encouraged.

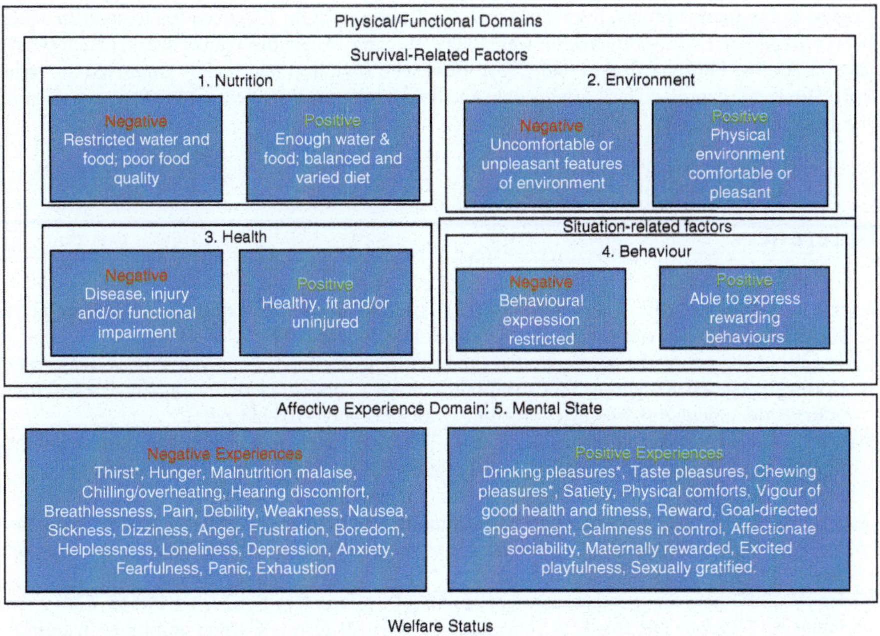

Fig. 8.2 An abbreviated version of the Five Domains Model (after Mellor 2015). Some elements are marked with an *asterisk* to indicate that they may not apply to cetaceans. *Image credit*: Mark Simmonds

8.6 Conclusions

The remarkable penta-radial symmetry of starfish sets them apart from many life forms, especially from bisymmetrical vertebrates like us. They are rather 'alien' really, lacking even a head or a brain, but I am not convinced that they cannot suffer—through, for example, an experience of 'distress' at being physically damaged.

When it comes to cetaceans, despite their somewhat fishy forms and great variety, it should be easier for us to conceptualise suffering in these intelligent and typically highly social marine mammal. This should enable us to evaluate with more empathy how their welfare is being impacted by human actions and to strive to respond appropriately and compassionately. Climate change gives us an enormous challenge in this (and for human kind more generally), but we are big-brained too, and, hopefully, wisdom and compassion will prevail.

I will conclude with the last part of the modern fable that I started this chapter with:

The old man made his way to the child and asked her what she was doing.

"I'm saving the starfish," she replied.

"You are wasting your time. You can't save them all, so what does it matter?"

Without pause, the child picked up another starfish. She tossed it back into the water. "It matters to this one," she said quietly but emphatically. The old man slowly bent down and started to help her.

Acknowledgements Firstly, my thanks to Andy Butterworth for the invitation to contribute here and for his patience whilst this chapter was slowly 'stewing'. Secondly, thanks to Andy Butterworth, Laetitia Nunny, Philippa Brakes and Claire Bass who reviewed and greatly improved an earlier draft. The views expressed here are my own and do not necessarily reflect those of any organisation that I am affiliated with.

References

Alter SE, Simmonds MP, Brandon JR (2010) Forecasting the consequences of climate-driven shifts in human behavior on cetaceans. Mar Policy 34:943–954

Anon (2011) Report of the whale welfare and ethics workshop 22–23 Mar 2011 Eden project, Cornwall, United Kingdom. IWC/63/WKM&AWI 4. https://awionline.org/sites/default/files/uploads/documents/ml-whalewelfareethicsworkshopreport-101111.pdf

Brakes P, Butterworth A, Simmonds MP (2004) Troubled waters: a review of the welfare implications of modern whaling activities. World Society for the Protection of Animals, London

Eiseley L (1978) The star thrower. Times Books (Random House), New York

Gaskin D (1982) The ecology of whales and dolphins. Heinemann, London

IWC (2016) Report of the IWC workshop on animal welfare held in the Kruger national park. Available from the IWC Secretariat

Jordan B (2005) Science-based assessment of animal welfare. Rev Sci Tech 4(2):515–528

Kaschner K, Tittensor DP, Ready J, Gerrodette T, Worm B (2011) Current and future patterns of global marine mammal biodiversity. PLoS One 6:e19653

Laidre KL, Stirling I, Lowry LF, Wiig Ø, Heide-Jørgensen MP, Ferguson SH (2008) Quantifying the sensitivity of arctic marine mammals to climate-induced habitat change. Ecol Appl 18:S97–S125

MacLeod CD, Bannon SM et al (2005) Climate change and the cetacean community of north-west Scotland. Biol Conserv 124(4):477–483

Macleod CD, Weir CR et al (2008) Temperature-based summer habitat partitioning between white-beaked and common dolphins around the United Kingdom and Republic of Ireland. J Mar Biol Assoc UK 88(6):1193–1198

Mellor DJ (2015) Updating animal welfare thinking: moving beyond the "five freedoms" towards "a life worth living". Animals 6. doi:10.3390/ani6030021

Mellor DJ, Reid CSW (1994) Concepts of animal well-being and predicting the impact of procedures on experimental animals. In: Improving the well-being of animals in the research environment. Australian and New Zealand Council for the Care of Animals in Research and Teaching, Glen Osmond, pp 3–18

Murray L, Simmonds MP (1998) The Atlantic Frontier; Britain's last ocean wilderness. Greenpeace and the Whale and Dolphin Conservation Society report

Nunny L, Simmonds MP (2016) Climate change and marine mammals: an assessment of the scientific literature. Paper submitted to the Scientific Committee of the International Whaling Commission. SC/66b/E/05, 9pp. https://iwc.int/sc66bdocs

Ramp C, Delarue J, Palsbøll PJ, Sears R, Hammond PS (2015) Adapting to a Warmer Ocean—Seasonal Shift of Baleen Whale Movements over Three Decades. PLoS ONE 10(3): e0121374. https://doi.org/10.1371/journal.pone.0121374

Reeves RR, Ewins PJ et al (2014) Distribution of endemic cetaceans in relation to hydrocarbon development and commercial shipping in a warming Arctic. Mar Policy 44:375–389

Reggente MAL, Alves F, Nicolau C, Freitas L, Cagnazzi D, Baird RW and Galli P (2016) Nurturant behavior toward dead conspecifics in free-ranging mammals: new records for odontocetes and a general review. J Mammal. doi:http://dx.doi.org/10.1093/jmammal/gyw089. First published online: 24 May 2016

Schumann N, Gales NJ, Harcourt RG, Arnould JPY (2013) Impacts of climate change on Australian marine mammals. Aust J Zool 61:146–159

Simmonds MP (2006) Into the brains of whales. Appl Anim Behav Sci 100(1–2):103–106. doi:10.1016/j.applanim.2006.04.015

Simmonds MP (2016) Cetaceans and global warming. In: Laffoley D, Baxter JM (eds). Explaining ocean warming: Causes, scale, effects and consequences. Full report. Gland, Switzerland: IUCN. pp. 303–320

Simmonds MP, Eliott WJ (2009) Climate change and cetaceans: concerns and recent developments. J Mar Biol Assoc UK 89(1):203–210

Whitehead H, McGill B, Worm B (2008) Diversity of deep-water cetaceans in relation to temperature: implications for ocean warming. Ecol Lett 408:1–5

Chapter 9
Managing the Welfare of Marine Mammals at Mass Strandings in Golden Bay, New Zealand

Mike Ogle

Abstract In this chapter, issues of marine mammal welfare are illustrated through recounting three mass stranding events of long-finned pilot whales which occurred in Golden Bay, New Zealand. For two of the mass strandings discussed, both were reported soon after the whales stranded and had good access and high numbers of volunteers assisting Department of Conservation (DOC) staff. One of these strandings had a high refloating success rate (89% of 345 whales), the other a moderate success rate (39% of 198 whales). This contrasted with the third stranding (comprising of 105 whales) which occurred in a remote location with difficult access and was first observed from an aircraft, 1 or possibly 2 days after the initial stranding. When DOC staff arrived at this remote site, less than one quarter of the pod was still alive, and these were suffering considerably. Given the whales' poor condition, high degree of suffering and low chance of survival, they were euthanised following DOC guidelines. These three mass strandings were relatively large and if combined accounted for approximately one third of the nearly 2000 cetaceans that stranded in Golden Bay between 1990 and 2016. New Zealand has a relatively high occurrence of strandings, with an average of 300 cetaceans stranded annually in the last 26 years. Stranding events are recorded on the New Zealand Whale and Dolphin Stranding Database, which is maintained by the DOC. This government organisation has statutory responsibility for management of marine mammals under the Marine Mammal Protection Act. Its role, obligations under the Treaty of Waitangi and use of volunteers at mass strandings are briefly described.

M. Ogle
Department of Conservation, Takaka, New Zealand
e-mail: mogle@doc.govt.nz

A. Butterworth (ed.), *Marine Mammal Welfare*, Animal Welfare 17,
DOI 10.1007/978-3-319-46994-2_9

9.1 Introduction

Marine mammal strandings are a regular occurrence in New Zealand; between 1990 and 2016 an average of 300 stranded annually (New Zealand Whale and Dolphin Stranding Database, accessed 6th April 2016). Over this period, there have been 308 mass strandings, with 39 of these mass stranding events involving 50 or more cetaceans. One definition of a mass stranding is a stranding involving more than one cetacean that is not a mother-calf pair (Gercai and Lounsbury in Jepson et al. 2013). Forty-one species have been recorded as stranding in New Zealand; the commonest species to strand is the long-finned pilot whale (*Globicephala melas*, Traill 1809). Long-finned pilot whales are classified as not threatened in New Zealand by Baker et al. (2016) and globally as data deficient in the IUCN red list (Taylor et al. 2008). The largest recorded stranding in New Zealand of 1000 pilot whales (*Globicephala* sp.) occurred on Chatham Island in 1918.

The Department of Conservation (DOC) is the central government organisation charged with promoting conservation of the natural and historic heritage of New Zealand. DOC has 1637 staff and 60 offices distributed around New Zealand, including offshore islands (DOC 2015a). DOC has the statutory responsibility of marine mammal management under the Marine Mammals Protection Act 1978 (MMPA). Also there is a statutory responsibility on DOC in the Conservation Act 1987 (the founding legislation of DOC) to give effect to the principles of the Treaty of Waitangi. The Treaty of Waitangi is an agreement between Maori (the indigenous people of New Zealand) and the Crown (i.e. government) signed in 1840. In practical terms this means that, at a marine mammal stranding, major decisions are made in partnership between DOC and the local iwi/tribe. It is an offence under the MMPA to herd or disturb marine mammals without permission from DOC. However, the penalties do not apply to anyone providing humane care to stranded, sick or injured marine mammals. While DOC is responsible for marine mammal strandings, assistance is often provided by large numbers of volunteers, some affiliated with nongovernment organisations, in particular Project Jonah. Project Jonah has 2200 volunteers trained to assist at marine mammal strandings, and many of these volunteers can be mobilised at short notice (Daren Grover, general manager, Project Jonah, pers. comm. April 2016). A service level agreement exists between DOC and Project Jonah, under which Project Jonah has agreed to provide assistance to DOC and to train people for marine mammal strandings (DOC 2015b). The organisational structure used by DOC during marine mammal strandings follows the Coordinated Incident Management System (CIMS) model (NZFS 1998) which is adaptable to small and large emergency events.

There are five locations in New Zealand where mass strandings have occurred in high numbers: Northland Region, Mahia Peninsula, Golden Bay, Chatham Islands and Stewart Island. These five locations account for 84% of cetaceans involved in mass strandings. Since 1990 nearly 2000 cetaceans have mass stranded in Golden Bay; this is the highest total of these five locations. New Zealand's third largest mass stranding, 345 pilot whales, occurred in Golden Bay in January 1991. Golden Bay

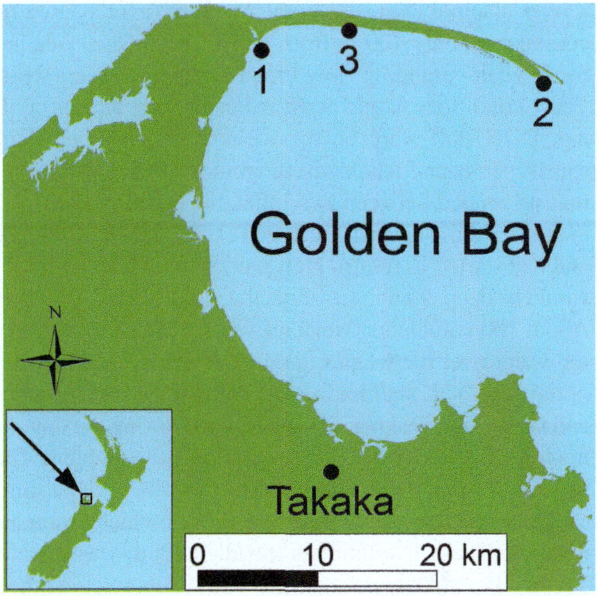

Fig. 9.1 Map of Golden Bay. Numbers *1*, *2*, and *3* are locations of strandings discussed in text as follows: (1) Puponga, 1991; (2) Bush End Point, 2009; (3) Farewell Spit, 2015. *Inset figure* shows location of main map in New Zealand (*Image credit*: Mike Ogle)

(40.6°S, 172.8°E) lies in the north-west corner of the South Island, New Zealand. Golden Bay has a population of just under 5000 permanent residents (Statistics New Zealand 2013), but is boosted in summer by seasonal residents and tourists. This semicircular bay faces east into the South Taranaki Bight, and the entrance (between the end of Farewell Spit and Separation Point) is 25 km across (Fig. 9.1). At this broad entrance, the maximum depth is around 35 m, and from this the seafloor gradually slopes up to the shoreline (LINZ 1999). Most of the 90 km shore is comprised of sandy gently sloping beaches with occasional rocky headlands. However, on the southern coastline, rocky headlands dominate, separated by small sandy bays. Along most of the shore, large tidal flats are exposed at low tide. The most extensive tidal flats are at Farewell Spit; here tidal flats are present along the entire 26 km length of the spit and at their widest can extend more than 7 km out from the high tide mark. The maximum difference between low and high tide is 4.5 m (LINZ 2015). The purpose of this chapter is to illustrate issues of marine mammal welfare at stranding events through discussion of stranding events in the Golden Bay area, New Zealand.

9.2 Puponga, January 1991

At 8:30 am on January 24, 1991, the Golden Bay DOC office received a report from a local tour operator that a pod of whales had stranded near Puponga (Stark 1991). Twenty minutes later it was confirmed that an estimated 200–300 whales had stranded. The whales had stranded on the tidal flat directly south of Puponga Point, adjacent to a river channel (Fig. 9.1 Point 1). Access for people here is relatively easy, with a coastal road only 400 m from the stranding site. Weather conditions were favourable for a whale stranding: low cloud and rain, strong wind and very cold temperature. These conditions would assist in keeping the whales cool and their skin from desiccating. DOC staff were on the scene from 10:40 am with rescue materials (buckets, sheets, slings and whale rescue pontoons). Whale rescue pontoons consist of a lifting mat suspended between two inflatable pontoons and are designed to lift whales of up to about 2 tons (Project Jonah 2012). The tide had reached its lowest point at 10:30 am and had started to return. However, it would not be until mid-afternoon that the water would be deep enough to refloat the whales, and high tide was forecast to be at 4:49 pm. About 300 volunteers (Nelson Evening Mail 24/1/1991) tended the whales by bucketing water over the whales and covering the whales with wet sheets. At the beginning of the day, DOC staff had assessed the whales to be in good condition, but despite this and favourable weather, 20 whales died during the day. As the tide came in and whales floated, they were guided into a group by the rescuers. The pod was released at approximately 3:30 pm, with almost all of the whales departing as one group.

The exception to this was five whales which swam away before the main pod was released. These five animals travelled south-west parallel to the coast for 2.5 km to Taupata Point. Despite attempts with a boat to guide the whales away from shore, the whales could not be stopped from restranding. Shortly after this, another 40 whales also stranded at Taupata Point. It was thought that the earlier five whales were responsible for luring the other 40 in to strand (Stark 1991). Rescuers were sent to these whales, but with the tide now receding, 13 whales could not be moved. To increase the probability of successfully refloating the other 32 whales that lay in deeper water, those whales that could not be moved were euthanised. By 6 pm the remaining 32 whales had been guided out beyond the low tide and half an hour later swam out to sea.

The next day during an early morning search by helicopter, 26 whales were observed stranded at Ferry Point, 15 km south-west of the initial stranding. They were spread out in two groups, 1 km apart from each other. The whales were kept wet and cool through the day by DOC staff and volunteers. Refloating begun at 4 pm and the two groups were brought together. For the next 45 min, people made a human barrier between the whales and the shore, after which the whales appeared to orientate themselves then headed out to sea.

In the initial stranding on the first day, 345 whales stranded; of these 325 were refloated and 20 died. Shortly after this first refloating, 45 of these whales restranded, 13 of which died and 32 were refloated. With another restranding of 26 whales the following day, five more whales died. Of the initial 345 whales stranded, 38 died and 307 (89%) were successfully refloated.

9.3 Bush End Point, December 2009

From a chartered light aeroplane, on the morning of December 26, 2009, a large pod of stranded whales were seen at the far eastern end of Farewell Spit at an area known as Bush End Point (Fig. 9.1, Point 2). The pilot contacted the local air control staff who forwarded the report to the DOC emergency duty phone. Normally two DOC staff members would be dispatched to the site to make an initial assessment. However, the site of this stranding was remote and difficult to access; there was no road, and vehicles could only be driven to the site along the 22 km of beach during the hours either side of low tide. The condition of the whales at this stage was unknown. Planning for the worst-case scenario, two additional staff were included in the initial assessment team to assist with logistics, and two rifles were taken in case euthanasia was determined to be the best course of action that would result in the least suffering. On the drive to the site, the four staff members discussed possible scenarios and the logistical issues these scenarios presented. If the whales were in suitable condition for refloating, the most difficult logistical consideration would be getting enough volunteers quickly to the site. A local tour company did have buses which regularly travelled along the beach to this far end of the spit. However, given that the next high tide was at 5:40 pm (and it would not be possible to drive along the beach 1–2 h before this), it was highly unlikely that enough volunteers could be transported to the stranding site in time to attempt refloating the whales on that evening's high tide. While there was limited accommodation associated with the lighthouse at the eastern end of the spit, it would not be enough for the anticipated number of volunteers that would be required to undertake a successful refloating of the pod. If volunteers were taken to the stranding site to attempt a refloat, they would need to be completely self-sufficient, including food, water, hygiene and shelter.

The initial assessment team, including the author, arrived on site at approximately 11 am. The pod was scattered over an area of approximately 4 km × 1 km, across a broad expanse of sandy tidal flat. The sunny, warm and windy weather conditions were not favourable for stranded whales. Without regular wetting, cetacean skin in these conditions soon desiccates, blisters and then peels (Fig. 9.2). At

Fig. 9.2 Without regular wetting, stranded cetacean skin in exposed warm, sunny and windy conditions soon desiccates, blisters and then peels. Dead pilot whales, from a pod of 105, Bush End Point 28/12/2009. *Image credit*: Greg Napp/DOC

this time, an automated weather station, 2 km from the stranding site, recorded an air temperature of 22 °C and a wind speed of 32 km/h, and no rain had fallen in the past 5 days (NIWA 2016). The first few pilot whales approached were already deceased. Carcasses were in the early stages of decomposition, some with up to a third of the skin dried and peeled off. However, some were still alive, but in very poor condition with blistered and/or peeling skin and showing signs of distress. Given the physical state of the whales, it was likely they had been stranded for at least 1 day, possibly 2. Experience from previous strandings led the team to conclude that it was unlikely the surviving whales would live much longer. The decision was made to euthanise the surviving 26 whales to avoid the whales enduring a slow and painful death. This was carried out following DOC guidelines (Boren 2012) by experienced staff using the rifle. A total of 105 pilot whales had stranded, ranging in size from 1.97 to 5.9 m in length. Two years later 21 pilot whales stranded at the same site; when they were eventually discovered, they were all already dead.

9.4 Farewell Spit, February 2015

At 10:50 am Friday morning of February 13, 2016, a staff member of the cafe near the base of Farewell Spit phoned the Golden Bay DOC office to report seeing a pod of over 30 whales or dolphins. He said the pod was 3–4 km away, stranded on the tidal flats of the inner beach (DOC 2015c). At this time, heat haze and distortion across the exposed tidal flats at that distance would have made the stranded cetaceans difficult to see and hence difficult to count accurately.

Two DOC staff members, including the author, were dispatched to assess the situation and arrived at the stranding site an hour after receiving the report. The distance from the base of the spit (and also the end of the road) to where whales were stranded was 6 km (Fig. 9.1, Point 3). A large pod of pilot whales was spread out in a nearly 1 km long strip of animals, orientated parallel to shore and about 500 m out from the high tide mark in the tidal sand flat. An estimate was made from the top of a high dune of 143 whales, but there were several dense groupings, making an accurate count difficult. This information was communicated to the DOC office in Takaka, where preparations had already begun. By 1 pm an accurate count was made while walking through the pod; the revised (and final) total was now 198 whales, of which 24 were dead. The skin of some of the live whales had already formed blisters, due to desiccation from wind, sun and heat. More DOC staff soon arrived, and the local Farewell Spit tour company delivered the first bus load of 34 volunteers at 2:20 pm. Three more bus loads of volunteers arrived over the next 2 h, and a few volunteers had walked the 6 km along the beach from the road end. High tide was predicted to occur at 5:20 pm and to reach a peak tide level at the same height as that morning's high tide. The incoming tide reached the first whales at about 2:45 pm, and by 4:30 pm about 75% of the pod was floating. At this stage there were approximately 100 volunteers and DOC staff on-site and around 170 live whales. Ideally, at this phase of a stranding, two people wearing wetsuits would be required per whale, to guide and hold the whales in chest deep water for up to an hour until the whales have recovered sufficiently from the stresses of stranding to regroup as a pod and propel themselves back out towards sea. However, at this stage, not only were there not enough people, but some of the volunteers had already become cold and exhausted whilst providing initial care to the stranded whales and so could not safely stay in the water for long. At 5:20 pm the tide started to recede, and while a good number of whales had been directed out away from shore, many were still stranded or had restranded. A count at 7 pm gave a total of 88 dead and 12 stranded live whales. Of these live whales, two were in very poor condition, having more than one third of their skin peeled off as a result of desiccation and abrasion; both these animals were euthanised that evening. Two more died overnight, and the remaining eight were in such poor condition by Saturday morning that these were also euthanised.

At 9:45 pm that Friday night, 81 stranded whales were found by a Project Jonah volunteer, 6 km west of the initial stranding site and only 1 km from the road end. This group of whales was most likely from those that had been refloated a few hours earlier. In the past severe injuries (e.g. broken thigh bone, knocked unconscious) have occurred to people working at night around stranded whales in Golden Bay. Since then it has been the policy to not work around stranded whales at night. So at first light, Saturday morning, people began tending to the whales, keeping them wet and 'up-righting' them. By this time 14 of these whales had died, and one more was to die later in the day. High tide was not expected until 6 pm and the whales needed to be kept cool and wet until the tide could reach them at about 4 pm. By 9:25 am, there were about 150 volunteers tending the whales, and this increased to well over 200 volunteers by 11:30 am (Fig. 9.3).

Fig. 9.3 Volunteers keeping stranded long-finned pilot whale (*Globicephala melas*) wet and cool at Farewell Spit, New Zealand, February 2015 (*Image credit*: Murray Hedwig)

The numbers of volunteers continued to increase through the day with over 400 on-site by 1:45 pm (Daren Grover, Project Jonah, pers. comm.). About 100 volunteers in wetsuits were briefed at 3:30 pm on the refloating phase; following this they proceeded to the whales, where the incoming tide had just arrived. When whales are first beginning to refloat, people will often endeavour to keep calves paired with the adult whale it was stranded next to. However, genetic and spatial analysis of stranded whales has shown that calves are often separated from their mothers at strandings (Oremus et al. 2013). Half an hour before the forecast high tide time, many of the larger whales were still not floating and so unable to be moved to deeper water. As high tides vary with many factors and do not always occur at the height or exact time predicted, there was no certainty that the tide would rise any further. Therefore, rather than the usual orderly process of herding whales together then releasing as one group, whales were urgently moved (some using slings and whale rescue pontoons on the larger whales) to a nearby narrow shallow channel that led out to sea (Fig. 9.4).

The end result was that many of the pods were released individually, rather than one large group. Earlier, one whale had been moved in a whale rescue pontoon out to a boat and was used as a 'lure' for the other whales. To the staff on the boat, there was no clear indication whether this 'lead' whale had any effect on the remaining whales or not. The boat stayed with the whales until 7:45 pm, when the whales were last seen 'swimming well' and heading out to sea. The total number of whales

Fig. 9.4 Volunteers guide a pilot whale to deeper water, Farewell Spit 4 pm 14/2/2015 (*Image credit*: Nadia Steenhouwer, Project Jonah NZ)

successfully refloated on this occasion was 66. From the previous evening 12 whales were unaccounted for and also assumed to have successfully refloated. Out of 198 whales of the initial stranding and over 2 days, a total of 78 whales (39%) were assumed to have been successfully refloated and swum back out to sea.

9.5 Discussion

There are many theories for the causes of marine mammal strandings; for examples see those referenced in Evans et al. (2005), Oremus et al. (2013) and Jepson et al. (2013). Many of these theories are difficult to prove and may or may not be relevant for Golden Bay. Nearshore topography is often mentioned as a possible cause of mass stranding. The key topographical features of Golden Bay are its semicircular shape and gently sloping seafloor. This has often resulted in Golden Bay being called a 'whale trap'. Another possibility may be that, because the bay is sheltered from large ocean swells and has gently sloping beaches, sick whales intentionally come here to rest (or die) and are followed in by their pod which then strands.

As illustrated by the stranding survival rates in the three Golden Bay examples detailed in this chapter, the proportion of whales that survive a stranding can be highly variable. Survival rates at mass strandings for all of New Zealand (1990–2016) also tend to extremes, with 56% of strandings having no survivors (100% mortality) and 15% of strandings in which animals all survive (0% mortality). The survival rates for all documented strandings are spread almost evenly between the two extremes, indicating that many factors are likely at play in determining the 'outcome' for a stranded animal. One key factor influencing mass stranding survival rate in many cases is likely to be the time elapsed between when whales first strand and when people start providing care (i.e. wetting and cooling). This factor may account for the lower survival rates recorded for locations where the human population is low and access is difficult (e.g. Chatham Islands, Stewart Island and Fiordland), compared to the higher survival rates near well-populated and accessible areas.

Maximising the survival rate of stranded marine mammals is a key focus for those DOC staff involved in strandings. Current plans to improve stranding survival rates include trialling a purpose-designed wheeled gantry, built by A-Ward Attachments Ltd (Auckland, New Zealand), for lifting stranded whales and transporting them across tidal flats. A protocol is in place with Massey University, to trial the use of 'on the beach' blood analysis of stranded pilot whales (similar to what has been done for dolphins at Cape Cod (Sharp et al. 2014)), to aid in health assessment and triage of the stranded individuals, and proposals to satellite track refloated pilot whales to confirm post release survival have recently been discussed. However, the key factor in maximising mass stranding survival rate is likely to be early detection, followed by rapid deployment of large numbers of volunteers to keep the whales or dolphins wet and cool.

Acknowledgements We would like to thank Kay Stark for the assistance with the 1991 stranding information and the Takaka DOC staff who have so thoroughly recorded communications in the whale incident management log books, from which much of the 2015 stranding material was referenced. The quality of this text was improved by comments from Greg Napp (DOC, Golden Bay) and two other anonymous reviewers.

References

Baker CS, Chilvers BL, Childerhouse S, Constantine R, Currey R, Mattlin R, van Helden A, Hitchmough R, Rolfe J (2016) Conservation status of New Zealand marine mammals, 2013, New Zealand threat classification series 14. Department of Conservation, Wellington

Boren L (2012). Area operational plan for marine mammal incidents. Guidelines. Department of Conservation, Wellington. DOC-776969

Department of Conservation (2015a) Department of Conservation annual report for year ended June 2015. Department of Conservation, Wellington

Department of Conservation (2015b) Service level agreement between Project Jonah and Department of Conservation. Department of Conservation, Wellington, New Zealand. DOC-2213663. Unpublished record

Department of Conservation (2015c) Whale incident management log book, vol 3. Department of Conservation, Takaka. Unpublished record

Evans K, Thresher R, Warneke RM, Bradshaw CJA, Pook M, Thiele D, Hindell MA (2005) Periodic variability in cetacean strandings: links to large-scale climate events. Biol Lett 1(2):147–150. doi:10.1098/rsbl.2005.0313

Jepson PD, Deaville R, Acevedo-Whitehouse K, Barnett J, Brownlow A, Brownell RL Jr et al (2013) What caused the UK's largest common dolphin (*Delphinus delphis*) mass stranding event? PLoS One 8(4):e60953. doi:10.1371/journal.pone.0060953

Land Information New Zealand (1999) New Zealand nautical chart NZ61 Karamea River to Stephens Island. Land Information New Zealand

Land Information New Zealand (2015) Tide predictions, Tarakohe 2015 CSV file. Land Information New Zealand. http://www.linz.govt.nz/sea/tides/tide-predictions. Accessed 19 July 2016

National Institute of Water and Atmosphere (2016) CliFlo: NIWA's national climate database on the web. http://cliflo.niwa.co.nz/. Accessed 17 July 2016

New Zealand Fire Service Commission (1998) The New Zealand Coordinated Incident Management System (CIMS): teamwork in emergency management. New Zealand Fire Service Commission, Wellington. ISBN:0-908920-32-6

Oremus M, Gales R, Kettles H, Baker CS (2013) Genetic evidence of multiple matrilines and spatial disruption of kinship bonds in mass strandings of long-finned pilot whales, *Globicephala melas*. J Hered 104(3):301–311

Project Jonah (2012) Marine mammal medic training handbook. Project Jonah, Auckland

Sharp SM, Knoll JS, Moore MJ, Moore KM, Harry CT, Hoppe JM, Niemeyer ME, Robinson I, Rose KS, Sharp WB, Rotstein D (2014) Hematological, biochemical, and morphological parameters as prognostic indicators for stranded common dolphins (*Delphinus delphis*) from Cape Cod, Massachusetts, U.S.A. Mar Mamm Sci 30:864–887. doi:10.1111/mms.12093

Stark K (1991) Whale stranding, Golden Bay 24–25 January 1991. Department of Conservation, Takaka. NMG-1 NHS-09-02-06. Unpublished file note

Statistics New Zealand (2013) 2013 Census population and dwelling tables—Tasman District. Statistics New Zealand. http://www.stats.govt.nz/~/media/Statistics/Census/2013%20Census/data-tables/population-dwelling/tasman.xls. Accessed 18 July 2016

Taylor BL, Baird R, Barlow J, Dawson SM, Ford J, Mead JG, Notarbartolo di Sciara G, Wade P, Pitman RL (2008) *Globicephala melas*. The IUCN red list of threatened species 2008. http://dx.doi.org/10.2305/IUCN.UK.2008.RLTS.T9250A12975001.en. Accessed 19 June 2016

Chapter 10
Social Change in Cetacean Populations Resulting from Human Influences

Philippa Brakes

Abstract Group living has a number of potential ecological and animal welfare benefits. The social environment of the 90 or so species (http://www.iucn-csg.org/index.php/status-of-the-worlds-cetaceans/) of cetaceans is highly diverse, ranging from the complex third-order alliances of male bottlenose dolphins (*Tursiops* sp.), to the matrilineal societies of pilot whales (*Globicephala* sp.), to the apparently less social beaked whale species. Nevertheless, even for some beaked whales, there is evidence of stable group associations. For larger, long-lived or wide-ranging species, such as blue whales (*Balaenoptera musculus*), there are also important spatio-temporal considerations for interpretation of behaviour and associations. As a result of the differing social structures and the opportunity for the transmission of social information, the relationship between sociality and welfare in this order of mammals is multifaceted. Sociality and social dynamics have the potential to influence individual and group welfare in both a positive and negative manner, and there are complex relationships between sociality, the impacts of human-induced rapid environmental change and the welfare of cetaceans. E.O. Wilson listed ten 'qualities' of sociality. Although used to classify animal societies according to their degree of sociality, some of these features also provide a useful roadmap for evaluating the importance of sociality for individual and group welfare. They are used here to examine the interplay between sociality, welfare and environmental change. The importance of the transmission of social information, culture and specific behaviours, such as play, is also explored within the context of environmental change and cetacean welfare. It is concluded that a more comprehensive understanding of the social mechanisms operating within and between cetacean social groups will enable a fuller understanding of the welfare

P. Brakes
Whale and Dolphin Conservation, 38 St. Paul Street, Chippenham, Wiltshire, UK

Centre for Ecology and Conservation, University of Exeter, Penryn Campus, Cornwall, UK
e-mail: philippa.brakes@whales.org

A. Butterworth (ed.), *Marine Mammal Welfare*, Animal Welfare 17,
DOI 10.1007/978-3-319-46994-2_10

implications of human-induced rapid environmental change. Alongside more traditional measures of welfare, such as body condition and disease, aspects of sociality may also provide important indicators for establishing welfare condition in these highly social species.

10.1 Introduction

The behaviour of cetaceans is influenced by both extrinsic and intrinsic factors. An important extrinsic influence, which potentially affects their welfare, is the social environment. The social environment has been described as being comprised of non-random and heterogeneous social interactions (Krause and Ruxton 2002; Croft et al. 2008). The social environment of the 90 or so species (IUCN 2016) of cetaceans is highly diverse, ranging from the complex third-order alliances of male bottlenose dolphins (*Tursiops* sp.) (Connor and Krützen 2015), to the matrilineal societies of pilot whales (*Globicephala* sp.) (Amos et al. 1993; de Stephanis et al. 2008), to the apparently less social beaked whale species. Nevertheless, even for some beaked whales, there is evidence of stable group associations (Fedutin et al. 2015). For larger, long-lived or wide-ranging species, such as blue whales (*Balaenoptera musculus*), there are also important spatio-temporal considerations for interpretation of behaviour and associations (Lomac-Macnair and Smultea 2016).

As a result of the differing social structures and the opportunity for the transmission of social information, the relationship between sociality and welfare in this order of mammals is multifaceted. This chapter will first examine how sociality and social dynamics have the potential to influence individual and group welfare in both positive and negative ways and then will explore the relationships between sociality, the impacts of human-induced rapid environmental change (HIREC) (Sih et al. 2011) and the welfare of cetaceans.

10.2 Benefits of Group Living

Group living has a number of potential ecological and animal welfare benefits, which include: predator defence, cooperative foraging, mating opportunities and reduced vulnerability to infanticide (Silk 2007). But living in groups can also incur costs that are important to welfare considerations: by facilitating the spread of disease, or increased parasite burdens, or through increased conspicuousness to predators and competition for resources (Krause and Ruxton 2002). Whilst the interplay between behavioural ecology and conservation of marine mammals is a developing field (Brakes and Dall 2016), beyond spread of disease or parasites, the importance of sociality is often overlooked when developing welfare metrics for wild marine mammal populations.

Group living, and in particular social structure and social information use, can influence fitness, gene flow and spatial distribution (Wilson 1975; Dall et al. 2005), all of which may have welfare implications. E.O. Wilson listed ten 'qualities' of sociality: group size, demographic distribution, cohesiveness, patterns of connectedness through communication, permeability of movement between social groups, the extent to which the population contains distinct social units, differentiation into social roles, integration of behaviour, information flow and the percentage of time devoted to social behaviour. Whilst these characteristics have been used to classify animal societies according to their degree of sociality, some of these features also provide a useful roadmap or framework for evaluating the importance of sociality for individual and group welfare (see Table 10.1).

Table 10.1 Wilson's ten 'qualities' of sociality (Wilson 1975) and welfare considerations for cetaceans

Quality	Welfare considerations
Group size	Welfare benefits may include predator defence, cooperative foraging, mating opportunities and reduced vulnerability to infanticide (after Silk 2007). Here 'group' is defined as 'animals that actively achieve or maintain spatiotemporal proximity' (after Whitehead 2008). However, some of these welfare benefits may also be obtained in aggregations, which are not the result of social interaction but instead result from patchy resource distribution
Demographic distribution	Populations and social groups may to some extent be robust to fluctuations in demographic distribution (from a welfare perspective), but this may depend on the extent and duration of parental and alloparental care and the social role of older individuals in predator defence or resource acquisition (Johnstone and Cant 2010; Whitehead 2015)
Cohesiveness	Wilson (1975) suggested that the proximity of individuals may be used as an index of sociality. Today the more common measure is the rate of interactions (Whitehead 2008). If the rate of interactions correlates with social behaviours, such as cooperative foraging, then it may follow that successful feeding could be correlated with interaction rate
Patterns of connectedness through communication	Cetaceans live in an aqueous medium, communicating through sound and touch, with some species using echolocation. As a result they interpret their world principally through sound. Communication is central to sociality, and as well as communicating vocally and through touch, there is some evidence that dolphins may eavesdrop on the echolocation of others (Gregg et al. 2007). Communication may also be important for the transmission of information, which may in turn be relevant to welfare, specifically where this relates to resource acquisition
Permeability of movement between social groups	May be important in relation to the spread of information between social groups, which may also be relevant to welfare in terms of information about predators and resource acquisition

(continued)

Table 10.1 (continued)

Quality	Welfare considerations
Distinct social units	Potentially relevant to the emergence of unique social behaviours and cultures. Resilience to environmental change may depend on the degree of behavioural plasticity exhibited within distinct social units and how plastic or conservative these social units are (in terms of information and social structure) in the face of external pressures (CMS 2014)
Differentiation into social roles	Female pilot and killer whales exhibit a post-reproductive phase, indicating an import role within their social groups (Johnstone and Cant 2010). This is supported by evidence that post-reproductive female orcas boost the fitness of kin (Brent et al. 2015). The removal of individuals with key social roles may have welfare repercussions for their social group (Williams and Lusseau 2006)
Integration of behaviour	Whitehead (2008) argues that measuring synchrony may be one way to examine integration of behaviour. The welfare implications of synchronous behaviour in cetaceans have not yet been extensively examined, but synchrony likely influences energy expenditure whilst travelling and hunting
	It may also be useful to examine how fluctuating asymmetry (FA) (Tomkins and Andrews 2001; Swaddle 2003) varies in relation to synchronous and other integrated behaviours. For example, FA may provide some insights in the distribution of personality types (Fink et al. 2005) within a social group, with potential consequences for individual and group welfare
Information flow	May be relevant to resilience, particularly in relation to innovative foraging techniques, resources patches and safe habitat (McNamara and Dall 2010; CMS 2014)
Time devoted to social behaviour	The welfare implications of the proportion of time devoted to social behaviour depend on the cost and benefits to the individual of spending time exhibiting that behaviour, which may be contingent on the other qualities of sociality identified by Wilson (1975)

For example, female pilot and killer whales exhibit a post-reproductive phase—a developmental stage extremely rare in mammals—which indicates that these older females have an important role within their social group (Johnstone and Cant 2010). This is supported by evidence that post-reproductive female killer whales boost the fitness of kin (possibly through the transfer of ecological knowledge, for example about foraging) (Brent et al. 2015). There is also complex interplay between cetacean social structure and the transmission of social knowledge (Cantor and Whitehead 2013). We do not yet understand the short- and long-term effects on individual welfare of disrupting such complex social systems.

Nevertheless, the buffering effects of social support are well recognised in our own species and, it is argued, are also relevant to farm animal welfare (Rault 2012). Since sociality itself evolved in response to various environmental challenges (Wilson 1975), it is perhaps to be expected that social support would be relevant to the welfare of many other species.

10.3 Social Learning and Social Information Use

For species that learn in social groups, the spread of information through social learning may be a key determinant of some responses to HIREC (Sih et al. 2011; Sih 2013). However, the importance of social information and social learning in relation to animal welfare has received little attention. Nevertheless, social learning is important in many mammalian species (Thornton and Clutton-Brock 2011), and it is likely that both opportunities for innovation and transmission of social information in the wild have implications for welfare, by enabling resilience to ecological and anthropogenic stressors. An example is the diversification of foraging strategies, through the use of tools, such as the sponges used by some bottlenose dolphins to assist foraging (Krützen et al. 2014).

As well as providing alternative prey items and potentially a more diverse diet, diverse foraging strategies could potentially provide latent resilience in the event that a particular prey type becomes unavailable. In contrast, the southern resident population of killer whales (*Orcinus orca*), which have developed a feeding specialisation on chinook salmon (*Oncorhynchus tshawytscha*), may be less resilient to fluctuations in prey abundance as a result of their cultural conservatism (Whitehead 2010).

Nevertheless, whilst social learning acts much faster than the intergeneration process of natural selection of genes and may play an important role in species' response to rapidly changing environments, other factors may also be at play. It is important to distinguish these other influences, such as local enhancement. The transmission of information in response to HIREC may also lead to maladaptive behaviours (Tuomainen and Candolin 2011). For example, sperm whale (*Physeter microcephalus*) depredation of sablefish from demersal longlines in the Alaskan fishery may be the result of social learning, but the fishing vessels may themselves be providing an acoustic cue (Thode et al. 2014). Whilst the welfare outcomes for these whales remain the same (potential entanglement), understanding the mechanisms associated with the behaviour can provide important insights for mitigation.

10.4 Culture

Social learning can result in persistent behavioural traits within a social group. It has been argued by Whitehead and Rendell (2015) that culture arising from social learning is widespread across many cetacean species. They define culture as: information or behaviours—shared within a community—which are acquired from conspecifics through some form of social learning.

The relationship between culture and individual and group welfare may be complex. There is good evidence that in some cetacean species, such as sperm and killer whales, cultural behaviour helps to shape both social relationships and social structure (Cantor et al. 2015) and potentially helps to develop symbolic markers of group identity (Whitehead and Rendell 2015). The welfare implications of disrupting these social systems are not yet well understood.

Although stranding behaviour may have various causes, it has also been suggested that there may be a link between conformist cultures and mass stranding in some odontocete (toothed whale) species (Rendell and Whitehead 2001). In the short-term, maladaptive cultural behaviour can evolve much more rapidly that genetic selection can counter it (Richerson and Boyd 2005; Whitehead 2010). The transmission of maladaptive behaviour arising as the result of human activities may have consequences for both individual and group welfare through lack of capacity to adapt rapidly to human-induced change.

10.5 Human Activities, Cetacean Sociality and Welfare

HIREC is now a widely acknowledged phenomenon (Sih et al. 2011), and there are few terrestrial or marine habitats that are not in some way affected by human activities. There are many ways that human activities can influence habitat quality, from climate change and ocean acidification, pollution, degradation of habitats, to depletion of prey by fisheries. One challenge for evaluating the animal welfare implications of HIREC is determining which aspects of sociality may be the most vulnerable to human-induced change and which anthropogenic threats are likely to impact sociality in the most significant ways. The behavioural domains which may be influenced by HIREC include: communication, foraging, migration and habitat use. The disruption of any of these behaviours has potential welfare implications.

Acute, direct threats to individual welfare, such as hunting and by-catch, may result in the removal of individuals that have a specific social role (Williams and Lusseau 2006). The impact of indirect threats, such as noise pollution, may influence the welfare of entire social groups in a chronic manner. However, welfare outcomes may be contingent on the 'strength' of the threat, since intense anthropogenic noise can result in acute, as well as chronic, welfare issues for cetaceans, and hunting may have long-term impacts beyond the loss of individual animals. As a result, there are no rules to follow when considering the impacts of HIREC on cetacean sociality, and each case must be evaluated independently. Nevertheless, there are some specific issues worth considering.

10.6 Noise and Sociality

One of the most important ways in which sociality, habitat and HIREC intersect for cetaceans is through the increase of anthropogenic noise in the oceans. Noise may have particularly profound effects on welfare if it has an influence on habitat selection or communication. Although ocean noise is a natural phenomenon, there is evidence that humpback whales (*Megaptera novaeangliae*) may not be able to cope with an increase in anthropogenic noise in the same way that they seem able to offset fluctuations in natural noise (Dunlop 2016). There is also strong evidence for auditory masking in some marine mammal populations (Erbe et al. 2015), which may result in changes in call rate or frequency. Interference with communication has unknown welfare implications, but it is conceivable that this type of disruption could influence cooperative behaviour (such as cooperative feeding strategies) or limit the ability to find a suitable mate or warn others about the presence of predators.

10.7 Removals Through Hunting and By-Catch

The welfare issues associated with hunting of cetaceans and incidental by-catch have been explored extensively (Bass and Brakes 2013). However, there may also be more subtle effects for social groups which should also be considered. For example, the removal of individuals that act as repositories of social knowledge may have welfare implications for their social group, beyond simple individual removals, which may have repercussions for subsequent generations, as evidenced in elephant social groups (McComb et al. 2001; Shannon et al. 2013).

There is good evidence from mitochondrial DNA that the calves of some species of baleen whale learn migration routes from their mothers (Carroll et al. 2011; Baker et al. 2013; Carroll et al. 2014). This social learning is thought to happen at a key stage in calf development. If this process is interrupted by significant removals through hunting, this may result in the loss of cultural knowledge, potentially an impediment to range recovery following exploitation (Clapham et al. 2008). In addition to the conservation issues associated with loss of cultural knowledge (CMS 2014), there are also related welfare concerns, particularly if knowledge about critical feeding habitat or safe havens from predators is lost.

In addition to the potential to remove key individuals who act as repositories for cultural knowledge, removing individuals from social groups has other welfare implications, such as leaving behind dependents. The plight of suckling calves whose mothers have been killed is often considered a welfare priority, but there may be other dependents within social groups which rely on social support, such as elderly or injured individuals, although these individuals may be difficult to identify. In addition, for those species that exhibit alloparental care (Best et al. 2015; Sakai et al. 2016), the welfare consequences of disruption to social networks are not yet well understood.

10.8 Pollution

The oceans are now replete with xenobiotic compounds (non-biological, chemical compounds foreign to living organisms). Residing at the top, or near the top of food chains, cetaceans are particularly vulnerable to bioaccumulation of chemical pollutants (for a full discussion of anthropogenic threats to cetaceans, see Brakes and

Simmonds 2013). Marine debris is also an increasing area of concern, although the full extent of the threat from ingested plastics and other marine debris remains unclear (Simmonds 2012). Some welfare implications are evident, through entanglement and ingestion, but the implications of pollution for sociality, and resulting impact on welfare, are unknown. We do know that pollution and marine debris can degrade habitat quality which may have implications for how whales and dolphins use space, and it has been argued that such effects deprive wildlife of 'life-sustaining habitat' (Paquet and Darimont 2010).

In addition, there is some evidence that certain types of chemical pollution may cause immunosuppression and reproductive impairment (Reijnders 1996; Houde et al. 2005), which may have implications for the social transmission of disease within and between social groups and potentially, reproductive behaviour.

10.9 Disturbance and Anthropo-Dependence

Persistent disturbance from vessel activity can lead to habituation or displacement from critical habitat. Research with North Atlantic right whales has indicated that whilst they showed little or no behavioural response to approaching vessels, they reacted mildly to the vocalisations of conspecifics and strongly to an experimental alert signal (Nowacek et al. 2004).

On the other hand, HIREC can also provide novel foraging opportunities for cetaceans, e.g. through provisioning (i.e. supplying opportunities for foraging, such as from trawler discards) (Mann and Kemps 2003) or cooperative fishing (Daura-Jorge et al. 2012). Such activities can create what has been termed anthropo-dependence (CMS 2014). Over time these human influences could change social structure and behaviour (Donaldson et al. 2012; Ansmann et al. 2012) or potentially create vulnerability within the social group involved, if the human activity were to change or cease.

10.10 Human Influences on Play Behaviour

Play is a widespread behaviour in mammals and there are many examples of object play among wild odontocetes (Kuczaj and Eskelinen 2014). Play has an important role in ontogeny (an organism's development). It is essential for social development and for learning motor skills that will be important for survival (Janik 2015). It is also important for developing social relationships and understanding social dynamics and may be an important aspect of alliance formation. As a result, although play frequency may decrease with age, play may have a role in ensuring good welfare of an individual throughout their life cycle (Held and Špinka 2011).

Fig. 10.1 Tail-walking behaviour in a bottlenose dolphin from Port Adelaide, Australia, learnt during rehabilitation in captivity, and transmitted some years later through a social group when the individual was returned to the wild. The behaviour may be play, but its transmission could potentially develop into an ethnic marker within the social group. *Image credit*: Mike Bossley

It is easy to misinterpret apparently exuberant behaviour in dolphins as play, but some behaviours, such as bow riding or creating bubble rings, in the absence of alternative explanations do strongly point towards play (Janik 2015) (Fig. 10.1).

Whilst there are few empirical studies on how play behaviour in cetaceans may be disrupted in the wild by human activities, disturbance and harassment are some of the activities most likely to have long-term effects on play behaviour. Marine mammals exhibit play behaviour under a wide range of natural circumstances, but where baseline data are available, cases of chronic disturbance should be examined for influence on diversity and frequency of play among all age cohorts.

10.11 Morality, Sociality and Welfare

It has been argued cogently that species other than our own have the capacity for moral behaviour and that morality has the potential for evolutionary advantage (Broom 2006). Moral behaviours include behaviours such as cooperation, empathetic and justice behaviours, where moral behaviour is defined as 'a suite of interrelated other-regarding behaviours that cultivate and regulate complex interactions within social groups' (Bekoff and Pierce 2009; Pierce and Bekoff 2012). The manner in which sperm whales moderate the use of their echolocation clicks (the most powerful sonar in the animal kingdom) has been suggested as an example of moral agency in cetaceans. In some regions these whales hunt in fairly compact groups, and it is thought that the potential risk of harming a conspecific through this intense sound production, specifically damaging others hearing, may be mitigated by socially learnt 'codes of conduct' within the group (for more examples of potentially moral behaviour in cetaceans, see Whitehead and Rendell 2015).

On first inspection, the topic of animal morality seems tangential to the issue of animal sociality and animal welfare. Nevertheless, a closer examination reveals some important intersections. Since social relationships and networks can add productivity to certain activities (Corning 2012), it follows that such productivity may have welfare implications and may be contingent on individuals within a network sticking to some rules of engagement. It is also conceivable that a threshold population size would be required for a particular moral framework to develop and be maintained within animal societies. Arguably, there is some reciprocal adaptive advantage between moral behaviours and living in social groups, in that sociality is required for the development of moral codes and moral codes themselves can assist social living and individual welfare.

Hal Whitehead argues in relation to mass stranding of some odontocete species that 'cultural group conformity in movement patterns may override an individual's survival instincts when the group gets into trouble' (further described in Brakes and Simmonds 2013). This suggestion that social cohesion may override individual interests is conceptually challenging and certainly very difficult to test empirically. But such intense cohesion could potentially have adaptive advantage in other circumstances, and if accurate, this could also have far-reaching welfare implications in terms of social group size and composition.

10.12 Cognition, Human Influences and Welfare

It is argued that the responses of wildlife to human influence, and human environmental change, are governed by cognitive processes ranging from perceptual processes to learnt behaviour. A better understanding of these cognitive processes can be utilised to reduce human impacts on wildlife (Greggor et al. 2014). By the same rationale, if cognition underlies a behaviour that is relevant to welfare, then understanding these cognitive processes may help in achieving better welfare outcomes for wild cetaceans.

Greggor et al. (2014) argue that mitigation methods used to stop birds colliding with human structures, such as wind farms, will only be effective if they are reliably perceived by the birds and that this perception is rapidly learnt (both of which are cognitive processes). Their argument is that 'cognitive theory can thus help predict how best to manipulate and exploit attentional biases, innate responses, and learning tendencies to enhance conservation efforts'. The question then, from the perspective of the social aspects of cetacean welfare, is: how can a better understanding of cetacean cognitive processes inform mitigation efforts from human impacts such as entanglement, or in areas where there is a high risk of ship strikes, so that nets, approaching boats or hazardous habitat can be reliably perceived, and the associated danger rapidly learnt by individuals, reacted to and then possibly transmitted?

10.13 Conclusions

Sociality is important for evaluating the impact of human activities on animal welfare (see Table 10.2). For highly social species, which demonstrate high-order cognitive capacities, such as cetaceans, this may be a particularly important consideration. Knowledge about sociality may lead to better understanding of their cognitive processes, and this may in turn facilitate better mitigation methods for protecting individual and group welfare.

There is an indisputable and important link between animal welfare and conservation outcomes. Collaboration between these two fields and incorporation of some of the emerging knowledge on cetacean sociality are likely to enable better outcomes for individuals, social groups and potentially populations.

Broom argues that the term humane can be applied where the treatment of animals is such that 'their welfare is good to a certain high degree' (Broom 2013). With our developing understanding of the complexity of some cetacean societies and the importance of certain associations for individual and group welfare, human activities in the ocean cannot be determined as 'humane' (using Broom's definition) if they interfere with social structure and processes. A better understanding of how human activities influence the welfare of cetaceans will only be achieved when aspects of sociality are also taken into consideration, alongside the range of more traditional welfare indicators such as body condition, wounding and disease.

Table 10.2 Summary of importance of social living for cetacean welfare

Positive	Negative
Resilience to change and adaptation through social learning	Conservative cultures may hinder adaptation and resilience
	Potential for spread of maladaptive behaviour through social learning
Individuals may act as repositories of social knowledge for the social group	Vulnerability from removal of repositories of social knowledge or individuals with specific role
Potential for alloparental care in some species	Vulnerability of some cohorts and dependents if 'carers' are removed
Predator defence and foraging cooperation	Foraging competition
	Disease transmission

References

Amos B, Schlötterer C, Tautz D (1993) Social structure of pilot whales revealed by analytical DNA profiling. Science 260:670–672

Ansmann IC, Parra GJ, Chilvers BL, Lanyon JM (2012) Dolphins restructure social system after reduction of commercial fisheries. Anim Behav 84:575–581. doi:10.1016/j.anbehav.2012.06.009

Baker C, Steel D, Calambokidis J et al (2013) Strong maternal fidelity and natal philopatry shape genetic structure in North Pacific humpback whales. Mar Ecol Prog Ser 494:291–306. doi:10.3354/meps10508

Bass CL, Brakes P (2013) Whaling and whale killing methods. Anim Welf 22:109–111

Bekoff M, Pierce J (2009) Wild justice: the moral lives of animals. University of Chicago Press, Chicago, IL

Best PB, Elwen SH, Palsboll PJ et al (2015) Possible non-offspring nursing in the southern right whale, *Eubalaena australis*. J Mammal 96:405–416. doi:10.1093/jmammal/gyv042

Brakes P, Dall SRX (2016) Marine mammal behavior: a review of conservation implications. Front Mar Sci 3:1–15. doi:10.3389/fmars.2016.00087

Brakes P, Simmonds MP (2013) Whales and dolphins: cognition, culture, conservation and human perceptions. Routledge, London

Brent LJN, Franks DW, Cant MA et al (2015) Ecological knowledge, leadership, and the evolution of menopause in killer whales. Curr Biol 25:1–5. doi:10.1016/j.cub.2015.01.037

Broom DM (2006) The evolution of morality. Appl Anim Behav Sci 100:20–28. doi:10.1016/j.applanim.2006.04.008

Broom DM (2013) The science of animal welfare and its relevance to whales. Anim Welf 22:123–126. doi:10.7120/09627286.22.1.123

Cantor M, Whitehead H (2013) The interplay between social networks and culture: theoretically and among whales and dolphins. Philos Trans R Soc Lond Ser B Biol Sci 368:20120340. doi:10.1098/rstb.2012.0340

Cantor M, Shoemaker LG, Cabral RB et al (2015) Multilevel animal societies can emerge from cultural transmission. Nat Commun 6:8091. doi:10.1038/ncomms9091

Carroll E, Patenaude N, Alexander A et al (2011) Population structure and individual movement of southern right whales around New Zealand and Australia. Mar Ecol Prog Ser 432:257–268. doi:10.3354/meps09145

Carroll EL, Rayment WJ, Alexander AM et al (2014) Reestablishment of former wintering grounds by New Zealand southern right whales. Mar Mamm Sci 30:206–220. doi:10.1111/mms.12031

Clapham PJ, Aguilar A, Hatch LT (2008) Determining spatial and temporal scales for management: lessons from whaling. Mar Mamm Sci 24:183–201

CMS (2014) Report of the CMS Scientific Council workshop on the conservation implications of cetacean culture

Connor RC, Krützen M (2015) Male dolphin alliances in Shark Bay: changing perspectives in a 30-year study. Anim Behav:1–13. doi:10.1016/j.anbehav.2015.02.019

Corning P (2012) The fair society: the science of human nature and the pursuit of social justice. University of Chicago Press, London

Croft DP, James R, Krause DJ (2008) Exploring animal social networks. Princeton University Press, Princeton

Dall SRX, Giraldeau LA, Olsson O et al (2005) Information and its use by animals in evolutionary ecology. Trends Ecol Evol 20:187–193. doi:10.1016/j.tree.2005.01.010

Daura-Jorge FG, Cantor M, Ingram SN et al (2012) The structure of a bottlenose dolphin society is coupled to a unique foraging cooperation with artisanal fishermen. Biol Lett 8:702–705. doi:10.1098/rsbl.2012.0174

Donaldson R, Finn H, Bejder L et al (2012) The social side of human-wildlife interaction: wildlife can learn harmful behaviours from each other. Anim Conserv 15:427–435. doi:10.1111/j.1469-1795.2012.00548.x

Dunlop RA (2016) The effect of vessel noise on humpback whale, *Megaptera novaeangliae*, communication behaviour. Anim Behav 111:13–21. doi:10.1016/j.anbehav.2015.10.002

Erbe C, Reichmuth C, Cunningham K et al (2015) Communication masking in marine mammals: a review and research strategy. Mar Pollut Bull. doi:10.1016/j.marpolbul.2015.12.007

Fedutin ID, Filatova OA, Mamaev EG et al (2015) Occurrence and social structure of Baird's beaked whales, *Berardius bairdii*, in the Commander Islands, Russia. Mar Mamm Sci 31:853–865. doi:10.1111/mms.12204

Fink B, Neave N, Manning JT, Grammer K (2005) Facial symmetry and the "big-five" personality factors. Personal Individ Differ 39:523–529. doi:10.1016/j.paid.2005.02.002

Gregg JD, Project DC, Dudzinski KM, Smith HV (2007) Do dolphins eavesdrop on the echolocation signals of conspecifics? Int J Comp Psychol 20:65–88

Greggor AL, Clayton NS, Phalan B, Thornton A (2014) Comparative cognition for conservationists. Trends Ecol Evol 29:489–495. doi:10.1016/j.tree.2014.06.004

Held SDE, Špinka M (2011) Animal play and animal welfare. Anim Behav 81:891–899

Houde M, Hoekstra PF, Solomon KR, Muir DCG (2005) Organohalogen contaminants in delphinoid cetaceans. Rev Environ Contam Toxicol 184:1–57. doi:10.1007/0-387-27565-7_1

IUCN (2016) Status of the World's Cetaceans I IUCN SSC – Cetacean Specialist Group. http://www.iucn-csg.org/index.php/status-of-the-worlds-cetaceans/. Accessed 9 Mar 2016

Janik VM (2015) Play in dolphins. Curr Biol 25:R7–R8. doi:10.1016/j.cub.2014.09.010

Johnstone R, Cant M (2010) The evolution of menopause in cetaceans and humans: the role of demography. Proc Biol Sci 277:3765–3771. doi:10.1098/rspb.2010.0988

Krause J, Ruxton GD (2002) Living in groups. OUP, Oxford

Krützen M, Kreicker S, Macleod CD et al (2014) Cultural transmission of tool use by Indo-Pacific bottlenose dolphins (*Tursiops* sp.) provides access to a novel foraging niche. Proc Biol Sci 281:20140374. doi:10.1098/rspb.2014.0374

Kuczaj SA, Eskelinen HC (2014) Why do dolphins play? Anim Behav Cogn 2:113. doi:10.12966/abc.05.03.2014

Lomac-Macnair K, Smultea MA (2016) Blue Whale (*Balaenoptera musculus*) behavior and group dynamics as observed from an aircraft off southern California. Anim Behav Cogn 3:1–21. doi:10.12966/abc.02.01.2016

Mann J, Kemps C (2003) The effects of provisioning on maternal care in wild bottlenose dolphins, Shark Bay, Australia. In: Gales N, Hindell M, Kirkwood R (eds) Marine mammals and humans: towards a sustainable balance. CSIRO Publishing, Clayton, pp 304–320

McComb K, Moss C, Durant SM et al (2001) Matriarchs as repositories of social knowledge in African elephants. Science 292:491–494. doi:10.1126/science.1057895

McNamara JM, Dall SRX (2010) Information is a fitness enhancing resource. Oikos 119:231–236. doi:10.1111/j.1600-0706.2009.17509.x

Nowacek DP, Johnson MP, Tyack PL (2004) North Atlantic right whales (*Eubalaena glacialis*) ignore ships but respond to alerting stimuli. Proc Biol Sci 271:227–231. doi:10.1098/rspb.2003.2570

Paquet PCD, Darimont CT (2010) Wildlife conservation and animal welfare: two sides of the same coin? Anim Welf 19:177–190

Pierce J, Bekoff M (2012) Wild justice redux: what we know about social justice in animals and why it matters. Soc Justice Res 25:122–139. doi:10.1007/s11211-012-0154-y

Rault JL (2012) Friends with benefits: social support and its relevance for farm animal welfare. Appl Anim Behav Sci 136:1–14. doi:10.1016/j.applanim.2011.10.002

Reijnders PJH (1996) Organohalogen and heavy metal contamination in cetaceans: observed effects, potential impacts and future prospects. In: Simmonds MP, Huchinson JD (eds) The conservation of whales and dolphins: science and practice. Wiley, Chichester, pp 205–217

Rendell LE, Whitehead H (2001) Culture in whales and dolphins. Behavioral and Brain Sciences 24:309–382.

Richerson PJ, Boyd R (2005) Not by genes alone: how culture transformed human evolution. University of Chicago Press, Chicago, IL

Sakai M, Kita YF, Kogi K et al (2016) A wild Indo-Pacific bottlenose dolphin adopts a socially and genetically distant neonate. Sci Rep 6:23902. doi:10.1038/srep23902

Shannon G, Slotow R, Durant SM et al (2013) Effects of social disruption in elephants persist decades after culling. Front Zool 10:62. doi:10.1186/1742-9994-10-62

Sih A (2013) Understanding variation in behavioural responses to human-induced rapid environmental change: a conceptual overview. Anim Behav 85:1077–1088. doi:10.1016/j.anbehav.2013.02.017

Sih A, Ferrari MCO, Harris DJ (2011) Evolution and behavioural responses to human-induced rapid environmental change. Evol Appl 4:367–387. doi:10.1111/j.1752-4571.2010.00166.x

Silk JB (2007) The adaptive value of sociality in mammalian groups. Philos Trans R Soc Lond B Biol Sci 362:539–559. doi:10.1098/rstb.2006.1994

Simmonds MP (2012) Cetaceans and marine debris: the great unknown. J Mar Biol 2012:1–8. doi:10.1155/2012/684279

de Stephanis R, Verborgh P, Pérez S et al (2008) Long-term social structure of long-finned pilot whales (*Globicephala melas*) in the Strait of Gibraltar. Acta Ethol 11:81–94. doi:10.1007/s10211-008-0045-2

Swaddle JP (2003) Fluctuating asymmetry, animal behavior, and evolution. Adv Study Behav 32:169–205. doi:10.1016/S0065-3454(03)01004-0

Thode AM, Wild L, Mathias D et al (2014) A comparison of acoustic and visual metrics of sperm whale longline depredation. J Acoust Soc Am 135:3086–3100. doi:10.1121/1.4869853

Thornton A, Clutton-Brock T (2011) Social learning and the development of individual and group behaviour in mammal societies. Philos Trans R Soc Lond B Biol Sci 366:978–987. doi:10.1098/rstb.2010.0312

Tomkins JL, Andrews S (2001) Fluctuating asymmetry. Enc Life Sci:1–5. doi:10.1055/s-0031-1275301

Tuomainen U, Candolin U (2011) Behavioural responses to human-induced environmental change. Biol Rev Camb Philos Soc 86:640–657. doi:10.1111/j.1469-185X.2010.00164.x

Whitehead H (2008) Analyzing animal societies: quantitative methods for vertebrate social analysis. University of Chicago Press, Chicago, IL

Whitehead H (2010) Conserving and managing animals that learn socially and share cultures. Learn Behav 38:329–336. doi:10.3758/LB.38.3.329

Whitehead H (2015) Life history evolution: what does a menopausal killer whale do? Curr Biol 25:R225–R227. doi:10.1016/j.cub.2015.02.002

Whitehead H, Rendell L (2015) The cultural lives of whales and dolphins. University of Chicago Press, Chicago, IL

Williams R, Lusseau D (2006) A killer whale social network is vulnerable to targeted removals. Biol Lett 2:497–500. doi:10.1098/rsbl.2006.0510

Wilson EO (1975) Sociobiology: the new synthesis. Belknap Press, Cambridge, MA

Chapter 11
Cetaceans in Captivity

Rob Lott and Cathy Williamson

Abstract Several species (and over 3000 individuals) of small cetacean are held in captivity around the world, primarily for public display and entertainment. Scientific evidence strongly supports concerns about individual animals' welfare, including mental and physical health. Conditions in captivity cannot meet an individual's biological needs, and restricted space, a limited social environment, artificial surroundings and behavioural restrictions all contribute to stress and early mortality. Wild cetacean populations in some countries are targeted by live captures to supply the public display industry, presenting a risk to conservation as well as welfare. Public opinion is shifting on cetacean captivity and may signal a change in the way cetaceans are held in captivity in the future.

11.1 Introduction

The capture and confinement of cetaceans presents a challenge to marine mammal welfare. The most commonly held cetaceans in captivity are belugas (*Delphinapterus leucas*), bottlenose dolphins (*Tursiops truncatus* or *Tursiops aduncus*) and orcas or killer whales (*Orcinus orca*). Other species, including finless porpoises (*Neophocaena phocaenoides*), harbour porpoises (*Phocoena phocoena*), Indo-Pacific humpbacked dolphins (*Sousa chinensis*), Irrawaddy dolphins (*Orcaella brevirostris*), Pacific white-sided dolphins (*Lagenorhynchus obliquidens*), Risso's dolphins (*Grampus griseus*) and short-finned pilot whales (*Globicephala macrorhynchus*), are among those species which are also held in captivity (Couquiaud 2005; Ceta-Base 2016a).

R. Lott (✉) • C. Williamson
Whale and Dolphin Conservation, 38, St. Paul Street, Chippenham, SN15 1LJ UK
e-mail: rob.lott@whales.org; cathy.williamson@whales.org

A. Butterworth (ed.), *Marine Mammal Welfare*, Animal Welfare 17,
DOI 10.1007/978-3-319-46994-2_11

While a few of these small cetacean species are held for research or even military purposes, the vast majority are held for public display and entertainment in stand-alone commercial facilities ('dolphinaria') or as exhibits in zoos or aquaria (Reeves and Fisher 2005). The majority of such facilities use the cetaceans in circus-style shows featuring tricks that bear little resemblance to the types of behaviour seen in wild cetaceans or which present a trained or choreographed version of 'wild-type' behaviours. An increasing number of facilities also offer interaction programmes where members of the public feed, touch or enter the cetacean enclosure to wade or swim with the individuals held (Whale and Dolphin Conservation 2015).

At least 3000 individual cetaceans are held in more than 50 countries around the world and in well over 300 facilities (Ceta-Base 2016a). In most countries where cetaceans are held, there is no official reporting system, and so the exact numbers of captive animals are hard to determine (Whale and Dolphin Conservation 2015; Whale and Dolphin Conservation Society 2009).

Welfare science is a growing field, and it is only recently that the first steps have been taken to quantify and systematically measure welfare among captive cetaceans (Clegg et al. 2015). Nevertheless, scientific evidence strongly supports a number of concerns relating to mental and physical health among captive cetaceans, all of which can have potentially negative impacts on an individual's welfare or wellbeing and, ultimately, on the animal's health and mortality (Rose 2014; Waples and Gales 2002; Maas 2000; Small and DeMaster 1995a). Ill health is difficult to diagnose in captive cetaceans (Rose et al. 2009) as clinical signs are often very subtle or are masked. It is not uncommon for dolphinarium staff to find an individual who is initially lacking in appetite, dying one or two days later and before any cause can be determined or treatment administered (Blake 2012). Furthermore, the lack of collated available data on captive cetaceans, and of their physiological, behavioural, survival and reproductive data (Whale and Dolphin Conservation 2015), currently makes effective welfare assessment problematic.

The primary threat to a cetacean's welfare in captivity is the zoo or aquarium's inability to provide a species-specific environment that meets an individual animal's biological needs (Whale and Dolphin Conservation 2015). Restricted space, a limited social environment, artificial surroundings and behavioural restrictions all contribute risk factors for stress, may contribute to abnormal behavioural changes, affect the health of the animals, necessitate the use of tranquilisers and result in early mortality in some animals (Maas 2000; Noda et al. 2007; Knight 2013).

11.2 Restrictive Space

In the wild, cetaceans are almost always in motion, even when resting. Many travel great distances every day, in search of food and for other activities. This is natural behaviour, for which they have adapted physically and behaviourally. Captive facilities provide only a fraction of the space across which a cetacean would travel in the wild (Tyack 2009). Bottlenose dolphins can travel tens of kilometres a day

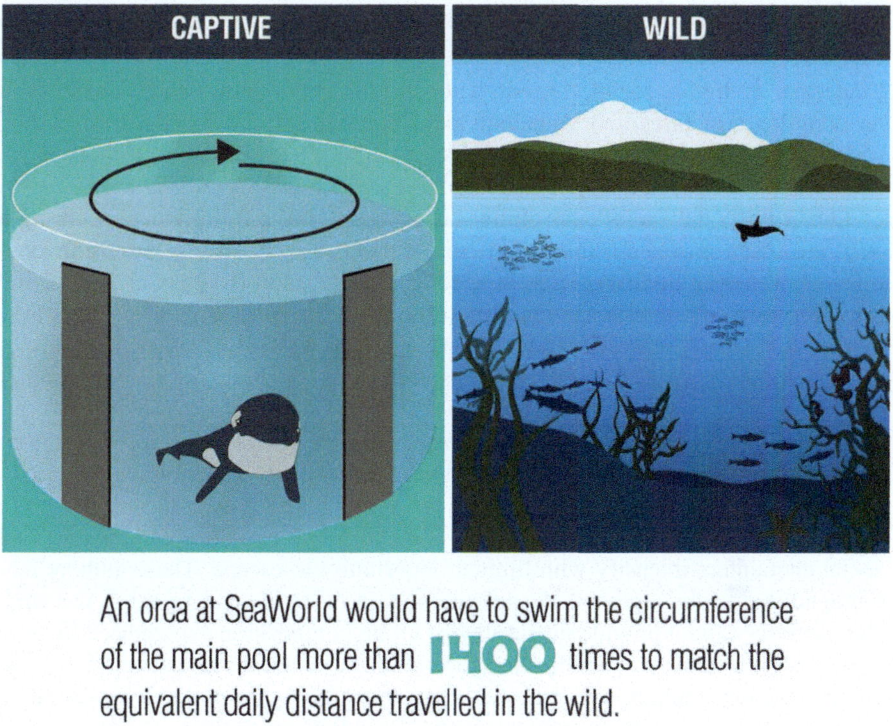

Fig. 11.1 The equivalent daily distance travelled in the wild by an orca cannot be achieved in a captive facility. *Image credit*: Kimberley Palfi for Whale and Dolphin Conservation

(Mate et al. 1995) with home ranges often exceeding 100 km^2 (Sprogis et al. 2015). A wild orca pod can cover over 160 linear kilometres a day, foraging and socialising (Baird 2000).

Even in the largest facilities, such as those at SeaWorld parks in the United States, a captive orca, for example, would need to swim around the perimeter of its tank 1400 times each day to cover the distance of its wild counterpart (Fig. 11.1). Pool depth is also severely restricted, as is the ability to swim at high speed. A common feature of captive orca 'society' (the social arrangements of a group kept in captivity) is the presence of dominant, often aggressive, hierarchies (Hargrove 2015). Restrictive enclosures offer no opportunities for subordinate whales to escape any given situation in order to diffuse an altercation. Similar threats are also known to exist for bottlenose dolphins and belugas in captivity (Waples and Gales 2002; Evans 2015).

Space may be further limited in captivity by the introduction of visitors to the cetacean's environment in 'swimming with dolphins' and other interaction programmes. Close contact between cetacean and human individuals in these programmes has potential to lead to the transmission of disease (Couquiaud 2005; Geraci and Ridgway 1991; Waltzek et al. 2012; Buck and Schroeder 1990; Hunt et al. 2008).

11.3 Limited Social Environment

Cetaceans are highly social, forming wide-ranging communities and societies built on complex structures and with individual interdependence formed from strong social bonds (King and Janik 2015; Krasnova et al. 2014; Blasi and Boitani 2014; Cantor and Whitehead 2013; Whitehead 2011).

In captivity, the social environment is severely limited. Individuals sharing a pool are often unrelated, may have been collected from widely different locations or may even be from different species or subspecies (Rose et al. 2009), and these mixtures of animals may not, therefore, share a common dialect. This may hinder their ability to exchange information and, as a result, limit social bonding, as individuals may not recognise the sounds or signals made by one another (Fig. 11.2).

Waples and Gales (2002) noted that psychological stressors in captive dolphins can be linked to social interactions between individuals, and this can result in aggression, injury, illness and impacts on the ability to rear calves and even result in death, where, in a limited physical environment, social pressures can escalate and social encounters intensify with limited opportunity to escape. These authors recommend that group structure in captivity should resemble that found in the wild. But captivity cannot provide the fluidity of group composition experienced by wild cetacean populations or provide the space to allow cetaceans to disperse from one another during conflict, avoidance mechanisms which are probably essential to reduce stress and violent encounters (Frohoff and Packard 1995).

Fig. 11.2 Individuals sharing a pool are often unrelated, perhaps hindering their ability to exchange information with one another. *Image credit*: Lee Harrison

Research has shown that orca societies have developed strong bonds between group members with individuals rarely spending more than a few hours apart from one another (Bigg et al. 1990). In this respect, they may be more highly bonded than humans. Orcas, like other dolphin species, can recognise themselves in a mirror (Delfour and Marten 2001), a trait that researchers attribute with being self-aware. This indicates that these animals probably have complex knowledge of themselves and their environment and so are likely to have thoughts about themselves and the world around them (Reiss and Marino 2001; Butterworth et al. 2013). Cetaceans are also considered to possess unique 'dialects' and have evolved a rich culture which is passed down through generations (Ford 1989).

The historic reputation of orcas as ruthless killers (Hoyt 1990) has given way to a greater appreciation of a creature that scientists now believe may be second only to humans in terms of behavioural, linguistic and ecological diversity and complexity (Rose 2014). Life in a small tank removes huge portions of the animals' capacity to make decisions, to judge situations focussed on feeding, social interaction or mobility, and profoundly limits 'choice' for these complex, sentient beings. They are denied key life strategies such as the ability to hunt, to explore and to migrate.

11.4 Aggression

A striking feature of orca society is the virtual absence of overt aggression within and between pods and ecotypes and also existence of a culture of cooperation and team work that prevails among groups of animals (Spong and Symonds 2000). The only recorded incident of a wild orca attacking a human occurred in 1972 when a Californian surfer, possibly mistaken for a seal, was bitten by an orca before being rapidly released (Lodi News-Sentinel 1996). The last 50 years have generated a long catalogue of aggressive acts by captive orcas towards each other and their trainers (Kirby 2012). Tilikum and Keto, male orcas held by SeaWorld, were implicated in the deaths of four humans (including three trainers) as documented in Gabriela Cowperthwaite's powerful, ground-breaking film, *Blackfish* (Fig. 11.3).

In 2015, a beluga died at a SeaWorld park after developing an infection in his jaw that was fractured during what was described as an 'interaction' with two other whales (Evans 2015). Visitors are also at risk of cetacean aggression. In 2008, three tourists were injured while swimming with dolphins in Curaçao after a bottlenose dolphin breached on top of them, seemingly deliberately (Rose et al. 2009; Marine Connection 2008). A number of such incidents have been reported in the media around the world, with many others likely going unreported (Vail 2012).

Fig. 11.3 There are no accounts of orcas deliberately attacking humans in the wild. In captivity, there are many recorded incidents of aggression by captive orcas towards each other and their trainers, some fatal. *Image credit*: Kimberly Palfi for Whale and Dolphin Conservation

11.5 Early Pregnancy and Calf Separation

In captivity, because of the artificial nature of the environment and the fact that calves of a number of cetacean species held in captivity are often separated from their mothers at a young age, whales and dolphins cannot learn the skills important to survival or essential nursing skills necessary to care for their own young (Rose et al. 2009). High rates of neonatal mortality are considered a major problem in captivity (Van Lint et al. 2006).

In the wild, orcas typically have their first calf at around 14 years of age and subsequent calves at intervals of approximately five years (Olesiuk et al. 2005). In captivity, however, orcas have routinely become pregnant—including via artificial insemination—much earlier (Hargrove 2015). At Loro Parque in the Canary Islands, Kohana, a female orca, became pregnant at just seven years of age and gave birth to Adan, a male, in 2010. In 2012, Kohana became pregnant again by the same male and gave birth to a female, Vicky. This young orca mother had two calves by the age of 10 and rejected them both. One theory as to why she did this is that she had no idea what to do with them as she herself was removed far too early from her own mother. This 'de maternalisation' is likely to be due to failure of one generation to 'teach' the next generation maternal skills. The calves had to be hand-reared by trainers, but, tragically, Vicky died in 2013 at just 10 months of age (Batt 2012).

The captive orca industry has a shallow gene pool (a limited number of reproductive animals), leading to many accounts of inbreeding. The father of both of Kohana's calves was in fact her *uncle.* Both Adan and Vicky were blood related to over 80% of the orcas held at SeaWorld, and Vicky and her mother shared the same grandfather (Batt 2012).

Wild orca offspring in the most studied populations stay with their mothers for life, with some matrilines consisting of four generations (Ford et al. 2000). SeaWorld has removed 19 orca calves from their mothers, including one at 10 months, one at 20 months and one at 24 months; and only two of these removals were on medical grounds (Hargrove 2015). Jett and Ventre (2015) demonstrated that captive orcas face the highest risk of dying between the ages of two and six and speculate that avoiding the separation of mothers and calves may reduce this figure.

11.6 Environmental Quality and Complexity

Orcas are the most widely distributed cetacean on the planet and probably the most widely distributed large mammal in the world after humans (Rice 1998). Clans of orcas roam every ocean of the world and most seas. They range from the polar ice edges to the tropics and from the shoreline to the deep, open ocean. These opportunistic predators have evolved sophisticated strategies to thrive in most marine ecosystems (Baird 2000). Scientists now recognise several different ecotypes of orcas around the world (Bigg et al. 1990; Pitman and Ensor 2003).

Bottlenose dolphins, comprising more than one species, are also widely distributed, consuming a large variety of different food and inhabiting a range of environmentally complex environments (Wells and Scott 2009).

Belugas inhabit Arctic or sub-Arctic environments and have adapted ecologically and behaviourally to these extreme conditions (O'corry-Crowe 2009).

A man-made tank can never replicate the complexity, expanse, choice and range of habitats in the ocean environment nor meet the full range of an individual cetacean's biological capacities and the range of exposure to the physiological adaptations with which the animal is equipped. In captivity, cetaceans cannot be provided with an environment that simulates their natural habitat. Water is chemically treated, often with chlorine, which prevents the placing of live fish (feeding of live prey such as fish is unlawful in many countries) and plants into their tanks and can also present health problems if used excessively or incorrectly (Couquiaud 2005). Tank water is also filtered to prevent the build-up of excrement and other waste, and most of the tanks holding cetaceans are smooth sided, small and empty of stimuli, perhaps to facilitate cleaning. Tanks lack species-specific enrichments (Couquiaud 2005) such as sand, rock, plants and changes in surface texture and depth, and, with nothing to use their anatomical and physiological adaptations on, many of the features which make cetaceans unique (their telos) become redundant, including the capacity to fully utilise their natural use of sound through echolocation (Au 2009). Some dolphinaria also provide only indoor facilities, lacking exposure to natural light and to natural daylight hours or daylight light patterns.

Captive cetaceans are often kept in climates to which they are not adapted (Couquiaud 2005), even to the extent of belugas, an Arctic species, being held in sea pens in the naturally warm seawater off the Turkish coast (Williamson 2008). Sea pens, while potentially offering greater environmental diversity and therefore a more enriched environment (Ruiz et al. 2009; Ugaz et al. 2013), have often been located in water that is too shallow, too warm and subject to tropical storms and in areas where pollution is a problem (Rose et al. 2009). Water quality can also be a problem in indoor tank environments, and many countries which regulate captive cetacean facilities include a number of water quality parameters that must be followed to comply with the law (Williamson 2006).

11.7 Noise

Cetaceans are highly adapted acoustic animals, living with the capacity to make sense of the complex auditory world of the ocean. Noise in the captive environment can have a potentially dramatic impact on their behaviour and physiology, in some cases causing them to refuse to eat (Couquiaud 2005). Noise is carried faster in water (Wright et al. 2007), and the loud music of shows and adjacent rides in facilities located in theme parks adds to, and contributes to, the noise of pumps and filters (Couquiaud 2005). The European Association for Aquatic Mammals (2009) recommends that mechanical equipment that produces sound in close proximity to dolphins should be isolated acoustically.

11.8 Behavioural Restrictions

In captivity, many of the choices available to individuals in the wild are removed. Food, shelter and medical care are provided, and breeding is usually controlled by the holding facility (Couquiaud 2005). Stereotypic behaviours, behavioural evidence of stress and high rates of infection and poor health are common among wide-ranging carnivores when they are denied sufficient space to carry out natural behaviour (Clubb and Mason 2003).

Orcas in the wild display a whole range of different adaptive behaviours, from 'spy hopping' to tail slaps and breaching (Jefferson et al. 2008). They are also one of the fastest moving creatures in the ocean, capable of swimming at speeds of over 20 km/h (10.8 knots) (Ford 1989). Orcas possess one of the largest brains by volume in the animal world and have developed some highly complex and sophisticated hunting strategies, which vary from region to region and also in the approaches taken to the targeted prey (Ford 2009). Perhaps the most spectacular behaviour is that witnessed among orcas in Antarctica, where certain populations use a hunting technique known as 'wave washing', in which the orcas work cooperatively to create a wave to flush a seal off an ice floe (Visser et al. 2008; Pitman and Durban 2012). Another hunting technique known as 'carousel feeding' has been perfected by orcas off the coast of Norway. This technique involves orcas cooperatively herding schools of herring into a tight ball and driving them towards the surface, then picking off individual fish that have been stunned by tail slapping (Similä and Ugarte 1993).

Bottlenose dolphins show a high capacity for problem-solving and tool use (Whale and Dolphin Conservation 2016a). Some members of a population in Australia have been documented carrying sponges on their beaks to protect them from sea urchins when foraging on the sea floor (Whale and Dolphin Conservation 2016b).

Belugas have a sophisticated sonar system, which helps them move around in shallow water, and are one of the most vocal of cetaceans. They sometimes travel hundreds of miles upstream in rivers to reach their summer calving grounds (Whale and Dolphin Conservation 2016c).

The one-dimensional caricature of cetacean behaviours which is demonstrated to the public in marine parks around the world, where all choice and decision-making has been removed, pays a great disservice to these cognitively outstanding creatures. Dysfunctional, socially disparate cetacean groupings are coupled with a lack of space, low environmental stimuli, no capacity to hunt or forage in a realistic way and combined with the spectacle of stereotypical behaviours such as jaw popping, bar chewing, repetitive swimming and motionless logging at the pool surface (Jett and Ventre 2011; Frohoff 2005).

11.9 Stress

Stress is reported to severely affect the health of cetaceans in captivity. Symptoms which are associated with stress include weight loss, lack of appetite, anti-social behaviour (including aggression), self-destructive behaviour, reduced breeding success, arteriosclerosis, stomach ulcers, blood cell count changes and increased susceptibility to diseases and increased mortality rates (Rose et al. 2009; Romero and Butler 2007; Frohoff 2004; Schmitt et al. 2010; Fair and Becker 2000; St. Aubin and Dierauf 2001).

Handling, restraint, confinement, transport, isolation or crowding and an artificial diet are risk factors for stress in captive cetaceans and, ultimately, lead to measurable reductions in their life expectancy (Maas 2000; Noda et al. 2007; Thomson and Geraci 1986). Waples and Gales (2002) describe three cases of illness or death in the space of one year among a group of captive bottlenose dolphins in Western Australia. These animals were most likely suffering from stress as a result of changes in social relationships, aggression from other dolphins and loss of social support. Schmitt et al. (2010) found that stress hormones (concentrations of plasma adrenocorticotropic hormone (ACTH), cortisol, and aldosterone) increased significantly in captive belugas during routine physical examination, and similar effects have been recorded in captive porpoises (Desportes et al. 2007).

11.10 Use of Tranquilisers

Psychotropic drugs are often used in the care of captive cetaceans (Knight 2013). The most commonly used is Diazepam (Valium® and generics), a benzodiazepine drug which veterinary staff use to facilitate the handling of whales and dolphins for certain procedures, such as clinical diagnostic tests (including bacteriological swabbing and blood sampling) and transport. Depending on the dose, benzodiazepines can be used to reduce anxiety and excitability and also to control stereotypical behaviours (Knight 2013).

Marine parks such as SeaWorld report that drugs such as benzodiazepines are used by the facility veterinarians for the care and treatment of the marine mammals they hold (Cornell 2011). At the Rimini dolphinarium in Italy, irregularities in the administration of tranquilisers were cited as one of the factors which resulted in the permanent closure of the facility by the public authorities (Cronin 2014).

As voluntary breathers, cetaceans must be conscious and awake to breathe (Lyamin et al. 2008). Diazepam can decrease the responsiveness of the respiratory system (Khan 2014), and so this possible side effect in whales and dolphins is of particular concern. Diazepam is also used to encourage feeding in some captive animals, as it appears to act by enhancing the taste and flavour of food (Dowling 2015). Its use on captive dolphins, however, is questionable, as research indicates they can only taste salt (Zhu et al. 2014).

11.11 Early Mortality

Female orcas in the wild can live to an estimated maximum of 90 years with a mean expectancy of 46 years. Male orcas live an estimated maximum of 70 years with a mean of 31 years (Olesiuk et al. 2005). Bottlenose dolphins can live for up to 50 years in the wild (NOAA Fisheries 2016).

Small and DeMaster (1995b) found that mortality rates of captured bottlenose dolphins increased by six times immediately after capture and that this mortality rate did not drop down to the 'base captive mortality rate' for up to 35–45 days. Two studies from the 1990s (Small and DeMaster 1995a; Woodley et al. 1997) demonstrate higher annual mortality rates for bottlenose dolphins (5.6 and 5.7% annually) and orcas (6.2% annually) in captivity than in the wild (bottlenose dolphins 3.9% and orcas 2.3% annually).

In a 2015 study by Jett and Ventre, looking at captive orca mortality on a global scale since 1961, it was found that nearly two-thirds of orca deaths occurred in the first five years of a whale's captivity. Orcas in US facilities fared better than facilities in other countries, with a median survival rate of 12 years, and since 1985, captive orca survival has improved but still lags far behind their wild counterparts.

Data is lacking to enable a clear comparison in mortality rates between wild and captive belugas, although Woodley et al. (1997) indicated that there was increased mortality in captivity. Re-evaluation of ageing techniques in belugas from the wild has put the maximum life span of belugas at 60 years (Stewart et al. 2006). In captivity, belugas routinely die before the age of 30 (Rose et al. 2009).

Considering that, in captivity, cetaceans receive veterinary care if they are found to be sick, do not have to hunt for food, are not exposed to pollution in the natural marine environment (but may be exposed to long term chemical exposure in tank water) and are protected from predators; it seems probable that other factors are playing a role in reducing the annual survival rates for cetaceans in captivity.

11.12 Threats to Wild Populations

Cetaceans rely on well-organised groupings for, inter alia, foraging, defence against predators and transmission of specialised behaviour between generations (Whitehead et al. 2004). The capture of cetaceans from wild populations for live display in captivity currently occurs in only a handful of places around the world, including Russia and Japan (IUCN 2015; International Whaling Commission 2014, 2015).

The removal of key individual cetaceans, animals crucial to social cohesion in cetacean populations, may have long-term implications for population viability (Lusseau and Newman 2004; Williams and Lusseau 2006). Reeves et al. (2003) noted that live removals are equivalent to killing, as the individuals brought into captivity can no longer help maintain the genetic inputs to their wild populations.

Live capture operations for public display typically target young female cetaceans whose temperament makes them easier to handle in aquaria (Rose et al. 2009). The bias in wild populations which results from taking into captivity young females is another cause for conservation concern (Williams and Lusseau 2006).

11.13 Beluga Captures

In the Russian Far East, belugas are captured in the Sakhalin–Amur region in the Sea of Okhotsk under a quota set by the Russian government (Shpak and Glazov 2013), capture being for display in aquaria in Russia and overseas. The belugas are targeted as they congregate in the relatively warm coastal waters during the summer months where they breed, forage and moult (Shpak et al. 2010), and selected animals are taken from a population estimated at just under 4000 individuals (Shpak and Glazov 2014).

In 2013, 81 beluga individuals were captured and transported to holding facilities in Russia prior to onward transfers to national and international aquaria. Thirty-four whales are believed to have died during capture, seven died at the holding facilities and three considered to be at risk of death were released (Shpak and Glazov 2014). Based on available knowledge, and noting that more research was needed, an independent scientific review panel looking at proposed removals of belugas from this population calculated the sustainable annual removal to be 29 individuals, way below current capture levels (Reeves et al. 2011). Concerns continue to be raised by local and international beluga scientists that the captures are unsustainable (International Whaling Commission 2014, 2015).

During capture, belugas are approached in shallow waters by the capture team in boats, encircled using seine nets while surrounded by further boats. Once within the confines of the net, any belugas deemed at risk of entanglement are wrapped in the net and held at the surface or tied to the side of one of the boats. The net (and the belugas trapped inside it) is then pulled to shore (Georgia Aquarium 2012). The stress involved in this process for these self-aware and socially aware whales, approached by boats, trapped in nets and pulled to shore, is reported to be severe (St. Aubin and Geraci 1992; Curry 1999; Butterworth et al. 2013). Footage of beluga captures in Russian waters from the late 1990s showed very crude methods of capture and transport that put the individuals targeted at considerable risk of injury or death (Woodyer 2012).

11.14 Box Out Case Study: Georgia Aquarium Application to Import Wild-Caught Belugas

In 2012, Georgia Aquarium applied to the US National Marine Fisheries Service (NMFS) to import 18 belugas captured in Sakhalin Bay for public display. While the application requested ownership of the belugas by Georgia Aquarium, the Aquarium planned for 15 whales to undergo immediate transfer to other US facilities, including three SeaWorld parks, under 'breeding loans' (Georgia Aquarium 2012). Following a public comment period, in which members of the public were invited to submit their views on the proposed import, and which resulted in approximately 9000 responses (NOAA Fisheries 2015), the NMFS denied Georgia

Fig. 11.4 Belugas captured in Russia are held in inadequate holding pen conditions. *Image credit*: Lloyd Hannemann

Aquarium its import proposal. The agency based its decision on the impact of live captures on the population, its belief that allowing the import would contribute to demand to capture further belugas for the United States and worldwide and its determination that five of the belugas proposed for import were potentially still nursing young and not yet independent at the time of their capture (NOAA Fisheries 2015). The Aquarium appealed this decision, but it was upheld in the US District Court of Atlanta (Georgia Aquarium Inc. vs. Penny Pritzker, 2015). The Sakhalin–Amur population of belugas has subsequently been included in the US Marine Mammal Protection Act as a 'depleted' population, now well below 60% of its historic abundance, which means imports are now prohibited (Whale and Dolphin Conservation 2016d). Meanwhile, captures for aquaria in China and other countries continue, with individuals exported from Russia to be held in wholly inadequate conditions (Fig. 11.4).

11.15 Japanese Drive Hunts

In Japan, annual quotas are given by the Japanese government for the killing and live capture of over 2000 small cetaceans in what are known as 'drive hunts' (Butterworth et al. 2013). Individuals are herded out at sea with small fishing vessels, and through the use of underwater noise, these groups of animals are driven towards the shore, where they are netted off and then removed alive for display in aquaria or killed for meat or other products (Butterworth et al. 2013; Vail 2015). Several species are targeted for live capture from the hunts, including bottlenose dolphins, false killer whales, Pacific white-sided dolphins, Risso's dolphins and short-finned pilot whales (Ceta-Base 2016b).

The prolonged and stressful process involved in the drive hunts during the herding offshore, dragging by the tail fluke alongside the capture boats, confinement in the netted-off cove and removal from the water and their pod mates (many of whom may go on to be killed), sometimes over many hours or even days, is likely to be have profoundly severe welfare impacts (Butterworth et al. 2013; Connor 2007).

Concerns regarding the sustainability of the drive hunts in Japan have been expressed by the International Whaling Commission and other scientific bodies (International Whaling Commission 1993; Kishiro and Kasuya 1993; IUCN 2015; Wells 2012; Marsh 2013). In 2014, the International Whaling Commission's Scientific Committee reported that the issue of total removals in the drive hunts needed to be more critically examined and incorporated into population assessments. It also noted that there was a lack of current accurate data on both stock identity and size for the bottlenose dolphins in the waters off Taiji, where the hunts take place (International Whaling Commission 2015).

11.16 Orca Captures

The first orca captures occurred in the Pacific Northwest of America in the early 1960s and continued until the mid-1970s when this practice was banned under state law (Pollard 2014). During this early capture period, 55 orcas were taken for display in marine parks. In 1976, the capture teams turned their attention towards Iceland, where 54 whales were taken over the next 13 years (Williams 2001). During the 1980s and 1990s, Japan was also active in supplying orcas to its marine parks—none of the 20 captured orcas taken during this time have survived (Jacobs 2006). The Russian government issues annual catch quotas for orcas (up to 10 per annum) for both the domestic market and export overseas (FEROP 2016), and today Russia remains the only country in the world where wild orca captures continue for the aquarium trade.

11.17 Conclusions: The Future for Captive Cetacean Welfare

Public opinion is shifting on cetacean captivity. Evidence of poor cetacean welfare has been brought to the attention of the millions of viewers of documentaries such as *Blackfish* and *The Cove*, and a majority of young Americans opposing cetacean captivity (Racanelli 2016). Concern for captive orcas has led this quantum shift in perspective (Whale and Dolphin Conservation 2014). In March 2016, in response to what he referred to as the changing mind-set of society and a shrinking customer base, SeaWorld's Chief Executive Officer, Joel Manby, announced an end to orca breeding at SeaWorld (Munarriz 2016).

Research reveals concerns for other species too, with a majority of UK holiday-makers indicating opposition to seeing whale and dolphin shows (Payne 2014) and discomfort about dolphin welfare expressed by people who had swum with them in captivity (Curtin and Wilkes 2007).

The number of facilities holding cetaceans in some parts of the world, including Europe, is declining (Whale and Dolphin Conservation 2015). However, in other parts of the world, including China and the Caribbean, it is increasing (China Cetacean Alliance 2015; Vail 2014).

Discussion is now focused on what alternatives exist for the thousands of individual bottlenose dolphins, orcas, belugas and other species currently in captivity. While a return to the wild under strict criteria may be possible for some (Williamson 2014), others may be too physically or mentally altered by long term captivity to survive without human care. Plans are now underway to create cetacean sanctuaries, offering individuals the chance to live out the remainder of their lives in enclosures in a natural cove or bay, protected from storms and pollution, where their health and welfare needs can be taken care of in a more naturalistic environment, without performing in shows, and with public observation strictly controlled or from a distance (Williamson 2016). This may be the future for cetaceans currently in captivity, a future which has the potential to address many of the threats to cetacean welfare presented by their current confinement in captivity.

For further information on captive cetacean welfare, news stories, blogs and up-to-date statistics, please visit whales.org/captivity.

References

Au WWL (2009) Echolocation. In: Perrin WF, Würsig B, Thewissen JGM (eds) Encyclopedia of marine mammals, 2nd edn. Academic Press, San Diego, pp 348–357

Baird RW (2000) The killer whale. In: Mann J, Connor RC, Tyack PL, Whitehead H (eds) Cetacean societies. University of Chicago Press, Chicago, pp 127–153

Batt E (2012) Op-Ed: Loro Parque in Tenerife welcomes second baby orca rejected by mom. Available via Digital Journal. http://www.digitaljournal.com/article/331012. Accessed 20 Apr 2016

Bigg MA, Olesiuk PF, Ellis GM, Ford JKB, Balcomb KC (1990) Social organization and genealogy of resident killer whales (*Orcinus orca*) in the coastal waters of British Columbia and Washington State. Report of the International Whaling Commission (Special Issue 12), pp 383–405

Blake L (2012) New dolphin death adds to puzzle at Minnesota Zoo. Available via Star Tribune. http://www.startribune.com/new-dolphin-death-adds-to-puzzle-at-minnesota-zoo/138853389/. Accessed 20 Apr 2016

Blasi MF, Boitani L (2014) Complex social structure of an endangered population of bottlenose dolphins (*Tursiops truncatus*) in the Aeolian Archipelago (Italy). PLoS One 9(12):e114849. doi:10.1371/journal.pone.0114849

Buck CD, Schroeder JP (1990) Public health significance of marine mammal disease. In: Dierauf LA (ed) CRC handbook of marine mammal medicine: health, disease and rehabilitation. CRC Press Inc., Boston

Butterworth A, Brakes P, Vail CS, Reiss D (2013) A veterinary and behavioral analysis of dolphin killing methods currently used in the "drive hunt" in Taiji, Japan. J Appl Anim Welf Sci 16(2):184–204
Cantor M, Whitehead H (2013) The interplay between social networks and culture: theoretically and among whales and dolphins. Philos Trans R Soc Lond B Biol Sci 368:20120340
Ceta-Base (2016a) Captive cetacean database. Living population. Available via Ceta-Base. http://www.ceta-base.org/captive/cetacean/. Accessed 20 Apr 2016
Ceta-Base (2016b) Captive cetacean database. Drive hunt results. Available via Ceta-Base. http://www.ceta-base.org/taiji/drive/results.php. Accessed 20 Apr 2016
China Cetacean Alliance (2015) Ocean theme parks: a look inside China's growing captive cetacean industry. Available via China Cetacean Alliance. http://chinacetaceanalliance.org/wpcontent/ uploads/2016/02/CCA-Report-Web.pdf. Accessed 6 Apr 2017
Clegg ILK, Borger-Turner JL, Eskelinen HC (2015) C-Well: the development of a welfare assessment index for captive bottlenose dolphins (*Tursiops truncatus*). Anim Welf 24:267–282
Clubb R, Mason G (2003) Captivity effects on wide-ranging carnivores: animals that roam over a large territory in the wild do not take kindly to being confined. Nature 425:473–474
Connor RC (2007) Complex alliance relationships in bottlenose dolphins and a consideration of selective environments for extreme brain size evolution in mammals. Philos Trans R Soc Lond B Biol Sci 362:587–602
Cornell L (2011) Sworn Affidavit of Dr Lanny Cornell. Seaworld Parks & Entertainment LLC and Marineland of Canada Inc. Ontario Superior Court of Justice. Court File No. 52783/11. Available via Scribd. https://www.scribd.com/doc/215567388/Seaworld-v-Marineland-Aff-of-Lanny-Cornell. Accessed 6 Apr 2017
Couquiaud L (2005) A survey of the environments of cetaceans in human care. Aquat Mamm 31(3):279–385
Cronin AM (2014) Abusive dolphinarium in Italy shut down for good! Available via One Green Planet. http://www.onegreenplanet.org/news/abusive-dolphinarium-in-italy-shut-down-for-good/. Accessed 20 Apr 2016
Curry B (1999) Stress in mammals: the potential influence of fishery-induced stress on dolphins in the eastern tropical Pacific Ocean. NOAA Technical Memorandum NMFS-SWFSC-260, p 121
Curtin S, Wilkes K (2007) Swimming with captive dolphins: current debates and post-experience dissonance. Int J Tour Res 9:131–146
Delfour F, Marten K (2001) Mirror image processing in three marine mammal species: killer whales (*Orcinus orca*), false killer whales (*Pseudorca crassidens*) and California sea lions (*Zalophus californianus*). Behav Process 53(3):181–190
Desportes G, Buholzer L, Anderson-Hansen K, Blanchet M-A, Acquarone M, Shephard G, Brando S, Vossen A, Siebert U (2007) Decrease stress; train your animals: the effect of handling methods on cortisol levels in harbour porpoises (*Phocoena phocoena*) under human care. Aquat Mamm 33(3):286–292
Dowling P (2015) Drugs affecting appetite (monogastric). Available via Merck Veterinary Manual. http://www.merckvetmanual.com/mvm/pharmacology/systemic_pharmacotherapeutics_of_the_digestive_system/drugs_affecting_appetite_monogastrichtml. Accessed 20 Apr 2016
European Association for Aquatic Mammals (E.A.A.M.) (2009) Standards and guidelines for the management of bottlenose dolphins (*Tursiops* sp.) under human care. Available via EAAM. http://www.eaam.org/jdownloads/Documents%20and%20Guidelines/eaam_standards_and_guidelines_for_the_management_of_bottlenose_dolphins_under_human_care_sept_2009.pdf. Accessed 20 Apr 2016
Evans SJ (2015) Nanuq the beluga whale dies at under-fire SeaWorld Orlando after fracturing his jaw and contracting infection while on loan. Available via Mail Online. http://www.dailymail.co.uk/news/article-2963937/Nanuq-beluga-whale-dies-fire-SeaWorld-Orlando-fracturing-jaw-contracting-infection-loan.html. Accessed 20 Apr 2016
Fair PA, Becker PR (2000) Review of stress in marine mammals. J Aquat Ecosyst Stress Recover 7:335–354

FEROP (2016) Russian orcas homepage. Available via Russian Orcas. http://www.russianorca.com/index.php?lang=en&mode=. Accessed 20 Apr 2016

Ford JKB (1989) Acoustic behaviour of resident killer whales (*Orcinus orca*) off Vancouver Island, British Columbia. Can J Zool 67:727–745

Ford JKB (2009) Killer whale *Orcinus orca*. In: Perrin WF, Würsig B, Thewissen JGM (eds) Encyclopedia of marine mammals, 2nd edn. Academic Press, San Diego, pp 650–657

Ford JKB, Ellis GM, Balcomb KC (2000) Killer whales: the natural history and genealogy of *Orcinus orca* in British Columbia and Washington state, 2nd edn. UBC Press, Vancouver

Frohoff TG (2004) Stress in dolphins. In: Behoff M (ed) Encyclopedia of animal behaviour. Greenwood Press, Westport, pp 1158–1164

Frohoff TG (2005) Report on observations and preliminary assessment at Boudewijn Seapark dolphinarium in Brugge, Belgium. Report to Global Action in the Interest of Animals

Frohoff TG, Packard JM (1995) Human interactions with free-ranging and captive bottlenose dolphins. Anthrozoos 8(1):44–53

Georgia Aquarium (2012) Application for a permit to import certain marine mammals for public display under the Marine Mammal Protection Act. Available via NOAA Fisheries. http://www.nmfs.noaa.gov/pr/permits/sci_res_pdfs/17324_final_application.pdf. Accessed 6 Apr 2017

Georgia Aquarium Inc. vs. Penny Pritzker, Secretary of Commerce, National Oceanic and Atmospheric Administration and National Marine Fisheries Service (2015) Case 1:13-cv-03241 2015 AT-103. Filed 28 Sept 2015

Geraci JR, Ridgway SH (1991) On disease transmission between cetaceans and humans. Mar Mamm Sci 7(2):191–194

Hargrove J (2015) Beneath the surface. Palgrave MacMillan, New York

Hoyt E (1990) Orca: the whale called killer, 3rd edn. Camden House Publishing, North York

Hunt TD, Ziccardi MH, Gulland FMD, Yochem PK, Hird DW, Rowles T, Mazet JAK (2008) Health risks for marine mammal workers. Dis Aquat Org 81:81–92

International Whaling Commission (1993) 1992 Scientific Committee report. Available via International Whaling Commission. http://iwc.int/scientifc-committee-reports. Accessed 20 Apr 2016

International Whaling Commission (2014) 2013 Scientific Committee report. Available via International Whaling Commission. http://iwc.int/scientifc-committee-reports. Accessed 20 Apr 2016

International Whaling Commission (2015) 2014 Scientific Committee report. Available via International Whaling Commission. http://iwc.int/scientifc-committee-reports. Accessed 20 Apr 2016

IUCN (2015) IUCN red list of threatened species. *Tursiops truncatus*. Available via the IUCN Red List of Threatened Species. http://www.iucnredlist.org/details/22563/0. Accessed 20 Apr 2016

Jacobs S (2006) Orcas deceased in captivity. Available via Orcahome. http://www.orcahome.de/orcadead.htm. Accessed 20 Apr 2016

Jefferson TA, Webber MA, Pitman RL (2008) Marine mammals of the world: a comprehensive guide to their identification. Academic Press, London

Jett J, Ventre J (2011) Keto and Tilikum express the stress of orca captivity. Manuscript for the orca project. Available via The Orca Project. https://theorcaproject.wordpress.com/2011/01/20/keto-tilikum-express-stress-of-orca-captivity/. Accessed 20 Apr 2016

Jett J, Ventre J (2015) Captive killer whale (*Orcinus orca*) survival. Mar Mamm Sci 31(4):1362–1377

Khan SA (2014) Tranquilizers, antidepressants, sleep aids and anticonvulsants (toxicity). Available via Merck Veterinary Manual. http://www.merckvetmanual.com/mvm/toxicology/toxicities_from_human_drugs/tranquilizers_antidepressants_sleep_aids_and_anticonvulsants_toxicity.html. Accessed 20 Apr 2016

King SL, Janik VM (2015) Come dine with me: food-associated social signalling in wild bottlenose dolphins (*Tursiops truncatus*). Anim Cogn 18:969–974

Kirby D (2012) Death at SeaWorld. St Martin's Press, New York

Kishiro T, Kasuya T (1993) Review of Japanese dolphin drive fisheries and their status. Report of the International Whaling Commission, vol 43
Knight JA (2013) Diazepam—its use in captive bottlenose dolphins (*Tursiops truncatus*). Personal communication. 20 Sept
Krasnova VV, Chernetsky AD, Zheludkova AI, Bel'kovich VM (2014) Parental behavior of the beluga whale (*Delphinapterus leucas*) in natural environment. Biol Bull 41(4):349–356
Lodi News-Sentinel (1996). Keiko reminds man of whale attack. Available via Google news. https://news.google.com/newspapers? nid=2245&dat=19960117&id=GIo1AAAAIBAJ&sjid=QiEGAAAAIBAJ&pg=3872,1646286. Accessed 6 Apr 2017
Lusseau D, Newman MEJ (2004) Identifying the role that animals play in their social networks. Proc Biol Sci 271(Suppl 6):S477–S481
Lyamin OI, Manger PR, Ridgway SH, Mukhametov LM, Siegel JM (2008) Cetacean sleep: an unusual form of mammalian sleep. Neurosci Biobehav Rev 32:1451–1484
Maas B (2000) Prepared and shipped: a multidisciplinary review of the effects of capture, handling, housing and transport on morbidity and mortality. A report for the Royal Society for the Protection of Animals, Horsham, UK
Marine Connection (2008) Dolphin "accident" at Dolphin Academy, Curacao. Available via You Tube. https://www.youtube.com/watch?v=rjUwL1I1YCc. Accessed 20 Apr 2016
Marsh H (2013) Additional letter to Japanese Government regarding dolphin and small whale hunts. Presidential Letters, The Society for Marine Mammalogy. Available via The Society for Marine Mammalogy. https://www.marinemammalscience.org/letters/additional-letter-to-japanese-government-regarding-dolphin-and-small-whale-hunts/. Accessed 20 Apr 2016
Mate BR, Rossbach KA, Nieukirk SL, Wells RS, Irvine AB, Scott MD, Read AJ (1995) Satellite-monitored movements and dive behavior of a bottlenose dolphin (*Tursiops truncatus*) in Tampa Bay, Florida. Mar Mamm Sci 11(4):452–463
Munarriz R (2016) This should be a great summer for SeaWorld. Available via The Motley Fool. http://www.fool.com/investing/general/2016/04/05/this-should-be-a-great-summer-for-seaworld.aspx. Accessed 20 Apr 2016
NOAA Fisheries (2015) Georgia Aquarium application to import 18 beluga whales (File No. 17324)
NOAA Fisheries (2016) Bottlenose dolphins. Available via Alaska Fisheries Science Center. http://www.afsc.noaa.gov/nmml/education/cetaceans/bottlenose.php. Accessed 20 Apr 2016
Noda K, Akiyoshi H, Aoki M, Shimada T, Ohashi F (2007) Relationship between transportation stress and polymorphonuclear cell functions of bottlenose dolphins, *Tursiops truncatus*. J Vet Med Sci 69(4):379–383
O'corry-Crowe GM (2009) Beluga whale *Delphinapterus leucas*. In: Perrin WF, Würsig B, Thewissen JGM (eds) Encyclopedia of marine mammals, 2nd edn. Academic Press, San Diego, pp 108–112
Olesiuk P, Ellis GM, Ford JKB (2005) Life history and population dynamics of northern resident killer whales (*Orcinus orca*) in British Columbia. Canadian Science Advisory Secretariat Research Document 2005/045, p 75
Payne E (2014) Free Willy! Eighty-six per cent of tourists no longer want to watch killer whales and dolphins performing tricks in captivity. Available via Mail Online. http://www.dailymail.co.uk/travel/article-2638686/Free-Willy-Tourists-no-longer-want-whales-dolphins-performing-tricks-captivity-finds-new-survey.html. Accessed 20 Apr 2016
Pitman RL, Durban JW (2012) Cooperative hunting behavior, prey selectivity, and prey handling by pack ice killer whales (*Orcinus orca*), type B, in Antarctic Peninsula waters. Mar Mamm Sci 28:16–36
Pitman RL, Ensor P (2003) Three forms of killer whales (*Orcinus orca*) in Antarctic waters. J Cetacean Res Manag 5(2):131–139
Pollard S (2014) Puget Sound whales for sale. Arcadia Publishing, Charleston
Racanelli J (2016) To conserve is to care: how an emphasis on animal welfare is transforming zoos and aquariums. Available via Centre for Humans and Nature. http://www.humansandnature.org/to-conserve-is-to-care. Accessed 20 Apr 2016

Reeves RR, Fisher SJ (2005) The global trade in live cetaceans: implications for conservation. J Int Wildl Law Policy 8:315–340

Reeves RR, Smith BD, Crespo EA, Notarbartolo di Sciara G (2003) Dolphins, whales and porpoises: 2002–2010 conservation action plan for the world's cetaceans. IUCN/SSC Cetacean Specialist Group. IUCN, Gland, Switzerland

Reeves RR, Brownell Jr RR, Burkanov V, Kingsley MCS, Lowry LF, Taylor BL (2011) Sustainability assessment of beluga (*Delphinapterus leucas*) live-capture removals in the Sakhalin–Amur region, Okhotsk Sea, Russia. Report of an independent scientific review panel. Occasional paper of the Species Survival Commission No. 44. IUCN, Gland, Switzerland, 34pp

Reiss D, Marino L (2001) Mirror self-recognition in the bottlenose dolphin: a case of cognitive convergence. PNAS 98(10):5937–5942

Rice DW (1998) Marine mammals of the world: systematics and distribution. Society for Marine Mammalogy, Spec Pub 4, Lawrence, KS

Romero LM, Butler LK (2007) Endocrinology of stress. Int J Comp Psychol 20:89–95

Rose, NA (2014) Killer controversy: why orcas should no longer be kept in captivity. Animal Welfare Institute, Washington, 16pp

Rose NA, Parsons ECM, Farinato R (2009) The case against marine mammals in captivity, 4th edn. The Humane Society of the United States and the World Society for the Protection of Animals, Washington

Ruiz CU, Sánchez A, Maldonado FG (2009) Social and individual behavior of a group of bottlenose dolphins (*Tursiops truncatus*) in open and closed facilities. Vet Méx 40(4):381–387

Schmitt TL, St. Aubin DJ, Schaefer AM, Dunn JL (2010) Baseline, diurnal variations, and stress induced changes of stress hormones in three captive beluga whales, *Delphinapterus leucas*. Mar Mamm Sci 26(3):635–647

Shpak O, Glazov D (2013) Review of the recent scientific data on the Okhotsk Sea white whale (*Delphinapterus leucas*) population structure and its application to management. J Cetacean Res Manag. SC/65a/SM23. 19pp

Shpak OV, Glazov DM (2014) Update report on the white whale (*Delphinapterus leucas*) live-captures in the Okhotsk Sea, Russia. International Whaling Commission Scientific Committee report SC/65b/SM14

Shpak OV, Andrews RD, Glazov DM, Litova DI, Hobbs RC, Mukhametov LM (2010) Seasonal migrations of Sea of Okhotsk beluga whales (*Delphinapterus leucas*) of the Sakhalin-Amur summer aggregation. Russ J Mar Biol 36(1):56–62

Similä T, Ugarte F (1993) Surface and underwater observation of cooperatively feeding killer whales in northern Norway. Can J Zool 71:1494–1499

Small RJ, DeMaster DP (1995a) Survival of five species of captive marine mammals. Mar Mamm Sci 11(2):209–226

Small RJ, DeMaster DP (1995b) Acclimation to captivity: a quantitative estimate based on survival of bottlenose dolphins and California sea lions. Mar Mamm Sci 11(4):510–519

Spong P, Symonds H (2000) Orcas. Available via OrcaLab. http://orcalab.org/orcas/. Accessed 20 Apr 2016

Sprogis KR, Raudino HC, Rankin R, Macleod CD, Bejder L (2015) Home range size of adult Indo-Pacific bottlenose dolphins (*Tursiops aduncus*) in a coastal and estuarine system is habitat and sex-specific. Mar Mamm Sci 32(1):287–308

St. Aubin DJ, Dierauf LA (2001) Stress and marine mammals. In: Dierauf LA, Gulland FM (eds) CRC handbook of marine mammal medicine, 2nd edn. CRC Press, Boca Raton, pp 253–269

St. Aubin DJ, Geraci JR (1992) Thyroid hormone balance in beluga whales, *Delphinapterus leucas*: dynamics after capture and influence of thyrotropin. Can J Vet Res 56:1–5

Stewart REA, Campana SE, Jones CM, Stewart BE (2006) Bomb radiocarbon dating calibrates beluga (*Delphinapterus leucas*) age estimates. Can J Zool 84:1840–1852

Thomson CA, Geraci JR (1986) Cortisol, aldosterone, and leucocytes in the stress response of bottlenose dolphins, *Tursiops truncatus*. Can J Fish Aquat Sci 43:1010–1016

Tyack P (2009) Behavior, overview. In: Perrin WF, Würsig B, Thewissen JGM (eds) Encyclopedia of marine mammals, 2nd edn. Academic Press, San Diego, pp 101–108

Ugaz C, Valdez RA, Romano MC, Galindo F (2013) Behavior and salivary cortisol of captive dolphins (*Tursiops truncatus*) kept in open and closed facilities. J Vet Behav 8:285–290

Vail CS (2012) The mixed messages of captivity. Available via Whale and Dolphin Conservation. http://uk.whales.org/blog/2012/12/mixed-messages-of-captivity. Accessed 20 Apr 2016

Vail CS (2014) One step at a time: victory for dolphins in the Bahamas. Available via Whale and Dolphin Conservation. http://uk.whales.org/blog/2014/07/one-step-at-time-victory-for-dolphins-in-bahamas. Accessed 20 Apr 2016

Vail CS (2015) Black Market. Can the World Association of Zoos and Aquariums prevent the trade in Taiji-caught dolphins? Whale & Dolphin (71)

Van Lint W, de Man D, Garn K, Hiddinga B, Brouwer K (2006) EAZA yearbook 2004. EAZA Executive Office, Amsterdam

Visser IN, Smith TG, Bullock ID, Green GD, Carlsson OGL, Imberti S (2008) Antarctic peninsula killer whales (*Orcinus orca*) hunt seals and a penguin on floating ice. Mar Mamm Sci 24:225–234

Waltzek TB, Cortes-Hinojosa G, Wellehan JFX Jr, Gray GC (2012) Marine mammal zoonoses: a review of disease manifestations. Zoonoses Public Health 59:521–535. doi:10.1111/j.1863-2378.2012.01492.x

Waples EA, Gales NJ (2002) Evaluating and minimising social stress in the care of captive bottlenose dolphins (*Tursiops aduncus*). Zoo Biol 21:5–26

Wells R (2012) Letter to Japanese Government regarding dolphin and small whale hunts. Presidential Letters, The Society for Marine Mammalogy. Available via The Society for Marine Mammalogy. https://www.marinemammalscience.org/letters/letter-to-japanese-government-regarding-dolphin-and-small-whale-hunts/. Accessed 20 Apr 2016

Wells RS, Scott MD (2009) Common bottlenose dolphin *Tursiops truncatus*. In: Perrin WF, Würsig B, Thewissen JGM (eds) Encyclopedia of marine mammals, 2nd edn. Academic Press, San Diego, pp 249–255

Whale and Dolphin Conservation (2014) Poll shows big jump in numbers of Americans opposed orca captivity. Available via Whale and Dolphin Conservation. http://uk.whales.org/news/2014/06/poll-shows-big-jump-in-numbers-of-americans-opposed-orca-captivity. Accessed 20 Apr 2016

Whale and Dolphin Conservation (2015) EU Zoo Inquiry. Dolphinaria. A review of the keeping of whales and dolphins in captivity in the European Union and EC Directive 1999/22, relating to the keeping of wild animals in zoos, 2nd edn. A report for the European coalitions ENDCAP and Dolphinaria-Free Europe in association with the Born Free Foundation

Whale and Dolphin Conservation (2016a) Common bottlenose dolphin *Tursiops truncatus*. Available via Whale and Dolphin Conservation. http://uk.whales.org/species-guide/common-bottlenose-dolphin Accessed 20 Apr 2016

Whale and Dolphin Conservation (2016b) Indo-Pacific bottlenose dolphin *Tursiops aduncus*. Available via Whale and Dolphin Conservation. http://uk.whales.org/species-guide/indo-pacific-bottlenose-dolphin. Accessed 20 Apr 2016

Whale and Dolphin Conservation (2016c) Beluga whale *Delphinapterus leucas*. Available via Whale and Dolphin Conservation. http://uk.whales.org/species-guide/beluga-whale. Accessed 20 Apr 2016

Whale and Dolphin Conservation (2016d) NMFS' designation of Russian beluga whales as depleted makes their import into US illegal. Available via Whale and Dolphin Conservation. http://uk.whales.org/news/2016/10/nmfs-designation-of-russian-beluga-whales-depleted-makes-their-import-into-us-illegal. Accessed 6 Apr 2017

Whale and Dolphin Conservation Society (2009) Dolphin protection and the Caribbean cruise industry: an overview of issues, recommendations, and opportunities for collaboration. Prepared for the Florida Caribbean Cruise Association

Whitehead H (2011) The cultures of whales and dolphins. In: Brakes P, Simmonds MP (eds) Whales and dolphins—cognition, culture, conservation and human perceptions. Earthscan, London, pp 149–165

Whitehead H, Rendell L, Osborne RW, Würsig B (2004) Culture and conservation of non-humans with reference to whales and dolphins: review and new directions. Biol Conserv 120:431–441

Williams R, Lusseau D (2006) A killer whale social network is vulnerable to targeted removals. Biol Lett 2:497–500. The Royal Society

Williams V (2001) Captive orcas: dying to entertain you. A report for the Whale and Dolphin Conservation Society

Williamson C (2006) Marine mammals—guidelines and criteria associated with captivity. Document produced as by the Whale and Dolphin Conservation Society, WDCS in collaboration with the SPAW/RAC and the CEP Secretariat for the Protocol Concerning Specially Protected Areas and Wildlife (SPAW) in the Wider Caribbean. Available via CAR.SPAW.RAC. http://www.car-spaw-rac.org/IMG/pdf/OVERVIEW_CAPTIVITY_MARINE_MAMMALS_WCR.pdf. Accessed 20 Apr 2016

Williamson C (2008) Dolphin captures in the agreement area. FINS. Newsletter of ACCOBAMS, Agreement on the Conservation of Cetaceans of the Black Sea, Mediterranean Sea and Contiguous Atlantic Area 4(1):10

Williamson C (2014) Can captive whales or dolphins be returned to the wild? Whale & Dolphin (65)

Williamson C (2016) SeaWorld SHOULD support sanctuaries. Available via Whale and Dolphin Conservation. http://uk.whales.org/blog/2016/04/seaworld-should-support-sanctuaries. Accessed 20 Apr 2016

Woodley TH, Hannah JL, Lavigne DM (1997) A comparison of survival rates for captive and free ranging bottlenose dolphins (*Tursiops truncatus*), killer whales (*Orcinus orca*) and beluga whales (*Delphinapterus leucas*). International Marine Mammal Association Inc. Draft technical report 93-01

Woodyer J (2012) Russian beluga captures 1999. Available via You Tube. https://www.youtube.com/watch?v=TZyRT1jRxD4. Accessed 20 Apr 2016

Wright AJ, Aguilar Soto N, Baldwin AL, Bateson M, Beale CM, Charlotte C, Deak T, Edwards EF, Fernández A, Godinho A, Hatch LT, Kakuschke A, Lusseau D, Martineau D, Romero LM, Weilgart LS, Wintle BA, Notarbartolo-di-Sciara G, Martin V (2007) Do marine mammals experience stress related to anthropogenic noise? Int J Comp Psychol 20:274–316

Zhu K, Zhou X, Xu S, Sun D, Ren W, Zhou K, Yang G (2014) The loss of taste genes in cetaceans. BMC Evolutionary Biology 14:218–227

Chapter 12
Assessing the Welfare of Cetacea

Isabella L.K. Clegg and Andy Butterworth

Abstract Most of the species from the order Cetacea appear to possess advanced cognitive abilities and close social networks and are also likely to experience different affective states comprising of more than just basic emotions. Welfare describes a balance of positive and negative affective states experienced by an individual, and this balance is a good indicator of how it perceives the surrounding environment. In this chapter, we discuss how the first steps in cetacean welfare science are being taken to establish this as a discipline. We discuss the pertinent areas of cetology that merit investigation to form the basis of possible cetacean welfare measures. In this arena of welfare assessment, much of the existing work comes from farm animal science, and this previous experience offers potential tools and techniques which could be adapted for cetaceans. We review these sources of information, make suggestions for relevant investigations and discuss how assessment of cetacean welfare might be accomplished.

12.1 Introduction

Increasing our understanding of cetacean welfare is essential if we want to progress in our research, conservation, care and protection of these animals. The available research suggests that most of the species from the order Cetacea appear to possess advanced cognitive abilities and close social networks and thus are also likely to experience different affective states comprising of more than just basic emotions. Overall welfare is a balance of positive and negative affective states where affective states are made up of moods and emotions formulated in response to events in the

I.L.K. Clegg (✉)
Laboratoire d'Ethologie Expérimentale et Comparée E.A. 4443 (LEEC), Université Paris 13, Sorbonne Paris Cité, Villetaneuse 93430, France
e-mail: izziclegg@hotmail.co.uk

A. Butterworth
Clinical Veterinary Science, University of Bristol, Langford, Bristol, UK

A. Butterworth (ed.), *Marine Mammal Welfare*, Animal Welfare 17,
DOI 10.1007/978-3-319-46994-2_12

surrounding environment. In previous chapters, we have been presented with the many environmental threats experienced by wild cetaceans and aspects of public display environments that have been called into question for captive cetaceans. Assessments of welfare will be necessary to understand the impacts of such contexts, from the animals' perspective, and hence to work towards reduction of these impacts. Although cetacean welfare science is not yet an established discipline, there are pertinent areas of cetology that can form the basis of possible welfare measures. In this arena of welfare assessment, much of the existing work comes from farm animal science, offering potential tools and techniques which could and can be adapted for cetaceans. In this chapter, we review these sources of information, make suggestions for potentially relevant future investigations and discuss how assessment of cetacean welfare might be accomplished.

12.2 The Science of Assessing Animal Welfare

Before we delve into the topic of cetacean welfare, and how we might assess it, we must understand the progression of terrestrial welfare measurement and discuss the difficulties associated with the assessment of an animal's welfare. The study of animal welfare has now developed into a multidisciplinary science (Ohl and van der Staay 2012; Webster 2005), with species-specific measures that comprise comprehensive welfare assessments (Blokhuis 2008; Botreau et al. 2007; Webster 2005). Welfare assessments have developed to the point where they now attempt to measure the quality of the animal's life in its surroundings, and so, although wild animal welfare assessments are not yet regularly applied, there exist assessment methods which could certainly make measurement of wild animal welfare feasible and could offer the possibility to provide insights into the animals' capabilities to adapt to changing environments.

12.3 Welfare Criteria and Measures

Welfare assessments exist in many forms and are capable of providing varying levels of detail, dependent on the time allocated for assessment and the welfare model the framework is based on. The 'Five Freedoms' principle (Farm Animal Welfare Council [FAWC] 1992) has been commonly used to assess welfare, and recent efforts follow this basic structure by expanding the categories used (e.g. WelfareQuality®, Blokhuis 2008). In the WelfareQuality® assessment system, to maximise the potential for standardisation between different species and methods of assessment, it was designed so that all assessments would use an agreed set of criteria, which then contained a number of species-specific measures. The criteria were devised to be 'minimal, exhaustive, independent, and agreed by stakeholders' (Botreau et al. 2007). The work of WelfareQuality®, and other similar research, has encouraged a transition towards the use of 'animal-based measures' and has stimulated cross-species application of welfare assessment frameworks (Veissier and Miele 2015), with a recent increase in studies measuring zoo animal welfare using assessment techniques originally created to assess farmed animals (Barber 2009; Whitham and Wielebnowski 2013; Clegg et al. 2015).

When validating welfare indicators within assessments, i.e. confirming that they are 'measuring what we want them to', there is widespread agreement that an appropriate approach is to cross-correlate results from behavioural and physiological measures, and if possible cognitive measures, to affirm that the true welfare state is being described (Désiré et al. 2002; Webster 2005; Boissy et al. 2007; Whitham and Wielebnowski 2013). Previously, only indicators of negative welfare were present in assessments, and the absence of these 'negative' indicators was taken to mean the animal had good welfare (Yeates and Main 2008). However, positive experiences and states are now recognised to be (at least) as important as negative states in their contribution to overall well-being (Fraser and Duncan 1998; Boissy et al. 2007; Yeates and Main 2008), and thus it is considered important (where they exist) to assess indicators of positive affect (see reviews by Boissy et al. 2007; Yeates and Main 2008).

12.4 Animal-Based Measures

Increased importance is now given to animal-based measures over resource-based measures (Whitham and Wielebnowski 2013). Animal-based measures, also called outcome measures, are parameters measurable directly from the actual animal, such as certain behaviours, physiological measures or disease incidence levels. Resource-based measures, or input measures, are those which provide information on the resources or procedures applied to the animals, for example, the space within the enclosure, quality of bedding, husbandry procedures or type of food provided (Webster 2005; Veissier et al. 2008). Animal-based measures are generally considered to be valid welfare indicators as they reflect the 'outcome' of what has actually happened to the animals, rather than judging the 'likely outcome' of provision of a certain resource. However, animal-based or outcome measures are sometimes less easy to carry out—they may require close observation of the animals over a period of time, or examination of, for example, lesions or changed disease levels in the animals; therefore, outcome measures have been considered by some to be less 'feasible' than resource-based measures. Currently, existing assessment protocols which aim to focus on animal-based measures use these wherever possible and supplement them with resource-based measures in order to comprehensively fulfil the overall set of criteria (Rushen et al. 2011; Whitham and Wielebnowski 2013).

12.5 Zoo Animal Welfare Assessment

This chapter is concerned with the welfare of wild animal species, whether maintained in the wild or in captivity, and the zoo welfare literature is likely to be more relevant to this area than that which comes from the assessment of domestic or laboratory animals. Zoological institutions and aquaria (hereafter zoos) are only recently embracing welfare science and its applications, although they have been responding to increased public concern by aiming to exceed minimum standards and by publishing their experience in the management of the many species they maintain (see review by Whitham and Wielebnowski 2013). It has been acknowledged that there is much potential for adapting farm animal welfare measures to zoo animals (Swaisgood 2007; Hill and Broom 2009; Mason and Veasey 2010). As a means to achieve this, especially considering the huge species diversity seen in captivity, welfare scientists are starting to be employed by zoos or permanently associated with them (Barber 2009; Maple 2007). So far zoo welfare research has focussed predominantly on elephants (Veasey 2006; Maple 2007; Mason and Veasey 2010), using cortisol as a stress measure (Pifarré et al. 2012; Shepherdson et al. 2013; Ugaz et al. 2013), as well as studies on the effects of enrichment and its impacts on behaviour (Carlstead and Shepherdson 2000). The only proposal for a comprehensive welfare assessment for captive zoo animals has been for bottlenose dolphins (*Tursiops truncatus*), where Clegg et al. (2015) adapted a welfare assessment framework originally devised for farm animals (WelfareQuality®) to this species. We will expand upon the details of this project and the other few existing studies on cetacean welfare in the following sections.

12.6 Existing Cetacean Welfare Studies

12.6.1 Studies Focussed on Cetacean Welfare in Captivity

Studies expressly investigating welfare in captive or wild cetaceans are scarce (Clark 2013; Ugaz et al. 2013; Clegg et al. 2015). Although the welfare of captive cetaceans seems to be an important topic of debate in public and the media (Grimm 2011; Jett and Ventre 2015), this interest has not stimulated an equivalent level of research thus far. Gygax (1993), Frohoff and Packard (1995) and Galhardo et al. (1996) were the first to look at the behaviour of dolphins in terms of what it might indicate about welfare, but conclusions were often contradictory to each other in terms of the function, and welfare implications, of certain behaviours. Waples and Gales (2002) provided useful observations correlated with physiological data from Indian Ocean bottlenose dolphins (*Tursiops aduncus*) that were experiencing social stress, and these authors proposed some initial indicators of poor welfare, although they did not label them as such. Although the behavioural findings of these studies were presented anecdotally, they described how inappetence, the incidence and severity of rake marks (superficial wounds from teeth of another dolphin, Scott et al. 2005), aggressive behaviours and social isolation were the main indicators accompanying severe social stress. Furthermore, these authors recorded changes in white-blood cell profiles (eosinopenia, lymphopenia and neutrophilia), weight loss and gastric ulceration (two out of three subjects) in their study animals. Collection of salivary cortisol has recently been confirmed as a useful sampling method with dolphins (Pedernera-Romano et al. 2006), and while Ugaz et al. (2013) correlated levels of salivary cortisol with potential behavioural welfare indicators in bottlenose dolphins, the choice of behaviours studied by these authors may not have been the most relevant to welfare (e.g. swimming depth, swimming direction). One investigation proposed that decreased vocalisation rate could be a measure of poor welfare in belugas (*Delphinapterus leucas*), although this study only examined two subjects (Castellote and Fossa 2006).

For captive killer whales (*Orcinus orca*), the small number of peer-reviewed, published studies related to welfare focus on mortality rates as opposed to behavioural indicators, and there is a marked lack of research on the behaviour of killer whales in captivity in general. Two recent studies on killer whale longevity in captivity when compared to the wild presented findings in direct contrast with each other (Jett and Ventre 2015; Robeck et al. 2015). In a study by Jett and Ventre (2012), it was suggested that increased surface behaviour of captive killer whales compared to wild counterparts put them at higher risks of mosquito-borne diseases, but argued this using personal observations, and unfortunately did not present behavioural data. Otherwise, there are no published studies investigating behavioural or physiological indicators of stress, welfare or affective states in killer whales.

Recently a comprehensive welfare assessment for a cetacean species was proposed: the C-Well® assessment for bottlenose dolphins (Clegg et al. 2015). The authors adapted the well-established farm animal assessment WelfareQuality®, to dolphins, using literature reviews, veterinary and professional expertise, and testing of the practical application of these assessment methods on 20 animals in three different facilities. The result was that 36 measures (Table 12.1) were proposed, 21 (58%) of which were animal-based and were capable of yielding individual welfare scores, comparable in many different ways (e.g. by measure, by criteria, when compared by pool, sex, age class and similarity of facilities). The C-Well® assessment may be seen as an initial step in the area of captive cetacean welfare assessment, with validation occurring through solicitation of expert opinion. The measures proposed are (as yet) unweighted, i.e. they are all given the same degree of ‘impact’, and the early use of these measures may stimulate research questions for future studies and prompt other researchers to contribute to work in this area.

Table 12.1 Criteria and measures of the Welfare Quality® framework adapted to establish the C-Well® assessment, taken from Clegg et al. (2015): 36 welfare measures are organised into 11 criteria which are part of four general principles

	Welfare Quality® criterion		Welfare Quality® measure[a]	C-Well® criterion		C-Well® measure
Good feeding	1	Absence of prolonged hunger	Body condition score	1	Absence of prolonged hunger	1.1. Body condition score
						1.2 Frequency of weight measurements
						1.3 Dietary records
	2	Absence of prolonged thirst	Water provision, cleanliness of water points, number of animals using water points	2	Absence of prolonged thirst	2.1 Capillary refill time
						2.2 Hydration protocol
Good housing	3	Comfort around resting	Time needed to lie down, cleanliness of animals	3	Resting comfort	3.1 Time budget
	4	Thermal comfort	No current measure	4	Thermal comfort	4.1 Frequency of water temperature testing
						4.2 Water temperature and diet
						4.3 Shade
	5	Ease of movement	Pen features according to live weight, access to outdoor loafing area in pasture	5	Appropriate environment	5.1 Topography
						5.1.1 Echolocation
						5.1.2 Complexity of enclosure
						5.2 Ability to exhibit complex movements
						5.2.1 Swim speed
						5.2.2 Aerials
						5.3 Water quality
						5.3.1 Salinity
						5.3.2 Coliform
						5.3.3 pH
						5.3.4 Chlorine
						5.3.5 Frequency of water quality testing
						5.4 Enrichment
						5.4.1 Application of enrichment

Good health	6	Absence of injuries	Lameness, integument alterations	6	Absence of injuries	6.1 Total wound threshold
						6.2 Wounds from enclosure
	7	Absence of disease	Coughing, nasal discharge, ocular discharge, hampered respiration, diarrhoea, bloated rumen, mortality	7	Absence of disease	7.1 Respiratory system
						7.1.1 Frequency of coughing
						7.1.2 Inhalation duration
						7.2 Eye diseases
						7.2.1 Discoloration
						7.2.2 Squinting
						7.3 Skin diseases
						7.3.1 Skin abnormalities
						7.3.2 Mouth abnormalities
						7.4 Blood parameters
						7.4.1 Blood sampling protocol
	8	Absence of pain induced by management procedures	Disbudding/dehorning, tail docking, castration	8	Absence of pain induced by management procedures	8.1 Blood draw
						8.2 Gastric tubing
						8.3 Voluntary restraint
						8.4 Emergency containment training
Appropriate behaviour	9	Expression of social behaviours	Agnostic behaviours, cohesive behaviours	9	Expression of social behaviours	9.1 Presence of social behaviours
	10	Expression of other behaviours	Access to pasture	10	Absence of abnormal behaviours	10.1 Stereotypic behaviour
	11	Good human–animal relationships	Avoidance distance	11	Positive human–animal relationship	11.1 Response to trainer while not under stimulus control
						11.2 Non-food tactile interactions
	12	Positive emotional state	Qualitative behaviour assessment			

12.6.2 Studies on Cetacean Welfare in the Wild

As mentioned previously, research on the welfare of wild animal species is quite sparse, and this is likely due to a combination of feasibility problems such as access, repeatable sampling and also perhaps a perceived lack of human 'responsibility' for welfare issues in wild animals, despite the suggestion that human impact on wild animal populations is increasing very significantly (Kirkwood et al. 1994). Wild cetaceans are no exception in the paucity of published welfare information. A common, direct, interaction between wild cetaceans and humans is whale-watching tourism, and researchers have endeavoured to quantify the impact of tourist boat activity on the animals, often finding that multiple cetacean species show mild to strong avoidance responses (e.g. killer whales: Williams and Ashe 2007; bottlenose dolphins: Bejder et al. 2006; minke whales (*Balaenoptera acutorostrata*): Christiansen and Lusseau 2015). Demonstrating further that this avoidance is salient and directly related to boat activity, a number of studies showed that population-level indicators, including reproductive rate and abundance, declined in both mysticete and odontocete species exposed to high levels of whale-watching activities (Bejder et al. 2006; Christiansen and Lusseau 2014). However, as yet, researchers have not focussed on applying direct, animal-based welfare measures in these contexts, although discussions supporting this approach are starting to take place (Swaisgood 2007; Ohl and van der Staay 2012).

Unfortunately, humans also interact with wild cetaceans in more lethal ways and these situations at least have spurred studies focussed on assessing the animals' welfare, albeit at the point of death. Butterworth et al. (2013) empirically evaluated dolphin welfare in the Taiji drive hunts, and several studies have argued against whaling mysticete species on the grounds of animal welfare concerns (e.g. Kestin 1995; Gales et al. 2008). Entanglement in marine debris can have many and varied lethal and non-lethal effects, and Butterworth et al. (2012) applied the Five Freedoms principles to discuss how entanglement affects individual animal welfare in cetaceans among other species. Cassoff et al. (2011) reviewed the causes and effects in cases of mysticete entanglement, where the majority of animals suffered protracted deaths, and concluded that entanglement may be one of the 'worst forms of human-caused mortality in any wild animal'.

Given the lack of published, validated animal-based measures of welfare, in the next two sections, we discuss the potential areas where such indicators may be found for both captive and wild cetaceans.

12.7 Animal-Based Welfare Measures for Cetaceans in Captivity

12.7.1 Inputs and Outcomes in the Captive Environment

We will start this part of the chapter by looking at what kinds of resources and management protocols are provided in the cetaceans' environment in captivity, i.e. assessed by resource-based measures, in order to highlight what kinds of outcomes are seen in the animal, i.e. the possible animal-based measures.

Legislation from around the world regulates the environment of cetaceans maintained for public display. This legislation is based on grouped resource-based requirements, with associated thresholds above which animal welfare is assumed to be good/acceptable (e.g. Animal Welfare Act (AWA 1966) in the USA and Council Directive 1999/22/EC in Europe). These laws dictate aspects such as the space required, water quality and facility hygiene levels, but the requirements are viewed as minimum standards by many institutions (McBain 1999; Joseph and Antrim 2010). Although animal-based measures are more reflective of actual welfare, resource-based measures are still likely to be a very important way to assess animal facilities, since the large variation in facility types in which cetaceans are kept worldwide is likely to differentially and very significantly impact their welfare (Joseph and Antrim 2010). However, animal-based measures could, and perhaps should, now be considered for inclusion in recommendations and codes of conduct, if not in the legislation as well, and we propose some possible categories of outcome measures in Fig. 12.1. It is worth noting that this large variation in display facilities, as well as the diversity of cetacean species kept in captivity, means that much research with large numbers of animals is needed to start to explain the variation and develop effective animal-based measures. Application of animal-based measures of cetacean welfare could improve the accuracy of resource-based measures, e.g. space provided, where animals' behavioural and physiological responses to different pool sizes can aid in establishing accurate thresholds. Animal-based indicators of welfare can be either behavioural, health-related or cognitive, and correlating data from these three categories can help to validate measures during the development process of a welfare framework (based on the 'triangulation' principle, Webster 2005). Furthermore, including measures from these categories confirms the comprehensive nature of the assessment where the hypothetical aim is to cover all aspects of the animal's life within the measures. Correlating such measures gives us an indication of the emotional responses of the animal which make up its affective states (Webster 2005; Boissy et al. 2007), with overall welfare being a balance of positive and negative affective states (Spruijt et al. 2001; Watters 2014).

Fig. 12.1 Diagram showing examples of input (resource based) and outcome (animal based) measures of welfare which could constitute a welfare assessment for cetaceans in captivity (*Image credit*: Isabella Clegg)

12.7.2 Behavioural Measures of Welfare

Behaviour is likely to be one of the most direct and significant expressions of welfare state in cetacean species (Waples and Gales 2002; Joseph and Antrim 2010), and regular behavioural monitoring is considered as crucial when aiming to understand welfare status of animals kept in zoos (Maple 2007; Watters 2014). Since cetaceans are generally highly social animals, social behaviour will likely be an important animal-based measure of welfare (Clegg et al. 2015). Social group composition in captivity is somewhat artificial, as this is decided by the zoo staff and management (Clegg et al. 2015), and the importance of social groupings is often overlooked in captive animal regulations, although efforts in some facilities are being made to model compositions on wild groups (Wells 2009). Being highly social means that there are many opportunities for poor welfare from social stress, as has been shown in studies with bottlenose dolphins (Waples and Gales 2002). Conversely, successful grouping of animals may have benefits and possibilities for good welfare through social bonding and development of relationships. Research has already suggested how affiliative contact behaviour in dolphins such as contact swimming and pectoral fin rubbing might be related to positive emotions, as well as describing how they might be quantified (Dudzinski 2010; Kuczaj et al. 2013). Play behaviour is common in both juveniles and adults in dolphin species and likely indicates positive affective state (Kuczaj and Eskelinen 2014). On the whole play is considered a measure of positive welfare, but difficulties with its quantification have impeded its use in other species' welfare assessments thus far (Boissy et al. 2007; Held and Špinka 2011). Cetacean species under human care may experience positive welfare states through provision of enrichment, mainly due to their cognitive abilities (Herman 2012), their tendency to play and their innovative behaviours (Kuczaj and Eskelinen 2014). Recently it was shown that captive dolphins increased certain vocalisations during a cognitive task only when cooperating and succeeding (Eskelinen et al. 2016), suggesting that vocal behaviour could be a source of welfare measures as with other animals (Manteuffel et al. 2004). Future captive welfare studies could maintain focus on inputs and outcomes as described in Fig. 12.1: to give an example, regarding enrichment, input measures would be the type/quantity/frequency of enrichment provided, and the outcome measures would be whether the dolphins are motivated to interact with the objects and whether the interaction is comprised of positive, calm and affiliative behaviours, or otherwise (Delfour and Beyer 2012; Hoy et al. 2010).

Indicators of social stress in captive cetaceans have been reported, such as social isolation (excluded from/avoiding the group), inappetence, inactivity, high rates of aggression and abnormal repetitive behaviour (ARB) (Galhardo et al. 1996; Waples and Gales 2002; Clegg et al. 2015), and thus behaviours representative of these states are all potential animal-based welfare measures. Stereotypic behaviour, a type of ARB, has been proposed as a welfare measure for cetaceans, but the lack of published research on the frequency and context of such behaviours means more work is needed before the measure can be established with an agreed value (Clark 2013; Clegg et al. 2015), especially as terrestrial research has found that in some cases animals with stereotypies have better welfare since they use it as a coping response (Rushen and Mason 2008). The quantification of rake marks on the body as an indirect measure of aggression levels was proposed by Clegg et al. (2015) as a welfare indicator, based on the premise that bodily damage is a common animal-based measure for farm species (e.g. WelfareQuality® 2009a, b, c). Figure 12.2 shows the grid

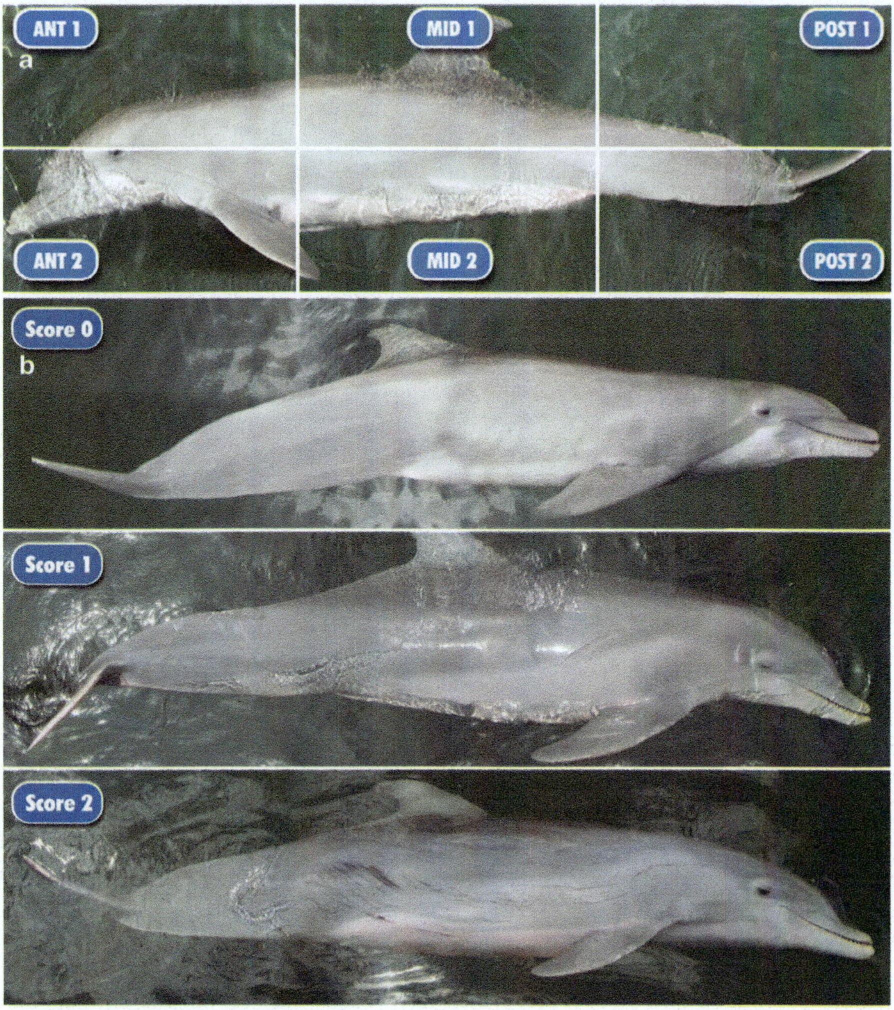

Fig. 12.2 Photographs showing the process of quantifying wounds and rake marks for use in the C-Well® assessment for bottlenose dolphins (*Tursiops truncatus*); descriptions taken from Clegg et al. (2015). (**a**) The grid indicates the six body regions utilised in the calculation, including *ANT1* anterior dorsal, *ANT2* anterior ventral, *MD1* mid-dorsal, *MID2* mid-ventral, *POST1* posterior dorsal and *POST2* posterior ventral; (**b**) the wound percent cover range and relevant welfare score for the dolphin in each image. Score 0, <15% new wounds and <30% old wounds; score 1, <10–15% new wounds and >30% old wounds; and score 2, >15% new wounds

used in the analysis of wounds and rake marks (estimates of old and new marks are made within further subdivisions of the grid) and the corresponding welfare scores (demonstrated by examples). Given that space is inherently limited in captivity, the link between aggression, rake marks and social stress may be different than that in wild contexts (Scott et al. 2005), but repeatedly quantifying rake marks over time and in conjunction with aggressive behaviour frequency and social isolation would, it is assumed, reveal the appropriate thresholds linked to welfare.

It is important to briefly discuss the role of 'natural' behaviour in welfare assessment, which is usually defined as the repertoire seen in free-ranging animals in the wild. Although a traditional view is that an animal not seen to be performing natural behaviours has poor welfare, as many authors have pointed out, absence of such behaviour should not be a welfare measure per se, since wild and captive animals do not experience the same environmental stimuli (e.g. predation, lack of food, environmental opportunities) which cause the performance of these behaviours (Dawkins 1980; Veasey 2006; Webster 2005). However, research is needed into whether there are natural behaviours in cetaceans which represent 'behavioural needs', i.e. those which are stimulated even in absence of the associated stimuli, and preference and motivation tests could perhaps answer these questions (Veasey 2006). An alternative would be to use outcome-based measures related to the richness of the environment and to measure the diversity of behaviours (Carlstead and Shepherdson 2000; Swaisgood 2007)—measures which would indicate whether the surroundings are stimulating for the animal.

12.7.3 Health-Related Measures of Welfare

In his triangulation model of animal welfare, Webster (2005) originally proposed physiological welfare measures as one of the key areas to be measured, due to their potential for assessing emotional responses (Boissy et al. 2007), but in this current chapter, we discuss health and associated physiological measures of dolphin welfare, since many previous studies have considered health as an important aspect of an animal's overall welfare (e.g. Dawkins 2006; Hill and Broom 2009; Mason and Veasey 2010; Held and Špinka 2011). Furthermore, health measures may have increased species-specific relevance to cetaceans, since most species appear to mask or conceal symptoms of pain and disease as much as possible (Castellote and Fossa 2006; Waples and Gales 2002). For this reason, the use of health measures may reveal potential welfare problems which have been 'masked' by the animals and may have the potential to detect problems before changes are seen in behavioural parameters.

The calculation of Body Condition Scores (BCS) employs measurements of fat cover on animals' bodies and is often used in farm animal welfare assessments (Welfare Quality® 2009a, b, c; Roche et al. 2009). In the past, BCS of cetaceans has been conducted in a few studies using descriptive scales and without reference to photographic scales or use of standardised graphics (e.g. North Atlantic right whales

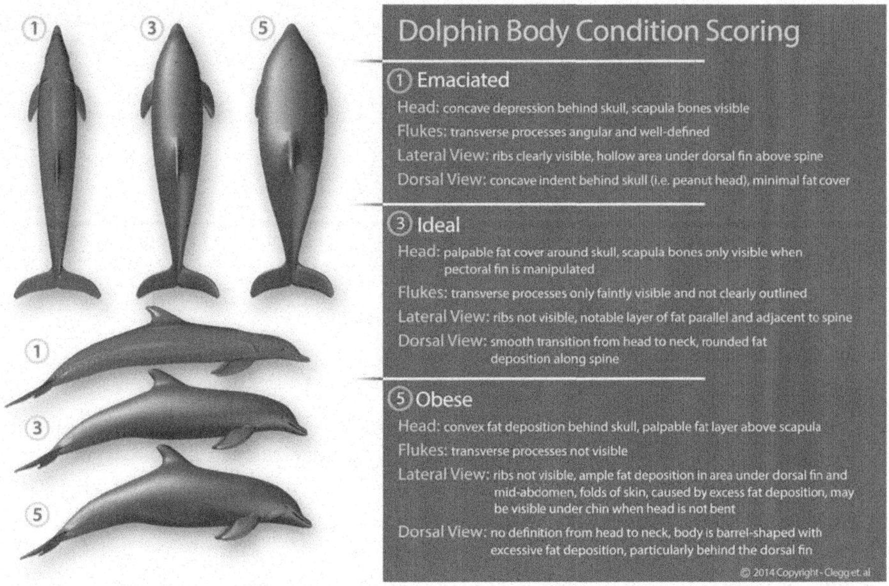

Fig. 12.3 Body condition score graphic for bottlenose dolphins (*Tursiops truncatus*) developed by Clegg et al. (2015)

(*Eubalaena glacialis*), Pettis et al. 2004). However, recently, the first two standardised protocols for BCS were published for short-beaked common dolphins (*Delphinus delphis*) (Joblon et al. 2014) and bottlenose dolphins (Clegg et al. 2015, Fig. 12.3). Further investigation into the degree of change in BCS that correlates to changes in welfare is needed, similar to research with cows clarifying the link (Roche et al. 2009). The link between other health measures and welfare state in cetaceans similarly needs to be clarified before their use in assessments. One approach would be to consider that only departures from health that cause a change in affective state, i.e. how the animal feels, can be considered as sufficient to cause poor welfare, in line with 'feelings-based' welfare definitions of authors including Fraser et al. (1997), Spruijt et al. (2001) and Mason and Veasey (2010). When an animal performs 'sickness behaviours', which in cetaceans may appear as lethargy, logging (remaining stationary at the surface), deliberate social isolation or inappetence (Joseph et al. 1986; Sweeney and Ridgway 1975), the illness or disease can be related to a decrease in welfare. Haematological indices can be used to measure some parameters of disease, and are used in human and animal health studies and in clinical diagnosis as 'markers' of disease progression. However, haematological measures are difficult to use in welfare evaluations due to the invasive requirement to collect blood samples, which is likely to interfere with unimpeded behaviour assessment, and also as a result of high inter- and intra-individual variation. Nevertheless, baseline and disease haematological profiles for both wild and captive cetaceans continue to be published (Dierauf and Gulland 2001; Thomson and Geraci 1986; Wells 2009), and if associated with accompanying behavioural data, it is possible that blood sampling could

Fig. 12.4 Photographic reference scale for assessment of squinting in bottlenose dolphins (*Tursiops truncatus*), as part of a welfare-related measure of eye disease developed by Clegg et al. (2015)

contribute to welfare information in some specific circumstances—perhaps where animals have had blood collected for clinical diagnostic purposes.

In farm animal studies, welfare measures related to health include presence of diarrhoea, skin inflammation, skin lesions, respiratory rate, lameness, lying time, eye conditions and coughing (Welfare Quality® 2009a, b, c). Clegg et al. (2015) adapted some of these measures for use in *T. truncatus* welfare assessment (skin and eye condition, coughing). Standardised measurement of eye condition and squinting in captive cetaceans (Clegg et al. 2015, Fig. 12.4) in conjunction with water quality parameters (resource-based measures) could confirm whether a relationship between the outcome (health measure) and the input (water quality) exists; this finding could be of real value in steps towards the prevention of disease and the improvement of welfare conditions (Joseph and Antrim 2010).

In welfare studies of other species, cortisol and corticosterone have been used as physiological indicators of stress and affective state (Webster 2005). Serum cortisol levels are often measured in cetaceans as part of blood analyses (Suzuki et al. 2003), and measuring cortisol from saliva or 'blow' (expired air) samples is now also being investigated, with the obvious advantage that the potential confounding factor of stress during sampling is diminished (Ugaz et al. 2013). These sampling techniques are still in the early stages of validation, with a major problem being water contamination, but varying collection protocols and analytical techniques holds much promise for combatting this issue (Atkinson et al. 2015). However, a recent review (Atkinson et al. 2015) warned that terrestrial animal stress models might not always be applicable to marine mammals, and these authors presented evidence to suggest that neuroendocrine hormones may be regulated very differently in cetaceans and called for increased research effort on this topic.

12.7.4 Cognitive Measures of Welfare

Cognitive measures of welfare are the least well-established category within welfare assessments, perhaps due to the high degree of experimental control and involvement required to conduct cognition studies (Mendl and Paul 2004; Paul et al. 2005; Mendl et al. 2009). Cognitive measures aim to show how behavioural and physiological responses to environmental stimuli are situated within cognitive processes, which are known to regulate, as well as be regulated by, emotions (Désiré et al. 2002; Paul et al. 2005; Boissy et al. 2007). The study of these phenomena appeared to experience a breakthrough with successful methodologies being established to assess judgement bias in animals, and Mendl et al. (2009) review the results from species tested using these methods to date. On the whole, animals experiencing environmentally induced reductions in welfare are more likely to judge ambiguous stimuli negatively, whereas animals in good or enriched environments tend to judge more optimistically (Mendl et al. 2009). The protocols used in these studies are generally physically non-invasive, but do require the animal to be trained (for possibilities of training see a review on the progress with marine mammals in captivity, Brando 2010), and so this area of cognition and welfare research does hold potential for cetaceans. Apart from measuring welfare, cognitive tests of these kinds may further our knowledge of animal consciousness, which represents only a minor aspect of most current welfare research (Mendl and Paul 2004; Herman 2012).

Anticipatory behaviour preceding food consumption is closely related to dopaminergic activity and has been shown to increase the positive value of the following consummatory (consumption) event (Spruijt et al. 2001; Boissy et al. 2007). Anticipatory behaviour reflects the intensity of 'wanting' the food, which is separate to the present-moment pleasure taken in consuming the food, and this phenomenon (anticipation of consumption) been proposed as a measure of welfare (Boissy et al. 2007), and may have promise for zoo animal assessment (Watters 2014). An initial study on anticipatory behaviour in bottlenose dolphins found that they display significant anticipatory behaviour before feeding sessions (Jensen et al. 2013), and a recent study of false killer whales (*Pseudorca crassidens*) indicated that certain vocalisations increased in conjunction with anticipatory behaviours (Platto et al. 2015).

12.7.5 The Human–Animal Relationship (HAR) and Welfare

One understudied aspect of cetaceans' lives in captivity merits discussion here in terms of potential positive and negative welfare impacts and the potential for welfare measures: the Human–Animal Relationship (HAR). We are just starting to understand how much the HAR affects animals: farm species studies show clearly that a positive HAR is likely to increase welfare and a negative HAR decreases it (see review by Waiblinger et al. 2006), and this seems also to be the case with zoo animals (Hosey 2008; Whitham and Wielebnowski 2013). A positive HAR between owners and dogs has been shown to invoke oxytocin release in both parties and forms part of a feedback loop which may have facilitated the co-evolution of human–dog bonds (Nagasawa et al. 2015).

12.7.5.1 Potential for Negative and Positive Welfare

Cetaceans under human care are fed their daily food intake in multiple sessions of operant conditioning (Brando 2010; Jensen et al. 2013), and therefore a routine part of their day is spent with those who conduct the sessions: the animal care staff/trainers. This HAR may be salient for the animals, in terms of the extended time the humans and animals spend in close contact, and the fact that captive cetaceans are considered undomesticated (Brando 2012). There is likely to be a strong association between type of training and HAR (Brando 2010). The majority of cetacean facilities use positive reinforcement training to teach (human) desired behaviours (Brando 2012; Jensen et al. 2013), and the technique of positive reinforcement training pioneered in cetacean facilities has now spread to other animal industries (Laule et al. 2003). This type of training may be conducive to good welfare (and/or to reductions in adverse welfare impacts) through establishment of a positive bond between the trainer and animal, providing a stimulating and safe environment, and ensuring voluntary cooperation from the animal (Laule et al. 2003; Brando 2012). There is evidence that training sessions, especially those using positive reinforcement techniques, are also cognitively enriching (e.g. with other animals: Laule et al. 2003; Bassett and Buchanan-Smith 2007; Dorey et al. 2015; cetaceans: Brando 2012; Clark 2013). In the only study of its kind, Perelberg and Schuster (2009) showed that outside of feeding sessions, their group of bottlenose dolphins was willing to perform behaviours asked by the trainers, where the only reinforcement was human applause and furthermore that they approached humans voluntarily to receive rubs and petting. Cetacean species are known to be very tactile in their interactions with each other (Dudzinski 2010; Kuczaj et al. 2013), and thus this might be a significant aspect of HARs they form in captivity. Some studies investigating dolphins' behavioural repertoires in response to guest–dolphin interactions found an increase in play and behavioural diversity following the sessions, which the authors tentatively took to mean that the animals did not view the interaction negatively (Trone et al. 2005; Miller et al. 2011). Schmitt et al. (2010) found that stress hormones in belugas showed no change between levels before or after wade-contact guest interactions or from baseline levels.

However, there are some captive facilities worldwide that continue to use negative reinforcement or positive punishment training (see Brando 2012 for definitions), and here we discuss the potential impacts on welfare. Other types of training are centred on forced cooperation, and although there are no published data for cetaceans, we may assume that the invoked feelings of fear, frustration and aggression might be similar to those of farm and laboratory animals who interact with humans in this way (Waiblinger et al. 2006; Bassett and Buchanan-Smith 2007; Brando 2012). Inappetence in captive cetaceans may be an extreme consequence of a poor HAR or disinterest in training sessions and is a strong indicator of poor welfare for cetaceans (although it has many other causal factors including health and social reasons, Waples and Gales 2002). The presence of unfamiliar humans in the cetaceans' environment is an additional type of HAR that may influence welfare: Hosey

(2008) provides a useful matrix and model developed for zoo animals describing how various HARs with familiar and unfamiliar humans can coexist (Hosey 2008). Studies focussing on cetaceans' responses to unfamiliar humans have found mixed results, where some found negative behaviours such as avoidance and aggression to increase during training sessions (Samuels and Spradlin 1995; Kyngdon et al. 2003; for positive behaviours see Trone et al. 2005; Miller et al. 2011).

12.7.5.2 Human–Animal Relationship-Based Welfare Measures

Increased research into measuring cetacean HARs in captivity would help to clarify mixed findings from past studies and provide insight about the significance of HAR to welfare, results of which may also translate to wild contexts, such as tourists swimming with cetaceans. Clegg et al. (2015) proposed a simple approach-avoidance test, adapted from farm animal assessments, for measuring the HAR of bottlenose dolphins where the response to a trainer (outside of training sessions) was tested. This test and the study protocol described by Perelberg and Schuster (2009) are types of preference tests (similar to Dorey et al. (2015) with *Canis lupus* spp.) and could be developed further to include measures of motivation to interact with humans. Anticipatory behaviour before sessions could give insights to cetaceans' perception of HARs, but studies would need to disentangle this from the desire for food acquisition. Future studies might develop ethograms of cetacean approach, avoidance and frustration behaviours in HAR contexts (Waiblinger et al. 2006), and potential confounding factors could be controlled, for example, interspecific social behaviour during sessions, behaviour of guests and type of guest interaction may influence HAR, as well as the duration and frequency of sessions. Standardised, detailed recordings of appetence levels and accompanying behaviours during training sessions could be conducted at all facilities, since inappetence is a potential measure of a poor HAR, social stress or disease: all strong negative welfare indicators. Whitham and Wielebnowski (2009) review how keeper ratings of animal behaviour and well-being may be used as assessment tools, and since most cetacean facilities already take these kind of records on a daily basis, efforts could be made to standardise and utilise this kind of routinely collected data.

12.8 Animal-Based Welfare Measures for Cetaceans in the Wild

In this section, we refer to Fig. 12.1 as a guide when discussing potential outcome-based measures of welfare for wild cetaceans. However, identifying specific inputs and outcomes in the wild setting is more difficult since the quality of observations, the number of repeated observations per individual, the use of physiological samples and human interaction with the animals are some of the variables in practice, and controlling for these variables is unlikely to be fully feasible in many situations (Dudzinski 2010). Long-term studies generally collect individual behavioural and life history data, with physiological and genetics parameters if possible (Wells 2009), and thus should be a focal point for past studies and future wild cetacean welfare research.

12.8.1 Behavioural, Health-Related and Cognitive Measures of Welfare

Behavioural measures of cetacean welfare are, in principle, non-invasive, feasible in several species and less 'expensive' to conduct (as they require little in the way of equipment or laboratory diagnostic tests, although are fairly labour intensive): thus, they have the potential to become important in evaluating the welfare of wild cetaceans. As with their captive counterparts, social behaviour is likely to provide indicators of welfare state, and social network analysis has been well used in studies to evaluate the bonds within cetacean groups (Lusseau 2003). Stanton and Mann (2012) showed how the development of early social networks can predict survivability, where fewer social bonds in early life resulted in a decrease in fitness in bottlenose dolphins, suggesting that the number and quality of social bonds an individual has may be linked to longer-term welfare. Other, shorter-term measures of social behaviour described earlier in Sect. 12.7.2 could be applicable for measuring positive and negative welfare of wild cetaceans (e.g. contact swimming, rubbing, aggressive behaviours), and similar validation protocols could be used to correlate these behaviours with other indices. The rake mark quantification protocol developed by Clegg et al. (2015), based on the fact that rake mark cover can be used as a proxy for aggression levels (Scott et al. 2005; Orbach et al. 2015), could be applied as a welfare measure for wild cetaceans. Since many odontocete species maintain social hierarchies which are partially sustained through play and aggression, and in which conspecifics are 'raked' (MacLeod 1998; Visser 1998), a high level of rake marks could indicate those animals experiencing social stress or a reduction in fitness (Waples and Gales 2002; Orbach et al. 2015; see Sect. 12.7.2 for potential link with welfare), and the quantification method proposed for captive dolphins by Clegg et al. (2015) (Fig. 12.2) could be applicable to wild dolphins.

Epidemiological parameters have potential as health-related welfare measures for wild cetaceans, since they are likely to represent the outcome of chronic, long-term and population-driven welfare states. However, such parameters are likely to have only limited feasibility as shorter-term assessments for individuals. Population measures such as longevity and reproductive rate would be most useful for welfare assessments if they were used in conjunction with other data (Swaisgood 2007; Barber 2009): examples from the farm animal literature show that assessment of reproductive success alone can be very misleading if used alone to imply good welfare (Dawkins 1980). A recent study (Christiansen and

Lusseau 2015) examined body condition, behaviour and foetal growth rate in minke whales, *Balaenoptera acutorostrata*, and used these population measures to assess the impact of whale watching, and although welfare was not expressly discussed, their conclusions provide support for assessing the long-term affective states of wild cetaceans. Data on individual or population-level disease parameters, already being collected in wild cetacean health assessments (e.g. Reif et al. 2008; Schwacke et al. 2014), could be correlated to behaviours in future studies to identify sickness behaviours and poor welfare indicators.

Pack (2010) discusses the progress of cognition research in wild cetaceans and emphasises how in the future collaborations between wild and captive fields will maximise our understanding and the value of the conclusions attributable to the data, as has been the case with cognition studies in non-cetacean species. Cognitive welfare measures for wild cetaceans might seem to be unlikely in terms of practicality, and certainly controlled, experimental settings are not as readily possible as they are in captivity, but nevertheless these are promising areas which merit further research. One of these is the occurrence of lateralised behaviours, a result of the differential processing of information by the two hemispheres of the brain (Rogers 2010; Leliveld et al. 2013). Thus far, studies in other species indicate that when stressed, the right hemisphere may be used preferentially (Rogers 2010). Leliveld et al. (2013) suggest that the right hemisphere processes negative emotions, including anxiety and fight or flight responses, and the left deals with positive emotions. Although the link between lateralised behaviours and welfare is not yet clear (Rogers 2010), this phenomenon could be a source relevant information about affective states and furthermore is relatively easy to measure in wild cetaceans: examples can be found in humpback whales (*Megaptera novaeangliae*), grey whales (*Eschrichtius robustus*), belugas, killer whales and bottlenose dolphins during various behaviours (Kasuya and Rice 1970; Silber and Fertl 1995; Clapham et al. 1995; Sakai et al. 2006; Karenina et al. 2010, 2013). In the research with the closest links to affective states, Karenina et al. (2010, 2013) found that during nonthreatening situations, belugas and killer whales (*Orcinus orca*) positioned calves on their right side, with killer whales favouring the left side as the situation became increasingly threatening, and Sakai et al. (2006) found that in affiliative flipper rubbing behaviour, the left pectoral fin and eye were used more by *Tursiops aduncus*. In this arena of research, there are indications that laterality in certain behaviours may be linked to affective state and welfare, but further work is needed to be able to unravel the potential environmental and evolutionary causal factors.

12.8.2 Welfare Measures Related to Human Interactions

There is a need to monitor the welfare of wild cetaceans specifically as result of human interactions or their related activities (Butterworth et al. 2012). We have already reviewed some of the studies that have investigated the disturbance or pain inflicted on cetaceans by humans (e.g. Cassoff et al. 2011; Butterworth et al. 2013), but in this section of the chapter, we emphasise the need for researchers to tackle the problem of assessing the welfare of animals affected by acute human interactions. Welfare is an all-encompassing, multidimensional aspect and thus incorporates many different areas of science (Dawkins 2006), and, although human-impact studies on cetacean do examine some of these elements separately, bringing them all together in a 'holistic' or integrated way could aid in understanding the effects of the issue in real time (Butterworth et al. 2012). Ohl and van der Staay (2012) explain how welfare assessments of wild animals need to be more dynamic and flexible to really answer questions about wild animal welfare and that the adaptive value of welfare states must be considered. We discuss a few examples below: situations where welfare frameworks could aid in the management and conservation efforts related to human-inflicted welfare issues.

There are a small number of studies which assess welfare as a by-product of their main questions of interest. Christiansen and Lusseau (2015) correlated health-, behavioural- and population-level welfare measures to assess non-lethal impacts of whale watching. Similar approaches have been seen in the research conducted to assess the effects of the provisioning (daily feeding) of a wild population of bottlenose dolphins in Australia. In these studies, behaviour and population parameters were combined with long-term ontogenetic data, to suggest that reproductive rate is lowered in provisioned females and that mother–calf behaviour is significantly different (Mann et al. 2000; Mann and Kemps 2003). In situations such as entanglement, tourists swimming with wild cetaceans, and anthropogenic noise, measures of poor welfare could be identified, and behavioural indicators of frustration, distress and fear might be used. Collaborations between captive and wild animal researchers may help to reveal potential welfare indicators for some species, with an example being Dudzinski et al.'s (2012) study on pectoral fin rubbing frequency—applicable in both wild and captive settings, although the focus in the study as presented was exploration of function of the behaviour, as opposed to use of the behaviour as a measure of welfare.

12.8.3 Welfare Evaluations During Strandings

Strandings may be caused by humans' agency, or by other factors, and when humans try and rehabilitate stranded animals, welfare questions are inevitably raised. With single strandings, the first question usually concerns the animal's chances of survival and whether or not efforts to treat it are futile (Butterworth et al. 2004). In this case, monitoring indicators of system functioning, alertness and vital signs are crucial for decisions made concerning the animal's welfare, and some of this data has been collected during past stranding events (e.g. Greenwood and Taylor 1980). Butterworth et al. (2004) examined 12 indicators of sensibility and 6 of vitality in multiple captive cetacean species, to be used as the basis for assessment of viability in stranding situations. Measuring hearing capabilities in stranded cetaceans has been recently advised to assess suitability for rehabilitation and discover the potential reason for stranding (André et al. 2007). In mass stranding events, resources are limited, and care is often allocated after a triage process. In the triage assessment, rapid analysis of each animal's 'need versus probability of success' is made. Proficiency in conducting physiological, behavioural and anatomical animal-based measures, by the multiple team members present, would potentially be useful and may save animals' lives. Other welfare indicators such as body condition, wound severity or rake mark cover could provide information about the health state prior to stranding, which may also indicate the likelihood of survival post-refloating or release and, also potentially, information on the cause of stranding (Joblon et al. 2014). Some standardised protocols for these measures have been proposed (Joblon et al. 2014; Clegg et al. 2015) and could be adapted for a wider range of species. When applied in stranding situations, welfare assessment tools may aid in ethical decision-making, helping to ensure that resources are used efficiently and animal suffering is minimalised.

12.9 Recommendations for Developing Measures of Cetacean Welfare

Research towards developing cetacean welfare measures must, because of their nature, be conducted in situ (Dawkins 2006), i.e. in the facilities or environments inhabited by the animals, and must look for measures and approaches specific to the species in question (Botreau et al. 2007; Hill and Broom 2009). The layout and organisation of zoological institutions mean they might already be well prepared to conduct in situ assessments of the animals in the captive environment. The animals are often visible and identifiable in their enclosures, there are multiple daily interactions with the keepers, and there are systems of individualised care and record-keeping (Barber 2009). Underwater windows in captive cetacean facilities are useful for behavioural observations (Dudzinski 2010), and therefore welfare assessment and thus extra considerations must be made for facilities without windows. Behavioural monitoring of captive cetaceans will likely be of utmost importance in welfare evaluations and should be conducted regularly and thoroughly (Waples and Gales 2002; Clegg et al. 2015). In the wild, conditions are more difficult for measuring welfare, but long-term studies of cetacean populations could more readily answer some welfare-focussed questions. Such studies of wild populations often have access to individual animals' life history, genetics, past behaviour, photographic records and, sometimes, physiological records as well (Wells 2009). Inclusion of welfare assessment protocols as part of ongoing research in well-studied wild populations might form the best starting point for investigation of welfare indicators in wild cetaceans.

In the previous sections, we have suggested research areas for potential behavioural, health-related and cognitive welfare measures, and we now give our final considerations to how best to validate measures once they have been identified. This is a challenging process, since when researching welfare in a relatively understudied arena there are not any validated measures to correlate with newly proposed measures (Boissy et al. 2007). Using lessons drawn from farm animal welfare research, in certain contexts, it is possible to explore the effects of perceived increases or decreases in welfare, which occur 'naturally' within the day-to-day environment of the animals (Désiré et al. 2002). Farm animal research has created a specifically focussed set of welfare measures for use during farm animal transport (e.g. Bradshaw et al. 1996), and this approach could also be considered for cetacean studies. Castellote and Fossa (2006) used transport of cetaceans between facilities, which occurs for breeding purposes or social group changes, to study vocal activity as a potential welfare measure and also to examine other potential measures of welfare status, including salivary cortisol and respiratory rate. Other welfare-altering and impacting situations which captive cetaceans may experience include medical examinations (Schmitt et al. 2010), mixing new groups together (Waples and Gales 2002) and situations where the animal is showing inappetence (Waples and Gales 2002).

Conversely, it may also be possible to conduct measurements in contexts where welfare is likely increased, for example, during feeding events (Platto et al. 2015), times of enrichment provision (Boissy et al. 2007; Clark 2013) and tactile interactions with humans, if these can be further confirmed to offer positive welfare potential (Perelberg and Schuster 2009). As mentioned in earlier sections of this chapter, research could be conducted into cognitive measures such as tests of judgement bias, to determine whether the biases in dolphins vary with welfare state, as has been demonstrated in other species. If this proves to be the case, then cognitive assessment methods may have strong potential to validate other measures taken in conjunction and could be utilised in the validation of a range of measures (Mendl et al. 2009). Contexts for validating wild cetacean welfare measures could be anthropogenic disturbances (e.g. Butterworth et al. 2012; Christiansen and Lusseau 2015), periods of social stress (e.g. hierarchy change; see review in Waples and Gales (2002)) and situations of environmental change. Valid welfare measures from either the wild or captivity could, and perhaps should, be applied during collaborative studies between the two settings, and the potential synergy may strengthen the meaning of outcome measures of welfare (Dudzinski 2010; Pack 2010).

Another approach to validation perhaps only relevant to captivity is to apply a range of non-validated assessments at different animal establishments and to investigate whether the results correlate to other welfare-related information about the facility (e.g. mortality rates, reproductive rates, government resource-based assessments, rate of serious human/animal incidences). This approach is starting to occur with farm animals (e.g. Temple et al. 2011) and would be feasible for cetaceans using assessment systems such as that described by Clegg et al. (2015).

Cetacean welfare assessments, especially in captivity, have potential to benefit the animals themselves if they can highlight situations and practices correlated to positive welfare states (Clegg et al. 2015) and can help to promote change to reduce negative impacts. As has been suggested with other species kept in zoos (Veasey 2006; Whitham and Wielebnowski 2013), objective data on cetaceans' welfare in captivity could result in improved future regulations, or in some cases, prohibiting of some practices or management methods, and might help the public to make informed consumer decisions about whether (or not) to visit and support captive cetaceans. Wild cetaceans are facing increasingly significant anthropogenic threats, and validated measures of cetacean welfare may aid in revealing poor welfare of individual animals (e.g. entanglement, Cassoff et al. 2011), or threats of poor welfare to populations, and potentially stimulate public support for conservation efforts. Since welfare encompasses many different areas of science, cetacean welfare research based on objective, animal-based assessment methods could advance our knowledge and understanding of these animals in both the wild and captivity.

12.10 Conclusions

Animal welfare science can provide tools and frameworks which can aid in the assessment of cetacean welfare. There are only a few studies that have considered measuring cetacean welfare in captivity, and the concept is not yet discussed for those in the wild. Combining behavioural, health-related and cognitive measures of welfare is likely to be the best way to reveal valid indicators, and we may find that social behaviour, anticipatory behaviour, cortisol and cognitive bias experiments could yield the first welfare measures of captive cetaceans. Social behaviour, assessment of rake marks, population parameters and visual laterality have potential for use as wild animal indicators. Collaborations between wild and captive researchers would increase the chances of identifying welfare indicators that are meaningful and considered by a wide group of stakeholders to be valid. The next logical steps in this progressive process are that, after identifying an initial group of potential cetacean welfare indicators, they are validated through practical application, and the correlations between measures explored. The implications of an agreed and established set of measures of cetacean welfare would be widespread, with the potential for direct benefits to the animals themselves, more accurate information for the public, greater support for conservation and the reviewing and improvement of regulations.

Acknowledgements We would like to thank Fabienne Delfour and Niels van Elk for their stimulating discussions and input related to this subject.

References

André M, Delory E, Degollada E, Alonso JM (2007) Identifying cetacean hearing impairment at stranding sites. Aquat Mamm 33:100

Atkinson S, Crocker D, Houser D, Mashburn K (2015) Stress physiology in marine mammals: how well do they fit the terrestrial model? J Comp Physiol B 185:463–486

AWA (1966) Animal Welfare Act. Public Law 89–544. In: 89th congress, H. R. 13881, USDA, Washington, DC, 24 August 1966

Barber JC (2009) Programmatic approaches to assessing and improving animal welfare in zoos and aquariums. Zoo Biol 28:519–530

Bassett L, Buchanan-Smith HM (2007) Effects of predictability on the welfare of captive animals. Appl Anim Behav Sci 102:223–245

Bejder L, Samuels AMY, Whitehead HAL, Gales N, Mann J, Connor R, Heithaus M, Watson-Capps JANA, Flaherty C, Kruetzen M (2006) Decline in relative abundance of bottlenose dolphins exposed to long-term disturbance. Conserv Biol 20:1791–1798

Blokhuis HJ (2008) International cooperation in animal welfare: the Welfare Quality® project. Acta Vet Scand 50:S10

Boissy A, Manteuffel G, Jensen MB, Moe RO, Spruijt B, Keeling LJ, Winckler C, Forkman B, Dimitrov I, Langbein J, Bakken M (2007) Assessment of positive emotions in animals to improve their welfare. Physiol Behav 92:375–397

Botreau R, Veissier I, Butterworth A, Bracke MBM, Keeling LJ (2007) Definition of criteria for overall assessment of animal welfare. Anim Welf 16:225

Bradshaw RH, Parrott RF, Goode JA, Lloyd DM, Rodway RG, Broom DM (1996) Behavioural and hormonal responses of pigs during transport: effect of mixing and duration of journey. Anim Sci 62:547–554

Brando SICA (2010) Advances in husbandry training in marine mammal care programs. Int J Comp Psychol 23:777–791

Brando SICA (2012) Animal learning and training: implications for animal welfare. Vet Clin North Am Exot Anim Pract 15:387–398

Butterworth A, Kestin SC, McBain JF (2004) Evaluation of baseline indices of sensibility in captive cetaceans. Vet Rec 155:513–518

Butterworth A, Clegg I, Bass C (2012) Untangled–marine debris: a global picture of the impact on animal welfare and of animal-focused solutions. World Society for the Protection of Animals International, London

Butterworth A, Brakes P, Vail CS, Reiss D (2013) A veterinary and behavioral analysis of dolphin killing methods currently used in the "drive hunt" in Taiji, Japan. J Appl Anim Welf Sci 16:184–204

Carlstead K, Shepherdson D (2000) Alleviating stress in zoo animals with environmental enrichment. In: Moberg GP, Mench JA (eds) The biology of animal stress: basic principles and implications for animal welfare. CABI Publishing, Wallingford, pp 337–354

Cassoff RM, Moore KM, McLellan WA, Barco SG, Rotstein DS, Moore MJ (2011) Lethal entanglement in baleen whales. Dis Aquat Org 96:175–185

Castellote M, Fossa F (2006) Measuring acoustic activity as a method to evaluate welfare in captive beluga whales (*Delphinapterus leucas*). Aquat Mamm 32:325

Christiansen F, Lusseau D (2014) Understanding the ecological effects of whale-watching on cetaceans. In: Whale watching sustainable tourism and ecological management. Cambridge University Press, Cambridge, pp 177–192

Christiansen F, Lusseau D (2015) Linking behavior to vital rates to measure the effects of non-lethal disturbance on wildlife. Conserv Lett 8:424–431

Clapham PJ, Leimkuhler E, Gray BK (1995) Do humpback whales exhibit lateralized behaviour? Anim Behav 50:73–82

Clark FE (2013) Marine mammal cognition and captive care: a proposal for cognitive enrichment in zoos and aquariums. J Zoo Aquar Res 1:1–6

Clegg ILK, Borger-Turner JL, Eskelinen HC (2015) C-Well: the development of a welfare assessment index for captive bottlenose dolphins (*Tursiops truncatus*). Anim Welf 24:267–282

Council Directive (1999) Council Directive 1999/22/EC of 29th March 1999 relating to the keeping of wild animals in zoos. http://eur-lex.europa.eu/legal-content/EN/TXT/?uri=celex:31999L0022. Accessed 21 March 2016

Dawkins MS (1980) Animal suffering: the science of animal welfare. Chapman and Hall, London

Dawkins MS (2006) A user's guide to animal welfare science. Trends Ecol Evol 21:77–82
Delfour F, Beyer H (2012) Assessing the effectiveness of environmental enrichment in bottlenose dolphins (*Tursiops truncatus*). Zoo Biol 31:137–150
Désiré L, Boissy A, Veissier I (2002) Emotions in farm animals: a new approach to animal welfare in applied ethology. Behav Process 60:165–180
Dierauf L, Gulland FM (eds) (2001) CRC handbook of marine mammal medicine, 2nd edn. CRC Press, Boca Raton
Dorey NR, Mehrkam LR, Tacey J (2015) A method to assess relative preference for training and environmental enrichment in captive wolves (*Canis lupus* and *Canis lupus arctos*). Zoo Biol 34:513–517
Dudzinski KM (2010) Overlap between information gained from complementary and comparative studies of captive and wild dolphins. Int J Comp Psychol 23:566–586
Dudzinski KM, Gregg J, Melillo-Sweeting K, Seay B, Levengood A, Kuczaj SA II (2012) Tactile contact exchanges between dolphins: self-rubbing versus inter-individual contact in three species from three geographies. Int J Comp Psychol 25:21–43
Eskelinen HC, Winship KA, Jones BL, Ames AE, Kuczaj II SA (2016). Acoustic behavior associated with cooperative task success in bottlenose dolphins (*Tursiops truncatus*). Animal Cognition. doi:10.1007/s10071-016-0978-1. Published online 29 March 2016
Farm Animal Welfare Council, FAWC (1992) FAWC updates the five freedoms. Vet Rec 17:357
Fraser D, Duncan IJ (1998) 'Pleasures', 'pains' and animal welfare: toward a natural history of affect. Anim Welf 7:383–396
Fraser D, Weary DM, Pajor EA, Milligan BN (1997) A scientific conception of animal welfare that reflects ethical concerns. Anim Welf 6:187–205
Frohoff TG, Packard JM (1995) Human interactions with free-ranging and captive bottlenose dolphins. Anthrozoös 8:44–53
Gales N, Leaper R, Papastavrou V (2008) Is Japan's whaling humane? Mar Policy 32:408–412
Galhardo L, Appleby MC, Waran NK, Dos Santos ME (1996) Spontaneous activities of captive performing bottlenose dolphins (*Tursiops truncatus*). Anim Welf 5:373–389
Greenwood AG, Taylor DC (1980) Humane handling of stranded cetaceans. Vet Rec 106:345–345
Grimm D (2011) Are dolphins too smart for captivity? Science 332:526–529
Gygax L (1993) Spatial movement patterns and behaviour of two captive bottlenose dolphins (*Tursiops truncatus*): absence of stereotyped behaviour or lack of definition? Appl Anim Behav Sci 38:337–344
Held SD, Špinka M (2011) Animal play and animal welfare. Anim Behav 81:891–899
Herman LM (2012) Body and self in dolphins. Conscious Cogn 21:526–545
Hill SP, Broom DM (2009) Measuring zoo animal welfare: theory and practice. Zoo Biol 28:531–544
Hosey G (2008) A preliminary model of human–animal relationships in the zoo. Appl Anim Behav Sci 109:105–127
Hoy JM, Murray PJ, Tribe A (2010) Thirty years later: enrichment practices for captive mammals. Zoo Biol 29:303–316
Jensen A-L, Delfour F, Carter T (2013) Anticipatory behaviour in captive bottlenose dolphins (*Tursiops truncatus*)—a preliminary study. Zoo Biol 32:436–444
Jett J, Ventre J (2012) Orca (*Orcinus orca*) captivity and vulnerability to mosquito-transmitted viruses. J Mar Anim Ecol 5:9–16
Jett J, Ventre J (2015) Captive killer whale (*Orcinus orca*) survival. Mar Mamm Sci 31:1362–1377
Joblon MJ, Pokras MA, Morse B, Harry CT, Rose KS, Sharp SM, Niemeyer ME, Patchett KM, Sharp WB, Moore MJ (2014) Body condition scoring system for delphinids based on short-beaked common dolphins (*Delphinus delphis*). J Mar Anim Ecol 7:5–13

Joseph B, Antrim J (2010) Special considerations for the maintenance of marine mammals in captivity. In: Kleiman DG, Thompson KV, Kirk Baer C (eds) Wild mammals in captivity: principles and techniques for zoo management, 2nd edn. The University of Chicago Press, Chicago, pp 181–216

Joseph BE, Cornell LH, Simpson JG, Migaki G, Griner L (1986) Pulmonary aspergillosis in three species of dolphin. Zoo Biol 5:301–308

Karenina K, Giljov A, Baranov V, Osipova L, Krasnova V (2010) Visual laterality of calf–mother interactions in wild whales. PLoS One 5:e13787. doi:10.1371/journal.pone.0013787

Karenina K, Giljov A, Ivkovich T, Burdin A, Malashichev Y (2013) Lateralization of spatial relationships between wild mother and infant orcas, *Orcinus orca*. Anim Behav 86:1225–1231

Kasuya T, Rice DW (1970) Notes on baleen plates and on arrangement of parasitic barnacles of gray whales. Sci Rep Whales Res Inst 22:39–43

Kestin SC (1995) Welfare aspects of the commercial slaughter of whales. Anim Welf 4:11–27

Kirkwood JK, Sainsbury AW, Bennett PM (1994) The welfare of free-living wild animals: methods of assessment. Anim Welf 3:257–273

Kuczaj SA, Eskelinen HC (2014) Why do dolphins play. Anim Behav Cogn 1:113–127

Kuczaj SA, Highfill LE, Makecha RN, Byerly HC (2013) Why do dolphins smile? A comparative perspective on dolphin emotions and emotional expressions. In: Watanabe S, Kuczaj S (eds) Emotions of animals and humans: comparative perspectives. Springer, New York, pp 63–85

Kyngdon DJ, Minot EO, Stafford KJ (2003) Behavioural responses of captive common dolphins *Delphinus delphis* to a 'swim-with-dolphin' programme. Appl Anim Behav Sci 81:163–170

Laule GE, Bloomsmith MA, Schapiro SJ (2003) The use of positive reinforcement training techniques to enhance the care, management, and welfare of primates in the laboratory. J Appl Anim Welf Sci 6:163–173

Leliveld LM, Langbein J, Puppe B (2013) The emergence of emotional lateralization: evidence in non-human vertebrates and implications for farm animals. Appl Anim Behav Sci 145:1–14

Lusseau D (2003) The emergent properties of a dolphin social network. Proc R Soc Lond B Biol Sci 270:S186–S188

MacLeod CD (1998) Intraspecific scarring in odontocete cetaceans: an indicator of male 'quality' in aggressive social interactions? J Zool 244:71–77

Mann J, Kemps C (2003) The effects of provisioning on maternal care in wild bottlenose dolphins, Shark Bay, Australia. In Marine Mammals and Humans: Towards a sustainable balance, eds N. Gales, M. Hindell, and R. Kirkwood (Clayton: CSIRO Publishing), 304–320

Mann J, Connor RC, Barre LM, Heithaus MR (2000) Female reproductive success in bottlenose dolphins (*Tursiops* sp.): life history, habitat, provisioning, and group-size effects. Behav Ecol 11:210–219

Manteuffel G, Puppe B, Schön PC (2004) Vocalization of farm animals as a measure of welfare. Appl Anim Behav Sci 88:163–182

Maple TL (2007) Toward a science of welfare for animals in the zoo. J Appl Anim Welf Sci 10:63–70

Mason GJ, Veasey JS (2010) How should the psychological well-being of zoo elephants be objectively investigated? Zoo Biol 29:237–255

McBain JF (1999) Cetaceans in captivity: a discussion of welfare. J Am Vet Med Assoc 214:1170–1174

Mendl M, Paul ES (2004) Consciousness, emotion and animal welfare: insights from cognitive science. Anim Welf 13:17–25

Mendl M, Burman OH, Parker RM, Paul ES (2009) Cognitive bias as an indicator of animal emotion and welfare: emerging evidence and underlying mechanisms. Appl Anim Behav Sci 118:161–181

Miller SJ, Mellen J, Greer T, Kuczaj SA (2011) The effects of education programmes on Atlantic bottlenose dolphin (*Tursiops truncatus*) behaviour. Anim Welf 20:159–172

Nagasawa M, Mitsui S, En S, Ohtani N, Ohta M, Sakuma Y, Onaka T, Mogi K, Kikusui T (2015) Oxytocin-gaze positive loop and the coevolution of human-dog bonds. Science 348:333–336
Ohl F, van der Staay FJ (2012) Animal welfare: at the interface between science and society. Vet J 192:13–19
Orbach DN, Packard JM, Piwetz S, Würsig B (2015) Sex-specific variation in conspecific-acquired marking prevalence among dusky dolphins (*Lagenorhynchus obscurus*). Can J Zool 93:383–390
Pack AA (2010) The synergy of laboratory and field studies of dolphin behavior and cognition. Int J Comp Psychol 23:538–565
Paul ES, Harding EJ, Mendl M (2005) Measuring emotional processes in animals: the utility of a cognitive approach. Neurosci Biobehav Rev 29:469–491
Pedernera-Romano C, Valdez RA, Singh S, Chiappa X, Romano MC, Galindo F (2006) Salivary cortisol in captive dolphins (*Tursiops truncatus*): a non-invasive. Anim Welf 15:359–362
Perelberg A, Schuster R (2009) Bottlenose dolphins (*Tursiops truncatus*) prefer to cooperate when petted: integrating proximate and ultimate explanations II. J Comp Psychol 123:45
Pettis HM, Rolland RM, Hamilton PK, Brault S, Knowlton AR, Kraus SD (2004) Visual health assessment of North Atlantic right whales (*Eubalaena glacialis*) using photographs. Can J Zool 82:8–19
Pifarré M, Valdez R, González-Rebeles C, Vázquez C, Romano M, Galindo F (2012) The effect of zoo visitors on the behaviour and faecal cortisol of the Mexican wolf (*Canis lupus bailey*). Appl Anim Behav Sci 136:57–62
Platto S, Wang D, Wang K (2015) Variation in the emission rate of sounds in a captive group of false killer whales *Pseudorca crassidens* during feedings: possible food anticipatory vocal activity? Chin J Oceanol Limnol 34:1218–1237
Reif JS, Fair PA, Adams J, Joseph B, Kilpatrick DS, Sanchez R, Goldstein JD, Townsend FI Jr, McCulloch SD, Mazzoil M, Zolman ES, Hansen LJ, Bossart GD (2008) Evaluation and comparison of the health status of Atlantic bottlenose dolphins from the Indian River Lagoon, Florida, and Charleston, South Carolina. J Am Vet Med Assoc 233:299–307
Robeck TR, Willis K, Scarpuzzi MR, O'Brien JK (2015) Comparisons of life-history parameters between free-ranging and captive killer whale (*Orcinus orca*) populations for application toward species management. J Mammal 96:1055–1070
Roche JR, Friggens NC, Kay JK, Fisher MW, Stafford KJ, Berry DP (2009) Invited review: body condition score and its association with dairy cow productivity, health, and welfare. J Dairy Sci 92:5769–5801
Rogers LJ (2010) Relevance of brain and behavioural lateralization to animal welfare. Appl Anim Behav Sci 127:1–11
Rushen J, Mason G (2008) A decade-or-more's progress in understanding stereotypic behavior. In: Mason G, Rushen J (eds) Stereotypic animal behaviour: fundamentals and applications to welfare. CAB International, Wallingford, pp 1–18
Rushen J, Butterworth A, Swanson JC (2011) Animal behavior and well-being symposium: farm animal welfare assurance: science and application. J Anim Sci 89:1219–1228
Sakai M, Hishii T, Takeda S, Kohshima S (2006) Laterality of flipper rubbing behaviour in wild bottlenose dolphins (*Tursiops aduncus*): caused by asymmetry of eye use? Behav Brain Res 170:204–210
Samuels A, Spradlin T (1995) Quantitative behavioral study of bottlenose dolphins in swim-with-dolphin programs in the United States. Mar Mamm Sci 11:520–544
Schmitt TL, St Aubin DJ, Schaefer AM, Dunn JL (2010) Baseline, diurnal variations, and stress-induced changes of stress hormones in three captive beluga whales, *Delphinapterus leucas*. Mar Mamm Sci 26:635–647
Schwacke LH, Smith CR, Townsend FI, Wells RS, Hart LB, Balmer BC, Collier TK, De Guise S, Fry MM, Guillete LJ Jr, Lamb SV, Lane SM, McFee WE, Place NJ, Tumlin MC, Ylitalo GM, Zolman ES, Rowles TK (2014) Health of common bottlenose dolphins (*Tursiops truncatus*) in Barataria Bay, Louisiana, following the Deepwater Horizon oil spill. Environ Sci Technol 48:93–103

Scott EM, Mann J, Watson-Capps JJ, Sargeant BL, Connor RC (2005) Aggression in bottlenose dolphins: evidence for sexual coercion, male-male competition, and female tolerance through analysis of tooth-rake marks and behaviour. Behaviour 142:21–44

Shepherdson D, Lewis KD, Carlstead K, Bauman J, Perrin N (2013) Individual and environmental factors associated with stereotypic behavior and fecal glucocorticoid metabolite levels in zoo housed polar bears. Appl Anim Behav Sci 147:268–277

Silber GK, Fertl D (1995) Intentional beaching by bottlenose dolphins (*Tursiops truncatus*) in the Colorado River Delta, Mexico. Aquat Mamm 21:183–186

Spruijt BM, van den Bos R, Pijlman FT (2001) A concept of welfare based on reward evaluating mechanisms in the brain: anticipatory behaviour as an indicator for the state of reward systems. Appl Anim Behav Sci 72:145–171

Stanton MA, Mann J (2012) Early social networks predict survival in wild bottlenose dolphins. PLoS One 7:e47508

Suzuki M, Uchida S, Ueda K, Tobayama T, Katsumata E, Yoshioka M, Aida K (2003) Diurnal and annual changes in serum cortisol concentrations in Indo-Pacific bottlenose dolphins *Tursiops aduncus* and killer whales *Orcinus orca*. Gen Comp Endocrinol 132:427–433

Swaisgood RR (2007) Current status and future directions of applied behavioral research for animal welfare and conservation. Appl Anim Behav Sci 102:139–162

Sweeney JC, Ridgway SH (1975) Common diseases of small cetaceans. J Am Vet Med Assoc 167:533–545

Temple D, Dalmau A, de la Torre JLR, Manteca X, Velarde A (2011) Application of the Welfare Quality® protocol to assess growing pigs kept under intensive conditions in Spain. J Vet Behav Clin Appl Res 6:138–149

Thomson CA, Geraci JR (1986) Cortisol, aldosterone, and leucocytes in the stress response of bottlenose dolphins, *Tursiops truncatus*. Can J Fish Aquat Sci 43:1010–1016

Trone M, Kuczaj S, Solangi M (2005) Does participation in Dolphin–Human Interaction Programs affect bottlenose dolphin behaviour? Appl Anim Behav Sci 93:363–374

Ugaz C, Valdez RA, Romano MC, Galindo F (2013) Behavior and salivary cortisol of captive dolphins (*Tursiops truncatus*) kept in open and closed facilities. J Vet Behav Clin Appl Res 8:285–290

Veasey J (2006) Concepts in the care and welfare of captive elephants. Int Zoo Yearb 40:63–79

Veissier I, Miele M (2015) Short historical overview of animal welfare sciences: how a societal concern has become a transdisciplinary subject. INRA Prod Anim 28:399–409

Veissier I, Butterworth A, Bock B, Roe E (2008) European approaches to ensure good animal welfare. Appl Anim Behav Sci 113:279–297

Visser IN (1998) Prolific body scars and collapsing dorsal fins on killer whales (*Orcinus orca*) in New Zealand waters. Aquat Mamm 24:71–81

Waiblinger S, Boivin X, Pedersen V, Tosi MV, Janczak AM, Visser EK, Jones RB (2006) Assessing the human–animal relationship in farmed species: a critical review. Appl Anim Behav Sci 101:185–242

Waples KA, Gales NJ (2002) Evaluating and minimising social stress in the care of captive bottlenose dolphins (*Tursiops aduncus*). Zoo Biol 21:5–26

Watters JV (2014) Searching for behavioral indicators of welfare in zoos: uncovering anticipatory behavior. Zoo Biol 33:251–256

Webster J (2005) Animal welfare: limping towards Eden. Blackwell Publishing Ltd, Oxford

Welfare Quality® (2009a) Welfare Quality® assessment protocol for cattle (fattening cattle, dairy cows, veal calves). Welfare Quality® Consortium, Lelystad

Welfare Quality® (2009b) Welfare Quality® assessment protocol for pigs. Welfare Quality® Consortium, Lelystad

Welfare Quality® (2009c) Welfare Quality® assessment protocol for poultry. Welfare Quality® Consortium, Lelystad

Wells RS (2009) Learning from nature: bottlenose dolphin care and husbandry. Zoo Biol 28:635–651

Whitham JC, Wielebnowski N (2009) Animal-based welfare monitoring: using keeper ratings as an assessment tool. Zoo Biol 28:545–560

Whitham JC, Wielebnowski N (2013) New directions for zoo animal welfare science. Appl Anim Behav Sci 147:247–260

Williams R, Ashe E (2007) Killer whale evasive tactics vary with boat number. J Zool 272:390–397

Yeates JW, Main DCJ (2008) Assessment of positive welfare: a review. Vet J 175:293–300

Part II
Seals

Chapter 13
The Welfare Impact on Pinnipeds of Marine Debris and Fisheries

Andy Butterworth and Sue Sayer

Abstract Uncounted, and usually unobserved, numbers of the animals that live in the oceans find themselves snared, trapped or entangled in lost fishing gear, monofilament line, nets, rope, plastic packaging and packing bands from crates, or become hooked on discarded fishing gear, or ingest human marine debris. Seals, sea lions and walrus (the pinnipeds) seem particularly susceptible to entanglement in marine debris—their exploratory natures may make this more likely, or perhaps they come upon plastic waste and rope on the shoreline to a greater extent than the other fully aquatic mammals. Pinnipeds meeting with plastic, either in the sea or on the shoreline, may carry debris wrapped around themselves for long periods. They often die as a result, sometimes from major chronic wounds. Although a wide range of the global species of seals can be affected by marine debris, some species are much more significantly affected than others. The key seal species affected by entanglement are monk seals, fur seals and California sea lions. Seals which become entangled or who ingest marine debris may be subjected to distress, pain, trauma, infection, skin and muscle lesions and compromised ability to move, feed and carry out normal behaviour. For these reasons marine debris has the capacity to present a significant and global issue with respect to animal welfare, as well as to more immediately apparent concerns regarding habitats and the quality of the marine environment.

13.1 Introduction

Uncountable and unobserved in many cases, large number of pinnipeds are becoming tangled, or trapped, in discarded or lost fishing gear, net, rope, packaging and monofilament fishing line, or are hooked on fishing equipment (Convention on Biological Diversity 2012). Pinnipeds may be captured as by-catch and die (e.g. in

A. Butterworth (✉)
Department of Clinical Veterinary Science, University of Bristol, Langford, Bristol, UK
e-mail: andy.butterworth@bris.ac.uk

S. Sayer
Cornwall Seal Group Research Trust, Cornwall, UK

A. Butterworth (ed.), *Marine Mammal Welfare*, Animal Welfare 17,
DOI 10.1007/978-3-319-46994-2_13

live operational fishing gear—mostly gill and trawl nets) or become live entangled (mostly in storm damaged or discarded ghost fishing gear) with consequent welfare implications. The pinnipeds appear particularly susceptible to entanglement in this kind of marine waste; perhaps they encounter net, rope and waste on shorelines and in coastal waters more than the oceanic marine mammals. As highly intelligent mammals, pinnipeds appear curious about their environment and so likely to investigate materials floating in the water column, particularly juvenile animals when playing. Seals, walrus and sea lions meeting with waste or ghost fishing gear (*ghost gear is abandoned, lost or otherwise discarded fishing gear—ALDFG—which continues to 'fish' in an indiscriminate way*), in the shallow water of the coast or on the shore, may carry this material wrapped around them for a long period and sometimes die from the penetrating wounds caused by the rope and line. A broad range of species can be affected by entanglement, but some are much more commonly seen wrapped with rope, net or marine debris than others, especially monk seals, fur seals, grey seals and California sea lions. Seals wrapped or trapped in loops of marine debris may experience pain, fear, skin lesions and infection and sometimes deeply incised wounds from rope or line, which can amputate limbs, and cut down to bone (Fig. 13.1). The lines or fragments of net can interfere with their ability to move and perform natural behaviours—to keep up with conspecifics, to hunt, to forage, to mate and to move through the water at speed. Entanglement may also lead to complications such as oedema in pregnant females with the potential for reducing

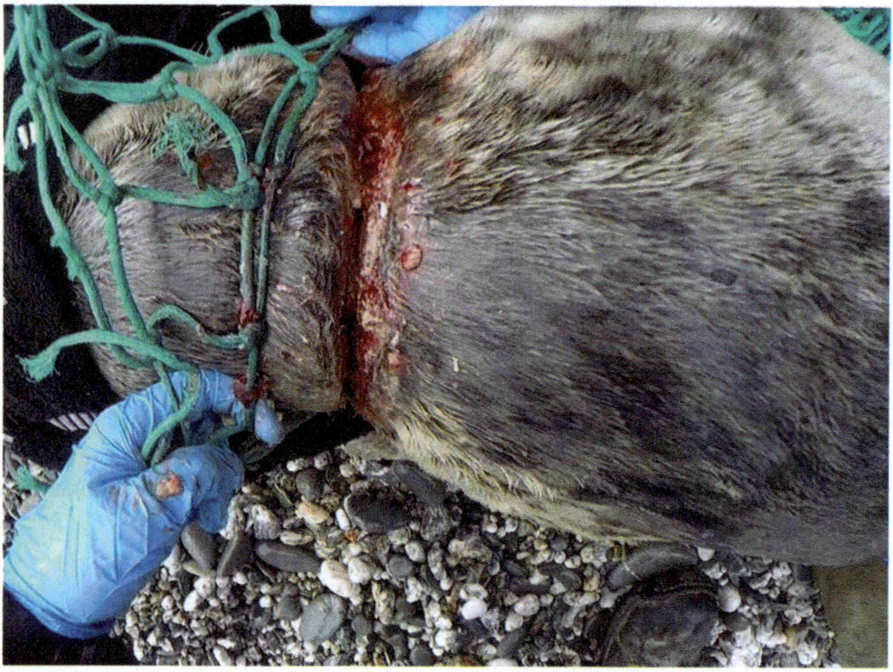

Fig. 13.1 Live juvenile grey seal with deep open wound from trailing trawl net freed by the British Divers Marine Life Rescue. *Image credit*: Sue Sayer, Cornwall Seal Group Research Trust

survival and fecundity. Entangled debris presents a global animal welfare concern, and the recent launch of the Global Ghost Gear Initiative by World Animal Protection (WAP 2016) is the first major organisational initiative with a direct focus on marine debris in relation to animal welfare. Debris can damage local habitats by smothering rock and seabed substrates, resulting in the need to animals to alter their feeding behaviour. Many pinniped species have telescopic necks that improve their ability to accelerate forward to snatch prey, and with many entanglements occurring around the neck and head area, this ability can be severely reduced. The natural panic reaction for some pinnipeds is to spin their bodies and this can further entangle them in fishing gear. Different types of entangling materials have different impacts. Monofilament net or line tends to incise deeply through skin and into flesh, caused by the animal's movement alone and then by subsequent growth. Multifilament net may be more prone to harbour bacteria and so likely result in infection—one grey seal was known to have died within 128 days of his last pre-entanglement sighting as a result of trawl net (Sayer et al. 2015), whilst others have been known to live over 14 years with presumed monofilament wounds. Post-mortems have shown that flesh and skin can completely regrow over the entangling material (Sayer et al. 2015).

Marine debris may also be a source of chemical pollutants in the sea; plastics may release plasticisers and additives, which can cause toxicity in top predators when these toxins accumulate in their marine food.

Marine debris is found in all corners of the oceanic world, but the reporting of the effects on pinnipeds is not uniform and is linked to the number of 'observers' who report entangled animals (Fig. 13.2). Perhaps because of the patchiness of

Fig. 13.2 Adult female grey entangled seal dead and decomposing, undetected whilst alive. *Image credit: Liz Clark, Cornwall Seal Group Research Trust*

reporting rates, there is almost no reported and published information on pinniped entanglement in some parts of the world. Moore et al. (2013) reported that post-mortems of dead and live stranded pinnipeds correlate with the distribution of human impacts including fishing gear entanglement, boat strikes and malicious gunshot wounding. Harcourt et al. (1994) suggest that published rates of entanglement are likely to be underestimates, because they report only animals seen when they come onto shore and do not report or detect those animals which die out at sea (Fig. 13.3). Both of these authors note that inaccessibility, delayed discovery and human safety concerns for access to places where these animals are found may limit the accurate reporting of the cause of death and so result in under-reporting of animals dying as a result of marine debris and entanglement. Assumptions are often made about similarly entangled animals being the same individual, and only detailed photo identification research can reveal the true extent of this issue.

There are big variations in the geographical spread of research into marine debris and its potential effects on animals. The 2012 Convention on Biological Diversity report (CBD 2012) identifies this imbalance and indicates the numbers of reports reviewed which concern entanglement in debris in a wide range of species (not only pinnipeds) from different oceans: They report 'Americas (North and South) (117), Australasia (56), Europe (52), Africa (12), Antarctic (7), Asia (6) and Arctic (5)'.

Fig. 13.3 Litter raft of mixed materials including lost fishing gear and a dead grey seal. *Image credit: Mike Stephens, Cornwall Seal Group Research Trust*

Estimates for animal entanglement and ingestion rates rely on reports of animals seen alive, or which have only recently died (otherwise the carcases become too decomposed for full analysis); therefore, the scale of this issue is likely to be seriously underestimated. If animals die unseen, as will be the case for many, possibly even the majority, of animals, then they will not be reported. Dead stranded animals with ghost fishing gear around their necks have been observed to decompose in such a way as leading to headless carcasses which further clouds accurate reporting. As Cole et al. (2006) say—'Our greatest concern remains the number of animals we never saw'.

Overall comments on the reporting variability for entanglements are made by Butterworth et al. (WSPA 2012). And even if regional reporting bias is taken into account, it is apparent that some areas produce higher risks of ingestion and entanglement than others, and so it is possible that highly targeted action in these areas of high risk might act to ameliorate localised marine debris impacts and that it may be worthwhile to focus resource and work to create improvements in these areas. The reported 'hotspots' for entanglement of pinnipeds are the western coast of the USA, sea lions and fur seals; the eastern coast of Australia, fur seals; the south African coast, fur seals; and the Celtic and North Seas, where the gulf stream is known to bring large amounts of debris, grey seals.

13.2 A Short History of Marine Debris

When the explorer Thor Heyerdahl crossed the Atlantic Ocean in 1970, he was so concerned about the marine debris that he observed on the oceans that he submitted a report to the United Nations 1972 Stockholm conference on the Human Environment (United Nations 1972). Marine litter is defined by the United Nations Environment Programme as 'any persistent, manufactured or processed solid material discarded, disposed or abandoned in the marine and coastal environment', and the United Nations Environment Programme of 2005 (United Nations 2005) estimated that 6.4 million tonnes of 'litter' end up in oceans every year. Estimates for the total amount of marine debris now present in the oceans vary, but, on average, around 300,000 items of litter and debris are estimated to be present per km^2 of ocean surface (NRC 2008). Marine waste and debris comprise plastics, metal, glass, rubber, paper and objects comprised of multiple man-made substances such as packaging boxes, bottles, fishing nets and floating accumulations of mixed waste material bound together into litter rafts (Fig. 13.3).

Plastic dominates marine litter because it is usually either neutrally buoyant or slightly denser than sea water, and because of its longevity. The top debris items collected between 1989 and 2007 were (ICC 2008)

> 'Cigarettes/cigarette filters: 24.6%, Bags (paper and plastic): 9.4%, Caps/lids: 9.1%, Food wrappers/containers: 8.9%, Cups/plates/forks/knives/spoons: 7.2%, Plastic cans: 4.6%, Straws/stirrers: 4.4%, Rope: 2.1%'.

Wilcox et al. (2016) listed the potential impacts of various forms of marine debris on marine mammals from the highest rank (risk) to the lowest rank: 'Buoys/traps/pots; Monofilament line; Fishing nets; plastic bags; Butts (cigarette butts); Plastic utensils; Balloons; Plastic caps; Food packaging; Plastic food lids; Straws/stirrers; Takeout containers; Hard plastic; Cans; Cups and plates; Glass bottles; Beverage bottles; Paper bags'.

'Plastics' are made from synthetic organic polymers—common forms of plastic include polyesters, polyethylene aramids and acrylics, polyethylene terephthalate (PET), polypropylene, nylon and high-density polyethylene (HDPE). Most rope, monofilament line and fishing net and a large proportion of packaging material are manufactured from plastic, sometimes woven, braided or plaited, to increase its strength as a fibre. Rope, monofilament line and net are specifically designed for use in the sea, are very strong, are resistant to degrading by saltwater and sunlight, and are resistant to abrasion. Plastics are usually neutrally dense or buoyant in the sea and float at the surface or sink only slowly in the water and can be carried by ocean currents. Nylon monofilament fishing line was first sold in 1939 (New World Encyclopedia 2016), and since that time monofilament plastic line has become much stronger, almost invisible in water (monofilament lines have low optical density) and extremely strong. Fishing lines are strong when related to their thickness, and this thin strength can result in extreme tissue damage when animals become entangled. Some plastics may last for up to 600 years in the sea, and because of their durability and longevity, abandoned, lost or otherwise discarded fishing gear (ALDFG) or derelict fishing gear (DFG), nets, lines, lost traps, floats with line, rope or net attached and monofilament fishing line (sometimes with hooks) are a particular concern for animal welfare.

'Packing bands' are extremely strong (by design) and are used to close containers and packages. They are usually made from polypropylene, nylon or polyester, often reinforced with other plastic fibres, and they are not only strong and resistant to degradation but are commonly formed into loops (around the original container), and these looped structures more commonly trap animals, particularly pinnipeds, than straight lines or ropes. Loops of packing band are seen in a wide range of loop size, and each type of loop may represent a particular hazard to a species or age group of seal or sea lion.

The US National Marine Debris Monitoring Program (Sheavly 2007) indicated that 17.7% of marine litter found on beaches came from ocean activity, with a large proportion of debris linked with fishing, including nets, fish baskets, fishing line, rope, buoys, floats, pots and traps. In the UK, fishing-derived marine debris includes nets, buoys, line and floats, and is the second largest source of marine debris after litter from beach visitors (Marine Conservation Society [MCS] 2007). Sayer and Williams (WAP, 2015) identified differences in the fishing gear found lost at sea (buoys and floats, 41%; trawl net, 17%; monofilament net, 14%; rope, 12%; others, 9%; pot related, 6%, and monofilament line, 1%) to that recorded on land (beaches) in the same area (monofilament line, 29%; rope, 26%, pot related, 11%; trawl net, 11%, others, 10%; buoys and floats, 8%; monofilament net, 5%)—this representing a kind of 'selection' for some types of fishing-related gear to be more likely to occur as lost in the sea. Marine debris comes from a wide range of other man-made sources; from intentional and unintentional waste tipping from shipping, including fishing vessels; from accidental or deliberate dumping of domestic, commercial or industrial waste into the sea from the land; from waste blown from shore or from boats; and from land-based debris or waste moving down rivers and into the sea after storms or floods.

The manufacturing origin (however, not the disposal location) of many objects can be determined from the barcode that the object carries (the initial three letters of the code indicate the manufacturing country). Santos et al. (2005) reported the source of debris found on beaches in Brazil and found that the country of origin of identifiable objects was 'USA 12.2%, Italy 7.6%, South Africa 6.4%, Argentina 6.0%, Germany 5.6%, United Kingdom 4.6%, Taiwan 4.4%, Singapore 3.6%, Spain 3.6%, Malaysia 3.1%, with 'others' 35.2% and 'unidentified' 7.6%'.

Barcode tracing for plastic debris shows that marine debris can be found 10 years later and 10,000 km from its origin (Barnes et al. 2009).

Marine litter in the ocean slowly breaks down into small particles, and these plastic pieces are now found in the water and marine sediments across the world. The Great Pacific Oceanic Gyre has debris estimated to have a mass of 100 million tonnes, and this is particularly concentrated into an area the size of France and Spain together (Sheavly 2007). Before the 1980s, relatively small quantities of marine litter reached the Southern Ocean. Today, there is now movement and accumulation of marine litter across the whole southern hemisphere, and significant amounts of marine debris have moved towards Antarctica (Barnes 2005). Plastic tends to break down rather slowly in the marine environment. Wang et al. (2016) report that the effects of UV-B radiation and exposure to oxygen, and autocatalytic degradation of plastic in the low temperatures of the sea is very slow when compared to degradation in the terrestrial environment. Zalasiewicz et al. (2016) state that degraded plastic is so widespread in ocean sediments that degraded plastic may become a key future geological indicator of the Anthropocene (current time, time of mankind).

One perceived route to reduction of marine debris, and hence having the potential to reduce wildlife entanglement, is through educational programmes. Pearson et al. (2014) report a survey used to assess the familiarity of the Australian public in coastal communities with an initiative called 'Seal the Loop'—an educational programme aimed at protecting seals from marine litter. A majority of the participants in the study were familiar with the education programme, but 32% of the participants were not able to explain what the risks of marine debris to wildlife actually were. The respondents also underestimated the actual impact on wildlife numbers, however, this study did conclude that 'learning something new about the impact of marine debris did change waste disposal behaviours'. A lost fishing gear recording scheme in Cornwall, UK, saw the removal of 50 tonnes of lost fishing gear recorded in a 12-month period, with an assessed reduction in serious risk posed to grey seals from 47 to 26% (Sayer and Williams 2015).

13.3 Plastic Waste Impacts on Animal Welfare Through the Entanglement of Pinnipeds

For many people, a description of an animal as having 'good welfare' might include the animal being 'well' (i.e. not unwell) and also that the animal had the potential for 'well-being'—or, at least, not subject to high levels of distress or high frequencies of interference. With regard to a state of 'good welfare', disease or physiological or anatomical damage, injury and trauma would provide potential welfare challenges. Sandoe and Simonsen (1992) used the term 'cost of coping' implying that emotional distress, pain or increased levels of physiological or disease-related challenge would have a 'cost' to the animal and that if this cost was great, or in some cases excessive, then the animal would be less likely to 'cope'. Prolonged failure to cope would probably result in suffering.

For wild animals, entanglement in a loop of rope, a discarded net or a packing band could represent a severe compromise to their ability to cope and so induce suffering. The entanglement could result in altered; feeding behaviours, use of food sources, social interactions and breeding patterns, hunting or foraging patterns and territorial or animal–human interactions.

For an individual animal, the capacity to cope (or not) would depend on the severity of the entanglement and whether the entanglement caused restriction of movement or, in some cases; trauma, skin lesions, wounds and an altered ability to swim, mate or feed. The size, locality, physiology, feeding habits, behaviours and types of marine debris found in the sea around different pinniped species will affect whether entanglement happens, how and when it takes place, at what age (linked to body size and inquisitive behaviour) and with what debris items. Entanglement could be 'acute', causing sudden and severe welfare problems such as asphyxiation, or trapping underwater, or 'chronic', in which the welfare impacts may increase over time through incisive wounds, susceptibility to infection and long-term restriction of behaviours.

A large number of seal and sea lion species are recorded to have been entangled, with 58% of all species of seals and sea lions reported by Boland and Donohue (2003). The incidence rate of entanglement for seal and sea lion species is reported to be from 0.001 to 5% annually of the local seal population, with notably high levels of entanglement of up to 7.9% in California sea lions from Mexico (Harcourt et al. 1994). Williams et al. (2011) report high entanglement rates for northern elephant seals (*Mirounga angustirostris*), Steller sea lion (*Eumetopias jubatus*) and harbour seal (*Phoca vitulina*) around the coast of British Columbia. A study of Bering Sea northern fur seals estimated that 40,000 seals were killed by marine debris entanglement each year (Derraik 2002). Rates of entanglement in grey seals in South West England are of a similar magnitude, averaging 3.1% between 2000 and 2013 (Sayer et al. 2015).

When seals become entangled, this can involve a ring of packing strap, or a fragment of fishing net, or a loop of monofilament line—which commonly forms a collar around the neck, or less commonly a loop around the central abdomen. The loop becomes tighter as the seal grows and may become deeply trapped in the skin. This is because the animal cannot remove it due to its tension or the directional hair of the coat (which is flattened against the body in the direction of least water resistance). If the seal is adult, the loop can cut into the tissues of the flipper or the neck and may become firmly embedded in the skin, subcutaneous fat or muscle and sometimes, finally, into bone. If the loop becomes deeply enmeshed or embedded, then it is unlikely that the seal can ever remove it. Most entanglements are in young animals, maybe because they are more curious, inquisitive and exploratory than adults, or perhaps because they are naïve feeders, less familiar with the hazards represented by fishing net fragments, or packing band loops. Young seals with severe constrictions may have feeding restricted to the point of starvation. Loop ligatures can cause amputation of the flippers, or create wounds open to infection, which limit the likelihood of survival. The constriction around the neck can embed in the tissues and finally cause strangulation as the animal grows into the noose. Because plastic-based rope, net and packaging bands are so durable, after death, the debris can returned to the sea, with the potential to entangle other animals (WSPA 2012).

Trailing entangling materials have a tendency to cause asymmetrical wounds as they catch under the animal's body during locomotion on land, causing deeply incised wounds at the back of the neck when the animal moves on land. Longer trailing materials can have a significant impact on survivorship, with longer trailing material lengths associated with poorer survival rates (Sayer et al. 2015). Entangled seals will experience increased drag during swimming (Boland and Donohue 2003). Derraik (2002) describe how northern fur seals (*Callorhinus ursinus*) entangled in even small net fragments of as little as 200 g in weight experience a fourfold increase in the energetic requirement to compensate for drag caused by altered water flow. This drag effect restricts movement and may ultimately lead to the exhaustion or drowning of the animal. Where stellar sea lions (*Eumetopias jubatus*) in Alaska and British Columbia ingest lost fishing line with hooks attached, the hooks and lures lodge in the animal or can damage the mouth and the digestive tract and reduce the animal's capacity to forage and feed effectively.

13.4 Severity Scoring for Pinniped Interactions with Marine Debris

In human medicine, scoring scales are used to describe wounds and to enable clinicians to gauge and communicate how the wounds are healing. The Red Cross has a classification of war wounds, used to describe wounds based on their visual appearance (not based on what caused them) (Coupland 1992). Work has been initiated on the assessment of entanglements in marine mammals. At the 2007 NOAA/NMFS (NOAA 2007) Serious Injury Technical Workshop, held in Seattle, a hierarchical descriptive scale for entanglement injuries to marine mammals was proposed:

"Serious—gear-related injury; ingestion of gear; trailing gear (e.g. flasher or lure), when it has the potential to anchor or drag, or when it is wrapped around the animal; gear attached to the body with the potential to wrap around flippers, body, or head; foreign bodies penetrating into a body cavity;

Multiple wraps around the body; missing flippers—front and back flipper (serious), for both otariids or phocids; deep external injuries."

"Non-serious—gear-related injuries; hooked in the lip; hooked in flipper, etc. with minimal trailing gear that does not have the potential to wrap around body parts, accumulate drag, or anchor; freely swimming animals encircled by purse seine nets."

"Grey area—gear-related injuries (less clear how serious the welfare impact is): hooked in head (serious injury could be assumed, but it depends on several factors, including where on the head the hooking took place, the depth of the hooking, the type of hook, etc.); animals stressed by being encircled or trapped (e.g., purse seine); animals released without gear following entanglement (this designation depends on the extent of the injury or how long the animal was submerged, how long the gear was on the animal, and the degree of restraint)."

Other impacts of interactions with humans were also discussed: 'Pinniped brought onto a vessel' (this was considered in this report to be 'non-serious') and the severity for the animal of being brought up onto a boat which depended on how the animal was brought up, e.g. in net, or a roller (a fishing boat net handling device), or through the power block (the powered device used to haul a net onto the deck).

Some scenario examples of 'serious scores' are provided to illustrate the possible welfare impacts, which could cause severe welfare insults, and based on descriptions of observed seal entanglements from Spraker and Lander (2010):

"Rope fragment wrapped around shoulder, strands had cut through the muscles of the right shoulder and halfway through the mid-portion of the humerus.

Material wrapped around upper neck, line had cut through the lower half of trachea.

Line wrapped around mid-neck, had cut through all dorsal muscles of the neck exposing the dorsal spinal processes of the cervical vertebrae."

Successful trials were conducted to assess the risks posed to marine life by lost fishing gear by Sayer and Williams (WAP, 2015). Firstly, risk was assessed in terms of likelihood of marine life interaction with the lost fishing gear—described as 'possible' (P) if seals/birds used the area routinely, 'likely' (L) if seals/birds were within 5 m of the item and 'witnessed' (W) if they were observed touching the item; otherwise the risk was assessed as 'unlikely' (U). Secondly, risk was assessed according to the likelihood of marine life entanglement in the lost fishing gear—described as 'possible' (P) if the item was looped/meshed or a balled mass, 'likely' (L) if they were within 5 m of a looped/meshed or balled item and 'witnessed' (W) if marine life was seen entangled; otherwise the risk was assessed as 'unlikely' (U). The two risk ratings were combined into the following categories: UU, UP, PU, PP, LU, LP, LL, LW, WL and WW. Categories PP + (PP, LP, LL, LW, WL or WW) were considered to pose a serious threat to marine life (especially seals and birds).

13.5 Fur Seals

Hofmeyr et al. (2002) recorded 101 fur seals (*Arctocephalus* spp.) and five southern elephant seals (*Mirounga leonina*) entangled over a period of 10 years on Marion Island in the Southern Ocean. These authors describe how 67% of the materials causing the entanglement came from the fishing industry. Polypropylene packaging straps (associated with the fishery) were the most common material causing entanglement, followed by fish trawl netting. These authors also noted longline hooks embedded in animals and that fishing line entanglements only started to be seen after longline fishing started in 1996 in this area. Hofmeyr et al. (2002) estimated that 0.24% of this population of fur seals were entangled each year. Hofmeyr et al. (2006), in a further study of Antarctic fur seals (*Arctocephalus gazella*) between 1996 and 2002 on Bouvetøya, an Antarctic island, reported entanglement rates from 0.024 to 0.059% and concluded that these rates were relatively low when compared to other pinniped populations because, they suggested, of the isolation of the site. This 2006 study found that more than two-thirds of materials causing entanglement were from fisheries sources.

Spraker and Lander (2010) estimated the causes of mortality in northern fur seals (*Callorhinus ursinus*) in the Alaskan St. Paul Islands. These authors describe combinations of the pathological effects of entanglement, with trauma and asphyxiation being caused by net fragments or packing band loops. In one case a heavily entangled living animal was dragging a decomposing seal in the same piece of entangling net.

Lawson and co-workers carried out a study on a series of beaches from the islands around the coast of Southern Australia, where there is an estimated Australian fur seal (*Arctocephalus pusillus doriferus*) population of about 30,000 animals (Lawson et al. 2015). Between 1997 and 2012, 138 entanglements were reported and the entangling debris was collected. In the debris, 50% ($n = 69$) of the objects were plastic twine or rope, which included trawl nets; 20% ($n = 27$) were packing straps, plastic bags and balloon strings; 17% ($n = 24$) were monofilament fishing line (which included gill nets); and 8% ($n = 11$) were rubber litter items. This study also recorded the characteristics of the entangling material; its 'type, colour, mesh size, overall mass, number of threads, whether the item was braided, twisted, knotted, if it was monofilament, and the number of strands for all entanglement items'. White plastic packaging straps were the most common (67%, $n = 6$) of the packing strap entanglements; 61% ($n = 43$) of rope entanglements were with green-coloured material, whilst grey- and white-coloured rope accounted for lower percentages of entangling material at 10% ($n = 7$) and 9% ($n = 6$), respectively. For the monofilament line entanglements, most of the monofilament was clear or green in colour (52% and 26%, respectively). Information on the location, date, age of the seal (pup, juvenile, adult) and the type and severity of the injury (whether the wound was cutting deep into tissue or was a surface wound) was also compiled. Analysis of this carefully collected data indicated that the majority 94% ($n = 46$) of entanglements involved pups or juvenile seals, with more pups (53%) than juveniles (41%) being entangled.

McIntosh and others, working at Seal Rocks, South-Eastern Australia, reported 359 entangled Australian fur seals and showed that the most common entanglement materials were from commercial fisheries and that entanglements were most frequent in pups and juveniles (McIntosh et al. 2015). Entanglement was most commonly observed from July to October, when the animals approached weaning. Using generalized additive mixed models (GAMMs), these authors estimated that 1.0% (CI = 0.6–1.7%) of the local population was entangled each year.

The loop diameters of entangling materials, which entangled Antarctic fur seals from a study at Bird Island, South Georgia, are described by Waluda and Staniland (2013). They reported material found entangling 90 animals, with loops from 11 to 69 cm in diameter (with a median diameter of 18 cm). These authors found that loop diameter was closely related to age class. Pups were more commonly entangled in small loops (median = 15.5 cm), and juveniles and adult females were entangled in loop diameters of about 17 cm (adult females = 17 cm, juveniles = 18 cm). Adult males were more likely to be snared in large loops (median = 34 cm). These authors report that juveniles were five times more likely to be snared than adult females. They propose that younger animals meet entangling material through inquisitive play. Adult males were least likely to become entangled, which may be because of the shape of their broad muscular necks and also their relatively small numbers within the total population and also potentially due to differences in their feeding and exploratory behaviours. This report notes also that if entanglement is fatal to a juvenile, then individuals prone to entanglement will possibly have been selected out of the population. This study also identifies that more 'very severe' entanglements occurred in the (Southern) winter, and these authors speculate that this may be due to changes in the ability to observe and report entangled animals, rather than a true alteration in entanglement rate. During winter, the animals are hauled out onto the shoreline and are thus more readily observed. This report also suggested that there has been a decline in the number of seals snared in packaging bands at Bird Island across the period of the study. In '1988/1999—58% of entanglements were with packing bands, between 1989 and 1994 this fell to 46%, and between 1994 and 2013 the proportion was 39%'. These authors suggest that the Commission for the Conservation of Antarctic Marine Living Resources (CCAMLR) ban on packaging bands, which began in 1995, may have started to have a reducing, but not complete eliminating, effect on the rate of packing band entanglements. Other studies suggest that the rate of entanglement of Antarctic fur seals (*Arctocephalus gazella*) halved over the 5-year period (1990–1994) after the introduction of MARPOL Annex V (in 1973, the International Maritime Organization IMO adopted the International Convention for the Prevention of Pollution from Ships, now known as MARPOL, which has been amended by the Protocols of 1978 and 1997 and kept updated with relevant amendments) (IMO 2016); however, polypropylene packing straps, synthetic fibre rope and fishing net fragments were still found to be common debris items which entangled seals in all the years of this study (Arnould and Croxall 1995).

Page et al. (2004) indicate that, in New Zealand, fur seals are most commonly entangled in loops of packing tape and pieces of trawl net originating from the rock lobster and trawl fisheries. These authors (Page et al. 2004) published entanglement rates for Australian sea lions and New Zealand fur seals in derelict fishing gear and in other marine debris. In 2002, the authors calculated that the Australian sea lion entanglement rate was 1.3% of the population annually, and the New Zealand Fur seal entanglement rate was 0.9%. Australian sea lions were commonly found to be entangled in monofilament line or net (rather than any other entangling materials), and these fishing materials appeared to be most likely derived from the local shark fishery.

On St. Paul Island, in the Alaskan Pribilof Islands, northern fur seal (*Callorhinus ursinus*) entanglement rates were studied by Fowler (1987), and various objects were found to entangle these animals around their necks, shoulders and flippers, with an estimated incidence rate of about 0.4% annually. The majority of these entanglements were with trawl nett fragments and plastic packing bands. This author noted that entanglement was more common in young animals, which were 'sometimes observed entangled together in groups attached to the same large items of debris'.

Shaughnessy (1980) reports entanglement in Cape fur seals (*Arctocephalus pusillus*), in the period 1972–1979. The majority of the entangling objects were around the seals' necks, with the incidence rate recorded at the Cape Cross colony of 0.56–0.66% per year. Animals were entangled with 'string, rope, fishing net, plastic straps, monofilament line and rubber O-rings', with a rate of entanglement estimated to be 0.4% annually of the population. These authors estimated 15,000 seal entanglements to take place each year and that 5700 of these animals would die as a result of their entanglement. Zavadil et al. (2007) reports northern fur seals (*Callorhinus ursinus*) on St. George Island to have an estimated entanglement rate of 0.06–0.08% annually for pups and with the maximum entanglement rate occurring in October with up to 0.11% of the population entangled just before weaning.

New Zealand fur seals (*Arctocephalus forsteri*) in the Kaikoura region of New Zealand breed close to a busy tourist and fishing area and become entangled in lost net and plastic waste (Boren et al. 2006). Entanglement rates are described in Boren et al.'s study as being in the range of 0.6–2.8% annually, with green trawl net pieces (42%) and plastic strapping bands (31%) being the most common entangling items. These authors also report that, perhaps due to the high density of 'observers' in this area, nearly half of the entangled seals were caught and released from their entanglement (43%) and that post-release monitoring has shown that the likelihood of an individual surviving is high, even after a significant entanglement wound.

Hanni and Pyle (2000) describe 914 California sea lions (*Zalophus californianus*), Steller sea lions (*Eumetopias jubatus*), Pacific harbour seals (*Phoca vitulina*), northern elephant seals (*Mirounga angustirostris*) and northern fur seals (*Callorhinus ursinus*) reported as entangled at South-east Farallon Island, in North California, between 1976 and 1998. The most common entangling materials were *monofilament line and net, heavy fishnet, other net, salmon fishing lure and line, fish hooks and line, packing straps, other miscellaneous marine debris and 'constriction'* (where no actual material could be seen, but material was presumed to be present, hidden in the fur or in a wound, with a circular indentation or wound present around the head, neck or torso).

13.6 Elephant Seals

Campagna et al. (2007) discuss the impact of entanglement on Southern elephant seals (*Mirounga leonina*) in relation to the characteristics of the wounds around the neck caused by monofilament fishing lines. In this study, entangled elephant seals were caught, and, where possible, the material was removed. The monofilament line found was typically 1.3–1.5 mm thick and was tied into a loop with a knot, presumably by the original fisherman. In some animals the entangling line still had lures or hooks attached, and the configuration of hooks and lures was typical of that found in gear used for local squid fishing. However, they do suggest this to be an underestimate, as observations were made at a time of year when juveniles were not present. These authors discuss how the monofilament line entanglement becomes a deep chronic wound associated with infection and note the severe consequences for the animals affected, and they judge, from the depth of the wounds, that entangled seals may have lived for months or even years with the monofilament line cutting into the neck tissues.

13.7 Sea Lions

In Australia, it is estimated that 1500 Australian sea lions (*Neophoca cinerea*) die annually from entanglement, mostly from snaring in monofilament gillnet from the shark fishery located where the sea lions forage (Page et al. 2003).

In California, Dau et al. (2009) report 1090 seal entanglements, of which 11.3% were related to fishing gear and with a particularly high incidence of fishing gear entanglement injury observed in the San Diego region. Zavala-González and Mellink (1997) report entanglement in California sea lion (*Zalophus californianus*) from a population which extends from British Columbia to Mazatlan in Mexico, including populations from the Gulf of California. The population of sea lions in the Mexican part of range area is estimated at 74,467 along the Pacific coast and 28,220 and in the Gulf of California, and these authors report annual entanglement rates in this region of 2.24% (which could equate to approximately 2300 animals annually).

A survey reported by the National Oceanic and Atmospheric Administration (NOAA 2012) indicates that packing bands cause more than 50% of neck entanglement in Steller sea lions (*Eumetopias jubatus*) in Alaska. A survey of 386 Steller sea lions (*Eumetopias jubatus*) in South-east Alaska and northern British Columbia reported an estimated incidence annual rate of entanglement of 0.26% (Raum-Suryan et al. 2009). These authors reported that the common materials causing entanglement were packing bands (54%), large rubber bands (rubber packing bands)

Fig. 13.4 California sea lion (known as Shammyrock) is seen here with an entanglement on March 16, 2014. *Image credit: The Marine Mammal Center*

(30%), pieces of net (7%), rope (7%) and monofilament fishing line (2%) (Fig. 13.4). This study also looked at the incidence of fishing gear ingestion or entanglement for these Steller sea lions and reports that 'salmon fishery flashers and lures (80%), long-line gear (12%), hooks and line (4%), spinners or spoons (2%), and bait hooks (2%)' comprised the major items found. Raum-Suryan et al. also describe a local education campaign—'Lose the Loop!'—which promoted cutting of entangling loops of fishing material and elimination of packing bands from local waste to help prevent entanglements.

13.8 Monk Seals

Donohue and Foley (2007) assess the influence of storm weather on monk seal entanglement in the North Pacific Ocean. They describe how, for the 23 years leading up to 2007, monk seal entanglement increased during episodes of severe weather associated with El Niño. They propose that ocean current processes linked with El Niño may contribute to changes in entanglement potentially because of introduction of new marine debris along with the changes in the ocean currents. The Hawaiian monk seal (*Neomonachus schauinslandi*) is an endangered species breeding only on six small islands and atolls in North-west Hawaii. Between 1996 and 2000, an initiative in this area aimed to reduce the amount of derelict fishing gear in the reefs close to the breeding sites for these seals (Boland and Donohue 2003) and a total of 195 tonnes of derelict fishing gear was removed from the area. Karamanlidis (2000) found that entanglement in abandoned nets was having a measurable effect on the population of monk seals (*Monachus monachus*) in the Mediterranean, and this author reported that the use of gillnets posed a significant threat to this endangered population of monk seals around the Desertas Islands off Madeira.

13.9 Grey Seals and Common Seals

Entanglements of grey seals (*Halichoerus grypus*) on the Dutch coast in the period between 1985 and 2010 are described by Hekman and Osinga (2010). They report that entanglement was relatively (relative to population size) more commonly observed in grey seals than common seals (*Phoca vitulina*) (about twice as often in the grey seal), and that in both species more of the entangled seals were males and that entanglement was more likely to occur in juveniles. ALDFG (lost fishing gear) was the most common entangling material, and the numbers of grey and common seals seen entangled and reported were believed to be only a small portion of the number of animals affected because of the animals assumed to be lost and undetected at sea.

Allen et al. (2012) report the physiological and anatomical effects of debris entanglement on grey seals (*Halichoerus grypus*) in Cornwall, UK, between 2000 and 2008. They describe how an under-reported aspect of entanglement is the effect of increased drag from trailing material and the increased foraging time required to feed because of the raised metabolic demands created by the entangling material. Allen et al. discussed the animal welfare impact of the entanglement injuries and report the types of injuries sustained by the animals to be '"constriction" (43%); "wound" (7%); "constriction and wound" (14%); "evident" (visible entanglement but wound type unclear, 36%)'. Allen et al. (2012) estimated that entangled seals form 8.7% of the seals recorded in the Cornish photo identification database (up to the end of 2011) and that of 58 seals showing evidence of entanglement in the database, 37 (64%) had visible lesions showing a constriction or an open wound, or both (Figs. 13.5, 13.6, 13.7, 13.8, 13.9, 13.10, and 13.11). These authors estimate entanglement rates in these seals to have declined from 5% (annually) of sightings in 2004 to 3% in 2011 and that entanglement had a significant impact on survivorship. A report (Sayer et al. 2015) extending and summarising this dataset obtained between 2000 and 2013 (262 animals) reported a mean annual rate of 3.1% of animals observed to be entangled. In contrast to other studies, most entangled animals were adult (62%), with an approximately even split between males and females. When visible, the entangling material was identified ($n = 92$), and all but one was fishery related with the majority being monofilament (72%) (Fig. 13.7) or trawl net

Fig. 13.5 Adult female grey seal severely injured with constricted open wound, Isles of Scilly. *Image credit: Rebecca Allen, Cornwall Seal Group Research Trust*

Fig. 13.6 Juvenile grey seal in a packing band with which she was observed playing. *Image credit: Dave McBride, Cornwall Seal Group Research Trust*

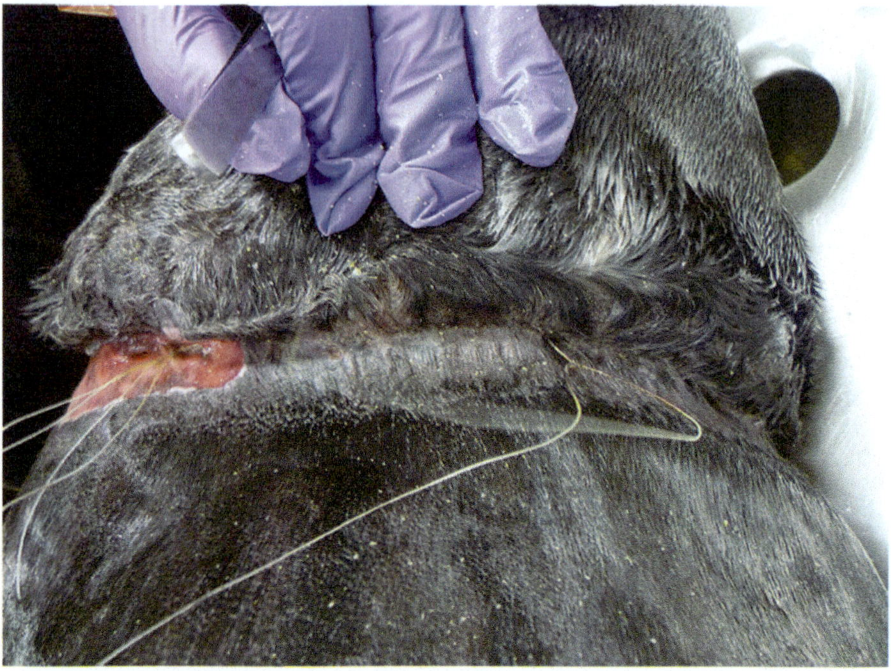

Fig. 13.7 (Post-mortem) monofilament lesion in an entangled grey seal. *Image credit: Sue Sayer, Cornwall Seal Group Research Trust and James Barnett, University of Exeter/Cornwall Wildlife Trust Marine Strandings Network*

Fig. 13.8 Juvenile grey seal entangled in plastic packing material— later successfully rescued. *Image credit: Simon Bone, Cornwall Seal Group Research Trust*

Fig. 13.9 Juvenile grey seal being rescued from trawl net by the British Divers Marine Life Rescue and the Cornish Seal Sanctuary. *Image credit: Sue Sayer, Cornwall Seal Group Research Trust*

Fig. 13.10 Adult male grey seal named 'Railway Arch' has lived with a partly healed entanglement wound for 13 years. *Image credit: Sue Sayer, Cornwall Seal Group Research Trust*

Fig. 13.11 Galapagos fur seal (*Arctocephalus galapagoensis*) with neck entanglement. *Image credit*: *Juan Pablo Muñoz*

(11%) (Figs. 13.1 and 13.9). Entanglements were observed around the neck (89%), body (2%), head (1%), mouth (2%), flipper (1%) and across multiple parts of the body (5%). The length of the trailing material and the presence of deeply constricted wounds were both significantly linked to reduced survivorship. Almost twice as many non-entangled seals survived over 10 years compared to those with deep constrictions. Rescues have routinely been performed successfully ($n = 30$) in this area in conjunction with the British Divers Marine Life Rescue and the Cornish Seal Sanctuary (Figs. 13.1 and 13.9). Post rescue photo identification of rescued, rehabilitated and released disentangled animals shows at the time of writing they can survive for long periods (up to 7 years is recorded by Sayer et al. 2015).

13.10 Conclusions

Pinnipeds are visible barometers of the spectrum of marine animals which can become snared, entangled, trapped or caught in marine debris. Seals are more visible than many marine animals because of their partial terrestrial habit. Marine plastic in the form of net, rope, monofilament line and packing bands can cause entanglement in a wide range of pinniped species, sometimes with severe consequences. There is the potential for severe acute welfare impacts on the individual animals through starvation and highly restrictive entanglement and some animals live for months or years (up to 16 years in one study of grey seals) with chronic deep incised wounds from net, packing band or monofilament line/net looped entanglement. Entanglement lesions can become chronic wounds, with deep infection that have debilitating consequences for the individual animal and leading to premature death in others.

Plastic is a 'new' challenge to these animals, man-made and entering the ocean in large quantities during the last century, and with an apparent dramatic rise in quantity, spread and effect particularly in the last 20 years. Plastic is probably very long lived in the sea (we don't yet know how long in practice), and there are plastic objects floating in the sea which have travelled thousands of kilometres. The effects of marine debris are not just aesthetic; marine debris has the potential to cause significant, widespread and 'hidden and unreported' animal suffering, through wounding, constriction, amputation, drag, infection, compromised feeding and ingestion. The pinniped species most likely to be affected by entanglement are fur seals, monk seals, California sea lions, grey seals, common seals and monk seals. Entanglement rates described in the literature range up to 7.9% of local populations annually (see

Table 13.1 Summary tabulation of reported entanglement rates for the pinniped species found in different ocean regions—the rate of entanglement (estimated % of population annually), the net, plastic and fishing line (% of reported entanglement cases for each category respectively) and the published source of the data

Ocean region	Species/subspecies	Rate of entanglement (%)	Net	Plastic	Fishing line	Published source
North-east Pacific	Steller sea lion	0.26	7	54	2	Raum-Suryan et al. (2009)
	Northern fur seal	0.4	65	19		Fowler (1987)
	Northern fur seal	0.08–0.35	39	37	9	Allen and Angliss (2014)
Eastern Central Pacific	Californian sea lion	0.08–0.22	19	25	14	Stewart and Yochem (1987)
	Californian sea lion	3.9–7.9	50		33	Harcourt et al. (1994)
	Northern elephant	0.15	19	36	33	Stewart and Yochem (1987)
	Harbour seal	0.09		33		Stewart and Yochem (1987)
	Northern fur seal	0.24	50			Stewart and Yochem (1987)
	Steller sea lion		4		4	Hanni and Pyle (2000)
Central Pacific	Hawaiian monk seal	0.7	32	8	28	Henderson (2001)
South-west Pacific	Kaikoura fur seal South	0.6–2.8	42	31		Boren et al. (2006)

(continued)

Table 13.1 (continued)

Ocean region	Species/subspecies	Rate of entanglement (%)	Net	Plastic	Fishing line	Published source
North-west Atlantic	Grey seal	3.1–5				Allen et al. (2012)
South-east Atlantic	Antarctic fur seal	0.024–0.059	48	18		Hofmeyr et al. (2002)
	Antarctic fur seal	0.4		46–52		Arnould and Croxall (1995)
	Cape fur seal	0.1–0.6		50		Shaughnessy (1980)
South-west Atlantic	Southern elephant seal	0.001–0.002		36	64	Campagna et al. (2007)
	Australian fur seal	1.9	40	30		Pemberton et al. (1992)
	New Zealand fur seal	0.9	29	30	3	Page et al. (2004)
	Australian sea lion	1.3	66	11	6	Page et al. (2004)
Western Indian Ocean	Antarctic and subantarctic fur seal	0.24	17	41	10	Hofmeyr et al. (2002)

Table 13.1)—with packing bands; fragments of lost net, rope, monofilament line and net; fishery flashers and lures; longline fishing gear, hooks and line; and bait hooks as the common and recurrent entangling materials in a number of seal and sea lion species.

The spread of plastic material in the ocean leaves seals entangled and, through entanglement and injury, sometimes results in their death through acute or chronic lesions, and this is a welfare concern. Entanglement results from human activity which was not anticipated or directly intentional, but which nonetheless is having a significant effect on animal welfare.

References

Allen BM, Angliss RP (2014) National Fur Seal (*Callorhinus ursinus*): Eastern Pacific Stock; Annual Human-Caused Mortality and Serious Injury National Oceanic and Atmospheric Administration. http://www.nmfs.noaa.gov/pr/sars/pdf/stocks/alaska/2014/ak2014_ northern-furseal- cp.pdf. Accessed 20 Oct 2016

Allen R, Jarvis D, Sayer S, Mills C (2012) Entanglement of grey seals *Halichoerus grypus* at a haul out site in Cornwall, UK. Mar Pollut Bull 64:2815–2819

Arnould JPY, Croxall JP (1995) Trends in entanglement of Antarctic fur seals (*Arctocephalus gazella*) in man-made debris at South Georgia. Mar Pollut Bull 30(11):707–712

Barnes DKA (2005) Remote islands reveal rapid rise of southern hemisphere, sea debris. Sci World J 5:915–921

Barnes DKA, Galgani F, Thompson RC, Barlaz M (2009) Accumulation and fragmentation of plastic debris in global environments. Philos Trans R Soc B 364(1526):1985–1998

Boland RC, Donohue MJ (2003) Marine debris accumulation in the nearshore marine habitat of the endangered Hawaiian monk seal, *Monachus schauinslandi* 1999–2001. Mar Pollut Bull 46:1385–1394

Boren LJ, Morrissey M, Muller CG, Gemmell NJ (2006) Entanglement of New Zealand fur seals in man-made debris at Kaikoura, New Zealand. Mar Pollut Bull 52:442–446

Campagna C, Falabella V, Lewis M (2007) Entanglement of southern elephant seals in squid fishing gear. Mar Mamm Sci 23(2):414–418

Cole T, Hartley D, Garron M (2006) Mortality and serious injury determinations for baleen whale stocks along the Eastern Seaboard of the United States, 2000–2004. Northeast Fisheries Science Center Reference Document 06-04

Convention on Biological Diversity (CBD) (2012) Impacts of marine debris on biodiversity: current status and potential solutions. Secretariat of the Convention on Biological Diversity and the Scientific and Technical Advisory Panel—GEF (2012), Montreal. Technical series no. 67. ISBN 92-9225-444-8

Coupland RM (1992) The Red Cross classification of war wounds: the E.X.C.F.V.M. scoring system. World J Surg 16(5):910–917

Dau BK, Gilardi KV, Gulland FM, Higgins A, Holcomb JB, Leger JS, Ziccardi MH (2009) Fishing gear-related injury in California marine wildlife. J Wildl Dis 45(2):355–362

Derraik JGB (2002) The pollution of the marine environment by plastic debris: a review. Mar Pollut Bull 44(9):842–852

Donohue MJ, Foley DG (2007) Remote sensing reveals links among the endangered Hawaiian monk seal, marine debris, and El Niño. Mar Mamm Sci 23(2):468–473

Fowler CW (1987) Marine debris and northern fur seals: a case study. Mar Pollut Bull 18(6b):326–335

Hanni KD, Pyle P (2000) Entanglement of pinnipeds in synthetic materials at south-east Farallon Island, California, 1976–1998. Mar Pollut Bull 40(12):1076–1081

Harcourt R, Aurioles D, Sanchez J (1994) Entanglement of California sea lions at Los Islotes, Baja California Sur, Mexico. Mar Mamm Sci 10:122–125

Hekman R, Osinga N (2010) Stichting Zeehondencreche Lenie t' Hart 2010. http://www.zeehondencreche.nl/wb/pages/wetensch.-onderzoek/wetenschappelijke-publicaties.php. Accessed 2/4/16

Henderson JR (2001) A pre- and post-MARPOL Annex V summary of Hawaiian monk seal entanglements and marine debris accumulation in the northwestern Hawaiian Islands, 1982–1998. Mar Pollut Bull 42:584–589

Hofmeyr GJG, De Maine M, Bester MN, Kirkman SP, Pistorius PA, Makhado AB (2002) Entanglement of pinnipeds at Marion Island, Southern Ocean: 1991–2001. Aust Mammal 24:141–146

Hofmeyr GJ, Bester M, Kirkman SP, Lydersen C, Kovacs KM (2006) Entanglement of Antarctic fur seals at Bouvetøya, Southern Ocean. Mar Pollut Bull 52:1077–1080

International Coastal Cleanup (ICC) (2008) Annual report—Ocean Conservancy 2008. http://www.oceanconservancy.org/our-work/marine-debris/2016-data-release/2016-data-release-1.pdf. Accessed 27 June 2016

International Maritime Organization (IMO) (2016) International convention for the prevention of pollution from ships. http://www.imo.org/en/OurWork/Environment/PollutionPrevention/Pages/Default.aspx. Accessed 18 Oct 2016

Karamanlidis AA (2000) Monitoring human and Mediterranean monk seal activity in the National Marine Park of Alonnissos and Northern Sporades, Greece. Monachus Guardian 3(1). http://www.monachus-guardian.org/mguard05/05covsto.htm. Accessed 3 Apr 17

Lawson TJ, Wilcox C, Johns K, Dann P, Hardesty BD (2015) Characteristics of marine debris that entangle Australian fur seals (*Arctocephalus pusillus doriferus*) in southern Australia. Mar Pollut Bull 98(2015):354–357

Marine Conservation Society (MCS) (2007) Beachwatch 2006. The 14th annual beach litter survey report. Marine Conservation Society, Ross on Wye. https://www.mcsuk.org/press/view/181 Accessed 6 Apr 16

MARPOL International Convention for the Prevention of Pollution from Ships (MARPOL) International Maritime Organization (1973). http://www.imo.org/en/About/Conventions/ListOfConventions/Pages/International-Convention-for-the-Prevention-of-Pollution-from-Ships-(MARPOL).aspx. Accessed 27 June 2016

McIntosh RR, Kirkwood R, Sutherland DR, Dann P (2015) Drivers and annual estimates of marine wildlife entanglement rates: a long-term case study with Australian fur seals. Mar Pollut Bull 101(2015):716–725

Moore MJ, van der Hoop J, Barco SG (2013) Criteria and case definitions for serious injury and death of pinnipeds and cetaceans caused by anthropogenic trauma. Dis Aquat Org 103(3):229–232

New World Encyclopedia (2016) History of fishing line. http://www.newworldencyclopedia.org/entry/Fishing_line. Accessed 18 Oct 2016

NOAA (2012) Marine mammal entanglement: pinniped. https://alaskafisheries.noaa.gov/pr/pinniped-entanglement. Accessed 12 July 2016

NOAA/NMFS (2007) Serious injury technical workshop, Seattle, 2007. Severity indexes page 7. http://www.nmfs.noa,a.gov/pr/pdfs/interactions/serious_injury_techmemo2008.pdf. Accessed 2 Apr 2016

NRC National Research Council (2008) Tackling marine debris in the 21st century. Committee on the effectiveness of international and national measures to prevent and reduce marine debris and its impacts. http://docs.lib.noaa.gov/noaa_documents/NOAA_related_docs/marine_debris_2008.pdf. Accessed 2 Apr 2016

Page B, Welling A, Chambellant M, Goldsworthy SD, Dorr T, van Veen R (2003) Population status and breeding season chronology of Heard Island fur seals. Polar Biol 26:219–224

Page B, McKenzie J, McIntosh R, Baylis A, Morrissey A, Calvert N, Haas T, Berris M, Dowie D, Shaughnessy PD, Goldsworthy SD (2004) Entanglement of Australian sea lions and New Zealand fur seals in lost fishing gear and other marine debris before and after Government and industry attempts to reduce the problem. Mar Pollut Bull 49:33–42

Pearson E, Mellish S, Sanders B, Litchfield C (2014) Marine wildlife entanglement: assessing knowledge, attitudes, and relevant behaviour in the Australian community. Mar Pollut Bull 89:136–148

Pemberton D, Brothers NP, Kirkwood R (1992) Entanglement of Australian fur seals in man-made debris in Tasmanian waters. Wildl Res 19:151–159

Raum-Suryan KL, Jemison LA, Pitcher KW (2009) Entanglement of Steller sea lions (*Eumetopias jubatus*) in marine debris: identifying causes and finding solutions. Mar Pollut Bull 58:1487–1495

Sandoe P, Simonsen HB (1992) Assessing animal welfare: where does science end and philosophy begin? Anim Welf 1:257–267

Santos IR, Friedrich AC, Barretto FP (2005) Overseas garbage pollution on beaches of northeast Brazil. Mar Pollut Bull 50(7):782–786

Sayer S, Williams K (2015) Ghost gear in Cornwall, UK 2014 to 2015. World Animal Protection Commissioned Report

Sayer S, Hockley K, Allen R (2015) Entanglement and its effects on grey seals (*Halichoerus grypus*) 2000 to 2013. World Animal Protection Commissioned Report, Cornwall and North Devon, UK

Shaughnessy PD (1980) Entanglement of cape fur seals with man-made objects. Mar Pollut Bull 11:332–336

Sheavly SB (2007) National Marine Debris Monitoring Program: Final Program Report, Data Analysis and Summary. Prepared for U.S. Environmental Protection Agency by Ocean Conservancy, Grant Number X83053401-02. 76 pp

Spraker TR, Lander ME (2010) Causes of mortality in northern fur seals (*Callorhinus ursinus*), St. Paul Island, Probilof Islands, Alaska, 1986–2006. J Wildl Dis 46(2):450–473

Stewart BS, Yochem PK (1987) Entanglement of pinnipeds in synthetic debris and fishing net and line fragments at San Nicolas and San Miguel Islands, California, 1978–1986. Mar Pollut Bull 18:336–339

United Nations (1972) Declaration of the United Nations conference on the human environment, Stockholm, 5–16 June 1972. http://www.unep.org/documents.multilingual/default.asp?documentid=97&articleid=1503. Accessed 18 Oct 2016

United Nations (2005) Environment programme, UNEP annual report 2005. http://www.unep.org/Documents.Multilingual/Default.asp?DocumentID=67&ArticleID=5125&l=en. Accessed 18 Oct 2016

Waluda CM, Staniland IJ (2013) Entanglement of Antarctic fur seals at Bird Island, South Georgia. Mar Pollut Bull 74:244–252

Wang J, Zhi Tan Z, Peng J, Qiu Q, Li M (2016) The behaviors of microplastics in the marine environment. Mar Environ Res 113:7–17

Wilcox C, Mallos NJ, Leonard GH, Rodriguez A, Hardesty BD (2016) Using expert elicitation to estimate the impacts of plastic pollution on marine wildlife. Mar Policy 65(2016):107–114

Williams R, Ashe E, O'Hara PD (2011) Marine mammals and debris in coastal waters of British Columbia, Canada. Mar Pollut Bull 62(6):1303–1316

World Animal Protection WAP (2016) Global ghost gear initiative. http://www.ghostgear.org. Accessed 18 Oct 2016

World Society for the Protection of Animals (WSPA) (2012) Marine debris: a global picture of the impact on animal welfare and of animal-focused solutions. In: Butterworth A, Clegg I, Bass C (eds). WSPA International, London. http://oceansandplastics.info/wp-content/uploads/2015/07/REP_WSPA_2012.pdf. Accessed 27 June 2016

Zalasiewicz J et al (2016) The geological cycle of plastics and their use as a stratigraphic indicator of the Anthropocene. Anthropocene 13:4–17. doi:10.1016/j.ancene.2016.01.002

Zavadil PA, Robson BW, Lestenkof AD, Holser R, Malavansky A (2007) Northern Fur Seal Entanglement Studies on the Pribilof Islands in 2006. For National Oceanic and Atmospheric Administration Prescott Stranding Grant Program and Alaska Regional Stranding Network Coordinator. http://doc.nprb.org/web/research/research%20pubs/outside_publications/733_entanglement.pdf. Accessed 2 Apr 2016

Zavala-González A, Mellink E (1997) Entanglement of California sea lions, *Zalophus californianus* californianus, in fishing gear in the central-northern part of the Gulf of California, Mexico. Fish Bull 95:180–184

Chapter 14
Loss of Habitat: Impacts on Pinnipeds and Their Welfare

Sheryl Fink

Abstract Pinnipeds around the world have been affected by habitat loss as a result of climate change and anthropomorphic activity, such as marine and coastal development. In addition to the physical reduction of available habitat, pinnipeds are impacted by secondary effects of habitat loss, such as disease and changes in prey availability. The impacts of global climate change are thought to be the most wide reaching, with changes in the availability and stability of sea and pack ice habitat expected to be most significant for at least 11 ice-associated species. Potential impacts on pinniped welfare occur as a result of changes in distribution and migration patterns, increased pup mortality, reduced foraging success, and decrease in body condition. Reductions in survival due to increased storm activity, increased exposure to disease and parasites, and human development have also been observed.

14.1 Introduction

While overexploitation is considered to have been the most important factor affecting the abundance and welfare of marine mammals historically, habitat destruction and fragmentation have become increasingly important threats to pinnipeds around the world.

Being relatively large and highly mobile marine species, pinnipeds are often thought to be less affected by habitat loss than many terrestrial animals. However, most have specific habitat needs for breeding or feeding. Identifying and quantifying habitat loss is challenging in marine environments, and understanding the impact on individual welfare is complex. Regardless, it seems clear that diminishing and deteriorating habitat are having a negative impact on the welfare and abundance of many pinniped species (Kovacs et al. 2011).

S. Fink
Director, Canadian Wildlife Campaigns, International Fund for Animal Welfare,
Guelph, Canada
e-mail: sfink@ifaw.org

A. Butterworth (ed.), *Marine Mammal Welfare*, Animal Welfare 17,
DOI 10.1007/978-3-319-46994-2_14

Of growing concern is the impact of climate change, which is resulting in a direct loss of ice habitat for many ice-associated pinniped species but may increase the availability of land-based habitats for others. Loss of ice and warming ocean temperatures occurring as a result of climate change will alter marine food webs, which will affect the distribution and availability of prey and result in changes to seals' foraging habitat and success. Ocean acidification, a result of increased carbon dioxide in the atmosphere, may also impact ice-associated seal survival and recruitment through disruption of food webs dependent on calcifying organisms (Kovacs et al. 2012). Rising sea levels are expected to result in reductions in available shoreline habitat for some pinniped species.

Other indirect impacts of climate change include induced habitat loss (Cooper et al. 2006), which will have a negative effect on pinniped welfare, increased disease and parasite risk (Karamanlidis et al. 2016), and increased impacts from human traffic and development in previously inaccessible areas (Skeate et al. 2012).

Apart from climate change, habitat loss continues to occur as a direct result of human activity, primarily coastal and marine development. Mineral, oil, and gas extraction, renewable energy development (Davis 2010; Tougaard et al. 2009), and practices such as aquaculture (Kemper et al. 2003) and the repeated use of mobile fishing gear have the potential to destroy or degrade areas of critical habitat (Skeate et al. 2012). Pinniped welfare may be compromised, and survival and reproductive rates may fall as a result of increased risk of entanglement in fishing gear (please see Chap. 13 this volume), exposure to chemicals that reduce immune system function or reproduction, and exposure to new pathogens or noise pollution (Harwood 2001). Even disturbance from tourism may pose a risk to seal welfare by forcing them to abandon preferred breeding or resting habitats (Johnson and Lavigne 1999b).

While it would be difficult to cover all sources and examples of pinniped habitat loss in this chapter, we address a few of the better known examples and species. Although little has been published on the welfare impacts of habitat loss, it is assumed that changes resulting in increased mortality, reduced pup survival, and decreased body condition will have negative welfare impacts for individual animals.

14.2 Climate-Driven Losses of Habitat

Global climate change is the most pervasive threat to pinnipeds worldwide, and climate-driven habitat losses are the most wide reaching in impact. The consequences of climate change on marine mammals have been increasingly documented in recent decades, with direct loss of sea ice habitat recognized as a prominent threat to Arctic marine mammals (Ragen et al. 2008, Tynan and DeMaster 1997; Simmonds and Isaac 2007, Laidre et al. 2008, 2015, Moore and Huntington 2008, Kovacs et al. 2011, 2012).

It is well documented that the earth's atmosphere is warming, causing regional adjustments in temperature, wind, ocean circulation, precipitation, ice cover and sea level, and pH balance, which are amplified in the polar regions (IPCC 2013). The extent of Arctic sea ice is now more than two million square kilometres less than it was in the late twentieth century (Kinnard et al. 2011). The declines in Arctic sea ice extent and thickness resulting from these changes are expected to continue into the future at increasing rates (IPCC 2007, 2013), with predictions suggesting that we are rapidly moving toward a seasonally ice-free Arctic (Overland and Wang 2013). Climate change is also affecting Antarctic habitat and ecosystems, although patterns have been very different in various sectors, with some areas experiencing increases in ice extent while others exhibiting a decrease. Central West Antarctica is one of the most rapidly warming regions on earth (Bromwich et al. 2013).

Sea ice habitats are unique in that they are spatially extensive, have few surface predators, and are virtually free of disease vectors (Kovacs et al. 2011). For ice-associated pinnipeds, ice acts as a physical platform, a marine ecosystem foundation, and a barrier to non-ice-adapted marine mammals and human commercial activities. Sea ice (and its loss) is an important factor in marine ecological dynamics, influencing productivity, species interactions, population mixing, gene flow, and pathogen and disease transmission (Post et al. 2013). The direct loss and deterioration of sea ice is the most obvious and immediate threat to ice-associated pinnipeds, particularly those in the Arctic.

14.3 Impacts of Climate Change-Induced Habitat Loss on Pinnipeds

Both Arctic and Antarctic seal species have evolved traits that depend on the larger-scale predictability of pack ice development, movements, persistence, and extent. Although accustomed to interannual fluctuations in ice and prey availability, ice-associated pinnipeds are vulnerable to a fast-changing environment and ill equipped to respond quickly to permanent or complete habitat loss as a result of climate change (Laidre et al. 2008; Moore and Huntington 2008).

Climate-driven habitat loss is expected to be most significant for ice-breeding pinniped species that require long periods of stable ice late in the spring season and specialist feeders who rely on prey species that are sensitive to changes in ice.

In the Arctic, seven pinniped species are considered ice associated, relying on ice to at least some extent for survival: ringed (*Pusa hispida*), bearded (*Erignathus barbatus*), spotted (*Phoca largha*), ribbon (*Histriophoca fasciata*), harp (*Pagophilus groenlandicus*) and hooded seals (*Cystophora cristata*), and walrus (*Odobenus rosmarus*). Of these, the walrus, ringed seal, and bearded seal are considered to be ice dependent and restricted to spatial and temporal domains influenced by sea ice.

In the Antarctic, four species of seal are closely tied to the presence of pack ice and require certain sea ice characteristics to complete their life cycle: the crabeater seal (*Lobodon carcinophagus*), Weddell seal (*Leptonychotes weddellii*), leopard seal (*Hydrurga leptonyx*), and Ross's seal (*Ommatophoca rossii*). These species rely on sea ice for critical portions of their life history and have demonstrated sensitivity to small changes in the sea ice environment. Climate change-driven impacts on prey distribution may also impact the ice-tolerant Antarctic fur seal (*Arctocephalus gazella*) and southern elephant seal (*Mirounga leonina L.*) (Siniff et al. 2008), who winter and forage in open water and marginal ice zones but reproduce on land.

Many pinniped species have already been affected by reductions in the geographic extent, seasonal duration, and stability of sea and pack ice (Siniff et al. 2008; Kovacs et al. 2011), and these impacts are expected to intensify. Changes in ice directly reduce the habitat available for seals that give birth and molt on sea ice, hide from predators or seek protection from inclement weather within ice fields, or eat ice-associated fish and other prey (Kovacs et al. 2011). This loss of available habitat can be expected to affect welfare through negative impacts on foraging success, breeding success, body condition, and mortality rates (Moore and Huntington 2008; Kovacs et al. 2011). Some of the impacts of ice habitat loss on pinnipeds that are likely to have welfare considerations include changes in distribution and migration patterns, increased pup mortality, decreased foraging success resulting from greater distances to food or changes to food webs, increased use of land-based haulouts, increased exposure to disease and contaminants, and increased exposure to human development and traffic (Tynan and Demaster 1997; Kovacs et al. 2011).

14.4 Changes in Distribution, Migration, and Abundance

Changes in the distribution and migration routes of some pinniped species are already occurring and are expected to continue, and this will ultimately alter population structure and genetic exchange rates. Ice habitat loss means many pinniped species will experience compression of their range concurrent with a loss of suitable breeding or foraging habitat, potentially resulting in population reductions. At the same time, subarctic and temperate pinniped species are likely to exhibit northward expansions of their ranges, which may place competitive pressure on endemic arctic species, further reducing their available habitat and putting them at greater risk of predation, disease, and parasite infections (Kovacs et al. 2011).

The manner and degree to which pinnipeds may adapt their behaviour, or relocate their breeding areas, in response to changing ice conditions, is still highly uncertain. Some species may be able to adapt; however species which are fixed in their traditional spatial and temporal cycles, and unable to shift, may be threatened with extirpation (removal or uprooting from a locality) or extinction (Kovacs et al. 2011). Major declines in abundance or pup production have already been documented for hooded seals in the Northeast Atlantic harp seals in the White Sea and ringed seals in Hudson Bay, which have largely been attributed to climate change impacts on ice conditions (Kovacs et al. 2012).

14.5 Increased Pup Mortality

In the complete absence of ice, female harps seals may move to find suitable ice outside of their historical pupping areas. However, if ice is present, females will give birth on ice insufficiently thick to persist throughout the nursing period, resulting in high levels of pup mortality (Stenson and Hammill 2014). There is no evidence that harp seal females have successfully adapted to give birth on land. In 2010 and 2011, poor ice conditions in the Gulf of St. Lawrence, Canada, resulted in mother harp seals giving birth on small, loose ice pans (Figs. 14.1 and 14.2), many barely able to hold the mother's weight. Pup mortality was extremely high, with large numbers of pups assumed to have drowned.

Fig. 14.1 Harp seal mother and pup in the Gulf of St. Lawrence, Canada, 2010. *Image credit: Sheryl Fink*

Fig. 14.2 A harp seal pup clings to a fragile pan of sea ice on March 25, 2006, in the midst of the annual commercial seal hunt. *Image credit: HSI/Frank Loftus*

Examination of dead and abandoned pups found on shore has demonstrated that causes of death included starvation, trauma from crushing by ice pans, pneumonia, and other infections that have not been observed among pups born on pack ice (Stenson and Hammill 2014). Predation by coyotes and eagles was also reported, species not present on pack ice. The type and extent of ice cover has also been found to be negatively correlated with the incidence of yearling harp seal strandings in the Northwest Atlantic (Soulen et al. 2013).

The ringed seal also depends almost exclusively on sea ice as a breeding habitat and haul-out platform. Ringed seals construct snow dens or subnivean lairs over breathing holes maintained in the ice, which are used for resting, and for females to give birth in the early spring. Both stable ice and sufficient snow to cover the lairs must be maintained long enough in the spring season to successfully complete the six-week period of nursing (Kovacs et al. 2011), in order to provide protection from predators and freezing. Spring rains or warm temperatures may cause the roofs of lairs to collapse, pushing pups out of the shelters and exposing them to predators such as polar bears, arctic fox, and ravens or gulls before they are able to survive in the open. In areas of less stable ice, pups tend to be smaller and may suffer higher mortality due to early separation from their mothers or an increased need to expend energy on thermoregulation. Ringed seals in some areas are already showing relatively long-term declines in reproductive rates and pup survival. With the disappearance of sea ice, many species such as walrus may be forced to rely on land-based haulouts. Terrestrial haul-out sites alone will not support the same number of walruses that the mixed seasonal use of sea ice and land has permitted in the past, and increasingly crowded conditions are likely to increase mortality among younger animals due to normal herd behaviours such as threat displays, fighting among bulls, manoeuvring for preferred positions within a herd, and general agonistic behaviour (Fay 1982; Jay et al. 2012; Kovacs et al. 2012).

Walruses on terrestrial haulouts are also at increased risk from polar bears, aircraft, boats, tourists/recreationists, hunters, feral dogs, etc. When threatened or disturbed, walrus stampede to the ocean which can lead to trampling and death of hundreds to thousands of animals each year, particularly young animals (Jay et al. 2011).

14.6 Nutritional Stress Due to Changes in Prey and Prey Availability

Decreases in prey abundance caused by a reduction of ice may impact pinnipeds' foraging success, body condition, reproductive rate, and pup survival. Availability of prey may also be reduced where reductions in ice habitat force pinniped species (e.g., bearded seal, walrus) to seek haul-out or whelping sites that provide poorer access to food. In the Antarctic, declines in ice extent have been correlated to reductions in krill, squid, and fish, which have been linked to decreased pup survival for southern elephant seals and crabeater seals (McMahon and Burton 2005). Conversely, increases in sea ice are thought to be related to reductions in phytoplankton blooms, which affect fish availability and lead to decreased foraging success for pregnant Weddell seal females (Siniff et al. 2008).

Reductions in sea ice have also been associated with declines in the clam populations that are critical prey for Pacific walruses, and the overall impact of less extensive seasonal ice coverage is expected to have a long-term negative impact on food resources of both walrus subspecies (Kovacs et al. 2015). Increased use of land haulouts, particularly by mothers and calves, could result in increased distances to food and greater energy expenditures from foraging trips and reduced access to preferred feeding grounds (Jay et al. 2011; Kovacs et al. 2011). Abandoned calves of Pacific walrus have been reported at sea, suggesting that females with dependent young may be experiencing nutritional stress with the retreat of their usual sea-ice resting platform separating them from feeding areas (Cooper et al. 2006).

14.7 Rising Sea Levels and Storms

Rising sea levels can mean a loss of habitat for seals and sea lions that rely on low-lying coastal areas for rest, moulting, giving birth, and pup rearing. Rising sea levels may submerge some of the low atolls, beaches, and small caves currently being used for giving birth by Mediterranean monk seals *(Monachus monachus),* and an evaluation of potential effects on the endangered Hawaiian monk seal (*Monachus schauinslandi*) in the Northwestern Hawaiian Islands found that maximum projected habitat loss ranged from 65–75% under modelled levels of sea level rise, which could increase extinction risk for these small isolated populations (Kovacs et al. 2012).

Increased storm activity associated with climate change is also likely to increase the risk of pinniped pups being separated from their mothers in beaches or caves and swept to sea and increase mortality of pups born on ice through trauma and crushing.

14.8 Disease and Contaminants

Although the causes are often difficult to pinpoint, pinniped mortality events have been linked to habitat degradation resulting from both climate change and coastal development. Individuals that are stressed or weakened are more susceptible to pathogens such as microparasites and morbilliviruses. Phycotoxins associated with algal blooms were linked to the mass die-offs of Mediterranean monk seals that occurred in the mid-1990s (Hernández et al. 1998).

14.9 Increased Human Development and Traffic

It is expected that the opening of previously inaccessible ice-covered areas will present increased threats from shipping and development (including oil and gas extraction) and spread from other oceanic areas of disease, parasite, and contaminant risks (e.g., Tynan and Demaster 1997, Ragen et al., 2008, Kovacs et al. 2011, Kovacs et al. 2012).

14.10 Habitat Loss Through Coastal and Marine Development

Pinnipeds, particularly those which haul out and give birth on land rather than ice, are also vulnerable to habitat loss by more 'traditional' causes of habitat loss: human activities such as coastal and marine development. About 44% of the world's population lives within 150 km of the coast (UN Atlas of Oceans 2016). As human population increases in coastal areas, pinniped species will feel the impacts of habitat loss, fragmentation, and change. Even if habitat is not completely lost, human disturbance can have a negative impact on resting and breeding behaviour if not properly managed.

Boating and shipping can impact marine mammals through direct boat strike injury and mortality, disturbance, and habitat destruction. While injury and mortality may appear to pose the greatest threats to welfare, behavioural changes associated with boat disturbance may affect energy budgets and general health and the well-being of individuals. Other pinniped species may be attracted to sounds of certain types of vessels or fishing activity, which can lead to boat impacts, entanglement, and death (Marsh et al. 2003).

14.11 Coastal Development

Coastal development has resulted in the reduction of suitable breeding habitat for a number of pinniped species, the most well known of which may be the Mediterranean monk seal (*Phoca monacus)* – currently the most endangered seal species (Aguilar 1999; Karamanlidis et al. 2016). Human persecution and disturbance has driven the Mediterranean monk seal into increasingly marginal habitat over centuries, with dramatic declines in both abundance and geographical range as a result of habitat deterioration, and with negative impacts on the welfare and conservation of the species (Johnson and Lavigne 1999a; Harwood et al. 1996, Karamanlidis et al. 2016).

Historically thought to haul out in colonies on open beaches and rocks that offered habitat to relatively large colonies of seals, Mediterranean monk seals have been displaced by centuries of development and human activity into progressively smaller and inaccessible marine caves to give birth and rear their pups (Johnson and

Lavigne 1999a). As space for colony and family group formation has disappeared, individual mothers with pups became the norm, forcing reduced gregarious behaviour, limited social interactions, and probably reduced mating and breeding success. Disturbance from construction and tourism may also force females to abandon their pups before weaning, resulting in death by starvation. Storm surges entering the caves can separate newborns from their mothers and sweep pups to their death when they drown or die from starvation (Gazo et al. 1999, 2000; Karamanlidis et al. 2016). In parts of the Eastern Mediterranean, seals have been discovered inhabiting caves that are little more than water-filled crevices with no internal beach or haul-out area, forcing animals to rest and sleep while floating in the water. It is unlikely that caves of this type can meet the essential biological or welfare needs of the species (Johnson and Lavigne 1999b).

14.12 Energy Exploration and Development

Anthropogenic pulsed sounds from activities such as seismic surveys, sonar, explosives, or pile driving are common in the marine environment and likely to increase, raising concerns about potential impacts on marine mammals. Ocean energy exploration and power generation presents a wide range of potential welfare concerns for pinnipeds. These include direct disturbance from increased human and vessel activity, changes in foraging conditions (which may be positive or negative), pollution, and the presence of noise which can cause direct physical auditory damage and affect patterns of distribution and abundance through behavioral responses (Tougaard et al. 2009; Skeate et al. 2012; Hastie et al. 2016).

Even renewable energy development may pose welfare threats. Many sites for wind farms are located on offshore sandbanks, which overlap with important pinniped habitats (Hastie et al. 2016). The extreme noise generated from pile-driving turbine bases may lead to the displacement of species such as harbour seals and increased competition with other species such as grey seals, which may be more tolerant of this activity (Skeate et al. 2012). There are concerns that the low-frequency anthropogenic noise emitted during the construction and maintenance processes of energy development may mask low-frequency underwater pinniped acoustic signals used for social communication, foraging, navigation, and mating. Potential collision and entanglement in mooring cables associated with wave energy or wind-float parks may also pose a welfare concern (Davis 2010).

14.13 Aquaculture

Marine aquaculture is the fastest growing world food industry and is rapidly occupying pinniped habitat in coastal and estuarine waters. Habitat degradation is caused not only by the loss of physical space but by the noise created by constant vessel traffic, the potential for local pollution, operational maintenance, and harvesting.

These impacts on pinnipeds are less studied (Wursig and Gailey 2003). Shooting, acoustic harassment devices, underwater explosives, exclusion nets, electric fencing, chasing, bright lights, and trapping and relocation are frequently used to deter pinnipeds from aquaculture sites and mitigate damage to equipment. Death and injury due to entanglement in antipredator nets is often reported (Kemper et al. 2003). Impacts of operational interactions with pinnipeds (with their associated welfare concerns of mortality and injury) have been well studied, but the effects of aquaculture on behaviour, range, demography, and ecology are far less explored.

14.14 Conclusions

Although it is clear that many pinniped species are being affected by various forms of habitat loss, little direct information on the welfare impacts of this loss currently exists. Habitat loss may have impacts on pup mortality, foraging success, reproductive success, and health, all of which will almost certainly affect the welfare of individual animals. There are numerous reported studies which examine the effects of climate change and human development on pinnipeds and their environments, but there is still great uncertainty in predicting future impacts and how species might adapt to these changes.

References

Aguilar A (1999) Status of Mediterranean monk seal populations RAC-SPA. United Nations Environment Program, Alos Editions, Tunis

Bromwich DH, Nicolas JP, Monaghan AJ, Lazzara MA, Keller LM, Weidner GA, Wilson AB (2013) Central West Antarctica among the most rapidly warming regions on earth. Nat Geosci 6:139–145

Cooper LW, Ashjian CJ, Smith SL, Codispoti LA, Grebmeier JM, Campbell RG, Sherr EB (2006) Rapid seasonal sea-ice retreat in the Arctic could be affecting Pacific walrus (*Odobenus rosmarus divergens*) recruitment. Aquat Mamm 3(1):98–102

Davis AE (2010) Report: potential impacts of ocean energy development on marine mammals in Oregon, College of Oceanic and Atmospheric Sciences, Oregon State University, Corvallis, OR 97330 https://ir.library.oregonstate.edu/xmlui/bitstream/handle/1957/19586/Amy%20Davis'%20review%20paper.pdf?sequence=1 Accessed 27 April 27, 2016

Fay FH (1982) Ecology and biology of the Pacific walrus, *Odobenus rosmarus divergens* Illiger. United States Fish and Wildlife Service, North American Fauna 74, Washington, DC, p 280

Gazo M, Layna JF, Aparicio F, Cedenilla MA, Gonzalez LM, Aguilar A (1999) Pupping season, perinatal sex ration and natality rates of the Mediterranean monk seal from the Cabo Blanco colony. J Zool 249:393–401

Gazo M, Gonzalez LM, Grau E (2000) Age at first parturition in a Mediterranean monk seal monitored long-term. Mar Mamm Sci 16:257–260

Harwood J (2001) Marine mammals and their environment in the twenty-first century. J Mammal 82(3):630–640

Hastie GD, Russell DJ, McConnell B, Thompson D, Janik VM (2016) Multiple-pulse sounds and seals: results of a harbor seal (*Phoca vitulina*) telemetry study during wind farm construction. Adv Exp Med Biol 875:425–430. doi:10.1007/978-1-4939-2981-8_50

Hernández M, Robinson I, Aguilar A, González LM, López-Jurado LF, Reyero MI, Cacho E, Franco J, López-Rodas V, Costas E (1998) Did algal toxins cause monk seal mortality? Nature 393:28–29

IPCC (2007) Chapter 4: climate change 2008 – the physical science basis. Observations: snow, ice and frozen ground. Cambridge University Press, Cambridge, pp 227–385

IPCC (2013) Climate Change 2013: The physical science basis. http://www.ipcc.ch/meetings/session36/p36_doc3_approved_spm.pdf Accessed 27 April 2016

Jay CV, Marcot BG, Douglas DC (2011) Projected status of the Pacific walrus (*Odobenus rosmarus divergens*) in the twenty-first century. Polar Biol 34:1065–1084

Jay CV, Fischbach AS, Kochnev AA (2012) Walrus areas of use in the Chukchi Sea during sparse sea ice cover. Mar Ecol Prog Ser 268:1–13

Johnson WM, Lavigne DM (1999a) Monk seals in antiquity. The Mediterranean monk seal (*Monachus monachus*) in ancient history and literature. Mededelingen 35:1–101

Johnson WM, Lavigne DM (1999b) Mass tourism and the Mediterranean monk seal. The role of mass tourism in the decline and possible future extinction of Europe's most endangered marine mammal, *Monachus monachus*. The Monachus Guardian. http://monachus-guardian.org/mguard04/04scien12.htm Accessed 28 April 2016

Karamanlidis AA, Dendrinos P, de Larrinoa PF, de Larrinoa PF, Gucu AC, Hohnson WM, Kirac CO, Pires R (2016) The Mediterranean monk seal *Monachus monachus*: status, biology, threats, and conservation priorities. Mammal Rev 46(2):92–105

Kemper CM, Pemberton D, Cawthorn M, Heinrich S, Mann J, Gales N, Hindell M, Kirkwood R (eds) (2003) Aquaculture and marine mammals: co-existence or conflict? Marine mammals: fisheries, tourism and management issues. CSIRO, Collingwood, VIC, Australia, pp 208–225

Kinnard C, Zdanowicz CM, Fisher DA, Isaksson E, de Vernal A, Thompson LG (2011) Reconstructed changes in Arctic Sea ice over the past 1450 years. Nature 479:509–512

Kovacs K, Lydersen C, Overland J, Moore E (2011) Impacts of changing sea-ice conditions on Arctic marine mammals. Mar Biodivers 41:181–194

Kovacs K, Aguilar A, Aurioles D, Bukarov V, Campagna C, Gales N, Gelatt T, Goldsworthy SD, Goodman SJ, Hofmeyr GJG, Harkonen T, Lowry L, Lydersen C, Schipper J, Sipila T, Southwell C, Stuart S, Thompson D, Trillmich F (2012) Global threats to pinnipeds. Mar Mamm Sci 28(2):414–436

Kovacs KM, Lemons P, MacCracken JG, Lydersen C (2015) Walruses in a time of climate change. Arctic Report Card: Update for 2015. http://www.arctic.noaa.gov/reportcard/walruses.html Accessed 27 April 2016

Laidre KL, Stirling I, Lowry LF, Wiig O, Heide-Jørgensen MP, Ferguson SH (2008) Quantifying the sensitivity of Arctic marine mammals to climate-induced habitat change. Ecol Appl 18(2):S97–S125

Laidre KL, Stern H, Kovacs KM, Lowry L, Moore SE, Regehr EV, Ferguson SH, Wiig Ø, Boveng P, Angliss RP, Born EW, Litovka D, Quakenbush L, Lydersen C, Vongraven D, Ugarte F (2015) Arctic marine mammal population status, sea ice habitat loss, and conservation recommendations for the 21st century. Conserv Biol 29(3):724–737

Marsh HD, Arnold P, Freeman M, Haynes D, Read A, Reynolds J, Kasuya T (2003) Strategies for conserving marine mammals. In: Gales N, Hindell M, Kirkwood R (eds) Marine mammals: fisheries tourism and management issues. CSIRO publishing, Collingwood, pp 1–19

McMahon CR, Burton HR (2005) Climate change and seal survival: evidence for environmentally mediated changes in elephant seal, *Mirounga leonina*, pup survival. Proc R Soc B 272: 923–928

Moore S, Huntington H (2008) Arctic marine mammals and climate change: impacts and resilience. Ecol Appl 18(2.) Supplement S157-S156

Overland JE, Wang M (2013) When will the summer Arctic be nearly sea ice free? Geophys Res Lett 40(10):2097–2101

Post E, Bhatt US, Bitz CM et al (2013) Ecological consequences of sea-ice decline. Science 341(6145):519–524

Ragen TJ, Huntington HP, Hovelsrud GK (2008) Conservation of arctic marine mammals faced with climate change. Ecological Applications (2 Suppl) S166–74

Simmonds MP, Isaac SJ (2007) The impacts of climate change on marine mammals: early signs of significant problems. Oryx 41(1):19–26
Siniff DB, Garrott RA, Rotella JJ, Fraser WR, Ainley DG (2008) Projecting the effects of environmental change on Antarctic seals. Antarct Sci 20(5):425–435
Skeate ER, Perrow MR, Gilfory JJ (2012) Likely effects of construction of Scroby Sands offshore wind farm on a mixed population of harbour *Phoca vitulina* and grey *Halichoerus grypus* seals. Mar Pollut Bull 64:872–881
Soulen BK, Cammen K, Schultz TF, Johnston DW (2013) Factors affecting harp seal (*Pagophilus groenlandicus*) strandings in the Northwest Atlantic. PLoS One 8(7):e68779. doi:10.1371/journal.pone.0068779
Stenson GB, Hammill MO (2014) Can ice breeding seals adapt to habitat loss in a time of climate change? ICES J Mar Sci 71(7):1977–1986
Tougaard J, Henriksen OD, Miller LA (2009) Underwater noised from three types of offshore wind turbines: estimation of impact zones for harbor porpoises and harbor seals. J Acoust Soc Am 125(6):3766–3773
Tynan CT, DeMaster DP (1997) Observations and predictions of Arctic climate change: potential effects on marine mammals. Arctic 50(4):308–322
UN Atlas of the Oceans (2016) Human Settlements on the Coasts. http://www.oceansatlas.org/servlet/CDSServlet?status=ND0xODc3JjY9ZW4mMzM9KiZwcmludGVyRnJpZW5kbHk9dHJ1ZSZzaG93Q2hpbGRyZW49dHJ1ZSYzNz1rb3M~ Accessed 28 April 2016
Wursig B, Gailey GA (2003) Marine mammals and aquaculture: conflicts and potential resolutions. In: Stickney RR, McVey JP (eds) Responsible marine aquaculture. CABI Publishing, Wallingford, pp 45–59

Chapter 15
Welfare Aspects of Commercial Hunting and Climate Change

Rebecca Aldworth

Abstract The ice-breeding seals that are the focus of the largest commercial kills of marine mammals on earth also face another human-caused welfare threat: climate change. The sea ice habitat that some pinniped populations rely on is melting at an alarming rate, causing increased mortality and associated poor welfare outcomes. At the same time, the lack of sea ice is changing the working environment of commercial sealers, exacerbating existing welfare problems and making it even more difficult, and at times impossible, to apply humane killing methods.

15.1 Introduction

Pinnipeds can experience some of the most severe direct impacts of human activity. Multiple species of seals have been, and continue to be, the targets of large-scale commercial hunts that have led to significant population declines as well as poor welfare outcomes (EFSA 2007; WTO 2013). Moreover, the ice-breeding seals that are the targets of large-scale commercial seal hunts are affected by climate change, which is fast destroying the seals' sea ice habitat and causing increased mortality by drowning and starvation. While these outcomes are discussed in detail by Fink (this volume), changes in sea ice conditions are also altering commercial sealing (Fig. 15.1) in ways that increase the risks of poor welfare which will be discussed later in Sects. 15.3, 15.3.1 and 15.3.2 in this chapter.

R. Aldworth
Humane Society International, Montreal, Quebec, Canada
e-mail: raldworth@hsi.org

A. Butterworth (ed.), *Marine Mammal Welfare*, Animal Welfare 17,
DOI 10.1007/978-3-319-46994-2_15

Fig. 15.1 A harp seal in the Gulf of St. Lawrence, Canada, watches approaching sealing boats at the commercial seal hunt. *Image credit: HSI/Kathy Milani*

15.2 What is a "Good Death?"

The word "euthanasia" is derived from the Greek terms "eu" (good) and "thanatos" (death). According to the American Veterinary Medical Association (Leary et al. 2013), "euthanasia" is generally used to describe ending the life of an individual animal in a way that minimizes or eliminates pain and distress. In other words, a "good death" can be equated to the humane ending of an animal's life.

Veterinary authorities around the world have contemplated the welfare implications of various methods of killing animals. Generally there is agreement that a "good death" involves the following processes:

- Stunning, using a humane stunning method, without causing unnecessary pain, fear or distress. There should be no need to repeat application of the stunning method, except in rare cases of mis-stunning. In these cases of mis-stunning, the kill is not considered to be humane.
- Monitoring to confirm unconsciousness *immediately* following stunning, using a series of indicators to confirm unconsciousness. Indicators of consciousness include corneal reflexes, righting reflexes, coordinated movements and rhythmic breathing.
- Killing, using an accepted method (such as bleeding), to be carried out *without delay* so that recovery of consciousness does not occur before death. The operator should be able to inspect and access the animal at all times during the bleeding process.

15.3 Commercial Sealing and Humane Slaughter

Canada's commercial seal hunt is the largest commercial slaughter of marine mammals in the world. Over the past two centuries, harp seals have been the primary targets of commercial sealers and, in that time, more than 50 million seals have been killed in Canada, with "allowable kills" of more than 350,000 in some recent years (DFO 2012).

Given its scale, it is unsurprising that the Canadian commercial seal hunt has been the subject of veterinary scrutiny for half a century. Recent reports on commercial sealing have consistently identified challenges to humane killing at the hunt and have made recommendations to reduce suffering (Burdon et al. 2001; Butterworth et al. 2007; Butterworth and Richardson 2013; Daoust and Caraguel 2012). Yet, despite resulting regulatory changes, there is evidence that poor welfare outcomes persist in commercial seal hunts (Butterworth and Richardson 2013). In response to animal welfare and conservation concerns, more than 35 countries have prohibited trade in products of commercial seal hunts, leading to a reduction in scale of commercial sealing. But while the number of seals killed has decreased and thus the overall welfare impacts of the commercial seal hunt have been diminished, there remain serious concerns with commercial sealing methods and their impacts on seals (Johnston et al. 2012; Butterworth and Richardson 2013).

In 2013, the World Trade Organization ruled on a challenge of the European Union prohibition on seal product trade by Canada and Norway. The panel, which evaluated considerable evidence submitted by all parties, stated in its reports (WTO 2013):

> …the circumstances and conditions of seal hunts present certain specific challenges to the humane killing of seals. Such challenges result in a risk in any given seal hunt that the targeted animals may suffer poor animal welfare outcomes of varying intensity and duration. Specifically, there are characteristics of the physical environment of seal hunting that affect the way seals are stunned and that can impact the degree of effectiveness of stunning attempts. We have also noted that attempts to strike or shoot a targeted seal more than once may not ameliorate the risks of ineffective stunning. Combined with the difficulties of assessing the consciousness of seals, seal hunting can present delays in carrying out the killing process and may pose specific animal welfare problems for seals that are struck and lost as well as for seals that are gaffed and hauled onto a sealing vessel while conscious. The challenge of reconciling the requirements of humane killing with the practical risks and difficulties of seal hunting, together with the potentially large territory of the hunt, poses an obstacle to monitoring and enforcement of the application of humane killing methods. Our assessment of the evidence taken together indicates that these risks to seal welfare are present in seal hunts in general.

The Canadian Marine Mammal Regulations permit two methods of stunning seals: clubbing (with a wooden bat or "hakapik") (Figs. 15.2, 15.3, 15.4) and shooting (with a shotgun or rifle) (Fig. 15.5). While it is possible that seals can be killed during the stunning process, the Regulations require that death be ensured via exsanguination (bleeding). The Marine Mammal Regulations do require the general steps involved in humane slaughter: stunning (by clubbing and shooting), monitoring (through observing for directed movement and/or skull palpation) and bleeding.

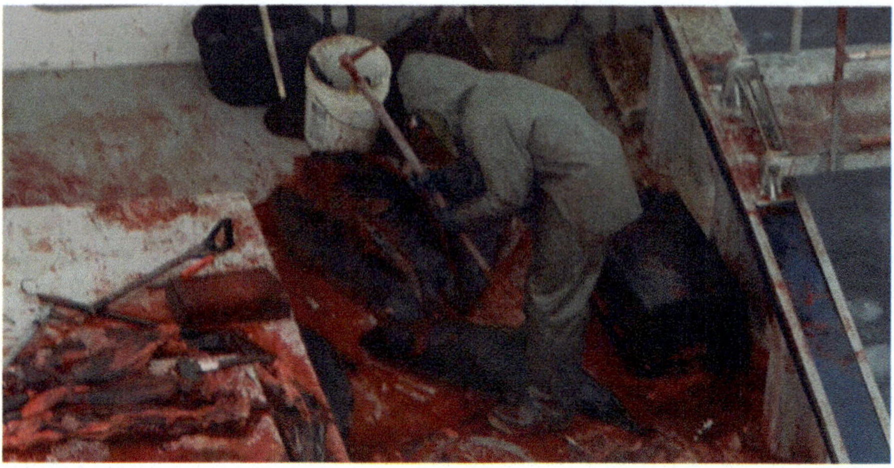

Fig. 15.2 A sealer clubs a seal that has been shot, impaled on a gaff and dragged onto the deck of the vessel on April 21, 2015. With diminishing ice conditions, this sequence during killing is very common as sealers are less able to disembark onto the ice. *Image credit: HSI/Michael Bernard*

Fig. 15.3 A sealer clubs a grey seal in the Canadian commercial seal hunt on February 23, 2011. *Image credit: HSI/Rebecca Aldworth*

Fig. 15.4 A sealer clubs a harp seal off Canada's East Coast on March 25, 2009. *Image credit: HSI /Jason Sanchez*

Fig. 15.5 A sealer aims his rifle at a seal during the commercial seal hunt in April 2011. *Image credit: HSI/Michael Bernard*

Yet, unlike the requirements in most recognized humane slaughter guidelines, it is not mandatory that these steps occur without delay, or prior to gaffing, dragging and moving the seal.

According to the WTO (2013):

> …we observe that delay between steps in the killing method can lead to prolonged suffering in seals and enhance the magnitude of poor welfare outcomes…Regarding actual delays in the killing process in seal hunts, evidence indicates that delays between steps in the killing process are an occurrence in seal hunts and that such delays can be attributable to pervasive characteristics of the hunts, including the physical conditions and the instruments used.

There is a good reason that Canada's sealing regulations do not actually prescribe a true humane slaughter process. The distance of vessels from the seals when they are shot, paired with the inability of sealers to work safely on the sea ice in many cases, makes application of humane slaughter steps in immediate succession a practical impossibility.

It is for this reason that many veterinary panels have sought to *improve* humane practice at the commercial seal hunt, while admitting that a low standard of welfare is unavoidable in commercial sealing. For example, Smith (2005), representing a panel of veterinarians who worked closely with the Canadian government on making improvements to the marine mammal regulations, stated:

> The Group recognizes that part of contributing to improved animal welfare and reduced suffering is to produce recommendations that are realistic in the context of the hunt, so that sealers will accept and implement them. There needs to be a realistic balance between ideal procedure and methodology, and what is practical and achievable.

As an example of this, Smith (2005) noted that while some members of the panel felt that bleeding should be a requirement of the Marine Mammal Regulations, others felt that "worker safety and the difficulties presented by the natural environment in which the hunt takes place were considerations that could make such a regulation difficult to apply, specifically in relation to hooking a seal." As a result, the report provided the following recommendation:

> Unless there is concern with sealer safety or possible loss of the seal, bleeding should take place immediately after checking and prior to hooking or skinning the animal.

Thus, Smith (2005) acknowledges that the welfare of seals can be secondary to safety and other concerns in commercial sealing. According to the American Veterinary Medical Association's Guidelines for the Euthanasia of Animals (Leary et al. 2013), "Selecting a method of euthanasia for free-ranging marine mammals can be a substantial challenge. Currently available euthanasia methods generally have significant limitations…." The Guidelines recognize "an inherent lack of control over free-ranging wildlife" and acknowledge that, "the quickest and most humane means of terminating the life of free-ranging wildlife in a given situation may not always meet all criteria established for euthanasia." In other words, the *most* humane way of killing free-ranging wildlife may not actually approach the standards that would be anticipated or expected for euthanasia in other species.

It appears that neither stunning/killing method for seals in Canada and Norway (clubbing and shooting) would be considered acceptable for general use in other commercial slaughter operations. Yet after decades of research, clubbing and shooting remain the only practical methods in commercial sealing.

15.3.1 Clubbing

> *Clubbing is a physical act, and the clubber must strike every blow with precision to ensure humane clubbing. It is probably impossible to invariably achieve this precision, given the cold and slippery conditions on the ice, the long hours, the pressure to work fast, and the possibility of a moving target.*
>
> *–Royal Commission on Seals and the Sealing Industry*

Leary et al. (2013) state that manually applied blunt force trauma to the head (clubbing) is "generally unacceptable for most species excluding piglets and small laboratory animals" and recommends to "replace, as much as possible, manually applied blunt force trauma to the head with alternate methods." The Council of the European Union states that stunning by a percussive blow shall not be used as a routine stunning method, but "only when there are no other methods available for stunning." In cases when stunning by manual percussive blow is necessary, the Regulation states that it should only be used for animals, including fur animals, up to 5 kg in weight.

Poor welfare outcomes associated with the widespread use of clubbing in commercial sealing have been reported for many years. Platt (1970) observed clubbing of seals and found that when seals were alarmed by the approach of a seal hunter, two distinct and completely separate reactions would result. The seal would either:

(1) Attempt to escape but after a short distance would stop and raise its neck and head into an almost vertical position.
(2) Freeze (play possum) with head withdrawn and protected by a large, loose fold of blubber that enveloped all but the anterior part of the face. The caudal limbs would be in a state of tension and crossed to protect the genitals.

In these cases, "examination of skull damage after skinning, invariably revealed damage entirely confined to the facial bones, or to the anterior aspect of the cranial bones, i.e. in the case of (1) destruction of the incisive or nasal bones and in (2) the upper aspect of the orbital cavities."

Platt's findings have been echoed in a number of other reports, which also showed a consistent lack of cranial injury during examinations of the skulls of seals clubbed by Canadian seal hunters (Hughes 1966, 1981; Simpson 1966; Scott 1971, 1977; Jordan 1978; Taylor 1979; Rowsell 1980; Burdon et al. 2001; Butterworth et al. 2007).

In 2001, Burdon et al. conducted random post-mortems on 76 seal carcasses abandoned on the ice. They found that in 17% of the seals, there were no apparent skull fractures, and in these cases, association with a level of unconsciousness would be highly improbable. A further 25% of the seals had minimal or moderate fractures. In cases of minimal fractures, unconsciousness would have been highly improbable, and, while moderate fractures would be more likely to be associated with a more significant decrease in consciousness, they would still not have a high level of probability to be associated with unconsciousness. Thus, Burdon et al. reported that in 42% of seals examined, there was not enough evidence of cranial injury to ensure that unconsciousness had occurred and concluded, "the current methods and competency of clubbing is significantly inaccurate in location, resulting in severe and unacceptable suffering."

In 2002, Daoust et al. reported on observations of the 2001 seal hunt in the Gulf of St. Lawrence. Dr. Pierre-Yves Daoust was stationed on board sealing vessels and observed 167 seals clubbed or shot. In 155 instances, the method of killing was observed. Though the report cautions that Dr. Daoust's "presence on board of sealing vessels may have incited sealers to hit the seals' skulls more vigorously," of the 100 seals he observed clubbed, 14% had skulls that were either not crushed or were only partially crushed. In 2% of cases, the seal was still alive after being brought onto the vessel and was subsequently clubbed. An additional of nine seals were wounded and then lost at sea.

Butterworth et al. (2007) performed *post-mortems* on 17 seals found abandoned on the ice. All of these had been clubbed and one had also been shot. In 47% of cases, the seals had been clubbed on the face or neck, and 82% had ocular damage. For 15 of the 17 animals, there were significant welfare concerns; most seals had multiple fracture sites, and a significant number (59%) had pre-mortem bleeding in the mouth or nostrils. Fifty-nine percent had damage to the face and lower jaw, suggesting inaccurate blows when clubbing the seal.

15.3.2 Shooting

Leary et al. (2013) stated, "For wildlife and other freely roaming animals, the preferred target area should be the head. It may, however, not be possible or appropriate to target the head when killing is attempted from large distances (missed shots may result in jaw fractures or other nonfatal injuries)."

Malouf (1986) explained the inherent obstacles to accuracy in shooting in commercial sealing operations, noting, "Many Canadian hunts take place, or have taken place, under conditions which make it impossible to obtain an acceptably high proportion of kills with head shots…The causes include long-range shooting, shooting from moving boats, and shooting at seals in the water." Additional potential obstacles to accuracy in shooting include high winds, extreme weather conditions, cold temperatures, ocean swells, low visibility and operator fatigue (Butterworth and Richardson 2013).

According to EFSA (2007):

> The main disadvantage with firearms is the risk of the targeted animal being hit with insufficient force and/or accuracy to cause instantaneous death or unconsciousness, and possibly escaping wounded. This may be caused by one or more of the following: poor marksman-

ship, excessive distance (e.g. shots fired over ranges > 50 m), unstable platforms (e.g. a boat or ice floe in rough weather conditions), unanticipated movement by the animal (e.g. a sudden movement of the head just before the hit) or inadequate firearms/ammunition. In any seal hunt, it is likely that a certain proportion of the animals will be only wounded, regardless of the power of the ammunition. Wounded seals may escape before they are re-shot, as there is no guarantee that the rifleman will be able to inflict a successful repeat shot immediately.

This echoes the findings of Malouf (1986), which found, "Shooting is clearly a humane way of killing if the animal is killed outright, but in any large-scale operation some proportion of seals will merely be wounded. These animals may recover or may die some time later; in either event the shooting inflicts a serious degree of suffering."

Multiple studies have recorded very high rates of wounding when seals are shot in commercial seal hunts (Butterworth and Richardson 2013) (Figs. 15.5–15.10). Burdon et al. (2001) observed 57 seals that were shot. Of these, three were struck and lost, 21 were struck multiple times, one was shot and not immediately recovered despite evidence of life, and 15 were shot and then impaled with hooks and dragged onto the vessels in absence of a test for unconsciousness. Of the latter group, one seal was observed being clubbed on board the boat.

In 2002, Daoust et al. reported on observations of the commercial seal hunt in the Gulf of St. Lawrence. The author was stationed on four different sealing boats and observed 167 seals that had been clubbed or shot and were brought on board the

Fig. 15.6 A seal that has been shot and wounded crawls across the ice, swims to a second ice pan, and then struggles onto it before a sealer arrives to club the seal on April 8, 2010. *Image credit: HSI/Gray Mitchell*

Fig. 15.7 A seal that has been shot and wounded crawls through her own blood while a sealing vessel approaches in April 2010. *Image credit: HSI*

Fig. 15.8 A seal is shot in the water at the commercial seal hunt on April14, 2014. *Image credit: HSI/Frank Loftus*

Fig. 15.9 A seal is shot at repeatedly and wounded in open water in April 2009. The sealers must impale the wounded seal on a gaff and drag the animal up and onto the deck before any effective monitoring for consciousness can occur. *Image credit: HSI*

vessels or lost in the ocean. His report indicates about half of the seals were clubbed and about half were shot. However, in 12 of these cases, he was unable to observe the method of killing or perform an examination of the carcass. Of the 43 seals observed shot, 93% were struck again with a club. For the vast majority (≥85%), Dr. Daoust recorded the interval between the shot and the blow(s) was less than or equal to 1 min. The report noted that, "A certain proportion of animals (3 of 8, in one instance where exact records were kept) were still alive during that interval, as shown by the conspicuous movements of their head."

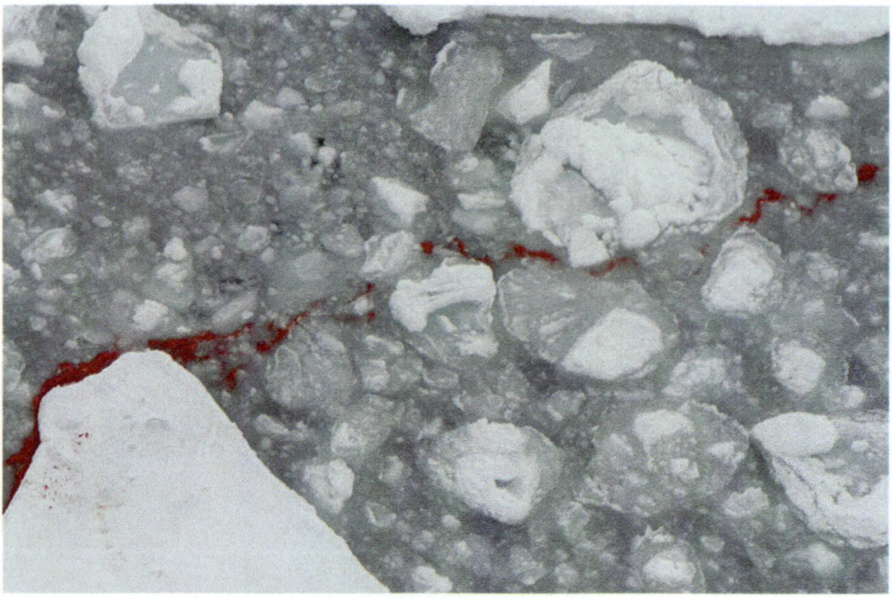

Fig. 15.10 A blood trail from a struck and lost seal at the commercial seal hunt in April 2007. The original site of the shot can be seen by the blood on the ice pan in the bottom left of the image. *Image credit: HSI*

Butterworth et al. (2007) evaluated 51 sequences in which seals were shot and the point of impact could be established. Of these, 59% were shot in regions of the body other than the head or were missed entirely. In total, 66% of the shot seals required multiple strikes, and a further 16% responded to stimuli after being shot but were not struck again. The mean duration of time from first shot to contact by the sealer was 48.8 ± 9.4 seconds, indicating a considerable period of potentially poor welfare (EFSA 2007).

Butterworth et al. (2007) concluded:

> We compare the welfare standards of the seal hunt with the expectations in commercial slaughterhouses in both the EU and Canada. The likelihood of a single effective shot or strike in dispatching seal pups is well below the corresponding standards achieved for animals in slaughterhouses. Since the number of affected animals is large, and the levels of wounding are high, the hunting method should be considered unacceptable.

15.4 Climate Change is Exacerbating Welfare Problems

Our climate is changing, and it seems to be happening at an unprecedented rate. Our planet is warming, sea ice is declining, and pollutants are accumulating within the environment and within organisms (Moore 2008). These changes are most certainly affecting marine ecosystems, and there is strong evidence that many marine mammal species will soon be, and already are, negatively impacted by climate change (McCarthy 2001). Ice-associated species such as harp seals, the primary focus of Canada's commercial seal hunt, provide examples of the conservation impacts of climate change on marine mammals. Like polar bears, these seals are ice-dependent animals, and they rely on the sea ice to give birth and to nurse their pups. Diminished sea ice cover can cause reduced reproductive success of adult females and higher rates of pup mortality (Johnston et al. 2005; Cote and Pigeon 2007; Department of Fisheries and Oceans 2010).

Unfortunately, climate change is causing sea ice to decline at an alarming rate (Moore 2008) and ice cover in seal whelping areas is expected to continue to decline, causing increased mortality for seals (Cote and Pigeon 2007; Department of Fisheries and Oceans 2010). While the conservation concerns associated with habitat loss for ice breeding seals has been reported in multiple studies, another negative outcome related to climate change has received less attention. The diminishing sea ice is compromising animal welfare in commercial sealing operations through changes in the ways in which the hunters interact with the seals (Butterworth and Richardson 2013).

15.4.1 Higher Wounding Rates

The reduction in sea ice cover associated with climate change is causing an increased reliance on long distance shooting from sealing vessels (WTO 2013). Very high wounding rates in seals that are shot have been recorded in multiple studies (Butterworth and Richardson 2013), as have the challenges and delays in effective monitoring for consciousness when shooting from vessels (WTO 2013). As sea ice continues to diminish, it is logical to anticipate an increase in the poor welfare outcomes related to shooting in the commercial seal hunt.

15.4.2 Increased Numbers of Open Water Shootings

Smith (2005) provided recommendations on improving welfare in commercial sealing operations. The Canadian House of Commons Standing Committee on Fisheries and Oceans (Keddy 2007) subsequently advised the Canadian government to implement these recommendations. While the Canadian government based some of its 2009 amendments to the Marine Mammal Regulations on Smith (2005), the recommendation that *"a seal should not be shot in the water, or in any circumstance when it is possible the carcass cannot be recovered"* was omitted. In its formal response to the Standing Committee, the Canadian government stated that the sealing industry and Department of Fisheries and Oceans would continue to consult with the authors of Smith (2005) on *"the practical application of some recommendations, such as a prohibition of shooting seals in the water which could have a detrimental effect on the fishery when poor ice conditions prevail."*

Ultimately, the Canadian government did not prohibit the shooting of seals in open water, and the practice is likely to increase in frequency as sea ice diminishes. Shooting seals in the water significantly increases the risk of poor welfare outcomes (Daoust and Caraguel 2012). Application of the humane slaughter method is impossible when shooting seals in the water because effective monitoring for consciousness is impossible prior to hooking and hoising onto the vessel (Butterworth and Richardson 2013). Shooting seals in open water also contributes to incidences of struck and loss (WTO 2013).

15.4.3 Higher Struck and Lost Rates

Loss rates are higher for sea hunts than for land hunts given the ability of marine mammals to dive beneath the surface of the water and remain there for prolonged periods of time and unconscious seals are in a negative state of buoyancy and can sink (NAMMCO 2006). According to Sjare and Stenson (2002), *"One of the major sources of unreported mortality during the commercial harp seal* (Pagophilus groenlandicus) *hunt in the Northwest Atlantic is the number of animals that are killed but not recovered or reported, commonly referred to as struck and lost."* Struck and lost seals can survive with injuries that profoundly impact their continued survival (WTO 2013).

Given the deteriorating ice conditions over the past decade, virtually all recent sealing now occurs amidst broken sea ice pans. If the recommendation by Smith (2005), to prohibit shooting in any circumstance in which it is possible the carcass may not be recovered, were implemented, the present-day seal hunt would effectively be ended. Keddy (2007) noted that a prohibition on shooting seals in open water would have a detrimental effect on the profitability of the seal hunt in poor ice years, indicating that sealers shoot more seals in open water when there is reduced sea ice cover, while WTO (2013) states that shooting seals in open water contributes to incidences of struck and lost.

15.4.4 Increased Risks of Conscious Seals Gaffed onto Vessels

Climate change has caused a significant deterioration in the sea ice cover that forms the working environment for sealers during the commercial sealing season (Johnston et al. 2012). The sea ice pans are smaller and more fragile and often will not support the weight of a sealer. This can make it impossible for sealers to retrieve wounded seals with any other means than hooking with a gaff (a long wooden pole with a metal hook at the end) or hakapik (a clubbing instrument with a hook at the end) (Figs. 15.11–15.13). WTO (2013) reports that recovering seals onto vessels with hooks and gaffs is consistent with the occurrence of unstable ice conditions and the related increase in the use of firearms.

Fig. 15.11 A 3-week-old seal pup is shot and wounded and left to suffer before being impaled on a gaff while visibly opening his mouth in April 2011. *Image credit: HSI/Michael Bernard*

Fig. 15.12 A 3-week-old seal pup is shot and wounded and crawls across the ice before being gaffed onto a sealing vessel on April 21, 2015. *Image credit: HSI/Michael Bernard*

Fig. 15.13 A wounded seal is gaffed onto solid ice. *Image credit: HSI/Michael Bernard*

EFSA (2007) recommended that unless they are in the water, seals should not be "moved, i.e. gaffed, hauled or moved from the position they have come to rest, until it has been confirmed that they are dead or irreversibly unconscious, or have been bled out." Yet, there are many recorded instances of hooking and gaffing of potentially conscious seals in evidence (WTO 2013).

In a 2008 interview with the National Post, the president of the Canadian Sealers Association explained that when sea ice is too thin for the sealer to stand on it, the seal can only be monitored for consciousness after being impaled on a hook and hoisted onto the vessel (Hanes 2008). WTO (2013) states:

> Physical conditions and concerns for sealers' safety may demand that the seal be hooked onto the boat if it cannot be checked on the ice. This may have potentially severe negative impacts on animal welfare if a seal is conscious and sensible to pain during this process… We find compelling evidence to show that the possibility of retrieving seals by hook/gaff is important to the feasibility of commercial seal hunting in Canada and Norway. Furthermore, given the difficulties of assessing the consciousness of the seal and the challenges of re-stunning by firearm, there is a possibility that some seals will be conscious when hooked or gaffed leading to severe negative consequences for animal welfare.

15.4.5 Increased Delays in Monitoring

In contrast to seals that are clubbed with a wooden bat or hakapik, when a seal is shot, there is no requirement that animals are monitored for unconsciousness immediately after stunning. Rather, the Regulations only require the sealer to perform the test *"as soon as possible."* As detailed in a previous section, as the sea ice deteriorates, reliance on long-distance shooting as a primary stunning method is increasing, and Smith (2005) confirms that when seals are shot, there is often a delay in sealers being able to monitor for consciousness. WTO (2013) echoes this observation, stating that, "Evidence specifically confirms that the likelihood of delay is greater when a firearm is the stunning instrument used because of the distance between the sealer and the seal." Butterworth et al. (2007) noted that, *"the mean duration from first shot to contact by the sealer was 48.8 ± 9.4 seconds, indicating a considerable period of potentially poor welfare."*

15.4.6 Increasing Inability of Sealers to Perform Humane Slaughter Techniques on the Ice

Leary et al. (2013) state that animals should be bled out immediately following confirmation of unconsciousness, while Smith (2005) states, "Bleeding of a seal will ideally be done on the ice immediately following the checking process." The World Organization for Animal Health (2016) states that it should be possible for the slaughterer "to observe, inspect and access the animals throughout the bleeding period," and that any animal showing signs of recovering consciousness should be re-stunned. Unfortunately, Canada's Marine Mammal Regulations, as amended in 2009 (Government of Canada 2009), do not mirror these requirements. Instead, the Regulations stipulate that, *"No person shall skin a seal until the cranium has been crushed and at least one minute has elapsed after the two axillary arteries of the seal located beneath its front flippers have been severed to bleed the seal."* Thus, bleeding can occur at any time after monitoring for unconsciousness and after the animal has been impaled with a gaff hook and dragged across the ice and onto a vessel.

Climate change is causing a reduction in sea ice in commercial sealing areas, which makes it increasingly difficult for sealers to operate on the ice. This, in turn, creates significant obstacles to bleeding out seals on the ice and remaining with them for the duration of the process.

15.5 Conclusions

The environment in which the Canadian commercial seal hunt occurs presents significant obstacles to consistent application of humane slaughter methods. This has resulted in poor welfare outcomes for seals, which have been reported in multiple studies over the past half century.

Climate change has caused sea ice cover off Canada's East Coast to diminish over the past decade, and the trend toward decreased sea ice formation is expected to continue in the coming ones. The conservation impacts of diminished sea ice on the ice-dependent harp seals are apparent, and reductions in sea ice are also impacting adversely on animal welfare in the context of the commercial seal hunt. As ice conditions worsen, the suffering witnessed during the hunt can be anticipated to increase.

References

Burdon RL, Gripper J, Longair JA, Robinson I, Ruehlmann D (2001) Veterinary Report, Canadian Commercial Seal Hunt, Prince Edward Island, March 2001. Unpublished report to the International Fund for Animal Welfare

Butterworth A, Gallego P, Gregory N, Harris S, Soulsbury C (2007) Welfare aspects of the Canadian seal hunt: Preliminary report, 30 pp and Final report, p 45. Document submitted to European Food Safety Authority

Butterworth A, Richardson M (2013) A review of animal welfare implications of the Canadian commercial seal hunt. Mar Policy 38:457–469

Cote F, Pigeon M (2007) Climate change and seal populations in Canada. Parliamentary information and research service. http://www.sealsandsealing.ca/resources/ClimateChange07.pdf Accessed 23 Oct 16

Daoust PY, Caraguel C (2012) The Canadian harp seal hunt: observations on the effectiveness of procedures to avoid poor animal welfare outcomes. Anim Welf 4:445

Daoust PY, Crook A, Bollinger TK, Campbell KG, Wong J (2002) Animal welfare and the harp seal hunt in Atlantic Canada. Can Vet J 43(9):687

Department of Fisheries and Oceans (DFO) (2010) A review of ice conditions and the harp seal total allowable catch (TAC) for 2010. Canadian Science Advisory Secretariat. Science response. 2010/004

Department of Fisheries and Oceans (DFO) (2012) Current Status of Northwest Atlantic Harp Seals, (Pagophilus groenlandicus). DFO Canadian Science Advisory Secretariat. Sci. Advis. Rep. 2011/070

European Food Safety Authority (EFSA) Animal Health and Animal Welfare Panel (2007) Animal Welfare aspects of the killing and skinning of seals - Scientific Opinion of the Panel on Animal Health and Welfare. p 165. doi:10.2903/j.efsa.2007.610

Government of Canada Marine Mammal Regulations SOR/93-56 Amended 2009. http://laws-lois.justice.gc.ca/eng/regulations/SOR-93-56/ Accessed 23 Oct 16

Hanes A (2008) More humane seal hunt takes to ice. National Post. http://bleudeterre.wordpress.com/2008/03/23/more-humane-seal-hunt-takes-to-ice/ Accessed 23 Oct 16

Hughes TI (1981) 1981 Sealing Report. Report for Commission on Seals and Sealing. Ministry of Supply and Services Canada ISBN 0 660 12127 1

Hughes TI (1966) Report of inspection of 1966 seal hunting in the Gulf of St Lawrence. Ontario Humane Society, Toronto

Johnston DW, Friedlaender AS, Torres LG, Lavigne DM (2005) Variation in sea ice cover on the east coast of Canada from 1969 to 2002: climate variability and implications for harp and hooded seals. Clim Res 29(3):209

Johnston DW, Bowers MT, Friedlaender AS, Lavigne DM (2012) The effects of climate change on harp seals (*Pagophilus groenlandicus*). PLoS One 7(1):e29158. doi:10.1371/journal.pone.0029158

Jordan WJ (1978) The killing of harp seal pups. Report for Royal Society for the Prevention of Cruelty to Animals

Keddy G (2007) Ensuring a sustainable and humane seal harvest. Report of the Standing Committee on Fisheries and Oceans. http://www.parl.gc.ca/HousePublications/Publication.aspx?Language=E&Mode=1&Parl=39&Ses=1&DocId=2872843&File=0 Accessed 23 Oct 16

Leary S, Underwood W, Anthony R, Cartner S, Corey D, Grandin T, Greenacre CB, Gwaltney-Bran S, McCrackin MA, Meyer R, Miller D (2013) AVMA American Veterinary Medical Association guidelines for the euthanasia of animals: 2013 edition

Malouf A (1986) Seals and sealing in Canada: report of the Royal Commission, Montréal: Royal Commission on seals and the sealing industry in Canada, vol 3. Ministry of Supply and Services Canada, Canada. ISBN 0 660 12127 1

McCarthy JJ (2001) Climate change 2001: impacts, adaptation, and vulnerability: contribution of working group II to the third assessment report of the intergovernmental panel on climate change. Cambridge University Press, UK

Moore SE (2008) Marine mammals as ecosystem sentinels. J Mammal 3:534–540

North Atlantic Marine Mammal Commission (NAMMCO) (2006) Report of the NAMMCO workshop to address the problems of "Struck and Lost" in seal, walrus and whale hunting. North Atlantic House. Copenhagen, Denmark; 14–16 November 2006. http://www.nammco.no/webcronize/images/Nammco/818.pdf Accessed 23 Oct 16

Platt C (1970) Harp seal Symptoms exhibited by seal pups in relation to stunning and killing with hakapik. Report for International Society for the Protection of Animals

Rowsell HC (1980) Observations on the harp seal hunt: the front – 1980. Royal Commission on seals and the sealing industry in Canada. Ministry of Supply and Services Canada, Canada. ISBN 0 660 12127 1

Scott TH (1971) Annual seal hunt – Northwest Atlantic: gulf of St. Lawrence area, 1971 season. Report: International Society for Protection of Animals, London

Scott TH (1977) Landsmen and the long liner seal hunt off the northeast coast of Newfoundland, April 1977. Report: International Society for the Protection of Animals, London

Simpson E (1966) Observations on the seal hunt in the Gulf of St. Lawrence, march 1966. Final report for New Brunswick Society for the Prevention of cruelty to animals. Fredericton, New Brunswick

Sjare B, Stenson GB (2002) Estimating struck and loss rates for harp seals (*Pagophilus groenlandicus*) in the Northwest Atlantic. Mar Mamm Sci 18(3):710–720

Smith B (2005) Improving humane practice in the Canadian harp seal hunt: A report of the Independent Veterinarians' Working Group on the Canadian harp seal hunt

Taylor GB (1979) Report on the seal cull off the Magdalen Islands. Report: International Society for the Protection of Animals, London

World Organization for Animal Health (OIE) (2016) Terrestrial Animal Health Code. 7.5.1. Slaughter of animals. http://www.oie.int/index.php?id=169&L=0&htmfile=chapitre_aw_slaughter.htm. Accessed 23 Oct 26

WTO World Trade Organization (Reports of the Panel) (2013) European Communities—Measures Prohibiting the Importation and Marketing of Seal Products. p 186 (WT/DS400/401)

Chapter 16
Assessing the Welfare of Pinnipeds

Isabella L.K. Clegg and Andy Butterworth

Abstract A recent increase in collaborative and independent studies on sea lions, seals and walruses has advanced our knowledge and interest in pinniped welfare. Nevertheless published discussions of the welfare of pinnipeds, and secondly of potential measures to assess their welfare, are, respectively, very few and non-existent. This chapter aims to make first steps in the discussion on assessing pinniped welfare, with the goal of stimulating future welfare investigations. Pinniped species are able to thrive in two opposing environments, the land/ice margin at the coast and in the sea, and these animals use these two 'domains' for different functions. Welfare measurement is concerned with the outcome of an animal's internal and external responses to its environment, and pinniped species' evolutionary biology may be especially important in this respect, in terms of our understanding of the animals' responses and interactions within their two domains. Pinnipeds are being directly impacted by serious anthropogenic disturbances in the wild, including human interference at established feeding and breeding grounds, hunting, entanglement and climate change, and are also often kept in captive collections. Feasible evaluations of welfare can therefore be assumed to have potential widespread utility, including applications benefitting the animals themselves.

16.1 Introduction

Although the literature on pinnipeds is not as extensive at that of the charismatic megafauna cetacean species, many collaborative as well as independent studies on sea lions, seals and walruses have advanced our knowledge and sustained the interest in these fascinating animals. Nevertheless published discussions of the welfare of pinnipeds, and secondly of potential measures to assess their welfare, are very

---

I.L.K. Clegg (✉)
Laboratoire d'Ethologie Expérimentale et Comparée,
E.A. 4443 (LEEC), Villetaneuse 93430, France
e-mail: izziclegg@hotmail.co.uk

A. Butterworth
Department of Clinical Veterinary Science, University of Bristol,
Langford, Bristol, UK

A. Butterworth (ed.), *Marine Mammal Welfare*, Animal Welfare 17,
DOI 10.1007/978-3-319-46994-2_16

few and non-existent. This chapter aims to make first steps in the discussion on assessing pinniped welfare and hopes to stimulate future welfare investigations. Pinniped species are able to thrive in two opposing environments, the land/ice margin at the coast and in the sea, and these animals use these two 'domains' for different functions. Welfare measurement is concerned with the outcome of an animal's internal and external responses to its environment, and pinniped species' evolutionary biology may be especially important in this respect, in terms of our understanding of the animals' responses and interactions within their two domains. Pinnipeds are being directly impacted by serious anthropogenic disturbances in the wild, including human interference at established feeding and breeding grounds, hunting, entanglement and climate change, and are also often kept in captive collections, so feasible evaluations of welfare could be assumed to have widespread utility, including applications benefitting the animals themselves.

16.2 Tools from Welfare Science Applicable to Pinnipeds

In this chapter, we will discuss the Otariidae (sea lions and fur seals), Phocidae (true seals) and Odobenidae (walruses) separately where possible, as these families have each followed a different evolutionary path to become adapted to both the land and the sea (Renouf 1991) and thus sometimes may merit differential consideration in terms of welfare. However, in many areas of the literature concerning wild and captive pinnipeds, there is a paucity of information on certain topics among the families and species, and in these cases our suggestions for potential welfare measures will be inter-specific and thus remain conservative. As we reviewed in Chap. 12, welfare science has established itself in regard to farm animal species, and scientists working in laboratory, zoo and companion animal welfare are adapting some of the tools and techniques first developed for farm animal assessment, to these species too (Barber 2009; Whitham and Wielebnoswki 2013). One notable disparity between welfare measurement for farm and zoo animals is that the former may, in some cases, have herd-based assessments, whereas the latter are usually afforded individual assessments. Pinnipeds are held in many zoological collections, where individual welfare assessment is logical because often the animals are observable, animals may have detailed discrete records, and come into contact with, and are often known individually by, the keepers (Barber 2009). Measuring the welfare of wild animals such as pinnipeds is possible but has been little investigated (see Chap. 12 for fuller discussion), and we suggest here how this might be accomplished and where measures for captive and wild pinnipeds might overlap. We start by reviewing the existing research on pinniped welfare since this provides context for our suggestions and discussions.

16.3 Existing Pinniped Welfare Research

There are very few studies discussing directly the welfare of pinnipeds in general and none which investigate how it might be measured comprehensively or 'holistically' (i.e. considering a wide range of risk factors and variables which may affect welfare).

Seal welfare has featured in legislative documents and guidance articles relating to the culling of seals for fur or population control practices (e.g. Fitzgerald 2011), but discussions are limited to the ethics of welfare states as perceived by humans. One recent report investigated the welfare implications of seal hunts for individual animals and in particular the qualities of the killing process itself where the authors highlight the poor welfare likely experienced by wounded animals which escape the sealers (Butterworth and Richardson 2013). Butterworth et al. (2012) reviewed how entanglement may impact pinniped welfare at the time of death or through debilitating wounds or increased energetic demands. From available data, the authors calculated that 0.24–2.21% of pinniped populations are currently entangled and discussed how the different species' behaviour and ecology alters the risk for each type of entanglement.

The only published research available which discusses multiple aspects of pinniped welfare (a more holistic approach to welfare assessment, rather than a focus on single welfare impacts) is a report on sea lion welfare in traveling circuses, where opinions on welfare were given by 20 experts (Hopster and de Jong 2014). The experts were in agreement on important risk factors (i.e. input measures) for pinnipeds in this situation (pool dimensions, space, social conditions, food and water quality), but overall there was a large degree of variation in the experts' opinions on the welfare significance, leading the authors to conclude that empirical data is urgently needed on the subject. Two other studies have looked at stereotypic behaviour as a single potential measure of pinniped welfare: Kastelein and Wiepkema (1988) found that conducting training sessions reduced Steller sea lions' (*Eumetopias jubata*) stereotypy frequency, and Smith and Litchfield (2010) showed that enrichment reduced stereotypic behaviour in Australian sea lions (*Neophoca cinerea*). Notably, in contrast to progress in cetology, emotion studies or discussions of affective states in pinnipeds are non-existent (e.g. these are not mentioned in the widely read book on pinniped behaviour by Renouf 1991). As a result and due to the paucity of welfare studies, our proposals for measures in this chapter must be partially based on extrapolation from other species' research and should be regarded conservatively as an initial 'pilot effort' aimed at focussing investigative effort on pinniped emotion and welfare.

16.4 Specific Considerations for Pinnipeds

Welfare measures for pinnipeds may be challenging to develop due to the two environments they inhabit, especially since the time spent in each of these two locations is likely to vary seasonally, interspecifically and also inter- and intra-individually (Riedman 1990; Renouf 1991). In general, pinnipeds use the water for foraging while the land is used for resting and reproduction, with social interactions occurring in both environments. Some species can rest and reproduce in the sea (Gulland et al. 2001; Walsh et al. 2001; Renouf 1991). To establish potential pinniped welfare evaluations, detailed species-specific knowledge of behavioural ecology must be gathered (see reviews by Riedman (1990) and Reynolds and Rommel (1999)). It has been noted that certain pinniped species recently brought into captivity will eat more readily in water than in land (Gulland et al. 2001). Pinnipeds do not echolocate and have highly developed and sensitive visual, tactile and passive listening systems, which vary among species and individuals depending on their environment (Schusterman et al. 2000). Mating systems differ greatly among species, and breeding seasons are variable, resulting in marked behavioural and physiological adaptations of the different pinniped species, and these are well documented (Boyd 1991; Thompson et al. 1994; Robeck et al. 2001). Understanding of the variability in pinniped behaviours, both in the wild and captivity, is likely to be an important consideration when developing welfare measures for these species.

16.5 Input and Outcome-Based Measures

Input or resource-based measures of welfare are those that assess the resources available to the animal, and outcome or animal-based measures focus on the multidimensional responses of the animal to the environment; outcome measures are considered more accurate measures of welfare (Webster 2005; Veissier et al. 2008). At the present time, legal requirements for baseline requirements for pinniped husbandry and welfare in captivity are based on inputs (resources), and examples of these resource requirements can be found in the Animal Welfare Act (AWA 1966) in the USA and Council Directive 1999/22/EC in Europe.

The pinnipeds' various ecological adaptations suggest that the input-outcome measurement frameworks (see Chap. 12 for full explanation) would be more complex to establish for pinnipeds than for cetaceans, but could nevertheless be useful tools for pinniped welfare research. The input measures of welfare would need to cover both land and water environments and might consider variables such as water quality, topography of the pool surface and haul-out site, diet and social group membership. Potential outcome-based measures might focus on aspects of the animal's behaviour, physiology and cognitive processes (Fig. 16.1).

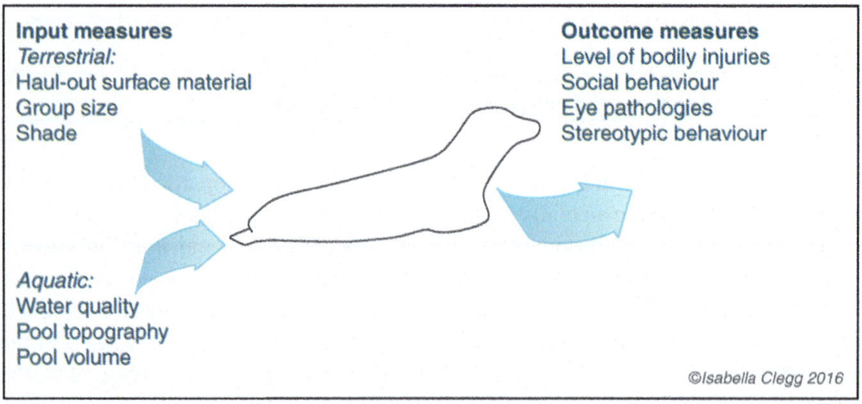

Fig. 16.1 Diagram showing some examples (not an exhaustive list) of potential input (resource-based) and outcome (animal-based) measures of welfare for pinniped species in captivity. (Image credit: Isabella Clegg)

Joseph and Antrim (2010) provide recommendations for the resources which should be provided to captive pinnipeds. In terms of designing observation methods and recording data for outcome, or animal-based, measures, the two concurrent environments inhabited by pinnipeds will mean that extra considerations must be made: for example, during behavioural observations, where animals are generally free to move between land and water, there might have to be two observers present, one watching each environment. Pinnipeds are observed to be 'intelligent' (responsive to their environment in adaptive ways) and easily trained (Schusterman et al. 2002), and so conducting physiological and cognitive measures should be no more difficult than with cetaceans, at least in captive situations. In a similar way to the model for cetacean welfare described in Chap. 12 of this book, the Triangulation principle could be used to study pinniped welfare, where welfare is assessed and the respective measures are validated by correlating against data from behavioural, health and cognitive measures (based on Webster 2005).

16.6 Animal-Based Welfare Measures for Pinnipeds in Captivity

16.6.1 Behavioural Measures of Welfare

Relatively little is known about captive pinniped behaviour, and very few published ethograms (definitions of behavioural repertoires) from the wild exist (Smith and Litchfield 2010). This makes it hard to incorporate information concerning normal behavioural patterns and seasonality of behaviours into welfare measurements (e.g. Fig. 16.2 shows a normal resting position for a wild Steller sea lion).

Behavioural measures of welfare are likely to be crucial in welfare evaluations if experience from other species is considered a template for use in pinnipeds. Behaviour can indicate how the surroundings are perceived by the animal (Gonyou 1994; Veasey 2006), which in the case of pinnipeds is their utilisation of two complex and contrasting environments which may be difficult to fully simulate in captivity. Behavioural measures of welfare can provide information about the adequacy of the captive environment and thus whether the animals' needs are likely being fulfilled (Veasey 2006). Recently, partial ethograms of behaviours in captivity were published for two pinniped species (Australian sea lions, Smith and Litchfield 2010; Cape fur seals (*Arctocephalus pusillus*), Wierucka et al. 2016). These ethograms suggest that regular behavioural monitoring for captive pinnipeds would aid in establishing 'normal' baselines (Maple 2007; Wierucka et al. 2016). For example, basic, long-term data on the topography and frequency of play, affiliative and agonistic behaviours in different species, and in varying contexts, would start to reveal the function and applicability of pinniped behaviours for welfare assessment.

There are no behavioural measures of welfare already validated for pinnipeds. Compared to cetaceans there are likely to be fewer potential measures of social

Fig. 16.2 Photo showing chosen resting position of a wild Steller sea lion (*Eumetopias jubata*): published data and observations of wild pinniped behaviour will aid in understanding of 'normal' behaviours in captivity. (Image credit: Isabella Clegg)

behaviour, due to the (generally) less complex social systems seen in pinnipeds when compared to, for example, cetaceans (Schusterman et al. 2002). Using knowledge of other species' welfare and indications from published pinniped research, some 'pilot' behavioural measures can be suggested:

a. **Inappetence** (poor or reduced appetite) has strong potential as a pinniped behavioural measure. Pinnipeds' appetite for food can be lost or diminished in response to a number of infectious and non-infectious diseases (Gulland et al. 2001), but decreases in response to environmental stress (Mellish et al. 2006) and in response to behavioural stressors, e.g. in the breeding season (Kastelein et al. 1995; Petrauskas and Atkinson 2006). Future studies should discriminate between these contexts and associated levels of inappetence before inappetence can be used as an indicator of poor welfare.
b. **Stereotypic behaviours** are perhaps the most studied potential welfare indicators in pinnipeds, although as with other species in captivity, it is not clear whether animals with stereotypies have worse welfare or improved welfare compared to non-stereotyping conspecifics due to the potential that the activity may function as a 'coping mechanism' in challenging environments (Rushen and Mason 2008). Walruses are a species for which stereotypic behaviour is often

reported (Mason 2010; Kastelein and Wiepkema 1989), with repetitive tusk rubbing being repeatedly cited. In addition to the potential detrimental impacts on the animal's affective state, this behaviour can cause secondary health problems such as tusk pulpitis (Walsh et al. 2001) and thus is likely to be valid measure of poor welfare. Other stereotypies include regurgitation, flipper chewing, pattern swimming, weaving and head shaking (Smith and Litchfield 2010; Hopster and de Jong 2014). As with cetaceans, pattern swimming would require meticulous observation before classification as a stereotypy, to be sure that there is really no function (i.e. that the behaviour is not adaptive or offers functional advantages) and that it does not vary (Clegg et al. 2015). Pattern swimming in pinnipeds can reach high frequencies, e.g. 45% of observed time in Smith and Litchfield's (2010) study, and thus may be detrimental to affective state through the prevention of other activities. Broom (1983) stated that an animal performing stereotypic behaviour for more than 10% of waking time has poor welfare, but this statistic may need to be established as having the same significance when applied to the pinnipeds. Multiple studies have found that provision of enrichment is effective in reducing pattern swimming frequency in pinnipeds (e.g. Grindrod and Cleaver 2001; Smith and Litchfield 2010; Hocking et al. 2015). The next step for this area of welfare research could be to record physiological and cognitive data and use it concurrently with data on stereotypy type and frequency, to explore how these behaviours impact emotions and affective states (Rushen and Mason 2008).

c. **Anticipatory behaviour** related to feeding sessions in captive animals has been suggested as an indicator of affective state (Spruijt et al. 2001), and the approach taken by a study with dolphins (Jensen et al. 2013), which involved assessment of recognised anticipatory behaviours, could be applied to captive pinnipeds, who are also likely to perform this behaviour. Anticipatory behaviour may signify positive expectations of the event to come and thus be a positive indicator of welfare (Spruijt et al. 2001), but concurrently high levels of such behaviours might represent prolonged fixation on the event and lack of other stimulation in the environment. More research in pinnipeds and other species is needed to understand how we might use this measure in welfare assessments.
d. **Agonistic behaviour** is used as a welfare measure in farm animal welfare assessment (WelfareQuality 2009a,b,c; Mononen et al. 2012). Aggressive behaviour has been documented in captive pinniped species, especially during the breeding seasons (Miller 1975; Robeck et al. 2001; Wartzok 1991). Due to the effects of seasonality and sexual dimorphism on aggressive behaviour, measurements of the type and frequency of agonistic behaviour may be of limited use as a welfare indicator; however, the secondary effects of aggressive interactions in terms of wounds (discussed in the next section on health) and social isolation from the group could be investigated as outcome measures of welfare.
e. **Affiliative behaviours**: In terms of behaviours which may indicate positive states of welfare, literature from other species indicates that play and affiliative behaviours may be good candidates as welfare assessment tools (Held and Špinka, 2011). As mentioned before, there are few published studies on play frequency in captive pinnipeds (see Wartzok 1991 for a review). Affiliative behaviours and their link to

emotional state are also very little studied, but there are some potential behaviours which may yield welfare measures, such as allogrooming and 'facial expressions'. Facial expressions have been linked to emotions in other species (Boissy et al. 2007; Waller and Micheletta 2013), and unlike cetaceans, pinniped species display many different facial expressions, some very canid-like (Miller 1975). Naso-nasal contact and the accompanying facial expression were proposed by Miller (1975) as an affiliative interaction for New Zealand fur seals (*Arctocephalus forsteri*) and Pacific walruses (*Odobenus rosmarus divergens*), and this author also suggested that the degree of erection of vibrissae could be used as an indicator of how relaxed the animals were, since for the New Zealand fur seals the vibrissae seemed to be more erect when naso-nasal contact was made with unfamiliar or more dominant animals. In captivity, where breeding is often controlled, and male-male competition limited or eliminated, the distance maintained between animals over time could be investigated as an indicator for the strength of social bonds (accounting for any temperature and space variations, Baldi et al. 1996).

16.6.2 Health-Related Measures of Welfare

Epidemiological measures of disease or pathology incidence can be used as indicators of health and welfare and can be broadly informative. Such parameters have been published for pinnipeds in captivity. Roberts and Demaster (2001) examined the annual survival rate of six pinniped species in 95 different captive facilities and found it to range from 0.957 (i.e. survival rate of 95.7% of the population per year) for the South American sea lion (*Otaria flavescens*) to 0.884 for the Northern fur seal (*Callorhinus ursinus*). Small and Demaster (1995) highlighted that while mortality rate for California sea lions (*Zalophus californianus*) had improved significantly over the years studied, it was significantly higher in the first 40 days after animals were taken from the wild or transferred between facilities (whether captive or wild born), leading them to conclude that homeostasis was significantly disturbed during this acclimation period and that it was the most stressful time for the animals. Mason (2010) used mortality rates and reproduction success in a captive and wild pinniped species to hypothesise that walruses were more likely to have 'relatively poor welfare' in captivity and grey seals (*Halicheorus gypus*) likely to have 'relatively good welfare'. While this approach provides a starting point for pinniped welfare research based on the blunt indicator of mortality, animal-based indicators measurable in real time are more likely to be accurate in determining day-to-day welfare impacts.

The relationship between health and welfare states is complex (Boissy et al. 2007; Mason and Veasey, 2010), and, as in Chap. 12 of this volume, we consider here how health and disease may cause changes in affective state. In many other species, health status data is used as a welfare measure (concurring with "feelings-based" welfare definitions, e.g. Fraser et al. 1997; Mason and Veasey 2010). Pinnipeds can contract a number of debilitating diseases and health problems in the wild and in captivity (wild, Bossart 2011; captive, Dierauf and Gulland 2001). One of the most revealing analyses for animals in relation to health status comes from records of haematological sampling, and several studies have published data for captive pinnipeds (Roletto 1993; Mellish et al. 2006; Trumble et al. 2006). Roletto (1993) compared profiles in healthy and "diseased" animals in 395 California sea lions, Northern elephant seals (*Mirounga angustirostris*) and Pacific harbour seals (*Phoca vitulina richardsi*) over 6 years. This author found that the animals experiencing a range of diseases had significantly higher red blood cell counts, haemoglobin values, haematocrit and blood urea nitrogen among others, but that the different white blood cells measured were highly variable and could not always be used to predict health status. The animals studied by Roletto (1993) were those being rehabilitated after stranding, and this is likely to have resulted in variation from 'normal' haematological baseline values. Considering that captive pinnipeds are intelligent, adaptable and trainable (Schusterman et al. 2002) and can, in some cases, voluntarily participate in medical husbandry procedures (Brando 2010), captive facilities could potentially publish the baseline data found in each species to facilitate the use of blood parameters in future welfare assessment of the welfare impact of disease states.

Eye pathologies in captive pinnipeds are common and have been attributed to pool design and water quality (Colitz et al. 2010; Gage 2011) (Fig. 16.3). Although cataracts and lens luxation are common age-related diseases in humans and other animals, additional risk factors for their incidence in captive pinnipeds were identified by Colitz et al. (2010). These risk factors include insufficient access to shade, a history of fighting and previous ocular disease. Access to shade (or deep water) is likely a crucial resource-based measure of welfare for pinnipeds and is stipulated in the AWA (1966), but could be more conservative, i.e. more rigorous, as proposed by Clegg et al. (2015) for dolphins. Although pool colour was not reported to be a significant factor in Colitz et al.'s (2010) study, the authors nevertheless suggest that colour should be considered for inclusion in welfare assessments. Individual clinical measures of eye disease could follow the standardised method suggested by Clegg et al. (2015) in which a photographic scale (reference scale) is used (see Chap. 12). Photographic reference scales of eye opacity and squinting in pinnipeds

Fig. 16.3 Cataracts and lens luxation are commonly seen in pinnipeds in captivity: the photo shows a severe case in a California sea lion (*Zalophus californianus*). (Image credit: Isabella Clegg)

could be developed, and these would allow facilities to assess the severity of eye pathologies. Consequently such data could be pooled between facilities and correlated to other behavioural and physiological data to help to understand how this problem impacts the animals' affective states.

Body condition scoring (BCS) is another useful tool which could be standardised to allow assessments of the deposition of tissue and body fat on an animal, as is widely used in welfare assessments for other species (WelfareQuality 2009a,b,c; Mononen et al. 2012; Clegg et al. 2015). Although the link between BCS and affective state is unclear (Roche et al. 2009), as with eye pathologies, a standardised scoring system could encourage widespread data collection. The pinniped literature provides a strong background for protocols to measure body condition which have been used in number of past health assessments (e.g. Castellini et al. 1993; Guinet et al. 1998; Trites and Donnelly 2003), and ultrasound imaging of tissue and fat reserves could be a useful validation tool for this measure (Mellish et al. 2004).

Interspecific aggression which leads to injuries is a prominent and common wild behaviour (Bartholomew 1970; Chilvers et al. 2005; Wartzok 1991) and in terms of captive animal welfare represents an interesting dilemma in so far that it warrants the question whether wild welfare states are the sole sought-after standard. Pinnipeds species are generally sexually dimorphic and some of the most polygynous of all animals (Riedman 1990): males fight during and around breeding seasons, subadult males practice fighting and mating skills and females defend their resources and pups, and as a result interspecific injuries are common, often debilitating, and sometimes fatal (Fig. 16.4). For example, 84% of females had permanent interspecific-injury scars in a New Zealand sea lion population (*Phocarctos hookeri*), 0.5% of the

Fig. 16.4 Two male Northern elephant seals (*Mirounga angustirostris*) fighting, a behaviour which is often linked to high rates of morbidity and mortality. (Image credit: Ari Friedlaender)

breeding females were killed by males each year (Chilvers et al. 2005) and around 6% of pups born each year in a northern elephant seal population were killed by males (Le Boeuf 1974). Body injuries are a valid assessment of welfare due to their direct and indirect effects on affective state (Broom 1991; Welfare Quality® 2009a,b,c; Mononen et al. 2012). Furthermore, injuries are evidence of interactions that are not always observable and thus are a proxy indicator of aggression (Scott et al. 2005). For these reasons cetologists have taken steps to try to quantify bodily injuries (Scott et al. 2005; Clegg et al. 2015). For pinnipeds, systematic reviews of past research are needed, since behavioural data and mortality rates are sometimes available concurrently, which might enable the creation of methods to quantify the level of injuries on the body and set thresholds for poor welfare. This method could be a simple accumulation of systematic estimations, such as that described for cetaceans in Clegg et al. (2015), in which old scars and new injuries were quantified separately. The data might then be correlated with levels of aggressive behaviour and inappetence. Monitoring of aggression levels in captivity will help us to understand if there is a threshold level of injuries above which welfare decreases rapidly.

Physiological welfare measures for captive pinnipeds provide information on animal emotions and thus can support and validate other measures (Boissy et al. 2007). Faecal glucocorticoid levels were measured in Steller sea lions in response to season (Petrauskas and Atkinson 2006) and medical and restraint procedures (Petrauskas et al. 2008). No significant differences resulted from these variables, but there was much individual variation found between the experimental groups. Kershaw and Hall (2016) found that wild harbour seal (*Phoca vitulina*) plasma cortisol spiked in response to a capture procedure, but that blubber cortisol did not, instead varying significantly with season and sex. Myers et al. (2010) investigated variance in serum cortisol levels in wild and captive Steller sea lions, which was elevated in breeding and annual moult months. Risk factors associated with stress at sampling could be significantly reduced if the most stressful elements of the procedure were understood, and of course could be removed entirely with the use of positive reinforcement training techniques. More theoretical and practical data is required to 'make sense' of the welfare picture as described by glucocorticoids measures, including accounting for the likely physiological differences between the action of terrestrial and marine mammal stress hormone (Atkinson et al. 2015).

16.6.3 Cognitive Measures of Welfare

Cognitive measures of welfare aim to assess the appraisal of emotions, whether conscious or unconscious (Mendl and Paul 2004; Paul et al. 2005). They are the least explored of the 'welfare assessment methods', when compared to the behavioural and health components of the Triangulation approach (Webster 2005), but there are promising new techniques derived from human and other species research (Mendl and Paul 2004; Rogers 2010) which could be applicable to pinnipeds.

Measures may be easier to apply to pinnipeds in captivity than their wild counterparts (Rosen 2009), and since cognitive tests are often based on experimental psychology and require repeated measures, past data and training of the animals, researchers in captive facilities could take the lead in this area of pinniped welfare.

Similar to our assertions for cetaceans in Chap. 12 of this volume, we suggest one of the most promising cognitive measures of pinniped welfare might be cognitive bias testing. Cognitive bias describes an individual's appraisal of ambiguous stimuli, which the individual animal may interpret positively or negatively, and has been shown with other animals to be closely linked to welfare. For example, animals kept in impoverished environments, with anxious dispositions, or subject to physical examinations, i.e. situations likely to induce poor welfare, judge the ambiguous stimuli more negatively (Mendl et al. 2009). The design of tasks created to test this has, in the past, included Go/No-Go or active response operant discrimination tasks, using spatial, visual or auditory cues, and these sorts of tests would certainly be feasible with captive pinnipeds. Schusterman et al. (2002) share their experiences of training cognitive tasks with sea lions, which could be used when designing such future studies. They found that animals retained language-representative signals for short periods of time when other distracting stimuli were present, but were able to successfully compare sample stimuli for extended periods, and thus seem to demonstrate easily disrupted short-term memory but a reliable, accurate long-term memory (Schusterman et al. 2002). If a condition of supposed poor or enhanced (e.g. addition of enrichment) welfare is imposed, cognitive bias results may confirm that these resources have impacts on welfare. Furthermore these results could be correlated to other animal-based measures of welfare to validate them.

Other cognitive measures with possible links to welfare include laterality and tests of preference and motivation. Brain and behavioural lateralisation is the differential processing of stimuli by the brain's left and right hemispheres, which can then translate into lateralised behaviours (Rogers 2010). A study of pinniped lateralised behaviour in the wild was conducted, with walruses seeming to prefer the right flipper for feeding (Levermann et al. 2003). A small number of studies were carried out in captivity: California sea lion populations showed directional swimming dependent on sex (Wells et al. 2006) and a right-ear preference for adults listening to conspecific calls (Böye et al. 2005), and directional swimming was recently reported in northern fur seals (Pryaslova et al. 2009). The link between lateralisation and animal welfare has been reviewed by Rogers (2010), and it appears that the left hemisphere deals with non-stressful and the right deals with stressful emotions (Leliveld et al. 2013). Although there exist examples of lateralisation in certain contexts, these traits have not been explored widely as species-specific welfare measures. Nevertheless further studies are encouraged in pinnipeds, and lateralisation data could be collected alongside other experimental data, as laterality can be relatively simple to measure. Preference and motivational tests can indicate which resources the animal seeks and how hard it is willing to work for them, and when different conditions are imposed, the internal appraisal of the resource in relation to external factors can be revealed (Gonyou 1994; Paul et al. 2005; Boissy et al. 2007). In terms of welfare, this can indicate those resources which the animals covet most, and as a next step, the provision of these resources could be used to test potential behavioural and physiological indicators of positive welfare.

16.7 Animal-Based Welfare Measures for Pinnipeds in the Wild

16.7.1 Behavioural Measures of Welfare

Pinniped species divide their time variably between land and sea and can spend from days to months at sea on foraging trips (Riedman 1990; Renouf 1991). Due to the difficulties of underwater observations of these highly mobile animals, behavioural measurements of wild pinnipeds have traditionally been on land (Renouf and Lawson 1987). Therefore we must remain conservative in our suggestions for animal-based welfare measures, since these are likely to be based on limited knowledge of the full behavioural repertoire of the animals.

First we discuss those behavioural measures proposed in the captive pinniped section which also merit investigation in wild populations. Examples of measures which may assess positive affective states include naso-nasal contacts, allogrooming, play and the level of contact between individuals. Renouf and Lawson (1987) found patterns of play over time in harbour seals, showing that adults played almost as much as juveniles, that solitary play was far more frequent than social play and that there were sex differences in frequency and type of play. These authors suggested that the high levels of adult play may represent reduced foraging and energetic pressures and thus would be an indicator for other needs having been satisfied. Wartzok (1991) pointed out that play behaviour may, to one of the participants, be stressful and injurious and still function as play for another, as is the case with older males play fighting or during sexual play with younger animals. Quantifying and defining play is the first step towards its use in welfare assessment but definition of what constitutes play is recognised as a difficult challenge (Held and Špinka, 2011). A recent study proposed an intensity index for play in captive elephants (Vicino and Marcacci 2015), an approach which could be applied to wild and captive pinnipeds. Miller (1975) and Renouf and Lawson (1987) among others have expressed the opinion that a significant amount of pinniped play occurs underwater, and thus future efforts could aim to document this important part of the behavioural repertoire. The distance between individuals could also be a measure of positive (or negative) welfare but would need much investigation first, as the effects of temperature and space available would have to be tested, and 'distance adoption behaviour' may only be a measure applicable outside of the breeding season. Nevertheless, Baldi et al.'s (1996) study on southern elephant seals (*Mirounga leonine*) inhabiting a stretch of coastline with an unusually large amount of available space showed that the females tended to keep a distance of between one and two body length from each other. Smaller-scale differences within this range of 'distance maintained' could be measured to explore whether this may be used as a sign of affiliation (and thus positive affective state), since existing research supports the positive aspects of social recognition and long-term bonds which are seen in pinnipeds (Wartzok 1991; Schusterman et al. 2002; Insley et al. 2003).

Aggression is common within wild pinniped groups and should be considered a potential welfare measure. Excessive levels of aggression have often been used as an indicator for poor welfare states (Swaisgood 2007; Welfare Quality® 2009b), whether caused directly by social interactions or as result of displaced aggression (Broom 1991; Gonyou 1994). Although the frequency of aggressive behaviours per se could not be used alone as an accurate welfare indicator without consideration of other factors, including seasonality and space available, measures such as position in the hierarchy and vigilance behaviour could indicate the level of social stress experienced by individuals. The literature on other species shows that the most stressful hierarchical position in a group can either be the most dominant or the most subordinate individuals and varies between species, with possibilities for intraspecific variability (Abbott et al. 2003; De Vries et al. 2003; Sapolsky 2005). Pinniped species differentially engage in territorial defence or dominance hierarchies to maintain harems of females, with phocid species especially favouring a hierarchical system (Riedman 1990). The dominance hierarchies of male grey seals (*Halichoerus grypus*) in three populations were quantified (Twiss et al. 1998), and further work in this species suggests that levels of aggression are more related to proximity and familiarity of the closest neighbour than to dominance (Bishop et al. 2015). Therefore the chronic social stress in such systems might result from disturbances to social stability and as a result of space restrictions (Bishop et al. 2015). Future studies might control these factors and enable measurement of the frequency of aggression indicators and other physiological indicators to reveal the effect on welfare.

Pinnipeds are one of the most vocal mammalian taxa (Schusterman et al. 2001). Vocal behaviour in other species is used as a measure of welfare, since vocalisation is likely to accompany a particular mood or emotion. However, in order for the emotional state to be indicated, there must be well-established understanding of the 'meaning' of the vocalisations (Manteuffel et al. 2004). Schusterman et al. (2001) concluded in their review paper that pinnipeds have consistent calls (some similar between species) for mothers reuniting with pups and during threats, alarm and aggressive behaviour. Miller (1975) noted that 'whimpering' in male New Zealand fur seals occurred in 'mild distress' situations: during territorial swimming, accompanying mild threat displays, and in interactions with females. Vocalisations linked to these contexts could be tested in relation to affective state (i.e. alongside other potential welfare measures), and more work is needed to define the vocal nuances for each species before welfare states can be assumed or alluded to. Captive pinniped research teams could focus their research efforts in this area (mapping of vocalisations), as data collection could build on the results of current studies (Schusterman et al. 2001) and through collaborations with wild animal studies.

16.7.2 Health-Related Measures of Welfare

Like cetaceans, pinnipeds are considered sentinels of ocean health since they are high in the food chain and mobile, gregarious animals, thereby able to provide us with biomarkers of the state of the environment (Bossart 2011). Reviews of the pathologies in pinniped diseases are found elsewhere (Dierauf and Gulland 2001; Bossart 2011), but here we are concerned with measures which correspond to the animals' affective state. Our suggestions discussed for captive pinnipeds (Sec. 16.3) for how blood analyses and cortisol measurement might relate to welfare which are potentially applicable to wild pinnipeds. Bringing together findings from both wild and captive studies could increase the efficiency of data use, adopting the 'three Rs' principles (refine, reduce, replace) and so lessen the stressors to wild and possible endangered animals (Nolen and Bishop 2001). 'Sickness behaviours' are caused by disease impact on affective state and include anorexia, lethargy, depression and antisocial behaviours (Broom 1991; Sneddon et al. 2014). Documentation recording the frequency of these behaviours and of any changes in accompanying behaviours or physiology in wild pinnipeds would greatly assist work in this area. Millman (2007) suggests that these sickness behaviours should be included within welfare assessments and explores the theory that sickness behaviour is a motivational state which competes with other needs dependent on the context.

In addition to contracting diseases, wild pinniped populations have been observed to undergo severe nutritional stress (Trites and Donnelly 2003; Rosen 2009). The health and physiological changes in pinnipeds resulting from starvation or shortage of food include metabolic depression, poor body condition, immunosuppression and a blood profile typical of the sequential starvation process: elevated blood glucose (carbohydrate utilisation), followed by elevated ketone and non-esterified fatty acid levels (fat utilisation), followed by elevated blood urea nitrogen which signifies protein metabolism in the last stage of starvation (Trites and Donnelly 2003). Epidemiological measures during times of nutritional stress in pinnipeds include reduced growth rate, reduced pup survivability, increased mortality rate and reduced reproduction success rate; these data can reveal the early stages of food deficiencies and could be used to select populations for validation of other animal-based welfare indicators. Rosen (2009) reviewed the results of captive studies conducted on nutritional stress and emphasised how collaborative efforts between researchers focusing on either the wild or captivity can help to identify the tangible and measurable effects of a lack of, or a reduction in, available food.

A standardised body condition scoring protocol and method of wound quantification, as discussed in the captive pinniped section (Sect. 16.6), would also aid welfare measurement of wild pinnipeds for the same reasons. The assessment of wounds on wild animals may sometimes be hindered by visibility issues, and a potential tool could be the use of thermal imaging cameras as presented in Walsh and Gaynor (2001), used by these authors to assess infection in conspecific wounds in captive California sea lions and walruses. This technique could be validated in captivity where the severity of wounds assessed through thermograms could be correlated with external appearance, haematology and behavioural indices and then used in wild pinniped studies.

16.7.3 Cognitive Measures of Welfare

Experimental cognitive studies are logistically harder to conduct in wild animals, and in the past, simple 'counts of cognition' per taxonomic species have been used as an indicator of intelligence level (Lefebvre 2010). Schusterman et al. (2002)

hypothesised that pinnipeds have advanced problem-solving abilities, as demonstrated by their raiding of fisherman's catches, and these authors' captive cognition studies showed that California sea lions classify stimuli successfully and can recall elements easily from long-term memory, skills useful for social interactions, foraging and vigilance behaviour in the wild. However, when looking for cognitive measures of welfare, the focus is more on how the animals process emotional experiences and how this affects the treatment of other information (Paul et al. 2005): for these topics, there are very few studies yet concerning any wild animal species.

Earlier in this chapter, we presented the advantages of cognitive bias tests as non-invasive measures of welfare which are capable of drawing together other multidimensional indicators. Although cognitive bias studies have previously not been conducted with any wild animals, there may be potential to assess such biases in a passive way (without the operant conditioning step) through using salient stimuli in the animals' environment which are already associated with positive or negative outcomes (Brilot et al. 2009). In addition to cognitive bias, Paul et al. (2005) also suggest attention and judgement biases as welfare measures, where affective state is shown to impact decision-making or attention to a task, and again such experiments could be possible with wild pinnipeds. With captive three-spined sticklebacks (*Gasterosteus aculeatus*), Purser and Radford (2011) show how such tests may reveal impacts on affective state which were being missed before: they found that although their imposed condition (noise) only induced a mild fear response, there were significant increases in food-handling errors and decreased food to non-food discrimination which led to overall reduced foraging efficiency. These studies could be attempted with wild pinnipeds, for which anthropogenic noise and disturbances might be good candidates for an assumed deleterious condition. Lastly, as reviewed earlier in this chapter, behavioural lateralisation could potentially reveal aspects of affective states and thus welfare which are otherwise not accessible (Rogers 2010). Studies could continue to build on existing laterality findings with pinnipeds (Levermann et al. 2003; Böye et al. 2005; Wells et al. 2006; Pryaslova et al. 2009), perhaps with more focus on collecting multidimensional data on potential emotional states at the same time.

16.8 Recommendations for Developing Measures of Pinniped Welfare

Although only one study thus far has explicitly aimed to assess pinniped welfare (Hopster and de Jong 2014), the literature reviewed in this chapter shows that there is enough reported knowledge to support initial investigations. There is very little 'emotion' research in pinnipeds, and since welfare is a balance of affective states made up of moods and emotions (Spruijt et al. 2001; Mendl et al. 2010), this topic should be established further in order to facilitate welfare research. Other species' emotion research can contribute to the foundations of measures applicable to pinnipeds: Mendl et al. (2010) offer theoretical and practical advice for advancing animal emotion research, and Boissy et al. (2007) present robust indicators of positive emotions in terrestrial mammals, e.g. play, inter-animal distance, and allogrooming. Désiré et al. (2002) stipulated that to investigate emotions (in their case with farm animals), fine-scale behaviours such as posture and body position must be studied as opposed to only the more obvious actions, and this is likely to also be true for pinnipeds.

With cetaceans, a welfare assessment has been proposed for bottlenose dolphins (*Tursiops truncatus*) since they are one of the most studied species and the most commonly kept in captivity (Clegg et al. 2015). The same approach could be used with pinnipeds, where California sea lions, Steller sea lions, northern fur seals and harbour seals might be the first to be assessed for the welfare impact of captivity. Welfare studies should always be conducted in situ in the environment the animals are inhabiting (Dawkins 2006). Initial welfare research on baseline behaviours for these well-studied species would be best conducted outside of the breeding season, since marked behavioural and physiological changes often occur (Riedman 1990; Robeck et al. 2001; Kershaw and Hall 2016) which may distort the data collected and mask the welfare significance from the parameters assessed. However, clearly these periods should not be ignored, as they are often the source of high levels of stress and injury for the animals (Le Boeuf 1974; Chilvers et al. 2005; Fig. 16.4), and thus could be used in the context of validation of measures used to assess 'poor welfare' (discussed earlier in this chapter, e.g. injury extent, excessive aggression, body condition).

Other contexts for validation of measures can be used, and for wild and captive pinnipeds, this could involve situations in which there is human interaction. Similar to our discussion in Chap. 12 for cetaceans, welfare data could be gathered during interactions and used to explore correlations among sets of measures. For example, a recent phenomenon has been the stranding of California sea lions along the central Californian coast, where the causes listed include leptospirosis (*Leptospira interrogans*) infection, domoic acid poisoning, malnutrition (sometimes with gastric ulcers) and human interactions (Greig et al. 2005). Data on measures such as body condition scoring, inappetence and play/vocalisation/activity level could be taken from animals with chronic infections and compared with those which stranded, but were clinically healthy (as occurred in 0.7% of strandings from 1991 to 2001, Greig et al. 2005). In wild pinnipeds, entanglement in marine debris can cause significant injury and debilitation and affect a measurable proportion of some pinniped populations (Butterworth et al. 2012) (also see Chap. 13 this volume): the proposed welfare measures discussed in this chapter could be applied to these animals as part of the assessment of the poor welfare caused by marine debris.

In captivity, pinnipeds usually receive their food within multiple training sessions, and thus there is potential for the human-animal relationship (HAR) to be a factor in enhancing or reducing welfare (as with other captive species, Waiblinger et al. 2006; Whitham and Wielebnowski 2013). California sea lions can use the pointing gestures of their trainers as referential cues, suggesting that through their habituation to humans, they have learned to exploit human body language (Malassis and Delfour 2015). Captive pinnipeds' behaviour towards trainers during training sessions could be measured to indicate whether they find this interaction aversive, neutral or positive. For example, vocalisations have been proposed as an emotional indicator in pinnipeds (Schusterman et al. 2001), and approach/avoidance tests could also be applied, as they have with dolphins (Clegg et al. 2015). Once basic indicators of pinniped emotional state are established, different human interactions in captivity might be used as either a situation where good welfare is likely (e.g. cognitively enriching training sessions, Clark 2013) or where poor welfare is likely (e.g. some travelling circuses, Hopster and de Jong 2014; when being transported between facilities, as with cetaceans, Castellote and Fossa 2006).

16.9 Conclusions

Past research has described the behaviour, health and cognition of captive and wild pinnipeds, and this foundation of published material can be used to progress the application of welfare research and welfare assessment methods for pinnipeds. There are no existing established pinniped welfare measures per se as yet, but proposals for indicators in cetaceans and other terrestrial mammals show some significant potential for creation of pinniped species-specific tests, tools and measures. There is very little emotion research thus far in pinnipeds, and this topic should be established further in order to facilitate welfare research and could include development of indicators of positive emotions—play, inter-animal distance, allogrooming and the use of fine-scale behaviours such as posture and body position. The breeding season should be carefully defined as a context but not ignored, as this is often a period of high levels of stress and injury for the animals. The first pinniped welfare investigations might benefit from focussing on those species most well studied and commonly kept in captivity. Human interaction with both captive and wild pinnipeds is likely to be important to their welfare: captive pinnipeds' behaviour towards trainers during training sessions could be measured to indicate whether they find this interaction aversive, neutral or positive. In wild pinniped populations, entanglement in marine debris negatively impacts many animals, and measures discussed in this chapter could be applied to these animals to assess the welfare impact of debris, as well as validating objective welfare indicators for use in other contexts.

References

Abbott DH, Keverne EB, Bercovitch FB, Shively CA, Mendoza SP, Saltzman W, Snowdon CT, Ziegler TE, Banjevic M, Garland T, Sapolsky RM (2003) Are subordinates always stressed? A comparative analysis of rank differences in cortisol levels among primates. Horm Behav 43:67–82

Atkinson S, Crocker D, Houser D, Mashburn K (2015) Stress physiology in marine mammals: how well do they fit the terrestrial model? J Comp Physiol B 185(5):463–486

AWA (1966) Animal welfare act. Public law 89–544. 89th congress, H. R. 13881, august 24, 1966. USDA, Washington DC
Baldi R, Campagna C, Pedraza S, Le Boeuf BJ (1996) Social effects of space availability on the breeding behaviour of elephant seals in Patagonia. Anim Behav 51:717–724
Barber JC (2009) Programmatic approaches to assessing and improving animal welfare in zoos and aquariums. Zoo Biol 28:519–530
Bartholomew GA (1970) A model for the evolution of pinniped polygyny. Evolution 24:546–559
Bishop AM, Pomeroy P, Twiss SD (2015) Variability in individual rates of aggression in wild gray seals: fine-scale analysis reveals importance of social and spatial stability. Behav Ecol Sociobiol 69:1663–1675
Boissy A, Manteuffel G, Jensen MB, Moe RO, Spruijt B, Keeling LJ, Winckler C, Forkman B, Dimitrov I, Langbein J, Bakken M (2007) Assessment of positive emotions in animals to improve their welfare. Physiol Behav 92:375–397
Bossart GD (2011) Marine mammals as sentinel species for oceans and human health. Vet Pathol 48:676-690
Boyd IL (1991) Environmental and physiological factors controlling the reproductive cycles of pinnipeds. Can J Zool 69:1135–1148
Böye M, Güntürkün O, Vauclair J (2005) Right ear advantage for conspecific calls in adults and subadults, but not infants, California Sea lions (*Zalophus californianus*): hemispheric specialization for communication? Eur J Neurosci 21:1727–1732
Brando SICA (2010) Advances in husbandry training in marine mammal care programs. Int J Comp Psychol 23:777–791
Brilot BO, Normandale CL, Parkin A, Bateson M (2009) Can we use starlings' aversion to eyespots as the basis for a novel 'cognitive bias' task? Appl Anim Behav Sci 118:182–190
Broom DM (1983) Stereotypies as animal welfare indicators. In: Smidt D (ed) Indicators relevant to farm animal welfare. Springer, Netherlands, pp 81–87
Broom DM (1991) Assessing welfare and suffering. Behav Process 25:117–123
Butterworth A, Clegg I, Bass C (2012) Untangled–marine debris: a global picture of the impact on animal welfare and of animal-focused solutions. World Society for the Protection of Animals International, London, UK
Butterworth A, Richardson M (2013) A review of animal welfare implications of the Canadian commercial seal hunt. Mar Policy 38:457–469
Council Directive (1999) Council Directive 1999/22/EC of 29th March 1999 relating to the keeping of wild animals in zoos. Retrieved from http://eur-lex.europa.eu/legal-content/EN/TXT/?uri=celex:31999L0022 on March 21st 2016
Castellini MA, Davis RW, Loughlin TR, Williams TM (1993) Blood chemistries and body condition of Steller Sea lion pups at Marmot Island, Alaska. Mar Mamm Sci 9:202–208
Castellote M, Fossa F (2006) Measuring acoustic activity as a method to evaluate welfare in captive beluga whales (*Delphinapterus leucas*). Aquat Mamm 32:325
Chilvers BL, Robertson BC, Wilkinson IS, Duignan PJ, Gemmell NJ (2005) Male harassment of female New Zealand Sea lions, *Phocarctos hookeri*: mortality, injury, and harassment avoidance. Can J Zool 83:642–648
Clark FE (2013) Marine mammal cognition and captive care: a proposal for cognitive enrichment in zoos and aquariums. J Zoo Wildl Med 1:1–6
Clegg ILK, Borger-Turner JL, Eskelinen HC (2015) C-well: the development of a welfare assessment index for captive bottlenose dolphins (*Tursiops truncatus*). Anim Welf 24:267–282
Colitz CM, Saville WJ, Renner MS, McBain JF, Reidarson TH, Schmitt TL, Nolan EC, Dugan SJ, Knightly F, Rodriguez MM, Mejia-Fava JC (2010) Risk factors associated with cataracts and lens luxations in captive pinnipeds in the United States and the Bahamas. J Am Vet Med Assoc 237:429–436
Dawkins MS (2006) A user's guide to animal welfare science. Trends Ecol Evol 21:77–82
DeVries AC, Glasper ER, Detillion CE (2003) Social modulation of stress responses. Physiol Behav 79:399–407

Désiré L, Boissy A, Veissier I (2002) Emotions in farm animals: a new approach to animal welfare in applied ethology. Behav Process 60:165–180
Dierauf L, Gulland FM (eds) (2001) CRC handbook of marine mammal medicine, 2nd edn. CRC press, Boca Raton, FL
Fitzgerald PL (2011) "morality" may not be enough to justify the EU seal products ban: animal welfare meets international trade law. J Int Wildl Law Pol 14:85–136
Fraser D, Weary DM, Pajor EA, Milligan BN (1997) A scientific conception of animal welfare that reflects ethical concerns. Anim Welf 6:187–205
Gage LJ (2011) Captive pinniped eye problems, we can do better. J Mar Anim Ecol 4:25–28
Gonyou HW (1994) Why the study of animal behavior is associated with the animal welfare issue. J Anim Sci 72:2171–2177
Greig DJ, Gulland FM, Kreuder C (2005) A decade of live California Sea lion (*Zalophus californianus*) strandings along the Central California coast: causes and trends, 1991-2000. Aquat Mamm 31:11–22
Grindrod JAE, Cleaver JA (2001) Environmental enrichment reduces the performance of stereotypic circling behavior in captive common seals (*Phoca vitulina*). Anim Welf 10:53–63
Guinet C, Roux JP, Bonnet M, Mison V (1998) Effect of body size, body mass, and body condition on reproduction of female south African fur seals (*Arctocephalus pusillus*) in Namibia. Can J Zool 76:1418–1424
Gulland FMD, Haulena M, Dierauf LA (2001) Seals and sea lions. In: FMD G, Dierauf LA (eds) CRC handbook of marine mammal medicine, 2nd edn. CRC Press, Boca Raton, FL, pp 907–926
Held SD, Špinka M (2011) Animal play and animal welfare. Anim Behav 81:891–899
Hocking DP, Salverson M, Evans AR (2015) Foraging-based enrichment promotes more varied behaviour in captive Australian fur feals (*Arctocephalus pusillus doriferus*). PLoS One 10:e0124615
Hopster H, de Jong IC (2014) Welfare of sea lions in travelling circuses, Research report, vol 770. Wageningen UR Livestock Research, Lelystad, The Netherlands, p 92
Insley S, Phillips AV, Charrier I (2003) A review of social recognition in pinnipeds. Aquat Mamm 29:181–201
Jensen A-L, Delfour F, Carter T (2013) Anticipatory behaviour in captive bottlenose dolphins (*Tursiops truncatus*) – a preliminary study. Zoo Biol 32:436–444
Joseph B, Antrim J (2010) Special considerations for the maintenance of marine mammals in captivity. In: Kleiman DG, Thompson KV, Kirk Baer C (eds) Wild mammals in captivity: principles and techniques for zoo management, 2nd edn. Chicago, The University of Chicago Press, pp 181–216
Kastelein RA, Wiepkema PR (1988) The significance of training for the behaviour of Steller Sea lions (*Eumetopias jubata*) in human care. Aquat Mamm 14:39–41
Kastelein RA, Wiepkema PR (1989) A digging trough as occupational therapy for Pacific walruses in human care. Aquat Mamm 15:9–17
Kastelein RA, Kershaw J, Berghout E, Wiepkema PR (1995) The food consumption of south American sea lions (*Otaria flaveseem*). Aquat Mamm 21:43–53
Kershaw JL, Hall AJ (2016) Seasonal variation in harbour seal (*Phoca vitulina*) blubber cortisol-a novel indicator of physiological state? Sci Rep 6:21889
Le Boeuf BJ (1974) Male-male competition and reproductive success in elephant seals. Am Zool 14:163–176
Lefebvre L (2010) Taxonomic counts of cognition in the wild. Biol Lett. doi:10.1098/rsbl.2010.0556
Leliveld LM, Langbein J, Puppe B (2013) The emergence of emotional lateralization: evidence in non-human vertebrates and implications for farm animals. Appl Anim Behav Sci 145:1–14
Levermann N, Galatius A, Ehlme G, Rysgaard S, Born EW (2003) Feeding behaviour of free-ranging walruses with notes on apparent dextrality of flipper use. BMC Ecol 3:1
Malassis R, Delfour F (2015) Sea lions' (*Zalophus californianus*) use of human pointing gestures as referential cues. Learn Behav 43:101–112

Manteuffel G, Puppe B, Schön PC (2004) Vocalization of farm animals as a measure of welfare. Appl Anim Behav Sci 88:163–182
Maple TL (2007) Toward a science of welfare for animals in the zoo. J Appl Anim Welf Sci 10:63–70
Mason GJ (2010) Species differences in responses to captivity: stress, welfare and the comparative method. Trends Ecol Evol 25:713–721
Mason GJ, Veasey JS (2010) How should the psychological well-being of zoo elephants be objectively investigated? Zoo Biol 29:237–255
Mellish JAE, Tuomi PA, Horning M (2004) Assessment of ultrasound imaging as a noninvasive measure of blubber thickness in pinnipeds. J Zoo Wildl Med 35:116–118
Mellish JAE, Calkins DG, Christen DR, Horning M, Rea LD, Atkinson SK (2006) Temporary captivity as a research tool: comprehensive study of wild pinnipeds under controlled conditions. Aquat Mamm 32:58–65
Mendl M, Paul ES (2004) Consciousness, emotion and animal welfare: insights from cognitive science. Anim Welf 13:17–25
Mendl M, Burman OH, Parker RM, Paul ES (2009) Cognitive bias as an indicator of animal emotion and welfare: emerging evidence and underlying mechanisms. Appl Anim Behav Sci 118:161–181
Mendl M, Burman OH, Paul ES (2010) An integrative and functional framework for the study of animal emotion and mood. Proc R Soc Lond B Biol Sci 277:2895–2904
Miller EH (1975) A comparative study of facial expressions of two species of pinnipeds. Behaviour 53:268–284
Millman ST (2007) Sickness behavior, and its relevance to animal welfare assessment at the group level. Anim Welf 16:123–125
Mononen J, Moller SH, Hansen SW, Hovland AL, Koistinen T, Lidfors L, Malkmvist J, Vinke CM, Ahola L (2012) The development of on-farm welfare assessment protocols for foxes and mink: the WelFur project. Anim Welf 21:363–371
Myers MJ, Litz B, Atkinson S (2010) The effects of age, sex, season and geographic region on circulating serum cortisol concentrations in threatened and endangered Steller Sea lions (*Eumetopias jubatus*). Gen Comp Endocrinol 165:72–77
Nolen AL, Bishop LJ (2001) Animals in research and education: ethical issues. Kennedy Inst Ethics J 11:91–112
Paul ES, Harding EJ, Mendl M (2005) Measuring emotional processes in animals: the utility of a cognitive approach. Neurosci Biobehav Rev 29:469–491
Petrauskas LR, Atkinson SK (2006) Variation of fecal corticosterone concentrations in captive Steller Sea lions (*Eumetopias jubatus*) in relation to season and behavior. Aquat Mamm 32:168
Petrauskas L, Atkinson S, Gulland F, Mellish J, Horning M (2008) Monitoring glucocorticoid response to rehabilitation and research procedures in California and Steller Sea lions. J Exp Zool 309:73–82
Purser J, Radford AN (2011) Acoustic noise induces attention shifts and reduces foraging performance in three-spined sticklebacks (*Gasterosteus aculeatus*). PLoS One 6:e17478
Pryaslova YP, Lyamin OI, Mukhametov LM (2009) Lateralized behavior in northern fur seals. Dokl Biol Sci 1:171–174
Renouf D, Lawson JW (1987) Quantitative aspects of harbour seal (*Phoca vitulina*) play. J Zool 212:267–273
Renouf D (1991) The behaviour of pinnipeds. Dordrecht, The Netherlands, Springer Science & Business Media
Reynolds JE, Rommel SA (eds) (1999) Biology of marine mammals. Smithsonian Institution Press, Washington, DC, p 578
Riedman M (1990) The pinnipeds: seals, sea lions, and walruses. University of California Press, Berkeley, California, p 439
Robeck TR, Atkinson AKC, Brook F (2001) Reproduction. In: FMD G, Dierauf LA (eds) CRC handbook of marine mammal medicine, 2nd edn. CRC Press, Boca Raton, FL, pp 193–236

Roberts SP, Demaster DP (2001) Pinniped survival in captivity: annual survival rates of six species. Mar Mamm Sci 17:381–387

Roche JR, Friggens NC, Kay JK, Fisher MW, Stafford KJ, Berry DP (2009) Invited review: body condition score and its association with dairy cow productivity, health, and welfare. J Dairy Sci 92:5769–5801

Rogers LJ (2010) Relevance of brain and behavioural lateralization to animal welfare. Appl Anim Behav Sci 127:1–11

Roletto J (1993) Hematology and serum chemistry values for clinically healthy and sick pinnipeds. J Zoo Wildl Med 24:145–157

Rosen DA (2009) Steller Sea lions *Eumetopias jubatus* and nutritional stress: evidence from captive studies. Mammal Rev 39:284–306

Rushen J, Mason G (2008) A decade-or-more's progress in understanding stereotypic behavior. In: Mason G, Rushen J (eds) Stereotypic animal behavior: fundamentals and applications to welfare. CAB International, UK, pp 1–18

Sapolsky RM (2005) The influence of social hierarchy on primate health. Science 308:648–652

Schusterman RJ, Kastak D, Levenson DH, Reichmuth CJ, Southall BL (2000) Why pinnipeds don't echolocate. J Acoust Soc Am 107:2256–2264

Schusterman RJ, Southall BL, Kastak D, Reichmuth Kastak C (2001) Pinniped vocal communication: Form and function. In: Proceedings of the 17th International Congress on Acoustics, Rome, Italy, pp 1-2)

Schusterman RJ, Kastak CR, Kastak D (2002) The cognitive sea lion: meaning and memory in the laboratory and in nature. The cognitive animal. Empirical and theoretical perspectives on animal cognition. The MIT Press, Cambridge, Massachusetts, pp 217–228

Scott EM, Mann J, Watson-Capps JJ, Sargeant BL, Connor RC (2005) Aggression in bottlenose dolphins: evidence for sexual coercion, male-male competition, and female tolerance through analysis of tooth-rake marks and behaviour. Behaviour 142:21–44

Small RJ, Demaster DP (1995) Survival of five species of captive marine mammals. Mar Mamm Sci 11:209–226

Smith BP, Litchfield CA (2010) An empirical case study examining effectiveness of environmental enrichment in two captive Australian sea lions (*Neophoca cinerea*). J Appl Anim Welf Sci 13:103–122

Sneddon LU, Elwood RW, Adamo SA, Leach MC (2014) Defining and assessing animal pain. Anim Behav 97:201–212

Spruijt BM, van den Bos R, Pijlman FT (2001) A concept of welfare based on reward evaluating mechanisms in the brain: anticipatory behaviour as an indicator for the state of reward systems. Appl Anim Behav Sci 72:145–171

Swaisgood RR (2007) Current status and future directions of applied behavioral research for animal welfare and conservation. Appl Anim Behav Sci 102:139–162

Thompson PM, Miller D, Cooper R, Hammond PS (1994) Changes in the distribution and activity of female harbour seals during the breeding season: implications for their lactation strategy and mating patterns. J Anim Ecol 63:24–30

Trites AW, Donnelly CP (2003) The decline of Steller Sea lions *Eumetopias jubatus* in Alaska: a review of the nutritional stress hypothesis. Mammal Rev 33:3–28

Trumble SJ, Castellini MA, Mau TL, Castellini JM (2006) Dietary and seasonal influences on blood chemistry and hematology in captive harbor seals. Mar Mamm Sci 22:104–123

Twiss SD, Anderson SS, Monaghan P (1998) Limited intra-specific variation in male grey seal (*Halichoerus grypus*) dominance relationships in relation to variation in male mating success and female availability. J Zool 246:259–267

Veasey J (2006) Concepts in the care and welfare of captive elephants. International Zoo Yearbook 40:63–79

Veissier I, Butterworth A, Bock B, Roe E (2008) European approaches to ensure good animal welfare. Appl Anim Behav Sci 113:279–297

Vicino GA, Marcacci ES (2015) Intensity of play behavior as a potential measure of welfare: a novel method for quantifying the integrated intensity of behavior in African elephants. Zoo Biol 34:492–496

Waiblinger S, Boivin X, Pedersen V, Tosi MV, Janczak AM, Visser EK, Jones RB (2006) Assessing the human–animal relationship in farmed species: a critical review. Appl Anim Behav Sci 101:185–242

Waller BM, Micheletta J (2013) Facial expression in nonhuman animals. Emot Rev 5:54–59

Walsh MT, Andrews BF, Antrim J (2001) Walruses. In: FMD G, Dierauf LA (eds) CRC handbook of marine mammal medicine, 2nd edn. CRC Press, Boca Raton, FL, pp 927–937

Walsh MT, Gaynor EV (2001) Thermal imaging of marine mammals. In: FMD G, Dierauf LA (eds) CRC handbook of marine mammal medicine, 2nd edn. CRC Press, Boca Raton, FL, pp 643–652

Wartzok D (1991) Physiology of behaviour in pinnipeds. In: Renouf D (ed) The behaviour of pinnipeds. Springer Science & Business Media, Dordrecht, NE, pp 236–299

Webster J (2005) Animal welfare: limping towards Eden. Blackwell Publishing Ltd, Oxford, UK

Welfare Quality® (2009a) Welfare quality® assessment protocol for cattle (fattening cattle, dairy cows, veal calves). Welfare Quality® Consortium, Lelystad, The Netherlands

Welfare Quality® (2009b) Welfare quality® assessment protocol for pigs. Welfare Quality® Consortium, Lelystad, The Netherlands

Welfare Quality® (2009c) Welfare quality® assessment protocol for poultry. Welfare Quality® Consortium, Lelystad, The Netherlands

Wells DL, Irwin RM, Hepper PG (2006) Lateralised swimming behaviour in the California Sea lion. Behav Process 73:121–123

Whitham JC, Wielebnowski N (2013) New directions for zoo animal welfare science. Appl Anim Behav Sci 147:247–260

Wierucka K, Siemianowska S, Woźniak M, Jasnosz K, Kieliszczyk M, Kozak P, Sergiel A (2016) Activity Budgets of Captive Cape Fur Seals (Arctocephalus pusillus) Under a Training Regime. Appl Anim Wel Sci 19:62–72

Part III
Sirenians

Chapter 17
Human Interactions with Sirenians (Manatees and Dugongs)

Robert K. Bonde and Mark Flint

Abstract There are three extant sirenian species of the Trichechidae family and one living Dugongidae family member. Given their close ties to coastal and often urbanized habitats, sirenians are exposed to many types of anthropogenic activities that result in challenges to their well-being, poor health, and even death. In the wild, they are exposed to direct and indirect local pressures as well as subject to large-scale stressors such as global climate change acting on regions or entire genetic stocks. In captivity, they are subject to husbandry and management practices based on our collective knowledge, or in some cases lack thereof, of their needs and welfare. It is therefore reasonable to consider that their current imperiled status is very closely linked to our actions. In this chapter, we identify and define human interactions that may impact dugongs and manatees, including hunting, fisheries, boat interactions, negative interactions with man-made structures, disease and contaminants, and global climate change. We examine techniques used to investigate these impacts and the influence of sirenian biology and of changing human behaviors on potential outcomes. We examine how this differs for dugongs and manatees in the wild and for those held in captivity. Finally, we provide possible mitigation strategies and ways to assess the efforts we are making to improve the welfare of individuals and to conserve these species. This chapter identifies how the welfare of these species is intrinsically linked to the human interactions these animals experience, and how the nature of these interactions has changed with societal shifts. We proffer suggested ways to minimize negative impacts. Current knowledge should be used to minimize negative human interactions and impacts, to promote positive impacts, and to protect these animals for the future.

R.K. Bonde (✉)
Wetland and Aquatic Research Center, U.S. Geological Survey,
7920 NW 71st Street, Gainesville, FL, USA
e-mail: rbonde@usgs.gov

M. Flint
School of Forest Resources and Conservation, University of Florida, Gainesville, FL, USA

School of Veterinary Science, The University of Queensland, St Lucia, QLD, Australia

A. Butterworth (ed.), *Marine Mammal Welfare*, Animal Welfare 17,
DOI 10.1007/978-3-319-46994-2_17

17.1 Introduction

There are three extant sirenian species of the Trichechidae family, which include *Trichechus manatus*, the West Indian manatee; *Trichechus inunguis*, the Amazonian manatee; and *Trichechus senegalensis*, the African manatee. The other living Sirenia family, the Dugongidae, is composed of the dugong (*Dugong dugon*) which ranges from East Africa to the north and western Pacific Ocean. All living species are listed as "vulnerable" under the International Union for Conservation of Nature, with the Steller's sea cow listed as "extinct." Given their close ties to coastal and often urbanized habitats, they are exposed to many types of anthropogenic activities that result in challenges to their well-being, poor health, and even death. In the wild, they are exposed to direct and indirect local pressures as well as subject to large-scale stressors such as global climate change acting on a region or entire genetic stocks. In captivity, they are subject to husbandry and management practices based on our collective knowledge, or in some cases lack thereof, of their needs and welfare. It is reasonable to consider that their current imperiled status is very closely linked to our actions.

Consequently, our long history of interacting with sirenian species as both a food source and collecting them as zoological specimens, and our recent western-world interest in conserving them, makes the welfare of wild and captive sirenians of particular interest at both local and global scales. Through the use of historic and new knowledge, it may be possible to propose solutions to population stabilization and recovery which we can proactively implement on a local scale or even on a global scale. However, can this be achieved in time to prevent maladaptive responses of these remaining populations to human-induced threats? Understanding the biological impact of anthropogenic threats to sirenian populations is likely to improve our adoption of country-specific recovery plans.

In this section, we identify human interactions that may impact dugongs and manatees, including hunting, fisheries, boat collisions, impact of man-made structures, disease and contaminants, and global climate change. We examine techniques used to investigate these impacts and the influence of sirenian biology and of changing human behaviors on potential outcomes. We examine how this differs for dugongs and manatees in the wild and for those held in captivity. Finally, we provide possible mitigation strategies and ways to assess the efforts we are making to improve the welfare of individuals and to conserve these species.

Many of these issues are first defined in this chapter and then discussed in greater detail in the subsequent sirenian chapters under their respective categories. This chapter aims to show how the welfare of these species is intrinsically linked to the human interactions these animals experience, and how the nature of these interactions has changed with societal shifts. We also suggest ways to minimize negative impacts.

17.2 Necropsy Examination

Strandings have provided an opportunity to examine the background to these events and to conduct detailed necropsies on carcasses (Fig. 17.1). Through this work, scientists have documented evidence for causes of mortality that are of concern. These concerns lie principally in our ability to identify anthropogenic sources of death in wildlife. Direct human interactions with marine animals often result in death, but survival can also be impacted by indirect detrimental changes to the ecosystem. Many countries have developed long-term salvage programs to determine the causes of deaths in their animals and to try to understand how to minimize this, and a spin-off of this is that the

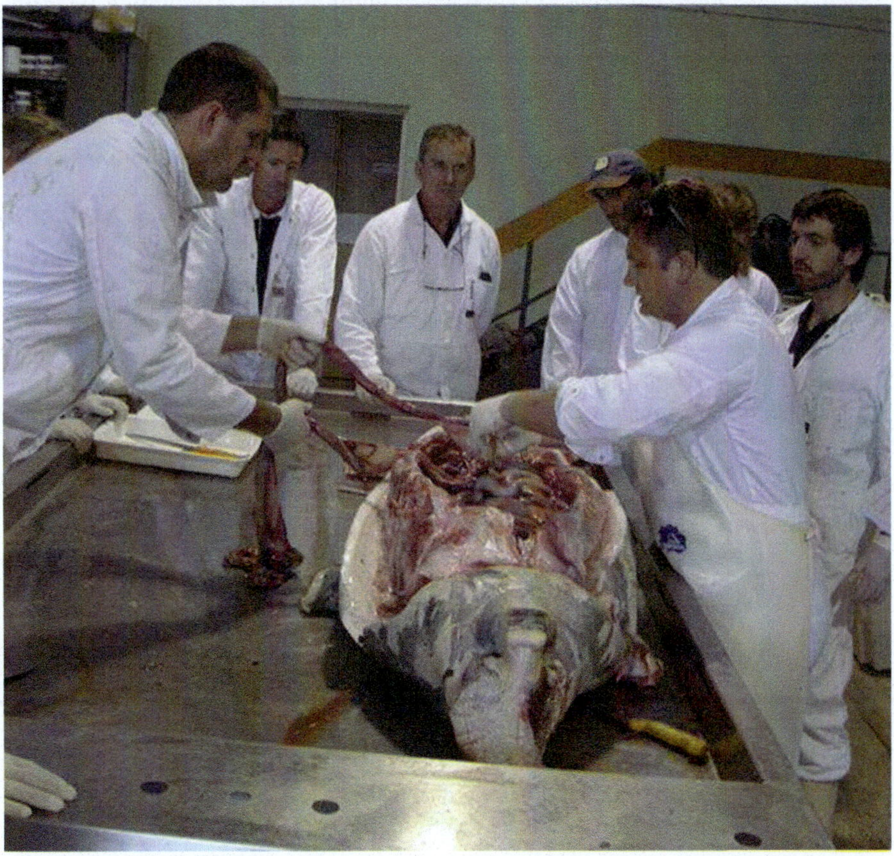

Fig. 17.1 Necropsy demonstration and training being performed on a dugong to show how to systematically examine a carcass in order to obtain the samples necessary to determine cause of death. By training multiple professionals, consistent data can be collected, and welfare-orientated management decisions can be made at the regional level (*Image credit: Mark Flint*)

positive benefits of such programs are being adopted in other countries. Establishing projects across the range of marine mammals in many countries allows decisions to be made regarding resourcing and where to direct monitoring efforts. This maximizes information available for managers tasked with recovery of imperiled species.

Since the early 1980s, useful manuals have been provided that help guide the pathologist or prosector through the necropsy of sirenians. These include necropsy manuals for the dugong (Eros et al. 2007) and manatees (Bonde et al. 1983). The latter manual has been translated into Spanish and French, and in addition an atlas on Amazonian manatee anatomy in Portuguese has recently been released (Marmontel et al. 2016). All these guides detail the descriptive observations that can be made to provide the best information from a postmortem examination—SOPs (standard operating procedures) encourage a uniformity of postmortem reporting that allows for comparison between cases, species, and regions. As such, there is now minimal baseline data that should always be recorded by the pathologist. This includes the date, time, location, which animals are involved, and mitigating circumstances that may help determine the actual cause of death. Along with written documentation, it is imperative to collect biological and voucher samples (tissues that can validate species and health condition and be used to reference-specific cause of death), as well as photographs. The quality of the detail recorded often relates directly to the data. Specifically, it is helpful to collect information that can be used to build a case that informs us about why marine mammals strand and what impact human activities may have had on this stranding.

Determining the causes of sirenian mortality and morbidity has been a useful tool in understanding the pressures these marine mammals face and on the conservation of the species. Population demographic assessments, mitigation strategies, and recovery plans have largely been based on the findings of these programs of data collection from specific areas, and these can be applied to global issues through rollout of the programs and sharing of data and of conservation planning and strategies.

17.3 Hunting

The approximately fifty-million-year evolutionary history of sirenians has been extensively documented through fossil evidence (for summary, see Marsh et al. 2011). The recent sirenians are classified into two families: the three extant species of manatee are in the family Trichechidae. The dugong and the extinct Steller's sea cow are in the family Dugongidae. These two lineages have been separate for 25–40 million years. Steller's sea cow was restricted to the coastal fringes of two islands near the Kamchatka Peninsula of Siberia when it was discovered by Europeans in the middle of the eighteenth century. It was quickly exploited for its meat and became extinct within 27 years of its discovery as a result of its restricted distribution, small population size, likely low rates of potential population growth, and overharvest (Marsh et al. 2011).

The coastal, estuarine, and riverine habitats of manatees and dugongs make them accessible to humans with small canoes and boats, and they have been hunted for millennia (Marsh et al. 2011). Hunting and commercial exploitation are believed to have contributed to decline of all four species in some locations (see Marsh et al. 2011 for references), and hunting is now banned in most countries, although in practice hunting still occurs in some places due to lack of enforcement or for reasons of cultural heritage (Castelblanco-Martinez et al. 2012). However, hunting can be sustainable if the sirenian population is large and largely inaccessible to hunters. Dugongs have been hunted in the Torres Strait Region between Australia and Papua New Guinea for at least 4000 years (Crouch et al. 2007), and the numbers hunted have been substantial for at least the last 400 years (McNiven and Bedingfield 2008). Nonetheless, several lines of evidence indicate that the modern harvest, which is permitted under an international treaty between the two countries, is sustainable (Marsh et al. 2015).

As discussed in Chap. 18, the animal welfare concerns associated with hunting are contentious and subject to campaigns by anti-hunting lobby groups, particularly in Australia (Watkin et al. 2016). In some countries, recent efforts have been directed at encouraging fishermen and hunters to take other species for subsistence or adopt forms of ecotourism using live sirenians by employing fishermen and aborigines as local tour guides to take paying visitors out to view them. The efficacy of this approach is location specific and depends on the logistics of the tourism operation, the clarity of the water, and the values associated with hunting. The approach is unlikely to succeed when sirenians are cultural keystone species (Butler et al. 2012) and the cultural values of hunting exceed the provisioning values of the meat (Delisle 2013).

17.4 Fisheries and Interactions

Sirenians may become the victim of accidental death due to human interactions such as fishing operations. These include entanglement in active or discarded fishing gear and drowning in fishing nets that can often lead to death (Adimey et al. 2014; Beck and Barros 1991). Not only is entanglement a mortal threat, it also raises concern for trapped animals struggling in a net that impedes their movement, causes injury, and ultimately death by drowning or through exhausted attempts to reach the surface to breathe. Some of these entanglement sources can be eliminated or reduced though successful regulation and cooperation with fishermen. For example, certain traditional netting practices have been reduced or banned once they were identified as a cause of harm to sirenians. Active sources of net entanglement often include gill nets, hoop-nets, or shark entanglement nets used to protect bathers on the beach. Discarded or unattended nets present the most dangerous situation for sirenian entanglement and drowning. One reduction strategy used in the Queensland Shark Control Program has been to integrate or replace shark nets with drumlines (large baited unmanned hooks suspended from a float and usually secured to the sea floor by long anchored lines; these are used to lure and capture sharks (Queensland Government 2016)) in high-density dugong areas to minimize the amount of net needed to protect beach bathers.

Fig. 17.2 A manatee which has lost its limb due to entanglement and subsequent strangulation and devitalization of the distal tissue (*Image credit: Sirenia Project—U.S. Geological Survey*)

Ingestion of debris has also been identified as a source of risk to sirenians (Adimey et al. 2014; Beck and Barros 1991). This includes plastic debris, discarded monofilament fishing line, or rope, which can lead to ingestion or entanglement. Sirenians ingest debris discarded by humans into the water mistaking it for a food source or when it becomes hidden among seagrass beds. Some reported ingestion cases have resulted in internal problems such as gut blockage, or intoxication from the chemicals released by the debris as it breaks down, which in some cases can lead to death. Externally, if line is bound up tightly around a flipper, this can result in strangulation, self-amputation, and impeded movement (Fig. 17.2).

The best prevention of this induced threat is education, cleaning up the environment, and proactive work to discontinue the discarding of trash into the water.

17.5 Watercraft Strikes

A significant negative interaction between humans and sirenians in their marine and aquatic environment is the threat of collisions with watercraft (Beck et al. 1982; Lightsey et al. 2006; O'Shea et al. 1985; Rommel et al. 2007). There have been numerous documented cases involving death due to impact with boats and ships, and this is more common in areas where people and sirenians share the same waterways. Further, welfare concerns exist concerning sublethal interactions where the dugong or manatee is injured or impeded by a boat strike, and then has a chronic decrease in their quality of life, predisposing them to secondary conditions and risks and removing their functionality in the population. Injuries are not always visible, and therefore some animals may avoid detection and the opportunity for medical care, but they continue to function in a compromised or impaired way as part of their herd. Sometimes there is convincing evidence that the impact was due to a spinning propeller, as the wounds inflicted of this are diagnostically evident—multiple parallel lacerations of similar size, depth, and spacing. However, in many cases of watercraft collision, blunt trauma or bruising to the sirenian alone (concussive and percussive injuries) is the primary cause of death with no external propeller lacerations. Therefore, watercraft-induced lesions can be propeller only, impact trauma only, or a combination of impact and propeller trauma. Thus, forensic pathology becomes an important tool in determining causes of injury and in developing subsequent mitigation strategies.

Many manatees in Florida carry signs of sublethal encounters with watercraft as evidenced by the scar patterns on their bodies, but appear to remain functional members of the population (Beck and Reid 1995). What is not known is the degree of social impediment (injury leading to inability to interact or loss of cohorts, such as nursing females) that occurs as a consequence of these. One old Florida manatee had 53 different scar patterns inflicted at different times during its life. Another known Florida manatee had its tail completely removed by a very large vessel impact, and this major lesion subsequently healed. He was observed actively participating in a mating herd in south Florida—an activity that requires use of both the forelimbs and the tail to maneuver the female into a mating position. While he could compete with the other suitors, it was not known if he could successfully mate. These scars are a tragic reminder of how human actions can impact the lives and survival of sirenians (Fig. 17.3).

Auditory studies have suggested that Florida manatees may not hear approaching boats well, or may have difficulty localizing them, potentially creating a risk for these animals in areas where boating activities occur. However, recent evidence suggests that these animals use the sensory hair on their body to aid in detecting objects in the water. These hairs are whisker-like in structure and may be used in a way akin to the lateral line of fish, to detect approaching objects in the water (Gaspard et al. 2013; Sarko et al. 2007). Despite some questions on their auditory ability, manatees are much attuned to stimuli in the water, and this, in general, does heighten their ability to detect and avoid threats. They have excellent communication ability in close quarters through their single-note calls (O'Shea and Poche 2006), whereas their vision is believed to be poor (Bauer et al. 2003). Taking these features of manatee behavior into account, managers have chosen to reduce the enforced speeds of watercraft in areas where manatees are found in large numbers to give them a greater opportunity to detect approaching boats and avoid collisions. This boat speed regulation in Florida has aided in reducing manatee deaths and encouraged recovery of

Fig. 17.3 A healthy dugong caught during population surveillance showing a healed wound that resulted in loss of over 60% of its forked tail. This injury was presumably caused by a boat strike. In addition, the kink at the base of the tail is a presumed healed fractured spine (*Image credit: Department of Environment and Heritage Protection StrandNet*)

the species (Calleson 2014). Additionally, acoustic ambient noise levels in the environment can have detrimental effects on marine mammals that rely on sound reception for behavioral cues.

Mitigation of the impacts of vessels is a dynamic area of regulation. Go-slow zones, manatee and dugong exclusion (for vessels) zones, the compulsory use of propeller guards in some commercial vessels, and the requirement for many commercial vessels to appoint a "spotter" whose job it is to search for species such as manatees, dugongs, and sea turtles in the path of the vessel are some current mitigation tactics employed around the world.

17.6 Crushed or Drowned in Flood Structures, Locks, and Dams

While dugongs are strictly marine herbivores, the remaining sirenian species all utilize fresh and salt water. As such, other perturbations confronting manatees in developed places such as Florida include animals being crushed or drowned in flood control structures, locks, or dams. The flood control structures mechanically pin and trap manatees, resulting in severe crushing trauma and drowning (Odell and Reynolds 1979). One case in point occurred when two of the flood gates in south Florida started reporting manatee-associated deaths over a period of few years in the late 1970s. At that time, these two structures pinned and drowned 22 manatees in just a few weeks (Odell and Reynolds 1979). When this was brought to the attention of management, the structures were modified with timers to open the gate to an opening of 4 feet (1.2 m), allowing water flow to sweep any resting manatees safely through, then after a short time, moving to the final desired opening of 4″ (0.1 m). This change resulted in resolution of this problem, with no manatee deaths until the mid-1990s when the timers were deactivated. Shortly after this, dead manatees were reported from these structures again, and the timers were restored, and no manatee deaths have occurred since that time. Managers continue to monitor the operation of the structures to ensure compliance.

Lock gates can also kill sirenians when they get caught behind the closing door mechanism during operation (O'Shea et al. 1985). In Florida, a simple solution was employed (pressure sensors and/or fencing to keep manatees from getting trapped behind the doors), and this type of death is now rare.

Dams are extremely detrimental to healthy populations as they cut off and isolate breeding individuals. Impediments to travel corridors created by locks and dams hamper the ability of animals to make behaviorally timed movements or migrations, and this can ultimately be detrimental to their ability to feed and mate and ultimately to the individual and population health.

Ongoing collaboration between lock and gate operators, freshwater managers, and conservationists is considered the best tool for controlling accidental death and/or impediment in these types of structures.

17.7 Infectious and Emerging Anthropogenic Diseases

We view sirenians as sentinel species that can alert us to imbalances in the environment (Bonde et al. 2004). With this philosophy, keeping the environment healthy for sirenians will ensure that it is healthy for humans. One benefit of creating this link is that people are more likely to care about the welfare of a sentinel species if they can see how the sentinel's health is going to reflect their own. Zoonotic diseases (diseases with the potential for animal to human transmission) which can cause human ailments are present among sirenians. Monitoring for sirenia-borne zoonotic viruses, bacteria, and parasites was proposed by (Forrester 1992) and can better prepare us to deal with this issue. Of special concern are *Cryptosporidium*, toxoplasmosis from cat feces, and leptospirosis from ruminants, feral pigs, and rodents (Borges et al. 2011; Sulzner et al. 2012), all diseases capable of infecting both humans and sirenia. Monitoring water quality should remain a high priority, with a diligent watch for *Vibrio*, *Clostridia*, *Aeromonas*, and *Pseudomonas* species of pathogens (Nielsen et al. 2013).

As the sirenians are mammals, they are especially vulnerable to red tide algal blooms (Bossart et al. 1998; O'Shea et al. 1991). These algae produce profoundly toxic neurologic agents which can result in loss of consciousness, and therefore rapid death, in an air breather like a sirenian in the water. Development of harmful algal blooms (HABs) can be influenced by a complex interaction of anthropogenic factors; but their increased occurrence over recent years is a direct result of humans' negative impact on our nearshore habitats. To counter this harmful growth, many places with known populations of sirenians have serial water monitoring programs. Monitoring for blooms can alert rescue teams and first responders to potential exposure. The success rate for rescued intoxicated individuals can be good if the animal can be accessed prior to losing consciousness, because the animals can clear the neurologic agent after removal from the contaminated environment. Secondary effects can present with introduction of tolerant invasive species, and other algae, such as *Lyngbya*. Infiltration of *Lyngbya* into systems can lead to skin irritation in sirenians, as well as retarded growth of natural aquatic plants (Harr et al. 2008). Disease surveillance programs in live and dead animals are the most effective tools in monitoring and minimizing the impacts of disease threats (Bonde et al. 2004).

17.8 Contaminants: Pollution and Toxins

Due to anthropogenic influences on our environment through exposure to contaminants, detailed studies have been conducted on levels of well-known toxins identified in sirenians. Primary toxicology studies have focused on copper (which can result in acquired zinc deficiency), which is heavily used in control of nuisance plants. For example, with high levels of copper reported in manatees in Crystal River in the early 1980s, herbicide applications were regulated to avoid direct exposure to the wintering population (O'Shea et al. 1984). Since that mitigation, copper levels have been reduced in the manatee population. Mercury is another metallic toxin found in the environment that can have devastating consequences for mammals (Driscoll et al. 2013). Many environmental pollutants are associated with suitable habitat occupied by sirenians. These include chemicals, compounds, fatty acids, pharmaceuticals, trace elements, toxins, and poisons (O'Shea and Tanabe 2003). Agencies are tasked with reducing discharge of these pollutants into the environment. This is often not easy to do, and many of these contaminants are sequestered for decades in sediments and may not get reintroduced into the ecosystems unless the sediments are disturbed, agitated, or exposed. Harmful algal blooms caused by *Lyngbya* and *Karenia brevis* are becoming more frequent, and this is impacting the health of sirenians. Ongoing environmental monitoring and regulation of agricultural runoffs is the current practice used around the world to control the negative impacts of natural contaminants (Meager and Limpus 2014). Disaster mitigation plans, such as those used after oil spills, are the primary response tool to a toxin-induced unusual mortality event or threat. Increased awareness has resulted in sirenians being the primary species of concern in many regions after such disasters (Geraci and Lounsbury 2005).

17.9 Global Climate Change

The major concerns for sirenia of climate changes are alteration in sea level, seasonal weather anomalies, global warming trends, flooding, and unpredictable increases in stochastic events (Wells et al. 2015). It is predicted that these events will result in loss of habitat and that this will relate directly to the carrying capacity of sirenians, with the potential to cause localized population isolation and potential extinctions. To prepare ourselves for these fundamental changes, proactive action is preferred to a reactive response which may be provided too late. Efforts could focus on reduction of global carbon emissions and modeling to predict how sirenians and humans will respond to changes.

17.10 Captive Populations

Holding dugongs or manatees in captivity occurs for two primary purposes—either for (1) treatment or assessment prior to reintroduction back to the wild or (2) for entertainment and education. Regardless of the purpose, both share risks and rewards. The placement of any wild animal into a captive environment for any period of time alters their behavior and ability to interact socially with their cohort. However, the benefit of allowing close experiences of the public with such large marine mammals may serve to educate and increase awareness and ecological responsibility in those who see these creatures at close quarters and become aware of the conservation and environmental issues which are affecting them, although some are distressed by seeing these animals in captivity. The risks associated with captivity must be weighed against the greater benefit to the individual and the species, and this balancing of factors raises ethical and societal questions about why and whether to keep large mammals in captivity or to protect them in the wild. Efforts to optimize the facilities and care provided for large mammals, including sirenia in captivity, could be prioritized to benefit all species of concern.

17.11 Social Structure

Monitoring of populations of sirenians has identified various human stressors that can have an impact on the social structure of the population. This has been reported through increases in the incidence of orphan rescues due to death or displacement of their mother (Marsh et al. 2011). Nursing mothers also exhibit nutritional stress (poor body condition) relative to the quality of the local environment and, if unhealthy, can pass on diseases by contact and even accumulated toxins through vertical transmission via milk during suckling (O'Shea et al. 1984). Human activity can also result in reduced reproductive success or physical separation of potential breeding individuals. This form of isolation can prohibit opportunities to mate, as well as affect genetic fitness in the population. This form of human-induced impact can come from overcrowding, encroachment, and physical tampering of sirenians or their habitats (Marsh et al. 2011).

17.12 Positive Human Interactions

Although much of what we report on our interactions with sirenians are long-standing issues, which may have contributed to their current threatened status, there are examples of human activity which has positively influenced these species or at least has reduced anthropogenic impacts. Examples will be given in the following chapters and include go-slow zones in known manatee and dugong habitats, changes to fishing and waterway control practices, and laws to limit hunting and exploitation (Fig. 17.4). However, there are also practices such as long-term investment in rearing orphaned calves for return to their natural habitat, the science and research dedicated to gaining a better understanding of sirenia, and the supply of heated water as a by-product in discharge canals of electric facilities providing a warm water refuge to thousands of manatees in Florida over winter (an anthropogenic side effect). These current practices can be seen as making a positive difference to sirenian populations around the world.

Fig. 17.4 A manatee enforcement zone that is active during the winter when manatees are known to frequent the area (*Image credit: Jaylene Flint*)

17.13 Conservation, Education, and Possible Solutions

All of the threats discussed in the section above not only impact the survivorship of the species but also the welfare of individuals. All of these threats can either be made worse if we do not continue with ongoing efforts to mitigate and reverse their effects, or they can be overcome if we work together to protect and conserve our aquatic resources. This may best be achieved through education, communication, and regulatory implementation. In order to reduce the human impact on sirenians, proactive efforts could continue to be employed. Reducing our footprint on the environment will likely improve conditions for sirenians to allow them to better coexist with people. We need to address our philosophies regarding the intrinsic value of sirenians. Can we coexist with them with minimal negative impact? And, if so, how do we do that? They play an important role in the ecosystem and can act as a sentinel species to help us gauge emerging problems. When we identify problems, we can respond with measures that will aid in recovery of the population, and we do that already to an extent with appropriate signs, protected areas and sanctuaries, and through education programs (Fig. 17.5). In the history of sirenian species recovery, we have come a long way. However, we need to continue to move ahead

Fig. 17.5 With the general public being able to access and watch wild marine mammals from the comfort of purpose-built viewing centers, there has been an increase in public awareness of the issues facing them and support toward reducing threats to their survival. Shown is an example of such a viewing center in Central Florida where manatees come to seek refuge in the industrially warmed waters and people can watch them from the shore (*Image credit: Stacie Koslovsky*)

with a proactive vision for strategic gathering of data and information to help us better understand complexities facing sirenians and their role in the coastal environment.

17.14 Conclusions

Sirenians are a resilient species that have survived and evolved over millions of years. They are seemingly well suited to cope with the perturbations and problems that surround them; but human society needs to realize that the impact of our footprint on the environment for sirenians is huge and will last a long time. We can have positive impacts on sirenians through creation of sanctuaries, no entry zones, watercraft speed zones, fishery exclusion areas by enforcement action to prevent illegal hunting and trading, and through increased care regarding what we put into and take out of our aquatic environment. These actions can be enforced through governmental agency regulation. Now is the time to use our current knowledge to minimize negative human interactions and impacts, to promote positive impacts, and to protect these animals for the future.

References

Adimey NM, Hudak CA, Powell JR, Bassos-Hull K, Foley A, Farmer NA, White L, Minch K (2014) Fishery gear interactions from stranded bottlenose dolphins, Florida manatees and sea turtles in Florida, USA. Mar Pollut Bull 81:103–115

Bauer GB, Colbert DE, Gaspard J, Littlefield B, Fellner W (2003) Underwater visual acuity of Florida manatees (*Trichechus manatus latirostris*). J Comp Psychol 16:130–142

Beck CA, Barros NB (1991) The impact of debris on the Florida manatee. Mar Pollut Bull 22:508–510

Beck CA, Bonde RK, Rathbun GB (1982) Analyses of propellor wounds on manatees in Florida. J Wildl Manag 46:531–535

Beck CA, Reid JP (1995) An automated photo-identification catalog for studies of the life history of the Florida manatee. In: Population Biology of the Florida Manatee (*Trichechus manatus latirostris*). National Biological Service, p 120–134

Bonde RK, Aguirre AA, Powell J (2004) Manatees as sentinels of marine ecosystem health: are they the 2000-pound canaries? EcoHealth 1:255–262

Bonde RK, O'Shea TJ, Beck CA (1983) Manual of procedures for the salvage and necropsy of carcasses of the West Indian Manatee (*Trichechus manatus*). USGS-Biological Resources Division, Florida, p 175

Borges JCG, Alves LC, Faustino MAD, Marmontel M (2011) Occurrence of cryptosporidium spp. in Antillean manatees (*Trichechus manatus*) and Amazonian manatees (*Trichechus inunguis*) from Brazil. J Zoo Wildl Med 42:593–596

Bossart GD, Baden DG, Ewing RY, Roberts B, Wright SD (1998) Brevetoxicosis in manatees (*Trichechus manatus latirostris*) from the 1996 epizootic: gross, histologic, and immunohistochemical features. Toxicol Pathol 26:276–282

Butler JRA, Tawake A, Skewes T, Tawake L, McGrath V (2012) Integrating traditional ecological knowledge and fisheries management in the Torres Strait, Australia: the catalytic role of turtles and dugong as cultural keystone species. Ecol Soc 17(4):34

Calleson CS (2014) Issues and opportunities associated with using manatee mortality data to evaluate the effectiveness of manatee protection efforts in Florida. Endanger Species Res 26:127–136

Castelblanco-Martinez DN, Barba E, Schmitter-Soto JJ, Hernandez-Arana HA, Morales-Vela B (2012) The trophic role of the endangered Caribbean manatee *Trichechus manatus* in an estuary with low abundance of seagrass. Estuar Coasts 35:60–77

Crouch J, McNiven IJ, David B, Rowe C, Weisler M (2007) Berberass: marine resource specialisation and environmental change in Torres Strait during the past 4000 years. Archaeol Ocean 42:49–64

Delisle A (2013) A socio-economic investigation of the Torres Strait indigenous dugong and turtle fisheries doctor of philosophy. James Cook University, Australia

Driscoll CT, Mason R, Chan HM, Jacob DJ, Pirrone N (2013) Mercury as a global pollutant: sources, pathways, and effects. Environ Sci Technol 47:4967–4983

Eros C, Marsh H, Bonde RK, O'Shea TJ, Beck CA, Recchia C, Dobbs K, Turner M, Lemm S, Pears R, Bowater RO (2007) Procedures for the salvage and necropsy of the dugong (*Dugong dugon*), Second edn. Great Barrier Reef Marine Park Authority, Australia

Forrester DJ (1992) Parasites and diseases of wild mammals in Florida. University Press of Florida, Gainesville

Gaspard JCI, Bauer GB, Reep RL, Dzuik K, Read L, Mann DA (2013) Detection of hydrodynamic stimuli by the Florida manatee (*Trichechus manatus latirostris*). J Comp Physiol A 199:441–450

Geraci JR, Lounsbury VJ (2005) Marine mammals ashore: a field guide for strandings, Second edn. National Aquarium in Baltimore, Baltimore

Harr KE, Szabo NJ, Cichra M, Phlips EJ (2008) Debromoaplysiatoxin in Lyngbya-dominated mats on manatees (*Trichechus manatus latirostris*) in the Florida King's Bay ecosystem. Toxicon 52:385–388

Lightsey JD, Rommel SA, Costidis AM, Pitchford TD (2006) Methods used during gross necropsy to determine watercraft-related mortality in the Florida manatee (*Trichechus manatus latirostris*). J Zoo Wildl Med 37:262–275

Marmontel M, Guimaraes JP, Mari RB, Beck C, Bonde RK (2016) Atlas de anatomia de peixe-boi Amazônico, *Trichechus inunguis* Natterer, 1883. Instituto de Desenolvimento Sustentavel Mamiraua, Brazil

Marsh H, Grayson J, Grech A, Hagihara R, Sobtzick S (2015) Re-evaluation of the sustainability of a marine mammal harvest by indigenous people using several lines of evidence. Biol Conserv 192:324–330

Marsh H, O'Shea T, Reynolds JE (2011) Behaviour and habitat use. In: Ecology and conservation of the Sirenia: dugongs and manatees. Cambridge University Press, New York, p 145–208.

McNiven IJ, Bedingfield AC (2008) Past and present marine mammal hunting rates and abundances: dugong (*Dugong dugon*) evidence from Dabangai bone mound, Torres Strait. J Archaeol Sci 35:505–515

Meager JJ, Limpus C (2014) Mortality of inshore marine mammals in eastern Australia is predicted by freshwater discharge and air temperature. PLoS One 9(4):e94849. doi:10.1371/journal.pone.0094849

Nielsen KA, Owen HC, Mills PC, Flint M, Gibson JS (2013) Bacteria isolated from dugongs (*Dugong dugong*) submitted for postmortem examination in Queensland, Australia, 2000-2011. J Zoo Wildl Med 44:35–41

O'Shea TJ, Beck CA, Bonde RK, Kochman HI, Odell DK (1985) An analysis of manatee mortality patterns in Florida, 1976-81. J Wildl Manag 49:1–11

O'Shea TJ, Moore JF, Kochman HI (1984) Contaminant concentrations in manatees in Florida. J Wildl Manag 48:741–748

O'Shea TJ, Poche LB (2006) Aspects of underwater sound communication in Florida manatees (*Trichechus manatus latirostris*). J Mammal 87:1061–1071

O'Shea TJ, Rathbun GB, Bonde RK, Buergelt CD, Odell DK (1991) An epizootic of Florida manatees associated with a dinoflagellate bloom. Mar Mamm Sci 7:165–179

O'Shea, TJ, Tanabe S (2003) Persistent ocean contaminants and marine mammals: a retrospective overview. In: Toxicology of marine mammals. Taylor and Francis, London, p 99–134.

Odell DK, Reynolds JE (1979) Observations on manatee mortality in South Florida. J Wildl Manag 43:572–577

Queensland Government (2016) Shark control equipment and locations. https://www.daf.qld.gov.au/fisheries/services/shark-control-program/shark-control-equipment-and-locations. Accessed 23 Dec 2016

Rommel SA, Costidis AM, Pitchford TD, Lightsey JD, Snyder RH, Haubold EM (2007) Forensic methods for characterizing watercraft from watercraft-induced wounds on the Florida manatee (*Trichechus manatus latirostris*). Mar Mamm Sci 23:110–132

Sarko DK, Reep RL, Mazurkiewicz JE, Rice FL (2007) Adaptations in the structure and innervation of follicle-sinus complexes to an aquatic environment as seen in the Florida manatee (*Trichechus manatus latirostris*). J Comp Neurol 504:217–237

Sulzner K, Johnson CK, Bonde RK, Gomez NA, Powell J, Nielsen K, Luttrell MP, Osterhaus A, Aguirre AA (2012) Health assessment and seroepidemiologic survey of potential pathogens in wild Antillean manatees (*Trichechus manatus manatus*). PLoS One 7. doi:10.1371/journal.pone.0044517

Watkin L, Kim F, Delisle MK, Stoeckl N, Marsh H (2016) Setting the table: indigenous engagement on environmental issues in a politicized context. Soc Nat Resour. doi:10.1080/08941920.2016.1150541

Wells ML, Trainer VL, Smayda TJ, Karlson BSO, Trick CG, Kudela RM, Ishihawa A, Bernard S, Wulff A, Anderson DM, Cochlan WP (2015) Harmful algal blooms and climate change: learning from the past and present to forecast the future. Harmful Algae 49:68–93

Chapter 18
Impacts of Marine Debris and Fisheries on Sirenians

Helen Owen, Jaylene Flint, and Mark Flint

Abstract Harmful marine debris includes land and ship-sourced waste and abandoned fishing gear from recreational and commercial fisheries; these forms of debris are making their way into waterways and oceans with increasing frequency. For sirenians, marine debris and fisheries pose a significant risk to their well-being through entanglement, ingestion and hunting, both legal and illegal, as well as through more indirect ways, such as changing social structures and creating orphans through loss of cohorts. This chapter addresses the welfare impacts of marine debris and fisheries on sirenians. It also explores the changes in attitude that are occurring in many of the stakeholders involved and how these are translating into positive outcomes.

18.1 Introduction

For sirenians, marine debris and fisheries pose a significant risk to their well-being through entanglement, ingestion and hunting, both legal and illegal, as well as through more indirect ways such as changing social structures and creating orphans through loss of cohorts (Australian Government 2009a).

H. Owen (✉)
University of Queensland, Gatton, QLD, Australia
e-mail: h.owen1@uq.edu.au

J. Flint
Veterinary Marine Animal Research, Teaching and Investigation Unit,
School of Veterinary Science, University of Queensland, Gatton, QLD, Australia

Center for Conservation, The Florida Aquarium, Tampa, FL, USA

M. Flint
Veterinary Marine Animal Research, Teaching and Investigation Unit,
School of Veterinary Science, University of Queensland, Gatton, QLD, Australia

A. Butterworth (ed.), *Marine Mammal Welfare*, Animal Welfare 17,
DOI 10.1007/978-3-319-46994-2_18

This chapter addresses the effects of harmful marine debris, which include land-sourced waste and garbage (such as bags, bottles, ropes, fibreglass, piping, insulation, paints and adhesives), abandoned fishing gear from recreational and commercial fisheries (e.g. strapping bands, synthetic ropes, derelict fishing nets, floats, hooks, fishing line and wire trace) and ship-sourced, solid, non-biodegradable floating materials disposed of at sea (e.g. fibreglass and insulation) (Threatened Species Scientific Committee 2003). Marine debris is any manufactured or processed solid waste material that enters the ocean environment. Many forms of debris are composed of plastic and thus are persistent and buoyant for years to decades after entering the ocean (Laist 1987; Ryan 1987; Hansen 1990; Goldberg 1995; Goldberg 1997; Derraik 2002). Such debris can be carried extensive distances on ocean currents, increasing the likelihood of exposure to wildlife (Sheavly and Register 2007; Laist 1987).

Fisheries may also serve as a hazard to marine animals during active use, as animals can become entrapped in deployed gear or affected indirectly when vessels and gear damage necessary habitat such as seagrass beds.

Both types of effects result in stress on resident animals. No family of marine animal is immune to marine debris or the threat of fishery interactions. Six hundred and ninety three species of marine organisms have been reported to have encountered marine debris (Gall and Thompson 2015). There are records of entanglement in or ingestion of plastic debris for 54% of the International Union for Conservation of Nature's Red Listed species (STAP 2011). All of these items can negatively impact the welfare of marine animals.

18.2 Marine Debris: Extent of Impact

Although the proportion of Florida manatees dying following interactions with marine debris or fishery equipment is generally reported to be low (Bossart et al. 2004; Buergelt et al. 1984; O'Shea et al. 1985), interactions with fishery equipment are likely to be of greater significance for Amazonian manatees, West Indian manatees in the wider Caribbean region and dugong (Hines and Reynolds 2012). For example, the regional use of shark nets in Queensland, Australia has had a significant impact on dugong in this area (Marsh et al. 2005; Heinsohn 1972), and the practice of soak net fisheries has been known to cause dugong mortalities under unusual circumstances. These include when animals are being forced to forage in novel areas post-flooding and when new fisheries are introduced in a region where dugong regularly forage. The animals that die as a result of interactions with marine debris and fishery equipment are likely to represent only a small proportion of animals affected (Laist 1987). More likely, many interactions will have a sublethal impact and cause chronic debility (Beck and Barros 1991; Laist 1987). These more protracted effects are of greatest welfare concern.

The increased and widespread use of plastics and other durable manufactured material has resulted in an increase in marine debris (Laist 1987). Marine debris can cause detrimental effects when animals are entangled or when material is ingested. This can result in rapid or more protracted death, usually involving tissue damage or starvation. In sublethal cases, it can hamper an animal's ability to eat, move, escape predators and participate in social interactions (Laist 1987).

18.3 Ingestion

Marine debris can be ingested accidentally or when it is mistaken for food, particularly when debris becomes entrapped in mats of vegetation. The shallow waters where sirenians feed often accumulate debris, making them particularly prone to ingestion (Attademo et al. 2015a; Beck and Barros 1991). Beck and Barros performed a study on stranded manatee carcasses in Florida. Debris was found in the gastrointestinal tract of 14% of the 439 examined animals. Pieces of monofilament fishing line and fish hooks or wire were the most commonly ingested debris, but a large assortment of other material including string, rope, paper, cellophane and rubber bands were also found (Beck and Barros 1991).

The potential sublethal and/or chronic impact of marine debris ingestion is demonstrated by a case series on Antillean manatees (Attademo et al. 2015a). These animals presented, alive or dead, with extreme weight loss and gastrointestinal tract debris composed of mostly plastic bags and knit fabric commonly used to make crab traps. In these cases, the debris probably resulted in weight loss by obstructing the gastrointestinal tract and causing a physical impedance to the passage of food and, less commonly, by reducing nutrient absorption or reducing intake by stimulating a feeling of satiety. Intake could also have been reduced secondary to discomfort, either through the mass effect of the material or if the material caused abrasion or erosion and subsequent ulceration of the mucosal surface of the gastrointestinal tract (Attademo et al. 2015a). Ingested debris may cause animals to become more buoyant preventing the ability to submerge and dive to access seagrass beds and avoid threats at the water's surface (Laist 1987; Bjorndal et al. 1994).

Debris ingestion can also result in acute death, when the gastrointestinal tract is punctured by sharp objects like hooks and wire or when intestinal rupture occurs, sometimes secondary to impaction. Linear debris such as monofilament line can cause intestinal plications (folding up of loops of the gut pulled together by the internal monofilament line), intussusceptions with subsequent intestinal necrosis, rupture and peritonitis (Beck and Barros 1991; Attademo et al. 2015a; Forrester et al. 1975).

Finally, the ingestion of debris can cause a toxicosis through ingestion of suspended microplastics or other constituents of the material (Mato et al. 2001). Microplastics are small plastic particles that are introduced to the marine environment through breakdown of larger plastic items or through discardment after their use as microbeads in abrasives such as 'sand' blasting or cosmetic materials. These plastics may contain polychlorinated biphenyls (PCBs) and dichlorodiphenyltrichloroethane (DDT), which are known to compromise immunity and cause infertility in animals, even at very low levels (Attademo et al. 2015b; Mato et al. 2001).

The welfare impact of ingestion of marine debris is that there is often a protracted and uncomfortable demise. Given the high proportion of populations affected by ingested marine debris and the difficulty in removing these items from the marine environment, efforts to mitigate the production and pollution of such products are immediately required.

18.4 Entanglement

Entanglement is another way in which marine debris can impact on sirenian welfare. Most reported cases of entanglement involve discarded remnants of fishery gear which can subsequently be carried by the animal for years (Adimey et al. 2014; Beck and Barros 1991; O'Shea et al. 1985); however any floating foreign object is a potential attractant and entrapment device for marine vertebrates. During the current authors' careers, we have removed, among other things, undergarments stuck on heads, crab pots wrapped around tails which were causing slow amputation and half-ingested fishing hooks with the lead line wrapped around the forelimb. In the case of the fish hooks, the line sawed through the mandible and oesophagus as the animal tried to swim while entangled. It is likely that sirenians become entangled passively, while foraging for food. For Florida manatees at least, their exploratory nature may further predispose them to entanglement (Adimey et al. 2014).

Entanglement can result in anything from impeded movement and restriction of normal behaviour to severe injury affecting deep tissue that can cause limb strangulation and amputation or mortality. Tissue injury can occur as a result of the abrasive and cutting nature of some materials, with the animals' swimming motion likely to further exacerbate the injury in many instances. In severe cases, multiple tissue layers can become necrotic and in some instances infected by opportunistic pathogens. Entanglement can also result in both mechanical trauma resulting in compromised function and pathological fractures resulting in complete loss of limb capacity (Adimey et al. 2014; Beck and Barros 1991; O'Shea et al. 1985; Walsh and Bossart 1999; Bossart 2001). Extrapolating from other species, it is probable that death in these chronic cases occurs over a period of months (Moore and van der Hoop 2012).

18.5 Changing Attitudes of Marine Debris

Recently the threat that marine debris poses to the marine environment has become recognised after being debated for decades (Stefatos et al. 1999; Derraik 2002; Laist 1987). This has led environmental and resource managers to an increased research focus in this area and the development of multiple legislations and conventions in an attempt to discover the impact and reduce the threat (Derraik 2002).

At an international level, the United Nations in 2011 made a resolution calling states to cooperate in the prevention, reduction and control of any sources of marine debris (STAP 2011). There is also the legal framework of the United Nations Convention of the Law of the Sea (UNCLOS) which allows for the protection of the entire marine environment from all sources and types of marine pollution (STAP 2011).

Such is the acknowledgement of the significance of marine debris in causing harm to marine animals, the Australian Federal Government under the *Environmental Protection and Biodiversity Conservation Act of 1999* commissioned threat abatement plans to be developed to contribute to the future prevention, removal, mitigation of impact and monitoring of marine debris within Australian waters (Australian Government 2009b).

In addition to governance, as awareness has grown there has been a shift in public perception of the effect of plastic bags, bottles and trash being dumped into our oceans with a collective conscience to repair the damage. Countries such as the United States and Australia have regular 'clean-up the ocean' campaigns with local groups volunteering their time to remove potential hazards and trash. Perhaps the most successful of these efforts is 'Clean Up Australia Day' which started in 1990 with 300,000 volunteers picking up rubbish around Australian land and water, expanding to become 'Clean Up the World' involving 120 countries and forty million participants 20 years later (Fig. 18.1).

Fig. 18.1 Staff and volunteers of The Florida Aquarium in the United States conduct a regular local clean-up day in Tampa Bay, Florida. Crews recover trash items ranging from mobile phones (inset) to nets and fishing line (*Image Credit: The Florida Aquarium*)

18.6 Fisheries: Fishery Gear

The nets, lines, hooks and traps used in commercial and recreational fisheries are potential sources of entanglement and ingestion for sirenians (Adimey et al. 2014). In a study performed by Adimey et al. (2014), hook, line and trap pot gear most commonly entangled the flippers, while fishing net entanglements were most commonly found on the body of Florida manatees, sea turtles and bottlenose dolphins.

Sirenians are mammals with dorsally rostral nostrils fitted with valves akin to their proboscid relatives, the elephants, which allow for breathing at the surface of the water with barely their nose exposed to the air and the rest of their bodies submerged. However, if entangled and submerged, even this clever adaptation is of no use to them if they cannot break the surface to draw in air. Unlike terrestrial mammals that drown by inhalation of water into the lungs, preventing the gaseous exchange of carbon dioxide and air at the alveoli, sirenians are thought to drown by asphyxiation that may occur without a drop of water entering their lungs. This mechanism occurs in several marine vertebrate species and is induced by laryngospasm and irreversible closure of the epiglottis preventing the breathing of water or air. This is likely to occur quickly, with reports of death occurring within 3 to 5 min of submersion (Woolford et al. 2015; Heinsohn et al. 1976). The consequence of this is that even if an animal can be rescued alive, once laryngospasm has occurred, they may not start spontaneously breathing again. Subsequently, following onset, drowning is very likely fatal.

Along with net fishing, the fixed shark nets used to protect bathers in highly urbanised regions such as Queensland are another source of entanglement for sirenians (Heinsohn 1972; Marsh et al. 2005). A relatively high number of dugong were caught in these shark nets when they were first introduced; however this number fell considerably, soon after modifications were introduced by the Queensland Shark Control Program. These modifications involved the addition of baited drumline hooks (large baited unmanned hook suspended from a float and usually secured to the sea floor used to lure and capture sharks) as a substitute for nets in some areas and the education and training of contractors (Marsh et al. 2005). This reduced the total linear footage of nets in the water and so reduced the risk to nearby marine life as well as increasing the responsible monitoring of nets and lines.

Education and increased awareness have also played a big role in the reduction of fishing gear in the environment. Many fishing piers have receptacles for discarded fishing hooks and line to conveniently allow fishermen to dispose of their gear. Further, special interest groups such as zoos and aquariums and community groups regularly hold clean-up days around piers and along waterways. Ghost nets (discarded nets in the ocean and on beaches) are harder to collect due to their remote locations (on land or at sea) and the sheer size of many of the nets. In order to reduce the impacts of ghost nets, organisations such as the World Wildlife Foundation working with state and federal government agencies in Australia make available documents such as *The Net Kit* (World Wildlife Fund 2016), which guides an individual on how to identify and report a discarded net as well as provides examples of known nets (purpose and origin) to give information that may aid management decisions. In Australia, ghost nets more commonly impact the northern part of the country, which is predominantly inhabited by seafaring indigenous communities. Their livelihoods are affected by these nets, as the ghost nets have the potential to reduce the population of dugong and sea turtles available for hunting. Consequently, programs have been put in place to employ indigenous communities to help report and remove these nets and also to repurpose the nets into saleable art and clothing, which in turn serves to increase awareness of this issue.

18.7 Vessel Injury

Boat strike is a relatively common cause of sirenian mortality in places around the world where there are populations of people and sirenians in close proximity (Bossart et al. 2004; Buergelt et al. 1984; Lightsey et al. 2006; Joly et al. 2009; Greenland and Limpus 2007; Owen et al. 2012; Eros et al. 2000). Injury can result from sharp, penetrating trauma (e.g. from boat propellers) or from blunt concussive and percussive trauma (e.g. boat hull or skeg impacts) and can result in acute death due to extensive haemorrhage and tissue injury (Fig. 18.2). It appears dugong are more prone to overt external injuries caused by propeller cuts and Florida manatees more to the percussive internal injuries; both requiring different diagnostic

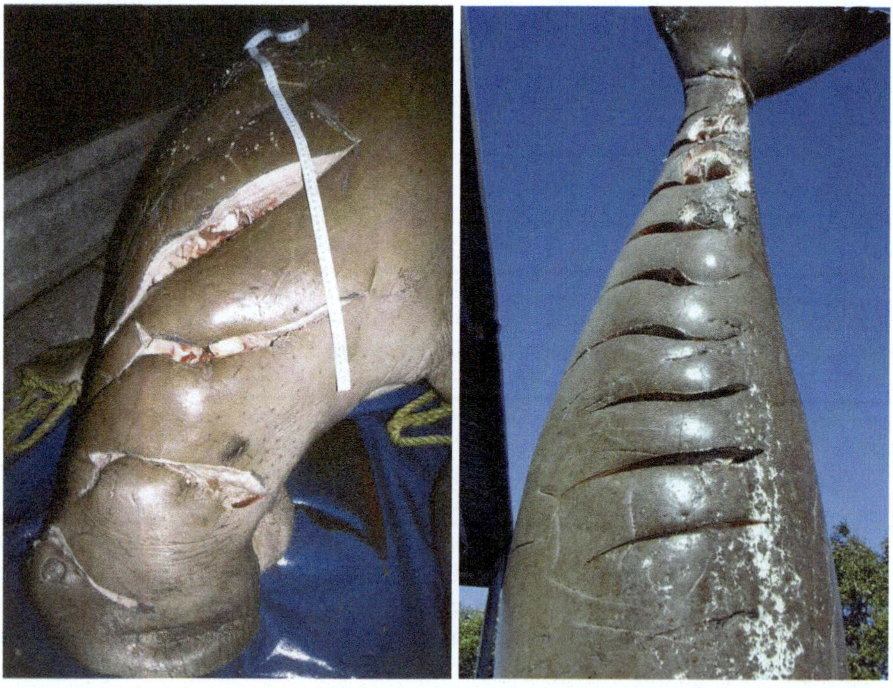

Fig. 18.2 Injuries caused by a propeller from a large vessel resulting in lethal trauma to adult dugong (*Image credit: Department of Environment and Heritage Protection StrandNet*)

approaches when determining the best course of monitoring, treatment and mitigation. Due to the dense nature of sirenian skin and bones, it is more common for boat strike to result in sublethal injuries (Yan et al. 2006a; Clifton et al. 2008a; Clifton et al. 2008b; Yan et al. 2006b). In dugong, injuries can be superficial but can lead to secondary infections and physical impediments. In Florida manatees, injuries often affect the lungs and bones (Lightsey et al. 2006; Bossart et al. 2004) producing potential lesions such as fractures, hydrothorax (fluid in the chest cavity), subcutaneous emphysema (air under the skin) and pneumothorax (air leaking from damaged lungs or externally into the thorax). While not always fatal, these injuries and the subsequent pathologies which result will all hamper the animal's ability to move, breathe, maintain a stable position in the water and dive. Secondary infections are common and include sepsis (septicaemia, circulating infection in the tissues and bloodstream), pyothorax (infection and pus in the thorax), pleuritis (inflammation of the lung surface and internal rib cage) and pleuropneumonia (infection in and around the lungs), peritonitis (infection of the tissues which surround the abdominal organs), osteomyelitis (infection within bone) and cellulitis (infections deep in the skin) (Harr et al. 2011; Buergelt et al. 1984; Bossart 2001; Lightsey et al. 2006; Walsh and Bossart 1999). These sequelae commonly result in chronic debility, and severe disease can take up to 18 months to resolve even with treatment in rehabilitation facilities (Bossart 2001). Given that not all boat strikes in dugong and manatees are visibly obvious to outside observers, welfare concerns arise. Individuals may evade detection and be left to suffer as they develop secondary conditions and are at risk of further injury, with reduced abilities to avoid vessels and to reduce negative interactions.

18.8 Indirect Effects of Fisheries

Fisheries and vessels do not necessarily have to have direct contact with sirenians to influence their well-being. It is possible that boat propellers, and some fishery practices such as bottom trawling, can result in damage to seagrass beds and destruction of habitat, causing overgrowth of unpalatable monocultures of aquatic vegetation and requiring animals to migrate to alternative sources of food (Marsh et al. 2005; Chin 2005; Hines 2012; Marsh et al. 2011; Spain and Heinsohn 1973; Marsh et al. 1982; Coles et al. 2007; Australian Institute of Marine Science 1996).

The noise associated with fishery (and recreational) vessels is also likely to adversely affect sirenian behaviour, with a high level of boat presence near a seagrass bed known to be grazed by sirenians being negatively correlated with seagrass bed usage (Miksis-Olds et al. 2007). A study by Miksis-Olds and Wagner (2011) demonstrated that Florida manatees spend more time feeding and less time milling and socialising when there are high levels of noise, possibly so they can reach their nutritional quota and leave the area as soon as possible. In general, these disruptions can lead to decreased socialisation, reduced feeding time in heavily trafficked areas, changes in diving and surfacing behaviour, changes to the habitats visited based on boat traffic and possible separation of mother and calf pairs (Deutsch et al. 2009; Marsh et al. 2011; Hodgson and Marsh 2007).

18.9 Sirenians as Target Species

Due to their coastal habitat preference, sirenians have been easily accessible to humans for hunting for thousands of years. This exploitation is thought to have contributed to the extinction of the Steller's sea cow in the 1700s, with fishermen taking advantage of the naïve and slow moving beasts as easy prey (Marsh et al. 2011). Globally, today this attitude has changed, and although manatees and dugong have historically been widely hunted for meat, skin, oil and bones, this practice is now largely prohibited by national governments. However, there are exceptions to this rule. Several countries still allow hunting of their sirenian species by indigenous communities as part of their traditional practices where the products are used for subsistence. In some countries, this provision extends past indigenous people. Current Brazilian legislation permits subsistence hunting by its citizens; but in theory, this is only for those who can demonstrate dire need (Hines and Reynolds 2012; Marsh et al. 2011).

Dugong hunting is of cultural significance to Australian Aboriginals and Torres Strait Islanders in certain geographical regions, so hunting of dugong for subsistence is currently allowed (Marsh et al. 1996; Heinsohn et al. 2004; Marsh et al. 1981) under the *Native Title Act*, *Environment Protection and Biodiversity Conservation Act* and the *Torres Strait Fisheries Act* (Fig. 18.3). Under these terms, subsistence hunting of dugong and manatees in many global regions can include subsistence use for food, medicine, poison, cooking, lamp fuel, aphrodisiacs, jewellery and charm production, carvings and leather products (Marsh et al. 2011).

Where sirenians are traditionally hunted, the welfare of the animal is usually a significant consideration by the hunters of any catch. For example, in the Torres Straits of Australia, the dugong is a totem animal for several island communities, and hunting is undertaken with a great deal of gravity and respect. The hunting of the dugong is described in detail by Marsh et al. (1981) and Marsh et al. (2011). In short, dinghies with outboard motors are used to locate a dugong. Once found, the engine is turned off, and the boat is allowed to drift towards the animal while it grazes in shallow water. Using only a hunting implement—a harpoon with a detachable pronged head called a "wap" (Marsh et al. 1981)—the animal is harpooned when it surfaces to breathe near the boat. While the dugong is surfacing, the harpooner uses his weight to drive the spear head into the animal's integument (Marsh et al. 1981). The dugong then usually swims rapidly away with the head of the harpoon inserted in its skin. The head of the harpoon is tethered by a light rope, which draws out from the spear shaft as the animal swims away. The harpoon is held firmly by the hunter so as to not lose the animal. The speared dugong usually becomes exhausted within about 5 min of rapid swimming. As it is drawn close to the boat, its tail is braced

Fig. 18.3 Dugong meat obtained from the legal and sustainable harvest of dugong from Torres Strait. This harvest provides important cultural and provisioning services to the traditional peoples of the region, and their right to hunt is protected by an international treaty between Australia and Papua New Guinea (*Image credit: Department of Environment and Heritage Protection StrandNet*)

against the gunwale so the head is held under the water until it drowns (Marsh et al. 1981). This usually takes about 3 to 5 min (Woolford et al. 2015). Once the carcass is returned to shore, the animal is butchered in a specific way as per each island's custom. This tradition is gradually passed from elders to younger apprentice hunters as they come of age. In these cultures, traditional dugong hunting is considered an honourable occupation. This practice has been reviewed by animal welfare specialists and considered the most suitable currently available method practicable in the light of human safety and concerns about firearms in small boats (Helene Marsh, pers. comm.). The method also avoids the 'struck and lost' issues associated with shooting untethered animals; as is seen in other species hunted recreationally or professionally (see Chap. 12, Human interactions with seals).

Unfortunately, illegal hunting and poaching of sirenians still occur in many areas. Harpoons are commonly used for illegal hunting in other parts of the world. Hunters have often identified and targeted sirenia feeding or mating sites and in some cases use more easily captured calves to lure their mothers close enough for the hunter to physically dispatch the animal. In some instances, once exhausted but not dead, the live animal is then brought to shore where it is killed either by stabbing or imparting blunt force to the head. Reports also describe some hunters in South America plugging the nostrils of manatees with wooden pegs to obstruct breathing. In parts of Africa, manatees are still baited alive into wooden traps where they remain alive until the hunter arrives and kills them. Gillnets and fishing traps are also used for illegal hunting practices in some regions, where the animals drown or are killed when they become entangled in the nets and alert the poacher (Hines and Reynolds 2012). These kinds of nontraditional hunting are likely to result in distress during restraint and may be associated with opportunistic predation whilst the animals are held in traps or nets, resulting in extensive tissue damage before death occurs.

18.10 Changing Attitudes of Fisheries

Along with the actual entanglement, dugong carcases have been found to have been mutilated post-mortem in what is believed to be an attempt by the fishermen to hide the carcases (Slater 1997). This type of activity is relatively common, and examples exist from around the world. The perception was that if fishermen were found to have caught dugong or manatees, they would be fined under various species protection laws and acts and have their enterprise closed down or be told to move to new sites, all of which represented a risk of lost time and money (Fig. 18.4). However,

Fig. 18.4 A dugong carcass that was suspected to have been be gutted and an anchor and weights tied to its peduncle to encourage sinking and avoid being detected by management authorities. It was suspected this animal was originally caught in a fishing net, although this was not confirmed (*Image credit: Department of Environment and Heritage Protection StrandNet*)

through open communication, these misperceptions have been addressed, and managers and fishermen now more freely work together.

Due to extreme weather events in Queensland, Australia, during 2011, commercial fisheries experienced changes in fish abundance and location compared with their normal catch sites. Understandably, to maintain their catch rate, fishermen relocated their nets to where the fish had moved. This change in location caused a change in the numbers of dugong exposed to fishery interactions, and over a short period of time, a dramatic increase in fishery-related captures and deaths occurred. These interactions were noticed through increased reports from fishermen, increased strandings of animals with signs of boat strike and increased rates of entanglement that were likely linked to the industry. As a result, the Queensland Government and Great Barrier Reef Marine Park Authority, in consultation with the fishery, set in place a moratorium and established temporary protection sites to prevent these interactions continuing to occur. Through this collaborative agreement, dugong were protected until the fish returned to their original location and the fisheries were saved the loss of time, product and income associated with sirenian interactions by knowing the locations where they could safely fish (Great Barrier Reef Marine Park Authority 2012).

Throughout the world there are zones in place around urbanised areas that are subject to recreational and commercial fishing and boating activities and are known sirenian habitats. These *Go Slow Zones* and *Green Zones* serve to either slow traffic down year-round, during certain seasons, or to exclude entry to an area or to certain activities, such as fishing, in an area—with the sole purpose of protecting dugong and manatees and other prominent sea creatures such as sea turtles and coral reefs. This form of habitat protection and zone enforcement has been shown to reduce the number of vessel strikes and negative fishery interactions (Calleson 2014).

Finally, although there has been much controversy over the harvesting of dugong, manatees and sea turtles by indigenous people throughout the world, in many regions, indigenous communities value sirenians and chelonids as totem animals. With this respect comes the responsibility to only take what is needed for survival and to operate within their communities' rules. However, with the advent of social media, aberrant behaviours, often by junior community members, have been posted and seen by the larger national community, and this has sometimes resulted in public outcry. One such event occurred in Australia in 2012 resulting in the changing of legislation, specifically Queensland's *Animal Care and Protection Act 2001*. This change in the law meant that the act of hunting by traditional methods exposed the hunter to potential prosecution. The affected indigenous communities responded well to these changes, by both re-educating their junior hunters in the traditional way and by working with the government and researchers to find a mutually acceptable alternative hunting protocol. Further, indigenous communities often create agreements to govern between themselves and other communities, to prevent other communities from entering their lands and waters with the intention of hunting their sirenians and turtles. In this capacity, indigenous communities can serve as protectors of their local dugong, manatee and sea turtle populations.

18.11 Conclusions

Human use and, sometimes, misuse of the marine environment has resulted in significant marine debris and fishery issues. While some of these issues are historic, and may be traced back thousands of years, others are recent and a consequence of our disposable, fast moving society producing abundant cheap plastics and other trash. However, societal awareness of the negative impacts associated with marine debris and unsustainable fishing practices is increasing, and the risks to sirenians are becoming more widely understood. Consequently, the attitudes of those who wish to change these practices to conserve the aquatic environment and its animals for future generations are becoming widely recognised as the only way forward.

References

Adimey NM, Hudak CA, Powell JR, Bassos-Hull K, Foley A, Farmer NA, White L, Minch K (2014) Fishery gear interactions from stranded bottlenose dolphins, Florida manatees and sea turtles in Florida, USA. Mar Pollut Bull 81(1):103–115. doi:10.1016/j.marpolbul.2014.02.008

Attademo F, Balensiefer D, Freire A, de Sousa G, da Cunha F, Luna F (2015a) Debris ingestion by the Antillean manatee (*Trichechus manatus manatus*). Mar Pollut Bull 101:284–287

Attademo FLN, Balensiefer DC, ACdB F, de Sousa GP, da FAGC C, FdO L (2015b) Debris ingestion by the Antillean manatee (*Trichechus manatus manatus*). Mar Pollut Bull 101(1):284–287. doi:10.1016/j.marpolbul.2015.09.040

Australian Government, Department of the Environment, Water, Heritage and the Arts (2009a) Background paper for the threat abatement plan for the impacts of marine debris on vertebrate marine life.

Australian Government, Department of the Environment, Water, Heritage and the Arts (2009b) Threat abatement plan for the impacts of marine debris on vertebrate marine life.

Australian Institute of Marine Science (1996) Data/AIMS ProjectNET - Index/Seagrasses/. Australian Government, Australian Institute of Marine Science. http://www.aims.gov.au/docs/projectnet/seagrasses.html. Accessed 21 Mar 2016

Beck CA, Barros NB (1991) The impact of debris on the Florida manatee. Mar Pollut Bull 22(10):508–510. doi:10.1016/0025-326x(91)90406-i

Bjorndal KA, Bolten AB, Lagueux CJ (1994) Ingestion of marine debris by juvenile sea turtles in coastal Florida habitats. Mar Pollut Bull 28(3):154–158. doi:10.1016/0025-326X(94)90391-3

Bossart G (2001) Manatees. In: Dierauf L, Gulland F (eds) CRC handbook of marine mammal medicine: health, disease and rehabilitation, Second edn. CRC Press, Boca Raton, FL, pp 939–960

Bossart G, Meisner R, Rommel S, Lightsey J, Varela R, Defran R (2004) Pathologic findings in Florida manatees (*Trichechus manatus latirostris*). Aquat Mamm 30(3):434–440

Buergelt CD, Bonde RK, Beck CA, O'Shea TJ (1984) Pathologic findings in manatees in Florida. J Am Vet Med Assoc 185(11):1331

Calleson CS (2014) Issues and opportunities associated with using manatee mortality data to evaluate the effectiveness of manatee protection efforts in Florida. Endanger Species Res 26(2):127–136. doi:10.3354/esr00638

Chin A (2005) Seagrasses. In: Chin A (ed) State of the great barrier reef on-line. Great Barrier Reef Marine Park Authority Townsville, Queensland. http://www.gbrmpa.gov.au/corp_site/info_services/publications/sotr/seagrasses/index.html

Clifton K, Yan J, Mecholsky J, Reep R (2008a) Material properties of manatee rib bone. J Zool 274(2):150–159
Clifton KB, Reep RL, Mecholsky JJ Jr (2008b) Quantitative fractography for estimating whole bone properties of manatee rib bones. J Mater Sci 43(6):2026–2034
Coles RG, McKenzie L, Rasheed M, Mellors J, Taylor H, Dew K, McKenna S, Sankey T, Carter A, Grech A (2007) Status and trends of seagrass habitats in the great barrier reef world heritage area. Department of Primary Industries and Fisheries, Australia
Derraik JGB (2002) The pollution of the marine environment by plastic debris: a review. Mar Pollut Bull 44(9):842–852. doi:10.1016/S0025-326X(02)00220-5
Deutsch CJ, Rycyk A, Barlas ME, Nowacek DP, Koslovsky SM, Frisch K (2009) Final progress report to Florida Fish and Wildlife Conservation Commission. Project contract no. 021426 to Florida State University.
Eros C, Marsh H, Bonde R, Beck C, O'Shea T, Recchia C, Dobbs K (2000) Procedures for the salvage and necropsy of the dugong (*Dugong dugon*). vol Reseach Publication No. 64.
Forrester DJ, White FH, Woodard JC (1975) Intussusception in a Florida manatee. J Wildl Dis 11:566–568
Gall SC, Thompson RC (2015) The impact of debris on marine life. Mar Pollut Bull 92(1–2):170–179. doi:10.1016/j.marpolbul.2014.12.041
Goldberg ED (1995) The health of the oceans - a 1994 update. Chem Ecol 10(1–2):3–8. doi:10.1080/02757549508035325
Goldberg ED (1997) Plasticizing the seafloor: an overview. Environ Technol 18(2):195–201. doi:10.1080/09593331808616527
Great Barrier Reef Marine Park Authority (2012) Australian Government. http://www.gbrmpa.gov.au/managing-the-reef/threats-to-the-reef/extreme-weather/ecosystem-impacts/dugong-and-turtle-strandings. Accessed 31 Mar 2016
Greenland J, Limpus C (2007) Marine wildlife stranding and mortality database annual report 2007 I. Dugong. Queensland Government. Environmental Protection Agency, Australia
Hansen J (1990) Draft position statement on plastic debris in marine environments. Editorial Page 15(3):16
Harr KE, Rember R, Ginn PE, Lightsey J, Keller M, Reid J, Bonde RK (2011) Serum amyloid a (SAA) as a biomarker of chronic infection due to boat strike trauma in a free-ranging Florida manatee (*Trichechus manatus latirostris*) with incidental polycystic kidneys. J Wildl Dis 47(4):1026–1031
Heinsohn G (1972) A study of dugongs (*Dugong dugon*) in northern Queensland, Australia. Biol Conserv 4:205–213
Heinsohn GE, Marsh H, Spain AV (1976) Extreme risk of mortality to dugongs (mammalia: sirenia) from netting operations. Wildl Res 3(2):117. doi:10.1071/WR9760117
Heinsohn R, Lacy RC, Lindenmayer DB, Marsh H, Kwan D, Lawler IR (2004) Unsustainable harvest of dugongs in Torres Strait and Cape York (Australia) waters: two case studies using population viability analysis. Anim Conserv 7:417–425. doi:10.1017/s1367943004001593
Hines E (2012) Dugongs in Asia. In: Hines E, Reynolds J, Mignucci-Giannoni A, Aragones L, Marmontel M (eds) Sirenian conservation: issues and strategies in developing countries. University Press of Florida, Gainsville
Hines E, Reynolds J III (2012) Sirenian conservation: issues and strategies in developing countries. University Press of Florida, Gainsville
Hodgson AJ, Marsh H (2007) Response of dugongs to boat traffic: the risk of disturbance and displacement. J Exp Mar Biol Ecol 340(1):50–61. doi:10.1016/j.jembe.2006.08.006
Joly DO, Heisey DM, Samuel MD, Ribic CA, Thomas NJ, Wright SD, Wrigh IE (2009) Estimating cause-specific mortality rates using recovered carcasses. J Wildl Dis 45(1):122–127
Laist DW (1987) Overview of the biological effects of lost and discarded plastic debris in the marine environment. Mar Pollut Bull 18(6, Supplement B):319–326. doi:10.1016/S0025-326X(87)80019-X

Lightsey JD, Rommel SA, Costidis AM, Pitchford TD (2006) Methods used during gross necropsy to determine watercraft-related mortality in the Florida manatee (*Trichechus manatus latirostris*). J Zoo Wildl Med 37(3):262–275. doi:10.1638/04-095.1

Marsh H, Channells P, Heinsohn G, Morrissey J (1982) Analysis of stomach contents of dugongs from Queensland. Wildl Res 9(1):55–67. doi:10.1071/WR9820055

Marsh H, Corkeron P, Lawler I, Preen A, Lanyon J (1996) The status of the dugong in the southern great barrier reef Marine Park, vol 41. Great Barrier Reef Marine Park Authority, Townsville

Marsh H, De'Ath G, Gribble N, Lane B (2005) Historical marine population estimates: triggers or targets for conservation? The dugong case study. Ecol Appl 15(2):481–492. doi:10.1890/04-0673

Marsh H, Gardner BR, Heinsohn GE (1981) Present-day hunting and distribution of dugongs in the Wellesley islands (Queensland) - implications for conservation. Biol Conserv 19(4):255–267. doi:10.1016/0006-3207(81)90002-1

Marsh H, O'Shea TJ, Reynolds JJI (2011) Threats. In: Ecology and conservation of the Sirenia. vol Conservation biology. Cambridge University Press, Cambridge, New York. doi:10.1017/CBO9781139013277

Mato Y, Isobe T, Takada H, Kanehiro H, Ohtake C, Kaminuma T (2001) Plastic resin pellets as a transport medium for toxic chemicals in the marine environment. Environ Sci Technol 35(2):318–324. doi:10.1021/es0010498

Miksis-Olds JL, Donaghay PL, Miller JH, Tyack PL, Nystuen JA (2007) Noise level correlates with manatee use of foraging habitats. J Acoust Soc Am 121(5):3011–3020. doi:10.1121/1.2713555

Miksis-Olds JL, Wagner T (2011) Behavioral response of manatees to variations in environmental sound levels. Marine Mammal Science 27(1):130–148.

Moore MJ, van der Hoop JM (2012) The painful side of trap and fixed net fisheries: chronic entanglement of large whales. J Mar Biotechnol 2012:4. doi:10.1155/2012/230653

O'Shea TJ, Beck CA, Bonde RK, Kochman HI, Odell DK (1985) An analysis of manatee mortality patterns in Florida, 1976-81. J Wildl Manag 49(1):1–11. doi:10.2307/3801830

Owen H, Gillespie A, Wilkie I (2012) Postmortem findings from dugong (*Dugong dugon*) submissions to the University of Queensland: 1997-2010. J Wildl Dis 48(4):962–970. doi:10.7589/2011-03-080

Ryan PG (1987) The origin and fate of artefacts stranded on islands in the African sector of the southern ocean. Environ Conserv 14(04):341–346. doi:10.1017/S0376892900016854

Sheavly S, Register K (2007) Marine debris & plastics: environmental concerns, sources, impacts and solutions. J Polym Environ 15(4):301–305

Slater J (1997) The legal and policy issues involved in protecting a population of dugongs (*Dugong dugon*) from Gill Netting in Shoalwater Bay, of the Great Barrier Reef World Heritage Area. Paper presented at the 16th Annual National Environmental Law Conference, Adelaide South Australia 3–5 April 1997.

Spain AV, Heinsohn GE (1973) Cyclone associated feeding changes in the dugong (Mammalia: sirenia). Mammalia 37:678–680

STAP (2011) Marine debris as a global environmental problem: introducing a solutions based framework focused on plastic. A STAP Information Document, Washington, DC

Stefatos A, Charalampakis M, Papatheodorou G, Ferentinos G (1999) Marine debris on the seafloor of the Mediterranean Sea: examples from two enclosed gulfs in Western Greece. Mar Pollut Bull 38(5):389–393. doi:10.1016/S0025-326X(98)00141-6

Threatened Species Scientific Committee (2003) Injury and fatality to vertebrate marine life caused by ingestion of, or entanglement in, harmful marine debris. Advice to the Minister for Environment and Heritage from the Threatened Species Scientific Committee on a public nomination of a Key Threatening Process under the Environment Protection and Biodiversity Conservation Act 1999. Australian Government, Department of the Environment. http://www.environment.gov.au/biodiversity/threatened/ktp/marine-debris.html. Accessed 17 March 2016

Walsh M, Bossart G (1999) Manatee medicine. In: Fowler M, Miller R (eds) Zoo and wild animal: current therapy. W.B. Saunders, Philadelphia, Pennsylvania, p 747

Woolford L, Franklin C, Whap T, Loban F, Lanyon JM (2015) Pathological findings in wild harvested dugongs *Dugong dugon* of Central Torres Strait, Australia. Dis Aquat Org 113(2):89–102. doi:10.3354/dao02825

World Wildlife Fund (2016) The Net Kit. http://awsassets.wwf.org.au/downloads/mo001_the_net_kit_1dec02.pdf. Accessed 17 March 2016

Yan J, Clifton KB, Mecholsky JJ, Reep RL (2006a) Fracture toughness of manatee rib and bovine femur using a chevron-notched beam test. J Biomech 39(6):1066–1074. doi:10.1016/j.jbiomech.2005.02.016

Yan J, Clifton KB, Reep RL, Mecholsky JJ (2006b) Application of fracture mechanics to failure in manatee rib bone. J Biomech Eng 128(3):281–289. doi:10.1115/1.2187044

Chapter 19
Impact of Climate Change and Loss of Habitat on Sirenians

Helene Marsh, Eduardo Moraes Arraut, Lucy Keith Diagne, Holly Edwards, and Miriam Marmontel

Abstract Although the impacts of climate change on the welfare of individual manatees and dugongs are still uncertain, the effects are likely to be through indirect interactions between meteorological and biotic factors and the human responses to climate change. We divided the potential impacts into (1) those that will potentially affect sirenians directly including temperature increases, sea-level rise, increased intensity of extreme weather events and changes in rainfall patterns and (2) indirect impacts that are likely to cause harm through habitat loss and change and the increase in the likelihood of harmful algal blooms and disease outbreaks. The habitat modification accompanying sea-level rise is likely to decrease the welfare of sirenians including increased mortality. Many species of tropical seagrasses live close to their thermal limits and will have to up-regulate their stress-response systems to tolerate the sublethal temperature increases caused by climate change. The capacity of seagrass species to evoke such responses is uncertain, as are the effects of elevated carbon dioxide on such acclimation responses. The increase in the intensity of extreme weather events associated with climate change is likely to decrease the welfare of sirenians through increased mortality from strandings, as well as habitat loss and change. These effects are likely to increase

H. Marsh (✉)
College of Science and Engineering, James Cook University, Townsville, QLD 4811, Australia
e-mail: helene.marsh@jcu.edu.au

E.M. Arraut
Remote Sensing Division, National Institute for Space Research, São José dos Campos, SP, Brazil

Wildlife Conservation Research Unit, The Recanati-Kaplan Centre, Oxford University, Tubney House, Abingdon Road, Tubney OX13 5QL, UK

L.K. Diagne
African Aquatic Conservation Fund, Chilmark, 2535 MA, USA

H. Edwards
Marine Mammal Department, Florida Fish and Wildlife Conservation Commission, St Petersburg, FL, USA

M. Marmontel
Instituto de Desenvolvimento Sustentável Mamirauá, Tefé, AM, Brazil

A. Butterworth (ed.), *Marine Mammal Welfare*, Animal Welfare 17,
DOI 10.1007/978-3-319-46994-2_19

the exposure of sirenians to disease and their vulnerability to predators, including human hunters. Climate-related hazards will also exacerbate other stressors, especially for people living in poverty. Thus the risks to sirenians from climate change are likely to be greatest for small populations of dugongs and manatees occurring in low-income countries. The African manatee will be particularly vulnerable because of the high levels of human poverty throughout most of its range resulting in competition for resources, including protein from manatee meat.

19.1 Introduction

The impacts of climate change on the welfare of individual manatees and dugongs have not been well explored and are still uncertain (Marsh et al. 2011). The effects are likely to be through indirect interactions between meteorological and biotic factors and the human responses to climate change. Consequently, the impacts of climate change and resultant shift in habitats are likely to be significant for sirenian species.

A simplistic view of the effects of human-induced climate change on sea levels and water temperatures is that they may be beneficial to sirenians. Increased water levels and temperatures could likely allow extension of the ranges of all extant species (e.g. see Lawler et al. 2007). With the exception of the extinct Steller's sea cow, sirenians have typically been restricted to tropical and subtropical waters throughout their evolutionary history (Marsh et al. 2011). Climate changes during the Pleistocene resulted in a northward expansion of manatees in North America (Marsh et al. 2011). However, our analysis of the available literature and data indicates that the impacts will be complex (Fig. 19.1), and on balance, we think climate

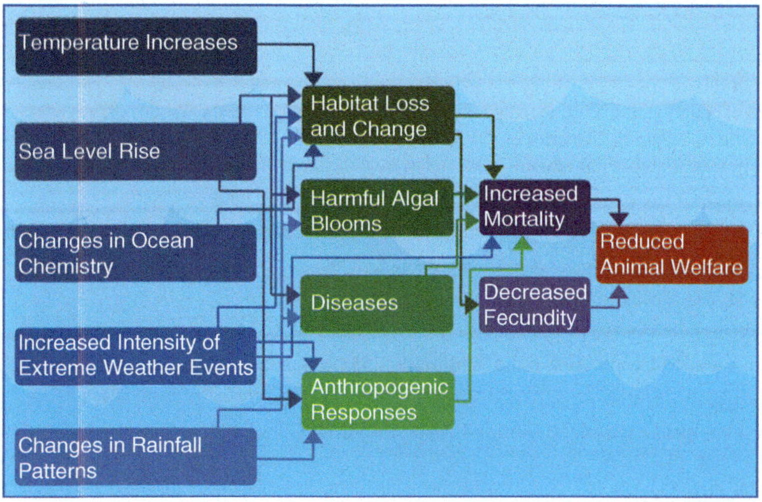

Fig. 19.1 Diagrammatic representation of the direct and indirect effects of climate change on the welfare of sirenians. The impact of these changes is likely to be spatially heterogeneous. In some areas, climate change could have a beneficial effect on sirenians (*Image credit: Elizabeth Collier and Helene Marsh*)

change is likely to have adverse effects on the welfare of manatees and dugongs by increasing mortality and reducing fecundity, hence increasing the likelihood of local extinctions.

In this brief review, we divide the potential impacts into (1) those that will potentially affect sirenians directly, including temperature increases, sea-level rise, increased intensity of extreme weather events and changes in rainfall patterns, and (2) indirect impacts that are likely to cause harm through habitat loss and change and the increase in the likelihood of harmful algal blooms and disease outbreaks. The anthropogenic responses to such changes will likely exacerbate these impacts on manatees and dugongs and have adverse effects on the welfare of individual animals (Fig. 19.1).

19.2 Direct Effects of Climate Change on Sirenians: Temperature Increases

According to the Intergovernmental Panel on Climate Change (IPCC, Church et al. 2013), each of the last three decades has been warmer than any preceding decade since 1850. The current estimates of globally averaged land and ocean surface temperature show a warming of 0.85 °C. Over the next 35 years, global mean surface temperatures will likely increase by 0.3 to 0.7 °C and are likely to exceed 1.5 °C by the end of the twenty-first century (Church et al. 2013). Warming of the oceans dominates the increase in energy stored in the climate system and accounts for more than 90% of the energy accumulated between 1971 and 2010 (Church et al. 2013). Worldwide, surface ocean waters have warmed by about 0.1 °C per decade over the last 45 years. Since the 1950s, areas of high surface salinity, where evaporation dominates, have become more saline and areas of low salinity, where precipitation dominates, have become fresher (Church et al. 2013). These regional trends in ocean salinity provide indirect evidence for changes in evaporation and precipitation over the oceans (IPCC 2014). These changes will in turn lead to alterations in ocean currents, mixing and ventilation (Doney et al. 2012). On a broad scale, rising sea surface temperatures are contributing to numerous changes, including sea-level rise, increased ocean stratification, altered patterns of ocean circulation, precipitation and freshwater input (Doney et al. 2012), resulting in loss of the habitat needed for the survival of manatees and dugongs.

The capacity of climate change models to predict changes at local scales is improving but is still limited and geographically uneven, a situation exacerbated by the variable longevity and quality of temperature records. For example, although long-term water temperature records are difficult to find for any region in Africa, it is clear that rising air temperatures and other factors associated with climate change such as deforestation leading to erosion will increase water temperatures as well. Temperatures in Africa are projected to rise faster than the global average increase during the twenty-first century (IPCC 2014). In the Sahel, air temperatures reach 42 °C at the hottest time of the year, and water temperatures of 33 °C have been recorded in African manatee areas (L. Keith Diagne unpublished data).

The temperatures of the freshwater ecosystems of the Amazon are also expected to increase with climate change (Killeen and Solórzano 2008). Higher temperatures stimulate algal growth, bacterial metabolism, nutrient cycling rates and macrophyte growth, resulting in an increase in trophic status. The release of nutrients from the death of this increased biomass, in turn, is expected to cause a feedback loop in which lake trophic status in general and macrophyte growth in particular are further increased (Ficke et al. 2007). Although it is unlikely that the current Amazonian manatee population is limited by food availability, the above scenario suggests that forage will be more abundant, widespread and, possibly, of better quality. The survival of Amazonian manatees might thus have the potential to improve, but the situation is likely to be very complex and geographically heterogeneous, as explained below.

Manatees and dugongs have limited morphological and physiological capacity to regulate their body temperatures without considerable metabolic cost (Elsner 1999). For example, Florida manatees (Irvine 1983), and likely other sirenians (but see Lanyon et al. 2006 for dugongs), have low metabolic rates. Dugongs have a limited blubber layer but well-developed dermis (Horgan et al. 2014). As a result of morphological and physiological limitations, the thermo-neutral zone (the temperature range within which the animal does not require to make regulatory changes in metabolic heat production or loss) of Amazonian manatees is estimated to be from only 23 to 32.5 °C (Gallivan et al. 1983), whilst the lower limit of the thermo-neutral zone of the Florida manatee is only 20 °C (Irvine 1983). The thermal-neutral zone of the other extant sirenians is not known, but their geographic distributions suggest that their thermal tolerances are also likely to be narrow and less than those of pinnipeds and cetaceans (Costa and Williams 1999).

Most of the work on the effects of water temperature on the welfare of sirenians has been on the effects of temperatures cooler than their thermo-neutral zone (see Marsh et al. 2011 for summary). Environmental temperatures that exceed the upper boundary of the thermo-neutral zone also result in an increase in metabolic rate as a consequence of the additional work required to rid the body of excess heat (Costa and Williams 1999). The direct effects of and risks from warmer water on the welfare of sirenians warrants further investigation. The indirect effects are considered below.

19.3 Sea-Level Rise

Sea levels are expected to rise by 8–16 mm/yr by the year 2100 as a result of ocean thermal expansion and glacier melting (IPCC 2014). It is virtually certain that sea-level rise will continue for many centuries past 2100. The capacity for sirenian habitats to expand with rising water levels, and at local scales, will depend on the coastal and riparian (the interface between land and river) infrastructure.

Many coastal low-lying regions that support manatees and dugongs are at risk. For example, Florida is extremely vulnerable to sea-level rise and could lose as much as 10% of its land area under future climate conditions. With a 1–2 m rise predicted in the next 88 years (Rahmstorf 2010; Parris et al. 2012), many of the thermal refuges (see Fig. 19.6) used by Florida manatees will be inundated, including six of the seven most widely used industrial sites, several smaller power plants and the largest passive thermal refuge. Many of these industrial sites are likely to be moved or have their operations limited in response to sea-level rise. Loss or reduction of the warm-water discharge produced at these sites as a byproduct of their cooling systems will inevitably reduce their carrying capacity for Florida manatees during the colder months when thousands of manatees seek refuge in these industrially warmed waters. This increased mortality due to loss of warm water could potentially limit the size and growth of the Florida manatee population (Edwards 2013).

Saltwater intrusion from sea-level rise, or from storm surges, can have significant negative impacts on sirenian species. Intrusion along with lower spring flows could reduce or eliminate the viability of natural springs as warm-water refuges for manatees. Nonetheless, some spring flows may increase due to sea-level rise and saltwater intrusion. In addition, lower stream spring flows could reduce the dilution of streams and rivers, concentrating salts and other pollutants (Bloetscher et al. 2009) or dry up spring runs altogether, isolating manatees from these much needed resources (Edwards 2013). In West Africa, rising sea levels would increase salinity intrusion in coastal lagoons and lakes, including very important manatee habitats in the extensive coastal lagoons of Gabon, Ivory Coast and Nigeria (Lucy Keith Diagne pers. comm.)

Arora and Boer (2001) predict a decrease in runoff and annual discharge of the Amazon River of approximately 34% by the end of this century. In association with sea-level rise, this decrease may cause saltwater intrusion and changes in water chemistry in the Amazon's lowland and delta regions (Ficke et al. 2007). Sea-level rise also increases the severity and area of impact of surges associated with severe coastal storms (Hamilton 2010). These changes would affect manatee distribution in the mouth of the Amazon, an area of sympatry (coexistence in a shared area) of Amazonian and West Indian manatees, and could result in habitat alteration, loss of food resources, episodes of separation of mother and calf pairs and eventual death of animals.

Thus we predict that the net effect of sea-level rise will decrease the welfare of sirenians through habitat modification leading to increased mortality.

19.4 Changes in Ocean Chemistry

The accumulation of carbon dioxide in the atmosphere is increasing carbon dioxide concentrations in seawater. Roughly half of all anthropogenic carbon dioxide emitted since the industrial revolution has dissolved into ocean water leading to a decrease in pH (acidification) and a change in the water chemistry (Errera et al. 2014). Koch et al. (2013) reviewed relevant literature and concluded that the photosynthetic and growth rates of seagrasses are likely to increase under elevated carbon dioxide. However, as discussed below, many species of tropical seagrasses live close to their thermal limits and will have to up-regulate their stress-response systems to tolerate the sublethal temperature increases caused through climate change. The effects of elevated carbon dioxide on the capacity of various species of seagrass to acclimate to increased sea temperatures are unknown. For these reasons, the impacts on the welfare of dugongs and manatees from the changes in ocean chemistry, and on seagrasses, resulting from climate change are, as yet, impossible to predict.

19.5 Increase in the Intensity of Extreme Weather Events

Global climate change is also likely to result in increased intensity of severe storms (Webster et al. 2005; Florida Oceans and Coastal Council 2010; Bender et al. 2010, Geophysical Fluid Dynamics Laboratory/NOAA 2015). Such storms are likely to be associated with higher rainfall rates than the present day, and these events may cause runoff into coastal regions smothering seagrasses, flushing toxins into waterways and altering the local habitat through increased water flow. These factors are all associated with creating stressors for sirenian species.

Records of Atlantic hurricane activity show that there is a correlation between local tropical Atlantic sea surface temperatures and the Power Dissipation Index (a measure of Atlantic hurricane activity which combines frequency, intensity and duration of hurricanes). Both sea surface temperatures and the Power Dissipation Index have risen sharply since the 1970s, and there is evidence that in recent years, the levels of the Power Dissipation Index are higher than in the previous active Atlantic hurricane era of the 1950s and 1960s (Geophysical Fluid Dynamics Laboratory/NOAA 2015). This evidence suggests that there may be a large anthropogenic influence on Atlantic hurricanes, which could have profound implications, since big increases in tropical Atlantic sea surface temperatures projected for the late twenty-first century would translate to substantial increases in hurricane destructiveness (~300% increase in the Power Dissipation Index by 2100; Geophysical Fluid Dynamics Laboratory/NOAA 2015).

As a result of climate change, the coastal United States is expected to experience more intense storms and possible changes in the El Niño cycle (Scavia et al. 2002; Emanuel 2005). Since 1996, the frequency of hurricane landfalls in the Southeastern United States has increased, and this pattern of elevated hurricane activity could continue (Goldenberg et al. 2001; Webster et al. 2005; Mallin and Corbett 2006). Under climate change, higher sea levels coupled with more intense storms could impact Florida manatee mortality both indirectly through impacts to habitats (see below) or directly through storm effects. Florida manatees have lower survival during years with intense storms or hurricanes (Langtimm and Beck 2006). The mechanisms that caused mortality are unknown but likely vary with timing, intensity and duration of the storms.

Global models indicate that similar patterns will be seen in other parts of the ranges of sirenians. It is believed that the increase of category 4–5 storms (Bender et al. 2010) will outweigh the effects of the overall decreasing number of storms, and it is estimated that there will be a 30% increase in potential damage. However, these climate change increases in storm intensity are not projected to occur for several decades. Coupling this change with sea-level rise, the impacts to coastal areas used by sirenians are expected to increase over time. Direct mortality of dugongs due to tropical cyclones has been documented (Heinsohn and Spain 1974) including strandings in receding high waters after a storm surge (Marsh 1989). Such direct impacts from extreme weather events are expected to increase along with the indirect impacts on the habitats of manatees and dugongs explained below.

Meager and Limpus (2014) point out that there are several lines of evidence to suggest that the health of dugongs (like Florida manatees) is limited by cooler temperatures at the southern limits of their range. Not only do dugongs move in response to water temperature (Allen et al. 2004; Sheppard et al. 2006), but cold stress syndrome has been described in dugongs in Moreton Bay (Owen et al. 2013), at the southern limit of their temperature range in Australia. Nonetheless, lagged minimum air temperature also explained mortality of dugongs in the tropical Townsville region suggesting that impacts on health may result from the associated impacts of freshwater discharge on seagrass rather than temperature per se (Meager and Limpus 2014), although the change in water temperature may be a contributing factor.

Some models also predict an increase or persistence in wintertime cold-air outbreaks with climate change (Easterling et al. 2000; Walsh et al. 2001). Manatees in Florida occur at the northern limit of the species' range in the temperate and subtropical regions of the Southeastern United States. Manatees are poorly adapted to the cold and have a relatively narrow thermo-neutral zone, as explained above. Chronic exposure to water less than ~18°–20°C can lead to cold stress, a condition that can, for example, result in emaciation, dehydration, skin lesions, gastrointestinal disorders and death (Bossart et al. 2003). Under current climatic conditions, water temperatures in much of Florida periodically drop below thresholds that manatees can tolerate for extended periods (Laist and Reynolds 2005). Climate scenarios for Florida predict that these events will continue in the future. A significant number of cold-related mortalities of Florida manatees have been documented (1130 deaths from 1986 to 2015; Florida Fish and Wildlife Conservation Commission unpublished data). The cold-related die-off in the winter of 2010 was the most notable, due to its geographic extent and the number of related manatee mortalities (Edwards 2013). Two-hundred and eighty (280) manatees died of cold stress during a 3-month period; for 197 others, the cause of death could not be determined, but the timing of the deaths and the location of the carcasses suggest that most were also a result of the cold. An additional 49 manatees were rescued for treatment of cold stress during that period (Barlas et al. 2011).

The Amazon basin experiences an extreme flood or drought at approximately 10-year intervals, and as recently as 2005 and 2010, two droughts heavily affected riverine life (Marengo et al. 2011). During extreme drought periods, river water levels decrease, lakes become shallow or even dry up, and manatees become more vulnerable

Fig. 19.2 Amazon basin in a severe drought showing the fragmentation of the water courses inhabited by manatees. Such fragmentation can result in an increase in hunting (*Image credit: Eduardo Arraut*)

to hunting (Miriam Marmontel and Eduardo Moraes Arraut, pers. obs.) as discussed below (Fig. 19.2). Under prolonged drought conditions, aquatic plants do not survive, and manatees resort to ingesting organic matter from the bottom, and consuming sediment in large quantities, a practice which may cause death by obstruction of the intestine (Miriam Marmontel pers. obs.). One of the scenarios for the hydrological cycle in the Amazon basin associated with climate change (Sorribas et al. 2016) predicts that extreme droughts will become more frequent and recurrent in the region. If this scenario prevails, not only will there be more frequent years in which manatee forage is extremely scarce, but manatees will also suffer from more frequent and prolonged periods of vulnerability to predation, especially hunting as discussed below.

Thus the increase in the intensity of extreme weather events associated with climate change is likely to decrease the welfare of sirenians through increased mortality.

19.6 Changes in Rainfall Patterns

Patterns of precipitation are expected to change in intensity and variability as a result of climate change. Some areas will get wetter, others drier, as explained by Christensen et al. (2013). The global monsoon is likely to strengthen in area and intensity, and the likelihood of precipitation extremes is likely to increase. The El

Niño-Southern Oscillation (ENSO) will remain the dominant mode of interannual variability, and the associated precipitation variability that comes with El Niño will likely intensify. Severe tropical storms are expected to increase in intensity as mentioned above, with an increase in precipitation rates.

About 20% of Florida, manatees use natural sites as sources of warm water in winter (e.g. natural springs and warm-water seeps). A major concern is that climatic variation may alter the hydrological cycle that replenishes the Florida aquifers that feed these natural springs (Bloetscher et al. 2009; Edwards 2013) causing them to cease operating.

The African manatee occupies the widest range of habitats of any sirenian. The waterways used by manatees range from lagoons within equatorial rainforests in Central Africa to rivers at the edge of the Sahara Desert, to coastal mangrove channels and islands in the Atlantic Ocean. Throughout their range, African manatees migrate seasonally, based upon changes in water levels, moving up flooded rivers during annual rainy seasons to exploit food resources that are unavailable during dry seasons (Powell 1996; Keith Diagne 2015). Climate change in West and Central Africa is predicted to make dry areas drier and wet areas wetter, with longer and more frequent dry periods (Christensen et al. 2013). Therefore, manatee habitat in Central Africa may increase, but habitat loss due to drying and desertification will likely occur in the northern part of the species' range. Manatee populations in Mali, Niger, Senegal, Mauritania, Chad and Northern Nigeria are likely to be the most affected by desertification (Figs. 19.3).

Fig. 19.3 The Senegal River is the boundary between Senegal, Mauritania and Mali, and its flow is now controlled by the Manantali Dam. This view of the river in the dry season shows the greatly reduced habitat for manatees that could become more severe with increasing desertification (*Image credit: Lucy Keith Diagne*)

There has been a well-documented decrease in rainfall in the Sahel since the late 1960s (IPCC 2014), coupled with an increasing human demand that draws more and more water from rivers. Over timescales measured in decades, rainfall in the Sahel is also affected by sea surface temperatures, with rising ocean temperatures leading to decreased rainfall (IPCC 2014). Climate models predict an increase 1.0–1.9 °C in mean air temperature by 2050, and evaporation rates across Africa are projected to increase 5–10% during the same timeframe (Gaye et al. 2013). Rainfall in West and Central Africa is also known to be influenced by the El Niño-Southern Oscillation (ENSO), the Atlantic Niño and the Mediterranean Sea (Wade et al. 2015). The impact of climate change on African manatees is expected to be exacerbated by the anthropogenic response to increasing human population and human activity changes resulting from climate change (as discussed below).

The main influence of changes in the Amazon's rainfall pattern on Amazonian manatees will come via effects on the dynamics of the Amazon River's flood pulse. Although predictions of such impacts are still controversial, some general trends are emerging. Overall, greater precipitation levels are foreseen for the northern part of the basin, whilst the southern part is expected to experience lower rainfall levels (IPCC 2014). According to Langerwisch et al. (2013), this exchange will result in an increase of about one third in total inundation area and in a decrease in the number of extremely dry years (and in the probability of occurrence of three consecutive extremely dry years). Moreover, though the total number of extremely wet years is not expected to change, the probability of three consecutive wet years decreases by 30% in the east and increases by 25% in the west. Although an alternative hydrological scenario on the impacts of climate change on the dynamics of flooding in the Amazon basin supports the abovementioned increase in duration and extent of the high-water season, it also predicts an increase in the frequency of extreme low-water seasons for Central and Eastern Amazon.

The changes in rainfall patterns associated with climate change are likely to decrease the welfare of sirenians through changes in habitat. The impacts are likely to be geographically uneven.

19.7 Indirect Effect of Climate Change on Sirenians: Habitat Loss and Change

Sirenians are mostly herbivores (Marsh et al. 2011), foraging on a variety of marine, estuarine and freshwater plant species, although all species will eat some animal matter. According to Domning (1981, 2001), seagrasses have played a key role in sirenian life history and evolution, and they are a major dietary component of both dugongs and Antillean and African manatees. Seagrasses are therefore intrinsically linked to the well-being and survival of sirenian species and are an essential component of any environment which they inhabit. Seagrass beds are already among the most threatened of the world's ecosystems (Waycott et al. 2009), with 29% of global seagrasses lost or degraded and with the rate of loss accelerating (at a rate of

110 km^2 yr^{-1}; Waycott et al. 2009). Impacts of climate change, including elevated atmospheric carbon dioxide, increased variability in sea temperatures, sea-level changes and alterations in levels of ultraviolet radiation, will alter the conditions for growth of seagrasses (Watson et al. 1996). These effects are likely to be geographically variable.

According to Short and Neckles (1999), the climate change impact that will have the greatest direct effect on seagrasses will be sea-level rise. Increased sea levels will impact the amount of light available to the plants for photosynthesis. A rise in sea level may alter the structure of some seagrass beds by reducing their distribution and productivity and ultimately reducing their functional value (Short and Neckles 1999).

Temperature is a critical factor controlling seagrass growth, survival and reproduction (Poloczanska et al. 2007). Seagrass tolerance to changes in temperature varies by species, and, for species living at the upper limit of their thermal tolerance, mean sea temperature increases of up to 2 °C may have a severe impact (Ralph 1998). Extreme or prolonged thermal stress (Collier and Waycott 2014) and the effect of poor water quality can be synergistic in causing seagrass mortality. Increasing the duration (i.e. more days in a row) of thermal events above 40 °C is likely to negatively impact the ecological function of tropical seagrass meadows (Collier and Waycott 2014). The loss of seagrass species and changes in species dominance as a result of climate change are likely to have significant ramifications for sirenians, as well as for the other species that use seagrass communities (Campbell et al. 2006).

In many parts of the world, severe storms have been a significant cause of disturbance to seagrasses (Steward et al. 2006), as a result of wind, erosion from wave action, storm surge, shifting substrates, torrential rains and changes in salinity, etc., along with increased water turbidity (Short and Neckles 1999; Fourqurean and Rutten 2004). Although not all storms cause damage to seagrass meadows, hurricanes, tropical cyclones and other major weather events have all resulted in the decline of seagrasses. Individual storm characteristics play a role in determining the type and extent of the damage that occurs (Greening et al. 2006).

The dugong is more dependent on seagrass than any other sirenian (Marsh et al. 2011). Loss of available seagrass reduces dugong abundance through temporary migration, increased mortality and negative effects on dugong body condition and female reproductive rates (Marsh and Kwan 2008; Meager and Limpus 2014; Fuentes et al. 2016). Meager and Limpus (2014) used a 17-year data set to investigate the drivers of natural dugong mortality across a subtropical-tropical latitudinal gradient of 13° (>2000 km of coastline). Peak mortality followed sustained periods of low air temperature (lagged by 3 months) and elevated freshwater discharge associated with extreme rainfall events (lagged by 9 months). At a regional scale, these results translated into a strong relationship between annual mortality rates and an index of El Niño-Southern Oscillation, mortality being higher following La Niña events (Meager and Limpus 2014). Meager and Limpus (2014) considered that the two most parsimonious functional explanations for the relationship between freshwater discharge and dugong mortality were (1) reduced food availability and (2) direct impacts on health. There is strong support for the first hypothesis for dugongs. Elevated flood-linked discharge reduces photosynthetic available radiation (PAR), exported sediment can smother seagrasses, and flood discharges can scour seagrass beds and seed banks (Campbell et al. 2006).

A study of Amazonian manatee feeding ecology identified 49 species of aquatic plants as food items, mostly from the Poaceae (a grass) family (Guterres-Pazin et al. 2014). As grasses with a C4 photosynthetic pathway tolerate high temperatures better than C3 plants (Hamilton 2010), global warming may result in lower plant diversity available to Amazonian manatees. Nutrient quality may also be affected, as C4 grasses have lower nitrogen contents (Hamilton 2010).

On the basis of the strong relationship between inundation duration and predominant vegetation type and structure class (Ferreira-Ferreira et al. 2014; Junk et al. 2012), the Amazonian floodplain in general, and particularly its western part, is expected to gradually shift towards larger areas being covered by vegetation classes associated with longer flooding times, such as herbaceous vegetation. Vegetation associated with shorter inundation times, such as high várzea (seasonal floodplain) forests, is expected to decrease under this scenario. Less forest would result in decreased filtering of the huge amounts of sediment that are brought from the Andes during the high-water season, which in turn would result in greater siltation of open water and herbaceous areas (Ficke et al. 2007). This situation would create a positive feedback loop, in which, as the lakes became shallower, the overall inundation extent would increase, further promoting an increase in the size of the areas dominated by open water and herbaceous vegetation. Overall, this scenario would, on the one hand, result in increased food availability (aquatic macrophytes) for manatees, but, on the other hand, would also increase their vulnerability to predators, including human hunters, as discussed below, owing to the decrease in the depth of floodplain lakes.

19.8 Algal Blooms

Species of phytoplankton are important components of the aquatic food chain and over 4000 species of phytoplankton inhabit our oceans (Davidson et al. 2014). Although most species are benign, some species can cause a variety of deleterious effects to the environment and to the organisms that live in it. Some algal species are capable of producing toxins that can kill or sicken birds, fish, humans and marine mammals including some sirenians. Other phytoplankton species can produce high biomass by blooming to excess (Davidson et al. 2014), and such blooms can result in oxygen depletion in bottom waters (hypoxia/anoxia) or can reduce the amount of light that reaches the benthos leading to the death of light-dependent benthic seagrass.

Globally, algal blooms appear to be increasing as a result of increased nutrient loads from human activity (Sellner et al. 2003; Gilbert and Burkholder 2006; Heisler et al. 2008). The relationship is complex, and the link is not universal, but regard-

Fig. 19.4 Algal blooms are capable of smothering underlying sea grasses and persisting for days to weeks under the right climatic conditions (*Image credit: Tom Reinert*)

less, algae blooms can critically alter ecosystems, disrupt food webs, stimulate the growth of pathogens and have other ecological consequences (Wells et al. 2015; Fig. 19.4). Increasingly there is concern that climate change will result in changes in the phytoplankton community, which will increase the prevalence and geographical spread of algae blooms (Wells et al. 2015). Some harmful species are likely to become more successful under future climate conditions, whilst other species may diminish (Hallegraeff 2010).

Several studies have indicated there may be a relationship between the frequency, duration and magnitude of algal blooms and climate (Moore et al. 2008). Temperature increases from climate change are predicted to increase the range and duration of blooms of some harmful algal species (Moore et al. 2008; Hallegraeff 2010). The warmer upper ocean temperatures predicted under climate change scenarios will reduce vertical mixing of the water column, which will impact phytoplankton growth (Doney 2006). As ocean water warms, blooms will begin earlier and last longer (Moore et al. 2008), and algal blooms will expand their range and the period of time they occur (Gessner and Middaugh 1995). Increases in ocean acidity are also likely to influence phytoplankton in ways that favour algal blooms. Studies show that there is generally a positive relationship between increasing pH and growth and toxin production in algae blooms; however, it is unknown if this relationship will be maintained under the more acidic ocean conditions that will result from climate change. Although scientists are still uncertain about the impact of climate change on algae bloom species in general, the potential for interaction between the two factors is high. The predicted impacts of climate change on tropical and subtropical habitats are likely to increase the probability of negative impacts on sirenian species from algae blooms as discussed below.

Climate changes can influence red tides over huge spatial scales. Large amounts of African dust are carried by wind to the western Atlantic, including Florida's coast and the Caribbean. The amount of dust being transported is highly negatively correlated with the amounts of rainfall in Africa (Prospero and Lamb 2003). Saharan minerals from this dust are thought to be used by diazotrophic, or nitrogen fixing, bacteria to fuel the nitrogen economy of red tides (Walsh and Steidinger 2001), thus triggering larger algal blooms. Local drier conditions also can create more favourable conditions for red tides nearshore. As the toxic dinoflagellate, *Karenia brevis*, favours areas of higher salinity, changes to Florida's climate, including more episodic storms and changes in the amount of precipitation and runoff, are likely to increase the probability of manatees' exposure to red tide which could increase future mortality of the species (Edwards 2013).

The impact of algal blooms on the welfare of sirenians is well documented for Florida manatees where poison from potent neurotoxins (brevetoxins) produced by *K. brevis* have resulted in the death or suspected death of over 669 manatees since 1984. Such brevetoxicosis can release inflammatory mediators that can result in fatal toxic shock (Bossart et al. 1998). Large die-offs of manatees from red tide occurred in 1996, 2002, 2003, 2005, 2007 and 2013 (highest on record at 277 deaths; Edwards 2013). Sublethal exposure to harmful algal blooms may have long-lasting impacts on a manatee's immune system, which could impact the overall health of a manatee population (Walsh et al. 2015).

According to Landsberg and Steidinger's review (1998), Florida manatees are at high risk of being impacted by red tide blooms in the late winter and spring when salinities are greatest, and when large numbers of manatees move through, or forage in, red tide-affected areas in Florida (Edwards 2013). Pathways of exposure to red tide can occur from consumption of toxin-sequestering seagrass and attached epiphytes, inhalation of toxic aerosols or ingestion of toxic seawater (Landsberg and Steidinger 1998). As the toxins can be retained in some organisms from weeks to months after a red tide has dissipated, exposure can be delayed, and manatees can be affected long after the bloom has subsided. In 2012–2013, an overgrowth of a protistan alga (*Resultor* sp.) caused unusually high manatee mortality in the Indian River Lagoon system in Florida (Edwards 2013). More than 100 manatees died, and more than 50% of seagrass in the lagoon was lost as a result of the bloom (Lori Morris pers. comm).

Most harmful algal species are dinoflagellates that are capable of moving through the water column. Since mobile forms are expected to prevail over other species, it is likely that blooms of motile alga like *K. brevis* will increase as a result of climate change. In addition, under predicted future climate conditions, blooms of *K. brevis* may have the potential to produce higher cell and toxin concentrations (Errera et al. 2014).

Although high mortality due to harmful algal blooms has to date been restricted to the Florida manatee, it is possible that global climate change could alter ecosystems in a way that could increase the negative impact on other sirenian populations. For example, toxic dinoflagellates, which make tumour-producing okadaic acid, and blooms of toxic *Lyngbya* are all found in dugong habitats (Marsh et al. 2011). Although no deaths have been attributed to their occurrence, changes to these ecosystems, from changes in temperature, salinity, light, carbon, nutrients and others, could increase the probability of negative impacts of harmful algal blooms on dugongs.

19.9 Disease

Host-parasite dynamics may be altered by climate change, through changes in transmission rates and host susceptibility (Ficke et al. 2007). Environmental and climatic changes may alter biogeochemical cycles, which may increase the incidence of infectious diseases, especially waterborne diseases, such as leptospirosis and cryptosporidiosis (Barcellos et al. 2009; Costello et al. 2009; Lau et al. 2010). Manatees under environmental stress become immunocompromised and susceptible to infectious agents (Bonde et al. 2004). Antibodies anti-*Toxoplasma gondii* and anti-*Leptospira* spp. have been identified in Amazonian manatee samples in Brazil and Peru (Mathews et al. 2012; Delgado et al. 2013, 2015). Freshwater discharge associated with the increased number of extreme rainfall events expected under climate change may also impact the health of dugongs and manatees by increasing their exposure to infectious pathogens such as toxoplasmosis and faecal coliform bacteria (derived from human and animal sewage) or to contaminants with immunosuppressive effects (see Marsh et al. 2011 for review).

Thus climate change is likely to decrease the welfare of sirenians by increasing their exposure to disease.

19.10 Anthropogenic Responses

Climate change is already inducing adaptive human responses, and such responses will increase as the biophysical impacts of climate change become more evident. Climate-related hazards exacerbate other stressors, especially for people living in poverty (IPCC 2014). The African manatee will be particularly vulnerable because of the high levels of human poverty throughout most of its range resulting in competition for resources, including protein from manatee meat. Thus the risks to sirenians from climate change are likely to be greatest for small populations of dugongs and manatees occurring in low-income countries (see Marsh et al. 2011 for details).

19.11 Fishing and Hunting Pressure

Climate change will increase food insecurity, which in turn is projected to increase pressure on the world's fisheries. As Marsh et al. (2011) point out, 21 of the 33 countries where the impacts of climate change on fisheries are projected to have the greatest national economic impacts (Allison et al. 2009) are in the ranges of dugongs and manatees. The potential national economic impacts of climate change on fisheries could not be assessed in a further 21 countries in the range of sirenians; 17 of these are classified as Least Developed Countries or Small Island States or both and are thus potentially especially vulnerable to climate change.

Both marine and freshwater fisheries in Africa are already in serious decline (Atta-Mills et al. 2004; Allan et al. 2005). The climate change effects of desertification in North Africa and potential shifts in oceanic species throughout Africa will further exacerbate the problem of overfishing (IPCC 2014). A negative correlation between the catches of subsistence fishers and 'wild meat' has already been established for West Africa (Brashares et al. 2004; Rowcliffe et al. 2005). The African manatee is already heavily hunted throughout its range, and its meat is more costly than fish. As fish stocks continue to decline, it is expected that hunting pressure on African manatees will increase, possibly leading to extinction in some parts of its range.

These projected changes in human population resulting from climate change are significant to the sustainability of manatee and dugong populations, because the population growth rate of all sirenians is highly sensitive to changes in adult survival rate (Marsh et al. 2011), and most local populations of sirenians cannot withstand a human-induced mortality of even a few animals per year. These changes are also likely to be important to the welfare of sirenians because the effectiveness of cultural norms and regulations designed to protect them are hampered by weak governance, as illustrated by the following example.

The Torres Strait region between Australia and Papua New Guinea supports the largest dugong population in the world (Marsh et al. 2011). Anthropologists have documented the cultural values of dugongs since the European colonisation of Torres Strait (see Haddon 1912; Nietschmann and Nietschmann 1981). Dugongs are fundamental in the culture of Torres Strait Islanders (Beckett 1987; Johannes and MacFarlane 1991; Marsh et al. 2011). Dugong meat has been an important component of the seafood harvested by the indigenous peoples of Torres Strait for at least 4000 years (Crouch et al. 2007). The Torres Strait dugong harvest, which currently appears to be sustainable (Marsh et al. 2015), has significant provisioning value, particularly for residents of the Western Province of Papua New Guinea. This region is one of the poorest regions in the world (Butler et al. 2015) and has significant food insecurity (Omot 2012).

Climate change is expected to alter the biological productivity of the Torres Strait marine environment and the adjacent terrestrial environments, particularly that of Papua New Guinea (Butler et al. 2015). In addition to the effects of climate changes on the dugong habitats of Torres Strait, the rate of sea-level rise is also expected to accelerate the impacts of inundation on the island communities (Duce et al. 2010). Some islanders may be forced to relocate by 2100. Much of the coastal area around the Papua New Guinea coastal villages is low lying and flat. Some of this region is also likely to be inundated, necessitating relocation. Nonetheless, the strategic deep-water port, Daru Island, may be deemed too important to evacuate and may be instead protected by sea walls.

Overall, the human population of the coastal areas of Western Province is expected to increase. Stoeckl et al. (in press) predict that this growing population, along with the impacts of extreme weather on subsistence gardens and the degradation of coastal habitat through mining and logging, will increase pressure on Torres Strait dugong stocks. Due to the poor governance systems in place in Papua New Guinea, these changes are likely to result in the increased use of unsustainable gear such as large mesh nets. This change in fishery practice is likely to be more significant than reductions in hunting that might otherwise come about from (i) the relocation of people from islands prone to inundation and (ii) the relocation of animals to deeper waters because of the loss of shallow-water seagrass. Regulation of the Torres Strait dugong harvest is likely to deteriorate wherever Papua New Guinea villagers take a greater proportion of catch, in areas with food shortages, and where regulatory effectiveness is hampered by weak governance or low levels of resourcing for enforcement. These factors are expected to be especially pertinent in regions experiencing limited capacity and the breakdown of cultural norms caused by migration. Thus the animal welfare impacts associated with the dugong harvest in this region are expected to increase.

Some of the climate scenarios described above for the Amazon basin predict lower water levels during the time of seasonal manatee migrations and extreme low-water seasons, trapping manatees between impassable stretches of low water. When trapped, manatees are usually slaughtered for food by local inhabitants. This was the case in the 2016 drought in Mahates wetland, Bolívar, northern Colombia, when an estimated 90 manatees became trapped in a 300 m long, 60 cm deep remnant bed of the river, with their food resources dried out. At least 25 of them were killed by hunters (El Espectador 2016). More extreme low-water seasons would result in shallower refuges, which, again, would likely be associated with increased slaughter of manatees by local inhabitants. This situation occurred, for example, in the 1996, 2005 and 2010 extreme droughts, when mass slaughter occurred in the *rias* Tefé, Manacapurú, Coari, Badajós, Monte Alegre and Jari areas of the Amazon basin (Thornback and Jenkins 1982; Arraut et al. 2010; and personal observations by Eduardo Moraes Arraut, and Miriam Marmontel).

19.12 Infrastructure

Forty-four percent (44%) of people live within 150 km of a coast (UN Atlas 2010), and many coastal marine ecosystems are already significantly degraded (Halpern et al. 2008). The impacts of climate change are likely to cause shoreline erosion, coastal flooding and water pollution and affect man-made infrastructure and coastal ecosystems (IPCC 2014). Addressing the additional stress of climate change is likely to result in new approaches to managing land, water, waste and ecosystems. The examples below illustrate some of the resultant likely problems for sirenians.

As discussed in an earlier chapter (Chap. 17), engineered and technological options are commonly implemented adaptive responses to climate change (IPCC 2014). Dams are a significant threat to the welfare of African manatees, both through entrapment and through crushing in gates (Keith Diagne 2015). Manatees in Sahelian countries are already permanently isolated in the Niger and Senegal Rivers by dams. Major hydroelectric and minor agricultural dams block the ability of manatees to move out of river systems as their habitat and food resources decrease. Three major new dams are already under construction on the Niger River in Guinea, Mali and Niger, and as water scarcity increases with climate change, further dams will trap riverine manatees into smaller and smaller habitat areas making it less likely that they will survive long term.

The threats from hydroelectric damming on Amazonian manatees will be significant if the current massive dam-building plan for the Amazon basin is realised (Arraut and Marmontel 2016). The plan is to build dams or dam cascades across all large- and most medium-sized Amazon rivers to provide energy for aluminium extraction in the region and for industries elsewhere (Castello and Macedo 2015; Brazil 2016; Winemiller et al. 2016). If implemented, this dam-building plan would partition the Amazonian manatee into many small, confined populations, each of which would suffer from inbreeding with resultant reduction in evolutionary potential (Frankham et al. 2014) and increased vulnerability to slaughter, especially during natural or dam-induced extreme droughts. As explained above, in addition to this direct population impact, manatee survival is also expected to decrease even further owing to the intense degradation of the floodplain habitat. In combination, these effects are likely to result in local extinctions of small and confined Amazonian manatee populations and a second species-level population collapse. (The first collapse was driven by commercial over-exploitation for meat and hide over two centuries (Domning 1982).) Local human socio-economic-environmental conditions are also expected to deteriorate further as a result of the dam building (Fearnside 2006, 2014, 2016a, 2016b), resulting in an increase in the hunting pressure on manatees. As manatees are slow-breeding animals, recovery from a second collapse associated with the expected increase in hunting pressure would be highly unlikely. These cumulative impacts have led to calls for the plan for extensive dam building to be reconsidered.

In Florida, manatees rely on both natural and industrial sources of warm water for refuge, when water temperatures drop below approximately 20 °C, as explained above. About 60% of Florida manatees, counted during statewide surveys conducted after cold fronts, were using an industrial source of warm water as a winter refuge (Fig. 19.5). Including ~3% of the total population which were using a single passive thermal refuge (Florida Fish and Wildlife Commission unpublished data). An estimated 93% of the state's human population relies on groundwater for their needs (SFWMD 2001). Florida's human population is putting ever-increasing

Fig. 19.5 Manatees overwintering at an industrial warm-water refuge in Florida (*Image credit: Stacie Koslovsky*)

stress on already compromised systems, and future water withdrawals are expected to increase by 21% by 2030 (Florida Department of Environmental Protection 2014). Loss of these refuge sites or reduction in their carrying capacity has been identified as a significant threat to the long-term viability of the population (Runge et al. 2007).

We conclude that the anthropogenic responses to climate change are likely to decrease the welfare of sirenians, through human responses to food insecurity and through modifications to rivers as a result of increased dam building.

19.13 Conclusions

This review indicates that climate change has considerable potential to negatively impact the welfare of dugongs and manatees. The effects are likely to be particularly serious in the low-income countries (IPCC 2014), which make up most of the ranges of all four sirenian species (Marsh et al. 2011). Given the uncertainties as to how climate change will affect sirenians and their habitats at local scales, the best precautionary actions are likely to be: (1) control factors known to increase mortality and reduce fecundity, as pointed out by Lawler et al. (2007) and Marsh et al. (2011), and (2) adopt an adaptive management approach as outlined in Fuentes et al. (2014). There is increasing global interest in a relatively newly appreciated ecosystem service provided by aquatic plants, including seagrasses, through their capacity to sequester carbon (Macreadie et al. 2014). The imperative to maintain wetlands as global greenhouse gas sinks has some potential to reduce the otherwise, mainly negative, impacts of climate change on the welfare of sirenians.

References

Allan JD, Abell R, Hogan Z et al (2005) Overfishing of inland waters. Bioscience 55:1041–1051

Allen S, Marsh H, Hodgson A (2004) Occurrence and conservation of the dugong Sirenia: Dugongidae in new South Wales. Proc Linnean Soc NSW 125:211–216

Allison EH, Perry AL, Badjec MC et al (2009) Vulnerability of national economies to the impacts of climate change on fisheries. Fish Fish 10:173–196. doi:10.1111/j.1467-2979.2008.00310.x/full

Arora VK, Boer GJ (2001) Effects of simulated climate change on the hydrology of major river basins. J Geophys Res 106(D4):3335–3348

Arraut EM, Marmontel M (2016) Amazonian manatee threatened with extinction by massive dam-building plan in the Amazon. Science. eLetter: http://science.sciencemag.org/content/351/6269/128.e-letters. Accessed 6 11 16

Arraut EM, Marmontel M, Mantovani JE et al (2010) The lesser of two evils: seasonal migrations of Amazonian manatees in the western Amazon. J Zool 280:247–256. doi:10.1111/j.1469-7998.2009.00655.x

Atta-Mills J, Alder J, Sumaila UR (2004) The decline of a regional fishing nation: the case of Ghana and West Africa. Nat Res Forum 28:13–21

Barcellos C, Monteiro AMV, Corvalán C, Gurgel C, Gurgel HC, Carvalho MS, Artaxo P, Hacon S, Ragoni V (2009) Mudanças climáticas e ambientais e as doenças infecciosas: cenários e incertezas para o Brasil. Epidemio Serv Saúde (Brasflia) 18(3): jul-set 2009

Barlas ME, Deutsch CJ, de Wit M et al. (2011) Florida manatee cold-related unusual mortality event, January–April 2010. Final report to USFWS (grant 40181AG037). Available via Florida Fish and Wildlife Conservation Commission http://myfwc.com/media/1536184/2010_Manatee_Cold_related_UME_Final.pdf. Accessed 30 April 2016

Beckett J (1987) Torres Strait islanders: custom and colonialism. Cambridge University Press, Cambridge

Bender MA, Knutson TR, Tuleya RE et al (2010) Modeled impact of anthropogenic warming on the frequency of intense Atlantic hurricanes. Science 327:454–458

Bloetscher F, Meeroff DH, Heimlich BN (2009) Improving the resilience of a municipal water utility against the likely impacts of climate change—a case study: city of pompano beach water utility. Florida Atlantic University, USA

Bonde RK, Aguirre AA, Powell J (2004) Manatees as sentinels of marine ecosystem health: are they the 2000-pound canaries? EcoHealth 1:255–262

Bossart GD, Baden DG, Ewing R et al (1998) Brevetoxicosis in manatees (*Trichechus manatus latirostris)* from the 1996 epizootic: gross, histologic and immunohistochemical features. Toxicol Pathol 26:276–282

Bossart GD, Meisner RA, Rommel SA et al (2003) Pathological features of the Florida manatee cold stress syndrome. Aquat Mamm 29:9–17

Brashares JS, Arcese P, Sam MK et al (2004) Bushmeat hunting, wildlife declines, and fish supply in West Africa. Science 306:1180–1183

Brazil (2016) Programa para Aceleração do Crescimento - PAC-II, Eixo Energia (Program for the Acceleration of Growth). In: Minist. Plan. http://www.pac.gov.br/infraestrutura-energetica. Accessed 30 April 2016

Butler JRA, Bohensky E, Skewes T, Maru Y et al (2015) Drivers of change in the Torres Strait region: status and trends. Report to National Environmental Research Program. Reef & Rainforest Research Centre Limited, Cairns, p 60

Campbell SJ, McKenzie LJ, Kerville SP (2006) Photosynthetic response to seven tropical seagrasses to elevated seawater temperature. J Exp Mar Biol Ecol 330:455–468

Castello L, Macedo MN (2015) Large-scale degradation of Amazonian freshwater ecosystems. Glob Chang Biol 3:990–1007. doi:10.1111/gcb.13173

Church JA, Clark PU, Cazenave A, Gregory JM et al (2013) Sea level change. In: Stocker TF et al (eds) Climate change 2013: the physical science basis. Contribution of working group I to the fifth assessment report of the intergovernmental panel on climate change. Cambridge University Press, Cambridge, pp 1137–1216

Christensen JH, Krishna Kumar K, Aldrian E et al (2013) Climate phenomena and their relevance for future regional climate change. In: Stocker TF et al (eds) Climate change 2013: the physical science basis. Contribution of working group I to the fifth assessment report of the intergovernmental panel on climate change. Cambridge University Press, Cambridge, pp 1217–1308

Collier CJ, Waycott M (2014) Temperature extremes reduce seagrass growth and induce mortality. Mar Pollut Bull 83:483–490

Costa DP, Williams TM (1999) Marine mammal energetics. In: Reynolds JE, Rommel SA (eds) Biology of marine mammals. Smithsonian Institution Press, USA, pp 218–286

Costello A, Abbas M, Allen A, Ball S, Bell S, Bellamy R, Friel S, Grace N, Johnson A, Kett M, Lee M, Levy C, Maslin M, McCoy D, McGuire B, Montgomery H, Napier D, Pagel C, Patel J, Oliveira JAP, Redclift N, Rees H, Rogger D, Scotte J, Stephenson J, Twigg J, Wolff J, Patterson C (2009) Managing the health effects of climate change. Lancet 373:1693–1733

Crouch J, McNiven IJ, David B et al (2007) Berberass: marine resource specialisation and environmental change in Torres Strait over the past 4000 years. Archaeol Ocean 42:49–64

Davidson K, Gowan RJ, Harrison PJ et al (2014) Anthropogenic nutrients and harmful algae in coastal waters. J Environ Manag 146:206–216

Delgado PM, Perea NS, Delgado JPM, Garcia CB, Malheiros AF, Davila CRG (2013) Detection of infection with toxoplasma gondii in manatees (*Trichechus inunguis*) of the Peruvian Amazon. Acta Biolgica Colombiana 18:211–216

Delgado PM, Perea NS, Garcia CB, Davila CRG (2015) Detection of infection with *Leptospira* spp. in manatees (*Trichechus inunguis*) of the Peruvian Amazon. Lat Am J Aquat Mamm 10:58–61

Domning DP (1981) Sea cows and sea grasses. Paleobiology 7:417–420

Domning DP (1982) Commercial exploitation of manatees *Trichechus* in Brazil c. 1785–1973. Biol Conserv 22:101–126

Domning DP (2001) Sirenians, seagrasses, and Cenozoic ecological change in the Caribbean. Palaeogeogr Palaeoclimatol Palaeoecol 166:27–50

Doney SC (2006) Plankton in a warmer world. Nature 444:695–696

Doney SC, Ruckelshaus M, Duffy JE et al (2012) Climate change impacts on marine ecosystems. Annu Rev Mar Sci 4:11–17

Duce SJ, Parnell KE, Smithers SG, McNamara KE (2010) A synthesis of climate change and coastal science to support adaptation in the communities of the Torres Strait. Synthesis report prepared for the Marine and Tropical Sciences Research Facility (MTSRF). Reef & Rainforest Research Centre Limited, Cairns, p 64

Easterling DR, Evans JL, Groisman PY et al (2000) Observed variability and trends in extreme climate events: a brief review. Bull Am Meteorol Soc 81:417–421

Edwards HH (2013) Potential impacts of climate change on warmwater megafauna: the Florida manatee example (*Trichechus manatus latirostris*). Clim Chang 121(4):727–738. doi:10.1007/s10584-013-0921-2

Elsner R (1999) Living in water: solutions to physiological problems. In: Reynolds JE, Rommel SA (eds) Biology of marine mammals. Smithsonian Institution Press, USA, pp 73–117

El Espectador (2016) Medio Ambiente 12 April 2016 Manatíes en ciénaga de Mahates, en riesgo por falta de alimento. http://www.elespectador.com/noticias/medio-ambiente/manaties-cienaga-de-mahates-riesgo-falta-de-alimento-video-626621. Accessed 6 11 16

Emanuel K (2005) Increasing destructiveness of tropical cyclones over the past 30 years. Nature 436:686–688

Errera RM, Yvon-Lewis S, Kessler JD et al (2014) Responses of the dinoflagellate *Karenia Brevis* to climate change: pCO_2 and sea surface temperatures. Harmful Algae 37:110–116

Fearnside PM (2006) Dams in the Amazon: Belo Monte and Brazil's hydroelectric development of the Xingu river basin. Environ Manag 38:16–27. doi:10.1007/s00267-005-0113-6

Fearnside PM (2014) Impacts of Brazil's Madeira River dams: unlearned lessons for hydroelectric development in Amazonia. Environ Sci Pol 38:164–172. doi:10.1016/j.envsci.2013.11.004

Fearnside PM (2016a) Environmental and social impacts of hydroelectric dams in Brazilian Amazonia: implications for the aluminum industry. World Dev 77:48–65. doi:10.1016/j.worlddev.2015.08.015

Fearnside PM (2016b) Tropical dams: to build or not to build? Science 351(80):456–457. doi:10.1126/science.351.6272.456-b

Ferreira-Ferreira J, Silva TSF, Streher AS et al (2014) Combining ALOS/PALSAR derived vegetation structure and inundation patterns to characterize major vegetation types in the Mamirauá sustainable development reserve, Central Amazon floodplain, Brazil. Wetl Ecol Manag 23:41–59. doi:10.1007/s11273-014-9359-1

Ficke AD, Myrick CA, Hansen LJ (2007) Potential impacts of global climate change on freshwater fisheries. Rev Fish Biol Fish 17:581–613

Florida Oceans and Coastal Council (2010) An update of the effects of climate change on Florida's ocean and coastal resources December2010. Available at: http://seagrantnoaagov/Portals/0/Documents/what_we_do/climate/Florida%20Report%20on%20Climate%20Change%20and%20SLRpdf. Accessed 30 April 2016

Florida Department of Environmental Protection (2014) Water use trends in Florida. Available via https://www.dep.state.fl.us/water/waterpolicy/docs/factsheets/wrfss-water-use-trends.pdf. Accessed 30 April 2016

Fourqurean JW, Rutten LM (2004) The impacts of hurricane Georges on the soft-bottom, back reef communities: site-and species-specific effects in South Florida seagrass beds. Bull Mar Sci 75:239–257

Frankham R, Bradshaw CJA, Brook BW (2014) Genetics in conservation management: revised recommendations for the 50/500 rules, red list criteria and population viability analyses. Biol Conserv 170:56–63. doi:10.1016/j.biocon.2013.12.036

Fuentes MMPB, Chambers LE, Chin A, Dann P, Dobbs K, Poloczanska E, Maison K, Turner M, Pressey RL, Marsh H (2014) Adaptive management of marine mega-fauna in a changing climate. Mitig Adapt Strateg Glob Chang. doi:10.1007/s11027-014-9590

Fuentes MMPB, Beatty B, Delean S, Grayson J, Lavender S, Logan M, Marsh H (2016) Spatial and temporal variation in the effects of climatic variables on dugong calf production. PLoS One, PONE-D-15-52097R1

Gallivan GJ, Best RC, Kanwisher JW (1983) Temperature regulation in the Amazonian manatee *Trichechus inunguis*. Physiol Zool 56:255–262

Gaye CB, Diaw M, Malou R (2013) Assessing the impacts of climate change on water resources of a west African trans-boundary river basin and its environmental consequences (Senegal River basin). Sci Cold Arid Reg 5:0140–0156

Geophysical Fluid Dynamics Laboratory/NOAA (2015) Global warming and hurricanes. http://www.gfdl.noaa.go.global-warming-and-hurricanes. Accessed 30 April 2016.

Gessner BD, Middaugh JP (1995) Paralytic shell fish poisoning in Alaska: a 20-year retrospective analysis. Am J Epidemiol 141:766–770

Gilbert PM, Burkholder JM (2006) The complex relationships between increasing fertilization of the earth, coastal eutrophication and proliferation of harmful algal blooms. In: Granéli E, Turner J (eds) Ecology of harmful algae. Springer, New York, pp 341–354

Goldenberg SB, Landsea CW, Mestas-Nuñez AM, Gray WM (2001) The recent increase in Atlantic hurricane activity: causes and implications. Science 293:474–479

Greening H, Doering P, Corbett C (2006) Hurricane impacts on coastal ecosystems. Estuar Coasts 29:877–879

Guterres-Pazin MG, Marmontel M, Rosas FCW, Pazin VFV, Venticinque EM (2014) Feeding ecology of the Amazonian manatee (*Trichechus inunguis*) in the Mamirauá and Amanã sustainable development reserves. Braz Aqua Mammal 40:139–149. doi:10.1578/AM.40.2.2014.139

Haddon AC (1912) Reports of the Cambridge anthropological expedition to Torres Strait. The University Press, Cambridge
Hallegraeff GM (2010) Ocean climate change, phytoplankton community response and harmful algal blooms: a formidable predictive challenge. J Phycol 46:220–235
Halpern BS, Walbridge S, Selkoe KA et al (2008) A global map of human impact on marine ecosystems. Science 319:948–952
Hamilton SK (2010) Biogeochemical implications of climate change for tropical rivers and floodplains. Hydrobiologia 657:19–35. doi:10.1007/ S10750-009-0086-1
Heinsohn GE, Spain AV (1974) Effects of a tropical cyclone on littoral and sub littoral biotic communities and on a population of dugongs (*Dugong dugon* [Müller]). Biol Conserv 6:143–152
Heisler J, Gilbert P, Burkholder J et al (2008) Eutrophication and harmful algal blooms: scientific consensus. Harmful Algae 8:3–13
Horgan P, Booth D, Nichols C, Lanyon JM (2014) Insulative capacity of the integument of the dugong (*Dugong dugon*): thermal conductivity, conductance and resistance measured by in vitro heat flux. Mar Biol 161:1395–1407. doi:10.1007/s00227-014-2428-4
Irvine AB (1983) Manatee metabolism and its influence on distribution in Florida. Biol Conserv 25:315–334
Junk WJ, Piedade MTF, Schöngart J et al (2012) A classification of major natural habitats of Amazonian white-water river floodplains (várzeas). Wetl Ecol Manag 20:461–475. doi:10.1007/s11273-012-9268-0
Johannes RE, Macfarlane JW (1991) Traditional fishing in the Torres Strait islands. CSIRO, Hobart
IPCC (2014) Climate change 2014: synthesis report contribution of working groups I, II and III to the fifth assessment report of the intergovernmental panel on climate change. IPCC, Geneva
Keith Diagne L (2015) *Trichechus senegalensis*. In: The IUCN Red List of Threatened Species 2015: e.T22104A81904980. http://dx.doi.org/10.2305/IUCN.UK.2015-4.RLTS.T22104A81904980.en. Accessed 15 April 2016
Killeen TJ, Solórzano LA (2008) Conservation strategies to mitigate impacts from climate change in Amazonia. Philos Trans R Soc B 363:1881–1888
Koch M, Bowes G, Ross C, Zhang X-H (2013) Climate change and ocean acidification effects on seagrasses and marine macroalgae. Glob Chang Biol 19:103–132. doi:10.1111/j.1365-2486.2012.02791.x
Laist DW, Reynolds JE III (2005) Florida manatees, warm-water refuges, and the uncertain future. Coast Manag 33:279–295
Landsberg JH, Steidinger KA (1998) A historical review of *Gymnodinium breve* red tide implicated in mass mortalities of the manatee (*Trichechus manatus latirostris*) in Florida, USA. In: Reguera B, Blanco J, Fernandez ML, Wyatt T (eds) Proceedings of the 8th international conference on harmful algae. Xunta de Galicia and Intergovernmental Oceanographic Commission of UNESCO, Paris, pp 97–100
Langerwisch F, Rost S, Gerten D, Poulter B, Rammig A, Cramer W (2013) Potential effects of climate change on inundation patterns in the Amazon Basin. Hydrol Earth Syst Sci 17:2247–2262
Langtimm CA, Beck CA (2006) Lower survival probabilities for adult Florida manatees in years with intense coastal storms. Ecol Appl 13:257–268
Lanyon JM, Newgrain K, Alli TSS (2006) Estimation of water turnover rate in captive dugongs (*Dugong dugon*). Aquat Mamm 32:103–108
Lau CL, Smythe LD, Craig SB, Weinstein P (2010) Climate change, flooding, urbanisation and leptospirosis: fuelling the fire? Trans R Soc Trop Med Hyg 104:631–638
Lawler IR, Parra G, Noad M (2007) Vulnerability of marine mammals in the great barrier reef to climate change. In: Johnson JE, Marshall PA (eds) Climate change and the great barrier Reef, a vulnerability assessment. Australia, Great Barrier Reef Marine Park Authority and Australian Greenhouse Office, pp 498–513
Macreadie PI, Baird ME, Trevathan-Tackett SM, Larkum AWD, Ralph PJ (2014) Quantifying and modelling the carbon sequestration capacity of seagrass meadows – a critical assessment. Mar Pollut Bull 83:430–439

Mallin MA, Corbett CA (2006) How hurricane attributes determine the extent of environmental effects: multiple hurricanes and different coastal systems. Estuar Coasts 29:1046–1061

Marengo JA, Tomasella J, Alves LM, Soares WR, Rodriguez DA (2011) The drought of 2010 in the context of historical droughts in the Amazon region. Geophys Res Lett 38:L12703. doi:10.1029/2011GL047436

Marsh H (1989) Mass stranding of dugongs by a tropical cyclone. Mar Mamm Sci 5:75–84

Marsh H, Kwan D (2008) Temporal variability in the life history and reproductive biology of female dugongs in Torres Strait: the likely role of sea grass dieback. Cont Shelf Res 28:2152–2159

Marsh H, O'Shea TJ, Reynolds JE III (2011) Ecology and conservation of the Sirenia: dugongs and manatees. Cambridge University Press, Cambridge

Marsh H, Grayson J, Grech A et al (2015) A re-evaluation of the sustainability of an indigenous marine mammal harvest using several lines of evidence. Biol Conserv 192:324–330

Mathews PD, da Silva VMF, Rosas FCW, d'Affonseca Neto JA, Lazzarini SM, Ribeiro DC, Dubey JP, Vasconcellos SA, Gennari SM (2012) Occurrence of antibodies to *Toxoplasma gondii* and *Lepstospira* spp. in manatees (*Trichechus inunguis*) of the Brazilian Amazon. J Zoo Wildl Med 43:85–88

Meager JJ, Limpus C (2014) Mortality of inshore marine mammals in eastern Australia is predicted by freshwater discharge and air temperature. PLoS One 9(4):e94849. doi:10.1371/journal.pone.0094849

Moore SK, Trainer VL, Mantua NJ et al (2008) Impacts of climate variability and future climate change on harmful algal blooms and human health. Environ Health 7:S4. doi:10.1186/1476-069X-7-S2-S

Nietschmann B, Nietschmann J (1981) Good dugong, bad dugong: bad turtle, good turtle. Natural History Magazine 90, New York

Omot N (2012) Food security in Papua New Guinea. AARES 56th Annual Conference – Fremantle, Western Australia, February 07–10, pp 17

Owen HC, Flint M, Limpus CJ, Palmieri C, Mills PC (2013) Evidence of sirenian cold stress syndrome in dugongs *Dugong dugon* from Southeast Queensland, Australia. Dis Aquat Org 103:1–7. doi:10.3354/dao02568

Parris A, Bromirski P, Burkett V *et al.* (2012) Global sea level rise scenarios for the United States National Climate Assessment. NOAA. Available via http://www.cpo.noaa.gov/sites/cpo/Reports/2012/NOAA_SLR_r3.pdf. Accessed 30 April 2016

Poloczanska ES, Babcock RC, Butler A et al (2007) Climate change and Australian marine life. Oceanogr Mar Biol Annu Rev 45:407–478

Powell JA (1996) The distribution and biology of the west African manatee (*Trichechus senegalensis*, link 1795). United Nations Environmental Program, Regional Seas Programme, Ocean and Coastal Areas, Nairobi, Kenya

Prospero JM, Lamb PJ (2003) African drought and dust transport to Caribbean: climate change implications. Science 302:1024–1027

Rahmstorf S (2010) A new view on sea level rise. Nat Rep Clim Chan 1004:44–45. doi:10.1038/climate.2010.2

Ralph PJ (1998) Photosynthetic response of laboratory-cultured *Halophila ovalis* to thermal stress. Mar Ecol Prog Ser 171:123–130

Rowcliffe JM, Milner-Gulland EJ, Cowlishaw G (2005) Do bushmeat consumers have other fish to fry? TREE 20:274–276

Runge MC, Sanders-Reed CA, Langtimm CA et al (2007) A quantitative threats analysis for the Florida manatee (*Trichechus manatus latirostris*). Available at U.S. Geological Survey https://pubs.er.usgs.gov/publication/ofr20071086. Accessed 30 April 2016

Scavia D, Field JC, Boesch DF et al (2002) Climate change impacts on U.S. coastal and marine ecosystems. Estuaries 25:149–164

Sellner KG, Doucette GJ, Kirkpatrick GJ (2003) Harmful algal blooms: causes, impacts and detection. J Ind Microbiol Biotechnol 30:383–406

SFWMD (Southwest Florida Water Management District) (2001) Hydrology and water quality of select springs in the Southwest Florida Water Management District. Available at Friends of mineral springs http://friendsofwarmmineralsprings.org/wp-content/uploads/2013/09/Hydrology-Water-quality-SW-Fla.pdf. Accessed 30 April 2016

Sheppard J, Preen AR, Marsh H, Lawler IR, Whiting S, Jones RE (2006) Movement heterogeneity of dugongs, *Dugong dugon* Müller over large spatial scales. J Exp Mar Biol Ecol 334:64–83

Short FT, Neckles HA (1999) The effects of global climate change on seagrasses. Aquat Bot 63:169–196

Sorribas MV, Paiva RCD, Melack JM et al (2016) Projections of climate change effects on discharge and inundation in the Amazon basin. Clim Chang. doi:10.1007/s10584-016-1640-2

Steward JS, Virnstein RW, Lasi MA et al (2006) The impacts of the 2004 hurricanes on the hydrology, water quality and seagrass in the Central Indian River lagoon, Florida. Estuar Coasts 29:954–965

Stoeckl N, Larson TM, Hicks C, Pascoe S, Marsh H (in press) Socioeconomic impacts of changes to marine fisheries and aquaculture that are brought about through climate change. In: Phillips B (ed) The impact of climate change on marine fisheries and aquaculture and their adaptations. Wiley, USA

Thornback J, Jenkins M (1982) The IUCN mammal red data book part 1: threatened mammalian taxa of the Americas and the Australasian zoogeographic region (excluding Cetacea). The International Union for Conservation of Nature (IUCN) and United Nations Environment Program (UNEP), Gland, Switzerland

UN Atlas (2010) UN Atlas: 44 percent of us live in coastal areas https://coastalchallenges.com/2010/01/31/un-atlas-60-of-us-live-in-the-coastal-areas/. Accessed April 30 2016

Wade M, Mignot J, Lazar A, Gaye AT, Carré M (2015) On the spatial coherence of rainfall over the Saloum delta (Senegal) from seasonal to decadal time scales. Front Earth Sci 3:30. doi:10.3389/feart.2015.00030

Walsh JJ, Steidinger KA (2001) Saharan dust and Florida red tides: the cyanophyte connection. J Geophy Res Oceans 106:11,597–11,612

Walsh JE, Phillips AS, Portis DH, Chapman WL (2001) Extreme cold outbreaks in the United States and Europe, 1948–99. J Clim 14:2642–2658

Walsh CJ, Butawan M, Yordy J et al (2015) Sublethal red tide toxin exposure in free-ranging manatees (*Trichechus manatus*) affects the immune system through reduced lymphocyte proliferation response, inflammation, and oxidative stress. Aquat Toxicol 161:73–84

Watson RT, Zinyowera MC, Moss RH (1996) Climate change 1995: impacts, adaptations, and mitigation of climate change. Scientific-technical analysis. Contribution of working group II to the second assessment report of the intergovernmental panel on climate change. Cambridge University Press, New York

Waycott M, Duarte CM, Carruthers TJB et al (2009) Accelerating loss of seagrasses across the globe threatens coastal ecosystems. Proc Natl Acad Sci U S A 106:12377–12381

Webster PJ, Holland GJ, Curry JA et al (2005) Changes in tropical cyclone number, duration, and intensity in a warming environment. Science 309:1844–1846

Wells ML, Trainer VL, Smayda TJ et al (2015) Harmful algal blooms and climate change: learning from the past and present to forecast the future. Harmful Algae 49:68–93

Winemiller KO, McIntyre PB, Castello L et al (2016) Balancing hydropower and biodiversity in the Amazon, Congo, and Mekong. Science 351(80):128–129. doi:10.1126/science.aac7082

Chapter 20
Sirenian Health and Well-Being in Managed Care

Michael T. Walsh and David J. Blyde

Abstract The recorded history of Sirenia species in managed care is short and quite variable with many areas of early efforts poorly documented with anecdotal material. The three extant Sirenia species of the Trichechidae family and the one extant species of the Dugongidae family are all listed as threatened by the IUCN (International Union for the Conservation of Nature). Initially hunted as a source of food in many locations, our understanding of their physiology, history, and role in the environment was slow to develop. Early literature on human interaction contributed by scientists, anatomists, and the curiosity of those who wished to share their involvement with the species was fragmented but important. Managed care of Sirenia in zoos and aquariums was initially catalyzed by a desire to show these strange animals to the public, but has morphed into a developing concern for the conservation of Sirenia populations. Public and scientific concern for the species led to protective measures in some of their ecosystems with improvements in our understanding of their biology, genetics, reproduction, disease challenges, and the influence of humans on their welfare. This evolution of public involvement led to rescue and rehabilitation efforts by aquariums, zoos, state, and federal agencies to intervene in individual animal health. Research into human mortality causes also supported better documentation of natural illnesses that effect the population's survival. The Florida manatee rehabilitation programs and Australian dugong efforts illustrate the intersection of science, medicine, and ecosystem health in advocating the needs of these unique animals and what is required to support their survival and encourage recovery. As we intersect with Sirenia in rehabilitation and exhibit

M.T. Walsh (✉)
Associate Professor in Aquatic Animal Health Program, Department of Comparative, Diagnostic and Population Medicine, College of Veterinary Medicine, University of Florida, Gainesville, FL, USA
e-mail: walshm@ufl.edu

D.J. Blyde
School of Agriculture and Food Sciences, University of Queensland, Gatton Campus, 4343 QLD, Australia
e-mail: David_Blyde@vrtp.com.au

A. Butterworth (ed.), *Marine Mammal Welfare*, Animal Welfare 17,
DOI 10.1007/978-3-319-46994-2_20

exposure for encouraging public support, it is important to provide suitable habitats for health and welfare and design their environments to their special needs while increasing protection of the wild habitats.

20.1 Introduction

The International Union for the Conservation of Nature (IUCN) is the world authority on the conservation of species and promotes a global approach to species management to coordinate efforts over subjective national borders crossed by wild species. Red lists are produced by political units to assess the risk of extinction to individual species (Deutsch et al. 2008; Keith Diagne 2015; Marmontel 2008; Marsh and Sobtzick 2015). The three extant sirenian species of the Trichechidae family and the one extant species of the Dugongidae family are all listed as threatened by the IUCN. The determination and listing of their biological status in the wild vary between organizations and is somewhat fragmented, with variations of protection in each country, and this variability can impact the efforts of facilities, biologists, and veterinarians involved in sirenian health care.

As an example of how a government can regulate a marine mammal wildlife species in the United States of America (USA), the classification of the manatee species (*Trichechus manatus latirostris*) as an endangered marine mammal places it under both the Marine *Mammal Protection Act of 1972* (MMPA) and the *Endangered Species Act of 1973* (ESA). The MMPA is a national policy with the National Marine Fisheries Service (NMFS) responsible for protecting whales, dolphins, porpoises, seals, and sea lions, whereas the US's Fish and Wildlife Service (USFWS; a division of the Department of the Interior) has responsibility for walrus, manatees, otters, and polar bears. The ESA, which is administered under the NMFS and USFWS, provides a program for the conservation of threatened and endangered species throughout all, or a significant portion, of their range and the conservation of the ecosystems on which they depend. Marine mammal species in captivity in the USA are regulated by the Animal and Plant Health Inspection Service (a division of the Department of Agriculture). These laws limit the extent of manatee facilities for exhibit and emphasize rescue and rehabilitation as the main focus.

Dugongs (*Dugong dugon*) in Australia are managed by the various state environmental organizations and federally by the Department of Environment under the Environment Protection and Biodiversity Act 1999. For example, it is the Queensland Department of Environment and Heritage Protection (EHP) which has responsibility, and in Western Australia it is the Department of Environment and Conservation. Various state bodies control the exhibition of animals in Australia such as the Department of Primary Industries in New South Wales through the Exhibited Animals Protection Act.

While these laws, adopted in these first world countries, provide general guidelines for the consideration of welfare of species with conservation concern such as sirenians, they do not outline detailed minimal standards or guidelines for their captive care. We discuss this in the following chapter.

This chapter will concentrate on the welfare considerations of sirenians that are taken into managed care for various reasons and discuss the past, current, and future considerations of care principals and species-appropriate standards for physical and mental wellness. It will briefly consider manatees and dugongs, so that the reader can see the similarities and differences in facilities from different parts of the globe.

20.2 History of Sirenian Captive Activity: Manatee

The information documenting the history of these species in captive care is piecemeal for both family groups, with only a few articles superficially describing early captive events. Early efforts in collection of many wild species included a wide range of approaches and effort geared to the animals' survival on the trip home, but with less concern to provide for physical or mental comfort. Relocation of these "intriguing" animals was attempted before knowledge was available on their transport, husbandry, nutrition, or habitat requirements. The initial "bring them back alive" imperative existed in captures by European explorers and naturalists who encountered these animals during their trips through the tropics. Western literature concerning manatees is sparse. In 1870, James Murie of the Zoological Society of London reviewed the information on the form and structure of the manatee based on two specimens that were necropsied after having failed to survive the trip to England (Murie 1872). These animals were obtained in Puerto Rico and Surinam (Dutch Guiana). He reviewed numerous anatomical accounts that dated back to 1818 and included information from Amsterdam, Germany, and Paris. At this time there was controversy as to the relationship of the manatee species to the other phylum of the animal kingdom, with some claiming it should be classed with cetaceans while others suggested the elephant was its closest relative. Manatees were freely eaten at this time by travelers and locals, and it was stated that the Catholic clergy in South America did not object to its being used on fast days (meatless Fridays were once part of the Catholic teaching) on the supposition of its being "allied to the fish tribe." Additional studies were performed on other specimens that were taken to Europe, and this resulted in detailed anatomic investigations and even in suggestions for improvements in manatee care and transportation, as it was understood at the time. These reports indicated potential causes of manatee mortality that included trauma, infection, starvation, hypothermia, and intestinal inflammation, all probably salient observations based on what we now know of these species and the likely conditions under which they were transported from the tropics to Europe. American institutions were also getting involved; Cornell in 1875 and the Zoological Gardens in Philadelphia attempted to maintain two animals in 1875, both of which died after only 90 days in captivity (Chapman 1875). By 1879, there was an apparent improvement in the care and management of captive sirenian with two manatees taken from Trinidad to the Brighton Aquarium in England. The female died after 7 months but the male lived for over 16 months. It was noted that he had "thrived on turnips and carrots," and this experience proved that manatees could be maintained (for a time) in captivity. He apparently died from hypothermia in January of the following year, 1880 (Crane 1881; Flower 1881).

While these historical accounts did suggest some improvements in the diet, transport, and other husbandry procedures, there would not have been any coordinated effort to bring this type of information together for over a century. Medical information was limited to anatomy and pathology of the animals necropsied after failed attempts to hold them in captivity, with little recording of the husbandry or veterinary information leading up to their death. In 1937, a manatee was born at the Miami Aquarium, and notes recorded by the owner were given to the *Journal of Mammalogy* (Barbour 1937). The calf lived for 18 months until it died when accidentally dropped. Subsequently, in 1975, a Florida manatee calf was born at the Miami Seaquarium (Clark 1976). A more detailed account of the birth of a manatee calf in the Netherlands with pre- and postnatal behavior notes was published shortly afterward (Dekker 1980). The mother originated from Guyana in 1972 and the father from Surinam in 1966. Since then there were additional births at Miami and SeaWorld, until the USFWS (US Fish and Wildlife Service) halted captive breeding in the 1990s, with the caveat that animals already housed in managed care prior to the enactment of protective legislation in the 1990s would still be allowed to breed. Other non-USA facilities with manatees have allowed breeding (European Endangered Species Preservation Program 2016 (EEP)).

20.3 Dugongs

Little is known about the history of holding and transporting dugongs. The level of care afforded to managed dugongs and manatees is regulated in developed regions, countries, and municipalities, but in developing countries, animals may be held in substandard conditions. This may include the use of physical restraint methods such as tail ropes to control and limit the animals' movement, similar to tethering a horse in a pasture. Occasionally when an artisanal fisherman accidently captures a live dugong (e.g., a cow with an attendant calf) in a net or *kelong* (fish trap), the fisherman will keep the animal in captivity in a sea cage (usually a fish trap) and charge the locals to see it (Fig. 20.1). Some animals are tethered, and there are recent reports of a tethered dugong and its calf being kept in separate cages off the remote island of Kokoya, Indonesia, for 7 years. Depending on the location of the holding cage, there may be attempts to feed the animal and its attendant calf. Veterinarians are typically not involved. Most of the reports of this practice have come from Malaysia and Indonesia, from regions where the dugong is rarely sighted and where it is considered a rarity (Helene Marsh pers. comm.).

While the domestic horse may be tolerate, through familiarization and domestication, a halter placed on the head, or a hobble on the leg, to allow for remote grazing, a wild sirenian should not be tail tethered under any circumstances. Pressure necrosis and tail deformation may result, with possible secondary infection of the

Fig. 20.1 Photos from video by Lim (2016) taken of a tethered, dugong in Indonesia. *Image credit: Delon Lim*

contact areas of the tail. The tailstock of sirenians is an appropriate attachment point only for the temporary attachment of satellite or radio collars that allow normal movement of the animal and which are designed to break away if the animal becomes entangled by the tether.

Several dugongs were held in captivity in India in the 1950s (Jones 1959). In the late 1960s and 1970s, Vic Oke, the then owner and manager of Cairns Oceanarium (which no longer exists), hand reared two orphaned female dugong calves—"Dolly" and "Daisy." They lived in captivity before succumbing to illness (Elliott et al. 1981). As far as we are aware, there are currently five dugongs held in captivity around the world. One is in Toba Aquarium, Japan; another in Underwater World, Singapore; and another in Sea World, Indonesia. The remaining two, a male and female, are held in the Sea Life Aquarium, Sydney, Australia. These two animals were orphans and were hand reared at Sea World, Gold Coast, Australia, before being transferred to Sydney in 2008.

20.4 Housing and Habitats

As it currently stands on a global level, the extent of managed care and the habitats provided to captive Sirenia depends on a number of factors that are heavily influenced by the experience and background of the care facility, the listing and legislative protection, and the government management of the species, all of which may vary by country and by national and local regulations. The three factors which most affect facility provision are (1) administration, (2) care personnel, and (3) veterinary staff. As described in the case discussed above, where a dugong was tethered by the tail in a make-shift enclosure for profit, the local methods used to keep the animal captive may mean that the animal's welfare needs are not met. Even in developed countries such as the USA and Australia, recognized and agreed standards for sirenian care are fully established.

20.5 Manatees

The agency responsible for the management of the manatee species in the USA is the US Fish and Wildlife Service (USFWS). Animals acquired prior to the ESA and the MMPA are not under the jurisdiction of the federal government and are considered "pre-act" with respect to how they are managed under captive care. This includes the right to display these animals for public interest and to allow reproduction. However, regardless of these exemptions, if manatees are maintained in a facility with public viewing or with public tours, the facility is regulated by the Animal Plant and Health Inspection Service under the US Department of Agriculture.

The minimal standards of care for manatees are currently under review (Michael Walsh, pers. comm.). These standards consider transport, water quality requirement, and housing, but do not address the full range of health requirements for the range of presentations of illness, injury, or age categories for other animal species that are commonly held in facilities for display, rehabilitation, or long-term care.

Enclosures: Regulatory space requirements in the USA only require a pool that is twice as long as the average length of an adult manatee with a depth one half the length (~8 × 2 m). The historical reasons for these restricted spaces may be that some of the original space requirement assumptions were partially based on rescue pools where sick animals were listless and did not exhibit normal swimming behavior. A list of known facilities holding manatees and dugongs is in Table 20.1. The facilities in the USA all exceed the minimal government standards as do most facilities worldwide. All manatees in the USA are maintained in artificial pools with a variation in the pool surface from smooth concrete to pseudo-natural habitats with rockwork and large variations in overall design. Other species such as fish may be incorporated in the habitat, and manatees will feed on food added to the environment for the fish (Fig. 20.2a).

Protection: Shade has not been a previous requirement for manatees, but should be provided in a portion of the environment if outdoors, and pool surfaces should be designed to decrease reflectivity from painted surfaces. Shore side or seaside pens are used in some facilities, and caretakers must be aware of water temperature ranges that are suitable for the animals. Manatee "interaction with guests" is present in some facilities (though not in the USA). Pens used for manatee interaction should

Table 20.1 Global sirenian facilities

Country	Institution	Summary of details	Number of animals	Website address (all accessed on 08/11/2016)	Species
Denmark	Zoo of Odense	One male, three females	4	http://www.odensezoo.dk/animals/south-america/vis/west-indian-manatee/?L=1)	*T. manatus*
Denmark	Randers Aquarium	One male, one female (Berthus & Gaffa)	2	http://www.regnskoven.dk/en	*T. manatus*
France	Beauval Zoo Park Aignan, División Académica de Ciencias Biológicas	Four or five males, two females "Herbert" (born Jan 21, 2009 in Nuremberg was sent there in 2015	6 or 7	http://www.zoobeauval.com	*T. manatus*
Netherlands	Burger's Zoo, Arnhem	One male, one female	2	http://www.burgerszoo.com	*T. manatus*
Spain	Faunia Madrid	Three males, one female	4	http://faunia.es/animales/manati-antillano-o-vaca-de-agua	*T. manatus*
Germany	Nuremberg Zoo	Two manatees, Mara (born 1994 in Nuremberg), Zorro (born 2009 in Odense)	2	http://tiergarten.nuernberg.de/en/discover/tiere-a-z/tier/nagelmanati.hT.manatus l	*T. manatus*
Germany	Tierpark Berlin	Two males, one female	3	http://www.tierpark-berlin.de/tierpark/tiere-wissenswertes.hT.manatusl	*T. manatus*
Italy	Aquarium of Genoa	Two males, one female	3	http://www.acquariodigenova.it/en/	
Poland	Breslau/Wroclaw	Two males, one female (Armstrong, Gumla, Theresa—came from Berlin in 2013 (?))	3	http://afrykarium.com.pl	*T. manatus*
France	Paris Zoo	One male Tinus	1	http://www.parczoologiquedeparis.fr/fr/animaux/animaux-amazonie-guyane/lamantin	*T. manatus*

(continued)

Table 20.1 (continued)

Country	Institution	Description	Number of animals	Website	Species
Manatee Facilities Asia					
Japan	Atagawa Tropical & Alligator Garden		1(?)	http://www4.i-younet.ne.jp/~wanien/index1.hT.manatus	Amazonian?
	Okinawa Churaumi Aquarium		3	http://oki-churaumi.jp/en/area/manatee-pool.hT.manatusl	Mexico *T. manatus*
	Toba Aquarium		3 1 Dugong	http://www.aquarium.co.jp	African 1 Dugong
Singapore	Singapore Zoo River Safari	Five males Six females	11	http://riversafari.com.sg	*T. manatus*
Taiwan	Hualien's Farglory Ocean Park	One female	1	https://taiwanaut.com	African
Seoul, S. Korea	Coex Aquarium	One female, Two males	3	www.coexaq.com	African
China	Hangzhou Underwater World	One male, one female	2	http://en.gotohz.com	African
	Nanning Zoo		2(?)		African
	Yantai Haichang Whale Shark Aquarium Yantai		2	http://www.oceanpark.com.hk	African
	Ocean Aquarium of Penglai Penglai,		4(?)		African
	Chimelong Ocean World Zhuhai City		7	http://zh.chimelong.com	African
Manatee facilities—Americas (excluding USA)					
Mexico	Dolphin Discovery—multiple sites		17 between parks	http://www.dolphindiscovery.com/manatees/	*T. manatus*
	Dolphin Discovery—Puerto Aventuras			http://www.dolphindiscovery.com	*T. manatus*
	Dolphin Discovery Chankanaab National Park—Cozumel			http://www.cozumelparks.com	*T. manatus*
	Xcaret—Quintana Roo		10	http://www.xcaret.com	*T. manatus*
	Acuario de Veracruz		6		*T. manatus*
	Universidad Veracruzana		1		*T. manatus*
	Aluxes Eco Park		4	http://aluxesecoaprque.tumblr.com	*T. manatus*

Table 20.1 (continued)

Country	Institution	Description	Number of animals	Website	Species
Belize	Manatee Rehabilitation Center			www.wildtracks.org	*T. manatus*
	Swallow Caye Wildlife Sanctuary—Caye Caulker			http://swallowcayemanatees.org/	*T. manatus*
Peru	Manatee Rescue Center—Quistococha			http://www.ayahumaamazontours.com	Amazonian
Guyana	Guyana Zoo—Georgetown			https://www.facebook.com/GuyanaZoo?sk=info&tab=page_info	*T. manatus*
Brazil	Aquasis—Caucaia Ceara			http://www.aquasis.org/subprograma-en.php?id_oquefazemos=8	*T. manatus*
	Centro Mamiferos Aquaticos—Pernambuco			http://mamiferosaquaticos.org.br	*T. manatus*
Dugong facilities—Australia					
Australia—Sydney	Sydney Aquarium	One male, one female	2	http://www.sydneyaquarium.com.au	Dugong
Manatee Facilities USA					
Florida critical care centers-exhibits	SeaWorld Orlando		Numbers vary with ill and injury presentations		*T. manatus*
	Lowry Park Zoo		Numbers vary with ill and injury presentations		*T. manatus*
	Miami Seaquarium		Numbers vary with ill and injury presentations		*T. manatus*
Florida rehabilitation holding facilities-exhibits	Homosassa Springs State Park		1 (?) Female		*T. manatus*
	Disney Animal Kingdom		2 injured males		*T. manatus*
	Mote Marine Lab-Sarasota		2 males		*T. manatus*

(continued)

Table 20.1 (continued)

Country	Institution	Description	Number of animals	Website	Species
Ohio holding facilities-exhibits	Cincinnati Zoo		2 older non-releasable females and orphan calf prerelease		*T. manatus*
	Columbus Zoo		2 older non-releasable females and orphan calf prerelease		*T. manatus*

Base data compiled by Cora Berchem at Save The Manatee Club (updated 09/13/2015). Additional data from Lucy Keith on African animals and data from the Mexican Studbook for manatees through 2016 from Roberto Sánchez Okrucky. Not all manatee facilities in Mexico are listed. The *Trichechus manatus* species is not broken down to subspecies *manatus* (Antillean) and *latirostris* (Florida) in this chart since the origin of some individuals is not fully known

Fig. 20.2 (**a**) Front view photo of the Singapore Zoo manatee exhibit. The largest facility built, but the post-build evaluation shows that newborn manatees intially may prefer the shallow water to the deeper areas. (**b**) Front view of a dugong enclosure in Australia that allows the animals to interact with visitors (as shown) but also to seek seclusion from other animals and the public. *Image credit: Singapore Zoo*

Fig. 20.2 (continued)

be large enough to allow the animal to avoid human interaction if needed. Sanctuary zones are made available to dolphins in human interaction programs in the USA. Manatees should be protected for excessive noise and poor air quality.

Water parameters: Water salinity may vary depending on the location, and the range seen in pools around the world ranges from freshwater to full salt water. Manatee facilities should allow for a range of temperatures from 26 °C (78 °F) to 30 °C (86 °F). Wild individuals do encounter temperatures outside of this range, but this published range is meant to represent a spread of tolerated seasonal fluctuations, and this temperature range may even not be ideal for manatee health, especially in compromised animals. Manatees may shiver from cold when temperatures drop below 21 °C (70 °F). A cold stress syndrome is suspected to be induced at temperatures below 20 °C, though the degree of effect or damage depends on the temperature and duration of the decrease. In 2010, wild temperatures dropped to as little as 5–9° (the 40's°F) in the Florida Bay and other shallow waters around Florida with animals dying acutely. In the winter, in Florida, freshwater springs provide relief from the surrounding cold water, with temperatures of spring water around 23 °C (73 °F), but these springs are not inhabited year round. In rehabilitation and managed care facilities, chlorine and ozone used for water disinfection can result in corneal damage if the chemical levels move out of the safe range for use, and so monitoring of oxidant levels must be strictly followed. Filtration must be enhanced to handle fecal loads.

It may be beneficial for manatees to have access to both freshwater and water of varying salinity, since it appears that the skin exfoliates after shifting between salt and freshwater, which may help to avoid retained skin becoming infected with fungal, algal, and bacterial growth. Manatees from saltwater areas drink freshwater to rehydrate after extended periods in salt water, whereas dugongs obtain all of their required fluids from the vegetation they eat. Manatees have been observed to "drink" from the overlaying freshwater layer above the saltwater layers where it is mixing, from springs and hoses, but also get much of their water from food.

20.6 Dugongs

There are no specific standards or requirements for the housing and exhibition of dugongs in any of the countries in which they are held. However, there are some fundamental standards that should be adhered to from a health and welfare perspective based on the experience of the authors.

Enclosures: Enclosures can be either natural or artificial. If artificial, the enclosure walls and floor should be durable, watertight, nonporous, nonabrasive, nontoxic, and easily cleaned and disinfected. Sharp projections and loose fittings should be avoided. Artificial pools have the advantage of being able to maintain water parameters better than natural pools. Natural pools have the advantage that they can be constructed to be much larger for less cost than artificial pools. The pool should be large enough to accommodate the natural behaviors of the animals and deep enough for the animals to dive and avoid the effects of ultraviolet radiation. Swim through and exhibit furniture should be included in the exhibit design to add some environmental enrichment for the captive animals. The minimum depth, length, and width for long-term housing of one adult animal are 4, 11, and 9 m, respectively, to allow for swimming and rolling without impeding natural behavior or causing collision with walls. If more than one animal is to be housed in the enclosure, then the length and the width should be increased by 50% per animal (Fig. 20.2b). Thought should go into the *exhibit furniture* in order that it does not present a hazard to the animal.

Protection: Animals should be protected from loud and/or continuous noise. Shade should be supplied, as well as shelter from any strong winds or rain.

Water parameters should be similar to those determined for captive dolphins and other marine mammals. Salinity should be within the range of 25–35 parts per thousand (ppt; specific gravity of 1.025–1.035). The temperature of the water should be maintained between 23 and 28 °C. If chemical treatment is used to maintain water quality and cleanliness, it should not cause harm to the animals. The bromine residual in the water should not exceed 1.0 mg/L and should be tested at least daily. The pH and ammonia should be maintained at 7.8–8.5 and below 0.7 mg/L, respectively, and should be tested daily. Fecal coliform counts should be <500 MPN (most probable number) per 100 mL water, measured weekly. If the coliform count exceeds this level, a water change or increased disinfection in the exhibit (noting requirement on maximum chemical treatment concentrations) may be necessary.

20.7 Captive Care: Manatees

The West Indian manatee accounts for the majority of sirenians maintained in the USA, with one Antillean manatee kept at an aquarium in Texas. Wild food availability is not assured, and reliance on non-wild foods like lettuce and other vegetables is common.

The number of sirenians in managed care is difficult to establish accurately, but a list of facilities is contained in Table 20.1. In the USA, at the time of publication, there are eight facilities accredited to hold manatees. In Florida, where the majority of manatees are resident, there are three critical care facilities (that also have manatee viewing for the public) and two zoological institutions that house animals prerelease, as part of the socialization for orphans or for non-releasable animals due to severe injury. In Florida, manatees are also maintained at three other holding facilities including a natural spring in Homosassa Springs state park. The history of manatees being held in facilities in the USA is fairly short, with Miami Seaquarium being the first facility in Florida which supported rescue and rehabilitation efforts as well as exhibiting animals to the public. The need for intensive care, due to human induced trauma and natural mortality causes, led to the rapid development of medical treatments and increased understanding of diseases common to this species. General anesthesia was not applied to manatees until the 1990s at SeaWorld, and prior to this these animals had been misidentified as voluntary breathers incapable of undergoing anesthetic procedures. Successful general anesthetic procedures have allowed a wider range of therapeutic approaches to be used on injured animals.

The focus within this chapter is not on sirenian diseases, but common mortality factors include death of neonatal animals, watercraft mortality, cold stress, known (red tide) and unknown biotoxins, and entanglement (Fig. 20.3). Some categories of mortality such as biotoxins and cold stress can vary dramatically in the number of animals killed each year with unusual mortality events causing epidemics. Less common disease includes other infectious organisms which result in sporadic

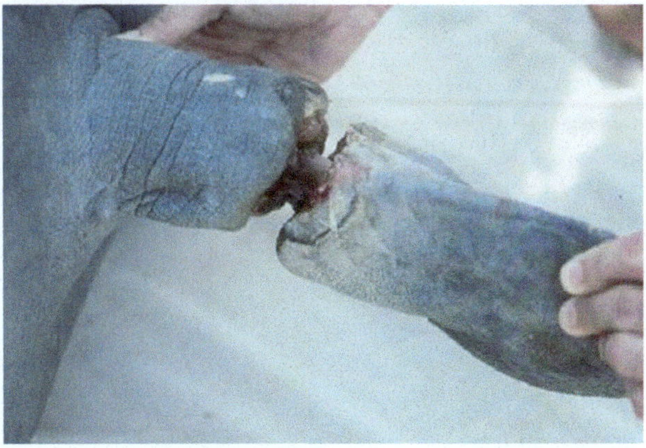

Fig. 20.3 An example of crab trap entanglement cutting through soft tissue and exposing the bone. In some instances, the limb may be saved, but often this type of human-induced injury requires amputation. Image credit: *Credit: Sea World Orlando*

diagnostic findings such as toxoplasmosis and mycobacterium. More detailed medical and husbandry information on manatee is available (Bossart 2001; Chittick et al. 2008; Murphy 2003; Walsh and Bossart 1999; Walsh and de Wit 2015).

20.8 Captive Care: Dugongs

Few dugongs have been maintained successfully in captivity, and this species has never bred in captivity. The composition of captive adult dugong diets includes seagrass, cos lettuce, snow pea sprouts, cabbage, endive, wheat grass, and spinach. Cos lettuce has been the predominant diet of some captive animals. A healthy adult dugong may consume up to 30 kg of cos lettuce per day, making the feeding of these animals very time consuming and expensive. Diets based on artificial ingredients such as cos lettuce may be deficient in essential vitamins and minerals, and supplementation may be necessary. In some places, captive dugongs have been fed seagrass obtained from local sources. A healthy adult dugong may need 20–25 kg of low-fiber seagrass per day to maintain bodyweight. Dugongs are notoriously inefficient eaters, and a lot of food is wasted, as observed in the wild, with trails of seagrass floating on the water's surface where dugongs have fed. Sourcing seagrass from the wild is very time consuming, and in certain jurisdictions, permits are required from the local conservation agency. In the wild, dugongs feed from the sea floor, and efforts should be made to feed the animals from the bottom of their enclosures in the captive situation. This requirement may require stacking the food into weighted trays and placing the trays on the bottom of the exhibit (Fig. 20.4).

Dugongs less than 1.5 m in length will be suckling, and those less than 1.8 m long are likely to still be deriving some of their nutrition from suckling (Marsh 1997). Attempts have been made to hand rear neonatal dugongs; however, little information is available on the process used (W. Blanshard, unpublished data), and

Fig. 20.4 An adult male, hand-reared dugong in Australia feeding on cos lettuce from feeding racks designed to sink to the bottom of the pool and mimic natural feeding behavior. *Image credit: Sea World Australia*

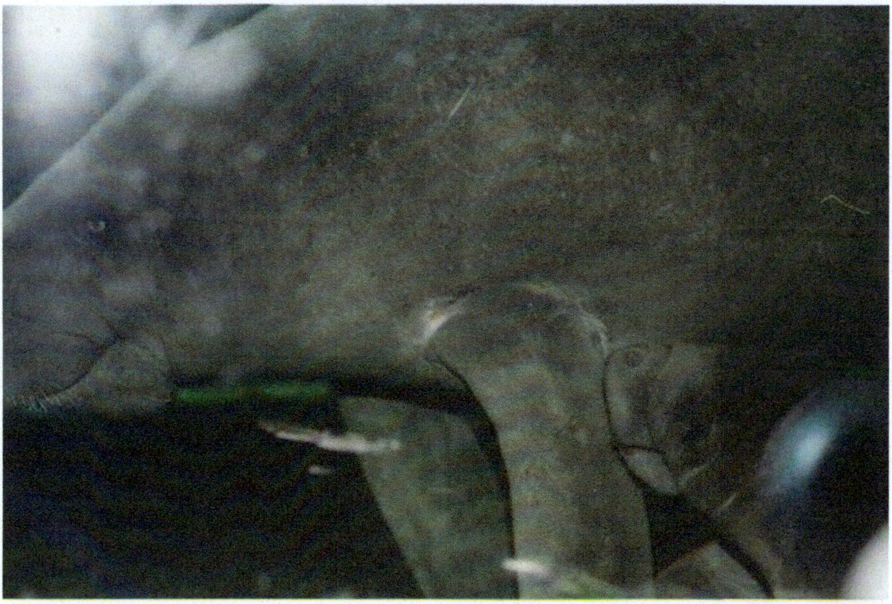

Fig. 20.5 Manatee calf nursing in Churami Aquarium in Okinawa facility. These births help biologists to understand manatee reproductive physiology. *Image credit Dr. Ueda*

early attempts were unsuccessful. In some manatee exhibits, natural feeding has been encouraged, and this is believed to be better for the health of the calf as well as to promote socialization (Fig. 20.5). Dugongs have been hand reared using milk diets based around Divetelact (Sharpe Laboratories Pty. Ltd., Ermington, New South Wales, Australia), powdered infant milk formula designed for multiple species of animals. Other ingredients such as coconut milk can be added to the Divetelact to increase the nutrient density. As with manatees, dugongs can be bottle-fed and generally nurse better while being cradled upside down in water, mimicking feeding from their mother (Fig. 20.5a). Hand rearing is a labor-intensive exercise as animals may need to be bottle-fed for up to 9 months before they begin to eat solid food (Fig. 20.6b, c).

In the wild, non-habituated dugongs are wary of noise and human activities such as boating, swimming, and diving. In captivity, possibly because most are hand reared to some extent or habituated to humans, they appear to be inquisitive and responsive to human interactions. Some captive facilities offer interactive experiences with captive animals, although it should be noted that hand-reared males have the potential to be aggressive to humans in the captive situation.

Dugongs can be conditioned to move through gates, swim onto platforms and stretchers, and present various parts of their body for inspection. Environmental enrichment and conditioning should be part of the husbandry of captive dugongs. Medical information on dugongs can be found in Woods et al. (2008).

Fig. 20.6 (**a**) Bottle feeding an orphaned neonatal manatee. (**b**, **c**) Modified feeding apparatus to allow bottle feeding of neonatal manatees *All image credits: Sirenia Project, United States Geological Survey*

20.9 Rescue and Rehabilitation

20.9.1 Manatees

Rescue and rehabilitation of manatees is the predominant captive holding activity related to manatees and is focused on treating the common reasons manatees present, disease, and trauma. The Florida Fish and Wildlife Conservation Commission (FWC) is the state agency responsible for manatee policy and also supports the unique Marine Mammal Pathobiology Laboratory where the majority of manatees are examined for cause of death. Information on agency and historic manatee mortality can be found at the FWC manatee web site (Florida Fish and Wildlife Conservation Commission 2016). A large portion of manatee deaths in Florida are a result of interactions with watercraft. The mortality causes seen at necropsy are reflected in the presentations to the rescue facilities with varying effort required to treat and hopefully return the animals to the wild. Manatee rescue and rehabilitation are extremely expensive and are supported by some state legislative funding. However, a large portion of the costs are absorbed by the facilities through the patronage of the public and through dedicated members of the facilities raising awareness and funds to support manatee rehabilitation. In addition, medical therapy is also very time consuming, and support for a sick individual can be challenging and dependent on the extent of damage present. Attention to support the "mental challenge" of adaption to a foreign environment, including the medical pool, can aid recovery and healing. Manatees may adapt better to a new environment when cohorts that are already eating are present, with food presented at the bottom of the pool and left overnight, and with restrictions on exposure to people. Antianxiety medication such as diazepam (7-chloro-1-methyl-5-phenyl-3H-1,4-benzodiazepin-2-one, a compound in the benzodiazepine family that typically produces a calming effect) has been used to help the animal to eat more quickly and help to adapt to the need for frequent treatment. This can have the effect of reducing the need for additional supplementation and decreasing additional injury. Physical restraint may need to be used, but with caution so as to assist, rather than hinder, the animal's recovery. Once stabilized the animal may go to a holding area for recuperation and if very young, will need to grow to an appropriate size before being considered for release. Holding area regulations are not under the guidance of APHIS (Animal and Plant Health Inspection Service of the United States Department of Agriculture) if the area does not involve the public, and currently guidelines are being developed to describe suitable standards for decision-making processes for animals being prepared for release.

20.9.2 Dugong

The rescue and rehabilitation of dugongs in Australia is usually conducted as a joint effort between the responsible government agency and the locally capable rehabilitation center. Without the same formal guidelines as the USA, all rescue efforts are undertaken on a case-by-case basis where the survivability of the animal, practicality of rescue, and welfare of the animal are taken into consideration by experts and an appropriate course of action is determined.

Traditionally, rehabilitation has been considered a challenge for dugong with low success rates in rearing calves. However, with medical advances many common ailments are routinely successfully treated and the animal released. Considerations for release above medical clearance include the animal's ability to find a herd it can join.

All strandings and mortalities are reported to the appropriate state agency and reflected in annual reports produced by each agency for the species.

20.10 Sirenian Welfare Expectations

The subjective determination of sirenian welfare principles is based upon a set of assumptions and expectations held by a number of groups and organizations. These may vary from country to country. The general tendency for these organizations is to "self-discover" their approach to sirenian welfare, which results in a wide variation in the rate of progress and the areas of importance which are emphasized.

There is a consistency in the principles of care between many manatee and dugong. These approaches are usually based on expectations and guidelines established by organizations dedicated to improving the care of managed animals. This is aided by open communication between facilities and through support of their governments to present a unified approach to sirenian care. The goals of some facilities are geared more to education and of others toward human interaction as the method of engaging the public. Perceived challenges to the health of wild manatees are listed in Table 20.2, and challenges seen in managed animals and rescued animals are listed in Table 20.3.

Table 20.2 List of common health challenges to wild manatees

1. Infectious diseases—parasitic, intestinal, toxoplasmosis (rare)
2. Trauma—watercraft, flood gates, chronic injury
3. Biotoxins—red tide, others undiagnosed
4. Pollutants
4. Algal blooms with food disruption—seagrass die-off
5. Cold stress—cold mortality
6. Neonatal mortality—inanition, maternal separation, maternal abandonment
7. Exuberant tourism—displacement from warm water sites, feeding interference
8. Loss or modification of habitat

Table 20.3 List of common health challenges to managed manatees

1. Obesity—appropriate wild diet availability
2. Repetitive hormonal cycling of females—hormone-related skin disease, in appetence, secondary illness
3. Male-related dominance behavior
4. Age-related reproductive tumor formation in old females
5. Chronic injury and infection secondary to initial traumatic injury

20.11 Species Specific Needs

Each species that has evolved in a specific environment has a genetically based set of needs that should be understood and addressed for the individuals, populations, and species to survive and thrive in a different habitat. Those needs include adequate and proper food availability, a suitable supportive habitat that is resilient in the face of change, access to the opposite gender for reproduction, and ability to vary their location to allow options for the acquisition of those needs. There are additional needs that may be linked to age for calf survival and growth, and others may be influenced by gender during breeding seasons. It is therefore reasonable to suggest that managed environments should strive as far as possible to replicate these provisions. Proper physical environments requiring movement for food or migration activity enhance the musculoskeletal or physical requirements. Though slow moving, the manatee moves during its foraging activity day to day and in some cases migrates great distances across the seasons to take advantage of other food sources located outside the normal range during warm weather. In addition to measured movement, both manatees and dugongs can swim rapidly for short periods of time to avoid danger.

One frequently encountered issue, at least for Florida manatees, is obesity. In facilities, and with restricted pool or lagoon volume, and reduced movement requirements, overfeeding can become a problem for long-term-managed animals. This is exacerbated for those that have access to excess food material when in multiage groups, especially if maintained in a group with juveniles who need extra food for growth. The solution to excess body weight is either diet control based on regular measurement of body weight or to increase exercise. Food restriction is difficult without separation of the different ages, and promoting increased activity is a challenge in that the animal's activity is based on foraging behavior or migration, and so providing adequate space and establishing multiple feeding stations to achieve this increased activity may not be practical. Nonetheless, obesity can be a problem, so working with input from a nutritionist may be helpful. Suggestions to use swim in place systems (powered water flows) within facilities designed to encourage movement seem logical but also difficult to implement and maintain.

20.12 Improving Sirenian Welfare with Managed Care Facilities

In the USA, there are often multiple organizations involved in the monitoring of both wild and managed sirenian welfare. Government agencies such as FWC are very active in the areas of rescue work (in collaboration with the facilities), manatee habitat, and cause of death research, through necropsy of all manatees that die. There are nongovernmental organizations such as the *Save The Manatee Club* that are involved in public education, legislative monitoring and action, as well as cooperation with facilities and manatee support organizations which advocate protection of the animal habitat and support for alliances between the public, universities, agencies, and facilities.

The immense pressures on all species due to human encroachment on their environments, physically, through habitat destruction, and through changes from indirect activity such as pollution, require that education of the general population becomes a priority for all animal species, whether wild or maintained. Though some people are opposed to keeping marine mammals in captivity, this rigid position is not seen by ourselves to be ultimately practical or sustainable and ignores the continual pressures placed on habitats by those who do not identify with the value of the animals or their environments. Education is the lynchpin for any theory of success related to wild animals. The goal of education is to identify with these species on an individual and "personal" level but also as part of the balance of the planet. This makes the education challenge more daunting since it is not limited to the animal alone, but how it fits into the ecosystem. Contrary to some beliefs, maintained animals can serve a function in this effort since a large percentage of people do not have a motivation to "care" unless there is a personal link to the animal. People's viewpoints vary, and efforts must address the wide range of learning styles. If sirenians are maintained in either man-made or "natural environments" (all environments are now to some degree modified by man), the requirements for their lives must include physically and mentally suitable living standards for all phases of life. The factors influencing successful managed care may include individual animal idiosyncrasy (personality) and maladaptation, deficiencies in husbandry, age, species maladaptation, facility, or environmental negatives such as noise, crowding, gender imbalance, lack of a sanctuary, or lack of adequate space.

Education should not be limited to the general public and must extend into the facilities involved with their care. There is a significant interest to develop metric systems for wellness (Clegg et al. 2015). While some caregivers are talented in their ability to identify with health challenges that the animal exhibits, it may be difficult for others to associate health status changes with environmental inadequacy or poor social structure. It can be challenging to align conservation of the species, public awareness, the growing need for ongoing improvement to facilities housing sirenians, and the need to better diagnose, care, treat, and release these animals.

20.13 Conclusions

To improve managed care, there should be an effort to develop universal species suitable standards of husbandry and care determined by bringing together all the current information regarding sirenian health parameters. This includes both physical and mental needs, which, while very difficult to assess, should start with a review of all facilities worldwide with an emphasis placed on the medical conditions and behavioral responses seen. In species that cannot vocalize their needs or express them very clearly through behaviors, then caretakers should be trained to recognize obvious, as well as subtle expressions of needs related to species suitable standards. Measurement and interpretation of animal welfare or wellness/well-being is dependent on the quality of the protocols in place which can detect insufficiencies in care or the expression of mental or physical health challenges.

References

Barbour T (1937) Birth of a manatee. J Mammal 18:106–107

Bossart G (2001) Manatees. In: CRC handbook of marine mammal medicine: health, disease and rehabiliation, 2nd edn. CRC Press, Boca Raton, pp 939–960

Chapman HC (1875) Observations on the structure of the manatee. Proc Acad Natl Sci Phila 27(3):452–462

Chittick EJ, Walsh MT, West G, Heard D, Caulkett N (2008) Sirenians (manatees and dugongs). In: Zoo animal and wildlife immobilization and anesthesia. Blackwell, Hoboken, pp 497–505

Clark B (1976) Florida manatee born at Miami Seaquarium on 3 May 1975. Int Zoo News 131:37–38

Clegg I, Borger-Turner J, Eskelinen H (2015) C-well: the development of a welfare assessment index for captive bottlenose dolphins (*Tursiops truncatus*). Anim Welf 24:267–282

Crane A (1881) Notes on the habits of the manatees (*Manatus australis*) in captivity in the Brighton aquarium. Proc Zool Soc London 49(2):456–460

Dekker D (1980) Pre-and postnatal behaviour in the manatee *Trichechus manatus* in captivity. Aquat Mamm 8:21–26

Deutsch CJ, Self-Sullivan C, Mignucci-Giannoni A (2008) *Trichechus manatus*. The IUCN Red List of Threatened Species 2008: e.T22103A9356917. http://www.iucnredlist.org/details/22103/0. Accessed 8 Nov 2016

Elliott H, Thomas A, Ladds P, Heinsohn G (1981) A fatal case of salmonellosis in a dugong. J Wildl Dis 17:203–208

European Endangered Species Preservation Program (EEP). Europaeisches Erhaltungszuchtprogram-Nuremberg coordinates the breeding program for manatees. http://www.eaza.net/activities/cp/Pages/EEPs.aspxlvfersen@odn.de. Accessed 8 Nov 2016

Florida Fish and Wildlife Conservation Commission (2016) 2015 Manatee Mortalities. http://myfwc.com/research/manatee/rescue-mortality-response/mortality-statistics/2015/. Accessed 8 Nov 2016

Flower WH (1881) Notes on the habits of the manatee. In: Proceedings of the Zoological Society of London, pp 453–456

Jones S (1959) On a pair of captive dugongs (*Dugong dugon*) (Erxleben). J Mar Biol Assoc India 1:198–202

Keith Diagne L (2015) *Trichechus senegalensis*. In: The IUCN Red List of Threatened Species 2015: e.T22104A81904980. http://dx.doi.org/10.2305/IUCN.UK.2015-4.RLTS.T22104A81904980.en. Accessed 8 Nov 2016

Lim, D (2016) Image of a tethered, dugong in Indonesia. http://www.dailymail.co.uk/news/article-3496001/Pictures-Indonesia-rare-dugong-mother-chained-caged-underwater.html. Accessed 12 Dec 2016

Marmontel M (2008) *Trichechus inunguis*. The IUCN Red List of Threatened Species 2008: e. T22102A9356406

Marsh H (1997) Dugongs. In: Stranded marine wildlife workshop hosted by Queensland Department of Encironment an Sea World Enterprises, Sea World Nara Resort, Main Beach, QLD

Marsh H, Sobtzick S (2015) *Dugong dugon*. The IUCN Red List of Threatened Species 2015: e. T6909A43792211. http://dx.doi.org/10.2305/IUCN.UK.2015-4.RLTS.T6909A43792211.en. Accessed 8 Nov 2006

Murie J (1872) On the form and structure of the Manatee (Manatus americanus). Trans Zool Soc Lond 8:127–220

Murphy D (2003) Sirenia. Zoo and wild animal medicine, 5th edn. Elsevier Science, St. Louis, pp 476–482

Walsh M, Bossart G (1999) Manatee medicine. In: Zoo and wild animal: current therapy. W.B. Saunders Company, Philadelphia, p 747

Walsh MT, de Wit M (2015) Sirenia. In: Fowler's zoo and wild animal medicine, vol 8. W.B. Saunders, St. Louis, pp 450–457

Woods R, Ladds P, Blyde D (2008) Dugongs. In: Medicine of Australian mammals. CSIRO Publishing, Collingwood, pp 615–627

Chapter 21
Assessing Welfare of Individual Sirenians in the Wild and in Captivity

Mark Flint and Robert K. Bonde

Abstract Assessing the welfare of wild populations of sirenians has required a "generalist" approach. The outcome has been a subjective decision as to whether what the observers are witnessing in an individual or group of animals is normal and whether that has positive or negative consequences. The understanding of sirenian welfare requirements, and a decision process for whether to support and maintain their natural habitats or to try to replicate it in a meaningful way in an artificial captive setting, is still in its early developmental stages and has dynamic qualities that are in need of urgent attention. In this chapter we use the knowledge and observations presented throughout the chapters on sirenians to outline a proposed standard approach for assessing welfare in individuals in wild populations, as well as guidelines for assessing captive groups of dugongs and manatees. In the wild, the suitability of the habitat and human impact on it, the limitations of carrying capacity, the dynamics of ecosystems, and the effects that the immediate environment will have on the known resident populations are examined. In captivity, we use the foundation of the *Five Freedoms*, based on experience derived from other captive species, and we combine this with experience from rehabilitating manatees in Europe and the United States and, more recently, dugongs in the Indo-Pacific, to identify requirements and to help us to assess the unique needs of these species when held in facilities. We present considerations and approaches to (1) holistically assess captive facilities and to assess the well-being of the individuals held in the facility, (2) derive a guideline for standard captive assessment, (3) determine if adequate welfare needs for the animals are being met, and (4) help to provide guidance on whether an animal is suitable for release after rehabilitation.

M. Flint (✉)
School of Forest Resources and Conservation, University of Florida,
The Florida Aquarium's Center for Conservation, Apollo Beach, Florida, USA

Veterinary-Marine Animal Research, Teaching and Investigation unit, School of Veterinary Science, The University of Queensland, Gatton, Queensland, Australia
e-mail: flintm@ufl.edu

R.K. Bonde
Wetland and Aquatic Research Center, U.S. Geological Survey, Gainesville, FL, USA
e-mail: rbonde@usgs.gov

A. Butterworth (ed.), *Marine Mammal Welfare*, Animal Welfare 17,
DOI 10.1007/978-3-319-46994-2_21

21.1 Introduction

Most vertebrate species, and especially the primate species, have detailed accounts of natural behavior and habitat usage, which has enabled a comprehensive baseline of collective knowledge to be used to assess their well-being in any given situation (Goodall 1986). Unlike these well-studied wild species, assessing the welfare of wild populations of sirenians has required a more "generalist" approach. Biologists, mariners, ecologists, veterinarians, pathologists, and modelers have collaborated to bring together segregated data and anecdotal information and to consider multiple environmental, populational, and individual characteristics which may correlate with certain specific behavioral responses. The outcome has been a subjective decision as to whether what the observers are witnessing in an individual or group of animals is normal and has positive or negative consequences for their welfare. In some ways the understanding of sirenian welfare requirements, and a decision process for whether to support and maintain their natural habitats or to try to replicate it in a meaningful way in an artificial captive setting, is still in its early developmental stages and has dynamic qualities that are in need of urgent attention.

In this chapter we use the knowledge and observations presented in the preceding chapters on sirenians to outline a proposed standard approach for assessing welfare of individuals in wild populations, as well as guidelines for assessing captive groups of dugongs and manatees.

In the wild, the suitability of the habitat and human impact on it, the limitations of carrying capacity, the dynamics of ecosystems, and the effects that the immediate environment will have on the known resident populations are examined.

In captivity, we use the foundation of the Five Freedoms (Farm Animal Welfare Council 1979), based on experience derived from other captive species, and we combine this with experience from several decades of rehabilitating manatees in Europe and the United States and, more recently, dugongs in the Indo-Pacific, to identify factors, judge their relative importance, and so help us to assess the unique needs of these species when held in captive facilities.

We present considerations and approaches to (1) holistically assess captive facilities and to assess the well-being of individuals held in the facility, (2) derive a guideline for standard captive assessment, (3) determine if adequate welfare needs for the animals are being met, and (4) provide guidance as to whether the animal is suitable for release after rehabilitation.

21.2 Wild: Indirect Assessment of Individuals in a Population

Indirect assessment of an individual can be carried out by assessing the features of the environment and then by applying a decision-making process on the suitability of those conditions relative to the carrying capacity for a particular group of animals that use or live in that area. This is not a straightforward approach, and many complex interactions, including those of competing species and humans, result in the need for best judgment as opposed to a set of rigid guidelines. However, it is quite possible to clearly determine the suitability of an environment to harbor dugongs and manatees, based on the assessment of known desirable characteristics.

High levels of human activity and occupation can deter use of a preferred area. Harmful algal blooms, such as red tide (*Karenia brevis*) exposure in Florida, can cause severe morbidity and mortality during blooms. A cyanobacterium (*Lyngbya majuscula*) can cause irritation, and its persistent presence in an area is not conducive with maintaining an ecosystem suitable for sirenians (Capper et al. 2013; Landsberg et al. 2009).

Global climate impacts are making site suitability a dynamic process. In Africa and South America, it cannot now be assumed that a body of water present during the rainy season will persist through the dry season and hence be perennially available. It is probable that this type of environmental unpredictability will become more prevalent as climate change impacts regions over the coming decades. Assessment of sites for long-term suitability should consider the possibility that some areas may become unsuitable habitats, and this should be considered alongside the human social implications of desertification discussed in the climate change chapter (Chap. 19).

Finally, increasing public awareness of the challenges faced by marine mammals over the last decade has resulted in a more informed public, and well-intentioned public now commonly directly alert authorities when they observe something which they perceive is amiss in the environment. This improved communication has allowed a greater capacity for those tasked with monitoring and assessing the well-being of sirenians, by enabling rapid collection of up-to-date information across huge geographical areas and through rapid reporting of "tip-offs" regarding potential issues or hotspots. This information directly assists efforts to manage imperiled sirenian stocks.

21.3 Wild Animals: Assessing Welfare at the Individual Level

Wild caught manatees and dugongs offer a rare direct insight into the health of members within the population (Figs. 21.1, 21.2, and 21.3). The assessment of the well-being of these individuals may allow decisions to be made based on welfare criteria which support responsible approaches to mitigation of problems.

The assessment of animals by radio tracking can offer useful information on the interaction of individuals within their environment (Flamm et al. 2005). The use of satellite telemetry and microchipping as part of a capture-mark-recapture program may yield vital data on (1) animal movement (distribution) and use of habitat or structures (seagrass beds, warm water refuges), (2) migratory patterns, (3) home range patterns, and (4) response to perturbations and anthropogenic activities (Fig. 21.4). Although considered to be of minimal risk to the animal and to present

Fig. 21.1 Manatee capture for health assessments. The animal is encircled by a net deployed from a boat (see behind manatee) and carefully captured by manually pulling the net to shore. *Image credit: Sirenia Project—United States Geological Survey*

Fig. 21.2 The animal is intensively monitored while medical and biological data is collected prior to release within approximately 30 min. *Image credit: Sirenia Project—United States Geological Survey*

Fig. 21.3 Dugong in a capture-mark-recapture study. These are usually performed by a "rodeo technique" where a jumper captures and secures the animal from a moving vessel. The animal is then supported either on or next to the vessel or by people in the water as shown in this image. *Image credit: Department of Environment and Heritage Protection StrandNet*

Fig. 21.4 Attaching a satellite telemetry tag to a dugong in southern Queensland, a region with a high urban population. This study determined the movements of dugong with respect to anthropogenic and climatic pressures. *Image credit: Department of Environment and Heritage Protection StrandNet*

only a short-term impediment, it is advised that any wild sirenian capture program includes documented welfare considerations or guidelines for reference during capture. Invasive methods are often used to collect samples from individuals for biological and health assessments, and this sampling is commonly incorporated into programs during tagging (Gerlach et al. 2015; Sulzner et al. 2012). This approach can yield detailed data on the functional health of the population, and allows determination of physiological parameters, and can guide responses to current environmental conditions or challenges. In addition to health (clinical pathology, body condition, abnormalities), other ancillary indicators of survivorship, such as nutritional state and reproductive status, can be determined during tagging procedures.

Guidelines could include assessment and minimization of the risk of death during handling and assessment of the ongoing effect of placement of any radio tag attachments or tissue or blood sampling used in research, which may have an adverse impact on the animal during the collection procedure or after release.

Live stranding of animals and the subsequent veterinary assessment and/or treatment of stranded individuals may help to identify that a part of the population is not performing well (Flint et al. 2010). These cohorts could be considered surrogates or proxies for assessing the health of the population. Furthermore, these individuals may highlight what the parameters of assessment should be for examining the health of the population. Frequently, the assessments carried out on stranded individuals include a baseline assessment of (1) body condition, body mass index, and/or nutritional status; (2) identification and diagnosis of disease; (3) behavioral response to capture and mentation or discussion of any indicators on the animal which may indicate a stress response and the current capacity of the animal to cope; and (4) the presence of injuries or external indicators of health status (e.g., parasites).

The collection of ancillary scientific data from these animals during rehabilitation has helped create physiological and anatomical baselines and expands our understanding of the requirements for the health and well-being of individuals. Combining this data with the environmental needs of the animal in a particular habitat, we can make better informed decisions on the requirements for maintaining a captive animal (if required) and what steps need to be taken prior to releasing an individual back into the wild.

21.4 Captive Animals: Assessing Welfare at the Facility Level

While many of the principles outlined for assessing individuals in the wild can be applied to assessing individuals at the facility level, there are additional considerations due to the potential for high stocking densities within facilities, resulting in low space/or water volume provided per animal when compared to the natural environment. In order to optimize an animal's artificial environment, assessment in captivity should take into consideration, as minimum, the *Five Freedoms*—standards derived for the holding of animals by the UK Farm Animal Welfare Council (1979).

The *Five Freedoms* simply outline that all animals held in captivity should have:

1. Freedom from hunger and thirst by ready access to freshwater and a diet to maintain full health and vigor
2. Freedom from discomfort by providing an appropriate environment including shelter and a comfortable resting area
3. Freedom from pain, injury, and disease by prevention or rapid diagnosis and treatment
4. Freedom to express normal behavior by providing sufficient space, proper facilities, and company of the animal's own kind
5. Freedom from fear and distress by ensuring conditions and treatment which avoid mental suffering

To achieve this, we strive to determine the necessary needs and wants of the species. Using our wild habitat knowledge for sirenians, we should provide:

1. Access to freshwater, appropriate seagrasses, and freshwater vegetation or substitutes located in a natural feeding location (i.e., on the floor for dugongs)
2. Protection from the sun, in the form of shade, replicate diurnal cycles when using indoor lighting, providing areas for isolation, and the ability to swim and turn unhindered
3. Available medical care and daily observation of health status
4. Tank design and size with adequate space per animal
5. Selection of structures, tools for enrichment development, and tank mates that allow social interaction in an environment safe from non-predatory hazards and one where areas are available for sanctuary from predators/competition/the public, while still achieving the general goals of the *Five Freedoms*

As is the case in many attempts to improve the welfare of facility inhabitants, the use of the Five Freedoms and creation of expected standards run the inherent risk of anthropomorphizing and the projection of personal human wants onto the animals. To avoid this wherever possible, reference to scientific data and analysis of the meaning of this data should form the basis of decision-making or as a reference. Robust data which can satisfy these needs can be derived from previous dugong and manatee captive experiences and our knowledge of their free-ranging behaviors and requirements.

There are additional considerations when artificially housing behaviorally complex species. The nature of the artificial housing may require adaptive behaviors that are otherwise not seen in the wild in order to maintain social structure and to support the well-being of individuals. For example, dominance structures exist in many hierarchical groups with a leader of each herd, flock, or group. This dominant position may be held by different members of the group for different activities, such as feeding, drinking, sleeping, or mating rights. With increased stocking densities and altered habitat, these activities may require the addition of complexities greater than observed in the wild or the creation of new dominance opportunities, like novel activities, to artificially develop and maintain social structure and help alleviate stress. This has been previously documented in other free-ranging species that become intensively managed (Flint and Murray 2001).

Conversely, natural dominance behaviors that exist in the wild may need to be curtailed in a captive environment. For example, for manatees, dominant (usually bigger) individuals may take more food than smaller individuals. This could lead to some smaller animals not getting the required minimal nutrition as a "behavioral expense" of being housed with larger individuals, if this is not monitored and controlled. Regarding the dugong, males can dominate other males, females, and juveniles by use of their tusks to rake (scar) herd mates into submission. If left unchecked in an enclosure, this could potentially lead to severe injury or morbidity and mortality.

For sirenians, natural and artificially induced behaviors are still an area of rapid learning for the animal keepers, with facility design requiring separate feeding areas and segregation of captive animals by size, sex, and breeding status as we progressively learn more about how to safely allow cohort interactions.

Building on the *Five Freedoms*, several principles have been proposed as being necessary to ensure animal welfare (Barnett and Hemsworth 2009):

1. Minimize stress.
2. Minimize negative emotions.
3. Maximize positive emotions.
4. Ensure adaptation.
5. Provide opportunity for normal or natural behaviors.
6. Provide natural environments.

One advantage afforded to manatees is that, as a Federally protected species, a consistency can be achieved through Federal standards being established for the care and management of sirenian undergoing rehabilitation in the United States. Unfortunately, this has not been achieved globally and would require many territories and countries to adopt the same guidelines or standards. The creation of US Federal standards effectively prescribes minimum standards against which all other needs may be assessed. In other parts of the world, outlined minimum requirements to optimize welfare tend to be based on each individual facility's best practices.

In the previous chapter (Chap. 20), Walsh and Blyde identified a range of health considerations and minimum standards which should be adopted in captive manatees and dugongs. Addressing Barnett and Hemsworth's (2009) principles, and common to all facilities, were space requirements, diet, and tank environmental factors. In some of the more resource-limited countries where standards are not as rigorously monitored, advocacy and assessment of the welfare of the dugongs or manatees may benefit from the input of special interest groups, and public reaction in these countries is starting to influence change and activity to support the safety, comfort, and health of these animals.

Finally, environmental enrichment has been a tool used in many aquatic and terrestrial species to combat negative behaviors and to provide a source of stimulation for the facility inhabitants (Anzolin et al. 2014). Sirenia are highly tactile animals, so the incorporation of novel objects in their enclosure such as flowing water or "toys" is believed to have beneficial effects on social development and cognitive abilities.

21.5 Captive Animals: Assessing Welfare at the Individual Level

When there is limited published data and all appropriate anecdotal knowledge has been employed, one mechanism to assess how effectively a captive enclosure is working is the behavioral response of the individuals within the facility. Two ways in which this can be achieved without the need to directly handle the animal are through the assessment of normal or stereotypical behaviors and through the public's response to the display and the animals in it.

Stereotypical behaviors have been reported in captive sirenians (Anzolin et al. 2014). One of the sirenians that the authors have dealt with had spent the majority of its 20-year life in captivity. While sirenians are usually docile animals in the wild, this individual was certainly not when people entered his tank to restrain him. He would quickly swim circles around the perimeter of the tank (the stereotypic part of the behavior) and repeatedly and rapidly head-butt anything in his path to avoid and prevent capture. This anticipation of an impending event posed a risk to any people in the pool but also illustrated the cognitive capacity of sirenians and their ability to respond to specific actions or perceived threats. Avoidance of these types of specific reaction, and steps to avoid the stimulus responsible or to prevent these types of behaviors, should be given consideration when dealing with individuals held in long-term captive situations.

With increased public awareness of the needs of captive animals, including sirenians, as well as the public demand for quality care in captivity, the comments made by guests about their experience of the animals are considered an important and rapid feedback mechanism to determine how well the animals are coping in captivity. This creates a two-way street for information, with facilities being an avenue to get the message out about conservation education and issues facing manatees and dugongs, but also producing a keen-eyed public that is informed and able to look for, and detect, signs of discontent. This public "eye," linked with the capacity for the public to be vocal (particularly through social media) about any anomalies or injustices they perceive, creates a situation in which public response to a situation is a very sensitive barometer for poor captive care. A case in point was a captive manatee in Venezuela that was being fed meat. The local zoo did not know that manatees were herbivores, but when this was observed and reported by a concerned citizen, the caretakers made amends and started feeding the malnourished manatee vegetation.

21.6 Guidelines for Use of Sirenians in Research

Research on both wild and captive sirenians follows the nationally adopted Institutional Animal Care and Use Committee (IACUC) standards based on the Animal Welfare Act in the United States, Animal Ethics Committee (AEC) standards based on the institute's state Animal Welfare Act in Australia, and systems such as Animal Research: Reporting In Vivo Experiments (ARRIVE) as part of the United Kingdom's 3R's (Replacement, Refinement, Reduction) to create standards to help maintain the well-being of these animals. These guidelines may also be used to provide information for intensive management.

21.7 Release of Sirenians Back into Their Natural Environment

In the United States and Australia, it is not legal to hold sirenians for public display. They can only be held in captivity for the purposes of medical treatment and rehabilitation, with the final intent of release back into the wild. Exceptions to this are if a panel of experts agree formally that the animal is non-releasable; that is, if the animal were to be returned to the wild, it would not be likely to survive or pose an imminent threat to the wild population. This is, in part, based on a determination of the animal's inability to thrive in the wild, or it is a health threat to the wild population.

When releasing sirenians back into their natural environment at the end of rehabilitation, there are several important considerations. In addition to requiring that the individual is free from disease and confirmation that they are suitably recuperated to survive unassisted in the wild, we must consider the environment and the resident population. Any release of animals must consider the genetic benefits and consequences of crossbreeding. This can occur through placing an animal from a different genetic population into a new area. Benefits might include the potential of hybrid vigor, but consequences might result from introducing a maladapted individual into a novel set of environmental conditions. In a similar way, the release of animals into inappropriate habitats, regardless of genetic effects, may be detrimental. The animals require "basic local knowledge" to meet the challenges for survival in a given area, such as finding access to freshwater (manatees), to food, to cohorts, to shelter (the location of warm water during winter in some populations), and to protection (from predators, human interactions, and loss of habitat).

Captive breeding programs have often been used as a final effort by recovery teams for critically endangered species. However, by the time a captive breeding program is implemented, it may be too late as there is already a low reproductive rate on captive bred animals. Most regional wildlife managers would prefer to allow sirenian species to breed successfully in the wild, but with some local populations at risk of going extinct, there is discussion about species reintroduction back into some extirpated areas.

21.8 Captive Assessment Guidelines

A tabulated score sheet (Table 21.1) is presented as an example of a potential approach to assessing the parameters we have discussed above and in the preceding chapters. It is by no means definitive or accurate for every facility, and any assessor could create their facility-specific own score sheet using this table as a guide.

Table 21.1 Assessing captive sirenian habitat suitability by the use of a cumulative weighted welfare parameter score

Facility assessment					Total
Parameter	Subparameter	Factor 1	Factor 2	Factor 3	
Housing	Cohorts in tank (#)	2 (10)	0 (5)	5+ (5)	
	Other species in tank	No (5)	Yes (0)		
	Predators	No (5)	Yes (−5)		
	Depth of tank	>2 m (5)	<2 m (0)		
	Area/animal	4 × length (10)	2 × length (5)	<2 × length (0)	
	Salinity-manatee	Freshwater (10)	Brackish (8)	Saltwater (4)	
	Salinity-dugong	Saltwater (10)	Brackish (6)	Freshwater (2)	
Enrichment	Included in tank	Yes (5)	No (0)		
	Form/structure used	Complex (10)	Cognitive (10)	Other objects (5)	
	Public contact allowed in tank?	No (5)	Yes (−10)		
Five Freedoms	Nutrition	Good (10)	Poor (−10)		
	Shelter	Adequate (10)	Absent (−10)		
	Normal Behavior	Yes (scored below)	No (0)		
	Freedom from fear	Yes (10)	No (−10)		
	Health care assessments	Daily (10)	Weekly (2)	None (−10)	
Individual Assessment (including answers based on those already housed)					
Behavior	Demeanor	Passive (5)	Aggressive (−10)		
	Eating	Well (5)	Small qty. (0)	No (−10)	
	Stereotypies (of others in tank)	No (10)	Yes (−25)		
	Average stay of animals in tank	<6 months (5)	>6 months (0)		
	Weight change in first 28 days	Gained weight (10)	Same weight (5)	Lost >5% weight (−20)	
	Medical issue (new since arrival)	No (10)	Yes (−20)		
Release	Date/timing of release	Known (20)	Unknown (−10)		
	Release site suitability	Known (20)	Unknown (−20)		
Total					

The premise of the assessment is that each parameter is assessed as accurately as is possible and the appropriate response is circled. Each response has a numeric value in parentheses next to it. From these values, each line is tallied (aggregated) to provide a subtotal. All assessments should be given a score, as some parameters may subtract from the final overall aggregated score. The subtotals are then tallied to provide a final score. In this example, a score of ≥100 is proposed to indicate a suitable facility, 80–99 may require further investigation and resource or other inputs prior to use of the facility, and <80 suggests the facility may not be suitable for the holding of sirenians.

21.9 Conclusions

The dugong and manatees are a group of species that is held in high regard by the public, and yet they are all threatened across the world. Through the efforts of many conservationists over the last few decades, it has been possible to collate a large amount of learned information about these animals and, to gain a greater understanding of their biology, habitat requirements and what is required to maintain these animals in healthy environments. In Chaps. 17–21 we have identified a range of human and natural threats that are creating stressed habitats and which are challenging sirenian survivorship. We have surveyed sirenian caretakers and the available literature to propose a standard set of requirements to assist successful care for individuals in rehabilitation. Finally, we used this collective knowledge to create and interpret assessments for wild and captive dugongs and manatees.

We hope that these chapters can improve the welfare of wild and captive sirenians and serve as a foundation from which to further fill in the gaps which exist in critical knowledge. With the current shift in public perception, and an increase in public awareness of the plight of numerous species, we hold hope that sirenians will become another recovered species that can be enjoyed for generations to come.

References

Anzolin DG, de Carvalho PSM, Viana PC, Normande IC, Souto AD (2014) Stereotypical behaviour in captive West Indian manatee (*Trichechus manatus*). J Mar Biol Assoc UK 94:1133–1137

Barnett JL, Hemsworth PH (2009) Welfare monitoring schemes: using research to safeguard welfare of animals on the farm. J Appl Anim Welf Sci 12:1–17

Capper A, Flewelling LJ, Arthur K (2013) Dietary exposure to harmful algal bloom (HAB) toxins in the endangered manatee (*Trichechus manatus latirostris*) and green sea turtle (*Chelonia mydas*) in Florida, USA. Harmful Algae 28:1–9

Flamm RO, Weigle BL, Wright IE, Ross M, Aglietti S (2005) Estimation of manatee (*Trichechus manatus latirostris*) places and movement corridors using telemetry data. Ecol Appl 15:1415–1426

Flint M, Murray PJ (2001) Lot fed goats- the advantages of environmental enrichment. Aust J Exp Agric 41:473–476

Flint M, Patterson-Kane JC, Limpus CJ, Mills PC (2010) Health surveillance of stranded green turtles in southern Queensland, Australia (2006–2009): an epidemiological analysis of causes of disease and mortality. Ecohealth 7:135–145

Gerlach TJ, Estrada AH, Sosa IS, Powell M, Lamb KE, Ball RL, de Wit M, Walsh MT (2015) Establishment of echocardiographic parameters of clinically healthy Florida Manatees (*Trichechus manatus latirostris*). J Zoo Wildl Med 46:205–212

Goodall J (1986) The chimpanzees of Gombe: patterns of behavior. The Belknap Press of Harvard University Press, Cambridge, p 673

Landsberg JH, Flewelling LJ, Naar J (2009) *Karenia brevis* red tides, brevetoxins in the food web, and impacts on natural resources: decadal advancements. Harmful Algae 8:598–607

Sulzner K, Johnson CK, Bonde RK, Gomez NA, Powell J, Nielsen K, Luttrell MP, Osterhaus A, Aguirre AA (2012) Health assessment and Seroepidemiologic survey of potential pathogens in wild Antillean manatees (*Trichechus manatus manatus*). PLoS One 7. doi:10.1371/journal.pone.0044517

UK Farm Animal Welfare Council (1979) The five freedoms. http://webarchive.nationalarchives.gov.uk/20121007104210/http:/www.fawc.org.uk/freedoms.htm. Accessed 21 Nov 2016

Part IV
Polar Bear

Chapter 22
Human–Polar Bear Interactions in a Changing Arctic: Existing and Emerging Concerns

Todd C. Atwood, Kristin Simac, Stewart W. Breck, Geoff York, and James Wilder

Abstract The behavior and sociality of polar bears (*Ursus maritimus*) have been shaped by evolved preferences for sea ice habitat and preying on marine mammals. However, human behavior is causing changes to the Arctic marine ecosystem through the influence of greenhouse gas emissions that drive long-term change in ecosystem processes and via the presence of in situ stressors associated with increasing human activities. These changes are making it more difficult for polar bears to reliably use their traditional habitats and maintain fitness. Here, we provide an overview of how human activities in the Arctic are likely to change a polar bear's behavior and to influence their resilience to environmental change. Developing a more thorough understanding of polar bear behavior and their capacity for flexibility in response to anthropogenic disturbances and subsequent mitigations may lead to successful near-term management interventions.

T.C. Atwood (✉) • K. Simac
Polar Bear Research Program, U.S. Geological Survey, Alaska Science Center, Anchorage, AK 99508, USA
e-mail: tatwood@usgs.gov; ksimac@usgs.gov

S.W. Breck
USDA–WS–National Wildlife Research Center, Fort Collins, CO, USA
e-mail: Stewart.W.Breck@aphis.usda.gov

G. York
Polar Bears International, Bozeman, MT, USA
e-mail: gyork@pbears.org

J. Wilder
Polar Bear Program, Marine Mammals Management, U.S. Fish & Wildlife Service, Anchorage, AK 99503, USA
e-mail: james_wilder@fws.gov

A. Butterworth (ed.), *Marine Mammal Welfare*, Animal Welfare 17,
DOI 10.1007/978-3-319-46994-2_22

22.1 Introduction

The polar bear (*Ursus maritimus*) is a conservation and cultural icon of the Arctic. Polar bears have a circumpolar distribution comprised of 19 subpopulations, which can be grouped into four ecoregions (Fig. 22.1; see Chap. 23 for a thorough description of the ecoregions) based on similarities between populations in life history and sea ice dynamics (Amstrup et al. 2008). Polar bears rely on sea ice to meet their key needs including foraging, searching for mates, and denning (Amstrup 2003). Their preferred habitat is first-year sea ice that occurs over biologically productive continental shelf (shallow) waters (Durner et al. 2009), which provide ready access to ringed (*Pusa hispida*) and bearded (*Erignathus barbatus*) seals (e.g., Smith 1980; Pilfold et al. 2014). Until recently, most populations—i.e., those occurring in the Polar Basin Convergent Ecoregion (PBCE), Polar Basin

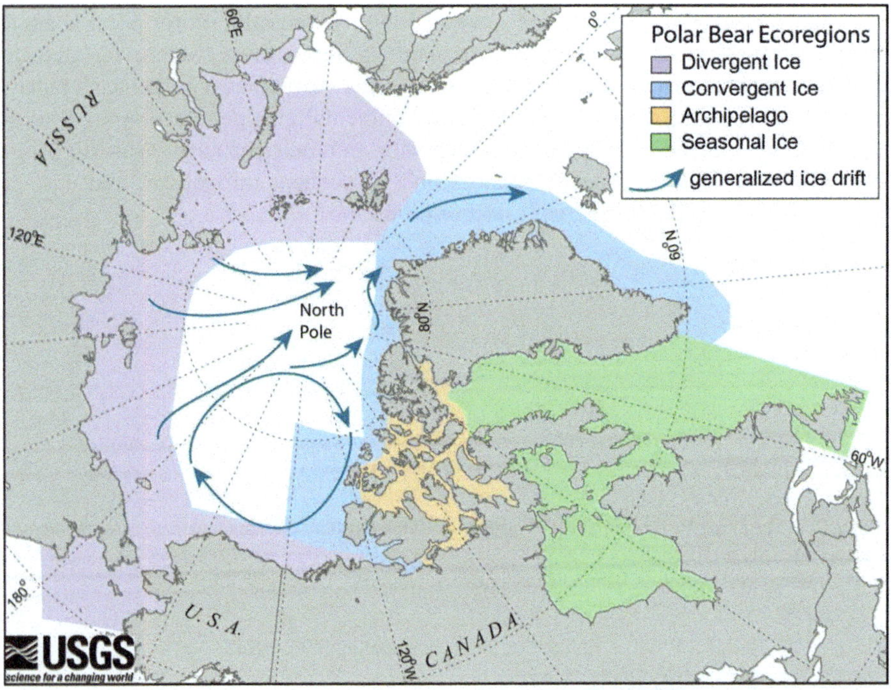

Fig. 22.1 Map of the four polar bear ecoregions defined by Amstrup et al. (2008). Ecoregion names are indicative of regional sea ice dynamics. The Polar Basin Divergent Ecoregion (PBDE) and Polar Basin Convergent Ecoregion (PBCE) occur within the Polar Basin; in the PBDE, annual sea ice advects toward the central Polar Basin, while in the PBCE, annual ice tends to converge along the western edge of Greenland and the Canadian Archipelago. In the Seasonal Ice Ecoregion (SIE), annual ice melts completely each summer, and the Archipelago Ecoregion (AE) is characterized by persistent multiyear ice. The map also includes a depiction of annual patterns of ice motion

Divergent Ecoregion (PBDE), and Archipelago Ecoregion (AE)—have been characterized by year-round availability of sea ice and bears that remained on the sea ice throughout the year, with the exception of some terrestrial denning. By contrast, populations in the Seasonal Ice Ecoregion (SIE) have long experienced the complete melt of sea ice in summer, which forces bears onto land where they rely on stored fat reserves for energy until the ice returns in the fall (Ramsay and Stirling 1988). The overwhelming reliance of polar bears on sea ice has mediated the types and extent of interactions with humans, but, with sea ice dynamics changing, the potential for increased human–polar bear interactions is becoming a management concern.

Like most large carnivore populations, there is poor understanding of how interaction with humans impacts polar bears and concomitantly how interactions impact humans living and working in polar bear range. Ultimate drivers (i.e., climate-induced loss of sea ice) will undoubtedly influence the nature of human–polar bear interactions, but these drivers are incredibly difficult to manage. Thus, identifying proximate factors leading to negative interactions (e.g., availability of anthropogenic resources) should be a high priority, given these factors should be "relatively easier" drivers to manage. Furthermore, as climate changes and anthropogenic activities increase in the Arctic, mitigating interaction and conflict between polar bears and humans is likely to be of growing importance. In this chapter, we first explore why understanding the effects of polar bear-human interactions is critical for their effective conservation in an increasingly human-dominated Arctic. We discuss the primary drivers of human–polar bear interactions and review what is known about how interactions with humans have affected polar bears. We then examine how climate-mediated changes to the environment are likely to influence the distribution and types of future interactions. We close by discussing how conservationists may be able to mitigate the likelihood of future interaction and conflict in the face of rapid environmental change.

22.2 Importance of Understanding Effects of Human Interactions with Wildlife

The expanding human population, and resultant growing demand for resources, has led to an increasingly pervasive reach of humans and infrastructure into wildlife habitats (e.g., Woodroffe et al. 2005). Over the last 100 years, the global human population has essentially quadrupled from 1.9 to 7 billion, and the corresponding anthropogenic footprint—defined as the measure of human impact on the Earth's ecosystems—has expanded commensurately (Roberts 2011). This has put extraordinary pressure on the integrity of terrestrial and marine ecosystems (e.g., Magnani et al. 2007; Halpern et al. 2008; Doney 2010). The expansion of human activities can impact wildlife by converting natural areas into altered land types (e.g., agricultural and urban) and also by altering the behavior of resident wildlife. These changes come with a number of attendant consequences, including a greater potential for human-wildlife interaction and conflict.

Interactions between humans and wildlife can have wide-ranging effects, including adversely impacting animal fitness and wildlife populations (Northrup and Wittemyer 2013). Indeed, most human-wildlife interactions are believed to be detrimental to wildlife: studies on a wide range of taxa show that human-wildlife interactions result in high mortality for many species (e.g., Moore and Seigel 2006 for reptiles; Müllner et al. 2004 for birds; Harrington and Veitch 1992 for mammals). In Africa and Asia, for example, large mammal population numbers, community diversity, and the geographical distribution of many species have been reduced due to hunting, habitat modification, and disease. Large carnivores are particularly vulnerable to exposure to human activities, primarily because their extensive home ranges are likely to increasingly overlap with human-occupied areas (e.g., Treves and Karanth 2003). Moreover, high dietary protein or fat requirements of large carnivores can lead to competition with humans for shared prey, particularly in regions where subsistence hunting is common (Dickman 2010). Competitive interaction over access to space or food resources often leads to conflict and lethal removal of wildlife, which ultimately may constitute a significant threat to the long-term persistence of focal wildlife species. For instance, in South Africa, where a majority of the total land area is now used for livestock and game farming (Department of Agriculture, Forestry, and Fisheries 2010), human-wildlife conflict is so frequent that the most recent national conservation assessment cites human persecution as the main threat to carnivores outside of protected areas (Friedmann and Daly 2004; Thorn et al. 2012).

Human interactions with wildlife can also impose a variety of economic and social costs to local communities including injury and loss of life, destruction of property, disease transmission to stock, companion animals or humans (Thirgood et al. 2005), and opportunity costs, where people forego economic or lifestyle choices due to impositions placed upon them by the presence of wild animals or conservation areas (Woodroffe et al. 2005). It is estimated that the economic cost to US agricultural producers from property damage and crop and livestock loss by wildlife exceeds $1 billion annually (NASS 2002). In both Asia and Africa, some communities may lose up to 15% of their total agricultural output to elephants (*Elephas maximus*, *Loxodonta* spp.) (Lamarque et al. 2009; Madhusudan and Sankaran 2010). In low-income countries including Mozambique and Namibia, over a hundred people are killed annually by crocodiles (Lamarque et al. 2009), while in India elephants kill more than one person every day (Rangarajan et al. 2010). Such losses, while seemingly insignificant at a national level, may give rise to high costs for the affected individuals and communities.

For polar bears, conflict with humans has been of little concern (except in isolated areas like Churchill, Manitoba, Canada) until recently. As a result, there is not a great deal of understanding about the potential impacts of conflict on polar bears and, concomitantly, their impact on people. However, for other bear species, conflict has been demonstrated to be an important consideration in population dynamics. For example, Schwartz et al. (2010) quantified the importance of conflict-related mortality for grizzly bears (*Ursus arctos*) inside and outside key protected areas. For black bears (*Ursus americanus*), Lewis et al. (2014) demonstrated how urban environments and climate-related factors can potentially act as either sources or sinks for bear populations depending upon how human-bear conflict is managed. These findings indicate that human–polar bear interactions that escalate into conflict could pose a challenge to conservation efforts, especially given that the Arctic marine environment is rapidly changing and the human footprint is growing.

22.3 Driving Forces of Human–Polar Bear Interactions

Loss of sea ice habitat due to a warming climate is the primary threat to the long-term persistence of polar bears (Amstrup et al. 2008; Atwood et al. 2016a) and, in many respects, the ultimate driver of contemporary and future human–polar bear interactions. The Arctic region is experiencing a warming trend that is driving pronounced changes in sea ice extent and structure. Since 1979, sea ice extent and volume during summer have declined at rates of approximately 14 and 28%/decade (Comiso 2012), respectively, with the greatest changes occurring most recently. Given the current trends in global greenhouse gas emissions (GHG) (IPCC 2014), and the lag times associated with global climate processes attaining equilibrium, Arctic warming will most likely continue for several decades even if aggressive mitigation were to occur immediately (Allen and Stocker 2013). Consequently, climate-induced loss of sea ice habitat will likely continue for several decades or longer if global GHG emissions are not reduced.

The primary way in which loss of sea ice habitat is believed to drive human–polar bear interactions is via the displacement of bears from preferred foraging habitat (i.e., sea ice over biologically productive shallow water; Durner et al. 2009). Sea ice is highly dynamic, with extent and volume varying seasonally as the eventual breakup of annual ice gives way to a minimum extent in mid-September, followed by the reformation of ice in fall (Barry et al. 1993). However, Arctic warming has caused the annual period of reduced ice availability (i.e., the open-water period: the time between ice breakup and freeze-up) to lengthen at a rate of nearly 13 days/decade since 1979, with a modest trend toward an earlier mean date of breakup and a pronounced trend toward later freeze-up (Stroeve et al. 2014). For subpopulations of the SIE, already accustomed to spending time onshore, the increasing duration of the annual open-water period has led to longer onshore residency and corresponding declines in body condition due to nutritional restriction (e.g., western Hudson Bay [WHB]; Regehr et al. 2007). In the PBDE, where sea ice historically was present year-round, the emergent and lengthening open-water period has caused some bears to displace from sea ice to land, with the occurrence and frequency of protracted use of terrestrial habitat increasing over time (Rode et al. 2015; Atwood et al. 2016b).

The diminishing Arctic sea ice is also driving changes in human behavior, most notably an escalation in development activities with the intent of providing new economic opportunities for Arctic countries but also increasing activities associated with recreation and research. Ocean resources become more readily available as the open-water period lengthens, and the Arctic is estimated to contain substantial reserves of extractable resources, including various rare earth elements. In Alaska, regulatory agencies have issued permits for low-density offshore oil and gas drilling over the continental shelf region of the southern Beaufort (SB) and Chukchi (CS) seas, and onshore exploration and drilling continues in the Prudhoe Bay and Kuparuk oil fields (National Petroleum Council 2015). Collectively, the increasing industrial and transport activities associated with resource extraction and changes to polar bear behavior have the potential to place humans and polar bears in closer proximity to each other for longer periods of time and escalate the likelihood of interaction and conflict.

22.4 Human–Polar Bear Interactions: Habituation and Conflict

Interactions between humans and polar bears occurred long before the establishment of present-day communities. The earliest recorded interactions were associated with sixteenth-century European exploration of the Arctic and early forms of commercial hunting (Lønø 1970). In the 1950s and 1960s, sport hunting reached a peak as commercially guided hunters began using airplanes to gain greater access to remote areas of the Arctic, and by the late 1960s, overhunting was considered the most significant threat to polar bear populations (Larsen 1975). Commercial and subsistence hunting occurred concurrently up to 1973 when the *Agreement on the Conservation of Polar Bears* was signed by the five polar bear range states (Canada, Denmark on behalf of Greenland, Norway, the then Soviet Union, and the USA) to, in part, reduce overharvest (e.g., DeMaster and Stirling 1983). Since implementation of the 1973 agreement, Norway and the Soviet Union, and then Russia, have not allowed the hunting of polar bears, while the USA and Greenland have only allowed hunting by indigenous peoples. Canada allows indigenous peoples to hunt polar bears or sell unused permits to sport hunters in some regions. Current estimates put the range-wide harvest at 3–4% of the total population of 20,000–25,000 bears, and harvest is believed to not represent a serious threat to long-term persistence (Atwood et al. 2016a).

Most interactions and conflict between humans and polar bears have occurred around settlements where bears were perceived to threaten life or property or were provoked by people (Stirling et al. 1977). The occurrence of animals near centers of anthropogenic activity is often attributed to habituation—animals have lost their natural wariness and grown tolerant of human presence (e.g., Immelman and Beer 1989). The prevailing thought is that habituation occurs when the benefits of not responding to a stimulus outweigh the perceived costs of response (Alcock 1988; Whittaker and Knight 1998). When polar bears are on land, they have little access to their marine mammal prey and spend the bulk of that time in a fasting state

(Ramsay and Hobson 1991; Rode et al. 2015). As with other ursids, polar bears on land are often attracted to novel scents and areas of human activities, including refuse dumps where they can gain access to human-related foods (Lunn and Stirling 1985), and may become conditioned to associate people or settlements with food rewards (e.g., Elfström et al. 2014). Habituation to humans and food conditioning, likely, play a prominent role in mediating the occurrence of bears around settlements and the resulting effects on bears and people.

Perhaps the best documented case history of past and contemporary human–polar bear interaction and conflict is from Churchill, on the western coast of Hudson Bay (within the boundaries of the WHB subpopulation and the SIE). Polar bears from this subpopulation move onto land when sea ice melts and spend the open-water period living off fat reserves accumulated prior to coming ashore (Derocher et al. 1993). Historically, a small proportion of the subpopulation routinely visited refuse dumps in the Churchill area to take advantage of the only reliable food resource available (Lunn and Stirling 1985). In 1969, a growing concern over the localization of bears around Churchill, and the consequent escalation of human-bear conflicts, prompted the closure of two dumps and the implementation of a program (i.e., Polar Bear Control Program, which became Polar Bear Alert Program after 1984), initiated to ensure public safety and reduce property damage (Kearney 1989). While this program has been successful in reducing threats to people and damage to property, human–polar bear interactions in the Churchill area have increased over time. From 1979 to 2013, the length of the open-water period in WHB increased at a rate of approximately 10 days/decade (Stroeve et al. 2014). During that time, the length of stay onshore by bears has increased commensurately, while the number of human–polar bear interactions nearly quadrupled (Towns et al. 2009)—despite a trend of declining population abundance (Regehr et al. 2007). In 2005, the last open refuse dump in the Churchill area was closed, but interaction and conflict continued to be a concern (Towns et al. 2009). In 2015, an access-controlled landfill was opened that should be effective in excluding bears. However, human–polar bear interactions are likely to increase in the future given the strong relationship between the timing of sea ice breakup and the nutritional condition of bears when they come ashore (Derocher et al. 2004).

Other forms of provisioned resources can habituate bears to human presence and foster potential for interaction and conflict. For example, there are three communities on Alaska's North Slope (Barrow, Nuiqsut, and Kaktovik) that harvest bowhead whales (*Balaena mysticetus*) from the SB in the fall. Unwanted remains from the harvest are aggregated at "bone piles" near Barrow, on Cross Island, and adjacent to Kaktovik, (Fig. 22.2), where they have served as focal attractors for polar bears since the late 1990s (Schliebe et al. 2008; Atwood et al. 2016b). The declining availability of sea ice over the SB's continental shelf has led to a marked increase in the proportion of the population coming ashore in summer and fall, with the greatest aggregation of bears occurring at the Kaktovik bone pile (Atwood et al. 2016b). Interactions between humans and bears are common during the process of butchering whales, which occurs onshore and often involves the diversionary feeding of bears

Fig. 22.2 A solitary adult (**a**) and family group (**b**) feeding at the bowhead whale bone pile located adjacent to Kaktovik, Alaska, September 2015. Unused remains from subsistence-harvested bowhead whales are occasionally aggregated at Point Barrow, and consistently at Cross Island (near Prudhoe Bay), and adjacent to Kaktovik on Barter Island following the cessation of the fall whaling season. The bone piles attract large numbers of bears that typically remain in the general area until the food is depleted. *Image credit: S.W. Breck*

to prevent them from approaching carcasses (T. Atwood, personal observation; Fig. 22.3). Interactions also occur at nearby camps used for subsistence activities (G. York, unpublished data). Unfortunately, documentation of human–polar bear interaction and conflict from this region is limited, but “defense of life” kills do

Fig. 22.3 An example of diversionary feeding of polar bears during the butchering of a subsistence-harvested bowhead whale in Alaska. Harvested whales are butchered onshore in fall, attracting polar bears that stay on the nearby barrier islands waiting for bone piles to be stocked with unused remains. Whalers often drag blocks of blubber away from the butchering site to feed bears and keep them at a safe distance. *Image credit: S.W. Breck*

occasionally occur when the bone piles become depleted and bears venture into town in search of food. As with WHB, the lengthening ice-free season is expected to result in bears from the SB spending more time onshore and becoming increasingly reliant on bowhead whale remains to ameliorate declines in body condition.

The examples from Churchill and Kaktovik highlight that polar bears motivated to obtain food appear more willing to risk interacting with humans. Fortunately, human–polar bear interactions rarely result in injury to humans, with only 73 documented attacks that resulted in 20 human fatalities and 63 injuries since 1870 (J. Wilder, unpublished data), numbers that are small in comparison to those of black bears (*Ursus americanus*), e.g., 63 fatalities from 1900 to 2009 (Herrero et al. 2011). A more frequent outcome of human–polar bear interactions is fatal injury to bears. For example, in Nunavut, Canada, 618 polar bears were killed in "defense of life or property" (a formal term that describes the lethal removal of an animal thought to pose an immediate threat to people or property) from 1970 to 2000, most of which (74%) occurred during the ice-free season at temporary camps established for subsistence activities (Dyck 2006). Similarly, Stenhouse et al. (1988) found that of the 265 polar bears killed as a result of conflict in the Northwest Territories, the majority (63%) were associated with activities on land and occurred during the open-water period. A common thread to interaction, conflict, and lethal removal of polar bears is the presence of attractants (most commonly human food waste) that bring land-bound bears into close proximity with humans. Projections of future sea ice conditions indicate that the ice-free season will continue to lengthen, increasing the amount of time bears must spend onshore and increasing the likelihood of nutritional restriction (Rode et al. 2015; Atwood et al. 2016b). Mitigating the resulting increased potential for human–polar bear conflict will require the proactive management of food attractants and human behaviors (e.g., supplemental feeding) that contribute to food conditioning and habituation.

22.5 Potential Threats of Increasing and Emerging Human Activities on Polar Bears

The continued decline of summer sea ice will allow greater human access to the Arctic Ocean, facilitating an increase in intensity and types of human activities. In 2013, the USA released the National Strategy for the Arctic Region, a document that acknowledges that the loss of summer sea ice provides an opportunity for increased economic development of the region (United States Government 2013), a notion recognized by other Arctic nations as well. Below, we identify activities that are likely to increase as the open-water period lengthens and discuss the implications of an increasingly ice-free Arctic for human–polar bear interactions.

Resource extraction—The Arctic is believed to contain about 30% of the world's undiscovered gas and 13% of the world's undiscovered oil supplies (Gautier et al. 2009), and the continued decline of summer sea ice will increase the economic viability of offshore resource extraction over time. Perhaps the greatest threat to polar bears from oil and gas development is the risk of exposure to oil spills and the direct effects that follow. Amstrup et al. (2006) modeled the effects of a hypothetical oil spill on polar bears in the SB and found that a modest spill of approximately 6000 barrels could directly impact a significant number of bears ($n = 74$; 8% of the estimated abundance of the SB subpopulation; Bromaghin et al. 2015). A large spill, like that from the Exxon Valdez in 1989 (over 250,000 barrels), would likely be catastrophic and potentially affect multiple subpopulations both directly and indirectly. Exposure to oil reduces the ability of bears to thermoregulate (Hurst and Øritsland 1982; Hurst et al. 1991), which could lead to reduced overwinter survival. Polar bears have also been observed consuming petroleum products (Derocher and Stirling 1991) or ingesting them through grooming (Øritsland et al. 1981) and have ingested oil when consuming contaminated prey (Stirling 1990). All routes of ingestion are a cause for concern given that even minimal consumption of oil can be lethal (St. Aubin 1990). In addition to oil, polar bears have consumed industrial chemicals that have resulted in mortality (Amstrup et al. 1989).

Exposure to infrastructure and activities associated with resource extraction (Fig. 22.4) are likely to indirectly affect polar bears by altering their behavior. For example, noise associated with extraction could cause polar bears and their prey to avoid areas adjacent to these activities (e.g., Geraci and Smith 1976; Kelly et al. 1988). Likewise, oil and gas development onshore could potentially affect polar bears by triggering the abandonment of maternal dens (Amstrup 1993) and reducing

Fig. 22.4 A polar bear family group resting near industrial infrastructure, Prudhoe Bay, Alaska, 2009. During summer and fall, polar bears are frequently observed along the coast of the bay. The bowhead whale bone pile on Cross Island is approximately 20 km north of the coast and attracts large numbers of bears when food is present. *Image credit: U.S. Geological Survey*

access to important terrestrial refugia and traveling habitat. By contrast, it is possible that bears could be attracted to onshore and offshore facilities. For instance, offshore facilities can alter ice conditions, creating semipermanent leads that would otherwise not exist (Stirling 1988) and attract seals at densities similar to that found away from development (Moulton et al. 2005). The presence of human foods could also attract bears to industrial facilities (Stenhouse et al. 1988). Given the limited information available to date, it is unclear if disturbances to polar bears and their prey would be short lived or lead to permanent changes in their behavior.

Resource extraction activities are believed to pose little threat to the long-term persistence of polar bear populations (Atwood et al. 2016a). However, as the industrial footprint expands, the likelihood of polar bears experiencing direct and indirect effects of associated activities will grow. Mitigating the effects of oil spills will prove challenging. The annual freeze-thaw cycle of sea ice will sequester and rerelease oil over multiple years. Oil also precipitates to the ocean floor where it can negatively affect the benthic ecosystem and lower levels of the food chain (Chen and Denison 2011) that polar bear prey species rely on. Additionally, the constant movement of sea ice may function to transport sequestered oil to regions of the Arctic beyond where spills originated. Managing the future risk to polar bears of exposure to resource extraction activities will require coordinated planning efforts between industry and natural resource managers across international boundaries.

Trans-Arctic and local shipping—As sea ice extent declines spatially and temporally, and industrial and recreational activities increase, it is natural that shipping through, between, and within Arctic regions will increase to improve efficiencies and meet local logistical demands. The best information available to assess the potential for increased shipping activity relies on climate model projections to determine potential navigability and season duration. For example, some predictions suggest that by the middle of the twenty-first century, changing sea ice conditions will enable expanded September navigability for common open-water ships crossing the Arctic along the Northern Sea Route (NSR), robust new routes for ice-strengthened ships, and new routes through the Northwest Passage (NP) (e.g., Smith and Stephenson 2013). Long-distance shipping routes that make use of the NSR or NP, and thus bypass the Suez Canal or Panama, could reduce transit distance by nearly 33%, providing a clear economic incentive for Arctic travel (Liu and Kronbak 2010). On a local scale, a survey of ship owners identified the lengthening ice-free season as likely to drive the emergence of a niche market for shipping services (supplanting more costly aerial services) that deliver goods to communities and industrial operations (Lasserre and Pelletier 2011). The need for local shipping services may be exacerbated by declines in accessibility of land-based supply routes if the utility of winter roads are impacted by milder winters (Stephenson et al. 2011).

Increased shipping, whether long-distance or local, could lead to direct disturbances of polar bears as well as to increased exposures to oil spills, novel pollutants, organisms, and pathogens. Direct disturbance is most likely to occur when ships bisect the path of swimming bears. Pagano et al. (2012) and Pilfold et al. (2016) reported increases in the frequency of long-distance swims (i.e., >50 km) by polar bears in the SB and WHB when extensive areas of open water were present. Most of the swims identified by Pagano et al. (2012) involved bears moving from the receding pack ice to land or from increasingly fragmented ice present in nearshore regions to pack ice, but bears also frequently swim short distances in the nearshore region of the SB as they travel between barrier islands (US Geological Survey, unpublished data). Additionally, polar bears have been observed swimming after both marine and terrestrial prey in open water in the Barents Sea (BS) (Stempniewicz et al. 2013; Stirling and van Meurs 2015). Collectively, these observations suggest the potential for shipping activity to interfere with seasonal migratory movement, routine travel, and foraging behavior.

A potential indirect effect of increased shipping activity is the unintentional discharge of pollutants and transport of invasive or pathogenic organisms and the cascading effects on bear and community health. Exposure of polar bears to contaminants/pollutants may occur from discharged ballast, bilge, or wastewater (Wilhelmsson et al. 2013). Additionally, vessels may vector new organisms to the Arctic that may, over time, alter food web dynamics or cause disease. For instance, estimates suggest that, at any one time, 7,000–10,000 different species are being carried in ships' ballast tanks around the world (Global Ballast Water Programme 2016). Given that polar bears appear to have a relatively naïve immune system (Weber et al. 2013), there is concern that they may be susceptible to the introduction of novel pathogens (Patyk et al. 2015). Such indirect effects arising from the discharge and transport by shipping vessels are likely to be clumped in space and time, with "hub" locations like those located near large communities and industrial developments likely to be disproportionately affected (Floerl et al. 2009). The spatiotemporal clustering of those effects provides an opportunity to put in place surveillance programs to monitor the frequency and nature of discharges and the occurrence of invasive and pathogenic organisms. While shipping activity is likely to increase over the coming decades, there will likely be considerable variability in the distribution and intensity of that activity based on shipping route availability. As a result, serious population-level consequences for polar bears are not expected (Atwood et al. 2016a).

Tourism and recreation—Declines in sea ice are also expected to result in increased access and opportunities for Arctic tourism and recreation. We define tourism and recreation as including activities centered on polar bears, such as commercial viewing, as well as recreational travel throughout polar bear habitat. It is difficult to estimate the intensity of recreational activities in the Arctic as there are few processes in place to capture that information. Yet in 2012, nature-related activities in the Arctic are believed to have generated $3.2 billion in Alaska and $40.4 billion in Canada (US Department of the Interior 2012), suggesting the opportunity exists to generate substantial revenues. In Churchill, commercial polar bear viewing is an important driver of the annual economy (Lemelin 2008). In Kaktovik, the opportunity to view polar bears visiting the bowhead whale bone pile has spurred the development of a nascent ecotourism industry (Fig. 22.5). Tour companies may in the future put pressure on governments to preserve their valued asset, the polar bears, and thus create a commercial value on the bears that could help to secure their

Fig. 22.5 Commercial polar bear viewing in Kaktovik, Alaska, September 2015. Bear viewing is done from either boats or vehicles during the fall, after the bowhead whale bone pile has been stocked with remains. *Image credit: U.S. Geological Survey*

future. While it is possible that increased tourism could lead to direct disturbances of polar bears—for example, causing heightened vigilance (Dyck et al. 2004)—it is believed that a relatively small proportion of polar bear subpopulations are regularly interacting with humans due to tourism and recreational activities. As a result, these activities are a less serious current and future threat than those described above.

Exposure to marine debris and fisheries—Compared to other marine species, there is little information on the threat posed to polar bears from exposure to marine debris. For example, while macrodebris ingestion and entanglement have been well documented in sea birds, turtles, and dolphins (Robards et al. 1995; Tomas et al. 2002; Denuncio et al. 2011), there is no documentation of ingestion by polar bears. Similarly, while ingestion of microplastics is a growing concern, there is no indication it represents a direct threat to polar bears. Marine debris could become a problem to individual animals if increased human activities result in increased occurrence and exposure to polar bears. But, at the subpopulation level, it is unlikely to pose a future risk.

As mentioned previously, the amount of commercial ship traffic in the Arctic is expected to increase as the duration of the open-water period lengthens, and it is possible that some of that traffic may include fishing vessels (Arctic Council 2009). There are unpublished reports of polar bears investigating ships (and videos can be found on the internet), and presumably they would be more likely to investigate fishing vessels due to the attraction of food odors. Additionally, there are two separate incidents from Norway (Spitsbergen) of land-based polar bears caught in lost commercial fishing nets (Spitsbergen News 2014). However, the extent to which the Arctic Ocean is capable of supporting a commercial fishing industry, and the risk it may pose to polar bears, is difficult to judge. While northward shifts in fish distribution in response to increasing ocean temperatures and reductions in sea ice have been recorded (Berge et al. 2015; Fossheim et al. 2015; Grebmeier et al. 2010), researchers are finding that a complex array of ecological factors need to be understood in order to predict successful range expansion by fish stocks into Arctic waters. For example, some of the increases in phytoplankton primary production that have accompanied reductions in sea ice (Arrigo and van Dijken 2015) may eventually be limited by nutrient availability as stratification due to fresh water inputs reduces nutrient mixing in the water column (Slagstad et al. 2015; Katlein et al. 2015). Limits on subarctic fish species range expansions, uncertainty over future levels of primary productivity, and the challenge of conducting fishing operations in a harsh remote environment all point to a low likelihood of a commercial fishing industry developing in Arctic waters in the near term (NOAA 2015; Berge et al. 2015; Slagstad et al. 2015). A more pressing concern may be the risk of human-bear interactions around subsistence fishing camps. For example, the polar bears congregating around Kaktovik, Alaska, to feed on bowhead whale remains have caused some community residents to reduce the use of fish camps due to safety concerns. Indeed, the number one priority identified at the recent Kaktovik Leadership meeting on human–polar bear conflicts was to restore the community's access to traditional fishing grounds (North Slope Borough 2015). With an increasing number of bears spending long periods of time on land during the open-water period, the potential for conflict associated with subsistence activities may increase.

Anthropogenic contaminants—Long-range atmospheric transport of little used anthropogenic contaminants (e.g., persistent organic pollutants [POP], polychlorinated biphenyls [PCB], organochlorine pesticides [OCP], polybrominated diphenyl ether [PBDE]) has concentrated those contaminants in the Arctic marine ecosystem (Muir et al. 1999). Accordingly, exposure to legacy contaminants remains a concern for wildlife and human health in the Arctic, where high lipid levels in the marine food web result in higher than expected concentrations in upper trophic level organisms (Borgå et al. 2001). Systematic studies of adverse health effects of pollution on polar bears began in earnest in the late 1990s via the implementation of circumpolar and national multi-species monitoring programs (e.g., Arctic Monitoring and Assessment Programme [AMAP]; AMAP 1998). Monitoring efforts have revealed a number of adverse effects on polar bear health resulting from exposure to contaminants, including impairment of liver and kidney function, as well as neurological alterations and decreased immune and endocrine system function (Sonne 2010; Bechshøft et al. 2011). Importantly, patterns of exposure have varied geographically, with the highest tissue concentrations occurring in polar bears from the Hudson Bay complex (McKinney et al. 2009), east Greenland (Sonne et al. 2012), Svalbard, and western Russia (McKinney et al. 2011), although exposure to many legacy contaminants has declined over time (Dietz et al. 2004; Verreault et al. 2005). Interestingly, exposure to new contaminants like perfluoroalkyls (PFC; Dietz et al. 2008) has been reported, sparking concern that environmental changes may be altering some contaminant pathways (Macdonald et al. 2005).

Climate change has already affected the distribution of select species (Hickling et al. 2006) and is expected to spur further changes in species distributions and migratory behavior that will expose Arctic endemics to new contaminant pathways (Muir and de Wit 2010). For example, climate-mediated changes to sea ice phenology (i.e., timing of annual sea ice formation and breakup) in WHB have led to a concurrent decline in polar bear consumption of ice-associated ringed and bearded seals, increased consumption of open-water-associated harbor (*Phoca vitulina*) and harp (*Pagophilus groenlandicus*) seals, and increased tissue concentrations of several chlorinated and brominated contaminants (McKinney et al. 2009). In addition to those contaminants already mentioned, there is concern over future exposure to heavy metals, such as mercury. Despite declining anthropogenic emissions, the Arctic ecosystem appears to be a sink for the accumulation of mercury for reasons that are not at all clear (Macdonald et al. 2005). One proposed mechanism by which mercury may be increasing in the Arctic is through a "reservoir effect," whereby methyl mercury is released from inventories accumulated in organic carbon phases of frozen ground as it thaws (Loseto et al. 2004). Similarly, it may be prescient to monitor exposure to artificial radionuclides (often released by the nuclear industry), given limited information available on potential effects, and concern that risk of exposure may be increasing (Lokas et al. 2014). Continued monitoring of polar bear exposure to anthropogenic contaminants (both legacy and new) remains a priority, particularly work that assesses the potential for synergistic effects between contaminants and environmental stressors (Patyk et al. 2015).

22.6 Conclusions

The Arctic sea ice ecosystem is changing rapidly and dramatically which, over time, will result in unprecedented levels of human activity (e.g., resource extraction, pollution, shipping, tourism, and recreation). Collectively, these changes pose risks to polar bears, though the type, severity, and consequence of the risk will differ geographically, as do processes that regulate sea ice and polar bear population dynamics (Amstrup et al. 2008; Rode et al. 2014). The behavioral plasticity that polar bears possess will determine how they respond to these myriad changes. But rapid behavioral adaptation is a tall order for a specialist predator with a long generation time. Currently, unabated rise in atmospheric GHG concentrations, and the resulting loss of sea ice habitat, is the primary threat to the long-term persistence of polar bears (Amstrup et al. 2008). Until further sea ice loss is stopped, management of other stressors, including those associated with human activities, may serve to slow the transition of populations to progressively worsened outcomes and improve the prospects for their long-term persistence (Atwood et al. 2016a). To that end, coexistence between humans and polar bears will require mitigation strategies that reduce the burden to local communities, as well as imposing mechanisms to manage further development. Ultimately, mitigating the effects of humans on polar bears will require an interdisciplinary approach that fosters circumpolar collaboration between affected communities, organizations, and regulatory agencies.

The myriad threats facing Arctic ecosystems also present a unique conservation opportunity. While ongoing and future habitat loss represents a clear and present threat to the long-term persistence of polar bears, they are still distributed across much of their historical range. This provides a unique opportunity to buoy and support populations facing likely decline by effectively managing factors such as conflict, habitat use (especially denning), and other forms of human disturbance or removal.

Mitigating human–polar bear conflict is also of mutual interest across communities, managers, researchers, and other stakeholders. Increasing community safety, enhancing social carrying capacity or tolerance for polar bears, and decreasing the need for unplanned and unwanted removals could act to address shared concerns from local to international scales. We have an opportunity to proactively address these issues before they become a problem in most parts of the Arctic. At various human-bear conflict meetings in Canada, the USA, and Norway, front line managers, community residents, and others working in polar bear habitat have expressed growing concern that polar bear encounters on land may be increasing and, given loss of habitat, will likely increase in the future.

These concerns have led to a variety of responses including use of local polar bear patrols or guards, regional approaches using existing government wildlife staff (conservation officers in Canada), national coordination of deterrence efforts by the US Fish and Wildlife Service in Alaska, and, internationally, the formation of the Conflict Working Group of the Polar Bear Range States. Efforts to date include improved international collation of human–polar bear interaction data, improved communication and coordination at all levels to share information, and work toward consistent messaging and training. Discussions have also highlighted the need for focal research on conflict issues to improve understanding of regional dynamics and to better quantify the effectiveness of deterrence tools.

References

Alcock J (1988) Animal behavior: an evolutionary approach, 4th edn. Sinauer Associates, Sunderland

Allen MR, Stocker TF (2013) Impact of delay in reducing carbon dioxide emissions. Nat Clim Chang 4:23–26. doi:10.1038/NCLIMATE2077

AMAP (1998) Arctic monitoring and assessment Programme: AMAP assessment report—Arctic pollution issues. AMAP, Oslo. http://www.amap.no

Amstrup SC (1993) Human disturbances of denning polar bears in Alaska. Arctic 46:246–250

Amstrup SC, Gardner C, Meyers KC et al (1989) Ethylene glycol (antifreeze) poisoning of a free-ranging polar bear. Vet Hum Toxicol 31:317–319

Amstrup SC (2003) Polar bear, Ursus maritimus. In: Feldhamer GA, Thompson BC, Chapman JA (eds) Wild mammals of North America: biology, management, and conservation. Johns Hopkins University Press, Baltimore, MD, pp 587–610

Amstrup SC, GM Durner, McDonald TL et al. (2006) Estimating potential effects of hypothetical oil spills on polar bears. U.S. Geological Survey, Alaska Science Center, Anchorage, AK

Amstrup SC, Marcot BG, Douglas DC (2008) A Bayesian network modeling approach to forecasting the 21st century worldwide status of polar bears. In: De Weaver ET, Bitz CM, Tremblay LB (eds) Arctic Sea ice decline: observations, projections, mechanisms, and implications. American Geophysical Union Geophysical Monograph No. 180, Washington, DC, pp 213–268

Arctic Council (2009) Arctic marine shipping assessment report. http://library.arcticportal.org/id/eprint/1400

Arrigo KR, van Dijken GL (2015) Continued increases in Arctic Ocean primary production. Prog Oceanogr 136:60–70

Atwood TC, Marcot BG, Douglas DC et al (2016a) Forecasting the relative influence of environmental and anthropogenic stressors on polar bears. Ecosphere 7(6):e01370. doi:10.1002/ecs2.1370

Atwood TC, Peacock E, McKinney M et al (2016b) Rapid environmental change drives increased land use by an Arctic marine predator. PLoS One. doi:10.1371/journal.pone.0155932

Barry RG, Serreze MC, Maslanik JA et al (1993) The Arctic Sea ice-climate system: observations and modeling. Rev Geophys 31:397–422

Bechshøft TØ, Sonne C, Dietz R et al (2011) Cortisol levels in hair of East Greenland polar bears. Sci Total Environ 409:831–834

Berge J, Heggland K, Lønne OJ et al (2015) First records of Atlantic mackerel (*Scomber scombrus*) from the Svalbard archipelago, Norway, with possible explanations for the extension of its distribution. Arctic 68:54–61

Borgå K, Gabrielsen GW, Skaare JU (2001) Biomagnification of organochlorines along a Barents Sea food chain. Environ Pollut 113:187–198

Bromaghin J, McDonald T, Stirling I et al (2015) Polar bears in the Beaufort Sea: population decline and stabilization in the 2000's. Ecol Appl 25:634–651

Chen J, Denison MS (2011) The Deepwater horizon oil spill: environmental fate of the oil and the toxicological effects on marine organisms. J Young Investig 6:84–95

Comiso JC (2012) Large decadal decline of the Arctic multiyear ice cover. J Climate 25:1176–1193

Denuncio P, Bastida R, Dassis M et al (2011) Plastic ingestion in Franciscana dolphins, *Pontoporia blainvillei* (Gervais and d'Orbigny, 1844), from Argentina. Mar Pollut Bull 62:1836–1841

DeMaster DP, Stirling I (1983) The estimation of survival and litter size of polar bear cubs. Int Conf Bear Biol Manag 5:260–263

Department of Agriculture Forestry and Fisheries (2010) Abstract of agricultural statistics. Department of Agriculture, Forestry and Fisheries, Pretoria

Derocher A, Stirling I (1991) Oil contamination of polar bears. Polar Rec 27:56–57

Derocher AE, Andriashek D, Stirling I (1993) Terrestrial foraging by polar bears during the ice-free period in western Hudson Bay. Arctic 52:251–254

Derocher AE, Lunn NJ, Stirling I (2004) Polar bears in a warming climate. Integr Comp Biol 44:163–176

Dickman AJ (2010) Complexities of conflict: the importance of considering social factors for effectively resolving human-wildlife conflict. Anim Conserv 13:458–466

Dietz R, Riget FF, Sonne C et al (2004) Seasonal and temporal trends in polychlorinated biphenyls and organochlorine pesticides in East Greenland polar bears (*Ursus maritimus*), 1990–2001. Sci Total Environ 331:107–124

Dietz R, Bossi R, Riget FF et al (2008) Increasing perfluoroalkyl contaminants in East Greenland polar bears (*Ursus maritimus*): a new toxic threat to the Arctic bears. Environ Sci Technol 42:2701–2707

Doney SC (2010) The growing human footprint on coastal and open-ocean biogeochemistry. Science 328:1512–1516

Durner GM, Douglas DC, Nielson RM et al (2009) Predicting 21st-century polar bear habitat distribution from global climate models. Ecol Monogr 79:25–58

Dyck MG, Baydack RK (2004) Vigilance behaviour of polar bears (Ursus maritimus) in the context of wildlifeviewing activities at Churchill, Manitoba, Canada. Biol Conserv 116:343–350

Dyck MG (2006) Characteristics of polar bears killed in defense of life and property in Nunavut, Canada, 1970–2000. Ursus 17:52–62

Elfström M, Zedrosser A, Stoen O-G et al (2014) Ultimate and proximate mechanisms underlying the occurrence of bears close to human settlements: review and management implications. Mammal Rev 44:5–18

Floerl O, Inglis GJ, Dey K et al (2009) The importance of transport hubs in stepping-stone invasions. J Appl Ecol 46:37–45

Fossheim M, Primicerio R, Johannesen E et al (2015) Recent warming leads to a rapid borealization of fish communities in the Arctic. Nat Clim Chang 5:673–677

Friedmann Y, Daly D (eds.) (2004) Red data book of the mammals of South Africa: a conservation assessment. CBSG Southern Africa, Conservation Breeding Specialist Group (SSC/IUCN) and The Endangered Wildlife Trust, Johannesburg, South Africa

Gautier DL, Bird KJ, Charpentier RR et al (2009) Assessment of undiscovered oil and gas in the Arctic. Science 324:1175–1179

Geraci JR, Smith TG (1976) Direct and indirect effects of oil on ringed seals (*Phoca hispida*) of the Beaufort Sea. J Fish Res Board Can 33:1976–1984

Global Ballast Water Management Programme (2016) The problem. http://globallast.imo.org/ballast-water-as-a-vector.htm. Accessed 10 Feb 2016

Grebmeier JM, Moore SE, Overland JE et al (2010) Biological response to recent Pacific Arctic Sea ice retreats. EOS Trans Am Geophys Union 91:161–162

Halpern BS, Walbridge S, Selkoe KA et al (2008) A global map of human impact on marine ecosystems. Science 319:948–952

Harrington FH, Veitch AM (1992) Calving success of woodland caribou exposed to low-level jet fighter overflights. Arctic 45:213–218

Herrero S, Higgins A, Cardoza JE et al (2011) Fatal attacks by American black bear on people: 1900–2009. J Wildl Manag 75:596–603

Hickling R, Roy DB, Hill JK et al (2006) The distributions of a wide range of taxonomic groups are expanding polewards. Glob Chang Biol 12:450–455

Hurst RJ, Øritsland NA (1982) Polar bear thermoregulation: effects of oil on the insulative properties of fur. J Therm Biol 7:201–208
Hurst RJ, Watts PD, Øritsland NA (1991) Metabolic compensation in oil-exposed polar bears. J Therm Biol 16:53–56
Immelman K, Beer CA (1989) A dictionary of ethology. Harvard University Press, Cambridge, MA
Intergovernmental Panel on Climate Change (IPCC) (2014) Climate change: impacts, adaptation, vulnerability. Part A: global and sectoral aspects. In: Contribution of the Working Group II to the Fifth Assessment Report of the Intergovernmental Panel on Climate Change. Cambridge University Press, Cambridge, UK
Katlein C, Fernández-Méndez M, Wenzhöfer F et al (2015) Distribution of algal aggregates under summer sea ice in the Central Arctic. Polar Biol 38:719–731
Kearney SR (1989) The Polar Bear Alert Program at Churchill, Manitoba. In: Bromely M (ed) Bear–people conflict: proceedings of a symposium on management strategies. Northwest Territories Department of Renewable Resources, Yellowknife, pp 83–92
Kelly BP, Burns JJ, Quakenbush LT (1988) Responses of ringed seals (*Phoca hispida*) to noise disturbance. In: Sackinger WM, Jeffries MO, Imm JL, Tracey SD (eds) Port and ocean engineering under Arctic conditions, Symposium on noise and marine mammals, vol 2. Geophysical Institute, University of Alaska, Fairbanks, pp 27–38
Lamarque F, Anderson J, Fergusson R et al. (2009) Human-wildlife conflict in Africa: causes, consequences and management strategies (No. 157). Food and Agriculture Organization of the United Nations (FAO), Rome
Larsen T (1975) Progress in polar bear research and conservation in the Arctic nations. B C Envtl Aff L Rev 4:295–308
Lasserre F, Pelletier S (2011) Polar super seaways? Maritimr transport in the Arctic: an analysis of shipowners intentions. J Transp Geogr 19:1465–1473
Lemelin RH (2008) Human-polar bear interactions in Churchill, Manitoba: the socio-ecological perspective. In: Higham J, Lück M (eds) Marine wildlife and tourism management: insights from the natural and social sciences. CAB International, Cambridge, MA, pp 91–108
Lewis DL, Breck SW, Wilson KR, Webb CT (2014) Modeling black bear population dynamics in a humandominated stochastic environment. Ecol Model 294:1–58
Liu M, Kronbak J (2010) The potential economic viability of using the Northern Sea route (NSR) as an alternative route between Asia and Europe. J Transp Geogr 18:434–444
Lokas E, Bartminski P, Wachniew P et al (2014) Sources and pathways of artificial radionuclides to soils as a high Arctic site. Environ Sci Pollut Res Int 21:12479–12493
Lønø O (1970) The polar bear (*Ursus maritimus Phipps*) in the Svalbard area. Nor Polarinst Skr 149:103
Loseto LL, Siciliano SD, Lean DR (2004) Methylmercury production in high Arctic wetlands. Environ Toxicol Chem 23:17–23
Lunn NJ, Stirling I (1985) The significance of supplemental food to polar bears during the ice-free period of Hudson Bay. Can J Zool 63:2291–2297
Macdonald RW, Harner T, Fyfe J (2005) Recent climate change in the Arctic and its impact on contaminant pathways and interpretation of temporal trend data. Sci Total Environ 342:5–86
Madhusudan MD, Sankaran P (2010) Seeing the elephant in the room: human-elephant conflict and the ETF report. Econ Pol Wkly 45:29–31
Magnani F, Mencuccini M, Borghetti M et al (2007) The human footprint in the carbon cycle of temperate and boreal forests. Nature 447:849–851
McKinney MA, Peacock E, Letcher RJ (2009) Sea ice-associated diet change increases the levels of chlorinated and brominated contaminants in polar bears. Environ Sci Technol 43:4334–4339
McKinney MA, Letcher RJ, Aars J et al (2011) Regional contamination versus regional dietary differences: understanding geographic variation in brominated and chlorinated contaminant levels in polar bears. Environ Sci Technol 45:896–902

Moore MJC, Seigel RA (2006) No place to nest or bask: effects of human disturbance on the nesting and basking habits of yellow-blotched map turtles (*Graptemys flavimaculata*). Biol Conserv 130:386–393

Moulton VD, Richardson WJ, Elliott RE et al (2005) Effects of an offshore oil development on local abundance and distribution of ringed seals (*Phoca hispida*) of the Alaskan Beaufort Sea. Mar Mamm Sci 21:217–242

Muir DC, de Wit CA (2010) Trends of legacy and new persistent organic pollutants in the circumpolar arctic: overview, conclusions, and recommendations. Sci Total Environ 408:3044–3051

Muir D, Braune B, de March B et al (1999) Spatial and temporal trends and effects of contaminants in the Canadian Arctic marine ecosystem: a review. Sci Total Environ 230:83–144

Müllner A, Linsenmair KE, Wikelski M (2004) Exposure to ecotourism reduces survival and affects stress response in hoatzin chicks, *Pisthocomus hoatzin*. Biol Conserv 118:549–558

National Agricultural Statistical Service (NASS) (2002) U.S. Wildlife Damage. U.S. Agricultural Service, Washington, DC

National Oceanic and Atmospheric Administration (NOAA) (2015) In: Third meeting of scientific experts on fish stocks in the Central Arctic Ocean. NOAA, Alaska Fisheries Science Center, Final report, Anchorage, AK, p 20

National Petroleum Council (2015) Arctic potential: realizing the promise of U.S. Arctic oil and gas resources. In: Committee on Arctic Research. National Petroleum Council, Washington DC, p 87. http://npcarcticpotentialreport.org/pdf/AR-Executive_Summary-Final.pdf. Accesssed 22 Dec 2016

North Slope Borough (2015) A report on the Kaktovik leadership meeting on human-polar bear conflict. North Slope Borough, Department of Wildlife Management, p 11

Northrup JM, Wittemyer G (2013) Characterising the impacts of emerging energy development on wildlife, with an eye towards mitigation. Ecol Lett 16:112–125

Øritsland NA, Engelhard FR, Juck FA et al. (1981) Effects of crude oil on polar bears. In: Environmental Studies No. 24. Northern Environmental Protection Branch, Indian and Northern Affairs, Ottawa, ON

Pagano AP, Durner GM, Amstrup SC, Simac KS, York GS (2012) Long-distance swimming by polar bears (*Ursus maritimus*) of the southern Beaufort Sea during years of extensive open water. Can J Zool 90:663–676

Patyk K, Duncan C, Nol P, Sonne C, Laidre K, Obbard M, Wiig Ø, Aars J, Regehr E, Gustafson L, Atwood TC (2015) Establishing a definition of polar bear health to guide research and management activities. Sci Total Environ 514:371–378

Pilfold NW, Derocher AE, Richardson E (2014) Influence of intraspecific competition on the distribution of a wideranging, non-territorial carnivore. Glob Ecol Biogeogr 23:425–435

Pilfold NW, McCall A, Derocher AE, Lunn NJ, Richardson E (2016) Migratory response of polar bears to sea ice loss: to swim or not to swim. Ecography. doi:10.1111/ecog.02109

Ramsay MA, Hobson KA (1991) Polar bears make little use of terrestrial food webs: evidence from stable-carbon isotope analysis. Oecologia 86:98–600

Ramsay MA, Stirling I (1988) Reproductive biology and ecology of female polar bears (*Ursus maritimus*). J Zool 214:601–634

Rangarajan M, Desai A, Sukumar R et al (2010) Securing the future for elephants in India. The report of the elephant task force. Ministry of Environment and Forests, New Delhi

Regehr EV, Lunn NJ, Amstrup SC et al (2007) Effects of earlier sea ice breakup on survival and population size of polar bears in western Hudson Bay. J Wild Manag 71:2673–2683

Robards MD, Piatt JF, Wohl KD (1995) Increasing frequency of plastic particles ingested by seabirds in the subarctic North Pacific. Mar Pollut Bull 30:151–157

Roberts L (2011) 9 billion? Science 333:540–543

Rode KD, Regehr EV, Douglas DC et al (2014) Variation in the response of an Arctic top predator experiencing habitat loss: feeding and reproductive ecology of two polar bear populations. Glob Change Biol 20:76–88

Rode KD, Wilson RR, Regehr EV et al (2015) Increased land use by Chukchi Sea polar bears in relation to changing sea ice conditions. PLoS One 10(11):e0142213. doi:10.1371/journal.pone.0142213

Schliebe S, Rode KD, Gleason JS, Wilder J, Proffitt K, Evans TJ, Miller S (2008) Effects of sea ice extent and food availability on spatial and temporal distribution of polar bears during the fall open-water period in the Southern Beaufort Sea. Polar Biol 31:999–1010

Schwartz CC, Haroldson MA, White GC (2010) Hazards affecting grizzly bear survival in the greater Yellowstone ecosystem. J Wildl Manag 74:654–667

Slagstad D, Wassmann PF, Ellingsen I (2015) Physical constrains and productivity in the future Arctic Ocean. Front Mar Sci 2:85. doi:10.3389/fmars.2015.00085

Smith TG (1980) Polar bear predation of ringed and bearded seals in the land-fast sea ice habitat. Can J Zool 58:2201–2209

Smith LC, Stephenson SR (2013) New trans-Arctic shipping routes navigable by midcentury. Proc Natl Acad Sci 110:4871–4872

Sonne C (2010) Health effects from long-range transported contaminants in Arctic top predators: an integrated review based on studies of polar bears and relevant model species. Environ Int 36:461–491

Sonne C, Letcher RJ, Bechshøft TØ et al (2012) Two decades of biomonitoring polar bear health in Greenland: a review. Acta Vet Scand 54:S15

Spitsbergen News (2014) Polar bear caught in fishing net. http://www.spitsbergen-svalbard.com/2014/08/03/polar-bear-caught-fishing-net.html. Accessed 22 Dec 2016

St. Aubin DJ (1990) Physiologic and toxic effects on polar bears. In: Geraci JR, St. Aubin DJ (eds) Sea mammals and oil: confronting the risks. Academic, San Diego, pp 235–240

Stempniewicz L, Kidawa D, Barcikowski M et al (2013) Unusual hunting and feeding behaviour of polar bears on Spitsbergen. Polar Rec 50:216–219

Stenhouse GB, Lee LJ, Poole KG (1988) Some characteristics of polar bears killed during conflicts with humans in the Northwest Territories, 1976–1986. Arctic 41:275–278

Stephenson SR, Smith LC, Agnew JA (2011) Divergent long-term trajectories of human access to the Arctic. Nat Clim Chang 1:156–160. doi:10.1038/NCLIMATE1120

Stirling I, Archibald WR (1977) Aspects of predation of seals by polar bears. J Fish Board Can 34:1126–1129

Stirling I (1988) Attraction of polar bears *Ursus maritimus* to off-shore drilling sites in the eastern Beaufort Sea. Polar Rec 24:1–8

Stirling I (1990) Polar bears and oil: ecological perspectives. In: Geraci JR, St. Aubin DJ (eds) Sea mammals and oil: confronting the risks. Academic, San Diego, pp 223–234

Stirling I, van Meurs R (2015) Longest recorded underwater dive by a polar bear. Polar Biol 38:1301–1304

Stroeve JC, Markus T, Boisvert L et al (2014) Changes in Arctic melt season and implications for sea ice loss. Geophys Res Lett 41:1216–1225

Thirgood S, Woodroffe R, Rabinowitz A (2005) The impact of human–wildlife conflict on human lives and livelihoods. In: Woodroffe R, Thirgood S, Rabinowitz A (eds) People and wildlife: conflict or coexistence? Cambridge University Press, Cambridge, pp 13–26

Thorn M, Green M, Dalerum F et al (2012) What drives human–carnivore conflict in the North West Province of South Africa? Biol Conserv 150:23–32

Tomas J, Guitart R, Mateo R et al (2002) Marine debris ingestion in loggerhead sea turtles, *Caretta caretta*, from the Western Mediterranean. Mar Pollut Bull 44:211–216

Towns L, Derocher AE, Stirling I et al (2009) Spatial and temporal patterns of problem polar bears in Churchill, Manitoba. Polar Biol 32:1529–1537

Treves A, Karanth KU (2003) Human–carnivore conflict and perspectives on carnivore management worldwide. Conserv Biol 17:1491–1499

United States Department of the Interior, U.S. Fish & Wildlife Service, and U.S. Department of Commerce, Bureau of the Census (2012) 2011 National survey of fishing, hunting and wildlife-associated recreation, Alaska

United States Government (2013) National Strategy for the Arctic Region. https://www.whitehouse.gov/sites/default/files/docs/nat_arctic_strategy.pdf. Accessed 22 Dec 2016

Verreault J, Muir DC, Norstrom RJ et al (2005) Chlorinated hydrocarbon contaminants and metabolites in polar bears (*Ursus maritimus*) from Alaska, Canada, East Greenland, and Svalbard: 1996–2002. Sci Total Environ 351:369–390

Weber DS, Van Coeverden De Groot PJ, Peacock E et al (2013) Low MHC variation in the polar bear: implications in the face of Arctic warming? Anim Conserv 16:671–683

Whittaker D, Knight RL (1998) Understanding wildlife responses to humans. Wildl Soc Bull 26:312–317

Wilhelmsson D, Thompson RC, Holmstrom K et al (2013) Marine pollution. In: Noone KJ, Sumaila UR, Diaz EJ (eds) Managing ocean environments in a changing climate: sustainability and economic perspectives. Elsevier, Burlington, pp 127–157

Woodroffe R, Thirgood S, Rabinowitz A (eds) (2005) People and wildlife: conflict or coexistence? Cambridge University Press, Cambridge, UK

Chapter 23
Polar Bears and Sea Ice Habitat Change

George M. Durner and Todd C. Atwood

Abstract The polar bear (*Ursus maritimus*) is an obligate apex predator of Arctic sea ice and as such can be affected by climate warming-induced changes in the extent and composition of pack ice and its impacts on their seal prey. Sea ice declines have negatively impacted some polar bear subpopulations through reduced energy input because of loss of hunting habitats, higher energy costs due to greater ice drift, ice fracturing and open water, and ultimately greater challenges to recruit young. Projections made from the output of global climate models suggest that polar bears in peripheral Arctic and sub-Arctic seas will be reduced in numbers or become extirpated by the end of the twenty-first century if the rate of climate warming continues on its present trajectory. The same projections also suggest that polar bears may persist in the high-latitude Arctic where heavy multiyear sea ice that has been typical in that region is being replaced by thinner annual ice. Underlying physical and biological oceanography provides clues as to why polar bear in some regions are negatively impacted, while bears in other regions have shown no apparent changes. However, continued declines in sea ice will eventually challenge the survival of polar bears and efforts to conserve them in all regions of the Arctic.

23.1 Introduction

The evolution of extant Arctic marine mammals is tightly linked to climatic factors that influenced the formation and development of sea ice, and as such current climate factors may threaten their persistence. The Arctic Ocean became seasonally ice covered due to a cooling environment beginning in the mid-Eocene (~45 million

G.M. Durner (✉)
U.S. Geological Survey, Alaska Science Center, Anchorage, AK, USA
e-mail: gdurner@usgs.gov

T.C. Atwood
Polar Bear Research Program, U.S. Geological Survey, Alaska Science Center, Anchorage, AK, USA
e-mail: tatwood@usgs.gov

A. Butterworth (ed.), *Marine Mammal Welfare*, Animal Welfare 17,
DOI 10.1007/978-3-319-46994-2_23

years ago, mya; Moran et al. 2006) with further cooling leading to the development of perennial sea ice ~14 mya (Darby 2008; Moran et al. 2006). This new sea ice habitat that persisted throughout the annual cycle presented ecological opportunities for the radiation of Arctic marine mammals. Phocid seals, which likely originated in southern latitudes ~20 mya, expanded their range to northern Atlantic waters at least 15 mya (Harington 2008)—largely coinciding with the appearance of perennial sea ice. This resulted in some northern hemisphere phocids becoming sea ice specialists, isolating them from their southern counterparts and causing their radiation into six species that fill sea ice niches today (Harington 2008). Arctic phocids are widespread and abundant, with at least one species comprising up to two million individuals (i.e., ringed seals, *Pusa hispida*; Kelly et al. 2010). The early adaptation and success by these seals to utilize Arctic sea ice presented opportunities for a new apex predator, the polar bear (*Ursus maritimus*).

The polar bear evolved as a specialist predator of ice-adapted seals (Fig. 23.1), primarily ringed seals but also bearded seals (*Erignathus barbatus*; Stirling and Archibald 1977; Stirling and Øritsland 1995). Polar bears are believed to have diverged from a brown bearlike ancestor sometime from as recently as 160 kya to as long ago as 5 mya (Hailer et al. 2012; Lindqvist et al. 2010; Miller et al. 2012). Regardless of which divergence estimate is used, the ancestor of present-day polar bears clearly entered a sea ice environment with abundant resources and few, if any, competitors. Despite periodic hybridization with brown bears (*U. arctos*; Bidon et al. 2014; Edwards et al. 2011; Miller et al. 2012), selective drivers in the Arctic have resulted in the genotypic and phenotypic traits in modern polar bears that are largely absent of brown bear ancestry (Cahill et al. 2015).

The present-day distribution of polar bears is concordant with the extent of Arctic and sub-Arctic marine waters that are normally covered by sea ice for at least 7–8 months of the year. The Arctic undergoes large seasonal fluctuations in sea ice extent, from a winter maximum of ~15.5 million km^2 to a summer minimum of ~6.3 million km^2 (1981–2010 average; National Snow and Ice Data Center (2016a, b). As a result, polar bears in much of their range change their distribution with the seasonal changes in sea ice. Despite the necessity for polar bears to adjust their distribution to this labile substrate, and the continuity of Arctic sea ice during most of the year, individuals show such strong philopatry (Amstrup et al. 2004; Paetkau et al. 1999) to regions that the entire world's population can be divided into 19 relatively discrete subpopulations (Fig. 23.2; Obbard et al. 2010).

These subpopulations, however, may be grouped into four ecoregions (please see Chap. 22; Fig. 23.2), each of which has distinct seasonal composition and dynamics of sea ice, underlying oceanography, and influence of adjacent land masses. Following the convention put forth by Amstrup et al. (2008), these include:

1. Seasonal Ice Ecoregion—sea ice typically melts completely and is absent 3–4 months during summer (subpopulations: Baffin Bay, Davis Strait, Foxe Basin, Southern Hudson Bay, and Western Hudson Bay [WH]). In the Seasonal Ice Ecoregion, polar bears are forced to summer on land where food consumption is negligible and activity is reduced.

Fig. 23.1 Typical hunting modes used by polar bears on the spring sea ice in the Beaufort Sea. (**a**) An adult male polar bear still hunting at a seal hole (15 April 2009) and (**b**) an adult female with her 2-year-old young at a ringed seal lair adjacent to a pressure ridge, where they successfully captured an adult seal (12 April 2000). *Image credits:* (**a**) *Michael Lockhart, United States Geological Survey (USGS);* (**b**) *George Durner, USGS*

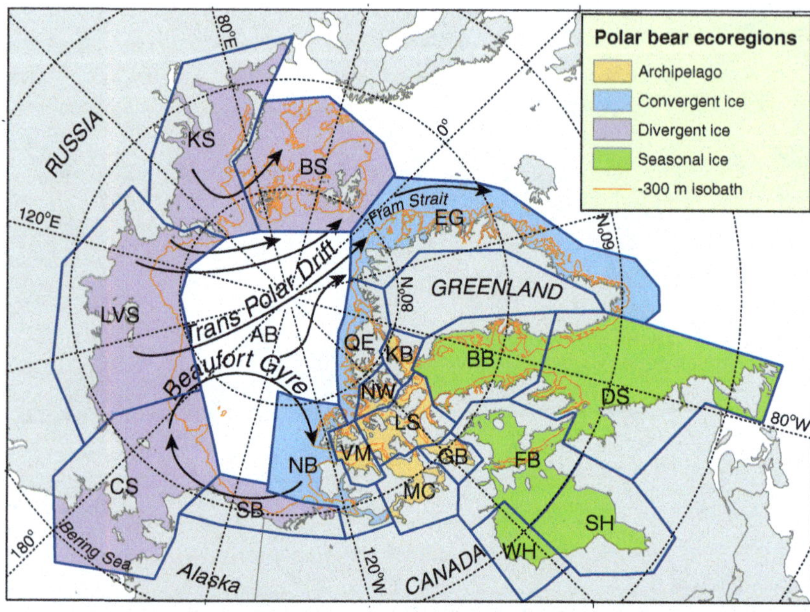

Fig. 23.2 Distribution of 19 polar bear subpopulations and four ecoregions. Southern Beaufort Sea (SB), Chukchi Sea (CS), Laptev Sea (LVS), Kara Sea (KS), Barents Sea (BS), East Greenland (EG), Queen Elizabeth (QE), Northern Beaufort Sea (NB), Southern Hudson Bay (SH), Western Hudson Bay (WH), Foxe Basin (FB), Davis Strait (DS), Baffin Bay (BB), Gulf of Boothia (GB), M'Clintock Channel (MC), Lancaster Sound (LS), Viscount-Melville Sound (VM), Norwegian Bay (NW), and Kane Basin (KB). Note the location of subpopulations relative to the 300 m bathymetric contour and ice drift patterns. *Black arrows* indicate generalized path of primary currents (*Source*: Amstrup et al. (2008) *Geophysical Monograph 180:213–268*)

2. Archipelago Ecoregion—the channels and bays within the island complex of northern Canada (subpopulations: Gulf of Boothia, Kane Basin, Lancaster Sound, M'Clintock Channel, Norwegian Bay, and Viscount Melville Sound). In the Archipelago Ecoregion, some sea ice survives the summer melt, resulting in a high proportion of ice >1 year old (i.e., multiyear ice) in subsequent seasons. Because of the proximity of islands and larger land masses, a high proportion of ice in the Archipelago Ecoregion does not drift as it is frozen to land (i.e., land-fast ice) from autumn to spring. Polar bears in the Archipelago Ecoregion have access to sea ice throughout the year as summer ice concentration remains relatively high (>50%) (Canadian Ice Service 2016a, b).
3. Divergent Ice Ecoregion—seas within the Arctic Ocean proper, primarily over the continental shelves of northern Eurasia, northern Alaska, and mainland northwest Canada (subpopulations: Barents Sea, Chukchi Sea, Kara Sea, Laptev Sea, and Southern Beaufort Sea [SB]). Historically, some sea ice over continental shelves was retained during summer and persisted into autumn freeze-up, providing hunting habitat for polar bears throughout the year. More recently, sea

ice over continental shelves in the Divergent Ice Ecoregion melts nearly to completion or is advected to other regions by currents of the Transpolar Drift and the Beaufort Gyre (Fig. 23.2). Recent patterns of sea ice melt and freeze now force most bears in this ecoregion to follow sea ice into the deep waters of the Arctic Ocean, or, to a lesser extent, onto land.

4. Convergent Ice Ecoregion—spanning northern and eastern Greenland and the northern edge of the Canadian Archipelago (subpopulations: Northern Beaufort Sea, Queen Elizabeth, and East Greenland). In the Convergent Ice Ecoregion, summer sea ice melt is minimal, and ice is received via the Transpolar Drift and the Beaufort Gyre from the Divergent Ice Ecoregion. Entrainment of ice by the Beaufort Gyre in the Northern Beaufort Sea and Queen Elizabeth subpopulations results in those waters consisting of the oldest and thickest sea ice in the Arctic.

There is commonality among ecoregions because polar bears are an ice obligate species that depend on sea ice for fulfilling life history requirements (e.g., hunting seals, traveling, seeking mates). However, variation in the composition, distribution, and pattern in the annual formation of sea ice influences polar bear distribution, life history, and demography. Climate-mediated displacement from preferred sea ice habitat, as is considered later, can have significant consequences for polar bear subpopulations.

23.2 Sea Ice Selected by Polar Bears Throughout Their Range

To appreciate how climate change is influencing polar bear habitat, it is helpful to understand how polar bears distribute themselves relative to sea ice extent and composition. Research on polar bear habitat use has been conducted on most of the 19 subpopulations. With few exceptions (Stirling et al. 1993; Pilfold et al. 2014), the bulk of these studies became possible through the advent of satellite radiotelemetry (Fig. 23.3; Fancy et al. 1988), which has provided location data of individual polar bears across the annual sea ice cycle (e.g., Amstrup et al. 2004). Satellite-derived locations of polar bears, when coupled with environmental data (e.g., sea ice attributes) and analyzed with statistical models (i.e., Resource Selection Functions, RSFs; Manly et al. 2002), have revealed broad-scale spatial and temporal characteristics of sea ice that are important for polar bears—aspects which researchers were largely unable to identify and quantify due to difficulty of making visual observations of bears across their range and throughout the year. In addition to pinpointing important habitat characteristics, RSFs have the added benefit, due to the wide spatial extent and frequent collection of satellite-collected environmental data, of providing information on the distribution of optimal sea ice habitats of polar bears across most of their range. The development of RSFs to describe important polar bear habitat has revealed how optimal habitat changes seasonally and across years and is likely to change decades beyond the present (Fig. 23.4; Durner et al. 2009).

Fig. 23.3 Placing a satellite radio collar on an adult female polar bear in the southern Beaufort Sea, 12 April 2005. *Image credit: Eric Regehr, USGS*

23.2.1 The Importance of Sea Ice Concentration and Composition

In subpopulations where RSF models have been developed, sea ice concentration (SIC; i.e., the areal extent of ice-covered versus ice-free waters within a defined area) can be one of the most important environmental variables affecting polar bear distribution. In other words, the mere presence or absence of sea ice is insufficient for predicting the distribution of optimal habitats and likely distribution of bears. The reasons for this are that polar bears require sea ice with characteristics that enable them to hunt efficiently, and which provides security against inclement weather. With the possible exception of the Archipelago Ecoregion, a stable ice platform is necessary for bears to avoid being swept into open ocean (Mauritzen et al. 2003). For this discussion we focus on polar bear habitat patterns modeled from SIC, as estimated largely from passive microwave (PMW) imagery. Since 1978, PMW estimates of Arctic SIC and extent (Cavalieri et al. 1996) have been disseminated as daily and monthly means in raster format (25 × 25 km pixel size) from the National Snow and Ice Data Center (NSIDC; http://nsidc.org/). These data provide a consistent long-term data source to measure changes in polar bear habitat.

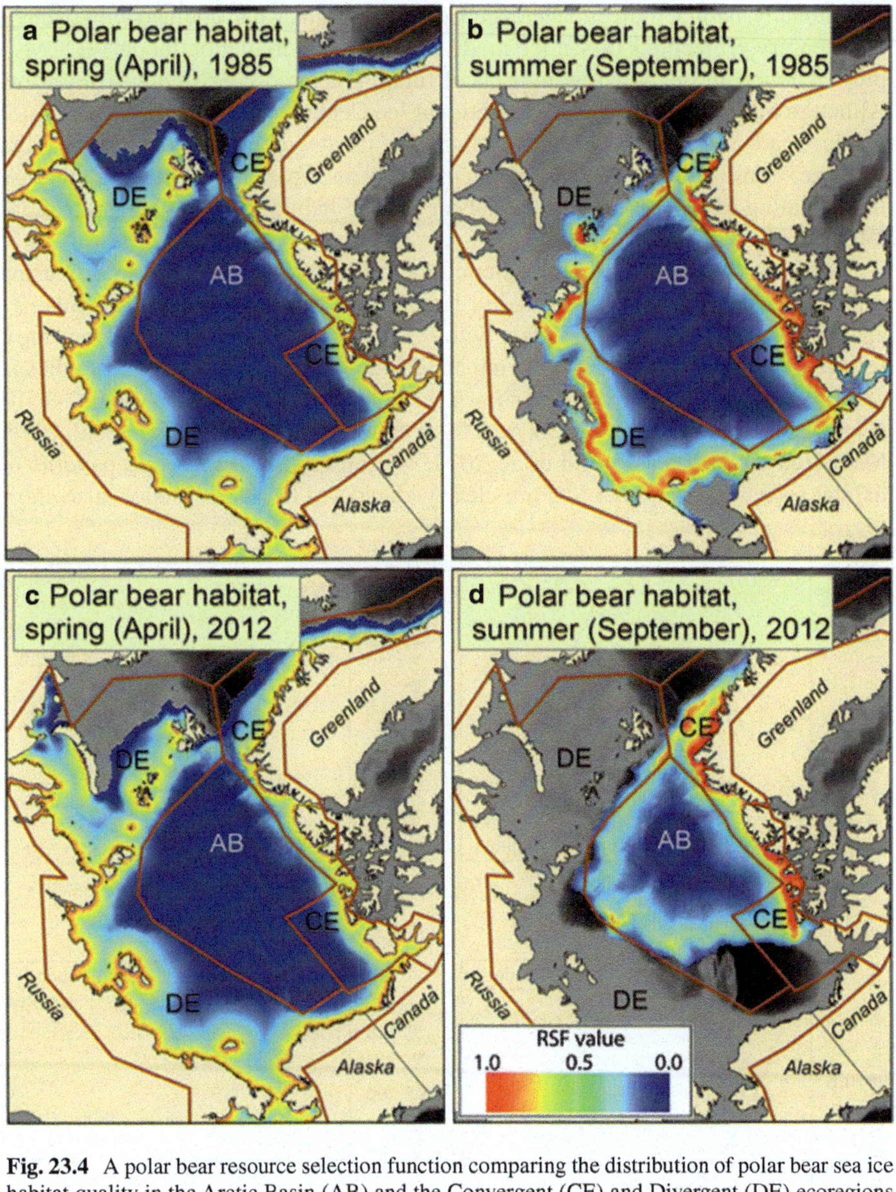

Fig. 23.4 A polar bear resource selection function comparing the distribution of polar bear sea ice habitat quality in the Arctic Basin (AB) and the Convergent (CE) and Divergent (DE) ecoregions (see Fig. 23.2) (1985 and 2012) during spring (**a** and **c**) and summer (**b** and **d**), based on Durner et al. (2009). Habitat quality ranges from poor (*blue tone* RSF) to optimal (i.e., the upper 20% of RSF-valued habitat, indicated by *yellow-red tone* RSF). *Gray tones* indicate ocean depth, where *light gray* indicates continental shelves and *dark gray* to *black* indicates deep polar basin waters. *Source*: Durner et al. (2009)

Additionally, habitat indices developed from PMW data, when coupled with twenty-first-century projections of sea ice made with general circulation models (GCMs), present a view of the potential impacts of future greenhouse gas-induced warming. Although other sea ice charts (e.g., National Ice Center 2016; Canadian Ice Service 2016a, b) are available that provide estimates of ice stage (thickness) and form (e.g., floe size), those data are more limited temporally and spatially, precluding their use in estimating decadal trends in polar bear sea ice habitat and making projections into the twenty-first century.

Arthur et al. (1996) first demonstrated the response of polar bears to SIC using satellite telemetry data from five adult female polar bears in the Chukchi Sea. By analyzing telemetry and PMW data with RSFs, they found polar bears were most selective of 51–75% SIC during spring and 21–50% SIC during summer. Studies in several other subpopulations have shown selection by polar bears for ~50–90% SIC (Fig. 23.5; Durner et al. 2004, 2006, 2009; Ferguson et al. 2000; Laidre et al. 2015; Mauritzen et al. 2003; Pilfold et al. 2014; Wilson et al. 2014). While patterns of habitat selection relative to SIC are clearly elucidated by RSFs, models also show

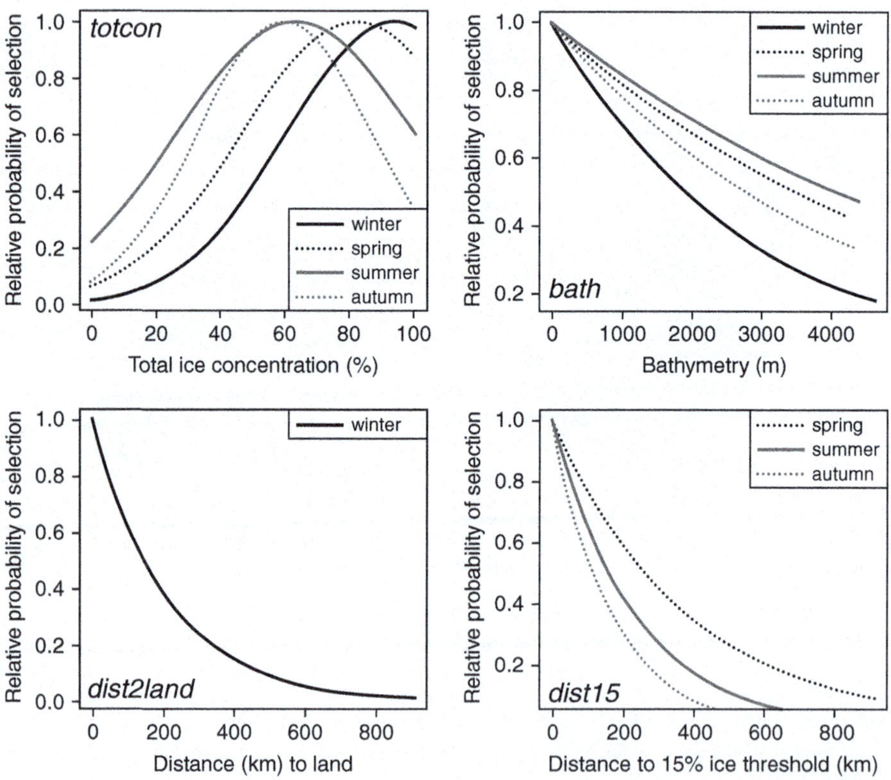

Fig. 23.5 Resource selection function results for the response of polar bears to habitat variables in the polar basin (1985–1995). Total ice concentration is the percent of ocean surface covered by sea ice. Bathymetry is ocean depth. *Source*: Durner et al. (2009)

that polar bears select relatively high SIC in close proximity to areas with low SIC (i.e., <15–50%) and near land (Fig. 23.5; Durner et al. 2009; Laidre et al. 2015; Pilfold et al. 2014). However, habitat selection is seasonally dependent (Durner et al. 2004, 2009; Ferguson et al. 2000; Laidre et al. 2015; Wilson et al. 2014), with lower concentrations of sea ice being selected during spring breakup (i.e., time when sea ice begins to fragment) and summer (Fig. 23.5; Durner et al. 2004, 2009; Ferguson et al. 2000; Laidre et al. 2015; Wilson et al. 2014). But in subpopulations whose sea ice is composed of a high proportion of landfast ice (i.e., sea ice attached to land, as in the Canadian Archipelago) or where sea ice converges (i.e., east Greenland Sea), concentrations near 100% are highly selected by polar bears during winter months (Ferguson et al. 2000; Laidre et al. 2015).

The distribution of sea ice relative to the continental shelves—Aside from the reliance on moderate to high SIC adjacent to low SIC, the spatial and temporal distribution of sea ice is also a determinant of habitat quality for polar bears. In general, the sea ice that we described in the prior paragraph can occur anywhere in the Arctic during some point in the annual cycle. However, the most biologically rich regions across the range of polar bears are in the Arctic's peripheral seas and over the continental shelves (<300 m deep). Several factors contribute to the relative richness of shelf waters. As pack ice is a dynamic substrate that drifts due to winds and currents (Spreen et al. 2011), the motion of ice next to landfast ice or shorelines creates regions of stress in the pack ice (i.e., shear zones) where fracturing of ice creates long cracks in the ice that remain open for hours to days (i.e., leads) and larger areas of open water (i.e., polynyas; Stirling 1997). Shelf waters are adjacent to basins with depths up to 3000 m (Jakobsson et al. 2008). This combination of shelf and basin results in a unique distribution of upwellings that mix deep-origin nutrient-rich waters with current-influenced waters near and over the continental shelf (Carmack and Wassmann 2006; Christensen 2008; Horner and Schrader 1982; Piatt and Springer 2003; Sigler et al. 2011). Seals remain in these productive shelf waters, even as the summer melt reduces or completely removes sea ice (Harwood and Stirling 1992; Harwood et al. 2012). Hence, biologically rich waters near and over the continental shelf coincide with an abundance of leads and polynyas to provide necessary habitat for several marine mammals, including polar bears (Stirling 1997).

RSFs for polar bears indicate a distribution of optimal sea ice habitat that is consistent with our understanding of primary productivity and prey habitat use. Arctic marine waters, with drifting ice of mid-to-high concentration over continental shelves, provide the nexus for optimal polar bear habitat (Fig. 23.4; Durner et al. 2009; Pilfold et al. 2014; Wilson et al. 2014).

23.2.2 Observed Changes in Optimal Habitat

Arctic sea ice extent has been substantially reduced and is projected to continue to decline during the twenty-first century (Comiso 2012; Meier et al. 2007; Overland and Wang 2013; Stroeve et al. 2007). PMW estimates of sea ice extent show multidecadal declines in sea ice both during the winter maximum and the summer minimum (Fig. 23.6). Likewise, polar bear habitat in most of the Arctic basin subpopulations and in the Seasonal Ice Ecoregion declined during the decade following 1995 (Durner et al. 2009; Stirling et al. 1999; Stirling and Parkinson 2006). Between the years 1985–1995 and 1996–2006, the total number of months of optimal sea ice habitat (i.e., the upper 20% of RSF-valued habitat) declined ~14% in the Barents Sea, ~12% in the Chukchi Sea, ~10% in the east Greenland Sea, and ~6% in the southern Beaufort Sea (Fig. 23.7; Durner et al. 2009). The greatest declines in

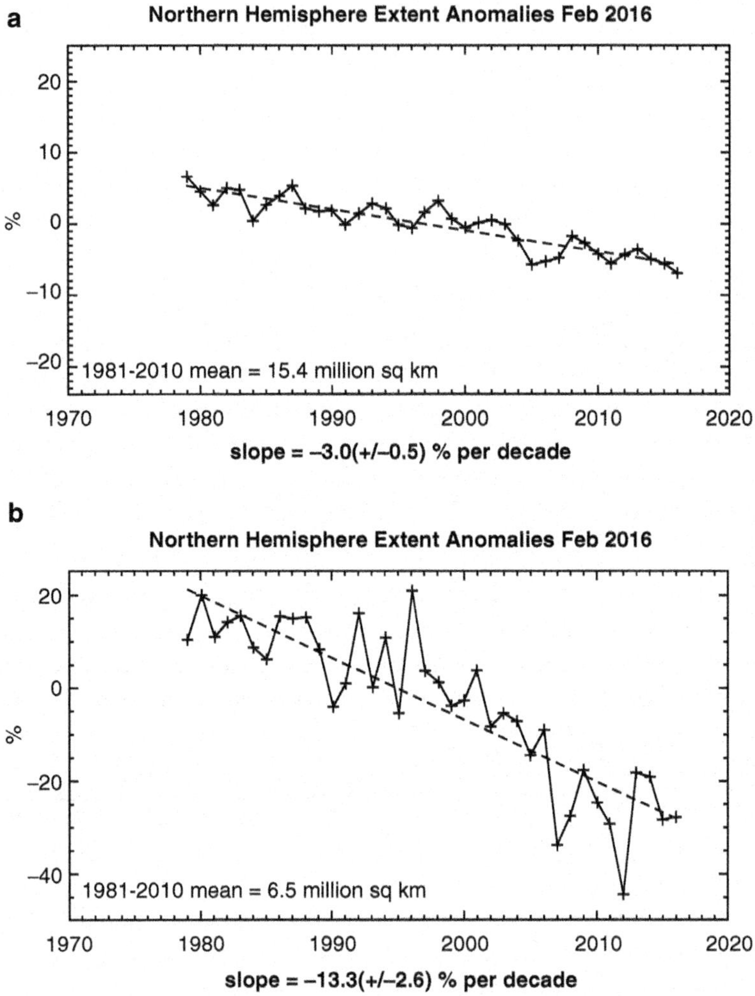

Fig. 23.6 Sea ice anomalies for (**a**) February (winter) and (**b**) September (summer) 1979–2015, relative to the 1981–2010 mean sea ice extent for the respective month (February, 15.4 million km^2; September, 6.5 million km^2). Arctic sea ice extent has been declining 3.0% decade^{-1} (winter maximum) and 13.4% decade^{-1} (summer minimum). *Source*: *NSIDC, ftp://sidads.colorado.edu/DATASETS/NOAA/G02135/, accessed 3 December 2016*

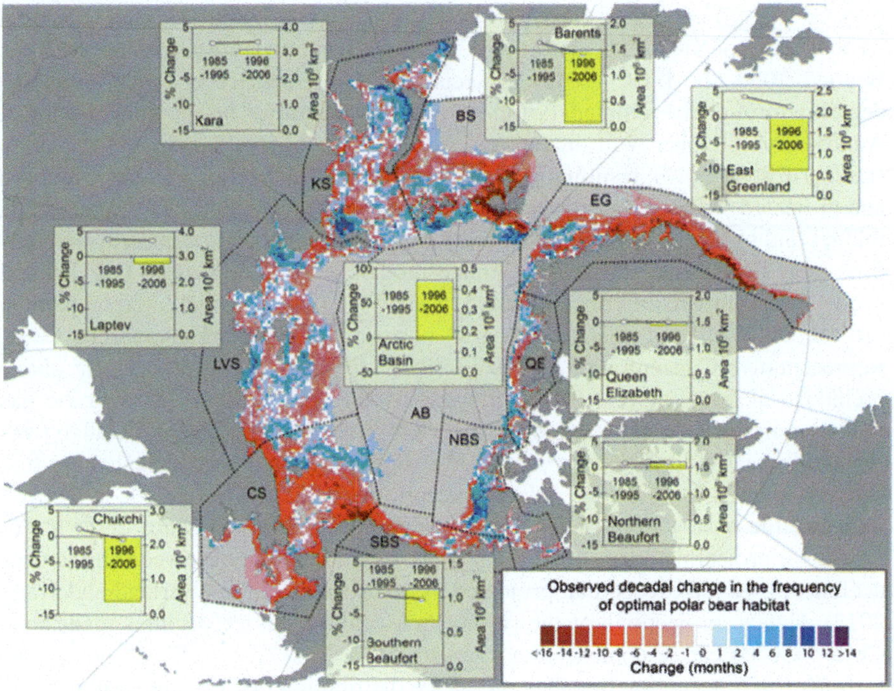

Fig. 23.7 Observed changes in polar bear optimal habitat (i.e., the upper 20% of RSF-valued habitat) in the polar basin, comparing the total number of months of optimal habitat in 1985–1995 compared to 1996–2006. Loss (gain) in optimal habitat is indicated by *red* (*blue*) colors. *Gray* indicates no optimal habitat was present in either decade. *Source*: Durner et al. (2009)

optimal habitat occurred during spring breakup and summer—largely as a result of sea ice melting beyond continental shelves and into the deep waters of the Arctic Ocean (Fig. 23.4d). Within the Seasonal Ice Ecoregion, a region that typically loses sea ice every summer, increasing duration of the ice-free season has displaced bears from preferred sea ice habitat to land for longer periods (Stirling et al. 1999; Stirling and Parkinson 2006). This is illustrated in Hudson Bay where from 1971 to 2003, annual breakup and freeze-up dates became 40 days earlier and 18 days later, respectively (Gagnon and Gough 2005).

23.2.3 *Projections of Future Polar Bear Optimal Sea Ice Habitat*

Current and future climate change is "extremely likely" to be linked to human activities, most notably from industrial emissions of carbon dioxide and methane (Intergovernmental Panel on Climate Change, IPCC; IPCC 2013). The IPCC (2013) projects that ≥5 consecutive years of nearly ice-free conditions (i.e., sea ice extent <1 × 10^6 km^2 of SIC >15%) in the Arctic during September are likely before 2050. Beyond 2050, the IPCC (2013) projects that Arctic sea ice is "very likely" to decrease through the remainder of the twenty-first century. The shoulder months of August and October will likely see Arctic-wide ice-free conditions by 2070 (Laliberté et al. 2016), resulting in an earlier breakup and later freeze-up. Observed and projected sea ice loss is driven in a large part by an ice cover that is becoming thinner and by the increased vulnerability of this thinning ice to complete summertime melt (Holland et al. 2010). Climate models (i.e., GCMs) from the Coupled Model Intercomparison Project Phase 5 (CMIP5) (Coupled Model Intercomparison Project (2016) utilized emission scenarios based on different radiative forcings (i.e., Representative Concentration Pathways (RCP), expressed as watts per m^2 above preindustrial levels at year 2100) imposed by different levels of greenhouse gas (GHG) emissions. As expected, high GHG emissions result in higher radiative forcing and increased likelihood of sea ice loss during the latter part of the twenty-first century (Fig. 23.8). With one notable exception, hindcasts from GCMs agree well with satellite observations of sea ice extent, and this lends confidence in the use of GCMs for projecting future conditions. The exception is that observed summertime declines in sea ice exceed that projected by GCMs (Fig. 23.8; Overland and Wang 2013). Hence, projections of the twenty-first-century sea ice declines made with CMIP5 GCMs can be considered conservative.

Since 1979, when satellite data on sea ice distribution first became available, six of the lowest September sea ice minimums have occurred after 2006, resulting in ice-free waters in most regions of the Divergent Ice Ecoregion, the entire Seasonal Ice Ecoregion, and parts of the Archipelago Ecoregion during those years (National Snow and Ice Data Center 2012). Of interest to the future status of polar bears is how the ice-free period may become longer in future years and how this may vary among regions. By using CMIP5 twenty-first-century projections of sea ice, Laliberté et al. (2016) demonstrated the advancement of the ice-free season to earlier months for the coming decades varies by region. Arctic seas adjacent to northern Asia (i.e., Kara Sea and Laptev Sea) may experience ice-free conditions in July as early as 2050. Near Alaska, projections of ice-free waters in the Chukchi and Beaufort seas during July are not expected until ~2080 and 2095, respectively, likely due to advection of thick multiyear ice from the central Arctic (Laliberté et al. 2016). At the extremes, ice-free conditions in the central Arctic Ocean and the Canadian Archipelago are not expected to occur until ~2070 during August; however, an ice-free Hudson Bay for July has been possible since 2010. The study by Laliberté et al. (2016) provides additional evidence that polar bear response to habitat loss will be regionally specific (Atwood et al. 2016).

Projections of the twenty-first-century polar bear habitat have included all ecoregions (Castro de la Guardia et al. 2013; Durner et al. 2009; Hamilton et al. 2014). In Hudson Bay 30–50% SIC has been identified as a critical threshold below which bears begin moving on to land during spring breakup, or 10% SIC, above which bears move from land to back to the sea ice during autumn freeze-up (Cherry et al. 2013; Stirling et al. 1999). This presents an important distinction between the standard of <15% SIC used by geophysicists (see, e.g., National Snow and Ice Data

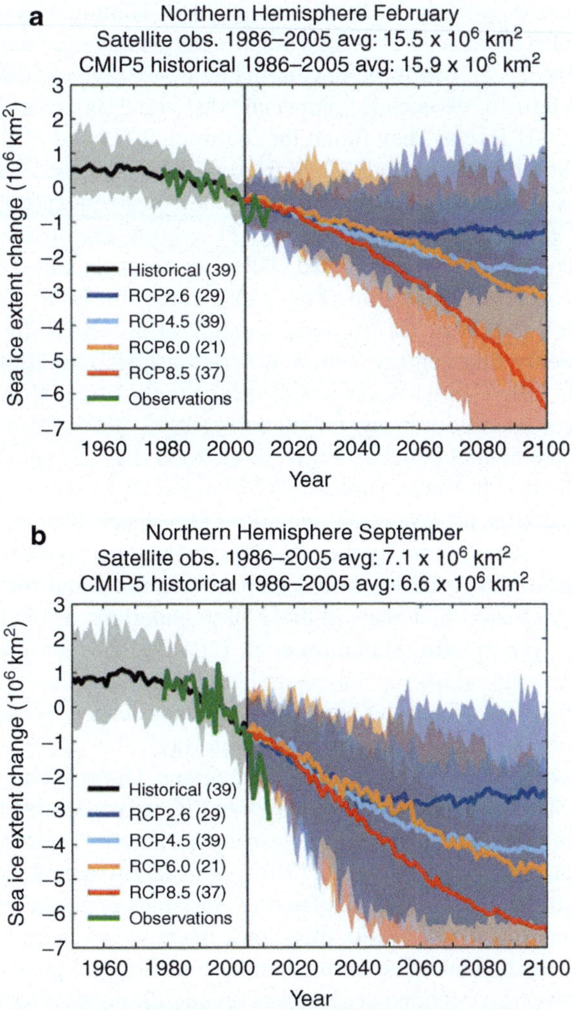

Fig. 23.8 The late twentieth- and twenty-first-century Arctic sea ice extent change (defined by concentration >15%) simulated by CMIP5 (Coupled Model Intercomparison Project) ensembles under Representative Concentration Pathways (RCP)2.6, RCP4.5, RCP6.0, and RCP8.5 for (**a**) February and (**b**) September, relative to conditions during 1986–2005. Solid lines represent multi-model means, and shading indicates the 5–95% range of the ensemble. Sample sizes for each RCP are indicated in the legend. Observational data of sea ice extent (1979–2012) are indicated by the *green solid line*. Note the greater observed loss of sea ice relative to that projected by GCMs after 2005. *Source*: *Fig. 12.28 in Collins et al. (2013). Chap. 12. Long-term climate change: projections, commitments and irreversibility (IPCC* 2013*)*

Center 2016a, b) to denote ice-free conditions and what are effectively ice-free conditions for polar bears. In projections of the twenty-first-century polar bear habitat in the Seasonal Ice and Archipelago Ecoregions, ice-free conditions were assumed to include the period when SIC reaches <30–50% in the spring to when SIC became

>10% in autumn (Castro de la Guardia et al. 2013; Hamilton et al. 2014). Using IPCC (2000) GHG emission scenarios for projections of the twenty-first-century WH sea ice (Fig. 23.2), Castro de la Guardia et al. (2013) projected spring SIC to decline from 1.0 to 13.5% decade^{-1} between 2001 and 2100, resulting in 2100 SIC at 20–84% of 2001 levels. They found that between 2035 and 2100, WH trends in breakup dates and ice-free periods range from 1.7 to 13.0 days decade^{-1} earlier and 2.2 to 20.7 days decade^{-1} longer, respectively. The overall effect projected by Castro de la Guardia et al. (2013) was for the 2100 ice-free season in WH to be 4.5–18.7 weeks greater than it would be had GHG emissions remained at 2001 levels.

For the Archipelago Ecoregion (Fig. 23.2), Hamilton et al. (2014) adopted a similar approach to that of Castro de la Guardia et al. (2013) by projecting SIC through the twenty-first century with a worst-case GHG emission scenario (i.e., RCP 8.5; IPCC 2013). In the seven polar bear subpopulations that they examined, none had ice-free conditions in any month during 1992–2005. Even with substantial GHG forcing, the annual pattern of SIC 2040–2060 was largely similar to 1992–2005 levels, albeit with lower summer SIC minima. However, during the last two decades of the century, all seven subpopulations are expected to experience multiple months of ice-free conditions. By the end of the twenty-first century, four subpopulations were projected to be ice-free for up to 5 months and the other three for 2–5 months. This means that bears in those subpopulations would be forced to use land during ice-free months. Hamilton et al. (2014) also show that this threshold SIC required for optimal habitat may be reduced to only 6 months a year, which is below the annual ice-covered duration currently experienced by the most southern polar bear subpopulation (i.e., Southern Hudson Bay).

For the Divergent and Convergent Ice Ecoregions, Durner et al. (2009) extrapolated RSFs to GCM projections of Arctic Ocean SIC to predict trends in the twenty-first-century optimal sea ice habitat for nine subpopulations. Their approach differed from that of Castro de la Guardia et al. (2013) and Hamilton et al. (2014) in that the RSF included, in addition to SIC, explanatory variables of ocean depth, distance to land, and distance to the 15% SIC threshold. Because ocean depth greatly influenced the RSF, the habitat value for a given level of SIC and proximity to the 15% SIC threshold over the continental shelf was greatly diminished for otherwise identical ice characteristics over deep waters beyond the shelf (Fig. 23.4). Because GCMs project that the majority of the twenty-first-century sea ice declines occur between spring and autumn, and that ice continues to return throughout the Arctic Ocean each winter, extrapolation of RSFs to GCM projections reveals high seasonal variability in habitat loss. Whereas Arctic Ocean winter habitat value decreased <10% by the end of the twenty-first century, the habitat value decreased >50% for summer (Durner et al. 2009). Likewise, there was considerable regional variation in the twenty-first-century habitat change, with most showing that the decadal cumulative frequency of optimal habitat occurring in any particular month would decline by between 25 and 50% by 2041–2050 (Fig. 23.9). Only the Arctic Basin, Queen Elizabeth, and the northern-most part of the Northern Beaufort subpopulations are likely to experience an increase in optimal sea ice habitat. Otherwise, the frequency of optimal habitat is projected to decrease in regions that are currently most important for polar bears (Fig. 23.9).

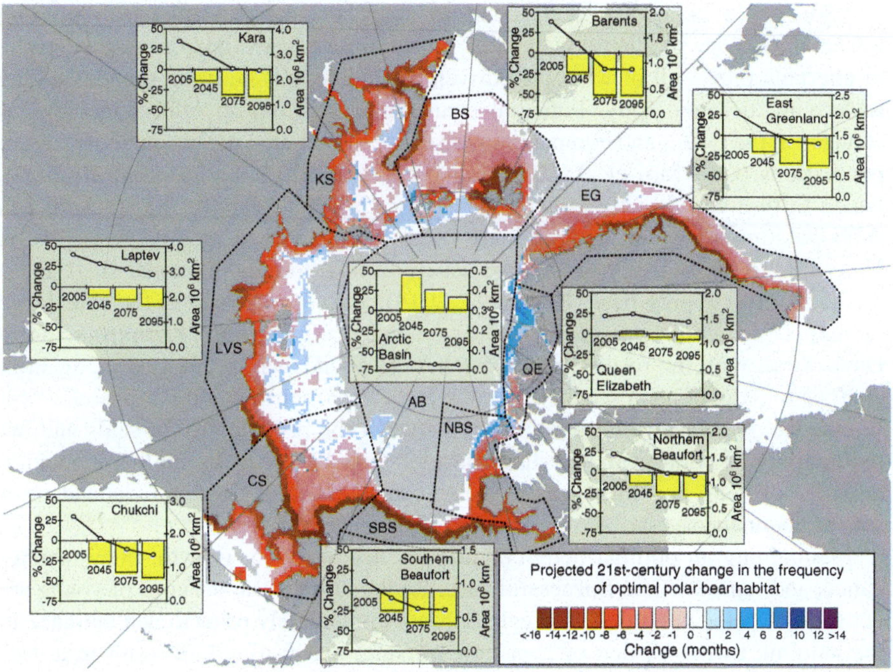

Fig. 23.9 Projected changes in polar bear optimal habitat (i.e., the upper 20% of RSF-valued habitat) in the polar basin, comparing the total number of months of optimal habitat in 2001–2010 compared to 2041–2050. Loss (gain) in optimal habitat is indicated by *red (blue) colors*. *Gray* indicates no optimal habitat was present in either decade. *Source*: Durner et al. (2009) *Ecol Monogr 79:25–58*

23.3 Consequences of Habitat Loss to Polar Bear Health, Reproduction, and Populations

Links between sea ice conditions and polar bear body condition (i.e., an index of health derived from body mass, linear body measurements, fatness estimates, or blood chemistry) have been drawn for several subpopulations from polar bears harvested by indigenous people for subsistence and from bears captured and released by scientists. Within the Seasonal Ice Ecoregion, correlations between declines in polar bear body condition and declining sea ice have been made for the WH (Lunn et al. 2016; Stirling et al. 1999), Southern Hudson Bay (Obbard et al. 2016), Davis Strait, and Baffin Bay (Rode et al. 2012) subpopulations. In the Divergent Ice Ecoregion, polar bears of the SB (Rode et al. 2010, 2014) and Barents Sea (Derocher 2005) subpopulations also appear to be showing declines in body condition that may be related to climatic variation. These studies largely point to a hypothesis of reduced accessibility to seal prey as the primary driver for declines in body condition, although density-dependent effects may have been a contributor in some subpopulations (Derocher 2005; Peacock et al. 2013; Rode et al. 2012).

Variation in polar bear body condition may also be attributed to bottom-up effects of primary productivity, population response of prey, and/or the appearance of alternate prey—all of which are driven by interactions between sea ice and the underlying oceanography. Potential examples of bottom-up effects include (1) changes in the extent and composition of sea ice influencing the productivity of seal populations (Crawford et al. 2015; Harwood et al. 2015; Ferguson et al. 2005), which in turn can directly affect polar bears; (2) polar bears of the Chukchi Sea subpopulation show unchanging body condition, despite sea ice declines, that is likely due to biologically productive waters over a broad continental shelf and a relatively diverse prey base (Rode et al. 2014); and (3) increases in the abundance of harp (*Pagophilus groenlandicus*) and hooded (*Cystophora cristata*) seals in Davis Strait with the resulting increase in bear abundance in that region, again in contrast to reductions in sea ice (Peacock et al. 2013; Stirling and Parkinson 2006). In essence, there is a tight relationship between the productivity of seals and the productivity of polar bears (Stirling and Øritsland 1995), such that the population response of polar bears to climate change can be, to some degree, independent of sea ice abundance.

Maintaining an optimal body condition is necessary for growth, body maintenance, and reproduction. Successful parturition and first-year survival of young are dependent on the condition of pregnant bears immediately prior to den entrance in the autumn, as the burden of gestation, nursing, and fasting can result in a den-bound adult losing 43% of its body mass during the 4–5 months of den tenure (Fig. 23.10; Atkinson and Ramsay 1995). Female polar bears exhibit delayed implantation, and there is evidence that bears <189 kg are unlikely to successfully reproduce (Derocher et al. 1992). To achieve body reserves sufficient for reproduction, female polar bears are dependent on a period of hyperphagia (i.e., near-constant, "super eating," over eating) (Ramsay and Stirling 1988) during late spring and early summer, when newly weaned naïve juvenile seals are abundant and sea ice is present in extent and composition that allows efficient hunting (Stirling et al. 1999). This reaches an extreme within the Seasonal Ice Ecoregion, as pregnant polar bears spend summer on land for 3–5 months, enter maternal dens in autumn, and do not resume hunting until the following March—a potential fasting duration of 8 months (Atkinson and Ramsay 1995; Stirling et al. 1999). For polar bears in other ecoregions, activity data collected from satellite telemetry suggests a similar pattern of hyperphagia during late spring and early summer (high activity) and fasting during summer (low activity) when preys are less available (Ferguson et al. 2001; Messier et al. 1992; Whiteman et al. 2015). Energetic models further support the necessity to polar bears for late spring and early summer sea ice, as advancement of breakup by only 1 month could result in 40–73% of female bears failing to reproduce and a reduction in litter size by 22–67% (Molnár et al. 2011).

Not only is the ability of adult female bears to successfully raise and wean young compromised in an Arctic with diminished sea ice, but other sex and age groups are vulnerable. Subadult bears, because they are likely to still be developing hunting skills, may be especially vulnerable to sea ice habitat loss (Bromaghin et al. 2015; Regehr et al. 2007). But even prime-age adult polar bears, which are the most resilient

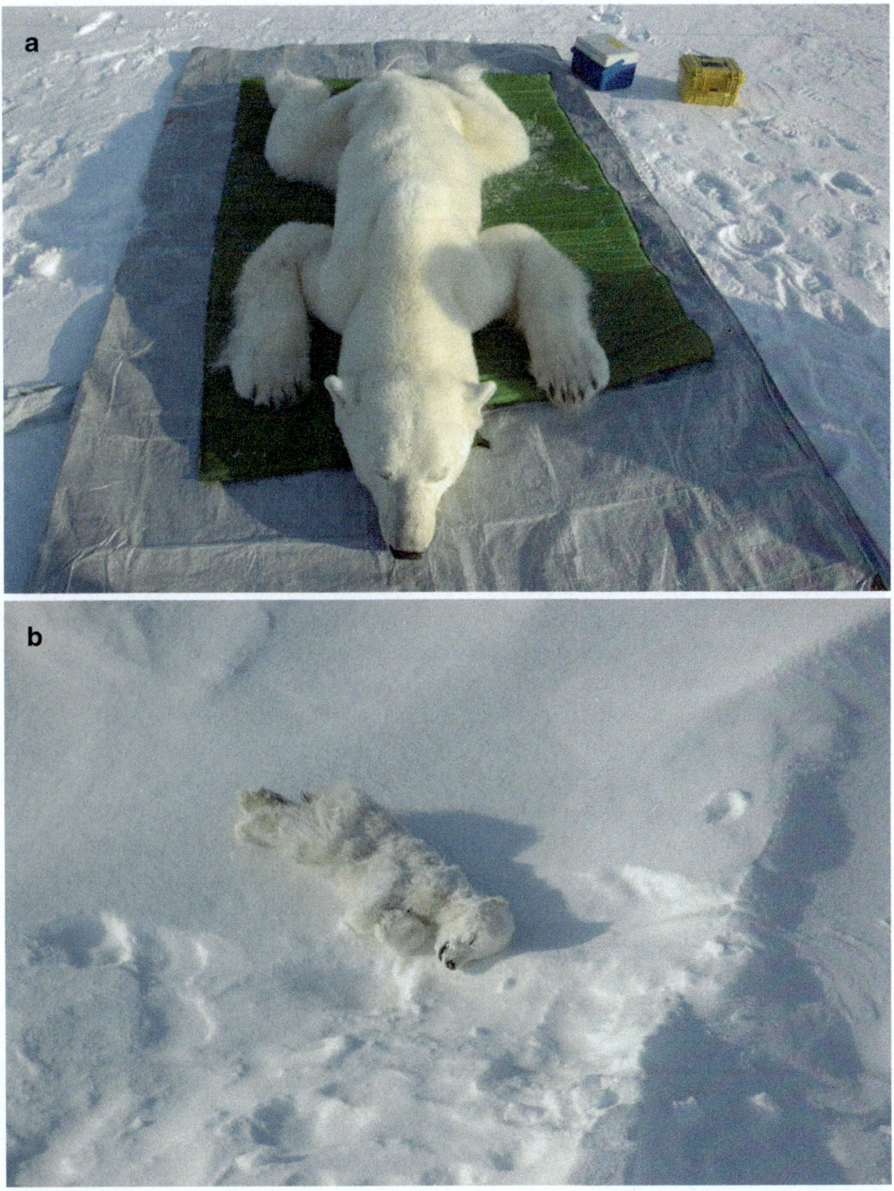

Fig. 23.10 (**a**) A starving adult female polar bear (137 kg) captured in the Beaufort Sea and (**b**) its emaciated dead cub (9 April 2009). Not shown is the surviving sibling of the dead cub. The adult entered a maternal den in the previous autumn with enough body fat for successful parturition and nursing in the den, but insufficient fat to continue nursing after den departure. Both the adult and surviving cub were observed roaming the pack ice on the following day. Subsequent genetic samples collected at a fur trap in Barrow, Alaska, in February 2011 confirmed survival of the adult. The cub has not been reobserved. *Image credit: George Durner, USGS*

members to environmental perturbations (Regehr et al. 2007), are unlikely to weather consecutive years of sea ice loss. Using a dynamic energy budget model, Molnár et al. (2010, 2014) showed that a 4-month absence of sea ice would result in 2–3% of adult males dying from starvation, but 9–21% could succumb to starvation if the duration of absence were to increase to 6 months. The relationship between adult survival and sea ice conditions is also supported by empirical models: in studies of the SB subpopulation, adult female survival declined precipitously when the number of ice-free days over the continental shelf increased from 3.5 to 4.5 months (Bromaghin et al. 2015; Regehr et al. 2010). For the WH subpopulation, Regehr et al. (2007) found a relationship between early ice breakup and reduced subadult survival but not for adult survival. Even in the northern Beaufort Sea, a region where sea ice has remained stable for three decades, polar bear survival was dependent on the condition of sea ice habitat (Stirling et al. 2011).

Increased displacement from optimal sea ice habitats during late spring and early summer during the hyperphagic feeding period (i.e., periods of very high food intake) has been hypothesized to be the reason for documented downward trends in body condition since the early 1980s (Obbard et al. 2016; Rode et al. 2010, 2012; Stirling et al. 1999). However, the energetic cost of changing sea ice dynamics for polar bears goes beyond displacement from optimal hunting habitat. Simply traveling across a sea ice substrate that, due to thinning, has become more fractured and vulnerable to drifting is likely to increase energy costs for polar bears as they attempt to search for and occupy the most optimal habitat for hunting seals and searching for mates (Castro de la Guardia et al. 2013; Sahanatien and Derocher 2012). Also, because there is a greater duration and extent of ice-free waters in the Arctic during summer, polar bears are increasingly engaging in open-ocean swims (Fig. 23.11) that can involve several hundred kilometers over as much as 9 days (Durner et al. 2011; Pagano et al. 2012; Pilfold et al. 2016). Although polar bears swim well in cold Arctic waters, swimming distances >100 km likely imposes high energetic costs, may result in mortality of the young (Durner et al. 2011), and increases their vulnerability to outright drowning (Monnett and Gleason 2006).

A changing sea ice platform can also affect the ability of polar bears to access suitable denning habitat (Stirling and Derocher 2012). Pregnant polar bears are reliant on

Fig. 23.11 A male polar bear swimming between ice floes in the Beaufort Sea (26 April 2009). *Image credit: Michael Lockhart, USGS*

Fig. 23.12 A female polar bear emerges from her maternal den next to a coastal bluff near Prudhoe Bay, Alaska (April 2009). *Image credit: Rusty Robinson, Brigham Young University*

dens of snow to provide a relatively warm and constant environment for parturition and growth of neonates, and the den must occur on a substrate that will remain stable for 3–5 months during the winter (Amstrup and Gardner 1994). Throughout the Arctic, most denning occurs on land (Fig. 23.12), although denning on sea ice occurs in the Beaufort Sea (Amstrup and Gardner 1994). For bears summering on pack ice, the greater extent of summertime open water and thinning sea ice platform that has been observed in recent years is increasing the necessity for pregnant bears to swim long distances or walk over a fractured surface (Pilfold et al. 2016; Sahanatien and Derocher 2012), and these conditions have prevented bears from reaching land traditionally used for denning (Derocher et al. 2011). SB polar bears represent the only subpopulation know to den on pack ice to a large degree (Amstrup and Gardner 1994; Fischbach et al. 2007). However, the suitability of sea ice as a denning substrate has declined since 1998 as Arctic Ocean pack ice has become thinner, younger and less stable (Fischbach et al. 2007). As a result, SB polar bears are increasingly using land for maternal denning (Fischbach et al. 2007), and, hence, an increasing proportion of the population must contend with unfavorable sea ice conditions to reach denning habitat.

Relating habitat loss to changes in population demography has only been possible for a few subpopulations which have received intensive capture-recapture studies. In regions that have high biological productivity and prey abundance

(Rode et al. 2014), have shown little loss of sea ice (Stirling et al. 2011), or have abundant alternate prey (Stirling and Parkinson 2006), the respective subpopulations have shown little apparent change in response to an otherwise warming Arctic. In the future Arctic, high-latitude marine waters including the northern Canadian archipelago and northern Greenland may experience an increase in habitat suitability as thick multiyear sea ice is replaced by thinner annual ice (Durner et al. 2009). This may have benefited some northern subpopulations of polar bears over the past three decades of declining sea ice (Stirling et al. 2011), and the relative persistence of sea ice in northern regions may serve as a refugium for a remnant population during the latter years of the twenty-first century (Atwood et al. 2016; Peacock et al. 2015). However, subpopulation sizes within the peripheral regions of the current circumpolar range of polar bears will likely decline, possibly to the point of extirpation, as sea ice is reduced in both its temporal and spatial extent (Amstrup et al. 2008; Atwood et al. 2016). Indeed, measureable population declines appears to have already begun in some subpopulations, such as the SB (Bromaghin et al. 2015; Regehr et al. 2010) and WH (Regehr et al. 2007).

23.4 Conclusions

Rapid loss of sea ice habitat in the twenty-first century brought on by anthropogenic greenhouse gas-driven climate warming presents the greatest stressor on all polar bear subpopulations (Amstrup et al. 2008; Atwood et al. 2016; Stirling and Derocher 2012). The impact of sea ice habitat declines on polar bears is multifaceted and will largely be a combined effect of reduced energy intake through reductions in prey availability; increased energetic costs to bears because of greater ice drift, greater open-water extent and duration, and an increasingly fractured icescape; and interference with the behavior of polar bears that is necessary for reproduction. Ultimately, these factors will reduce the availability of energy polar bears require to successfully reproduce. During the past three decades of warming global temperatures and concomitant declines in sea ice thickness and extent, some polar bear subpopulations have shown no apparent change and, indeed, may have even benefitted from ameliorating multiyear sea ice conditions as primary productivity and prey have increased. However, even those regions with currently stable subpopulations are vulnerable to the impacts of the projected twenty-first-century sea ice loss (Hamilton et al. 2014; Stirling et al. 2011). Continued climate warming-caused habitat loss will negatively impact all polar bear subpopulations by the later decades of the twenty-first century and currently represents the most significant conservation threat to the species (Amstrup et al. 2008; Atwood et al. 2016; Stirling and Derocher 2012).

Acknowledgments We thank Evan Richardson (Environment and Climate Change Canada) and Martyn Obbard (Ontario Ministry of Natural Resources) whose input on earlier versions greatly improved this chapter.

References

Amstrup SC, Gardner C (1994) Polar bear maternity denning in the Beaufort Sea. J Wildl Manag 58:1–10

Amstrup SC, McDonald TL, Durner GM (2004) Using satellite radiotelemetry data to delineate and manage wildlife populations. Wildl Soc Bull 32:661–679

Amstrup SC, Marcot BG, Douglas DC (2008) A Bayesian network modeling approach to forecasting the 21st century worldwide status of polar bears. In: De Weaver ET, Bitz CM, Tremblay LB (eds) Arctic Sea ice decline: observations, projections, mechanisms, and implications. American Geophysical Union Geophysical Monograph No. 180, Washington, DC, pp 213–268

Arthur SM, Manly FJ, McDonald LL, Garner GW (1996) Assessing habitat selection when availability changes. Ecology 77:215–227

Atkinson SN, Ramsay MA (1995) The effects of prolonged fasting of the body composition and reproductive success of female polar bears (*Ursus maritimus*). Funct Ecol 9:559–567

Atwood TC, Marcot BG, Douglas DC et al (2016) Forecasting the relative influence of environmental and anthropogenic stressors on polar bears. Ecosphere. doi:10.1002/ecs2.1370

Bidon T, Janke A, Fain SR et al (2014) Brown and polar bear Y chromosomes reveal extensive male-biased gene flow within brother lineages. Mol Biol Evol 31(6):1353–1363

Bromaghin J, McDonald T, Stirling I et al (2015) Polar bear population dynamics in the southern Beaufort Sea during a period of sea ice decline. Ecol Appl 25:634–651. doi:10.1890/14-1129.1

Cahill JA, Stirling I, Kistler L et al (2015) Genomic evidence of geographically widespread effect of gene flow from polar bears into brown bears. Mol Ecol 24(6):1205–1217

Canadian Ice Service (2016a) Historical total accumulated ice coverage for the Arctic archipelago. https://www.ec.gc.ca/glaces-ice/default.asp?lang=En&n=31AEB6F1-1. Accessed 22 Nov 2016

Canadian Ice Service (2016b) Sea ice charts. https://www.ec.gc.ca/glaces-ice. Accessed 22 Nov 2016

Carmack E, Wassmann P (2006) Food webs and physical-biological coupling on pan-Arctic shelves: unifying concepts and comprehensive perspectives. Prog Oceanogr 71:446–477

Castro de la Guardia L, Derocher AE, Myers PG et al (2013) Future Sea ice conditions in western Hudson Bay and consequences for polar bears in the 21st century. Glob Chang Biol 19(9):2675–2687

Cavalieri DJ, Parkinson CL, Gloersen P, Zwally HJ (1996) (updated yearly) Sea ice concentrations from Nimbus-7 SMMR and DMSP SSM/I-SSMIS passive microwave data, version 1. Boulder, CO, USA. NASA National Snow and Ice Data Center Distributed Active Archive Center. doi:10.5067/8GQ8LZQVL0VL

Cherry SG, Derocher AE, Thiemann GW et al (2013) Migration phenology and seasonal fidelity of an Arctic marine predator in relation to sea ice dynamics. J Anim Ecol 82:912–921

Christensen JP (2008) Sedimentary carbon oxidation and denitrification on the shelf break of the Alaskan Beaufort and Chukchi seas. Open Oceanogr J 2:6–17

Comiso JC (2012) Large decadal decline of the Arctic multiyear ice cover. J Climatol 25:1176–1193

Coupled Model Inter-comparison Project CMIP5 (2016) World Climate Research Programme. http://cmip-pcmdi.llnl.gov/cmip5. Accessed 22 Dec 2016

Crawford JA, Quakenbush LT, Citta JJ (2015) A comparison of ringed and bearded seal diet, condition and productivity between historical (1975–1984) and recent (2003–2012) periods in the Alaskan Bering and Chukchi seas. Prog Oceanogr 136:133–150

Darby DA (2008) Arctic perennial ice cover over the last 14 million years. Paleooceanography 23. doi:10.1029/2007PA001479

Derocher AE (2005) Population ecology of polar bears in Svalbard, Norway. Popul Ecol 47:276–275

Derocher AE, Stirling I, Andriashek D (1992) Pregnancy rates and serum progesterone levels of polar bears in western Hudson Bay. Can J Zool 70:561–566

Derocher AE, Andersen M, Wiig Ø et al (2011) Sea ice and polar bear den ecology at Hopen Island, Svalbard. Mar Ecol Prog Ser 441:273–279

Durner GM, Amstrup SC, Nielson R, McDonald T (2004) Using discrete choice modeling to generate resource selection functions for female polar bears in the Beaufort Sea. In: Huzurbazar S (ed) Resource selection methods and applications, Proceedings of the first international conference on resource selection, 13–15 January 2003, Laramie, Wyoming, USA, pp 107–120

Durner GM, Douglas DC, Nielson RM Amstrup SC (2006) A model for autumn pelagic distribution of adult female polar bears in the Chukchi Sea, 1987–1994. USGS Alaska Science Center, Contract Completion Report 70181-5-N240, Anchorage, 67 p

Durner GM, Douglas DC, Nielson RM et al (2009) Predicting 21st-century polar bear habitat distribution from global climate models. Ecol Monogr 79:25–58. doi:10.1890/07-2089.1

Durner GM, Whiteman JP, Harlow HJ et al (2011) Consequences of long-distance swimming and travel over deep-water pack ice for a female polar bear during a year of extreme sea ice retreat. Polar Biol 34:975–984

Edwards CJ, Suchard MA, Lemey P et al (2011) Ancient hybridization and an Irish origin for the modern polar bear matriline. Curr Biol 21(15):1251–1258

Fancy SG, Pank LF, Douglas DC et al. (1988) Satellite telemetry: a new tool for wildlife research and management. United States Department of the Interior, Fish and Wildlife Service, Resource Publication 172

Ferguson SH, Taylor MK, Messier F (2000) Influence of sea ice dynamics on habitat selection by polar bears. Ecology 81:761–772

Ferguson SH, Taylor MK, Born EW, Rosing-Asvid A, Messier F (2001) Activity and movement patterns of polar bear inhabiting consolidated versus active pack ice. Arctic 54:49–54

Ferguson SH, Stirling I, McLoughlin P (2005) Climate change and ringed seal (*Phoca hispida*) recruitment in western Hudson Bay. Mar Mamm Sci 21:121–135

Fischbach AS, Amstrup SC, Douglas DC (2007) Landward and eastward shift of Alaskan polar bear denning associated with recent sea ice changes. Polar Biol 30:1395–1409

Gagnon AS, Gough WA (2005) Trends in the dates of ice freeze-up and breakup over Hudson Bay, Canada. Arctic 58:370–382

Hailer F, Kutschera VE, Hallström BM et al (2012) Nuclear genomic sequences reveal that polar bears are an old and distinct bear lineage. Science 336:344–347

Hamilton SG, Castro del al Guardia L, Derocher AE et al (2014) Projected polar bear sea ice habitat in the Canadian Arctic archipelago. PLoS One 9(11):e113746. doi:10.1371/journal.pone.0113746

Harington CR (2008) The evolution of Arctic marine mammals. Ecol Appl 18(2 Suppl):S23–S40

Harwood LA, Stirling I (1992) Distribution of ringed seals in the southeastern Beaufort Sea during late summer. Can J Zool 70:891–900

Harwood LA, Smith TG, Auld JC (2012) Fall migration of ringed seals (*Phoca hispida*) through the Beaufort and Chukchi seas, 2001–02. Arctic 65:35–44

Harwood LA, Smith TG, George JC et al (2015) Change in the Beaufort Sea ecosystem: diverging trends in body condition and/or production in five marine vertebrate species. Prog Oceanogr 136:263–273

Holland MM, Serreze MC, Stroeve J (2010) The sea ice mass budget of the Arctic and its future change as simulated by coupled climate models. Clim Dyn 34:185–200

Horner R, Schrader GC (1982) Relative contributions of ice algae, phytoplankton, and benthic microalgae to primary production in the nearshore regions of the Beaufort Sea. Arctic 35:485–503

IPCC (2000) Special report on emission scenarios, Intergovernmental Panel onClimate Change. ISBN: 92-9169113-5. http://www.ipcc.ch/pdf/special-reports/spm/sres-en.pdf. Accessed 16 Mar 2016

IPCC (2013) Climate change 2013: the physical science basis. In: Stocker TF, Qin D, Plattner G-K, Tignor M, Allen SK, Boschung J, Nauels A, Xia Y, Bex V, Midgley PM (eds) Contribution of working group I to the fifth assessment report of the intergovernmental panel on climate change. Cambridge University Press, Cambridge, UK. , 1535 pp

Jakobsson M, Macnab RL, Mayer L et al (2008) An improved bathymetric portrayal of the Arctic Ocean: implications for ocean modeling and geological, geophysical and oceanographic analyses. Geophys Res Lett 35:L07602. doi:10.1029/2008GL033520

Kelly BP, Bengtson JL, Boveng PL et al. (2010) Status review of the ringed seal (*Phoca hispida*). U.S. Dep. Commer., NOAA Tech. Memo. NMFS-AFSC-212, 250 p

Laidre KL, Born EW, Heagerty P et al (2015) Shifts in female polar bear (*Ursus maritimus*) habitat use in East Greenland. Polar Biol 38:879–893

Laliberté F, Howell SEL, Kushner PJ (2016) Regional variability of a projected sea ice-free Arctic during the summer months. Geophys Res Lett 43:256–263

Lindqvist C, Schuster SC, Sun Y et al (2010) Complete mitochondrial genome of a Pleistocene jawbone unveils the origin of polar bear. Proc Natl Acad Sci 107(11):5053–5057

Lunn NJ, Servanty S, Regehr EV et al (2016) Demography of an apex predator at the edge of its range—impacts of changing sea ice on polar bears in Hudson Bay. Ecol Appl. doi:10.1890/15-1256

Manly BFJ, McDonald LL, Thomas DL et al (2002) Resource selection by animals. Kluwer Academic, Dordrecht

Mauritzen M, Belikov SE, Boltunov AN et al (2003) Functional responses in polar bear habitat selection. Oikos 100:112–124

Meier WN, Stroeve J, Fetterer F (2007) Whither Arctic Sea ice? A clear signal of decline regionally, seasonally and extending beyond the satellite record. Ann Glaciol 46:428–434

Messier F, Taylor MK, Ramsay MA (1992) Seasonal activity patterns of female polar bears (*Ursus maritimus*) in the Canadian Arctic as revealed by satellite telemetry. J Zool 226:219–229

Miller W, Schuster SC, Welch AJ et al (2012) Polar and brown bear genomes reveal ancient admixture and demographic footprints of past climate change. Proc Natl Acad Sci 109(36): E2382–E2390

Molnár PK, Derocher AE, Thiemann GW et al (2010) Predicting survival, reproduction and abundance of polar bears under climate change. Biol Conserv 143:1612–1622

Molnár PK, Derocher AE, Klanjscek T, Lewis MA (2011) Predicting climate change impacts on polar bear litter size. Nat Commun 2:186. doi:10.1038/ncomms1183

Molnár PK, Derocher AE, Thiemann GW et al (2014) Corrigendum to "predicting survival, reproduction and abundance of polar bears under climate change". Biol Conserv 143:1612–1622. doi:10.1016/j.biocon.2014.07.001

Monnett C, Gleason JS (2006) Observations of mortality associated with extended open-water swimming by polar bears in the Alaskan Beaufort Sea. Polar Biol 29:681–687

Moran K, Backman J, Brinkhuis H et al (2006) The Cenozoic palaeoenvironment of the Arctic Ocean. Nature 441:601–605

National Ice Center (2016) Sea ice charts. http://www.natice.noaa.gov. Accessed 22 Nov 16

National Snow and Ice Data Center (2012) Arctic sea ice extent settles at record seasonal minimum.http://nsidc.org/arcticseaicenews/2012/09/arctic-sea-ice-extent-settles-at-record-seasonal-minimum. Accessed 22 Nov 16

National Snow and Ice Data Center (2016a) 1981–2010 average. http://nsidc.org/arcticseaicenews/charctic-interactive-sea-ice-graph/. Accessed 3 March 2016

National Snow and Ice Data Center (2016b) Homepage. http://nsidc.org. Accessed 22 Nov 16

Obbard ME, Thiemann GW, Peacock E, Debruyn TD (2010) Polar bears: proceedings of the 15th working meeting of the IUCN/SSC Polar Bear Specialist Group, Copenhagen, Denmark, 29 June–3 July 2009. IUCN, Gland

Obbard ME, Cattet MR, Howe EJ et al (2016) Trends in body condition in polar bears (*Ursus maritimus*) from the southern Hudson Bay subpopulation in relation to changes in sea ice. Arch Sci. doi:10.1139/AS-2015-0027

Overland JE, Wang M (2013) When will the Arctic be nearly ice free? Geophys Res Lett 40:2097–2101

Paetkau D, Amstrup SC, Born EW et al (1999) Genetic structure of the world's polar bear populations. Mol Ecol 8:1571–1584

Pagano AM, Durner GM, Amstrup SC et al (2012) Long-distance swimming by polar bears (*Ursus maritimus*) of the southern Beaufort Sea during years of extensive open water. Can J Zool 90:663–676
Peacock E, Taylor MK, Laake J, Stirling I (2013) Population ecology of polar bears in Davis Strait, Canada and Greenland. J Wildl Manag 77:463–476
Peacock E, Sonsthagen SA, Obbard ME et al (2015) Implications of the circumpolar genetic structure of polar bears for their conservation in a rapidly warming Arctic. PLoS One 10(1):e112021. doi:10.1371/journal.pone.0112021
Piatt JF, Springer AM (2003) Advection, pelagic food webs and the biogeography of Beringia. Mar Ornithol 31:141–154
Pilfold NW, Derocher AE, Richardson R (2014) Influence of intraspecific competition on the distribution of a wide-ranging, non-territorial carnivore. Glob Ecol Biogeogr 23:425–435
Pilfold NW, McCall A, Derocher AE, Lunn NJ, Richardson E (2016) Migratory response of polar bears to sea ice loss: to swim or not to swim. Ecography 39. doi:10.1111/ecog.02109
Ramsay MA, Stirling I (1988) Reproductive biology and ecology of female polar bears (*Ursus maritimus*). J Zool 214:601–633
Regehr EV, Lunn NJ, Amstrup SC et al (2007) Effects of earlier sea ice breakup on survival and population size of polar bears in western Hudson Bay. J Wildl Manag 71:2673–2683
Regehr EV, Hunter CM, Caswell H et al (2010) Survival and breeding of polar bears in the southern Beaufort Sea in relation to sea ice. J Anim Ecol 79:117–127
Rode KD, Amstrup SC, Regehr EV (2010) Reduced body size and cub recruitment in polar bears associated with sea ice decline. Ecol Appl 20:768–782
Rode KD, Peacock E, Taylor M et al (2012) A tale of two polar bear populations: ice habitat, harvest, and body condition. Popul Ecol 54:3–18
Rode KD, Regehr EV, Douglas DC et al (2014) Variation in the response of an Arctic top predator experiencing habitat loss: feeding and reproductive ecology of two polar bear populations. Glob Chang Biol 20:76–88
Sahanatien V, Derocher AE (2012) Monitoring sea ice habitat fragmentation for polar bear conservation. Anim Conserv 15:397–406
Sigler MF, Renner M, Danielson SL et al (2011) Fluxes, fins, and feathers. Relationships among Bering, Chukchi, and Beaufort seas in a time of climate change. Oceanography 24:112–127
Spreen G, Kwok R, Menemenlis D (2011) Trends in Arctic Sea ice drift and role of wind forcing: 1992–2009. Geophys Res Lett 38. doi:10.1029/2011GL048970
Stirling I (1997) The importance of polynyas, ice edges, and leads to marine mammals and birds. J Mar Syst 10:9–21
Stirling I, Archibald WR (1977) Aspects of predation of seals by polar bears. J Fish Res Board Can 34:1126–1129
Stirling I, Derocher AE (2012) Effects of climate warming on polar bears: a review of the evidence. Glob Chang Biol 18:2694–2706
Stirling I, Øritsland NA (1995) Relationships between estimates of ringed seal (*Phoca hispida*) and polar bear (*Ursus maritimus*) populations in the Canadian Arctic. Can J Fish Aquat Sci 52:2594–2612
Stirling I, Parkinson CL (2006) Possible effects of climate warming on selected populations of polar bears (*Ursus maritimus*) in the Canadian Arctic. Arctic 59:261–275
Stirling I, Andriashek D, Calvert W (1993) Habitat preferences of polar bears in the western Canadian Arctic in late winter and spring. Polar Rec 29:13–24
Stirling I, Lunn NJ, Iacozza J (1999) Long-term trends in the population ecology of polar bears in western Hudson Bay in relation to climatic change. Arctic 52:294–306
Stirling I, McDonald TL, Richardson ES et al (2011) Polar bear population status in the northern Beaufort Sea, Canada, 1971–2006. Ecol Appl 21:869–876
Stroeve J, Holland MM, Meier W et al (2007) Arctic Sea ice decline: faster than forecast. Geophys Res Lett 34. doi:10.1029/2007GL029703

Whiteman JP, Harlow HJ, Durner GM et al (2015) Summer declines in activity and body temperature offer polar bears limited energy savings. Science 349:295–298

Wilson RR, Horne JS, Rode KD et al (2014) Identifying polar bear resource selection patterns to inform offshore development in a dynamic and changing Arctic. Ecosphere 5:136. doi:10.1890/ES14-00193.1

Chapter 24
Implications of Rapid Environmental Change for Polar Bear Behavior and Sociality

Todd C. Atwood

Abstract Historically, the Arctic sea ice has functioned as a structural barrier that has limited the nature and extent of interactions between humans and polar bears (*Ursus maritimus*). However, declining sea ice extent, brought about by global climate change, is increasing the potential for human-polar bear interactions. Loss of sea ice habitat is driving changes to both human and polar bear behavior—it is facilitating increases in human activities (e.g., offshore oil and gas exploration and extraction, trans-Arctic shipping, recreation), while also causing the displacement of bears from preferred foraging habitat (i.e., sea ice over biologically productive shallow) to land in some portions of their range. The end result of these changes is that polar bears are spending greater amounts of time in close proximity to people. Coexistence between humans and polar bears will require imposing mechanisms to manage further development, as well as mitigation strategies that reduce the burden to local communities.

24.1 Introduction

The capacity for environmental change to influence human and wildlife behaviors is nothing new, neither as a theoretical construct nor an empirical reality. The resource heterogeneities and uncertainties arising from changing environmental conditions, in some cases mediated by human activities, have acted as significant stimuli throughout history for behavioral and social change. In some instances, humans and wildlife have responded similarly to environmental change. For example, both have relied on a variety of migration strategies to cope with the seasonality of precipitation patterns, primary productivity, and long-term drought (e.g., Leimgruber et al. 2001; Gereta et al. 2004; Turney and Brown 2007). In other cases, the responses of

T.C. Atwood
Polar Bear Research Program, U.S. Geological Survey, Alaska Science Center, Anchorage, AK, USA
e-mail: tatwood@usgs.gov

A. Butterworth (ed.), *Marine Mammal Welfare*, Animal Welfare 17,
DOI 10.1007/978-3-319-46994-2_24

human and wildlife to environmental change have diverged. For instance, in parts of Central and South America, the expansion of transportation infrastructure into previously inaccessible areas has led to deforestation and eventual conversion of land to agricultural practices which, in turn, has greatly decreased the diversity of local floral and faunal communities (Brooks et al. 2002; Mosandl et al. 2008). These examples demonstrate ways in which human and wildlife behaviors and coexistence have been shaped by interactions with the environment and helped biologists identify the relationship between disturbances and patterns of response within communities.

In recent years, biologists have identified key life-history and behavioral traits that influence how species respond to environmental changes, such as habitat fragmentation and loss. Life-history traits, such as body size, govern perceptual ability (e.g., ability to perceive landscape heterogeneity) and determine the spatial scale over which organisms access various resources (e.g., prey, refugia, mates; Wiens 1989). Behavioral traits are usually consistent among populations of a species, are often easier to assess than population demographic variables, and are associated with various environmental variables of interest. Behavioral traits, like plasticity, determine the breadth of a species repertoire for responding to environmental cues, including stressors (Sih 2013). Collectively, these sorts of traits help determine the sensitivity of species to environmental change. As a logical corollary, identifying such traits may allow for greater insight into species responses to rapid environmental changes, like those brought about by a changing climate.

Life-history traits of polar bears (*Ursus maritimus*) are well documented. Polar bears are solitary apex predators that range over vast expanses of sea ice, specialize in preying on ice seals (primarily ringed (*Pusa hispida*) and bearded (*Erignathus barbatus*) seals), are long-lived, and are characterized by long generation times and interbirth intervals and extended care of dependent young (e.g., Amstrup 2003). By contrast, behavioral traits of polar bears are less well documented. The intent of this chapter is not to provide an extensive review of polar bear behavior, but to provide an overview of how human activities in the Arctic are likely to change behavioral traits that may influence polar bear sensitivity and, possibly, resilience to increasing human activities.

24.2 A Brief Primer on Polar Bear Sociality

Polar bears are non-territorial, ranging over large areas, and males tend to have larger home ranges than females (Amstrup et al. 2001; Laidre et al. 2013). Like most large carnivores, they have a relatively flexible social system that can vary seasonally and geographically. For example, while on the sea ice, males and females are mostly solitary except for the breeding season (April to May), when males adjust their spatial behavior to minimize encounters with other males and maximize encounters with females in estrus (Laidre et al. 2013). Females with dependent young tend to segregate from adult males while raising cubs, and the family group will remain together for approximately 2.5 years, with separation occurring when the adult female breeds again. In subpopulations where land use occurs during summer, bears have been observed aggregating into loose groups. For example, in western Hudson Bay (WHB), polar bears on land in summer will spatially segregate based on age and sex (Lunn and Stirling 1985; Derocher and Stirling 1990). Males generally remain closer to the coast and will sometimes form small "bachelor groups" of ≥2 individuals and engage in bouts of play (e.g., sparring; Latour 1981). Conversely, family groups and pregnant females will typically move further inland from the coast to be closer to denning areas (Lunn and Stirling 1985).

Similar to grizzly bears (*Ursus arctos*), polar bears have been observed to be fairly tolerant of each other around food resources. Near Churchill, Manitoba, Canada (WHB subpopulation), polar bears routinely visited local garbage dumps prior to their closure in 2005 (Towns et al. 2009), and while females with cubs could be intolerant of other bears, interactions between sex and age classes were generally nonaggressive (Lunn 1986). In Kaktovik, Alaska, USA (southern Beaufort Sea subpopulation [SB]), large numbers of polar bears congregate near a site ("bone pile") where the unused remains of subsistence-harvested bowhead whales (*Balaena mysticetus*) are dumped (Miller et al. 2006; Atwood et al. 2015). In general, incidents of interspecific strife appear to be rare at or near the bone pile, and mixed family groups often rest in close proximity to each other (Miller et al. 2015; Fig. 24.1).

Polar bear social behavior is more varied and nuanced than described above, but knowledge of the full scope is limited by the challenges associated with being able to routinely observe behavior. Key behaviors such as interspecific interactions and spatial aggregations (grouping together) are readily detectable and are sensitive to

Fig. 24.1 Polar bear family groups resting on a bluff near the bowhead whale bone pile adjacent to the community of Kaktovik, Alaska, USA, September 2015. Family groups often temporally partition use of the bone pile to avoid adult males and aggression between family groups is rare. *Image credit: S.W. Breck*

changes in environmental conditions and the availability/accessibility of prey. Derocher and Stirling (1990) posited that aggregating behavior (including the social behavior reported by Latour [1981]) in male polar bears from the WHB subpopulation may function to promote familiarity and thereby reduce the intensity of competition when individuals encounter each other on the sea ice. If correct, and applicable to other regions of the Arctic, aggregating behavior may help attenuate the potential for interspecific strife as climate-mediated accessibility of prey declines over time.

24.3 Stressors Likely to Induce Changes in Bear Behavior

24.3.1 Loss of Sea Ice Habitat

Polar regions are experiencing the most acute effects of climate change. Observations over the past 50 years show a pronounced decline in Arctic sea ice extent throughout the year, with the most prominent retreat in summer (Fig. 24.2; Serreze et al. 2007; Comiso 2012). Arctic-wide, the duration of the open-water season (i.e., the period of time between the breakup of sea ice in summer and freeze-up in fall) has increased at a rate of approximately 5 days/decade since 1979 (Stroeve et al. 2014), and some analysts have suggested that the Arctic may be ice-free (i.e., <1 million km^2 of sea ice) in September by 2030 or possibly earlier (Wang and Overland 2009; Overland and Wang 2013). Because of the phenology of sea ice (i.e., timing of annual formation and breakup), the duration of the open-water period is typically greatest for biologically productive (i.e., shallow) waters.

Loss of sea ice over biologically productive waters represents a loss of habitat for polar bears (see Chap. 23 for a full discussion of polar bear-habitat relationships). Durner et al. (2009) used general circulation models (GCMs), and polar bear habitat-use data to project that mean summer optimal habitat within the Polar Basin

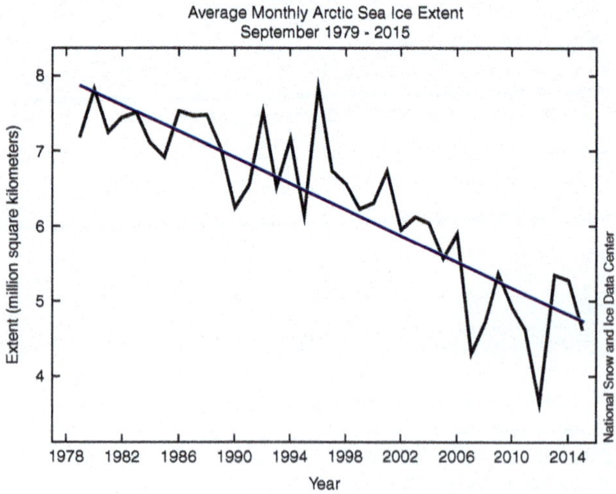

Fig. 24.2 Arctic-wide monthly September ice extent for 1979–2015, showing a decline of 13.4% per decade relative to the 1981–2010 average. Source: National Snow and Ice Data Center (NSIDC, Boulder, Colorado, USA). (Please see also further information on sea ice change described in Chap. 23)

(approximately 50% of the most northerly portion of polar bear range) could decrease by 68% by the end of the twenty-first century. Changes to sea ice phenology and habitat quality and distribution have influenced subpopulation vital rates (see Chap. 6.1) and are predicted to also influence subpopulation connectivity and the degree of genetic isolation. The latter are cause for concern given their potential to ultimately influence range-wide population viability.

24.3.2 *Industrialization*

The melting of Arctic sea ice will effectively unlock the Arctic Ocean, leaving it increasingly open to expanding human infrastructure and activities. Arguably, the human activity that generates the greatest concern for the polar bear is resource extraction, which is largely facilitated by increased maritime access. By mid-century, projections of oceanic shipping activity suggest that the five polar bear range state nations will gain increased maritime access to their currently designated exclusive economic zones (i.e., the region for each maritime nation that extends 370 nautical kilometers from the coast; Fig. 24.3), with the greatest increase in access

Fig. 24.3 A map of the exclusive economic zones (EEZ) of coastal Arctic countries. All countries are entitled to claim an EEZ extending up to 200 miles (*red dotted line*) (370 nautical km) from their coast. The "agreed borders" line delineates the border between each countries claimed EEZ. Where a maritime boundary has yet to be agreed, a line ("unsettled borders") equidistant between the countries is shown. *Source: http://www1.american.edu/ted/ICE/lomonosov.html*

occurring for Greenland (28% increase), followed by Canada (19%), Russia (16%), the USA (5%), and Norway (1%) (Stephenson et al. 2011). In turn, increased transiting of Arctic waters will make accessing and transporting the substantive untapped reserves of oil and gas more cost efficient. Projections suggest that the offshore region may contain up to 13% of the global mean estimate of undiscovered oil and up to 30% of the global mean estimate of undiscovered gas (Gautier et al. 2009), which provides a substantial economic incentive for Arctic nations to invest in infrastructure (roads, ports, security) to support industrialization.

Increased maritime access and the likely presence of enormous oil and gas reserves will spur the development of increased infrastructure and human activities in the Arctic, and the impacts of this will be greatest along the coastal and nearshore regions, areas that include valuable foraging (nearshore sea ice) and refugia (coastal lands) habitats for polar bears (Durner et al. 2009; Atwood et al. 2016a, b). In the USA, oil and gas development has largely been restricted to the coastal plain of Alaska's North Slope since 1977 (though nearshore extraction occurs from four artificial islands within 9 km of the coast; Fig. 24.4). However, additional offshore development plans are in place that would greatly expand the existing industrial footprint. For example, the Chukchi Sea Planning Area (CSPA) for offshore oil and

Fig. 24.4 Aerial view of an oil rig and infrastructure located in Prudhoe Bay, southern Beaufort Sea, Alaska. The oil rig is on an artificial island located approximately 8 km from the coast. A pipeline (right side of figure) runs from the rig to the mainland. *Image credit: US Geological Survey*

gas development lies approximately 50 km off the coast of northwestern Alaska. There are 487 active leases in the CSPA that cover an area of 225,000 km^2 (Bureau of Ocean Energy Management 2011), and seven exploration wells have been drilled since the early 1990s, with the most recent effort concluding in 2015. Recent analyses indicate that the CSPA may include between 15 and 50% of the total high-value sea ice habitat for the Chukchi Sea (CS) subpopulation of polar bears, depending on the season (Wilson et al. 2014).

In the adjacent southern Beaufort Sea Planning Area (BSPA), there are 183 active leases that cover an area of approximately 4000 km^2, with production occurring at three sites (Bureau of Ocean Energy Management 2011). Anticipated activities (within the next 10 years) for both planning areas include exploration—with seismic and ancillary surveys when sea ice is present—and eventual extraction (Bureau of Ocean Energy Management 2011). In other areas of the Arctic, there are also plans to conduct exploration and extraction activities (Harsem et al. 2011). Depending on the types of activities conducted, infrastructure could include icebreakers, drillships, multiple support vessels (e.g., for oil storage, spill response), and steel drilling caissons (platforms) as well as tugboat support. Accordingly, these activities could result in fragmentation of sea ice from icebreakers, noise disturbances, and the unintentional release of oil and industrial solvents/chemicals (Johnson et al. 2005; Amstrup et al. 2006; Moore et al. 2012).

To recapitulate, the primary threat to the long-term persistence of polar bears is climate-mediated loss of sea ice habitat (Amstrup et al. 2008; Atwood et al. 2016a); many of the other threats they face, like expanding human activities, are possible mainly because of the changing climate. Below, I discuss how loss of sea ice habitat and increasing anthropogenic activities are likely to drive changes in polar bear social structure and behavior.

24.4 Potential Changes in Social Structure and Behavior

24.4.1 Disruption of Gene Flow Between Subpopulations

Movement of individual polar bears between subpopulations has been clearly evident (Taylor et al. 2001; Mauritzen et al. 2002; Amstrup et al. 2004; Crompton et al. 2008) and led some to conclude that polar bears were panmictic (i.e., no mating restrictions throughout the entire range) and thus a single evolutionary unit (Paetkau et al. 1999). However, a recent circumpolar analysis of genetic structure and gene flow dynamics found that the 19 subpopulations studied clustered into four intermediate-scale units that corresponded to current ecological and oceanographic factors (e.g., patterns and timing of sea ice formation and melt, the convergence and divergence of ice from shore, predominant patterns of ice movement; Peacock et al. 2015). Additionally, this same work detected evidence of directional gene flow along a low-to-high latitudinal gradient for a subset of these clusters, which was considered to be generally consistent with a hypothesis of portions of the high Arctic serving as a refugium for polar bears as loss of sea ice habitat continues.

The concern is that continued reductions in the availability and connectivity of sea ice habitat will result in transient refugia and will trigger meta-population dynamics (Peacock et al. 2015) characterized by differential productivity and variable viability of groups of subpopulations. The mainland-island meta-population model, which involves the presence of suitable habitat patches located within dispersal distance from a large habitat patch that supports a permanent population of the species (Hanski and Simberloff 1997), fits polar bear dynamics that are likely to occur if loss of sea ice habitat continues. However, this model requires (1) density-dependent dispersal from mainland (source or refugia) subpopulations to island (sink) subpopulations and (2) enough habitat stability within sink subpopulations to allow year-round residency. The assumption of density-dependent dispersal rests on the notion that carrying capacity in refugia subpopulations is high enough to trigger density-dependent responses, which is not always the case (Betini et al. 2015). Also, while polar bears display high levels of vagility (i.e., ability to move over large areas), there is limited information available to gauge whether changing sea ice conditions may be impacting dispersal ability and frequency. For example, documentation of increased frequencies of long-distance swims (Pagano et al. 2012; Pilfold et al. 2016) suggests that bears are capable of traversing dispersal-like distances through water, but it is unknown if the increase in swimming behavior has compensated for decreased capacity to make long-distance movement over sea ice through loss of navigable ice. Finally, the assumption that there will be sufficient habitat to accommodate the year-round presence of individuals in sink subpopulations is tenuous (see Chap. 23). Projections of future sea ice extent generally show contraction of ice toward the pole, rather than fragmentation dynamics that create "islands" of sea ice habitat in an open-water matrix (IPCC 2014). Given the above, the more likely scenario for circumpolar social structure of polar bears is a coalescence around the Canadian Archipelago (projected to retain year-round sea ice the longest), with some subpopulations seasonally separated during the open-water period due to geographical features and transient use of a wider region of the Arctic during winter when sea ice is at its maximum extent.

24.4.2 Potential Impacts on Denning Behavior

A growing body of evidence indicates that polar bear denning behavior may be compromised by the continued loss of sea ice habitat. In most subpopulations, polar bears establish maternity dens either on land or landfast ice (Ramsay and Stirling 1990; Durner et al. 2003) and typically enter these dens by early November, after sea ice refreezes, providing a connective link between the pack ice and land. However, the increasing duration of the open-water period, and the protracted loss of connectivity between pack ice and land, may impede the ability of bears that mostly reside on sea ice year-round to safely access terrestrial denning habitat (Derocher et al. 2004) (bears in subpopulations that display a seasonal ice life history are onshore prior to denning on land and thus do not rely on connectivity of sea ice for accessing denning habitat). In the subpopulations of the Polar Basin (northern Beaufort Sea, SB, CS, Laptev Sea, and Kara Sea), the mean date of freeze-up has retreated at a rate approaching 13 days/decade since 1979 (Stroeve et al. 2014). If the pace of retreating freeze-up is maintained, connectivity between pack ice and terrestrial denning habitat could eventually be lost for some subpopulations by the middle of the century, if not sooner (Derocher et al. 2004; Stirling and Derocher 2012).

There does appear to be some capacity for plasticity in polar bear denning behavior. For example, in the SB, denning historically occurred mainly on the sea ice (Lentfer and Hensel 1980; Amstrup and Gardner 1994). However, between 1985 and 2005, a gradual shift in behavior occurred, marked by an increasing frequency of land-based dens such that by 2005, a majority of dens (67%) were located on land (Fischbach et al. 2007). This change in denning behavior occurred alongside declines in multi-year ice which could have affected sea ice stability for denning (Fischbach et al. 2007). The change also occurred concurrent with a greater frequency of long-distance swim events and increasing land use (Pagano et al. 2012; Atwood et al. 2016b), both of which may represent attempts by bears to cope with the increasing open-water period while maintaining access to denning habitat. Polar bear subpopulations that generally remain with the sea ice prior to moving to terrestrial habitats for denning are likely to face growing challenges of accessing land due to the lengthening open-water period (Bergen et al. 2007). It remains unknown whether polar bears have the capacity to shift from terrestrial denning back to a sea ice substrate. There may be a threshold for sea ice retreat, in which greater distances of pack ice from shore preclude bears from safely accessing land, and they choose, instead, to den on sea ice. However, that choice will not be feasible where inadequate pack ice is present.

24.4.3 Changes in Spatial Behavior

As mentioned above, in some subpopulations, bears are displaying divergent behavioral strategies for coping with the loss of sea ice habitat, while in others changing sea ice phenology is altering the timing of seasonal migratory movements. In the SB and CS subpopulations, polar bears have mostly remained on the sea ice year-round (with the exception of denning on land). However, over the last few decades, the protracted absence of sea ice from the continental shelf has driven growing proportions of the subpopulations to choose to summer on land rather than remain with the retreating sea ice. For example, in the SB, the proportion of radio-collared female bears on land between August and October increased from approximately 10% to 35% between historical (pre-2000) and contemporary periods (post-2000) (Atwood et al. 2016b), in the CS the proportion of radio-collared female bears on land increased from 20% (1986–1995) to 39% (2008–2013) (Rode et al. 2015). For both subpopulations, the average length of stay on land increased by approximately 30 days and was influenced by the lengthening open-water period. In WHB, where polar bears have historically retreated to land after sea ice melts completely in summer, changes in sea ice phenology have caused bears to adjust the timing of migration from sea ice to land and back. Cherry et al. (2013) examined the timing of arrival of polar bears onshore and departure back to see ice over two time periods, 1991–1997 and 2004–2009, and found the timing of migration showed trends of earlier arrival onshore in summer and later departure back to sea ice in fall, which closely tracked changes in the timing of sea ice breakup in spring and freeze-up in fall.

Changes to sea ice phenology and sea ice habitat connectivity is also believed to be responsible for the increased frequency of long-distance swims by polar bears, some of which are associated with seasonal migration from sea ice to land. In the SB and CS, Pagano et al. (2012) identified 50 long-distance swims (i.e., >50 km) by 20 bears over 6 years (2004–2009). Swim durations ranged from 1 to 10 days, distances ranged from 54 km to 688 km, and the frequency of swim events increased over time (Pagano et al. 2012). Pilfold et al. (2016) observed similar behavior in the Beaufort Sea and Hudson Bay from 2007 to 2012: swim duration ranged from 1 to 9 days, and distances ranged from 51 km to 404 km. In both studies, the majority of swims were associated with seasonal migrations between pack ice and land. Interestingly, Pilfold et al. (2016) found that the frequency of swims was similar for lone adult females and subadults, but lower for adult females with dependent young. This suggests that risk-averse individuals, such as adult females with cubs, may be less likely to undertake a long-distance swim given the energetic consequences and risk to cub survival. Indeed, Durner et al. (2011) describe an instance in which an adult female with a yearling cub made a 687-km swim shortly after being radio-collared. When the adult female was recaptured approximately 2 months later, she had lost 22% of her body mass and the yearling cub (Durner et al. 2011). These results support the hypothesis that long-distance swimming by polar bears is likely to occur more frequently as sea ice conditions change due to climate warming and represent an energetically costly and potentially dangerous mode of transiting fragmented habitat.

24.4.4 Changes to Foraging Behavior

Sea ice loss is increasingly limiting spring and summer hunting opportunities for polar bears (Stirling and Derocher 2012), thereby reducing energy stores available to maintain body condition during the coincident period of general food deprivation (Rode et al. 2010). One way in which bears may cope with declining foraging opportunities, and extended food deprivation, is to limit energy expenditure through reduced activity. For example, Whiteman et al. (2015) found that bears in the Beaufort Sea reduced levels of activity during the open-water season, regardless of whether they were on land or the retreating pack ice. This behavior could represent an attempt to reduce the energetic cost of continued searching for prey by increasing the amount of time spent inactive/resting until prey or other food resources become more accessible.

There is growing consensus that, in summer, polar bears that remain on the retreating pack ice are likely to have limited opportunities to encounter prey (Derocher et al. 2004; Stirling and Derocher 2012; Whiteman et al. 2015). Alternatively, polar bears that spend the summer on land may have greater opportunity to forage on a variety of food items, but the energetic benefit may not be sufficient to prevent declines in body condition in most cases (Rode et al. 2015). For example, in the Hudson Bay region, small numbers of polar bears have been documented consuming terrestrial foods (e.g., Russell 1975; Lunn and Stirling 1985; Ramsay and Hobson 1991; Rockwell and Gormezano 2009; Smith et al. 2010; Iversen et al. 2013), yet body condition and survival have generally declined (Stirling et al. 1999; Regehr et al. 2007). By contrast, polar bears from the SB that summer on land are able to scavenge on human-provisioned bowhead whale (*Balaena mysticetus*) remains (Atwood et al. 2016b; Fig. 24.5), and there is some evidence to suggest that these bears are in better condition than individuals that summer on the sea ice (US Geological Survey, unpublished data). However, it is important to note that a relatively small proportion (approximately 18%; Atwood et al. 2016b) of the SB subpopulation comes ashore during summer, so there is often limited demand for abundant bowhead whale remains. These findings corroborate the notion that the energy obtained from feeding on most terrestrial-based foods (i.e., protein- and carbohydrate-rich foods) is likely insufficient to offset daily energy expended. If polar bears eventually are forced to spend ≥5 months on land, it is unlikely that they will

Fig. 24.5 Polar bear family group (mother and cubs) at the bowhead whale "bone pile" in Kaktovik, Alaska, 2009. *Image credit: US Geological Survey*

be able to adjust their feeding behavior in a way sufficient enough to stave off declines in body mass (e.g., Molnár et al. 2014; Robbins et al. 2013).

24.4.5 Social Behavior While on Land

The supplemental feeding of polar bears that occurs along Alaska's North Slope (see Chap. 21 for further detail) is an interesting example of how a seemingly benign human activity has altered the behavior of both bears and humans. Briefly, there are three communities along the North Slope (SB region) that hunt bowhead whales in the fall. The unused whale remains are aggregated into bone piles that then attract large numbers of bears (e.g., up to ≈80 individuals around a single location; Atwood et al. 2016b), mostly individuals that were summering onshore, but occasionally bears will make a long-distance swim from pack ice to a bone pile location (US Geological Survey, unpublished data). Only one bone pile (adjacent to the community of Kaktovik) is both consistently used and accessible enough to allow polar bear viewing, and a fledgling commercial bear-viewing industry has developed as a result. During peak viewing (which typically occurs in October), it is not uncommon for 10–20 tourists and local residents/day to use vehicles and boats to view bears feeding at the bone pile and resting at nearby barrier islands, respectively.

The demographic characteristics of polar bears visiting bone piles reflects that of the SB subpopulation, but there is some indication of demographic groups partitioning the use of bone piles relative to the presence of people, conspecifics, and allospecifics. For example, Miller et al. (2015) observed temporal partitioning of the Kaktovik bone pile, with lone adult polar bears and grizzly bears mostly feeding at night, while polar bear family groups (adult females with dependent young) and subadults fed mostly during dawn and dusk, when more people were usually present. Females with dependent young may not be more tolerant of the presence of people; rather they may be more wary of the presence of adult males given the potential for infanticide (Rode et al. 2006; Derocher and Wiig 1999). Conversely, adult males may be more wary of people given that they likely experience the greatest hunting pressure, as evidenced by the disproportional harvest of adult males (Derocher et al., 1997; Rode et al. 2006; Molnar et al., 2008). The observation that adult male bears resting on land near Churchill were more vigilant in the presence of tourists corroborates the idea of their elevated wariness of people (Dyck and Baydack 2004). Importantly though, it is unknown if competitive interactions or wariness of people interferes with feeding behavior at bone piles.

24.4.6 Potential Impacts on Mating Success

Polar bears occur at low densities and range over vast areas, so they must rely on key behavioral traits to maximize the encounter of potential mates. As mentioned previously, movement data suggest that during the breeding season males minimize the likelihood of encountering each other, and maximize the likelihood of encountering females, by increasing the tortuosity of movement paths (Laidre et al. 2013). Additionally, ursids often rely on the deposition of scent marks on vertical substrates to communicate with conspecifics, but vertical features on sea ice are transient and make for poor substrates for chemical communication. Instead, there is evidence that polar bears may rely on pedal scent marking (via tracks left in snow) as a way to chemically communicate information (such as reproductive status) that may further facilitate encountering mates (Owen et al. 2015). However, the unpredictability of the Arctic sea ice, which has become more dynamic in spring as ice has thinned (Asplin et al. 2014), may impede the efficacy of these mate-searching behaviors by obstructing movement paths or disrupting scent trails. If conditions contributing to the impairment of mate-searching behaviors persist, it could increase the potential for Allee effects (i.e., a feature of small populations whereby low density limits individual fitness and population growth) in some subpopulations—especially those facing greater harvest pressure (in most subpopulations, males are harvested disproportionate to their abundance; see Chap. 25) (Molnar et al. 2008).

24.5 Conclusions

Increasingly, human behavior is causing changes to the Arctic marine ecosystem. These changes are occurring indirectly through the influence of greenhouse gas emissions on physical processes that drive long-term change in ecosystem processes and directly via in situ activities that can immediately alter the behavior and fitness of wildlife. Some wildlife species will be able to adapt to these changes, while others may not possess the plasticity necessary to persist (Van Hemert et al. 2015). Polar bears have evolved preferences for sea ice habitat and for preying on marine mammals, which have shaped their behavioral traits. But the rapid changes taking place in the Arctic are making it more difficult for polar bears and other ice-adapted species to reliably use their traditional habitats and maintain fitness (e.g., Stirling and Derocher 2012; Derocher et al. 2013). Behavioral plasticity is the initial response to dramatic environmental perturbations, followed by transmission of innovative behaviors within and across generations, eventually leading to evolution of the behavioral response over time (Tuomainen and Candolin 2011). However, behavioral plasticity may be an effective response by polar bears only if the rate of environmental change does not outpace transmission of behavioral innovations. In short, it remains to be seen whether polar bears (1) possess the plasticity to cope with these changes and (2) will have sufficient time to innovate and adapt.

In some subpopulations, the lengthening open-water season already has been linked to declines in fitness and survival (Stirling et al. 1999; Regehr et al. 2010; Rode et al. 2010; Obbard et al. 2016). One of the purported mechanisms for these declines is prolonged food deprivation brought about by a lack of access to prey when sea ice is unavailable (Stirling and Derocher 2012). Polar bears have long possessed a feast-and-famine lifestyle, relying on fat reserves accumulated in the spring to ameliorate drastic declines in body condition during times when seals are less available for capture. However, the plasticity of this lifestyle has limits. As mentioned earlier in this chapter, energy budget models indicate that an open-water period lasting >150 days could pose a substantial risk of reproductive failure and starvation. When ice is completely absent, bears will have no choice but to come ashore. The timing and pattern of sea ice melt and the presence of terrestrial-based food resources (e.g., Gormezano and Rockwell 2015; Rode et al. 2015) will largely determine when and where bears come ashore during the open-water period—some of these areas will be in close proximity to centers of human activities where food attractants are present. The eventual influx of a rising number of nutritionally stressed polar bears around coastal communities will likely result in increases in deaths from starvation, interspecific killing (e.g., Derocher and Wiig 1999), and human-bear conflict, all of which will challenge the limited capacities of managers.

Long-term projections of population status suggest that approximately two-thirds of the world's polar bears could be lost by the middle of this century, unless the global mean temperature is held to ≤2 °C above preindustrial levels (Amstrup et al. 2008; Amstrup et al. 2010). It is important to note that there will be temporal variation in the responses of subpopulations to sea ice loss. Some subpopulations already have experienced the effects of sea ice loss (e.g., Regehr et al. 2007, 2010; Bromaghin et al. 2015), while others have remained stable during an initial period of sea ice loss (e.g., Rode et al. 2014). However, ultimately the projections of sea ice loss are so dramatic that it is unclear how bears will be able to persist long-term in most parts of their current range. Until greenhouse gas emissions are stabilized and further sea ice loss is stopped, management of in situ stressors may serve to slow the transition of populations to progressively worsened outcomes (Atwood et al. 2016a), thereby buying much needed time for polar bears to adapt to a new stable state and hopefully improve the prospects for their long-term persistence. Developing a more thorough understanding of polar bear behavior and their capacity for flexibility in response to anthropogenic disturbances and subsequent mitigations may lead to successful near-term management interventions.

References

Amstrup SC, Gardner C (1994) Polar bear maternity denning in the Beaufort Sea. J Wildl Manag 58:1–10

Amstrup SC, Durner GM, McDonald TL et al (2001) Comparing movement patterns of satellite-tagged male and female polar bears. Can J Zool 79:2147–2158

Amstrup SC (2003) Polar bear, *Ursus maritimus*. In: Feldhamer GA, Thompson BC, Chapman JA (eds) Wild Mammals of North America: Biology, Management, and Conservation. Johns Hopkins University Press, Baltimore, Maryland, pp 587–610

Amstrup SC, McDonald TL, Durner GM (2004) Using satellite radiotelemetry data to delineate and manage wildlife populations. Wildl Soc Bull 32:661–679

Amstrup SC, Durner GM, McDonald TL et al (2006) Estimating potential effects of hypothetical oil spills on polar bears. U.S. Geological Survey, Alaska Science Center, Anchorage, Alaska

Amstrup SC, Marcot BG, Douglas DC (2008) A Bayesian network modeling approach to forecasting the 21st century worldwide status of polar bears. In: De Weaver ET, Bitz CM, Tremblay LB (eds) Arctic Sea Ice Decline: Observations, Projections, Mechanisms, and Implications. American Geophysical Union Geophysical Monograph No. 180, Washington, DC, pp 213–268

Amstrup SC, De Weaver ET, Douglas DC et al (2010) Greenhouse gas mitigation can reduce sea ice loss and increase polar bear persistence. Nature 468:955–960

Asplin MG, Scharien R, Else B et al (2014) J Geophys Res: Oceans 119:2327–2343

Atwood TC, Peacock E, McKinney MA et al (2015) Demographic composition and behavior of polar bears summering on shore in Alaska. U.S. Geological Survey, Administrative Report, Anchorage Alaska

Atwood TC, Marcot BG, Douglas DC et al (2016a) Forecasting the relative influence of environmental and anthropogenic stressors on polar bears. Ecosphere 7(6):e01370

Atwood TC, Peacock E, McKinney M et al (2016b) Rapid environmental change drives increased land use by an Arctic marine predator. PLoS One 11(6):e0155932

Bergen S, Durner GM, Douglas DC et al (2007) Predicting movements of female polar bears between summer sea ice foraging habitats and terrestrial denning habitats of Alaska in the 21st century: proposed methodology and pilot assessment. U.S. Geological Survey Administrative report, Anchorage, Alaska

Betini GS, Pardy A, Griswold CK et al (2015) The role of seasonality and nonlethal carry-over effects on density-dependent dispersal. Ecosphere 6:1–9

Bromaghin J, McDonald T, Stirling I et al (2015) Polar bears in the Beaufort Sea: population decline and stabilization in the 2000's. Ecol Appl 25:634–651

Brooks TM, Mittermeier RA, Mittermeier CG et al (2002) Habitat loss and extinction in the hotspots of biodiversity. Cons Biol 16:909–923

Bureau of Ocean Energy Management (BOEM) (2011) Biological evaluation: Oil and Gas Activities on the Beaufort and Chukchi Sea Planning Areas. Bureau of Ocean Energy Management, Anchorage, Alaska

Cherry SG, Derocher AE, Thiemann GW et al (2013) Migration phenology and seasonal fidelity of an Arctic marine predator in relation to sea ice dynamics. J Anim Ecol 82:912–921

Comiso JC (2012) Large decadal decline of the Arctic multiyear ice cover. J Clim 25:1176–1193

Crompton AE, Obbard ME, Petersen SD et al (2008) Population genetic structure in polar bears (*Ursus maritimus*) from Hudson Bay, Canada: Implications of future climate change. Biol Cons 141:2528–2539

Derocher AE, Stirling I (1990) Observations of aggregating behaviour in adult male polar bears (*Ursus maritimus*). Can J Zool 68:1390–1394

Derocher AE, Wiig Ø (1999) Infanticide and cannibalism of juvenile polar bears (*Ursus maritimus*) in Svalbard. Arctic 52:307–310

Derocher AE, Stirling I, Calvert W (1997) Male-biased harvesting of polar bears in Western Hudson Bay. J Wildl Manag 61:1075–1082

Derocher AE, Lunn NJ, Stirling I (2004) Polar bears in a warming climate. Integr Comp Biol 44:163–176

Derocher AE, Aars J, Amstrup SC et al (2013) Rapid ecosystem change and polar bear conservation. Conserv Lett 6:368–375

Durner GM, Amstrup SC, Fischbach AS (2003) Habitat characteristics of polar bear terrestrial maternal den sites in northern Alaska. Arctic 56:55–62

Durner GM, Douglas DC, Nielson RM et al (2009) Predicting 21st-century polar bear habitat distribution from global climate models. Ecol Monograph 79:25–58

Durner GM, Whiteman JP, Harlow HJ et al (2011) Consequences of long-distance swimming and travel over deep-water pack ice for a female polar bear during a year of extreme sea ice retreat. Polar Biol 34:975–984

Dyck MG, Baydack RK (2004) Vigilance behaviour of polar bears (*Ursus maritimus*) in the context of wildlife-viewing activities at Churchill, Manitoba, Canada. Biol Cons 116:343–350

Fischbach AS, Amstrup SC, Douglas DC (2007) Landward and eastward shift of Alaskan polar bear denning associated with recent sea ice changes. Polar Biol 30:1395–1405

Gautier DL, Bird KJ, Charpentier RR et al (2009) Assessment of undiscovered oil and gas in the arctic. Science 324:1175–1179

Gereta E, Meing'ataki GEO, Mduma S et al (2004) The role of wetlands in wildlife migration in the Tarangire ecosystem, Tanzania. Wetl Ecol Manag 12:285–299

Gormezano LJ, Rockwell RF (2015) The energetic value of land-based foods in Western Hudson Bay and their potential to alleviate energy deficits of starving adult male polar pears. PLoS One 10:e0128520. doi:10.1371/journal.pone.0128520

Hanski IA, Simberloff D (1997) The metapopulation approach, its history, conceptual domain, and application to conservation. In: Hanski IA, Gilpin ME (eds) Metapopulation Biology: Ecology, Genetics, and Evolution. Academic Press, San Diego, California, pp 5–26

Harsem Ø, Eide A, Heen K (2011) Factors influencing future oil and gas prospects in the Arctic. Energy Policy 39:8037–8045

Intergovernmental Panel on Climate Change (IPCC) (2014) Climate change: impacts, adaptation, vulnerability. Part A: global and sectoral aspects, Contribution of the Working Group II to the Fifth Assessment Report of the Intergovernmental Panel on Climate Change. Cambridge University Press, Cambridge, UK and NY

Iversen M, Aars J, Haug T et al (2013) The diet of polar bears (*Ursus maritimus*) from Svalbard, Norway, inferred from scat analysis. Polar Biol 36:561–571

Johnson CJ, Boyce MS, Case RL, Cluff HD, Gau RJ, Gunn A, Mulders R (2005) Cumulative effects of human developments on arctic wildlife. Wildl Monogr 160:1–36

Laidre KL, Born EW, Gurarie E et al (2013) Females roam while males patrol: divergence in breeding season movements of pack-ice polar bears (*Ursus maritimus*). Proc Royal Soc B 280: doi:10.1098/rspb.20122371

Latour PB (1981) Interactions between free-ranging, adult male polar bears (*Ursus maritimus* Phipps): a case of adult social play. Can J Zool 59:1775–1783

Leimgruber P, McShea WJ, Brookes CJ et al (2001) Spatial patterns in relative primary productivity and gazelle migration in the Eastern Steppes of Mongolia. Biol Cons 102:205–212

Lentfer JW, Hensel RJ (1980) Alaskan polar bear denning. Bears: Their Biol Manag 4:101–108

Lunn NJ, Stirling I (1985) The significance of supplemental food to polar bears during the ice-free period of Hudson Bay. Can J Zool 63:2291–2297

Lunn NJ (1986) Observations of nonaggressive behavior between polar bear family groups. Can J Zool 64:2035–2037

Mauritzen M, Derocher AE, Wiig Ø et al (2002) Using satellite telemetry to define spatial population structure in polar bears in the Norwegian and western Russian Arctic. J Appl Ecol 39:79–90

Miller S, Proffitt K, Schliebe S (2006) Demographics and behavior of polar bears feeding on bowhead whale carcasses at Barter and Cross Islands, Alaska, 2002–2004. Minerals management service technical report, Anchorage, Alaska, USA

Miller S, Wilder J, Wilson RR (2015) Polar bear-grizzly bear interactions during the autumn open-water period in Alaska. J Mammal 96:1317–1325

Molnar PK, Derocher AE, Lewis MA et al (2008) Modelling the mating system of polar bears: A mechanistic approach to the Allee effect. Proc R Soc B 275:217–226

Molnár PK, Derocher AE, Thiemann GW et al (2014) Corrigendum to "Predicting survival, reproduction and abundance of polar bears under climate change". Biol Cons 143:1612–1622

Moore SE, Reeves RR, Southall BL et al (2012) A new framework for assessing the effects of anthropogenic sound on marine mammals in a rapidly changing Arctic. Bioscience 62:289–295

Mosandl R, Günter S, Stimm B et al (2008) Ecuador suffers the highest deforestation rate in South America. In: Beck E, Bendix J, Kottke I, Makeschin F, Mosandl R (eds) Gradients in a tropical mountain ecosystem of Ecuador. Springer, Berlin Heidelberg, Germany, pp 37–40

Obbard ME, Cattet MRL, Howe EJ et al (2016) Trends in body condition in polar bears (*Ursus maritimus*) from the Southern Hudson Bay subpopulation in relation to changes in sea ice. Arct Sci 2:15–32

Overland JE, Wang M (2013) When will the Arctic be nearly ice free? Geophys Res Lett 40:2097–2101
Owen MA, Swaisgood RR, Slocomb C et al (2015) An experimental investigation of chemical communication in the polar bear. J Zool 295:36–43
Paetkau D, Amstrup SC, Born EW et al (1999) Genetic structure of the world's polar bear populations. Mol Ecol 8:1571–1584
Pagano AM, Durner GM, Amstrup SC et al (2012) Long-distance swimming by polar bears (*Ursus maritimus*) of the southern Beaufort Sea during years of extensive open water. Can J Zool 90:663–676
Peacock E, Sonsthagen SA, Obbard ME et al (2015) Implications of the circumpolar genetic structure of polar bears for their conservation in a rapidly warming Arctic. PLoS One 10(1):e112021
Pilfold NW, McCall A, Derocher AE et al (2016) Migratory response of polar bears to sea ice loss: To swim or not to swim. Ecography 40(1):189–199
Ramsay MA, Stirling I (1990) Fidelity of female polar bears to winter den sites. J Mammal 71:233–236
Ramsay MA, Hobson KA (1991) Polar bears make little use of terrestrial food webs: evidence from stable-carbon isotope analysis. Oecologia 86:598–600
Regehr EV, Lunn NJ, Amstrup SC et al (2007) Effects of earlier sea ice breakup on survival and population size of polar bears in western Hudson Bay. J Wildl Manag 71:2673–2683
Regehr EV, Hunter CM, Caswell H et al (2010) Survival and breeding of polar bears in the southern Beaufort Sea in relation to sea ice. J Anim Ecol 79:117–127
Robbins CT, Lopez-Alfaro C, Rode KD et al (2013) Hibernation and seasonal fasting in bears: the energetic costs and consequences for polar bears. J Mammal 93:1493–1503
Rockwell RF, Gormezano LJ (2009) The early bear gets the goose: climate change, polar bears and lesser snow geese in western Hudson Bay. Polar Biol 32:539–547
Rode KD, Farley SD, Robbins CT (2006) Sexual dimorphism, reproductive strategy, and human activities determine resource use by brown bears. Ecology 87:2636–2646
Rode KD, Amstrup SC, Regehr EV (2010) Reduced body size and cub recruitment in polar bears associated with sea ice decline. Ecol Appl 20:768–782
Rode KD, Robbins CT, Nelson L et al (2015) Can polar bears use terrestrial foods to offset lost ice-based hunting opportunities? Front Ecol Environ 13:138–145
Rode KD, Regehr EV, Douglas DC et al (2014) Variation in the response of an Arctic top predator experiencing habitat loss: feeding and reproductive ecology of two polar bear populations. Global Change Biol 20:76–88
Russell RH (1975) The food habits of polar bears of James Bay and southwest Hudson Bay in summer and autumn. Arctic 28:117–129
Sih A (2013) Understanding variation in behavioural responses to human-induced rapid environmental change: a conceptual overview. Anim Behav 85:1077–1088
Serreze MC, Holland MM, Stroeve J (2007) Perspectives on the Arctic's shrinking sea-ice cover. Science 315:1533–1536
Smith PA, Elliott KH, Gaston AJ et al (2010) Has early ice clearance increased predation on breeding birds by polar bears? Polar Biol 33:1149–1153
Stephenson SR, Smith LC, Agnew JA (2011) Divergent long-term trajectories of human access to the Arctic. Nat Clim Chang 1:156–160
Stirling I, Lunn NJ, Iacozza J (1999) Long-term trends in the population ecology of polar bears in western Hudson Bay in relation to climatic change. Arctic 52:294–306
Stirling I, Derocher AE (2012) Effects of climate warming on polar bears: a review of the evidence. Glob Chang Biol 18:2694–2706
Stroeve JC, Markus T, Boisvert L et al (2014) Changes in Arctic melt season and implications for sea ice loss. Geophys Res Lett 41:1216–1225
Taylor MK, Akeeagok S, Andriashek D et al (2001) Delineating Canadian and Greenland polar bear (*Ursus maritimus*) populations by cluster analysis of movements. Can J Zool 79:690–709

Towns L, Derocher AE, Stirling I et al (2009) Spatial and temporal patterns of problem polar bears in Churchill, Manitoba. Polar Biol 32:1529–1537

Tuomainen U, Candolin U (2011) Behavioral responses to human-induced environmental change. Biol Rev 86:640–657

Turney CSM, Brown H (2007) Catastrophic early Holocene sea level rise, human migration and the Neolithic transition in Europe. Quat Sci Rev 26:2036–2041

Van Hemert C, Flint P, Udevitz MS et al (2015) Forecasting wildlife response to rapid warming in the Alaskan Arctic. Bioscience 65:718–728

Wang M, Overland JE (2009) A sea ice free summer Arctic within 30 years? Geophys Res Lett 36:L07502. doi:10.1029/2009GL037820

Whiteman JP, Harlow HJ, Durner GM et al (2015) Summer declines in activity and body temperature offer polar bears limited energy savings. Science 349:295–298

Wiens JA (1989) Spatial scaling in ecology. Funct Ecol 3:385–397

Wilson RR, Horne JS, Rode KD et al (2014) Identifying polar bear resource selection patterns to inform offshore development in a dynamic and changing Arctic. Ecosphere 5:136. http://dx.doi.org/10.1890/ES14-00193.1

Chapter 25
Uncertainties in Forecasting the Response of Polar Bears to Global Climate Change

David C. Douglas and Todd C. Atwood

Abstract Several sources of uncertainty affect how precisely the future status of polar bears (*Ursus maritimus*) can be forecasted. Foremost are unknowns about the future levels of global greenhouse gas emissions, which could range from an unabated increase to an aggressively mitigated reduction. Uncertainties also arise because different climate models project different amounts and rates of future warming (and sea ice loss)—even for the same emission scenario. There are also uncertainties about how global warming could affect the Arctic Ocean's food web, so even if climate models project the presence of sea ice in the future, the availability of polar bear prey is not guaranteed. Under a worst-case emission scenario in which rates of greenhouse gas emissions continue to rise unabated to century's end, the uncertainties about polar bear status center on a potential for extinction. If the species were to persist, it would likely be restricted to a high-latitude refugium in northern Canada and Greenland—assuming a food web also existed with enough accessible prey to fuel weight gains for surviving onshore during the most extreme years of summer ice melt. On the other hand, if emissions were to be aggressively mitigated at the levels proposed in the Paris Climate Agreement, healthy polar bear populations would probably continue to occupy all but the most southern areas of their contemporary summer range. While polar bears have survived previous warming phases—which indicate some resiliency to the loss of sea ice habitat—what is certain is that the present pace of warming is unprecedented and will increasingly expose polar bears to historically novel stressors.

D.C. Douglas (✉)
U.S. Geological Survey, Alaska Science Center, Juneau, AK, USA
e-mail: ddouglas@usgs.gov

T.C. Atwood
U.S. Geological Survey, Alaska Science Center, Anchorage, AK, USA
e-mail: tatwood@usgs.gov

A. Butterworth (ed.), *Marine Mammal Welfare*, Animal Welfare 17,
DOI 10.1007/978-3-319-46994-2_25

25.1 Introduction

As discussed in the preceding chapters, polar bears (*Ursus maritimus*) have evolved preferences for sea ice habitat, which they rely on to meet a number of key life-history needs. However, global climate change, primarily caused by human activities that emit greenhouse gases, has caused the Arctic to warm at twice the rate of the rest of the planet (Overland et al. 2015). In turn, this warming has driven a multi-decadal reduction in sea ice extent that has been linked to declines in polar bear fitness and abundance in some subpopulations (e.g., Regehr et al. 2007; Rode et al. 2010; Obbard et al. 2016). As a result, continued loss of sea ice due to warming in the Arctic is considered the primary long-term threat to the persistence of polar bears (Atwood et al. 2016a).

If global greenhouse gas concentrations continue to rise unabated, most climate models project (Collins et al. 2013; Barnhart et al. 2015) that by century's end global mean surface air temperature will be ~4–5 °C above preindustrial levels and the Arctic Ocean will be ice-free in summer for as many as 5 months (Fig. 25.1, note RCP is defined below). If that occurs, polar bears could be forced ashore and food deprived (Rode et al. 2015a) for unsustainable periods across much of their current distribution, leading to widespread extirpation from many parts of their range (Amstrup et al. 2008; Robbins et al. 2012; Atwood et al. 2016a). On the other hand, with prompt and very aggressive mitigation of global emissions, most climate models project that earth's average air temperature would not rise more than 2 °C above preindustrial levels, and Arctic sea ice would persist all summer although at less than contemporary levels (Fig. 25.1). Such intervention in the current global warming trajectory, in conjunction with optimal management practices, would likely assure healthy polar bear populations could continue to occupy most of their historic range (Amstrup et al. 2010). These best- and worst-case global warming scenarios lead to distinctly different Arctic environments with distinctly different, but fairly certain, outcomes for polar bears. However, for intermediate emission

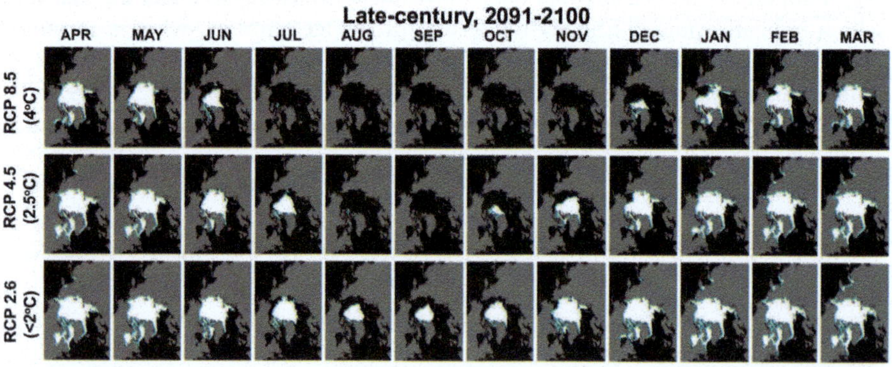

Fig. 25.1 Median monthly Arctic sea ice extent in the late twenty-first century based on projections by 13 general circulation models (GCM) from the Coupled Model Intercomparison Project Phase 5 (CMIP5) when forced with the "worst case" RCP 8.5 greenhouse gas emission scenario (*top row*), the "best case" representative concentration pathway (RCP) 2.6 scenario (*bottom row*), and an intermediate RCP 4.5 scenario (*middle row*). Maps view the North Pole (*center*) and show sea ice (*white*), ocean (*black*), and land (*gray*). *Source:* https://alaska.usgs.gov/science/biology/habitat_dynamics/sea_ice_future.php, accessed March 2016

scenarios, while greater global warming implies greater negative impacts on polar bears, the ability to precisely forecast their status Arctic-wide becomes less certain. In the sections that follow, we describe the primary sources of uncertainty associated with forecasting future polar bear status. We also illustrate how those uncertainties manifest in an analysis that asks where polar bears will be able to spend summer onshore at the end of the twenty-first century.

25.2 Uncertain Paths to the Future: Twenty-First-Century Emission Scenarios

The biggest contributors to uncertainties about the future status of polar bears are the presently unknown choices society may make regarding GHG emission pathways. In its Fifth Assessment Report (IPCC 2013), the United Nations Intergovernmental Panel on Climate Change (IPCC) evaluated general circulation model projections of twenty-first-century climate that were based on four different emission scenarios called representative concentration pathways (RCPs) (Fig. 25.2).

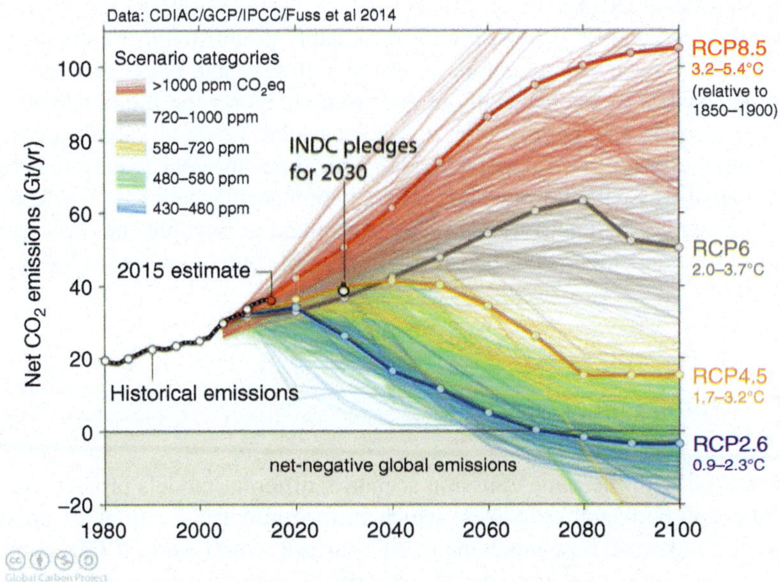

Fig. 25.2 Annual CO_2 emissions for most scenarios described by the IPCC Fifth Assessment Report Working Group 3 (*pale lines*) and four representative concentration pathways (RCPs) used for evaluating model projections of future climate changes by Working Group 1 (*bold lines*). Individual scenarios are grouped into five categories based on atmospheric concentrations of CO_2 equivalents in 2100. Historical CO_2 emissions are shown (*black*) with an estimated value for 2015 (*red dot*). Ranges of temperature increase for each RCP (*right*) refer to average warming in 2081–2100 relative to 1850–1900 (IPCC 2013). Emissions in 2030 are shown (*white dot*) assuming all countries meet their pledged (*nonbinding*) intended nationally determined contributions (INDCs) that were submitted under the Paris Climate Agreement in 2015. *Source: Global Carbon Project (http://www.globalcarbonproject.org), accessed March 2016*

RCPs are named by the approximate level of radiative forcing (above preindustrial levels) attained near the end of the century, expressed in units of watts per meter squared (W/m^2). The "warmest" scenario, RCP 8.5, represents a worst-case outcome of abandoning attempts to curtail global warming. It portrays a world with fast population growth (12 billion by 2100), little technological advancement, widespread poverty and slow economic growth, and high energy use (mostly from coal) and high emissions (van Vuuren et al. 2011). Models from this scenario project that by century's end average global temperature rise will climb upward of 4–5 °C (above preindustrial levels) and the Arctic Ocean will be ice-free for ~5 months during summer (Fig. 25.1).

The RCP 2.6 scenario represents a best-case outcome and portrays a world that keeps average global warming below 2 °C by promptly and aggressively reducing GHG emissions, even to the point of achieving negative emission rates (i.e., removing CO_2 from the atmosphere) by late-century. The aims of the Paris Climate Agreement, as adopted by 195 countries in 2015 (United Nations 2015), would be largely met if the RCP 2.6 was to be realized. At century's end under the RCP 2.6 scenario, most models project that summer sea ice will persist in the Arctic Ocean in all months (Fig. 25.1). Achieving an emission pathway like RCP 2.6 would require unprecedented global commitments and technological advances (Tollefson 2015; Smith et al. 2015).

The RCP 8.5 and RCP 2.6 scenarios reasonably establish upper and lower limits to all probable twenty-first-century emission pathways and upper and lower limits to the persistence of sea ice. Without question, the closer the future adheres to the RCP 2.6 scenario and its projection of sea ice availability in all months of the year, the better the prognosis for polar bears. With time, the real twenty-first-century emission pathway will play out, and the spread of plausible pathways for the remainder of the century will narrow. But today, the spread of possible emission scenarios remains broad and so too does the spread of possible outcomes for sea ice and polar bears.

25.3 Model Uncertainties

When forced with the same emission scenario, different models project somewhat different environmental outcomes, which in turn affect projections of polar bear status. It is unknown how much the earth's surface would warm if CO_2 concentration in the atmosphere were to double over the preindustrial era (termed the earth's climate sensitivity). Contemporary climate models differ in their estimates of the resultant warming, ranging between 2 and 4.5 °C (Knutti and Hegerl 2008). Similarly, different climate models project different estimates of when and by how much the sea ice will melt for any given level of greenhouse gas forcing. The uncertainties introduced by different model outputs are, however, expected and informative. Global climate models have been developed by various institutions worldwide where scientists have applied different strategies for approximating physical processes that occur at spatial and temporal resolutions beyond those of the model framework (Knutti 2008). For example, approximating the sub-grid-scale behaviors of clouds is among some of the most challenging and sensitive parameterizations. Since no best way exists to prescribe sub-grid-scale processes, and for other reasons, the spread of outcomes obtained from an ensemble of models reflects uncertainties attributable to the state of the art in global climate modeling.

25.4 Natural Climate Variability

The total amount of uncertainty in climate projections stems from three primary sources (Fig. 25.3): (1) differences between emission scenarios (i.e., RCP scenario spread), (2) differences between models (i.e., model spread), and (3) natural climate variability (i.e., internal variability). The relative contributions of these three sources change as a function of lead time (i.e., the length of time the forecast spans). Natural climate variability contributes a fairly constant level of uncertainty over all lead times, so it dominates uncertainty in short-term projections. Uncertainties associated with emissions and models increase with longer lead times. When projecting to mid-century, uncertainties owing to the spread among the RCP emission scenarios and the spread among contemporary models increase, and the amount of uncertainty

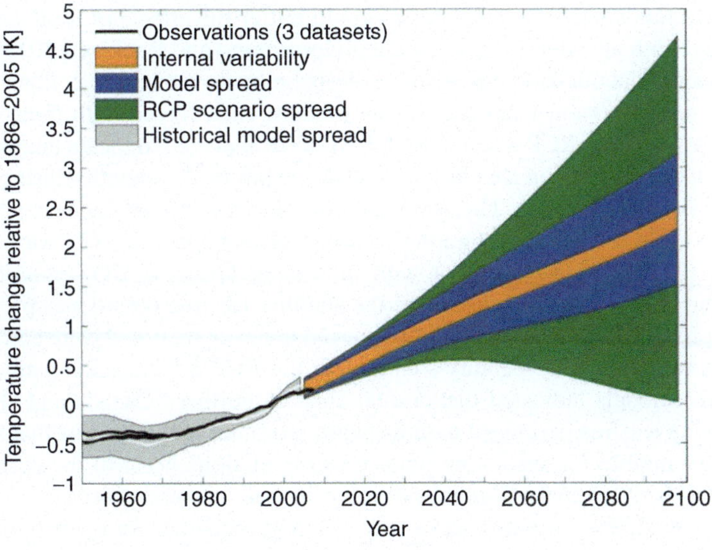

Fig. 25.3 Sources of uncertainty in climate projections as a function of lead time based on an analysis of CMIP5 results. Projections of global mean decadal mean surface air temperature to 2100 together with a quantification of the uncertainty arising from internal variability (*orange*), model spread (*blue*), and RCP scenario spread (*green*). *Reproduced from* Fig. 11.8 *in* Kirtman et al. (2013)

due to natural climate variability becomes proportionally less. During the second half of the century, the amount of uncertainty from today's broad spread of possible emission scenarios increasingly dominates over the uncertainties due to models. By the end of the century, the amount of uncertainty owing to the different emission scenarios is several times greater than that due to model spread, and uncertainty due to natural climate variability is inconsequential by comparison.

25.5 Forecasting Future Summer Habitat

How emissions and model uncertainties influence forecasts of polar bear status can be evaluated by asking: at the end of the twenty-first century, where can polar bears come ashore during summer without risk of undue stress from prolonged food deprivation? In some polar bear subpopulations, the longer open-water season (and thus the period of food deprivation) already has been linked to declines in fitness (Stirling et al. 1999; Obbard et al. 2016; Rode et al. 2010) and survival (Regehr et al. 2007, 2010). Moreover, energy budget models suggest that an open-water period lasting >150 days could result in a significant risk of reproductive failure and starvation (Molnár et al. 2010, 2014; Robbins et al. 2012), although that threshold likely has geographic dependencies due to variations in ocean productivity and prey accessibility that locally influence the nutritional condition of bears prior to their arrival on shore. Additionally, polar bears may develop a broader capacity to exploit alternative foods while on land, which could buffer the effects of food deprivation associated with an extended stay on land (Gormezano and Rockwell 2015).

To answer the question posed above, global climate model projections of future monthly sea ice extent from six climate models, each forced with three emission scenarios (RCP 2.6, RCP 4.5, and RCP 8.5), were analyzed to locate land where the minimum distance to sea ice did not exceed 200 km for ≥5 months during summer in every year, 2091–2100. The 200 km threshold was applied because adult polar bears are capable of swimming long distances (Pagano et al. 2012). Results of the analysis (Fig. 25.4) showed that with increasing levels of CO_2 emissions (i.e., increasing RCP), coastal areas where the summer ice-free period was projected to be no more than 4 months in duration occurred in fewer areas and were corroborated by fewer models. At the century's end under the RCP 8.5 emission pathway, only half of the models indicated that coastal areas in northern Canada and Greenland will have an ice-free period ≤4 months, while the other half indicated that all coasts will be unsuitable for sustaining polar bear populations because the entire Arctic Ocean will be ice-free for 5 months or more, at least in some years.

The three RCP scenarios lead to very different outcomes for polar bears. Under the RCP 8.5 scenario, the model spread raised uncertainty about whether polar bears will face extinction by the century's end or if they might persist in a refugium in northern Canada and Greenland. The RCP 2.6 and RCP 4.5 scenarios projected

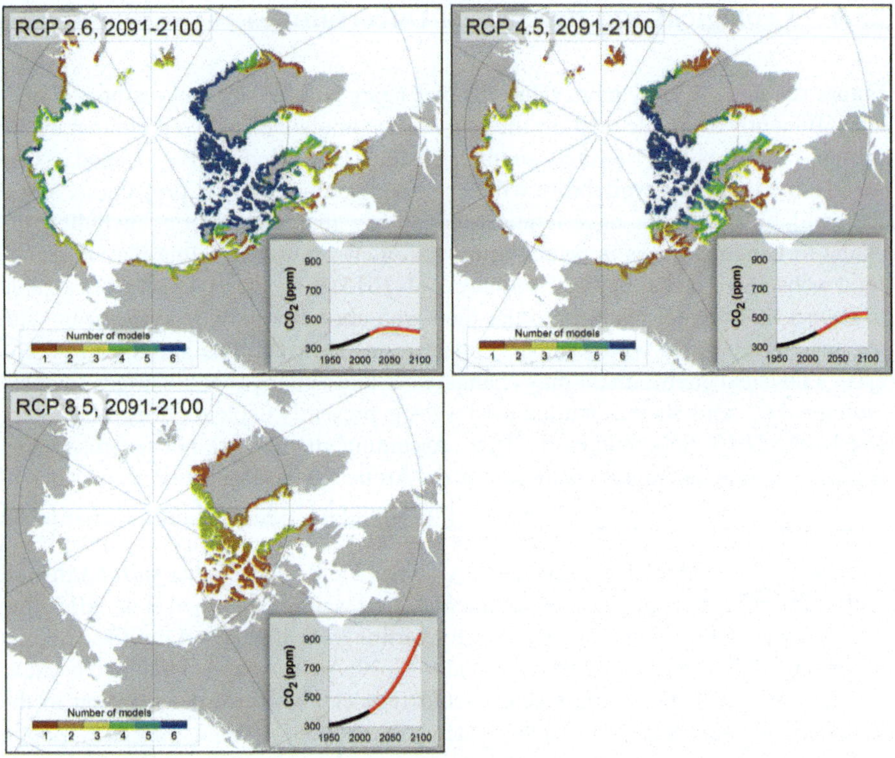

Fig. 25.4 Coastal areas where the summer ice-free period within 200 km of shore is projected to be 4 months or less in duration in each year, 2091–2100, as projected by six global climate models forced with three greenhouse gas emission scenarios (RCP 2.6, 4.5, and 8.5). Color shading along the coastline denotes the number of models in agreement. Inset shows the historic rise in atmospheric CO_2 concentration from 1950–2014 (*black line*) and the scenario-specific change from 2015–2100 (*red line*). (Six CMIP5 models included: CCSM4, CESM-CAM5, GFDL-CM3, HadGEM2-AO, HadGEM2-ES, and MPI-ESM-MR.)

very different outcomes compared to RCP 8.5, illustrating how the differences between emission scenarios inflate uncertainties in projections with long lead times (i.e., late-century). Under both the RCP 2.6 and RCP 4.5 scenarios, complete agreement among model projections in northern Canada and Greenland provided higher confidence that polar bears will be able to use those areas during summer at the century's end without being stranded onshore for ≥5 months (Fig. 25.4). Furthermore, under the RCP 2.6 scenario, a majority of models identified potential summer areas along the northern coast of Eurasia—but only half the models did so under the RCP 4.5 scenario. Hence, model uncertainties under RCP 4.5 introduced greater doubt about whether the Eurasian coast could support polar bears during summer by late-century, compared to the RCP 2.6 scenario.

25.6 Ecological and Behavioral Uncertainties

Future sea ice will only have value to polar bears before they come ashore if prey are sufficiently available to allow the bears to accumulate fat at levels comparable to present-day bears that routinely summer onshore (e.g., Hudson Bay, Canada). Can we assume that prey availability will accompany climate model projections of sea ice availability? Large uncertainties accompany that assumption, including the extent to which changes in primary production and nutrient cycling may influence food webs (Arrigo et al. 2008; Tremblay et al. 2015); however, we feel it reasonable to expect that greater changes to the food web are more likely for scenarios with greater warming. So while global climate models provide insights into how the earth's physical environment may change, how those changes could affect complex biological systems such as marine food webs is presently unclear (Hoegh-Guldberg and Bruno 2010; Schofield et al. 2010). Assuming changes are not severe, we can speculate how seals might redistribute as the Arctic sea ice ecosystem shrinks northward (Moore and Huntington 2008). For example, ringed (*Pusa hispida*) and bearded (*Erignathus barbatus*) seals, the primary prey of contemporary polar bears and the most ice-dependent seals in the Arctic, could be expected to shift northward with warming to occupy ice over continental shelf waters (Harwood et al. 2015) that has adequate stability and snow cover for birthing, weaning, and molting. Such a northward shift would likely be restricted to North America and Greenland because the deep basin of the Arctic Ocean would restrict a northward expansion in the Eurasian Arctic. Subarctic seals, such as spotted (*Phoca largha*), ribbon (*Histriophoca fasciata*), harp (*Pagophilus groenlandicus*), and hooded seals (*Cystophora cristata*), could expand their ranges northward into areas vacated by ringed and bearded seals. The net effects of these changes are unknown: many subarctic seals are adapted to extended bouts of pelagic behavior which may make them less available to polar bears, while some range shifts could introduce new prey that, if also available for capture, could improve conditions for polar bears (Peacock et al. 2013).

The faster the rate of ice loss, the more polar bears will be challenged by their low reproductive rate and long generation time and the likelihood that individual behaviors (such as where to spend the summer) are learned and possibly deep-rooted. As more summer ice melts in the future, more polar bears will likely come ashore (Rode et al. 2015b; Atwood et al. 2016b). If the future Arctic Ocean melts entirely in summer, then all polar bears will come ashore somewhere. Polar bears already possess a feast-and-famine lifestyle, in that they rely on fat reserves accumulated in spring to subsidize their energy requirements during the rest of the year when seals are less available for capture. The interplay between the amount of time spent onshore, the amount of fat reserves accumulated upon arrival, and the amount of available terrestrial food subsidies (Gormezano and Rockwell 2015; Rode et al. 2015a) will determine where oversummering is—and is not—possible. Some polar bears will perish when attempting to summer in marginal areas during years with extremely poor conditions, resulting in population-level selection pressure against the use of those areas in future summers. Under scenarios with greater warming and longer ice-free periods (e.g., Fig. 25.4), polar bear extirpation events will likely be more common and widespread because the overall rate of change will be faster, marginal areas will be more extensive, and extreme years will be more frequent.

25.7 Conclusions

We have described several sources of uncertainty that affect how precisely we can forecast the impacts of global warming on polar bear welfare. Under a worst-case emission scenario like RCP 8.5, it is uncertain if polar bears could survive as a species, and if they were able to persist, it would likely be in a high-latitude refugium in northern Canada and Greenland—assuming a food web also existed with enough accessible seals to fuel weight gains for surviving onshore during the most extreme years of summer ice melt. In all likelihood, such a refugium would be fragile and vulnerable, and ensuring its viability might be of little concern to a world grappling with more urgent ecological and humanitarian problems (Schneider 2009). On the other hand, if emissions could be aggressively mitigated like the RCP 2.6 scenario, healthy polar bear populations might continue to occupy all but the most southern areas of their contemporary summer range, while habitats in northern Canada and Greenland might even improve (Durner et al. 2009). The future for polar bears is yet to be determined, and many sources of uncertainty preclude our ability to precisely forecast their future status. The response of individual polar bear populations to a changing Arctic will likely vary based on the severity of future warming and the regional processes that regulate sea ice dynamics and biological productivity (Amstrup et al. 2008; Rode et al. 2014). Additionally, the extent of behavioral plasticity that polar bears possess may determine how well they respond to alterations in ecosystem structuring and to increasing human presence as the Arctic becomes more attractive to economic interests. However, time for ensuring the future of polar bears is running out. While polar bears have survived previous warming phases—which indicate some resiliency to the loss of sea ice habitat—what is certain is that the present pace of warming is unprecedented and will increasingly expose polar bears to historically novel stressors. The sooner global warming and sea ice loss are stopped, the better the long-term prognosis for the species.

References

Amstrup SC, Marcot BG, Douglas DC (2008) A Bayesian network modeling approach to forecasting the 21st century worldwide status of polar bears. In: DeWeaver ET, Bitz CM, Tremblay LB (eds) Arctic Sea ice decline: observations, projections, mechanisms, and implications, American Geophysical Union Geophysical Monograph No. 180, Washington, pp 213–268

Amstrup SC, DeWeaver ET, Douglas DC et al (2010) Greenhouse gas mitigation can reduce sea ice loss and increase polar bear persistence. Nature 468:955–960

Arrigo KR, van Dijken G, Pabi S (2008) Impact of a shrinking Arctic ice cover on marine primary production. Geophys Res Lett 35:L19603. doi:10.1029/2008GL035028

Atwood TC, Marcot BG, Douglas DC et al (2016a) Forecasting the relative influence of environmental and anthropogenic stressors on polar bears. Ecosphere 7(6):e01370. doi:10.1002/ecs2.1370

Atwood TC, Peacock E, McKinney M et al (2016b) Rapid environmental change drives increased land use by an Arctic marine predator. PLoS ONE 11(6):e0155932. doi: 10.1371/journal.pone.0155932

Barnhart KR, Miller CR, Overeem I et al (2015) Mapping the future expansion of Arctic open water. Nat Clim Chang 6:280–285

Collins M, Knutti R, Arblaster J et al (2013) Long-term climate change: projections, commitments and irreversibility. In: Stocker TF, Qin D, Plattner G-K, Tignor M, Allen SK, Doschung J, Nauels A, Xia Y, Bex V, Midgley PM (eds) Climate change 2013: the physical science basis. Contribution of working group I to the fifth assessment report of the intergovernmental panel on climate change. Cambridge University Press, Cambridge, UK, pp 1029–1136

Durner GM, Douglas DC, Nielson RM et al (2009) Predicting 21st-century polar bear habitat distribution from global climate models. Ecol Monogr 79:25–58

Gormezano LJ, Rockwell RF (2015) The energetic value of land-based foods in Western Hudson Bay and their potential to alleviate energy deficits of starving adult male polar pears. PLoS One 10:e0128520. doi:10.1371/journal.pone.0128520

Harwood LA, Smith TG, George JC et al (2015) Change in the Beaufort Sea ecosystem: diverging trends in body condition and/or production in five marine vertebrate species. Prog Oceanogr 136:263–273

Hoegh-Guldberg O, Bruno JF (2010) The impact of climate change on the world's marine ecosystems. Science 328:1523–1528

Intergovernmental Panel on Climate Change (IPCC) (2013) Climate change 2013: the physical science basis. Contribution of working group I to the fifth assessment report of the intergovernmental panel on climate change. Cambridge University Press, Cambridge, UK

Kirtman B, Power SB, Adedoyin JA et al (2013) Near-term climate change: projections and predictability. In: Climate Change 2013: The physical science basis. Contribution of working group I to the fifth assessment report of the intergovernmental panel on climate change. Cambridge University Press, Cambridge, UK, pp 953–1028

Knutti R (2008) Should we believe model predictions of future climate change? Philos Trans R Soc A 366:4647–4664

Knutti R, Hegerl GC (2008) The equilibrium sensitivity of the Earth's temperature to radiation changes. Nat Geosci 1:735–743

Molnár PK, Derocher AE, Thiemann GW et al (2010) Predicting survival, reproduction and abundance of polar bears under climate change. Biol Conserv 143:1612–1622

Molnár PK, Derocher AE, Thiemann GW (2014) Corrigendum to "predicting survival, reproduction and abundance of polar bears under climate change" [Biol Conserv 143 (2010) 1612–1622]. Biol Conserv 177:230–231. doi:10.1016/j.biocon.2014.07.001

Moore SE, Huntington HP (2008) Arctic marine mammals and climate change: impacts and resilience. Ecol Appl 18(sp2):S157–S165. doi:10.1890/06-0571.1

Obbard ME, Cattet MRL, Howe EJ et al (2016) Trends in body condition in polar bears (*Ursus maritimus*) from the Southern Hudson Bay subpopulation in relation to changes in sea ice. Arct Sci 2:15–32

Overland J, Hanna E, Hanssen-Bauer I et al (2015) Surface air temperature. In: Jeffries MO, Richter-Menge J, Overland JE (eds) Arctic Report Card 2015, pp 10–16. http://www.arctic.noaa.gov/reportcard. Accessed 23 Nov 2016

Pagano AM, Durner GM, Amstrup SC et al (2012) Long-distance swimming by polar bears (Ursus maritimus) of the southern Beaufort Sea during years of extensive open water. Can J Zool 90:663–676

Peacock E, Taylor MK, Laake J et al (2013) Population ecology of polar bears in Davis Strait, Canada and Greenland. J Wildl Manag 77:463–476
Regehr EV, Lunn NJ, Amstrup SC et al (2007) Effects of earlier sea ice breakup on survival and population size of polar bears in western Hudson Bay. J Wildl Manag 71:2673–2683
Regehr EV, Hunter CM, Caswell H et al (2010) Survival and breeding of polar bears in the southern Beaufort Sea in relation to sea ice. J Anim Ecol 79:117–127
Robbins CT, Lopez-Alfaro C, Rode KD et al (2012) Hibernation and seasonal fasting in bears: the energetic costs and consequences for polar bears. J Mammal 93:1493–1503
Rode KD, Amstrup SC, Regehr EV (2010) Reduced body size and cub recruitment in polar bears associated with sea ice decline. Ecol Appl 20:768–782
Rode KD, Regehr EV, Douglas DC et al (2014) Variation in the response of an Arctic top predator experiencing habitat loss: feeding and reproductive ecology of two polar bear populations. Glob Chang Biol 20:76–88
Rode KD, Robbins CT, Nelson L et al (2015a) Can polar bears use terrestrial foods to offset lost ice-based hunting opportunities? Front Ecol Environ 13:138–145
Rode KD, Wilson RR, Regehr EV et al (2015b) Increased land use by Chukchi Sea polar bears in relation to changing sea ice conditions. PLoS One 10:e0142213. doi:10.1371/journal.pone.0142213
Schneider S (2009) The worst-case scenario. Nature 458:1004–1005
Schofield O, Ducklow HW, Martinson DG et al (2010) How do polar marine ecosystems respond to rapid climate change? Science 328:1520–1523
Smith P, Davis SJ, Creutzig F et al (2015) Biophysical and economic limits to negative CO_2 emissions. Nat Clim Chang 6:42–50
Stirling I, Lunn NJ, Iacozza J (1999) Long-term trends in the population ecology of polar bears in western Hudson Bay in relation to climatic change. Arctic 52:294–306
Tollefson J (2015) Is the 2°C world a fantasy? Nature 527:436–438
Tremblay J-É, Anderson LG, Matrai P et al (2015) Global and regional drivers of nutrient supply, primary production and CO_2 drawdown in the changing Arctic Ocean. Prog Oceanogr 139:171–196
United Nations (2015) Adoption of the Paris Agreement, draft. Framework Convention on Climate Change, FCCC/CP/2015/L.9. https://unfccc.int/resource/docs/2015/cop21/eng/l09.pdf. Accessed 23 Nov 2016
van Vuuren DP, Edmonds J, Kainuma M et al (2011) The representative concentration pathways: an overview. Clim Chang 109:5–31

Chapter 26
The Harvest of Polar Bears Across the Circumpolar North

Elizabeth Peacock

Abstract Harvest of polar bears by aboriginal peoples has occurred for millennia across the circumpolar Arctic. While harvest for sport and the commercial fur trade increased dramatically as southerners expanded into the Arctic, the 1973 international Agreement for the Conservation of Polar Bears curtailed harvest largely to aboriginal peoples. This Agreement, catalyzed by global concern for declining polar bear populations, is a hallmark for international cooperation in conservation. In Russia, polar bear harvest has been illegal since 1957, although there are concerns of poaching by local people for food security and also for the black market fur trade. Norway banned all harvest with their ratification of the Agreement. The USA allows for polar bear harvest by the Inupiat of Alaska. Quotas for the two populations shared with the USA are determined by an international user-to-user agreement between aboriginal people of Alaska and Canada and an international agreement between the USA and Russia, respectively. In Greenland, polar bears are harvested by professional Inuit hunters under a quota system, currently based on historic numbers. In most of Canada, where two-thirds of the world's polar bears are harvested, anyone can harvest a polar bear but only within a quota system assigned to and managed by Inuit communities. This harvest is based not only on scientific information, but also on historic levels by treaty and local traditional ecological knowledge. Globally, polar bear harvest averages 798 (44 SD) per year. The vast majority is for subsistence, with 6% for sport (Canada), and a lesser proportion for defense of life and property. The legal international market for polar bear hides is supplied only by exports from Canada. Climate change poses a greater threat to polar bears than do the current levels of harvest. However, habitat change and harvest interact because of the increasing use of land by polar bears. Further, there are scientific and conservation questions about the appropriateness of harvesting polar bears, even for subsistence, from populations that are declining due to climate change.

E. Peacock
US Geological Survey, Anchorage 99508, AK, USA
e-mail: lpeacock@usgs.gov

A. Butterworth (ed.), *Marine Mammal Welfare*, Animal Welfare 17,
DOI 10.1007/978-3-319-46994-2_26

26.1 Introduction: The Historical Polar Bear Harvest

The Inuit and other aboriginal peoples of the circumpolar Arctic have hunted polar bears (*Ursus maritimus*), using their meat and fat for food and pelts for clothing and bedding, for thousands of years. They have also used the polar bear in trade as part of a subsistence hunting economy (Henri et al. 2010; Born et al. 2011). While polar bears constituted some nutritional and economic importance, their significance in the cultural and spiritual realm was perhaps greater (Henri et al. 2010). However, as explorers from the south and incoming missionaries and trappers expanded into the Subarctic and Arctic, the harvest and killing of polar bears for their hides increased, most dramatically with the advent of over-snow machines and non-aboriginal sport (trophy) hunting (Urquhart and Schweinsburg 1984). By the mid-twentieth century, unregulated hunting in Russia had increased around polar expedition bases, stations, and settlements, causing concern about population decline (Belikov and Boltunov 1998). Sport hunting peaked in the late 1960s at levels of several hundred per year (in addition to subsistence harvesting) in each of the USA, Canada, and Svalbard (Norway; Prestrud and Stirling 1994).

Concern over the level of polar bear harvest resulted in the first attempts to curtail harvest in the middle of the twentieth century. In 1949, Canada banned sport hunting by non-aboriginal people, although de facto it did not cease (Urquhart and Schweinsburg 1984). The Union of Soviet Socialist Republics established a complete ban on harvest in what is now the Russian Federation in 1956 (Anon. 2010). In the mid-1960s, the first international meeting of polar bear biologists and managers occurred in Fairbanks, Alaska, to discuss concerns about polar bear population decline (Flyger 1967). This meeting ultimately led to the international Agreement on the Conservation of Polar Bears in 1973 (hereafter, Agreement), in which the five nations that then hosted polar bears—USSR, Canada, the USA, Norway, and Denmark (Greenland)—agreed to restrict non-aboriginal polar bear harvest. Each member nation variously ratified the Agreement. The USA banned all non-Iñupiat harvest in Alaska. Norway ended all forms of polar bear harvest, including by traditional trappers. Denmark stopped non-aboriginal harvesting in Greenland (Born et al. 2011). With ratification of the Agreement, Canada formally allowed for continued non-aboriginal harvesting if Inuit communities chose to sell tags from their legal quotas to trophy hunters. These sport hunters were obligated to use a dog team and Inuit guide (Prestrud and Stirling 1994).

Thus with the signing of the Agreement, aboriginal harvesting of polar bears continued legally in three of the five nations. The territories and provinces of Canada, which are home to 13 of the world's 19 populations of polar bears, implemented the first population-based quotas in the late 1960s. In the Northwest Territories (including the modern territory of Nunavut), Manitoba, and Newfoundland and Labrador, quotas were explicit annual numbers based on traditional harvesting amounts that subsequently changed with updated science. Under aboriginal land claim agreements, Ontario allowed for continued Cree harvesting in Southern Hudson Bay at a set level, and Québec protected aboriginal harvest without quotas (Cooper 2015). In 2011, a user-to-user agreement regarding harvest in Southern Hudson Bay, a population shared by Nunavut, Québec, and Ontario, set the first voluntary quotas for aboriginal peoples in Québec. In the USA, in the Southern Beaufort Sea, which has a population of polar bears shared with Canada, harvest is regulated by a user-to-user agreement (Brower et al. 2002; Nageak et al. 1991) between the Iñupiat of Alaska and Inuit of the Yukon and Northwest Territories (the latter of whom are subject to government quotas). Until recently, in the Chukchi Sea of the USA (which is a population shared with Russia), harvest by the Iñupiat was unrestricted. In 2010 a shared quota was established in the Chukchi Sea, by the U.S.-Russia Polar Bear Commission, but the quota has not been implemented by the federal government of Russia out of deference to its 1956 ban. In Greenland, there were no quotas until 2006, but only traditional full-time hunters are allowed to harvest polar bears by non-mechanized means (Born et al. 2011).

The USA banned selling (unless processed into an art form) and exporting polar bear hides coincident with the signing of the international Agreement and their Marine Mammal Protection Act. Export of hides from Greenland continued until 2008 when Greenland failed to find a non-detriment finding (NDF; under their obligations to the Convention on International Trade in Endangered Species [CITES]) of international trade on polar bear population health. An NDF under CITES states that there are no detrimental effects of international trade of wildlife on the population. Greenland failed to find an NDF because they asserted that their research was at the time insufficient to know the status of their populations; therefore, Greenland was unable to know the effects of international trade of harvested animals on the health of their polar bear populations. This was a self-imposed voluntary and temporary ban. Therefore, currently, Canada is the only country from which export of hides is legal. Export of hides occurs as trophies from sport hunting but largely as pelts sold by aboriginal harvesters at international fur auctions. In 2008, Canada failed to find an NDF for export of hides from Baffin Bay, due to lack of current scientific data, and therefore export of hides from bears harvested in this population currently does not occur.

The curtailment of polar bear harvest since the 1970s has likely resulted in an increase in polar bear numbers globally, underscoring the efficacy of harvest management. However, it is difficult to compare modern population estimates with anecdotes or estimates derived by less reliable methods in the past. Nonetheless, there is demographic evidence that the population in the Barents Sea—shared between Norway and Russia—increased after the elimination of harvesting (Derocher 2005). Similar suggestions have been made for the polar bears in Davis Strait (Peacock et al. 2013), Southern Hudson Bay (Prevett and Kolenosky 1983), and the Southern Beaufort Sea (Amstrup et al. 1986) populations in which data have been available to evaluate the impact of harvest restrictions. It is also likely that curtailed sustainable, science-based quotas have allowed for modest growth or stability in Foxe Basin (Stapleton et al. 2015).

26.2 Current Polar Bear Harvest

The number of "human-caused removals" of polar bears globally averaged 798 (SD, 44) per year for the period 2009–2014 (PBSG 2014). Most of this figure comes from aboriginal harvest, although sport harvest and defense of life and property (DLP) kills are included. For example, in the last reported year (2014–2015), in the two jurisdictions of Canada where non-aboriginal sport harvest is allowed (the Nunavut Territory and the Northwest Territories), 40 polar bears were taken by sport hunters, representing 9% of the harvest (Anon. 2016; Dyck 2016). In Nunavut, specifically, over the last 15 years, sport harvest peaked in 2006–2007 at 24% (120 polar bears) and has remained at approximately 10% of Nunavut's harvest since this time (Dyck 2016).

DLP kills constitute a small fraction of global polar bear harvest. For example, in Nunavut and the Northwest Territories, DLP kills constituted 8% ($n = 37$) of the total harvest in 2014–2015 (Anon. 2016; Dyck 2016). However, DLP kills vary greatly among years. For example, in Nunavut DLP kills ranged between 17 and 87 per year over the last 15 years (Dyck 2016; Peacock et al. 2007). This variation may depend on freeze-up date, which influences time the bears spend on land during the summer and autumn, but also whether communities choose to self-report a kill as subsistence or DLP. DLP kills in Norway have averaged 2.0 per year (SD 1.8) for 1987–2015, with no evidence of increase (D. Vongraven, Norwegian Polar Institute, pers. comm.), and 1–3 per year in Russia (S. Belikov, All-Russian Research Institute for Nature Protection, pers. comm.). Greenland reports an average of 5.9 (SD 4.7) DLP kills per year for 2007–2015 (S. Erbs-Maibing, Government of Greenland, pers. comm.).

Recent legal international trade in polar bear hides (i.e., export of hides from Canada) increased between 2005 and 2013 from 266 to 400 per year but then dropped to 217 in 2014 (Cooper 2015) (Fig. 26.1). Cooper (2015) showed that the increase in exported hides largely reflects an increase in demand from China and argued that the demand has not generally increased harvest as there are many more polar bears killed per year than exported as hides. With the exception of a dramatic increase in polar bears harvested in Québec in the 2010–2011 harvest year (70% increase over the prior year's reported harvest), polar bears remain harvested below quotas set by jurisdictions. The 2011 harvest of polar bears from Québec was likely in response to an increase in demand and increased prices of furs at auction, although there may be some influence of better reporting mechanisms. More importantly, the manner in which ice formed in late 2010 in eastern Hudson Bay, which resulted in many bears being close to the Québec coast near the community of Inukjuak, provided ready access for hunters (M. Obbard, Ministry of Natural Resources, Ontario, pers. comm., and pers. comm. within Cooper 2015). Nonetheless, this much publicized single incident and the resulting public concern over a correlation between international trade and overharvest prompted a user-to-user voluntary agreement, resulting in the first polar bear quota for Québec, specifically for the Southern Hudson Bay population. In

Fig. 26.1 Polar bear hides drying in Resolute Bay, Nunavut. *Image credit*: *Mike Harte*

summary, Cooper's (2015) comprehensive review of harvest and trade data concluded that recent international trade has not affected polar bear harvest substantially over the last decade because, although demand is high, it does not surpass the legal, presumably sustainable, supply.

The legal means by which polar bears can be harvested differ among jurisdictions. The 1973 international Agreement prohibited the use of aircraft and large motorized vessels to hunt polar bears. In Greenland, polar bears must be harvested by professionally licensed hunters and only by dog sledge or boat (outboard motors allowed), but small over-the-snow motor machines are not permitted. In the USA, over-the-snow vehicles and small, motorized boats are permitted. There is no specific wording regarding weapons used in the USA, although the Russia-USA bilateral agreement dictates that poison and traps cannot be used in the harvest of polar bears in the shared Chukchi Sea (note, however, that polar bears are not legally harvested in Russia). In Canada, Inuit are able to use small over-the-snow machines and must harvest polar bears with high-caliber rifles, unless special permission has been granted by the various management bodies. In the early 2000s, an Inuit hunter sued the Government of Nunavut for the right to use a traditional spear/knife to harvest a polar bear. The court sided with the hunter, although this request was unique in modern times. Non-Inuit sport hunters in Canada must use an Inuit guide and must use a dog sledge for harvesting the bear. There are no restrictions on baiting or the use of dogs to pursue bears in any country.

26.3 The Science of Polar Bear Harvest Management

The most science-driven polar bear harvest management has occurred within Canada, specifically within the sophisticated quota systems of the Northwest Territories and Nunavut. Robust harvest management has been necessary, as Canada is home to two-thirds of the world's polar bears, and has active subsistence harvesting with the added economic incentive of the sport harvest and international trade. Since the establishment of the original quota levels, subsequent changes to quotas in Canada have been based largely on the best available science. As methods of polar bear capture and handling and scientific population modeling improved, it was suggested that polar bears could be harvested annually at 4.5% of their population size or 1.6% of the adult females in the population (Taylor et al. 1987a), reflecting the natural growth of a stable population. The 4.5% figure was derived from the birth and mortality rates measured from populations of polar bears in the High Arctic, which were assumed to be stable (Taylor et al. 1987b). As techniques became more sophisticated, each population's sustainable harvest rate could be estimated independently (e.g., Peacock et al. 2013; Taylor et al. 2002; Regehr et al. 2010). The common 4.5% guideline is no longer implemented, except in cases in which scientific data were unavailable or less robust. Non-numeric harvest regulations also became scientifically based. For example, Taylor et al. (1987a, b) established that sustainable harvest was most sensitive to adult female survival, so regulations called for protecting females with dependent young and dictated a 2:1 male/female ratio of the harvest. Furthermore, models started to incorporate stochastic variation of the recruitment and survival parameters, allowing for inherent variation in biological systems (Taylor et al. 1987a). Scientific modeling also evaluated the impact of trophy hunting and persistent male selectivity (McLoughlin et al. 2005; Molnar et al. 2007).

In regions where the harvest has been managed with science-based quotas, it has been assumed that the harvest of polar bears is ecologically compensatory. That is, the harvesting of individual polar bears compensates for natural mortality that would otherwise occur. Although this is unlikely at an individual level, given the bears that are hunted (predominately large males by sport harvesters or younger bears by Inuit), compensatory harvesting theoretically occurs in that the harvesting of bears necessarily reduces population density making it more favorable for other bears, thereby augmenting natural survival rates. A compensatory mechanism assumes that populations are stable and at carrying capacity, as theory and empirical evidence suggests that when large mammals exist close to the carrying capacity of their environment, they are largely naturally regulated by factors that are population density dependent (Fowler 1981). This theory, combined with evidence that harvesting polar bears at modest rates has not resulted in declines (Peacock et al. 2013; Stirling et al. 2011; Obbard et al. 2015; Stapleton et al. 2016), has led managers to assume that harvesting polar bears is not additive, i.e., that the harvest of polar bears ultimately reduces population size. Thus polar bear harvest management has relied on the premise that harvesting at a rate near population growth rate will result in a stable population, given that the habitat is stable (but see below, "Climate Change and Polar Bear Harvesting").

Science-based harvest management of polar bears is challenging when scientific data do not exist, are incomplete or not up-to-date. The estimation of demographic parameters to establish sustainable harvest based on population growth rate relies on capturing and marking bears. This requires extraordinary effort, expertise, and money and is only possible in populations with research access. For example, difficulties of access have complicated the estimation of population size for harvest management in the remote and expansive Chukchi Sea population (Rode et al. 2014). As another example, Nunavut requires research in each of its 12 populations every 15 years to ensure up-to-date scientific data; however, such frequent population inventories have yet to occur because of lack of resources. Further, assumptions required by the mathematical models are often challenged, bringing into question biases in the data. For example, a common challenge by local comanagement committees and within the scientific community is that the scientists sometimes do not cover the entire geographical area of the population; this can result in biased low population estimates because all bears are not susceptible to mark and recapture, a fundamental assumption of the mathematical models. This issue has been prominent in discussions around harvest in the Southern Beaufort Sea, Western Hudson Bay (Stapleton et al. 2014), and Baffin Bay (Dowsley and Wenzel 2008), populations which all had shown some evidence of decline (e.g., population size, body condition, reproductive parameters, etc.). Varying degrees of the robustness of data have complicated efforts to adjust harvest levels in these populations (Peacock et al. 2011).

The territory of Nunavut addresses the difficulty associated with frequent collection of scientific data in their community through comanagement plans, by allowing for Traditional Ecological Knowledge (TEK) to substitute for scientific data older than 7 years in the setting of harvest levels (Anon. 2005). Other jurisdictions also variously use local opinion and sometimes formalized TEK, via local comanagement boards, in the evaluation of scientific research and in the assessment of harvest levels. Recently, capture-based population demography work also requires permission by local comanagement authorities; in three recent attempts to collect robust scientific information for harvest management, alternative techniques (biopsy darting and aerial surveys) were used to collect data due to pressure from local Inuit groups to reduce the rate of capturing of bears (Foxe Basin, Baffin Bay, and Kane Basin; Stapleton et al. 2015; Peacock et al. 2011). As such, quotas in Canada, the USA, and Greenland are de facto based on varying degrees of scientific information.

The use of TEK to inform harvest management is not without controversy, because TEK is local in its nature, whereas populations are expansive. In addition, when economic incentives increase and harvest become more lucrative, expert-based methods for determining harvest levels may be put under pressure (see Foote and Wenzel 2009 for discussion of "conservation hunting"). Lastly, the international Agreement (international Agreement on the Conservation of Polar Bears in 1973) calls for scientific research to support harvest levels, and TEK is not globally accepted as a substitute for science-based harvest management, especially when populations are shared among international jurisdictions.

26.4 Climate Change and Polar Bear Harvesting

Previously, when establishing sustainable harvest levels, scientists have assumed that polar bear sea ice habitat was stable in quantity and quality. However, the sea ice habitat that polar bears use for foraging, mating, denning, and migration has declined in extent throughout the circumpolar Arctic since the 1990s as a result of global climate change (e.g., Durner et al. 2009; Hamilton et al. 2014; Sahanatien and Derocher 2012) (please also see Chaps. 23, 24, and 25 which discuss the effects of climate change). These reductions in sea ice habitat have been linked, in some populations, to declines variously in survival rates, recruitment, population size, and body condition (Stirling et al. 1999; Hunter et al. 2010; Regehr et al. 2010, 2007; Rode et al. 2010, 2012). Declines in demographic parameters have not always been detected following sea ice decline (Stapleton et al. 2015; Peacock et al. 2013; Rode et al. 2012, 2014; Obbard et al. 2015, 2016), perhaps due to other changes in the ecosystems, plasticity in bear behavior, or transient increase in habitat suitability. In addition to the presumption that population growth rates will ultimately decline as carry capacity declines with sea ice habitat, it is also speculated that as polar bears increase their time spent on land closer to human settlements (Atwood et al. 2016), defense kills (DLP) will increase, although this has yet to be documented. Regardless of the time scale and nature of ecosystem change, sea ice change should now be incorporated into evaluations of sustainable polar bear harvest (see the status table published by the IUCN/Polar Bear Specialist Group, http://pbsg.npolar.no/en/status/status-table.html).

A sustainable harvest implies that harvest will not be additive. Yet how can a population that is in decline as a result of reduced carrying capacity sustain a harvest? Indeed, Taylor et al. (1987b) unequivocally state that there “can be no sustainable harvest if population growth rate is $\leq$1.0%”. Does this mean that there is no basis for the harvest of polar bears in populations that are declining? The question involves scientific, conservation, ethical, and legal considerations (Derocher et al. 2013).

The conservation of animal species, not to be confused with preservation, allows for the wise use of resources, including hunting. With respect to hunting, the goals of conservation would be met through the assessment of sustainable harvest as defined above. In situations where population growth and/or the interaction between population productivity and future habitat change is unknown from a quantitative perspective (but see Hunter et al. 2010 for an example in which habitat change was incorporated in population modeling), conservation theory would advocate the precautionary principle. Even with science-based estimates of population growth rates, under a scenario of habitat change, these rates will have uncertain longevity. A precautionary approach, for example, would be to base harvest rates on the estimate of the lower confidence limit of population parameters, given the general assumption that reduced, or fragmented ice habitat, will result in lower population productivity. However, in most cases, empirical rates of population growth rate are simply unknown. In these situations, a precautionary approach could consider a wider variety of scientific information, such as relative population size, habitat metrics, body condition, recruitment indices, or index surveys. Because these data, although scientific, are proxies for population growth rate and may be less precise and/or accurate, the precautionary principle would suggest lowering harvest rates as uncertainty increases.

Southern (i.e., from non-Arctic peoples and governments) suggestions to reduce or eliminate harvest as a result of anthropogenic climate change are reminiscent of the history of imposition of outside perspectives and law on aboriginal communities (Peacock et al. 2011). Proposed reductions or elimination of polar bear harvest as a solution to counter the effects of climate change on polar bears is incongruous at best and at worst penalizes victims of climate change—Arctic aboriginal peoples—for a problem largely created by outside cultures and influences. Further, if habitat has declined to a lower carrying capacity, a reduced population is still theoretically harvestable at a lower level (Derocher et al. 2013). If indeed a population is not capable of a *sustainable* harvest, a shift in perspective to a lower, *cultural* harvest (i.e., aboriginal harvesting at a level to meet cultural needs) may be necessary. Finally, from a legal perspective, aboriginal harvest is protected under land claims agreements in both Canada (Peacock et al. 2011) and Greenland (Born et al. 2011) and will very likely continue in these countries in the future even in the scenario of population declines.

In the cases of continued aboriginal harvest, but likely declining polar bear populations, the precautionary principal could arguably call for the elimination of any real or perceived economic incentives to harvest additional animals. For example, in recent years we have seen local opposition to quota reductions in populations with active trophy hunting in the Western Hudson Bay (Regehr et al. 2007; Stirling et al. 1999) and Baffin Bay subpopulations (Aars et al. 2006). Further, as the cost of hides climbed in 2010 for various reasons, there was a rapid increase in harvest from Québec in Southern Hudson Bay. These phenomena argue against the notion that "conservation hunting" occurs (Freeman and Wenzel 2006) in Inuit communities. Further, it has been argued that trade of hides of legally harvested polar bears allows for the influx of pelts onto the international market from poached bears from Russia. Partly due to these concerns, Russia and the USA submitted proposals to CITES in 2009 and 2013 for the up-listing of the polar bear from Appendix II to Appendix I, which would have eliminated the commercial international trade of polar bear hides (Cooper 2015); these proposals were ultimately unsuccessful.

26.5 Conclusions: The Future of Polar Bear Harvesting

In decades past, the harvest of polar bears was implicated in population decline, but subsequent science-based management has been effective in restoring the numeric health of populations. Polar bear harvest is not thought to negatively affect polar bears at a global level, unlike climate change (Derocher et al. 2013). The future of polar bear harvests cannot be a unified approach, as climate change is differentially affecting polar bear populations, and the Arctic nations have various traditions and laws governing polar bear harvest. Continuing to rely on science, where available, to establish harvest levels, supports the 1973 international Agreement. When science is not available, evoking the precautionary principle is appropriate. Because it is difficult to establish robust quantitative assessments of population growth at frequent time intervals, expanding the tools used to assess population health (Pagano et al. 2014; Stapleton et al. 2016) warrants research attention. Realigning harvest to be not only based on sustainability but also in the interest of a continued cultural harvest is another consideration. Finally, the effects of international trade and economic incentives on polar bear harvest require vigilance. With these precautionary guidelines, aboriginal harvesting of polar bears will be able to continue for the foreseeable future.

References

Aars J, Lunn NJ, Derocher AE (2006) Polar bears. In: Proceedings of the 14th working meeting of the IUCN/SSC polar bear specialist group. Seattle, WA, USA, 20–24 June 2005. Occasional Paper of the IUCN Species Survival Commission No. 32

Amstrup SC, Stirling I, Lentfer JW (1986) Past and present status of polar bears in Alaska. Wildl Soc Bull 14:241–254

Anon. (2005) Polar bear management memoranda of understanding between community hunting and Trappers' organizations and the Department of Environment, Government of Nunavut, Iqaluit, Nunavut

Anon. (2010) Strategy for polar bear conservation in the Russian Federation. Ministry of Natural Resources and Environment of the Russian Federation, Russia. 59 pp

Anon. (2016) Inuvialuit settlement region submission to the polar bear technical committee meeting. Whitehorse, Yukon, Canada

Atwood T, Peacock E, McKinney MA, Lillie K, Wilson R, Douglas DC, Miller S, Terletzky P (2016) Rapid environmental change drives increased land use by an Arctic marine predator. PLoS One 11(6):e0155932. doi:10.1371/journal.pone.0155932

Belikov SE, Boltunov AN (1998) Problems with the conservation and sustainable use of polar bears in the Russian Arctic (International Association for Bear Research and Management). Ursus 10:119–127

Born EW, Heilmann A, Kielsen Holm L, Laidre KL (2011) Polar bears in Northwest Greenland—an interview survey about the catch and the climate. Monographs on Greenland, vol 351 and Man & Society, vol 41. Museum Tusculanum Press, University of Copenhagen, Copenhagen, Denmark

Brower CD, Carpenter A, Branigan ML, Calvert W, Evans T, Fischbach AS, Nagy JA, Schliebe S, Stirling I (2002) The polar bear management agreement for the southern Beaufort Sea: an evaluation of the first ten years of a unique conservation agreement. Arctic 55(4):362–372

Cooper EWT (2015) Review and analysis of Canadian trade in polar bears from 2005–2014: report prepared for the parties to the 1973 agreement on the conservation of polar bears. Environment Canada, Ottawa, Canada

Derocher AE (2005) Population ecology of polar bears at Svalbard, Norway. Popul Ecol 47(3):267–275

Derocher AE, Aars J, Amstrup SC, Cutting A, Lunn NJ, Molnar PK, Obbard ME, Stirling I, Thiemann GW, Vongraven D, Wiig O, York G (2013) Rapid ecosystem change and polar bear conservation. Conserv Lett 6(5):368–375. doi:10.1111/conl.12009

Dowsley M, Wenzel GW (2008) "The time of the most polar bears": a co-management conflict in Nunavut. Arctic 61(2):177–189

Durner GM, Douglas DC, Nielson RM, Amstrup SC, McDonald TL, Stirling I, Mauritzen M, Born EW, Wiig O, DeWeaver E, Serreze MC, Belikov SE, Holland MM, Maslanik J, Aars J, Bailey DA, Derocher AE (2009) Predicting 21st-century polar bear habitat distribution from global climate models. Ecol Monogr 79(1):25–58

Dyck MG (2016) Submission to the 45th annual Canadian federal/provincial/territorial polar bear technical committee. Whitehorse, Yukon, Canada

Flyger V (1967) The polar bear: a matter for international concern. Arctic 20(3):148–153

Foote L, Wenzel GW (2009) Polar bear conservation hunting in Canada: economics, culture and unintended consequences. In: Freeman MMR, Foote L (eds) Inuit, polar bears and sustainable use: local, national and international perspectives. CCI Press, Edmonton

Fowler CW (1981) Density dependence as related to life history strategy. Ecology 62:602–610

Freeman MMR, Wenzel GW (2006) The nature and significance of polar bear conservation hunting in the Canadian Arctic. Arctic 59(1):21–30

Hamilton SG, de la Guardia LC, Derocher AE, Sahanatien V, Tremblay B, Huard D (2014) Projected polar Bear Sea ice habitat in the Canadian Arctic archipelago. PLoS One 9(11). doi:10.1371/journal.pone.0113746

Henri D, Gilchrist HG, Peacock E (2010) Understanding and managing wildlife in Hudson Bay under a changing climate: some recent contributions from Inuit and Cree ecological knowledge. In: Loseto LL, Ferguson SH, Gilchrist HG (eds) A little less Arctic: top predators in the world's largest northern Inland Sea. Springer, Hudson Bay, pp 267–289

Hunter CM, Caswell H, Runge MC, Regehr EV, Amstrup SC, Stirling I (2010) Climate change threatens polar bear populations: a stochastic demographic analysis. Ecology 91(10): 2883–2897

McLoughlin PD, Taylor MK, Messier F (2005) Conservation risks of male-selective harvest for mammals with low reproductive potential. J Wildl Manag 69(4):1592–1600
Molnar PK, Derocher AE, Lewis MA, Taylor MK (2007) Modelling the mating system of polar bears: a mechanistic approach to the Allee effect. Proc R Soc B 275:217–226
Nageak BP, Brower CD, Schliebe SL (1991) Polar bear management in the southern Beaufort Sea: an agreement between the Inuvialuit game council and north slope borough fish and game committee. T N Am Wildl Nat Res 56:337–343
Obbard ME, Stapleton S, Middel KR, Thibault I, Brodeur V, Jutras C (2015) Estimating the abundance of the southern Hudson Bay polar bear subpopulation with aerial surveys. Polar Biol 38(10):1713–1725. doi:10.1007/s00300-015-1737-5
Obbard ME, Cattet MRL, Howe EJ, Middel KR, Newton EJ, Kolenosky GB, Abraham KF, Greenwood CJ (2016) Trends in body condition in polar bears (*Ursus maritimus*) from the southern Hudson Bay subpopulation in relation to changes in sea ice. Arctic Sci 2(1):15–32. doi:10.1139/as-2015-0027
Pagano AM, Peacock E, McKinney MA (2014) Remote biopsy darting of polar bears: implications for monitoring abundance, diet, and condition. Mar Mamm Sci 30(1):169–183
Peacock E, Dyck MG, Piugattuk F (2007) Submission to the polar bear technical committee. Department of Environment, Government of Nunavut, Igloolik, Nunavut
Peacock E, Derocher AE, Theimann GW, Stirling I (2011) Conservation and management of Canada's polar bears in a changing Arctic. Can J Zool 89:371–385
Peacock E, Taylor MK, Laake J, Stirling I (2013) Population ecology of polar bears in Davis Strait, Canada and Greenland. J Wildl Manag 77(3):463–476
Prestrud P, Stirling I (1994) The international polar bear agreement and the current status of polar bear conservation. Aquat Mamm 20(3):113–124
Prevett JP, Kolenosky GB (1983) Productivity and maternity denning of polar bears in Ontario. Int Conf Bear Res Manag 5:238–245
Regehr EV, Lunn NJ, Amstrup SC, Stirling I (2007) Effects of earlier sea ice breakup on survival and population size of polar bears in western Hudson Bay. J Wildl Manag 71(8):2673–2683
Regehr EV, Hunter CM, Caswell H, Amstrup SC, Stirling I (2010) Survival and breeding of polar bears in the southern Beaufort Sea in relation to sea ice. J Anim Ecol 79:117–127. doi:10.1111/j.1365-2656.2009.01603.x
Rode KD, Amstrup SC, Regehr EV (2010) Reduced body size and cub recruitment in polar bears associated with sea ice decline. Ecol Appl 20(3):768–782
Rode KD, Peacock E, Taylor MK, Stirling I, Born EW, Laidre KL, Wiig O (2012) A tale of two polar bear populations (*Ursus maritimus*): ice habitat, harvest, and body condition. Popul Ecol 54(1):3–18
Rode KD, Regehr EV, Douglas DC, Durner G, Derocher AE, Thiemann GW, Budge SM (2014) Variation in the response of an Arctic top predator experiencing habitat loss: feeding and reproductive ecology of two polar bear populations. Glob Chang Biol 20(1):76–88. doi:10.1111/gcb.12339
Sahanatien V, Derocher AE (2012) Monitoring sea ice habitat fragmentation for polar bear conservation. Anim Conserv 15(4):397–406
Stapleton S, Atkinson S, Hedman D, Garshelis D (2014) Revisiting western Hudson Bay: using aerial surveys to update polar bear abundance in a sentinel population. Biol Conserv 170:38–47. doi:10.1016/j.biocon.2013.12.040
Stapleton S, Peacock E, Garshelis D (2015) Aerial surveys suggest long-term stability in the seasonally ice-free Foxe Basin (Nunavut) polar bear population. Mar Mamm Sci 32(1):181–201. doi:10.1111/mms.12251
Stapleton S, Peacock E, Garshelis D (2016) Aerial surveys suggest long-term stability in the seasonally ice-free Foxe Basin (Nunavut) polar bear population. Mar Mamm Sci 32(1):181–201. doi:10.1111/mms.12251
Stirling I, Lunn NJ, Iacozza J (1999) Long-term trends in the population ecology of polar bears in western Hudson Bay in relation to climatic change. Arctic 52(3):294–306

Stirling I, McDonald TL, Richardson ES, Regehr EV, Amstrup SC (2011) Polar bear population status in the northern Beaufort Sea, Canada, 1971–2006. Ecol Appl 21(3):859–876

Taylor MK, Bunnell F, DeMaster D, Schweinsburg R, Smith J (1987a) ANURSUS: a population analysis system for polar bears (*Ursus maritimus*). Int Conf Bear Res Manag 7:117–125

Taylor MK, DeMaster DP, Bunnell FL, Schweinsburg RE (1987b) Modeling the sustainable harvest of polar bears. J Wildl Manag 51:811–820

Taylor MK, Laake J, Cluff HD, Ramsay M, Messier F (2002) Managing the risk from hunting for the viscount Melville sound polar bear population (International Association for Bear Research and Management). Ursus 13:185–202

Urquhart DR, Schweinsburg RE (1984) Polar bear: life history and known distribution of polar bear in the northwest Territories up to 1981. Department of Renewable Resources, Yellowknife, Northwest Territories

Chapter 27
Welfare of Captive Polar Bears and Their Value to In Situ Conservation Efforts

Randi Meyerson, Donald E. Moore, Sarah T. Long, and Judy Che-Castaldo

Abstract Polar bears (*Ursus maritimus*) have always been one of the most popular animals in zoos. Though their charismatic nature has made them a good flagship species for the Arctic habitat, there has been very little examination of the co-relationship or need for collaboration between the in situ and ex situ polar bear worlds. In the 1990s, polar bear populations in North American and European zoos were declining, and many zoos were closing their polar bear exhibits (Meyerson 2006; Linke 2015; Poirier and Lanthier 1995). Though still popular with the public, animal well-being concerns and increasing governmental regulations made it evident that in order to appropriately house and exhibit this large and intelligent species, significant financial resources would need to be invested. Given that the wild population numbers had rebounded as a result of the cooperative regulations enacted by the five Polar Bear Nations (i.e., the Range States: USA, Canada, Union of Soviet Socialist Republics, Norway and Denmark/Greenland) (United Nations Environment Program Register of International Treaties 1973), which addressed issues such as illegal harvest and environmental toxins, the conservation threat for the species was relatively low, and zoos were choosing to use their limited financial resources to build exhibits for species that had a greater conservation need.

R. Meyerson, DVM (✉)
Toledo Zoo and Aquarium, Toledo, OH, USA
e-mail: randi@toledozoo.org

D.E. Moore, PhD
Oregon Zoo, 4001 Southwest Canyon Road, Portland, OR 97221, USA
e-mail: don.moore@oregonzoo.org

S.T. Long • J. Che-Castaldo
Lincoln Park Zoo, 2001 N. Clark Street, Chicago, IL 60614, USA
e-mail: slong@lpzoo.org; jchecastaldo@lpzoo.org

A. Butterworth (ed.), *Marine Mammal Welfare*, Animal Welfare 17,
DOI 10.1007/978-3-319-46994-2_27

27.1 Introduction

Polar bears (*Ursus maritimus*) have always been one of the most popular animals in zoos. Though their charismatic nature has made them a good flagship species for the Arctic habitat, there has been very little examination of the co-relationship or need for collaboration between the in situ and ex situ polar bear worlds. In the 1990s, polar bear populations in North American and European zoos were declining, and many zoos were closing their polar bear exhibits (Meyerson 2006; Linke 2015; Poirier and Lanthier 1995). Though still popular with the public, animal well-being concerns and increasing governmental regulations made it evident that in order to appropriately house and exhibit this large and intelligent species, significant financial resources would need to be invested. Given that the wild population numbers had rebounded as a result of the cooperative regulations enacted by the five Polar Bear Nations (i.e., the Range States: USA, Canada, Union of Soviet Socialist Republics, Norway and Denmark/Greenland) (United Nations Environment Program Register of International Treaties 1973), which addressed issues such as illegal harvest and environmental toxins, the conservation threat for the species was relatively low, and zoos were choosing to use their limited financial resources to build exhibits for species that had a greater conservation need.

That was until climate change. In the early 2000s, when the threat to polar bears due to the effects of climate change and a warming Arctic became obvious, the link between wild and ex situ bear populations became evident. Polar bears in zoos started to be seen having the potential to play a vital role, both as conservation education ambassadors and as an ex situ research population to help address in situ questions. The Association of Zoos and Aquariums (AZA) based in the United States, the Canada's Accredited Zoos and Aquariums (CAZA), and the European Association of Zoos and Aquaria (EAZA) shifted their focus toward more intensive scientific and cooperative management of polar bears in the 2000s in order to address the needs of decreasing ex situ populations, as well as the increasing demand to more effectively educate the public about the threats climate change posed to the species (Meyerson 2015; Szánthó 2014; Szánthó and Spencer 2015). The reality is that in order to save polar bears in the wild you need to save the sea ice, their hunting and denning platforms (e.g., see Chaps. 23 and 25). In order to save the ice, greenhouse gas emissions and carbon use must decrease. This can only been done through changes in human activities. The public appeal of polar bears makes them the perfect ambassador for the effects of climate change in the Arctic. Zoos offer the ability to educate a group of people that are already engaged with animals. This captivated audience can have profound effects on the climate change issue by both changing their own personal habits, as well as helping to create motivations to form environmental policy through political pressure (Fig. 27.1). With over 180 million visitors a year, greater than the number of people attending all major professional sporting events combined in the

Fig. 27.1 Bears in our care are true ambassadors for their wild counters. They can inspire visitors to make changes in their own carbon use that can have global effects, thereby helping save the habitat of the bears in the wild. *Image credit: Toledo Zoo*

USA, Association of Zoos and Aquariums (AZA) accredited institutions have great public reach, with a focus on connecting people and animals. Being a founding member of the Network for Ocean and Climate Change Interpretation, AZA and its members work to change public understanding of the impact of a warming planet (Swim and Fraser 2013). Similarly, the EAZA, partnering with NGO Polar Bears International and Arctic Action, launched a major public climate change education program with their Pole to Pole Campaign from 2013–2015.

27.2 The Value of Ex Situ and In Situ Collaboration

With the increasingly apparent threat of climate change came new collaborative relationships between zoo professionals, nongovernmental organizations, and field scientists. Where once there was little interaction between the zoo community and polar bear field biologists, the potential value of that partnership was realized and fostered by the NGO Polar Bears International, who had working relationships with both groups. In addition to the great public reach offered by zoos for climate education, there also was a realization that polar bears in human care could act as an ex situ research population to help answer in situ questions, using, noninvasive methods.

Doing research in remote locations in harsh environments with limited repeated access to individuals is extremely difficult and expensive. Through researchers employing normal husbandry practices and positive reinforcement training techniques, bears in zoos have participated in studies on energetics, sensory perception, reproduction, and emerging diseases (Rode et al. 2016; Ware et al. 2015). Polar bears in zoos are being trained to wear accelerometers and to walk treadmills in order to help quantify energetic expenditure rates, with results helping determine the effects of increasingly prolonged periods of swimming with declining sea ice (A. Pagano, US Geological Survey, personal communication). Sensory perception research in zoos has included olfaction studies, looking at olfactory communication in polar bears by pedal secretions (bears secrete signal scents which can be detected in footpad imprints by other bears) of relevance to their increasingly fragmented environment (Owen et al. 2015). Auditory studies have been carried out to determine noise disturbance parameters, especially in maternal denning areas, relevant as anthropogenic activities are able to increase as the Arctic warms. Assisted reproductive techniques such as artificial insemination and ova and sperm rescue are being applied to captive polar bear husbandry (Curry et al. 2014; Curry and Roth 2016). These techniques potentially have future applications for maintaining genetic diversity in the wild population if population fragmentation or genetic bottlenecks occur. Having bears in human care may also help to predict future issues with emerging diseases from a warming Arctic. Studies are currently looking at the upregulation of genes in response to environmental stressors (Bowen et al. 2015), as well as the potential threat of diseases such as West Nile Virus, which was the cause of death of a male polar bear in 2006 (Dutton et al. 2010, and research in progress).

Collaborations between the ex situ and in situ communities are not only providing information for use in in situ situations. While originally developed as a tool for field researchers to consistently describe the body condition of wild polar bears, "The Polar Bear Score Card: A Standardised Fatness Index," is now used by zoos as well (Polar Bears International 2015) (Fig. 27.2).

27.3 Improving Standards of Care of Captive Polar Bears

Even before the concern of climate change, zoo animal care professionals realized that if polar bears were going to continue to be exhibited, that standards of their care needed to be better defined and facilities would need to be designed to

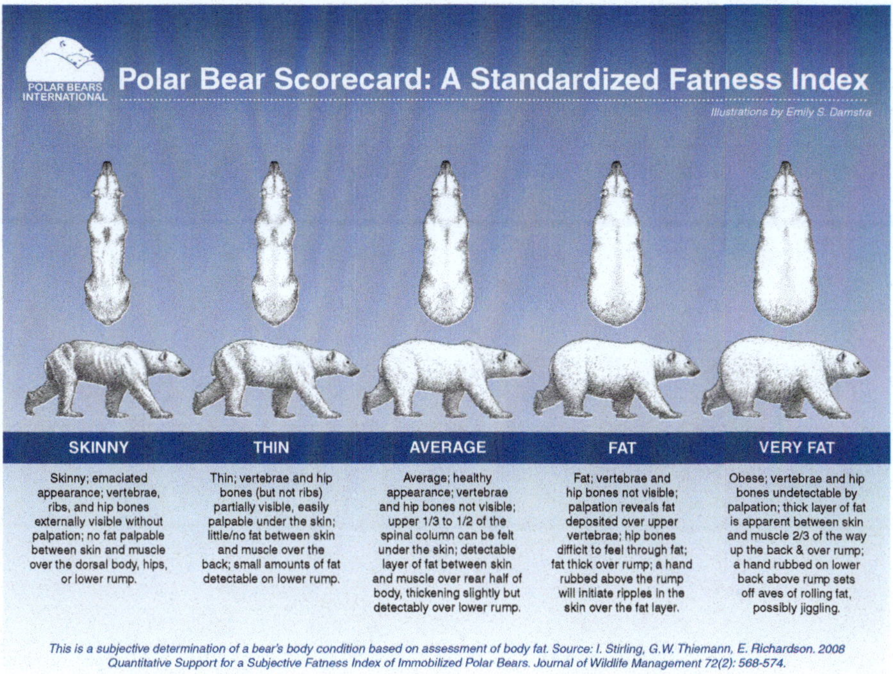

Fig. 27.2 "The Polar Bear Score Card: A Standardized Fatness Index," (Polar Bears International 2015). *Image credit: Polar Bears International*

address an increasingly complex understanding of the bears' welfare needs. Rightly so, there was and is a concern about housing an intelligent, large mammal whose natural range can be across an area of hundreds of square miles. Zoo professionals have worked hard to define exhibit and behavioral husbandry needs to provide good welfare for polar bears in our care. EAZA polar bear experts have developed the Ursid Husbandry Guidelines and AZA polar bear specialists have produced the Polar Bear Animal Care manual. Features of new exhibits include more space, larger pools, soft substrate, the ability to get out of view of conspecifics, elevated viewing areas for the bears, and the ability to make changes to exhibit furnishings (UHG 2007, PB ACM 2009; Shepherdson 2013, Fig. 27.3).

Simple husbandry practices that include allowing the bears a choice of access to a small indoor area during zoo open hours has been found to decrease pacing in bears, apparently because they can exercise more control and are able to see activities behind outdoor enclosure doors and inside holding areas where the bears' caretakers are. For instance, "Gus," the male polar bear at Central Park Zoo (New York City, Wildlife Conservation Society), decreased his swim-pacing behavior from 80% of the time during daytime hours with the most basic "enriched conditions," to consistently less than 25% during the same hours when training was

Fig. 27.3 The Toledo Zoo's new Polar Bear Exhibit, the Arctic Encounter opened in 2000. It's larger size and complexity, including varying topography, animal access to soft substrate and built in enrichment opportunities, allows the polar bears to exhibit a wider range of behaviors. *Image credits: Toledo Zoo*

Fig. 27.3 (continued)

increased and he was given continual access to an off-exhibit holding area. Additionally he chose to spend his time on-exhibit watching visitors and local birds (Moore, unpublished data 1996–2000). This kind of behavior change after receiving free choice as to whether they are in indoor holding or on-exhibit has been observed by other zoo scientists, including those at North Carolina, Chicago, and Toledo zoos (R. Meyerson, personal observation).

The training and enrichment component of the overall husbandry program is just as important as facility design for facilitating good welfare for polar bears in human care. Enrichment programs that facilitate an increased range of behaviors, problem solving, and choice also promote a healthy bear. This can be done by changing objects in their exhibit or offering novel stimuli. Effective enrichment programs provide stimuli in varying ways. Most common is the manipulation of how food items are offered. Common methods include scatter feeds, freezing in ice blocks, hiding them, or presenting them in a puzzle box where manipulation of an object is needed to get to the food. Other types of enrichment include auditory and olfactory stimulation, playing vocalizations, and offering different scents, spices or fur, and feces of other bears or species. Effective enrichment also promotes increased activity through the manipulation of objects. This can be provided both through exhibit design such as providing logs to claw, or mulch to dig through, or the provision of novel items such as 55 gallon plastic barrels, browse, PVC pipes, cardboard boxes, thick balls, and floating items. While enrichment is an important component of any polar bear husbandry program, novel items may also offer a safety risk, and for this reason good programs have an approval process that involves both animal and veterinary care staff.

Training polar bears for husbandry practices (e.g., shifting on and off exhibit) has been common, and recently zoos' training programs have been improved to include the ability to offer better veterinary care through enhanced husbandry training that has veterinary care goals. Behaviors, such as an "open mouth" for tooth exams, presenting appendages for injections, presenting feet for foot soaks, and most recently voluntary offering of feet for blood draws, allow for better care of polar bears; voluntary blood draws and other voluntary participation also facilitate the bears' participation in research that can increase our understanding and conservation of the species. In addition to these benefits, since polar bears are worked with only in protected contact where staff does not directly share space with them, this training participation is voluntary, and the time spent with the keepers can be seen as strengthening the keeper and bear bond (e.g., Fig. 27.4).

Fig. 27.4 This polar bear at the Oregon Zoo is being rewarded with fish snacks as staff perform a voluntary blood draw from the top of his rear foot. This type of cooperative training allows the veterinary staff to assess the bears' health without having to immobilize them, as well as enabling participation in research projects requiring multiple samples of small amounts of blood. *Image credit: Michael Durham/Oregon Zoo*

27.4 Positive Outcomes on Health and Well-Being of Captive Bears

With a better understanding of polar bear needs and behaviors, and better husbandry and veterinary techniques, positive effects on vital measures (mortality, fecundity, and longevity) are starting to be seen. Regional and international studbook databases have been used for decades to pool institutional records for all cooperatively managed bears—from the time of birth or import until death—and these studbooks can provide data on vital rates (fecundity, mortality) as well as cause of death. Studbook data from the European and American zoo associations' polar bear populations show mortality rates of the most vulnerable first age class (0–1 year old) decreasing in recent decades (from an average of 62% for females and 63% for males from 1970 to 1999 to 46 for females and 51% for males from 2000 to 2015) (Linke 2015). Causes of death recorded in the AZA regional studbook indicate there has been a significant reduction in intraspecific aggression (Meyerson 2015). From 1970 to 1999, injury from exhibit mates was the third most common cause of death (after euthanasia and unknown/other) of the 221 deaths recorded in the studbook for animals older than 1 year. However, in recent years (2000–2015), following the shift to more formal cooperative management as a Species Survival Plan, there were no deaths due to conspecific injuries, aggression, or self-inflicted injuries, and the number of deaths attributed to infection has decreased considerably (from 15% of deaths to 5%). Data from both North American and European accredited zoos indicate that animals are now living to older ages, with the mean age at death significantly higher in both regions in the more recent time period of modern management (during 2000–2015, on average 3.1 ± 0.61SD years older for AZA and on average 8.0 ± 0.57SD years older for EAZA) compared to prior management period (1970–1999) (Fig. 27.5). One of the driving factors of this increase is that proportionally fewer individuals died at younger ages during the more recent time period (Fig. 27.6).

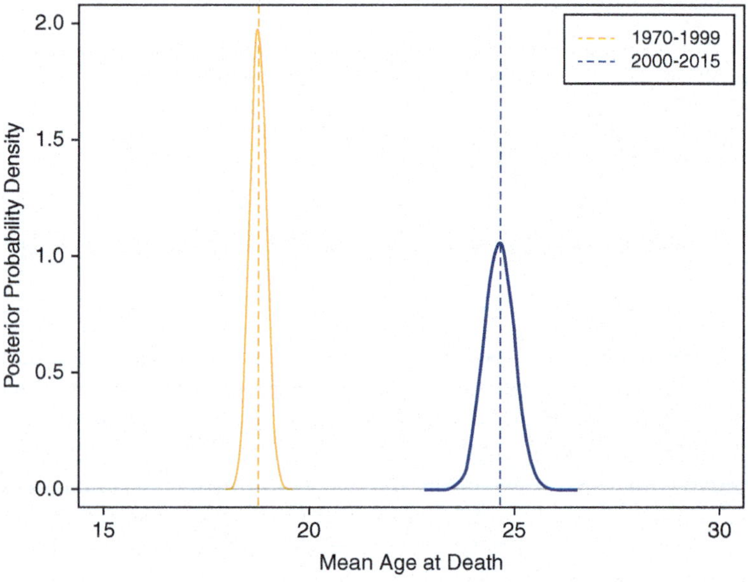

Fig. 27.5 A comparison of the mean age at death of polar bears in European and American zoos combined, between two time periods: 1970–1999 and 2000–2015. The expected ages at death are 18.8 (95% credible interval = 18.4–19.2) and 24.6 (23.9–25.4) years for the earlier and more recent time period, respectively (ages at death were modeled as a Poisson distributed variable in a Bayesian statistical model with non-informative priors. The dashed lines show the average age at death for each time period, and the curved lines show the range of likely ages or the posterior probability density of the model estimates)

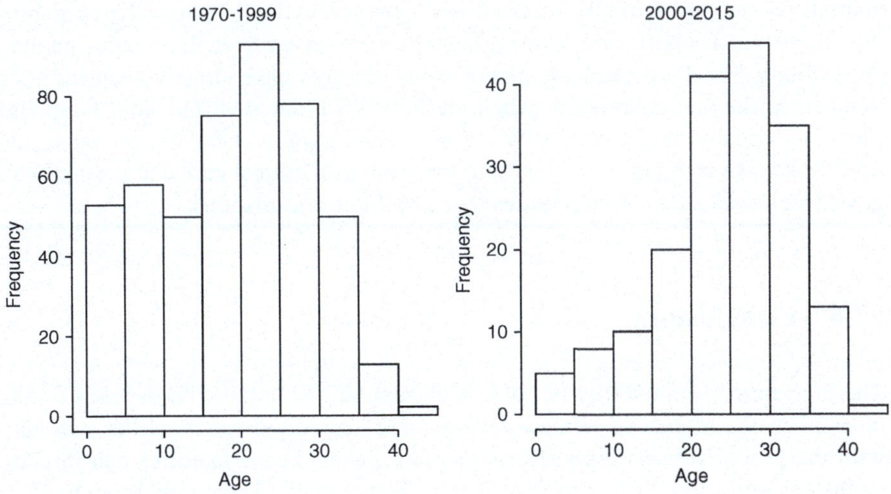

Fig. 27.6 Distribution of the ages at death for polar bears in European and American zoos during two time periods ($N = 472$ in 1970–1999, $N = 178$ in 2000–2016). In the more recent time period (2000–2015), proportionally fewer individuals died at younger ages and the median age at death is higher (20.0 for 1970–1999, 25.5 for 2000–2016)

27.5 Opportunities for Future Collaboration

Zoos work cooperatively through EAZA's European Endangered Species Programme (EEP) and AZA's Species Survival Plan (SSP) program to help to ensure that individual polar bears have appropriate facilities in which to live as both cubs and adults and that the populations remain genetically diverse and demographically stable (Ballou and Lacy 1995; Ballou et al. 2010). Governments also recognize the special requirements of the species and have enacted regulations to protect their welfare as well. As examples, in the United States, polar bears have their own regulations under the Animal Welfare Act (USDA 2013), and in Manitoba, Canada, captive polar bear management is regulated by the Polar Bear Protection Act (2002, 2008, and 2013). In 2012, the province also established the Leatherdale International Polar Bear Conservation Centre at the Assiniboine Park Zoo in Winnipeg. This center was established to transition orphaned polar bear cubs rescued in the province, for eventual placement in approved facilities. The center also contributes to understanding of the conservation of polar bears through its education and research programs.

In addition to having the skills to offer better welfare for the bears in our care, zoo professionals can offer additional skills that can directly help wild bears. With knowledge accumulated over time on how to house and handle polar bears, zoo professionals offer a large contingent of people experienced in working with live bears who can respond to environmental disasters like oil spills. Caring for the animals where they naturally occur allows them to remain there after their rehabilitation. For those whose care requires them to be removed from their native habitat, zoos offer a place for rescue. Collaborations like this have already occurred with input from the zoo community when the USFWS Polar Bear Oil Spill Response Plan was updated in 2015 (USFWS 2015). Additionally, AZA has an active Oil Spill Response program, working both to assist with animal care and partner with government and native communities (see http://aza.org/oilspill/).

27.6 Conclusions

The long-term sustainability of polar bear populations in the wild depends on the reversal of the effects of climate change, and accredited zoo facilities can help with the global conservation efforts of polar bears. These facilities can directly contribute education programs that seek to change public behaviors through effective education and interpretation—and these processes can affect hundreds of millions of visitors on-site and virtually. In addition, zoological facilities can continue to conduct ex situ basic biological and behavioral research with in situ applications, in developing assisted reproductive techniques in case of population bottlenecks, in researching emerging diseases and mitigation of disease and parasite effects, in executing contingency plans for the rescue of orphaned and compromised bears, and in understanding and responding to human-bear conflict and mitigating negative human-bear behaviors. The very real plight of the polar bear and the planet from the threat of climate change has facilitated partnerships between governments, zoos, researchers, nongovernmental organizations, and more recently with native communities. In order to save the bears, there is a need to continue to think outside the box on how to deal with these challenging and threatening issues.

References

Ballou JD, Lacy RD (1995) Identifying genetically important individuals for management of genetic diversity in pedigreed populations. In: Ballou JD, Foose TJ, Gilpin M (eds) Population management for survival and recovery. Columbia University Press, New York, NY, pp 76–111

Ballou JD, Lees C, Faust L, Long S, Lynch C, Bingaman-Lackey L, Foose T (2010) Demographic and genetic management of captive populations for conservation. In: Kleiman DG, Thompson KV, Baer CK (eds) Wild mammals in captivity: principles and techniques for zoo management, 2nd edn. University of Chicago Press, Chicago, IL, pp 219–252

Bowen L, Miles AK, Waters S, Meyerson R, Rode KD, Atwood TC (2015) Gene transcription in polar bears (*Ursus maritimus*) from disparate populations. Polar Biol 38(9):1413–1427. doi:10.1007/s00300-015-1705-0

Curry E, Roth TL (2016) A rapid, minimally-invasive method of collecting semen from polar bears. Reprod Fertil Dev 28:189. (abstract 118)

Curry E, Wyatt J, Sorel LJ, MacKinnon KM, Roth TL (2014) Ovulation induction and artificial insemination of a captive polar bear (*Ursus maritimus*) using fresh semen. J Zoo Wildl Med 45:645–649

Dutton CJ, Quinnell M, Lindsay R, DeLay J, Barker IK (2010) Paraparesis in a polar bear (*Ursus maritimus*) associated with West Nile virus infection. J Zoo Wildl Med 40:568–571

Swim JK, Fraser J (2013) Fostering hope in climate change educators. J Museum Edu. 38:286–297

Linke K (2015) World Zoo and Aquarium Association international polar bear studbook. Zoo Rostock, Rostock, Germany

Meyerson R (2006) Association of Zoo and Aquariums polar bear studbook. Toledo Zoo, Toledo, OH

Meyerson R (2015) Association of Zoo and Aquariums polar bear studbook. Toledo Zoo, Toledo, OH

Owen MA, Swaisgood RR, Slocomb C, Amstrup SC, Durner GM, Simac K (2015) An experimental investigation into chemical communication in the polar bear (*Ursus maritimus*). J Zool 295:36–43

Poirier P, Lanthier C (1995) North American regional polar bear studbook. Jardin Zoologique de Granby, Quebec, Canada

Polar Bear Animal Care Manual (2009) Association of Zoo and Aquariums. https://www.aza.org/uploadedfiles/animal_care_and_management/husbandry,_health,_and_welfare/husbandry_and_animal_care/polarbearcaremanual.pdf. Accessed 29 Nov 2016

Polar Bears International (2015) The polar bear score card: a standardized fatness index. http://www.polarbearsinternational.org/media/images/fatness-index-scorecard. Accessed 29 Nov 2016

Rode KD, Stricker CA, Erlenbach J, Robbins CT, Cherry SG, Newsome SD, Cutting A, Jensen S, Stenhouse G, Brooks M, Hash A, Nicassio N (2016) Isotopic incorporation and the effects of fasting and dietary lipid content on isotopic discrimination in large, carnivorous mammals. Physiol Biochem Zool 89(3):182–197

Shepherdson D, Lewis K, Carlstead K, Bauman J, Perrin N (2013) Individual and environmental factors associated with stereotypic behavior and fecal glucocorticoid metabolite levels in zoo housed polar bears. Appl Anim Behav Sci 147:268–277

Szánthó J (2014) European association of zoos and aquaria polar bear studbook. Artis Royal Zoo, Amsterdam, Netherlands

Szánthó J, Spencer W (2015) Zooquaria. European Association of Zoos and Aquaria News 89:18–19

The Polar Bear Protection Act (C.C.S.M. c. P94) (2002) Polar Bear Protection Regulation, Regulation 210/2002. Government of Canada

The Polar Bear Protection Act (C.C.S.M. c. P94) (2008) Polar Bear Protection Regulation, amendment (c. P94 de la C.P.L.M.) Regulation 150/2008. Government of Canada

United Nations Environment Program Register of International Treaties (UNEP) (1973) Agreement on Conservation of Polar Bears. The Governments of Canada, Denmark Norway, the Union of Soviet Socialist Republics and the United States of America. Oslo, 1973. http://sedac.ciesin.org/entri/texts/polar.bears.1973.html. Accessed 29 Nov 2016

United States Department of Agriculture (USDA) (2013) Animal Welfare Act and Animal Welfare Regulations

United States Fish and Wildlife Service (USFWS) (2015) Marine mammals management: oil spill response plan for polar bears in alaska

Ursid Husbandry Manual (2007) European Association of Zoos and Aquaria. https://www.aza.org/animal-care-manuals/. Accessed 29 Nov 2016

Ware JV, Rode KD, Pagano AM, Bromaghin J, Robbins CT, Erlenbach J, Jensen S, Cutting A, Nicassio-Hiskey N, Hash A, Owen M, Jansen HT (2015) Validation of mercury tip-switch and accelerometer activity sensors for identifying resting and active behavior in bears. (International Association for Bear Research and Management). Ursus 26:8–18

Chapter 28
Monitoring the Welfare of Polar Bear Populations in a Rapidly Changing Arctic

Todd C. Atwood, Colleen Duncan, Kelly A. Patyk, and Sarah A. Sonsthagen

Abstract Most programs for monitoring the welfare of wildlife populations support efforts aimed at reaching discrete management objectives, like mitigating conflict with humans. While such programs can be effective, their limited scope may preclude systemic evaluations needed for large-scale conservation initiatives, like the recovery of at-risk species. We discuss select categories of metrics that can be used to monitor how polar bears (*Ursus maritimus*) are responding to the primary threat to their long-term persistence—loss of sea ice habitat due to the unabated rise in atmospheric greenhouse gas (GHG; e.g., CO_2) concentrations—that can also provide information on ecosystem function and health. Monitoring key aspects of polar bear population dynamics, spatial behavior, health and resiliency can provide valuable insight into ecosystem state and function, and could be a powerful tool for achieving Arctic conservation objectives, particularly those that have transnational policy implications.

T.C. Atwood (✉)
Polar Bear Research Program, US Geological Survey, Alaska Science Center, Anchorage, AK, USA
e-mail: tatwood@usgs.gov

C. Duncan
Department of Microbiology, Immunology and Pathology, Colorado State University, Fort Collins, CO, USA
e-mail: Colleen.Duncan@colostate.edu

K.A. Patyk
USDA/APHIS/VS/STAS/Center for Epidemiology and Animal Health, United States Department of Agriculture, Fort Collins, CO, USA
e-mail: Kelly.A.Patyk@aphis.usda.gov

S.A. Sonsthagen
US Geological Survey, Alaska Science Center, Anchorage, AK, USA
e-mail: ssonsthagen@usgs.gov

A. Butterworth (ed.), *Marine Mammal Welfare*, Animal Welfare 17,
DOI 10.1007/978-3-319-46994-2_28

28.1 Introduction

The development of biologically meaningful and efficient approaches to monitoring wildlife populations has been a long-standing objective for researchers and managers. Most programs for monitoring the welfare of wildlife support efforts aimed at reaching distinct management objectives, like maintaining a sustainable harvest (Elmberg et al. 2006), conserving or improving habitat (Mawdsley et al. 2009), or mitigating conflict with humans (Treves and Karanth 2003). Such programs can be effective at reaching stated objectives but, because they are often limited in breadth and scope, may not allow for systemic evaluations needed for large-scale conservation initiatives like the recovery of threatened or endangered species.

In addition to providing information on population vital rates, monitoring programs geared toward ensuring the persistence of at-risk species should also provide feedback on ecosystem functioning and integrity (Lambeck 1997; Andelman and Fagan 2000). While challenging to implement, holistic monitoring programs have been used to track the distribution and abundance of threatened species relative to changing environmental conditions (Dawson et al. 2011), identify landscape characteristics that influence patterns of species richness (Niemi and Macdonald 2004), and monitor the influence of ecosystem stressors on population dynamics (Magurran et al. 2010). Ideally then, monitored species also could function as indicators to attributes characterizing the biophysical condition of an ecosystem, including changes in habitat, community composition (e.g., prey), and the influence of stressors (Landres et al. 1988).

The polar bear (*Ursus maritimus*) is a species of conservation concern given their vulnerability to climate-induced loss of sea ice habitat (e.g., Derocher et al. 2004; Amstrup et al. 2008; Stirling and Derocher 2012). There are numerous mechanisms (e.g., international and bilateral agreements, user-to-user agreements [subsistence users], memorandums of understanding, recovery plans) in place or under development to guide the conservation of polar bears. In most cases, these agreements either tacitly or directly rely on information collected through monitoring to guide decision-making. As a wide-ranging apex predator, polar bears are exposed to a broad array of ecosystem conditions and can be used as an indicator of attributes too difficult, inconvenient, or costly to measure directly. Given that, monitoring key aspects of polar bear population dynamics, spatial behavior, and health and resiliency can provide valuable insight into ecosystem state and function and could be a powerful tool for achieving Arctic conservation objectives, particularly those that have transnational policy implications.

28.2 The Monitoring Imperative

Polar bears range over the ice-covered seas of the circumpolar Arctic (Stirling and Derocher 2012) and most commonly occur in areas dominated by annual sea ice (i.e., sea ice that forms each year) covering shallow, biologically productive waters

(Durner et al. 2009). Currently, it is believed that there are approximately 20,000–25,000 polar bears distributed among 19 subpopulations (see Chap. 23), representing the majority of their historic range. However, monitoring circumpolar distribution and, particularly, abundance has been a long-standing challenge, and the magnitudes of change in status over time are uncertain for many subpopulations. For example, the most recent circumpolar subpopulation status assessment from the International Union for the Conservation of Nature/Polar Bear Specialist Group (IUCN/PBSG 2014) indicates three subpopulations are exhibiting a declining trend in abundance, six are considered stable, one is increasing, and nine are data deficient. For the data-deficient subpopulations, there are no abundance estimates for five subpopulations due, mostly, to logistical limitations that have precluded the collection of relevant data.

The necessity for monitoring polar bears has evolved over time. Monitoring began in earnest following the implementation (in 1976) of the 1973 *Agreement on the Conservation of Polar Bears*, which mandated that "the best available science" be used to manage polar bears and that nations should cooperate in the management of transborder subpopulations. In the 1970s and 1980s, a key role of monitoring was to ensure that polar bears were being harvested in a sustainable manner to allow recovery of some subpopulations from earlier overharvest in the 1950s and 1960s (e.g., Brower et al. 2002). Accordingly, stakeholders included agencies with management oversight, subsistence harvesters and communities within polar bear range, and sport-hunters (in those subpopulations where legal). In the mid-1990s, it became apparent that sea ice habitat was becoming less stable, and the prevailing management concern began to shift from sustainable harvest to the potential population-level consequences of sea ice decline (Stirling and Derocher 1993). As the loss of sea ice habitat accelerated, links to adverse effects on vital rates and fitness were detected (e.g., Stirling et al. 1999; Regehr et al. 2006, 2007, 2010; Rode et al. 2010), which helped inform the decision to list polar bears as "vulnerable" on the IUCN Red List in 2006 and "threatened" under the US Endangered Species Act (ESA) in 2008 (Schliebe et al., 2008; U.S. Fish and Wildlife Service 2008). Consequently, polar bears became a flagship species used to communicate to the public the perils of climate change, which broadened the group of stakeholders to include nongovernmental conservation organizations, industry, and the general public. In turn, the monitoring imperative shifted from harvest management to population viability in the face of global climate change (Derocher et al. 2004; Stirling and Derocher 2012; Amstrup et al. 2010).

Despite the recognition of the threat posed by climate-induced loss of sea ice habitat (see Chaps. 23 and 24), coordinated (i.e., multinational) holistic monitoring plans have been slow to develop. That said, the comprehensive effort put forth in Vongraven et al. (2012) to outline a framework for circumpolar monitoring has formed the foundation for the Circumpolar Action Plan developed under the aegis of the Polar Bear Range States (i.e., the parties signatory to the 1973 Agreement). Our intention here is not to reprise material covered by Vongraven et al. (2012). Rather, we focus on select categories of metrics that can be used to monitor how polar bears are responding to the primary threat to their long-term persistence—loss of sea ice habitat due to the unabated rise in atmospheric greenhouse gas (GHG; e.g., CO_2) concentrations (Amstrup et al. 2010; Atwood et al. 2016a). This can also provide information on ecosystem function and health. Ricke and Caldeira (2014) estimated that peak warming resulting from emissions occurs approximately 10 years after the CO_2 is added to the atmosphere. Further, sea ice stabilization is expected to require 20–30 years post peak warmth (Amstrup et al. 2010). This lag between mitigation of emissions and stabilization of sea ice habitat (sensu Allen and Stocker 2013) suggests that monitoring programs that capture long-term trends will be critical to measuring progress toward conservation objectives.

28.3 Metrics for Monitoring the Response of Polar Bears to Ecosystem Change Population Abundance and Vital Rates

Estimating the number of individuals within a given spatial unit is one of the most enduring and vexing challenges that wildlife researchers and managers face—yet, it also often constitutes the centerpiece of monitoring programs. As Vongraven et al. (2012) note, the question most often posed to polar bear biologists by policy makers and the public is some version of "how many bears are there?" Abundance and its trend are metrics that are readily translatable into an assessment of subpopulation status and, from a communication perspective, are easy for the public to understand—e.g., subpopulations are increasing, stable, or decreasing through time. From a monitoring perspective, tracking changes in abundance can provide periodic feedback on the efficacy of management activities and vulnerability to environmental conditions, provided data are collected in a manner sufficient to minimize uncertainties. However, the Arctic is a challenging environment to work in and aspects of polar bear life history make them difficult to sample. For example, like many large carnivore species, polar bears are mostly solitary and highly vagile and occur at low densities (Garner et al. 1990; Taylor and Lee 1995; Born et al. 1997; Amstrup et al. 2000, 2004). Polar bears easily transition from sea ice to open-water and, increasingly, terrestrial habitats when sea ice is absent. Collectively, these characteristics can present logistical and sampling (e.g., detection heterogeneities) challenges that may impair the ability to obtain regular and/or robust abundance estimates.

28.3.1 Abundance

Various methods have been used to estimate abundance, and each has its advantages and limitations. Briefly, the most commonly used quantitative methods rely on a physical mark-recapture (M-R) sampling design, where bears are captured and chemically immobilized from a helicopter, individuals are "marked" (e.g., via an individually identifiable tattoo, ear tag, or through obtaining a genetic identification) and released, and the process is repeated annually for a predetermined period of time (usually ≥3 years). Estimates of abundance are then derived based on the ratios of marked to unmarked individuals sampled (Amstrup et al. 2005). An alternative and less-invasive M-R method relies on the remote collection of genetic "marks" (i.e., adipose tissue samples) through biopsy darting from a helicopter (Pagano et al. 2014). While this method is also typically repeated annually for a minimum of 3 years, it does not require the capture and immobilization of individuals and thus can be more culturally acceptable in some northern communities (Peacock et al. 2011). Aerial surveys that incorporate line transect and distance sampling methodologies have become more common because, similar to remote M-R methods, they are less invasive than physical M-R and can also be used to estimate abundance from a single sampling occasion (Stapleton et al. 2014a). A fourth method is the use of high-resolution satellite imagery, which has been shown to be effective in estimating polar bear abundance under certain conditions (e.g., on land in summer), though significant analytical progress will be needed to optimize its widespread use to monitor abundance (Stapleton et al. 2014b; LaRue et al. 2015).

The methods described above differ in advantages and limitations. For example, while the costs associated with physical and remote M-R methods do not markedly differ, physical M-R allows for the collection of age data (beyond age classes), physical measurements, and biological samples, which can allow the monitoring of numerous other metrics discussed below (e.g., body condition, litter mass, exposure to infectious agents, and contaminants). Aerial surveys, depending on the intensity of sampling, can be as costly as physical M-R but only allow for the collection of information on age and sex classes (though determining the sex of subadults can be difficult and determining the sex of yearlings and cubs is not possible). The use of satellite imagery is substantially less costly than the other methods mentioned and allows for the monitoring of even the most remote regions of the Arctic, but sex and age class are indistinguishable from imagery and adverse weather conditions at the time of image acquisition can render imagery useless (Stapleton et al. 2014b).

The key to choosing a method for estimating abundance is to select one most appropriate to monitoring program objectives. For example, periodic abundance estimates using a multi-year physical M-R methodology may have been sufficient for monitoring subpopulations when sea ice conditions were stable, but, given the dramatic and rapid decline in the availability of sea ice habitat over the last 15 years, more frequent abundance estimates are probably warranted for some subpopulations. Since physical M-R is expensive and it may not be feasible to incur those expenses on a more regular timeframe, an aerial survey could be conducted to bridge the gap between periodic physical M-R studies to provide an up-to-date snapshot of subpopulation status. That said, data requirements of a monitoring program, may be such that only physical M-R, and the various data collection opportunities it provides, will suffice.

28.3.2 Survival

Like abundance, survival is a key metric used to monitor subpopulation status. And, like abundance, monitoring survival can be a costly and logistically challenging endeavor mainly because a large sample of marked individuals, and thus multiple years of monitoring, is required to detect demographic variation and trends (Coulson et al. 2005). The most common ways to monitor survival are via radiotelemetry and through M-R methods (both physical and remote [i.e., genetic] M-R; White and Burnham 1999; Murray and Patterson 2006; Schwartz et al. 2007), which can be leveraged to also monitor habitat use (in the case of radiotelemetry) and estimate abundance (based on M-R methodologies). Further, both methods enable the linking of mortality to causal factors, though only radiotelemetry allows for the reliable determination of specific forms of mortality (other than harvest) provided mortalities are investigated in a timely manner. Identifying causes of death (Fig. 28.1), while challenging, is valuable because it can provide important information on the nature and severity of threats to a population, as well as feedback on ecosystem health (e.g., presence of infectious agents, pollutants, prey availability/accessibility).

Fig. 28.1 Yearling polar bear found dead with its mother on a barrier island north of Prudhoe Bay, southern Beaufort Sea, Alaska, September 2012. Notice the *pink* coloration of exposed skin. Necropsy results indicated both bears ingested the commercial dye rhodamine B, though the cause of the death was unknown. *Image credit: U.S. Geological Survey*

Estimating sex- and age-specific survival (and mortality) rates can be particularly useful for monitoring population dynamics and changes in the nature of population and ecosystem stressors. For polar bears, mortalities have been linked to environmental conditions, natural causes, and anthropogenic activities, which have the potential to differentially affect demographic groups (e.g., Amstrup et al. 1989; Amstrup and Durner 1995; Derocher and Stirling 1996; Table 28.1) and drive popu-

Table 28.1 Published examples of confirmed and suspected sources of mortality (excluding harvest) for polar bears

Subpopulation	Mortality source[a]	Demographic group affected	Reference
Barents Sea	Adverse weather[s], intraspecific predation[d]	Cub-of-the-year (cubs born in the current year)	Larsen (1985)
	Intraspecific predation[d]	Cub-of-the-year, yearling	Derocher and Wiig (1999) and Stone and Derocher (2007)
Baffin Bay	Management action[b]	All	Dyck (2006)
Lancaster Sound	Management action[b]	All	Dyck (2006)
Western Hudson Bay	Starvation[s]	Cub-of-the-year, yearling	Derocher and Stirling (1996)
	Interspecific predation[s]	Adult	Lunn and Stenhouse (1985)
	Management action[b]	Subadult, adult	Lunn and Stirling (1985) and Stenhouse et al. (1988)
Southern Beaufort Sea	Natural causes[d]	Adult	Amstrup and Durner (1995)
	Poisoning[d]	Unspecified	Amstrup et al. (1989)
	Den collapse	Adult, cub-of-the-year	Clarkson and Irish (1991)
	Starvation[s]	Cub-of-the-year, yearling	Amstrup and Durner (1995)
	Adverse weather[s]	Unspecified	Monnett and Gleason (2006)
	Adverse weather[s] or starvation[s]	Yearling	Durner et al. (2011)
	Complications from capture (research)	Adult	Rode et al. (2014a, b)
Northern Beaufort Sea	Interspecific predation[s]	Cub-of-the-year	Richardson and Andriashek (2006)

Detection of mortalities and determination of causes are rare due to the difficulty of maintaining a monitoring effort of sufficient duration and intensity. For example, the death of satellite-collared individuals can be determined from location and sensor data, but biologists are often unable to reach the individual in a timely manner given the logistical difficulties of working in remote locations

[a]Source of mortality is coded with a superscript "d" when cause has been determined or a superscript "s" when cause is suspected

[b]Lethal removals of bears due resulting from human-bear interactions

lation dynamics. For example, juvenile survival, which determines recruitment, is often the most sensitive to limiting factors, whether caused by changes in population density or by environmental factors. Annual variation in juvenile survival can have multiple causes, including low birth weights, predation, and adverse environmental conditions (Amstrup and Durner 1995; Derocher and Stirling 1996; Rode et al. 2010). By contrast, prime-aged adult survival is buffered against most limiting factors (Wiig 1998; Regehr et al. 2015), and variability in survival typically has fewer causes (e.g., harvest). Persistently low adult survival rates may be indicative of a population experiencing a significant stress. However, survival of both adults and juveniles has been linked to changing sea ice conditions (Regehr et al. 2007, 2010; Bromaghin et al. 2015), indicating that monitoring survival (and select environmental covariates) can be a powerful approach for tracking effects of loss of sea ice habitat on population dynamics.

28.3.3 Body Condition

Body condition characterizes an individual's energy reserves and is often expressed as percent body fat (Cattet et al. 2002; Stevenson and Woods 2006). Variation in ursid body condition has been linked to multiple indicators of population and ecosystem health, including reproduction, cub survival, forage availability/quality, and environmental conditions (Atkinson and Ramsay 1995; Derocher and Stirling 1995; Zedrosser et al. 2006; Rode et al. 2010; McLellan 2011). In some polar bear subpopulations, declines in body condition have been associated with decreasing sea ice habitats and reduced access to their main prey, ice seals (Stirling et al. 1999; Obbard et al. 2006, 2016; Rode et al. 2010, 2012). These changes in condition preceded changes in subpopulation health, including reduction in litter mass, litter size, survival, and abundance (Derocher and Stirling 1994, 1996; Stirling et al. 1999; Regehr et al. 2007; Rode et al. 2010).

Various condition indices (CI) based on morphometric, biochemical, and physiological parameters have been used in polar bear monitoring. Most CI require the physical capture of individuals. For polar bears, body condition has mainly been reported using morphometric-based CI (Fig. 28.2) including body mass, length, skull size, girth, and derived measures including body mass index (BMI), body condition index (BCI), energy density, and lipid concentration of adipose tissue (Cattet et al. 2002; Stevenson and Woods 2006; McKinney et al. 2014). All of the morphometric-based CI, except the use of remotely collected adipose tissue biopsies (McKinney et al. 2014), require capture. By contrast, subjective fatness indices (FI; Stirling et al. 1989) (please see a body condition scale in Chap. 27) can be estimated from either physical examination (palpation) or visual (i.e., remotely via helicopter or other viewing platform) observation, though no work has been done to validate the accuracy of remotely collected visual FI measures.

Some CI perform consistently well in indicating biological responses to changing environmental conditions. Reductions in axillary girth have been associated with declines in summer sea ice extent in multiple subpopulations (Rode et al. 2010,

Fig. 28.2 Collection of morphometric data (skull width) from an immobilized polar bear, southern Beaufort Sea, Alaska, 2007. *Image credit: Daniel J. Cox*

2012). Reduced BCI values have been linked to the lengthening ice-free season (Obbard et al. 2007, 2008, 2016), and mean energy density of some demographic groups has varied among subpopulations characterized by different long-term changes in the availability of sea ice (Rode et al. 2014a). Likewise, FI of individuals have been shown to vary relative to disease status, contaminant exposure, and spatial distribution (Henriksen et al. 2001; Amstrup et al. 2006; Atwood et al. 2015). There may be opportunities to expand the CI monitoring tool chest by applying isotope dilution and bioelectrical impedance analysis (BIA) techniques to polar bears. While both have been validated against total body fat and measures of biological significance in other ursids (Farley and Robbins 1994; Robbins et al. 2012), they can exhibit high variability and/or subjectivity, which may preclude objective assessment of trends in body condition. In sum, body condition is an extremely useful and common metric that can be used to gain early insight into how subpopulations are responding to a rapidly changing Arctic.

28.3.4 Reproduction

Reproductive rates are expected to be highly sensitive to climate warming via declines in body condition of adult females (Molnar et al. 2011; Robbins et al. 2012). The reproductive biology of polar bears is well understood (Ramsay and Stirling 1988; Atkinson and Ramsay 1995). Like most long-lived mammals, they have a low reproductive rate. They enter estrus between March and June, and ovulation is induced by mating (Wimsatt 1963). Implantation of the fertilized egg is delayed until the fall, and the timing of implantation and birth is likely influenced by the nutritional condition of the female (Amstrup 2003). Pregnant females enter maternity dens around late October and emerge with cubs near the end of March. Females typically reach sexual maturity at approximately 5 years of age, generally give birth (typically twins) at 3-year intervals (Stirling et al. 1975; Lentfer et al. 1980), and reproductive senescence is believed to occur around 20 years of age (Ramsay and Stirling 1988). The interbirth interval (II) is marked by a period (≈2.5 years) of close association with cubs, which ends with weaning and the initiation of a new reproductive cycle (Amstrup 2003).

A number of metrics are used for monitoring polar bear reproduction, and many differ in their utility and effort required to collect data. For example, interbirth interval (II), the number of years between successive litters by adult females, provides indirect information on cub survival. If cubs die before weaning, females often have shorter IIs (i.e., <3 years; Vongraven et al. 2012) which can be indicative of declining recruitment of cubs to the subadult (3–4 years of age) age class. However, II can be difficult to estimate because it requires relatively long-term monitoring of known individuals and documentation of reproduction. Likewise, litter production rate, calculated from the number of females of a given age with cub-of-the-year (cubs born in the current year) litters divided by the total number of females of the same age, requires a large sample size of known-age adult females that can only be achieved through long-term monitoring. Similar to II, declines in litter production rate are often suggestive of subsequent declines in recruitment. Litter size and mass (following emergence of cubs-of-the-year from the den) are comparatively easier to collect and more commonly used. A single annual measure of litter size (e.g., each spring) is relatively uninformative with regards to population status: on average, most emerged litters are comprised of twins and variation in litter size is generally low (e.g., DeMaster and Stirling 1983; Larsen 1985; Ramsay and Stirling 1988; Rode et al. 2010). However, biannual (spring and fall) measures of litter size (expressed as ratios of cubs-to-adult female) can provide useful information on estimated cub survival over the critical ice-free season (e.g., DeMaster and Stirling 1983). Additionally, measuring litter mass may provide greater insight into population health given that heavier females have been shown to have heavier cubs (Derocher and Stirling 1994), which is believed to increase the likelihood of cub survival (Derocher and Stirling 1996).

Monitoring indices of reproduction can provide an early indication of changes in individual fitness and subpopulation health and environmental change and should be a key facet of monitoring programs. While most indices require the physical handling of bears, there are less-invasive (e.g., genetic-based measures of reproductive success; Zedrosser et al. 2007) and noninvasive options (e.g., aerial-based cub-to-adult female ratio). The selection of indices to monitor will depend on the quality of data needed to meet the objectives of the program.

28.4 Distribution and Habitats

For Arctic marine species, climate change is the most important driver of habitat loss and impaired access to critical resources (Wassmann et al. 2011). Declines in the extent and temporal availability of ice over shallow water areas reduce polar bear access to ringed (*Pusa hispida*) and bearded (*Erignathus barbatus*) seals, their preferred prey (Thiemann et al. 2008), and represent a loss of important foraging habitat. Since 1979, sea ice extent and volume during summer have declined at rates of −14%/ and −28%/decade (Comiso 2012), respectively, with the most pronounced change occurring over the last 15 years. A natural consequence of declining sea ice extent and volume has been a lengthening of the annual period of reduced ice availability (i.e., the open-water period), which has increased in duration at a rate of approximately 13 days/decade (Stroeve et al. 2014). Accordingly, there is an urgent need to understand the factors that influence the spatial behavior of polar bears.

Changes in the availability, distribution, and characteristics of sea ice habitat have forced polar bears in some subpopulations to modify their spatial behavior. For instance, in the southern Beaufort Sea (SB), the frequency of long-distance swimming events (i.e., continuous swim of >50 km) increased concurrent with the lengthening open-water period (Pagano et al. 2012; Fig. 28.3). In some cases, loss of sea

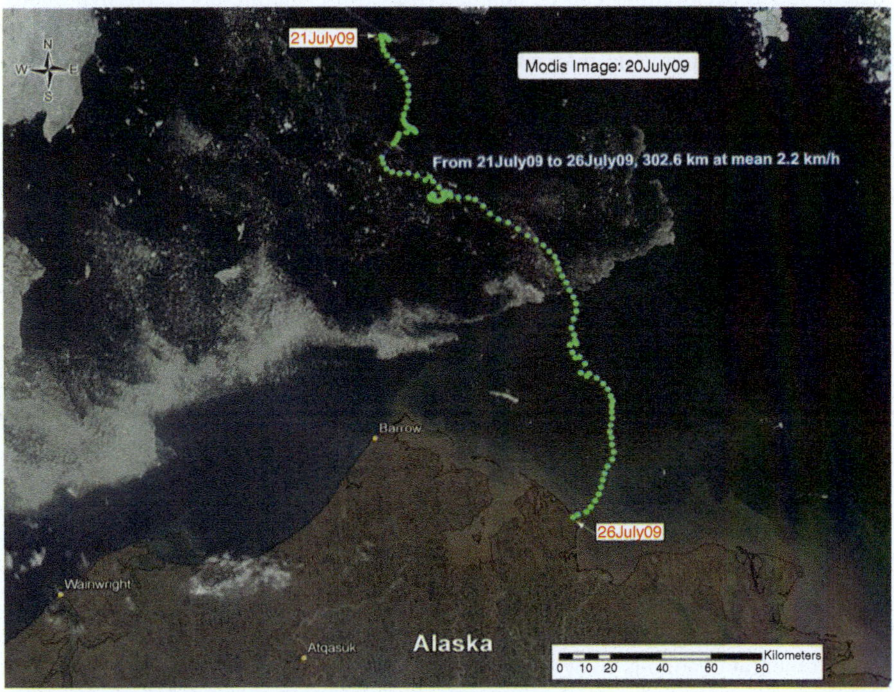

Fig. 28.3 An example of a long-distance swim by an adult female polar bear wearing a GPS-equipped satellite collar, July 2009. Over the course of 6 days, the bear swam 302 km from unconsolidated sea ice to the Alaska coast on 26 July 2009. *Image adapted from* Pagano et al. 2012

ice habitat has led polar bears to become increasingly reliant on land. For example, in western Hudson Bay (WHB), polar bears have historically displayed a pattern of displacing to land when sea ice in the bay melts completely each summer (Lunn and Stirling 1985). Over time, changing sea ice phenology has altered the timing of migration to land in summer, length of stay on land, and timing of departure back to ice in fall (Cherry et al. 2013). In the Chukchi Sea (CS) and SB, historic use of land has been rare and mainly restricted to denning, but recent observations indicate that the proportions of radio-collared bears coming ashore during summer have increased from 20% to 39% and 6% to 20%, respectively, over 15–20 years (Rode et al. 2015; Atwood et al. 2016b). As with WHB, the average length of stay on land for CS and SB bears has increased with the lengthening open-water period. In some polar bear subpopulations, increases in the length of stay on land have been linked to declines in body condition (Stirling et al. 1999; Obbard et al. 2006), and reproductive indices (Stirling and Parkinson 2006), and energy budget models suggest a 30-day increase in the length of stay on land could significantly increase the likelihood of reproductive failure and starvation (Molnár et al. 2010, 2014; Robbins et al. 2013).

Ideally, methods used to monitor distribution and habitats will be integrated, and there are a variety of tools available to suit that purpose. Polar bear location or occurrence data can be collected from satellite telemetry, direct observation (e.g., aerial survey), or noninvasive genetic methods. Coarse-scale habitat metrics (e.g., sea ice extent and concentration) can be characterized from satellite-borne passive microwave imagery (PMW), while finer-scale data can be gleaned from interpreted charts (provided in a geographic information system [GIS] format) of sea ice concentration, extent, and composition produced by the National Ice Center (NIC; Suitland, Maryland, USA; http://www.natice.noaa.gov/) and the Canadian Ice Service (CIS; Ottawa, Ontario, Canada; http://www.ec.gc.ca/glaces-ice/). The collection of location or occurrence data can be cost prohibitive——satellite telemetry collars average $3600 USD/per unit, and deployment requires helicopter-based capture. Collecting occurrence data from aerial surveys is slightly less costly than the deployment of collars, while the cost associated with land-based noninvasive genetic methods (e.g., hair-snags; Herreman and Peacock 2013) can be quite reasonable. Fortunately, daily PMW data (since 1979) and weekly to biweekly NIC (since 1997) and CIS data (since 1968) are freely available to users (Vongraven et al. 2012).

Data from polar bear space use and habitat characteristics have been used to identify patterns of resource selection (Mauritzen et al. 2003; Richardson et al. 2005; Durner et al. 2009), delineate subpopulation boundaries and characterize spatial structure (Amstrup et al., 2004; Sahanatien et al. 2015), quantify changes in the availability of high-quality sea ice habitats over time and project the distribution of high-quality habitat into the future (Durner et al. 2009; Durner 2014), and identify triggers of migratory behavior (Cherry et al. 2013; Rode et al. 2015; Atwood et al. 2016b). Monitoring spatial behavior and habitats has helped elucidate how future environmental change is likely to influence the distribution of polar bears, though linking patterns of resource selection to fitness has proven difficult. Monitoring programs that link estimates of habitat quality to measures of variation in demographic vital rates will further improve our understanding of how polar bears are likely to respond to the continued loss of sea ice habitat and may be critical in guiding reserve design and mitigating the impacts of human disturbance.

28.5 Monitoring Polar Bear Health

Health is one of the fundamental principles of animal welfare as defined by the World Organization for Animal Health (OIE 2015). However, there is not an agreed consensus on how animal health and disease are defined (Gunnarsson 2006; Nordenfelt 2011). To date, animal health monitoring has been largely disease centric, but it is slowly transitioning to a characterization of health in the context of vulnerability (Hanisch et al. 2012; Stephen 2014).

Recently, an expert group unanimously agreed that polar bear health can be applied at individual, species, and ecosystem levels and that some defining characteristics are whether a population can respond to factors in its environment and sustain itself long term (Patyk et al. 2015). The same group also agreed that there is a need for systematic and standardized monitoring and data collection to monitor health and identify early changes that may negatively impact subpopulations (Patyk et al. 2015). These, along with other needs and knowledge gaps to adequately assess and monitor health, are represented schematically in Fig. 28.4. While this list was developed by polar bear researchers with training and expertise in biology, ecology, and veterinary medicine, additional stakeholder involvement (e.g., subsistence users, Arctic villagers, physiologists, and climate scientists) will be necessary to

Fig. 28.4 Important concerns to polar bear health, grouped into three categories—biological, social, and ecological—depicted by the concentric circles. The eight themes listed within the figure are topics about concerns to polar bear health over the next 20 years. The means of survey scores are given in parentheses where respondents rated each concern on a scale from 1 to 10, where 1 = "little to no importance" and 10 = "utmost importance." The figure illustrates the multifaceted nature of health risks and the concerns to polar bear health across multiple levels and emphasizes the importance of integrated monitoring of health determinants across subpopulations. *Source* Patyk et al. 2015

further shape a polar bear health-monitoring program that could be adopted across all of the subpopulations. In order to do this, however, it will be necessary to collect and integrate health information in such a way that it can be used to address both current and future health concerns.

28.5.1 Monitoring Methods

Health information can be collected either actively or passively. Active (i.e., targeted) surveillance implies proactive data collection for a specific purpose and has been used in many of the pathogen and parasite investigations to date on polar bears (Fagre et al. 2015). In contrast, passive surveillance refers to data collected opportunistically, sampling that builds on existing monitoring programs, or the application of data from multiple sources. This type of data may include information collected during general examinations of bears captured as part of annual research programs, samples collected as a result of bears exhibiting abnormal behavior (e.g., hunter observation; Taylor et al. 1991), or data collected in association with a morbidity/mortality event (e.g., Atwood et al. 2015). Appropriate selection of animals is critical for the reduction of bias within the monitoring program.

Figure 28.5 depicts a set of metrics for monitoring health in polar bears and serves as a starting point for the development of more specific tests and scales of

Fig. 28.5 A word cloud depicting the most important metrics or critical indicators to monitor health in polar bears as identified by a panel of experts on polar bear research and management. The larger the word, the more frequently the metric or indicator was identified by the group of experts as described in Patyk et al. (2015). *Source: Patyk et al. (2015)*

measurement to ensure that results can be interpreted over time and across subpopulations. The nature of available biological samples (e.g., feces, hair, and blood) significantly influences, and often restricts, the type of testing that can be conducted on an animal (Fagre et al. 2015). There are many challenges associated with diagnostic tests in wild animals relative to their domestic counterparts for whom the assays are usually developed and validated (Stallknecht 2007; Ryser-Degiorgis 2013). These issues are exacerbated by environmental conditions that can complicate the collection and storage of many biological samples and restrict the potential field application of numerous biomedical tools (Fagre et al. 2015). Broadly speaking, direct detection methods for pathogens (i.e., bacterial culture) can be used in a wide range of species, while wildlife immunology has historically suffered from a lack of species-specific reagents and gaps in knowledge regarding the normal immune response in many free-ranging species (Gilbert et al. 2013). Fortunately, wildlife immunology, however, is an area in which rapid technological advances are being made and immune function assays are increasingly becoming more applicable to wildlife. For all diagnostic testing, the use of appropriate controls, metrics of test performance (i.e., sensitivity and specificity), and overall laboratory protocols (i.e., certification/accreditation and performance reviews) should also be considered.

Biological sample archives can be an extremely valuable resource to conduct retrospective investigations, e.g., to look at spatial or temporal trends of pathogens. However, care must be taken that samples are collected and stored in a method appropriate for future use (OIE 2013). The long-term costs associated with the storage and management of samples and data storage must also be considered.

28.5.2 Health Monitoring for the Future

The call to adopt a broader definition of polar bear health necessitates the recognition that health is more than a static biologic state or the absence of specific hazards and positions us to be more proactive in monitoring health for both known and unknown threats. Determinants of health are intertwined across biological, social, and ecological components (Fig. 28.5), and climate-driven changes to the marine ecosystem have prompted the need for a dynamic approach to monitoring health that is underscored by this broader conceptualization. To that end, a cumulative effect framework integrating multiple factors impacting on health over time has been identified as a critical need and, in some cases, has become an expectation of wildlife management agencies. The development of such a framework for polar bears is a critical first step. Additionally, given the limitations of biological sample collection on free-ranging animals, continual advancement of laboratory assays that can be conducted on currently available tissues is critical. As previously mentioned, wildlife immunology is an area of opportunity. The immune system is sensitive to both infectious and noninfectious stressors as well as both short- and long-term exposures, making it a very valuable system in which to study the cumulative effect of multiple stressors. A recent review highlighted systemic suppression in immune function of marine mammals exposed to contaminants and illustrated the potential population-level consequences that may arise when these species cannot appropriately respond to anthropogenic or infectious disease stressors (Desforges et al. 2016). Finally, in order to meaningfully advance polar bear health studies, there is a need for increased collaboration not only among polar bear biologists but also with disciplines not historically aligned with wildlife health programs (Patyk et al. 2015; Sleeman 2013). Individuals from a wide range of specialties have the ability to contribute significantly to the advancement of techniques and the interpretation of data related to polar bear health. Such groups should draw from both the biological (e.g., climate change and biomedical) and social (e.g., human dimensions and policy) sciences. As marine mammals have been proposed as a good sentinel for the health of marine ecosystems and human health hazards (Bossart 2011), the design and implementation of an innovative polar bear health-monitoring plan have the potential to impact other wildlife species both within the ecosystems they share and in other parts of the world where similar issues complicate effective and sustainable animal health-monitoring.

28.6 Assessing the Capacity for Resilience and Adaptation

Arguably, long-term persistence of polar bears will depend on their ability to either adapt or demonstrate plasticity to significant environmental stressors. However, to date, there has been limited work investigating the capacity for polar bears to be resilient to environmental change. Below, we briefly summarize those efforts and suggest additional monitoring opportunities that may prove useful in assessing the adaptive capacity of polar bears.

28.6.1 Diets

One way in which polar bears might cope with reduced opportunities to forage on ice seals is by switching to alternative food sources, particularly during the annual open-water period. Seasonal variation in the feeding ecology of polar bears is well documented—for example, there are observations, dating back several decades, of land-bound polar bears consuming terrestrial and freshwater food items (e.g., Russell 1975; Lunn and Stirling 1985; Ramsay and Hobson 1991). Moreover, recent evidence indicates that consumption of these foods is increasing in some subpopulations (Rockwell and Gormezano 2009; Smith et al. 2010; Iversen et al. 2013; Prop et al. 2015), and this has led to speculation that increased terrestrial feeding could mitigate the adverse energetic consequences associated with the lengthening open-water period (Gormezano and Rockwell 2013). While the geographic distribution of data on the occurrence of terrestrial feeding is limited, it is important to note that polar bear body condition and survival rates have declined in subpopulations where land use has increased and consumption of terrestrial-based foods has been documented (Rode et al. 2015).

There are a variety of analytical tools that can form the foundation of diet monitoring program, including the identification prey items from scat, but the most commonly used methods are stable isotope (SI) and fatty acid (FA) signature analyses (Thiemann 2008). Stable isotope-based estimates of diets can be derived using a variety of tissue types, including non-invasively collected hair samples, thus making it a relatively cost-effective method. By contrast, FA-based diet estimation relies on sample types (e.g., adipose tissue, milk, blood serum) that mostly require the physical handling of individuals (Iverson 1993; Cooper et al. 2005; Thiemann et al. 2008). As with many other monitoring tools, the selection of analytical technique and sample type should be guided by monitoring program objectives. Monitoring for changes in polar bear diets holds great potential for understanding the nutritional and energetic mechanisms by which body condition and reproduction are affected by sea ice loss and will aid in refining projections for the future status of polar bear populations as sea ice habitat continues to decline.

28.6.2 Genetics

Analyses of deeply sequenced genomes of polar, brown (*Ursus arctos*), and black bears (*Ursus americanus*) have allowed us to begin investigating the unique physiological adaptation of polar bear to Arctic environments (Miller et al. 2012; Liu et al. 2014; Welch et al. 2014). Deep genomic analyses uncovered regions in the ursid genome that appear to harbor genes and other functional elements that may be under positive selection, and a suite of 82 of these genes involving 20 known physiological pathways have been identified for polar bear (Miller et al. 2012). Among gene variants apparently under differential selection are those involved in muscle atrophy, protein conservation, and therefore, possibly, hibernation, the regulation of milk lipid droplet secretion, obesity, lean body mass, hyperphagia, synthesis and catabolism of high-density lipoproteins, and long-chain fatty acid synthesis and catabolism. Additionally, variation at genes related to the production of nitric oxide may allow polar bears to control trade-offs between energy production, oxygen consumption, and thermogenesis, providing a mechanism for polar bears to adaptively respond to Arctic extremes (Welch et al. 2014). Gene lineages within polar bears have been under stronger positive selection than in brown bears and may have restructured metabolic and cardiovascular function (Liu et al. 2014). As well, variants in a series of genes associated with skin and hair pigmentation differences among black, brown, and polar bears were also identified (Miller et al. 2012; Liu et al. 2014). It is not known yet whether these variants are fixed between species or whether polymorphism at these genes signals increased ability to adapt to changing conditions.

In some cases, the heritable variation in ecologically important traits for specialist species may be low (Kellermann et al. 2006; Williams et al. 2008). Species with low evolutionary potential that continue to reside in their current ranges can initially respond to changes in the environment by altering their phenotype (phenotypic plasticity; Bradshaw 1965; Przybylo et al. 2000) which, even in the absence of evolutionary potential, is an important mechanism for rapid species response because plastic characters can change within an individual's life span or a generation. However, while phenotypic plasticity provides a stop-gap measure for populations facing environmental change, there can be fitness associated costs (Chevin et al. 2010), and there are limits to plastic responses (Pigliucci 1996) in the absence of sufficient evolutionary potential. The degree of plasticity of traits depends on several factors such as phylogenetic constraints, degree of niche specialization, and physiological tolerance ranges (Nylin and Gotthard 1998; Williams et al. 2008). As global climate change proceeds, the optimal phenotypic changes required for timing of development, reproduction, and migration will exceed the limits of individual plasticity. Therefore, selection will likely alter traits, changing the genetic composition of populations and ultimately species. The ability of selection to alter traits is ultimately dependent upon the standing genetic variation within species and additional variation generated by mutation and immigration, as individuals cannot preadapt to future conditions (Lynch 1996; Bijlsma et al. 1997; Berteaux et al. 2004).

Genomics approaches are increasingly being used to link genetic variation to adaptively important trait variation, providing alternative measures of evolutionary potential (Hill 2012). Genomic information can provide value information regarding population persistence, inbreeding depression, outbreeding depression, hybridization, introgression, and adaptation (Allendorf et al. 2010; Angeloni et al. 2012). Identification of the genetic basis of adaptation to the Arctic in combination with ecological characterization of polar bears will provide new insights into the ability of polar bears to respond to a changing climate and development of conservation management programs (Eizaguirre and Baltazar-Soares 2014).

28.7 Conclusions

Despite the general consensus that polar bears are adversely affected by climatic warming, there is a great deal of uncertainty concerning short- and mid-term regional responses (e.g., Amstrup et al. 2008; Atwood et al. 2016a, b). In part, this is due to lack of recent (i.e., within the last 20 years) data on 7 of the 19 subpopulations (IUCN/PBSG 2014), such that basic vital rates for those subpopulations are presently unknown. However, ultimately the fate of polar bears will be determined by the fate of Arctic sea ice ecosystems. Localized adaptations may slow population declines and, in a best-case scenario, result in a remnant, vastly reduced "refuge" population. But the prospect of recovering a single remnant population to its former range is daunting if not unavailing, given the potential consequences of Allee effects, inbreeding depression, and bottlenecks, among others (e.g., Amos and Balmford 2001). The challenges to conservation are substantial and will require creative monitoring approaches to inform strategies for overcoming those challenges.

Clearly, what is needed to guide polar bear conservation planning is a circumpolar-based monitoring initiative that leverages the efforts of regional experts into range-wide inference regarding future trends and trajectories. The framework for such an approach is provided in great detail by Vongraven et al. (2012), and we reiterate some of those findings and provide additional suggestions in this chapter. Critical to the success of a program with that over-arching goal will be the identification of points (i.e., thresholds) at which changing sea ice conditions produce either pronounced population responses or substantially improve management outcomes (Samhouri et al. 2010). Inherent to this process is the notion that biologists have not become enamored with trying to identify and measure thresholds simply because they are scientifically interesting but rather because of their high conservation significance (Martin et al. 2009). In large part, the key to staying focused on the most relevant thresholds is to start by stating clear conservation objectives and potential management actions and then focusing on collecting information on metrics that characterize population response to those actions.

References

Allen MR, Stocker TF (2013) Impact of delay in reducing carbon dioxide emissions. Nat Clim Chang 4:23–26. doi:10.1038/NCLIMATE2077

Allendorf FW, Hohenlohe PA, Luikart G (2010) Genomics and the future of conservation genetics. Nat Rev Genet 11:697–709

Amos W, Balmford A (2001) When does conservation genetics matter? Heredity 87:257–265

Amstrup SC, Gardner C, Meyers KC et al (1989) Ethylene glycol (antifreeze) poisoning in a free-ranging polar bear. Vet Hum Toxicol 31:317–319

Amstrup SC, Durner GM (1995) Survival rates of radio-collared female polar bears and their dependent young. Can J Zool 73:1312–1322

Amstrup SC, Durner GM, Stirling I et al (2000) Movements and distribution of polar bears in the Beaufort Sea. Can J Zool 78:948–966

Amstrup SC (2003) Polar bear, *Ursus maritimus*. In: Feldhamer GA, Thompson BC, Chapman JA (eds) Wild mammals of North America: biology, management, and conservation. Johns Hopkins University Press, Baltimore, Maryland, pp 587–610

Amstrup SC, McDonald TL, Durner GM (2004) Using satellite radiotelemetry data to delineate and manage wildlife populations. Wildl Soc Bull 32:661–679

Amstrup SC, Durner GM, Stirling I et al (2005) Allocating harvests among polar bear stocks in the Beaufort Sea. Arctic 58:247–259

Amstrup SC, Stirling I, Smith TS et al (2006) Recent observations of intraspecific predation and cannibalism among polar bears in the southern Beaufort Sea. Polar Biol 29:997–1002

Amstrup SC, Marcot BG, Douglas DC (2008) A Bayesian network modeling approach to forecasting the 21st century worldwide status of polar bears. In: DeWeaver ET, Bitz CM, Tremblay LB (eds) Arctic Sea ice decline: observations, projections, mechanisms, and implications, American Geophysical Union Geophysical Monograph No. 180, Washington, DC, pp 213–268

Amstrup SC, DeWeaver ET, Douglas DC et al (2010) Greenhouse gas mitigation can reduce sea ice loss and increase polar bear persistence. Nature 468:955–960

Andelman SA, Fagan WF (2000) Umbrellas and flagships: efficient conservation surrogates or expensive mistakes? Proc Natl Acad Sci U S A 97:5954–5959

Angeloni F, Wagemaker N, Vergeer P et al (2012) Genomic toolboxes for conservation biologists. Evol Appl 5:130–143
Atkinson SN, Ramsay MA (1995) The effects of prolonged fasting of the body composition and reproductive success of female polar bears (*Ursus maritimus*). Funct Ecol 9:559–567
Atwood TC, Peacock E, Burek-Huntington K et al (2015) Prevalence and spatio-temporal variation of an alopecia syndrome detected in polar bears in the southern Beaufort Sea. J Wildl Dis 51:48–59
Atwood TC, Marcot BG, Douglas DC et al (2016a) Forecasting the relative influence of environmental and anthropogenic stressors on polar bears. Ecosphere 7(6):e01370
Atwood TC, Peacock E, McKinney M et al (2016b) Rapid environmental change drives increased land use by an Arctic marine predator. PLoS One 11(6):e0155932. doi:10.1371/journal.pone.0155932
Berteaux D, Réale D, McAdam AG et al (2004) Keeping pace with fast climate change: can arctic life count on evolution? Integr Comp Biol 44:140–151
Bijlsma R, Bundgaard J, Boerema AC et al (1997) Genetic and environmental stress, and the persistence of populations. In: Bijlsma R, Loeschcke V (eds) Environmental stress, adaptation and evolution. Birkhäuser-Verlag, Basel, Switzerland, pp 193–207
Born EW, Wiig Ø, Thomassen J (1997) Seasonal and annual movements of radio-collared polar bears (*Ursus maritimus*) in Northeast Greenland. J Mar Syst 10:67–77
Bossart GD (2011) Marine mammals as sentinel species for oceans and human health. Vet Pathol 48:676–690
Bradshaw AD (1965) Evolutionary significance of phenotypic plasticity in plants. Adv Genet 13:115–155
Bromaghin J, McDonald T, Stirling I et al (2015) Polar bears in the Beaufort Sea: population decline and stabilization in the 2000's. Ecol Appl 25:634–651
Brower CD, Carpenter A, Branigan ML et al (2002) The polar bear management agreement for the southern Beaufort Sea: an evaluation of the first ten years of a unique conservation agreement. Arctic 57:362–372
Cattet MRL, Caulkett NA, Obbard ME et al (2002) A body condition index for ursids. Can J Zool 80:1156–1161
Clarkson PL, Irish D (1991) Den collapse kills female polar bear and two newborn cubs. Arctic44:83–84.
Cherry SG, Derocher AE, Thiemann GW et al (2013) Migration phenology and seasonal fidelity of an Arctic marine predator in relation to sea ice dynamics. J Anim Ecol 82:912–921
Chevin LM, Lande R, Mace GM (2010) Adaptation, plasticity, and extinction in a changing environment: towards a predictive theory. PLoS Biol 8:e1000357. doi:10.1371/journal.pbio.1000357
Comiso JC (2012) Large decadal decline of the Arctic multiyear ice cover. J Climatol 25:1176–1193
Cooper MH, Iverson SJ, Heras H (2005) Dynamics of blood chylomicron fatty acids in a marine carnivore: implications for lipid metabolism and quantitative estimation of predator diets. J Comp Physiol 175:133–145
Coulson T, Gaillard J, Festa-Bianchet M (2005) Decomposing the variation in population growth into contributions from multiple demographic rates. J Anim Ecol 74:789–801
Dawson TP, Jackson ST, House JI et al (2011) Beyond predictions: biodiversity conservation in a changing climate. Science 332:53–58
DeMaster DP, Stirling I (1983) The estimation of survival and litter size of polar bear cubs. Int Conf Bear Biol Manag 5:260–263
Derocher AE, Stirling I (1994) Age-specific reproductive performance of female polar bears (*Ursus maritimus*). J Zool 234:527–536
Derocher AE, Stirling I (1995) Estimation of polar bear population size and survival in western Hudson Bay. J Wildl Manag 59:215–221
Derocher AE, Stirling I (1996) Aspects of survival in juvenile polar bears. Can J Zool 74:1246–1252

Derocher AE, Wiig Ø (1999) Infanticide and cannibalism of juvenile polar bears (*Ursus maritimus*) in Svalbard. Arctic 52:307–310

Derocher AE, Lunn NJ, Stirling I (2004) Polar bears in a warming climate. Integr Comp Biol 44:163–176

Desforges JPW, Sonne C, Levin M et al (2016) Immunotoxic effects of environmental pollutants in marine mammals. Environ Int 86:126–139

Durner GM, Douglas DC, Nielson RM et al (2009) Predicting 21st-century polar bear habitat distribution from global climate models. Ecol Monogr 79:25–58

Durner GM, Whiteman JP, Harlow HJ et al (2011) Consequences of long-distance swimming and travel over deep-water pack ice for a female polar bear during a year of extreme sea ice retreat. Polar Biol 34:975–984

Durner GM (2014) Impacts of a changing Arctic on habitat use and behavior of polar bears (*Ursus maritimus*). Dissertation, University of Wyoming, Laramie

Dyck MG (2006) Characteristics of polar bears killed in defense of life and property in Nunavut, Canada, 1970-2000. Ursus 17:52–62

Eizaguirre C, Baltazar-Soares M (2014) Evolutionary conservation—evaluating the adaptive potential of species. Evol Appl 7:963–967

Elmberg J, Nummi P, Pöysä H et al (2006) The scientific basis for new and sustainable management of migratory European ducks. Wildl Biol 12:121–127

Fagre A, Nol P, Atwood TC et al (2015) A review of infectious agents in polar bears (*Ursus maritimus*) and their long-term ecological relevance. EcoHealth. doi:10.1007/s10393-015-1023-6

Farley SD, Robbins CT (1994) Development of two methods to estimate body composition of bears. Can J Zool 72:220–226

Garner GW, Knick ST, Douglas DC (1990) Seasonal movements of adult female polar bears in the Bering and Chukchi Seas. 8th International Conference on Bear Research and Management 8:219–226.

Gilbert AT, Fooks AR, Hayman FT et al (2013) Deciphering serology to understand the ecology of infectious diseases in wildlife. EcoHealth 10:298–313

Gormezano LJ, Rockwell RF (2013) What to eat now? Shifts in polar bear diet during the ice-free season in western Hudson Bay. Ecol Evol 3:3509–3523

Gunnarsson S (2006) The conceptualisation of health and disease in veterinary medicine. Acta Vet Scand 47:71

Hanisch SL, Riley SJ, Nelson MP (2012) Promoting wildlife health or fighting wildlife disease: insights from history, philosophy, and science. Wildl Soc Bull 36:477–482

Henriksen EO, Wiig Ø, Skaare JU et al (2001) Monitoring PCBs in polar bears: lessons learned from Svalbard. J Environ Monit 3:493–498

Herreman J, Peacock E (2013) Polar bear use of a persistent food subsidy: insights from non-invasive genetic sampling in Alaska. Ursus 24:148–163

Hill WG (2012) Quantitative genetics in the genomics era. Curr Genet 13:196–206

Iverson SJ (1993) Milk secretion in marine mammals in relation to foraging: can milk fatty acids predict diet? Symp Zool Soc Lond 66:263–291

Iversen M, Aars J, Haug T et al (2013) The diet of polar bears (*Ursus maritimus*) from Svalbard, Norway, inferred from scat analysis. Polar Biol 36:561–571

IUCN/PBSG, International Union for the Conservation of Nature/Polar Bear Specialist Group (2014) Summary of polar bear population status per 2014. http://pbsg.npolar.no/en/status/status-table.html. Accessed 12 Dec 2016

Kellermann VM, Van Heerwaarden B, Hoffmann AA et al (2006) Very low additive genetic variance and evolutionary potential in multiple populations of two rainforest *Drosophila* species. Evolution 60:1104–1108

Lambeck RJ (1997) Focal species: a multi-species umbrella for nature conservation. Conserv Biol 11:849–856

Landres PB, Verner J, Thomas JW (1988) Ecological uses of vertebrate indicator species: a critique. Conserv Biol 2:316–328

Larsen T (1985) Polar bear denning and cub production in Svalbard, Norway. J Wildl Manag 49:320–326
LaRue M, Stapleton S, Porter C et al (2015) Expanding applications for using high-resolution satellite imagery to monitor polar bear abundance and distribution. Wildl Soc Bull 39:772–779
Lentfer JW, Hensel RJ, Gilbert JR et al (1980) Population characteristics of Alaskan polar bears. In: Martinka CJ, McArthur KL (eds) Bears: their biology and management. Fourth international conference on bear research and management. U.S. Government Printing Office, Washington, D.C., pp 102–115
Liu S, Lorenzen ED, Fumagalli M et al (2014) Population genomics reveal recent speciationand rapid evolutionary adaptation in polar bears. Cell 157:785–794
Lunn NJ, Stenhouse GB (1985) An observation of possible cannibalism by polar bears (*Ursus maritimus*). Can J Zool 63:1516–1517
Lunn NJ, Stirling I (1985) The significance of supplemental food to polar bears during the ice-free period of Hudson Bay. Can J Zool 63:2291–2297
Lynch M (1996) A quantitative-genetic perspective on conservation issues. In: Avise JC, Hamrick JL (eds) Conservation genetics. Chapman and Hall, New York, pp 471–501
Magurran AE, Baillie SR, Buckland ST et al (2010) Long-term datasets in biodiversity research and monitoring: assessing change in ecological communities through time. Trends Ecol Evol 25:574–582
Martin J, Runge MC, Nichols JD et al (2009) Structured decision making as a conceptual framework to identify thresholds for conservation and management. Ecol Appl 19:1079–1090
Mauritzen M, Belikov SE, Boltunov AN et al (2003) Functional responses in polar bear habitat selection. Oikos 100:112–124
Mawdsley JR, O'Malley R, Ojima DS (2009) A review of climate-change adaptation strategies for wildlife management and biodiversity conservation. Conserv Biol 23:1080–1089
McKinney M, Atwood TC, Dietz R et al (2014) Validation of adipose lipid content from biopsies as a body condition index using polar bears. Ecol Evol 4:516–527
McLellan BN (2011) Implications of a high-energy and low-protein diet on the body composition, fitness, and competitive abilities of black (*Ursus americanus*) and grizzly (*Ursus arctos*) bears. Can J Zool 89:546–558
Miller W, Schuster SC, Welch AJ et al (2012) Polar and Brown bear genomes reveal ancient admixture and demographic footprints of past climate change. Proc Natl Acad Sci U S A 109(36):e2382–e2390
Molnár PK, Derocher AE, Thiemann GW et al (2010) Predicting survival, reproduction and abundance of polar bears under climate change. Biol Conserv 143:1612–1622
Molnár PK, Derocher AE, Klanjscek T et al (2011) Predicting climate change impacts on polar bear litter size. Nat Commun 2:186
Molnár PK, Derocher AE, Thiemann GW et al (2014) Corrigendum to "Predicting survival, reproduction and abundance of polar bears under climate change" (Biol Conserv 143:1612–1622). Biol Conserv 177:230–231. doi:10.1016/j.biocon.2014.07.001
Monnett C, Gleason JS (2006) Observations of mortality associated with extended open-water swimming by polar bears in the Alaskan Beaufort Sea. Polar Biol 29:681–687
Murray DL, Patterson BR (2006) Wildlife survival estimation: recent advances and future directions. J Wildl Manag 70:1499–1503
Niemi GJ, McDonald ME (2004) Application ofgical indicators. Annu Rev Ecol Evol Syst 35:89–111
Nordenfelt L (2011) Health and welfare in animals and humans. Acta Biol 59:139–152
Nylin S, Gotthard K (1998) Plasticity in life history traits. Annu Rev Entomol 43:63–83
Obbard ME, Cattet MR, Moody T et al (2006) Temporal trends in the body condition of southern Hudson Bay polar bears. Applied Research and Development Branch, Ontario
Obbard ME, McDonald TL, Howe EJ et al (2007) Polar bear population status in southern Hudson Bay, Canada. Administrative Report. US Department of the Interior, US Geological Survey, Anchorage, AK

Obbard ME, Cattet MR, Moody T et al (2008) Temporal trends in the body condition of southern Hudson Bay polar bears. Research Information Note Number 3, Ontario Ministry of Natural Resources

Obbard ME, Cattet MR, Howe EJ et al (2016) Trends in body condition in polar bears (*Ursus maritimus*) from the southern Hudson Bay subpopulation in relation to changes in sea ice. Arct Sci. doi:10.1139/AS-2015-0027

Pagano AM, Durner GM, Amstrup SC et al (2012) Long-distance swimming by polar bears (*Ursus maritimus*) of the southern Beaufort Sea during years of extensive open water. Can J Zool 90:663–676

Pagano AM, Peacock E, McKinney MA (2014) Remote biopsy darting and marking for polar bears. Mar Mamm Sci 30:169–183

Patyk K, Duncan C, Nol P et al (2015) Establishing a definition of polar bear health to guide research and management activities. Sci Total Environ 514:371–378

Peacock E, Derocher AE, Thiemann GW et al (2011) Conservation and management of Canada's polar bears (*Ursus maritimus*) in a changing Arctic. Can J Zool 89:371–385

Pigliucci M (1996) How organisms respond to environmental changes: from phenotypes to molecules (and vice versa). Trends Ecol Evol 11:168–173

Prop J, Aars J, Bårdsen BJ et al (2015) Climate change and the increasing impact of polar bears on bird populations. Front Ecol Evol. doi:10.3389/fevo.2015.00033

Przybylo R, Sheldon BC, Merilä J (2000) Climatic effects on breeding and morphology: evidence for phenotypic plasticity. J Anim Ecol 69:395–403

Ramsay MA, Stirling I (1988) Reproductive biology and ecology of female polar bears (*Ursus maritimus*). J Zool 214:601–633

Ramsay MA, Hobson KA (1991) Polar bears make little use of terrestrial food webs: evidence from stable-carbon isotope analysis. Oecologia 86:598–600

Regehr EV, Amstrup SC, Stirling I (2006) Polar bear population status in the southern Beaufort Sea. U.S. Geological Survey, Open-File Report 2006–1337

Regehr EV, Lunn NJ, Amstrup SC et al (2007) Effects of earlier sea ice breakup on survival and population size of polar bears in western Hudson Bay. J Wildl Manag 71:2673–2683

Regehr EV, Hunter CM, Caswell H et al (2010) Survival and breeding of polar bears in the southern Beaufort Sea in relation to sea ice. J Anim Ecol 79:117–127

Regehr EV, Wilson RR, Rode KD et al (2015) Resilience and risk: a demographic model to inform conservation planning for polar bears. Open-File Report No. 2015–1029. US Geological Survey

Richardson E, Stirling I, Hik DS (2005) Polar bear (*Ursus maritimus*) maternity denning habitat in western Hudson Bay: a bottom-up approach to resource selection functions. Can J Zool 83:860–870

Richardson ES, Andriashek D (2006) Wolf (*Canis lupus*) predation of a polar bear (*Ursus maritimus*) cub on the sea ice off northwestern Banks Island, Northwest Territories, Canada. Arctic 59:322–324

Ricke KL, Caldeira K (2014) Maximum warming occurs about one decade after a carbon dioxide emission. Environ Res Lett 9:124002

Robbins CT, Ben David M, Fortin JK et al (2012) Maternal condition determines birth date and growth of newborn bear cubs. J Mammal 93:540–546

Robbins CT, Lopez-Alfaro C, Rode KD et al (2013) Hibernation and seasonal fasting in bears: the energetic costs and consequences for polar bears. J Mammal 93:1493–1503

Rockwell RF, Gormezano LJ (2009) The early bear gets the goose: climate change, polar bears and lesser snow geese in western Hudson Bay. Polar Biol 32:539–547

Rode KD, Amstrup SC, Regehr EV (2010) Reduced body size and cub recruitment in polar bears associated with sea ice decline. Ecol Appl 20:768–782

Rode KD, Peacock E, Taylor M et al (2012) A tale of two polar bear populations: ice habitat, harvest, and body condition. Popul Ecol 54:3–18

Rode KD, Regehr EV, Douglas DC et al (2014a) Variation in the response of an Arctic top predator experiencing habitat loss: feeding and reproductive ecology of two polar bear populations. Glob Chang Biol 20:76–88
Rode KD, Bromaghin J, Pagano A et al (2014b) Evaluating the potential for long- and short-term effects of capture on polar bears. Wildl Res 41:311–322
Rode KD, Robbins CT, Nelson L et al (2015) Can polar bears use terrestrial foods to offset lost ice-based hunting opportunities? Front Ecol Environ 13:138–145
Russell RH (1975) The food habits of polar bears of James Bay and Southwest Hudson Bay in summer and autumn. Arctic 28:117–129
Ryser-Degiorgis MP (2013) Wildlife health investigations: needs, challenges and recommendations. BMC Vet Res 9:1–17
Sahanatien V, Peacock E, Derocher AE (2015) Population substructure and space use of Foxe Basin polar bears. Ecol Evol 5:2851–2864
Samhouri JF, Levin PS, Ainsworth CH (2010) Identifying thresholds for ecosystem-based management. PLoS One 5:e8907
Schliebe S, Wiig Ø, Derocher A et al (2008) *Ursus maritimus*. The IUCN red list of threatened species 2008: e.T22823A9391171. The IUCN Red List of Threatened Species(tm) http://www.iucnredlist.org/details/22823/0. Accessed 12 Dec 2016
Schwartz MK, Luikart G, Waples RS (2007) Genetic monitoring as a promising tool for conservation and management. Trends Ecol Evol 22:25–33
Sleeman JM (2013) Has the time come for big science in wildlife health? EcoHealth 10:335–338
Smith PA, Elliott KH, Gaston AJ et al (2010) Has early ice clearance increased predation on breeding birds by polar bears? Polar Biol 33:1149–1153
Stallknecht DE (2007) Impediments to wildlife disease surveillance, research, and diagnostics. Curr Top Microbiol Immunol 315:445–461
Stapleton S, Atkinson S, Hedman D et al (2014a) Revisiting western Hudson Bay: using aerial surveys to update polar bear abundance in a sentinel population. Biol Conserv 170:38–47
Stapleton S, LaRue M, Lecomte N et al (2014b) Polar bears from space: assessing satellite imagery as a tool to monitor Arctic wildlife. PLoS One. doi:10.1371/journal.pone.0101513
Stenhouse GB, Lee LJ, Poole JG (1988) Some characteristics of polar bears killed during conflicts with humans in the Northwest Territories, 1976-86. Arctic 41:275–278
Stephen C (2014) Toward a modernized definition of wildlife health. J Wildl Dis 50:427–430
Stevenson RD, Woods WA (2006) Condition indices for conservation: new uses for evolving tools. Integr Comp Biol 46:1169–1190
Stirling I, Andriashek D, Latour P et al (1975) The distribution and abundance of polar bears in the eastern Beaufort Sea. Beaufort Sea technical report 2. Department of the Environment, Victoria, British Columbia
Stirling I, Spencer C, Andriashek D (1989) Immobilization of polar bears (*Ursus maritimus*) with Telazol® in the Canadian Arctic. J Wildl Dis 25:159–168
Stirling I, Derocher AE (1993) Possible impacts of climatic warming on polar bears. Arctic 46:240–245
Stirling I, Lunn NJ, Iacozza J (1999) Long-term trends in the population ecology of polar bears in western Hudson Bay in relation to climatic change. Arctic 52:294–306
Stirling I, Parkinson CL (2006) Possible effects of climate warming on selected populations of polar bears (*Ursus maritimus*) in the Canadian Arctic. Arctic 59:261–275
Stirling I, Derocher AE (2012) Effects of climate warming on polar bears: a review of the evidence. Glob Chang Biol 18:2694–2706
Stone IR, Derocher AE (2007) An incident of polar bear infanticide and cannibalism on Phippsøya, Svalbard. Polar Record 43:171–173
Stroeve JC, Markus T, Boisvert L et al (2014) Changes in Arctic melt season and implications for sea ice loss. Geophys Res Lett 41:1216–1225
Taylor M, Elkin B, Maier N et al (1991) Observation of a polar bear with rabies. J Wildl Dis 27:337–339

Taylor M, Lee J (1995) Distribution and abundance of Canadian polar bear populations: a management perspective. Arctic 48:147–154
Thiemann GW (2008) Using fatty acid signatures to study bear foraging: technical considerations and future applications. Ursus 19:59–72
Thiemann GW, Iverson SJ, Stirling I (2008) Polar bear diets and arctic marine food webs: insights from fatty acid analysis. Ecol Monogr 78:591–613
Treves A, Karanth KU (2003) Human-carnivore conflict and perspectives on carnivore management worldwide. Conserv Biol 17:1491–1499
U.S. Fish and Wildlife Service (2008) Endangered and threatened wildlife and plants; 12-month petition finding and proposed rule to list the polar bear (*Ursus maritimus*) as threatened throughout its range. Fed Reg 72:1064–1099
Vongraven D, Aars J, Amstrup S et al (2012) A circumpolar monitoring framework for polar bears. Ursus 23(sp2):1–66
Wassmann P, Duarte CM, Agusti S et al (2011) Footprints of climate change in the Arctic marine ecosystem. Glob Chang Biol 17:1235–1249
Welch AJ, Bedoya-Reina OC, Carretero-Paulet L et al (2014) Polar bears exhibit genome-wide signatures of bioenergetics adaptation to life in the Arctic environment. Genome Biol Evol 6:433–450
Williams SE, Shoo LP, Isaac JL et al (2008) Towards an integrated framework for assessing the vulnerability of species to climate change. PLoS Biol 6:2621–2626
Wimsatt WA (1963) Delayed implantation in the Ursidae, with particular reference to the black bear (*Ursus americanus* Pallas). In: Enders AC (ed) Delayed implantation. University of Chicago Press, IL, pp 49–76
White GC, Burnham KP (1999) Program MARK: survival estimation from populations of marked animals. Bird Study 46(S1):S120–S139
Wiig Ø (1998) Survival and reproductive rates for polar bears at Svalbard. Ursus 10:25–32
World Association of Animal Health (OIE) (2013) Collection, submission and storage of diagnostic specimens. Manual of diagnostic tests and vaccines for terrestrial animals 2015
World Association for Animal Health (OIE) (2015) Terrestrial Animal Health Code
Zedrosser A, Dahle B, Swenson JE (2006) Population density and food conditions determine adult female body size in brown bears. J Mammal 87:510–518
Zedrosser A, Bellemain E, Taberlet P et al (2007) Genetic estimates of annual reproductive success in male brown bears: the effects of body size, age, internal relatedness and population density. J Anim Ecol 76:368–375

Part V
Otters

Chapter 29
Effects on Otters of Pollution, Fisheries Equipment and Water-Borne Debris

Lesley Wright, Shawn Larson, Jan Reed-Smith, Nicole Duplaix, and Thomas Serfass

Abstract Welfare effects of human interactions on river and sea otters take many forms: trapping, fishing, water borne debris, toxic elements and captivity. Animal welfare issues associated with otter trapping, especially using leghold traps, are reviewed. Different trap models are described. Box traps or padded traps are used for the live capture of otters. Frequent trap checking correlates with minimising trap-related injuries. The impact of climate change on wetland habitats affects otter populations. Climate change may affect sea otter populations both negatively and positively. Ocean acidification is thought to have negative impacts on sea otter prey. Global warming may have a positive impact on available habitat for sea otter populations, opening new habitat as ocean ice retreats northwards. Our state of knowledge of husbandry techniques for river otters and sea otters in captivity has improved over the last 20 years. Poorly designed enclosures lead to chronic stress. Creating factual, documented activity budgets for individual otters is a first step to stimulating natural behaviours. Rehabilitation techniques for orphaned river and sea otter cubs have improved and successful releases have been documented.

L. Wright (✉)
Deputy Chair, IUCN/SSC Otter Specialist Group, Didcot, UK
e-mail: lc.wright@yahoo.co.uk

S. Larson
Seattle Aquarium, Seattle, USA
e-mail: s.larson@seattleaquarium.org

J. Reed-Smith
African Otter Outreach Project, Lake Odessa, MI 48849, USA
e-mail: jrsotter@iserv.net

N. Duplaix
Chair, IUCN/SSC Otter Specialist Group, Chair, IUCN/SSC Otter Specialist Group, Oregon State University, Corvallis, OR 97331-3803, USA

T. Serfass
Frostburg State University, Frostburg, MD 21532, USA
e-mail: tserfass@frostburg.edu

A. Butterworth (ed.), *Marine Mammal Welfare*, Animal Welfare 17,
DOI 10.1007/978-3-319-46994-2_29

29.1 Introduction

Otters are carnivores belonging to the subfamily Lutrinae (family Mustelidae). The 13 extant species are divided into semiaquatic (11 species) and marine (2 species). Otters have long, tubular bodies, muscular tails and short limbs, well suited for aquatic propulsion. They are distributed across all continents with the exception of Australasia and the poles. Of the 13 species, 5 are endangered, 5 threatened and 2 vulnerable (IUCN Red List 2015). Only one, the North American river otter, is considered out of danger. Clearly otters are at risk.

Human development has destroyed, transformed and fragmented otter habitats worldwide. Habitat-related stressors include pollution, sedimentation, drought, overfishing and invasive species. Only a limited number of unaltered rivers and wetland landscapes remain today. However, if some or all of these stressors are addressed and repaired, otters will return and re-establish stable populations. Such has been the case in areas of the USA (*Lontra canadensis*) and Europe (*Lutra lutra*) where protection and antipollution measures were enforced in the 1980s and otters were successfully reintroduced. Once considered critically endangered in the UK, the otter has returned to every county in the span of 40 years.

In the following chapters, we will focus on three aspects involving human activities which directly affect otter welfare: the hunting and trapping of otters, waterborne debris and poor otter husbandry in captivity.

29.1.1 Pollution: Oil

Oil is by far the most serious marine-borne threat to otters. For sea otters (*Enhydra lutris*) and marine otters (*Lontra felina*), a major oil spill has the potential to be a global population-threatening event.

On 24 March 1989, the T/V *Exxon Valdez* went aground in northwest Prince William Sound, Alaska, spilling an estimated 264,000 barrels of crude oil and contaminating more than 2000 km of shoreline (Bragg et al. 1994). The spill killed many species of marine wildlife, including several thousand sea otters (Ballachey et al. 1994). In addition to the acute mortalities over the decade following the spill, the chronic persistence of the oil in the nearshore intertidal areas continued to affect the ecosystem, causing delayed recovery of several wildlife species, including sea otters (Peterson et al. 2003).

As yet, no similar spill has occurred on the Pacific coast of South America, but should such an event happen off Peru or Chile, it could devastate the sparse, fragmented population of the endangered marine otter, *Lontra felina* (Valqui, J and Rheingantz, ML (2015). This exclusively coastal species is, like the sea otter, broadly dependent on kelp forest for prey and dens on rocky shores exposed to heavy seas—and oil spills.

Other otter species using coastal habitats have also been exposed to spilt oil, e.g. the Eurasian otter, *Lutra lutra* ('Erika' in 1999 off France; MV Braer, off Shetland in 1995), and the North American river otter, *Lontra canadensis* (Exxon Valdez also killed at least 39 river otters, though Bowyer et al. (2003) estimated that up to 80 animals may have been lost). These species are relatively abundant: the Eurasian otter is only Red List assessed as Near Threatened (Roos et al. 2015) and the river otter as Least Concern (Serfass et al. 2015). There is no major population threat, though local extirpation may occur.

The threat to coast-utilising African clawless otters, *Aonyx capensis*, posed by the frequent oil spills in the Niger delta (see, e.g. Snowden and Ekweozor 1987) has not received the global attention or research effort that tanker spills have received; nevertheless, it is possible that individual otters in this region are being exposed to oil.

Other otter species known to use coastal habitats also suffer from oil spills, such as the smooth-coated otter, *Lutrogale perspicillata*, and the Asian small-clawed otter, *Aonyx cinereus*. There has been no documented occurrence of otters (specifically) being affected by this, but spills, leaks and well blow-outs are impacting mangroves in Indonesia and Malaysia, which are otter habitat, such as the spill from the Lucky Lady off Java in 2004, which spilt 1200 tonnes of crude oil and impacted coastal mangroves (ITOPF 2015). Any coastal otter could be exposed to an oil spill, worldwide. What are the effects of this exposure?

Otters are affected physically, physiologically and psychologically by exposure to oil and, for some individuals, by capture, cleaning and rehabilitation either to the wild or into captivity. Unlike other marine mammals, otters do not have a layer of blubber under the skin for insulation—they rely on air bubbles trapped in their coats, warmed with body heat. Accordingly, they have a very strong urge to groom, distributing natural skin oils and replenishing the air layer. Sea otters are particularly assiduous in this activity and are frequently seen blowing air into their fur (Kenyon 1969). Young sea otter cubs have particularly fluffy coats, giving them buoyancy while they are too young to swim, relying on their mother maintaining the air content to stop them sinking and drowning.

An oiled otter will have a very strong drive to groom the oil from its fur, thus ingesting oil, inhaling the fumes and experiencing irritation of the eyes, nose, digestive tract and lungs. In addition, oil destroys the insulation of the fur, leading to both loss of buoyancy (especially in cubs) and hypothermia (Williams et al. 1988). Very many otters have died in 'horrendous suffering' as a result of contamination with oil (Kruuk 2006).

Oil spills often bring out a wave of compassion for wildlife, and in some countries, there is a well-organised network of responders ready to travel to spill sites as soon as an event is notified. Animals are rescued, assessed and, where possible, cleaned and either rehabilitated back to the wild or retained in captivity. The chronic effects of crude oil exposure then begin to be important. After capture, some otters will inevitably be judged unviable and euthanised. The rest will enter the cleaning process: after light sedation, the otter is washed (dish washing detergent liquid has been found to be effective) and then allowed to recuperate under veterinary care while the natural coat oils replenish.

After *Exxon Valdez*, this process was still largely experimental and variable, slow and expensive (estimated at $40,000–80,000 per otter), and some of the cleansing agents used were themselves toxic, adding to the otters' distress (Williams et al. 1988).

Of the 356 otters picked up alive, only 54% were subsequently released and 10% found homes in aquaria—34% died during the cleaning and rehabilitation process. Even perfectly healthy otters can suffer stress-induced mortality during capture, with rates of 5% and 10% (Waldichuck 1990; Estes 1992).

Since the *Exxon Valdez*, these loss and cost rates have been vastly improved in terms of cost (around ten times less), time spent in captivity and survival rates, mainly because of improved washing media, the provision of heated, softened, freshwater recuperation pools, infrared thermography (to assess body temperature) and subcutaneous temperature-sensitive passive integrated transponder tags (Jessup et al. 2012). With these modern protocols in place, the prognosis today for an oiled otter is far better than it was in 1989.

Even so, the otter which has been cleaned will have been exposed to a highly toxic substance. Inhalation of crude oil fumes are known, in the short term, to result in respiratory distress, but in the longer term can lead to chemical pneumonia, mucosal irritation, loss of coordination, irregular heartbeat, convulsions and coma (Goldstein 2010). Ingestion rapidly results in vomiting and diarrhoea, and the components of crude oil are known teratogens, as well as affecting the haemopoietic (blood-forming) system, damaging the liver and kidneys, and, in the case of benzene, are carcinogens, even at very low concentrations (Goldstein 2010). DeGrange (1990) notes that necroscopy of otters dying in the *Exxon Valdez* treatment centre showed pulmonary emphysema, subcutaneous emphysema, haemorrhagic enteritis and liver and kidney dysfunction. Waldichuck (1990) reports that many otters that survived cleaning exhibited liver damage, low blood sugar levels and low body temperature and died regardless of treatment. Baker et al. (1981), during necroscopies of Eurasian otters (*Lutra lutra*) following the Sullom Voe, Shetland Islands spill in 1978, saw tracheitis, excess fluid in the lungs, blood and mucus in the intestines, gastric contents in the trachea, lung congestion, oedema and areas of lung collapse, gastric erosion and kidney damage; one otter that survived cleaning died later with an ulcerated stomach wall and a gut full of clotted blood. Duffy et al. (1993), looking at river otters (*Lontra Canadensis*) affected by the Exxon Valdez spill, found haemolytic anaemia, causing 'Horrendous suffering' according to Kruuk (2006).

When any oiled otter, of any species, is initially recovered, there is no doubt that it is suffering greatly. Given the long-term effects of exposure, the prognosis for the individual animal must be carefully assessed: it may be better for that otter to be euthanised rather than put through the stress of cleaning and released back to the wild as a compromised individual. Against this, two other factors need to be considered.

First, in an endangered species with a small global population, every individual animal is precious, and a life in captivity, if contributing to the gene pool ark, may be preferred as a solution, since ongoing veterinary care will be available. With a more abundant species, the global impact of the loss of a single individual will be less profound.

Second, public opinion must be considered—it is very powerful! Not only do most conservation, rescue and rehabilitation organisations rely on donations from the public for their very existence, but public opinion can sway the policies of governments and spur the organisations responsible for spills into improving spill prevention, acting swiftly in the event of an accident and developing more effective containment and environmental cleaning methods. The public also provide the very large number of volunteers without whom, in most countries, there would be no rescue effort at all. The public's emotional need for a 'happy ending' is illustrated by the enduring popularity of 'Olive the Oiled Otter', who survived oiling in 2009, returned successfully to the wild, successfully raised cubs, had a presence on social media (California Dept of Fish and Wildlife 2016) and whose death from a shark bite 6 years later on 22 March 2015 was reported globally with a high level of attention, being frequently described as 'iconic'.

The effects of oil spills are evident decades after the event. The population of Alaskan sea otters is growing more slowly than expected (Peterson et al. 2003). Many otter prey items, such as clams and mussels, bioaccumulate hydrocarbons. Oil is still present in intertidal regions buried in sediment and is excavated by sea otters as they forage for clams (Bodkin et al. 2012), ingesting oil in prey and coming into direct contact. Duffy et al. (1993) noted that otters in oiled areas appeared smaller than those in pristine habitats.

29.1.2 Bioaccumulation of Heavy Metals and Organic Compounds

The effects of bioaccumulation of heavy metals and persistent organic pollutants have been well researched in Eurasian otters, *Lutra lutra*, where they are believed to have been responsible for the huge population declines and local extirpations of the second half of the twentieth century; none of this, however, involves a marine dimension as these animals inhabit inland waterways. There has, however, been work on sea otter populations, particularly in California, as a result of the near extinction of this subspecies.

Contaminants such as persistent organic pollutants (POPs), which include polycyclic aromatic hydrocarbons [PAHs], polyhalogenated biphenyls [PCBs and PBBs], brominated flame retardants, pharmaceuticals, heavy metals and plastics, all have the potential to affect the health and survival of sea otters directly and indirectly (Jessup et al. 2010; Crain et al. 2009; Cole et al. 2011; Bodkin et al. 2012; Ballachey and Bodkin 2015). Little work has been done to determine the levels of toxic contaminants in sea otters, but levels of some contaminants measured in California sea otters were reported as sufficiently high to be of concern and have been associated with risk of infectious disease and death (Kannan et al. 2006a, b; 2007; Nakata et al. 1998; Murata et al. 2008).

29.1.3 Parasites, Infections and Diseases

Sea otters have been diagnosed with, and are susceptible to, a wide range of infectious and noninfectious diseases. To date 14 bacterial and viral diseases within sea otter have been documented (Murray 2015). It is important to note that sea otter diseases don't follow the traditional land-based model; rather transmission of pathogens tends to be land to sea, a mechanism of transfer of infection that makes the study of sea otter diseases unique (Miller et al. 2010; Oates et al. 2012; Shapiro et al. 2012). Even though a number of diseases have been documented in sea otters, there is little evidence for unexpectedly significant negative consequences at a population level, except in the threatened California population where disease rates are found to be relatively high for a wildlife population (Miller et al. 2010; Murray 2015). Many factors affect the prevalence of disease within sea otter populations, and disease processes may be secondary to other population stressors such as food limitation, decreased habitat quality, increased levels of contaminants and limited genetic diversity, all of which may compound population losses from disease (Ballachey and Bodkin 2015).

The most frequently reported disease in sea otters is meningoencephalitis caused by *Toxoplasma gondii* (a protozoan parasite), in California and not seen in the Alaskan population. The transmission route was thought to be from domestic cats (a known host) via faeces flushed into the sanitary system and thus out to sea, where it may bioaccumulate in the filter-feeding molluscs that sea otters consume. This transmission hypothesis has been recently challenged by Lafferty (2015), who favours the theory that the transmission arises from land-based wildlife, and is considered to be a significant cause of mortality in this endangered population.

Other organisms found in sea otters include the marine *Brucella* bacteria and the fungus *Coccidioides immitis. C. immitis* infection causes the disease 'valley fever' in humans. For an individual otter, both of these infections cause significant disease experience; although some individuals may be rescued and rehabilitated, the majority of compromised individuals will simply die at sea—and sick individuals may also be more susceptible to predator attack, e.g. by sharks (Kreuder et al. 2003, p 501). Some infections can be treated with veterinary care, if the otter survives capture, but there is currently no effective treatment for *Toxoplasmosis* (Stoskopf 2015).

Toxic algal blooms can also incapacitate and kill sea otters: the population of the diatom *Pseudo-nitzschia* occasionally increases rapidly, secreting domoic acid, a neurotoxin, causing seizures and even death in otters who eat contaminated shellfish. Many otters develop inflammation of the heart muscles and heart failure after being exposed to domoic acid (Van Dolah et al. 2003). Otters have also been killed by cyanobacteria in river outflow biomagnified in bivalves (Miller et al. 2010).

29.1.4 Fishing Gear

Otters are opportunist predators and share prey preferences with humans engaging in fishing and aquaculture. As a result, it is inevitable that otters will explore the food opportunities presented by fishing equipment, and some individuals will be entangled or trapped and will drown. In population terms, this can cause sufficient loss of individuals to be enough to depress local populations, such as with Californian sea otters in the 1980s (Ballachey and Bodkin 2015).

By their nature, fish traps are designed to prevent the target species escaping once inside the trap, by using physical baffles, and backward-projecting obstacles.

Unfortunately, a curious otter can find itself unable to reverse out and, being an air-breathing mammal, will drown within a few minutes (sea otters may be able to hold their breath a little longer than other otter species, but this is still less than 10 min—much shorter than when compared to, for example, elephant seals (*Mirounga* sp.), which can dive for more than 2 h).

Various forms of amelioration have been put in place to reduce the impact of fish traps—either voluntarily or through legislation. In California in the 1980s, otters were found drowned in fish, crab and lobster traps (Hatfield et al. 2011); once this had been discovered, traps were restricted to waters deeper than sea otters dive—more than 60 m—and trap openings were modified to be made too small for otters to enter.

In Europe, otters have also been drowned in traps, particularly eel fykes (long, tapering cones of nets with baffles), which are typically set at depths of up to 5 m, well within foraging range (Jefferies et al. 1984), and in lobster creels, similarly shallowly placed (Vincent Wildlife Trust 1988). By law, in Europe, eel fykes must have otter guards or 'stop grids'—a mesh across the trap with an opening too small for an otter to pass through, but allowing the target species to pass freely. In addition, research is currently being conducted by Aktion Fischotterschutz eV in Germany on fish trap 'escape hatches'—weakened areas of net that will retain the target species in normal circumstances, but which an otter can break open to escape (Hans-Heinrich Krüger, pers. comm.). This proved successful in tests with live otters habituated to experimental procedures (Hans-Heinrich Krüger, pers. comm.). Otters can also become entangled in fish farm nets (Northridge 1988, describing this for *Lutra lutra* in the UK; Juan Valqui, pers. comm. describing this for *Lontra feline* in Peru). This often happens in shallower water, not observed by humans, and the otter may struggle to free itself for a very long time before succumbing to exhaustion or drowning. Various methods are employed to keep marine mammals away from nets, mainly because of the cost of replacing nets and/or escaped stock rather than from a welfare point of view, including the use of acoustic alarms, with variable success (Quick et al. 2002).

In some parts of the world, dynamite fishing is employed in habitat utilised by otters, e.g. along the coast of Peru, and the endangered marine otter, *Lontra felina*, is highly likely to be stunned or killed if in the proximity (Juan Valqui, pers. comm.).

In countries where conservation is deemed important, there has been research into mortalities caused by fishing gear, and amelioration has become enshrined in legislation. Despite legislation existing, the regulations may be flouted, and otters are dying in illegal gear—several otters each year are reported drowned in illegal fykes in the UK alone (Poole et al. 2007). In the rest of the world, however, otters will be dying in fish traps unregarded and may also be deliberately killed where they are perceived as competitors for fish (e.g. Sasekumar et al. 2012).

29.1.5 Marine-Borne Debris

Anthropogenic waste known as 'trash' or marine debris is increasing in the oceans and washing to shore, where any otter utilising the marine environment will come into contact with it. As yet, although unsightly, there is no evidence that it is posing a major population-level conservation challenge for otters ('... I've seen otters swimming in between plastic bags and their dens full of anthropogenic debris. Of course that should not be, and is very unpleasant to see, but I'm not really sure how to count that as a threat': Juan Valqui re Endangered Marine Otters, *Lontra felina*, Peru), though Ballachey and Bodkin (2015) report marine debris to be a significant and growing problem for sea otters in California.

However, individual otters suffer through ingestion or entrapment, often presenting with serious injuries. An example of this was an otter rescued on Skye in Scotland with a cable tie cut deep into his neck, with laboured breathing and a swollen head, and another a week later, on the same island, with fishing twine around his body, cutting in and causing pain and danger of infection. Fortunately both these individuals were spotted by concerned fishermen, and the otters were rescued, treated and rehabilitated successfully (IOSF 1997a, b) (see Chap. 13 for experience in pinnipeds of marine debris).

29.1.6 Emergent Threats: Microbeads

The development of new materials and their subsequent discharge into the oceans means that new threats are always emerging. Plastic microbeads, known as 'virgin plastic pellets' by the plastics supply industry, are used in soaps and shampoo products and have started appearing in otter spraint. The BBC's "Winter-Watch" program, in 2015, working at a relatively remote location on the River Dee in Scotland, found that 40% of the 39 otter spraints sampled contained microbeads (Klein-Aarts, 2015). However, their effect on the animal is unknown, but they appear to 'accumulate lipophilic compounds from the sea water and thereby transferring them up the food chain' (Arno Gutleb, pers. comm.). In the USA, the Microbead-Free Waters Act of 2015 (H.R. 1321) works to phase out production and sale of plastic microbeads in rinse-off cosmetics beginning in 2017, and this Act was unanimously passed by both House of Representatives and Senate and signed into law in 2015 by President Barack Obama (Monterey Bay Aquarium 2015). Canada may shortly follow suit (Pettipas et al. 2016). Other countries, however, continue to produce microplastic beads (Cole et al. 2011), with unknown consequences.

29.2 Social Changes in Otter Populations Resulting from Human Disturbances

Human disturbances take many forms, both direct and indirect. In areas with discreet human presence, otters may be diurnal and easy to observe at a distance such as in the Scottish Isles. If human density increases and the disturbances are loud, close and disruptive, otters tend to become nocturnal and often disappear altogether. Some species are more tolerant of human presence than others. The giant otter is diurnal, lives in groups and has become a favourite ecotourism target in Brazil and Peru, whereas the endangered hairy-nosed otter in Southeast Asia is rarely observed. In Singapore, a major city with canals and parks, several family groups of smooth-coated otter have become re-established and are popular with local residents.

If otters are molested, trapped or shot at, they become wary and avoid the area involved. In South America where otter trapping was intense up to the 1970s, otters disappeared from whole watersheds, and these were not repopulated for decades (Duplaix et al. 2015). However, if only a few otters were removed from a river, these were replaced in a few years as the otter population expanded again.

Sea otters are highly social mammals often forming relatively large groups of resting animals termed rafts (from 3 to over 1000 otters). In areas of high human density, people may try to approach these rafts of sea otters for pictures or a close encounter with wildlife. Wild sea otters tend to be wary of humans, and approaching boats may disturb sleeping animals, startling them and causing them to dive and swim away quickly and potentially harming individuals particularly pups. Ecotourism targeting sea otters in California in Monterey Bay had become a conservation issue that needed to be dealt with to comply with MMPA regulations. To address this, a programme called 'Be sea otter savvy' was developed by the Monterey Bay Aquarium, US Fish and Wildlife, California Department of Fish and Wildlife and Friends of the Sea Otter to educate tourists about their wildlife viewing actions potentially disturbing the sea otters normal behaviours and activities (Sea Otter Savvy Program 2016). Activity in Canada and Alaska by wildlife viewing boats and cruises aren't as numerous or as invasive as those in California, but speeding boats have been known to kill otters by inadvertently striking them.

29.3 Global Climate Change Effects on Otter Welfare

The impacts of climate change on biodiversity are being observed and documented in biomes around the world. The impact on freshwater and marine ecosystems and their hydrologic regimes make otters particularly vulnerable to climate change. Already threatened by habitat loss, pollution, persecution and a dwindling prey base in many countries, river otters and sea otters rank among the most vulnerable mammals to climate change. Extreme weather-related events such as droughts and flooding not only affect water quality but also seasonal events in the otter's life cycle such as cub-rearing when dens are flooded and the litter is lost. For instance, giant otters (*Pteronura brasiliensis*) live in ecosystems that have well-marked wet and dry seasons (Duplaix 1980; Duplaix et al. 2015). The birth season takes place at the beginning of the dry season when water levels are low and stable. Recently in the Pantanal and in the Amazon, the annual seasonal wet-dry cycle has become unpredictable with wet seasons of variable length. Reports of the otters skipping the breeding season altogether have become more frequent (Duplaix et al. 2015), which may seriously impact the long-term survival of this endangered species.

Climate change may affect sea otter populations both negatively and positively. Ocean acidification is thought to have negative impacts on sea otter prey. The effect of decreasing ocean pH may lead to a dramatic decrease in the shellfish populations upon which sea otters depend for food. As nearshore systems become more acidic, one anticipated response is that many shellfish such as clams, oysters, mussels and snails will have greater difficulties forming and maintaining their shells, resulting in increased mortality. However, sea otters tend to feed on a wide variety of prey and may be able to find and consume soft prey and fishes in addition to the limited shellfish available, which may allow them to survive in an increasingly acidic ocean (Ballachey and Bodkin 2015). On the other hand global warming may have a positive impact on available habitat for sea otter populations. Warming Pacific oceans may change the southern extent of winter pack ice in the Bering Sea. If this happens sea otters may be able to extend their range northward, into coastal areas of the Bering Sea, where there is abundant shallow habitat with abundant benthic prey populations. Today the only variable keeping sea otters out of this area is the winter pack ice (Ballachey and Bodkin 2015).

References

Baker JR, Jones AM, Jones TP, Watson HC (1981) Otter *Lutra lutra* mortality and marine oil pollution. Biol Conserv 20:311–321

Ballachey BE, Bodkin JL (2015) Challenges to sea otter recovery and conservation. In: Larson S, Bodkin JL, VanBlaricom GR (eds) Sea otter conservation. Academic, Boston, MA, pp 63–96. doi:10.1016/B978-0-12-801402-8.00004-4

Ballachey BE, Bodkin JL, DeGange AR (1994) An overview of sea otter studies. In: Loughlin TR (ed) Marine mammals and the Exxon Valdez. Academic, San Diego, CA, pp 47–59

Bodkin JL, Ballachey BE, Coletti HA, Esslinger GG, Kloecker KA, Rice SD, Reed JA, Monson DH (2012) Long-term effects of the Exxon Valdez oil spill: sea otter foraging in the intertidal as a pathway of exposure to lingering oil. Mar Ecol Prog Ser 447:273–287. doi:10.3354/meps09523

Bowyer RT, BlundellGM B-DM, Jewett SC, Dean TA, Duffy LK (2003) Effects of the Exxon Valdez oil spill on river otters: injury and recovery of a sentinel species. Wildl Monogr 153:1–54

Bragg JR, Prince RC, Harner EJ, Atlas RM (1994) Effectiveness of bioremediation for the Exxon Valdez oil spill. Nature 368(6470):413–418

California Dept of Fish and Wildlife (2016) Olive the oiled otter https://en-gb.facebook.com/OlivetheOiledOtter/. Accessed 5 Dec 2016

Cole M, Lindeque P, Halsband C, Galloway TS (2011) Microplastics as contaminants in the marine environment: a review. Mar Pollut Bull 62:2588–2597

Crain CM, Halpern BS, Beck MW, Kappel CV (2009) Understanding and managing human threats to the coastal marine environment. Ann N Y Acad Sci 1162:39–62

DeGrange AR (1990) The oil spill in Prince William Sound. Alaska IUCN Otter Species Group Bull 5:2–7

Duffy LK, Bowyer RT, Testa JW, Faro JB (1993) Differences in blood haptoglobin and length-mass relationships in river Otters (*Lutra canadensis*) from oiled and non-oiled areas of Prince William Sound, Alaska. J Wildl Dis 29(2):353–359

Duplaix N (1980) Observations on the ecology and behavior of the giant river otter (*Pteronura brasiliensis*) in Suriname. Rev Ecol (Terre Vie) 34:495–620

Duplaix N, Evangelista E, Rosas FCW (2015) Advances in the study of giant otter (*Pteronura brasiliensis*) ecology, behavior, and conservation: a review. Lat Am J Aquat Mamm 10(2):75–98

Estes JA (1992) Catastrophes and Conservation: Lessons from Sea Otters and the Exxon Valdez IUCN Otter Spec. Group Bull 7:33–35

Fallsons Angler (2015) Plastic microbeads in otter spraint. http://fallonsangler.net/little-plastic-balls/. Accessed 6 Dec 2016

Goldstein P (2010) effects of crude oil exposure to children, adults and pregnant mothers. American Medical Forensic Specialists (AMFS) Newsletter. http://www.amfs.com/resources/medical-legal-articles-by-our-experts/95/effects-of-crude-oil-exposure. Downloaded on 11 April 2016

Hatfield BB, Ames JA, Estes JA, Tinker MT, Johnson AB, Staedler MM, Harris MD (2011) Sea otter mortality in fish and shellfish traps: estimating potential impacts and exploring possible solutions. Endanger Species Res 13:219–229

IOSF (1997a). Some recent problems. International Otter Survival Fund, Otter News 13

IOSF (1997b). Litter causes more problems for Skye Otters; 3-day rescue. International Otter Survival Fund, Otter News 14 Archive. http://www.otter.org/Public/News.aspx. Accessed 8 Dec 2016

ITOPF (2015) Country profiles: Indonesia. International Tanker Owners Pollution Federation. http://www.itopf.com/fileadmin/data/Documents/Country_Profiles/indones.pdf. Downloaded on 11 April 2016

IUCN (2015). The IUCN Red List of Threatened Species. Version 2016-3. [http://www.iucnredlist.org. Downloaded 07 Dec 2016]

Jefferies DJ, Green J, Green R (1984) Commercial fish and crustacean traps: a serious cause of otter *Lutra lutra* (L.) mortality in Britain and Europe. Vincent Wildlife Trust, London, p 31

Jessup DA, Yeates LC, Toy-Choutka S, Casper D, Murray MJ, Ziccardi MH (2012) Washing oiled sea otters. Wildl Soc Bull 36:6–15. doi:10.1002/wsb.113

Jessup DA, Johnson CK, Estes J, Carlson-Bremer D, Jarman WM, Reese S (2010) Persistent organic pollutants in the blood of free-Ranging Sea otters (*Enhydra lutris ssp.) in* Alaska and California. J Wildl Dis 46(4):1214–1233

Kannan K, Agusa T, Perrotta E, Thomas NJ, Tanabe S (2006a) Comparison of trace element concentrations in livers of diseased, emaciated and non-diseased Southern Sea otters from the Californian coast. Chemosphere 65:2160–2167

Kannan K, Perrotta E, Thomas NJ (2006b) Association between perfluorinated compounds and pathological conditions in Southern Sea otters. Environ Sci Technol 40(16):4943–4948

Kannan K, Perrotta E, Thomas NJ, Aldous KM (2007) A comparative analysis of Polybrominated Diphenyl ethers and polychlorinated biphenyls in Southern Sea otters that died of infectious diseases and noninfectious causes. Arch Environ Contam Toxicol 53:293–302

Kenyon KW (1969) The sea otter in the eastern Pacific Ocean. North AM Fauna 68:1–352

Klein-Aarts, Lonneke > 5 Gyres. "Shocking result of BBC Winterwatch last night. In 40% of the 39 otter poo's around river Dee, Scotland they found plastic beads (from cosmetic origin)... Think the beads came there by the otters eating salmon. There are not that many hikers leaving cosmetical plastic beads that high up in the river Dee. Unfortunately BBC doesn't have easy access to former episodes for people living abroad. I think a large part of British population watched though. It might have created some awareness" Facebook 22 January 2015 [https://www.facebook.com/5gyres/posts/10152710426938920. Accessed 17 April 2015]

Kreuder C, Miller MA, Jessup DA (2003) Patterns of mortality in Southern Sea Otters (Enhydra lutris nereis) from 1998–2001. J Wildl Dis 39:495–509

Kruuk H (2006) Otters—ecology, behaviour and conservation. Oxford University Press, United Kingdom, p 265

Lafferty KD (2015) Sea otter health: challenging a pet hypothesis. Int J Parasitol Parasites Wildl 4(3):291–294. doi:10.1016/j.ijppaw.2015.05.005. http://www.ncbi.nlm.nih.gov/pmc/articles/PMC4491668/. Downloaded 12 April 2016

Miller MA, Kudela RM, Mekebri A, Crane D, Oates SC, Tinker MT (2010) Evidence for a novel marine harmful algal bloom: cyanotoxin (microcystin) transfer from land to sea otters. PLoS One 5(9):e12576

Monterey Bay Aquarium (2015) Microbeads banned! A victory over ocean plastic pollution. https://futureoftheoceanwordpresscom/2015/12/30/microbeads-banned-a-victory-over-ocean-plastic-pollution/. Downloaded 12 April 2016

Murata S, Takahashi S, Agusa T, Thomas NJ, Kannan K, Tanabe S (2008) Contamination status and accumulation profiles of Organotins in sea otters (*Enhydra lutris*) found dead along the coasts of California, Washington, Alaska (USA) and Kamchatka (Russia). Mar Pollut Bull 56:641–649

Murray MJ (2015) Veterinary medicine and sea otter conservation. In: Lawson SE, Bodkin JL, VanBlaricom GR (eds) Sea otter conservation. Academic, Monterey, CA, pp 160–197

Nakata H, Kannan K, Jing L, Thomas N, Tanabe S, Giesy JP (1998) Accumulation pattern of organochlorine pesticides and polychlorinated biphenyls in Southern Sea otters (*Enhydra lutris nereis*) found stranded along coastal California, USA. Environ Pollut 103(1):45–53

Northridge S (1988) Marine mammals and fisheries: a study of conflicts with fishing gear in British waters. Report to Wildlife Link Seals Group. Wildlife Link, London

Oates SC, Miller MA, Byrne BA, Chouicha N, Hardin D, Jessup D, Dominik C, Roug A, Schriewer A, Jang SS, Miller WA (2012) Epidemiology and potential land-sea transfer of enteric bacteria from terrestrial to marine species in the Monterey Bay region of California. J Wildl Dis 48(3):654–668

Peterson CH, Rice SD, Short JW, Esler D, Bodkin JL, Ballachey BE (2003) Long-term ecosystem response to the *Exxon Valdez* oil spill. Science 302(5653):2082–2086

Pettipas S, Bernier M, Walker TR (2016) A Canadian policy framework to mitigate plastic marine pollution. Mar Policy 68:117–122

Poole WR, Rogan G, Mullen A (2007) Investigation into the impact of fyke nets on otter populations in Ireland. In: Irish wildlife manuals, No. 27. National Parks and Wildlife Service, Department of Environment, Heritage and Local Government, Dublin, Ireland

Quick NJ, Middlemass SJ, Armstrong JD (2002) The use of anti-predator controls at Scottish marine Salmon farms. Fisheries research services report No 03/02. http://www.gov.scot/Uploads/Documents/0302CollCon.pdf. Accessed 12 April 2016

Roos A, Loy A, de Silva P, Hajkova P, Zemanová B (2015) *Lutra lutra*. The IUCN Red list of threatened species 2015: e.T12419A21935287. http://dx.doi.org/10.2305/IUCN.UK.2015-2.RLTS.T12419A21935287.en. Downloaded on 11 April 2016

Sasekumar A, Marshall A, Macintosh N (Eds) (2012). Ecology and conservation of southeast Asian marine and freshwater environments including wetlands. Springer Science & Business Media Netherlands 151–170.

Sea Otter Savvy Program (2016) Respect the Nap! http://www.seaotters.org/sea-otter-savvy-program/. Accessed 8 Dec 2016

Serfass T, Evans SS, Polechla P (2015) *Lontra canadensis*. The IUCN Red list of threatened species 2015: e.T12302A21936349. http://dx.doi.org/10.2305/IUCN.UK.2015-2.RLTS.T12302A21936349.en. Downloaded on 11 April 2016

Shapiro K, Miller M, Mazet J (2012) Temporal association between land-based runoff events and California Sea otter (*Enhydra lutris nereis*) protozoal mortalities. J Wildl Dis 48(2):394–404

Snowden RJ, Ekweozor IKE (1987) The impact of a minor oil spillage in the estuarine Niger Delta. Mar Pollut Bull 18(11):595–599

Stoskopf MK (2015) Parasitic diseases of marine mammals. Merck veterinary manual. http://www.merckvetmanual.com/mvm/exotic_and_laboratory_animals/marine_mammals/parasitic_diseases_of_marine_mammals.html. Downloaded 12 April 2016

Van Dolah FM, Doucette GJ, Gulland FMD, Rowles TL, Bosart GD (2003) Impacts of algal toxins on marine mammals. In: Vos JG, Bossart G, Fournier M, O'Shea T (eds) Toxicology of marine mammals. CRC, Boca Raton, pp 247–269

Vincent Wildlife Trust (1988) The effects of otter guards on the fishing efficiency of eel fyke nets. Vincent Wildlife Trust, London, 47pp

Williams TM, Kastelein RA, Davis RW, Thomas JA (1988) The effects of oil contamination and cleaning on sea otters (*Enhydra lutris*). I. . 66: 2776–2781

Chapter 30
Animal Welfare Issues Pertaining to the Trapping of Otters for Research, Conservation, and Fur

Thomas L. Serfass, Lesley Wright, Kelly Pearce, and Nicole Duplaix

30.1 Introduction

Legal trapping of otters is conducted for research (e.g., to equip individual animals with radio transmitters) and applied conservation (e.g., to obtain individuals for reintroduction projects) and for utilitarian purposes (i.e., the fur industry for some species). Until relatively recently, standards defining the most appropriate traps in relation to animal welfare for wildlife caught for utilitarian purposes (wildlife species killed for fur have become generically referred to as furbearers, a term that will be used hereafter) were poorly established. Trapping was usually subject to regulations imposed by individual wildlife management jurisdictions [e.g., state and provincial wildlife agencies in the United States of America (USA) and Canada, respectively]. Canada, Russia, the European Union (EU), and USA are involved in collaborative, ongoing efforts to develop and implement standards for what ostensibly constitutes "humane trapping." The motivation for developing trapping standards seems largely a response by Canada, Russia, and the USA (the three top wild fur-producing countries; Animal Legal and Historical Center 2010) to overcome

T.L. Serfass (✉) • K. Pearce
North American Coordinator, IUCN Otter Specialist Group, Department of Biology and Natural Resources, Frostburg State University, Frostburg, Maryland 21532, USA

Appalachian Laboratory, University of Maryland Center For Environmental Science, Frostburg, Maryland 21532, USA
e-mail: tserfass@frostburg.edu; kjpearce@frostburg.edu

L. Wright
Deputy Chair, IUCN/SSC Otter Specialist Group, Didcot, UK
e-mail: lc.wright@yahoo.co.uk

N. Duplaix
Chair, IUCN/SSC Otter Specialist Group, Fisheries and Wildlife Department, Oregon State University, Corvallis, Oregon 97331-3803, USA
e-mail: Nicole.Duplaix@oregonstate.edu

A. Butterworth (ed.), *Marine Mammal Welfare*, Animal Welfare 17,
DOI 10.1007/978-3-319-46994-2_30

legislation passed by the European Union in 1991 (Regulation 3254/91). This legislation bans the import of wild fur from countries allowing the use of leghold traps [now often referred to as "foothold trap," a semantical adjustment presumably adopted to depict trapping less harshly (i.e., more humanely) than "leghold trap." Leghold traps are banned in at least 80 countries (Fox 2004)].

Two agreements ratified by the EU council in 1998 [the first with Canada and Russia—"Agreement on International Humane Trapping Standards" (AIHTS)] and the other as a separate agreement with the USA [incorporating comparable standards as AIHTS, but in the form of "Best Management Practices for Trapping" (BMPs); Association of Fish and Wildlife Agencies (AFWA) 2006] through an agreed "Minute," which is nonbinding—i.e., apparently there are no penalties or enforcement to ensure standards are met) resulted in an exemption for Canada, Russia, and the USA. This enabled continued export of fur from wild-caught furbearers and use of leghold traps during an undefined evaluation period to assess humane issues pertaining to leghold and other traps (United States Department of Commerce 1997; Iossa et al. 2007; Proulx et al. 2012). These agreements brought about the first attempt to establish international standards [i.e., through the International Organization for Standardization (ISO)] to define what constitutes "humane" for traps within certain general trap-type categories (Harrop 2000; Princen 2004). Unanimity was not achieved on what constitutes key thresholds for traps regarding the extent of injuries caused by traps intended to restrain, but not kill, an animal and the time required for an animal to become unconscious when caught in traps designed for killing. However, a process was established to define performance of a trap (safety for the trapper and efficiency in capturing target species), to assess trauma related to physical injuries caused to animals caught in traps designed for restraint, and killing efficiency for traps designed to kill. Stress-induced trauma endured by a trapped animal currently is not a part of ISO welfare standards for trapping (Iossa et al. 2007). Fundamental to these agreements is that mandatory testing be conducted to determine if traps conform to standards established under AIHTS and BMPs for a particular species (i.e., become certified as acceptable under the agreement). Through the agreements, traps failing to meet agreed standards are expected to be phased from use. However, traps not meeting standards are permitted to remain in use if there are no alternative traps certified for the target species. This presumes that trap research continues with the intent of identifying a trap or traps that meet certification requirements. Trap standards are at various stages of completion (depending on species) (e.g., Fur Institute of Canada 2015), but design of trap testing protocols and evaluation of trap performance appear in some cases to be largely at the discretion of authorities responsible for managing furbearer trapping, with minimal external review. Specific details for outcomes of trap performance assessments are not readily available in the USA and have not been subjected to meaningful, external peer review. In contrast, Canada has published a variety of outcomes from trap testing and, along with Russia, has phased out the use of "traditional" leghold traps (Proulx 1999; AIHTS 2012).

The North American river otter (*Lontra canadensis*; hereafter river otter) serves particularly well for discussing traps and trapping systems in relation to animal welfare issues pertaining to otters in general for both research, and conservation and fur trapping—particularly in reference to populations in the USA. The river otter has received considerable conservation/research attention [predominantly in the USA where reintroduction projects involving live-trapping (i.e., the intention is for the trapped animal to be alive post-trapping event) and translocations of individuals from areas with viable populations have taken place in 22 states to restore extirpated populations]. The USA and Canada both kill substantial numbers of river otters each year for the fur trade, but Canadian populations did not suffer declines to the extent of those in the USA and have thus received less research/conservation attention based on live-trapping. The Eurasian otter (*Lutra lutra*) has received extensive research attention (see Kruuk 2006 for a review), but relatively few studies have been based on live-trapping (Fernandez-Moran et al. 2002; Ó Néill et al. 2007). Other species of otters generally have received little research attention or, as with the Eurasian otter, live-trapping has not been part of most studies. Paucity of live-trapping studies for otters outside of North America (NA) likely is related to greater concern for animal welfare regarding trapping and restrictions on the use of leghold traps. Hence, the following review of animal welfare issues pertaining to live-trapping for research and conservation focuses on the river otter in the USA, using examples from other species when applicable; those pertaining to fur trapping exclusively focus on the river otter in both the USA and Canada.

30.2 Types of Traps and Animal Welfare Standards

Traps considered for AIHTS agreements are placed in two general categories: (1) restraining traps and (2) killing traps. Restraining traps are designed to restrict a captured animal's movements and include leghold traps, modified leghold traps, powered and non-powered snares, and cage-type traps.

Among killing traps, rotating-jaw traps, which have spring-powered jaws that when triggered close forcibly across the body (the neck or chest is intended) of the trapped animal, have received considerable attention regarding animal welfare considerations pertaining to trapping (Proulx 1999; Proulx et al. 2012). However, restraining traps (leghold traps, and non-powered and powered snares) are also sometimes classified and used as killing traps, typically by setting the trap in a manner that will drown the captured animal (AFWA 2006). Drowning sets are typically used to kill semiaquatic mammals, including the river otter.

30.3 Restraining Traps

30.3.1 Leghold Traps

This type of trap is manufactured in a variety of configurations and sizes (Proulx 1999). The basic design of all leghold traps is the same, being comprised of two metal jaws that are held open at 180° by a triggering mechanism when set and clamp together (to grasp the trapped animal's limb) at 90° in reference to the set position

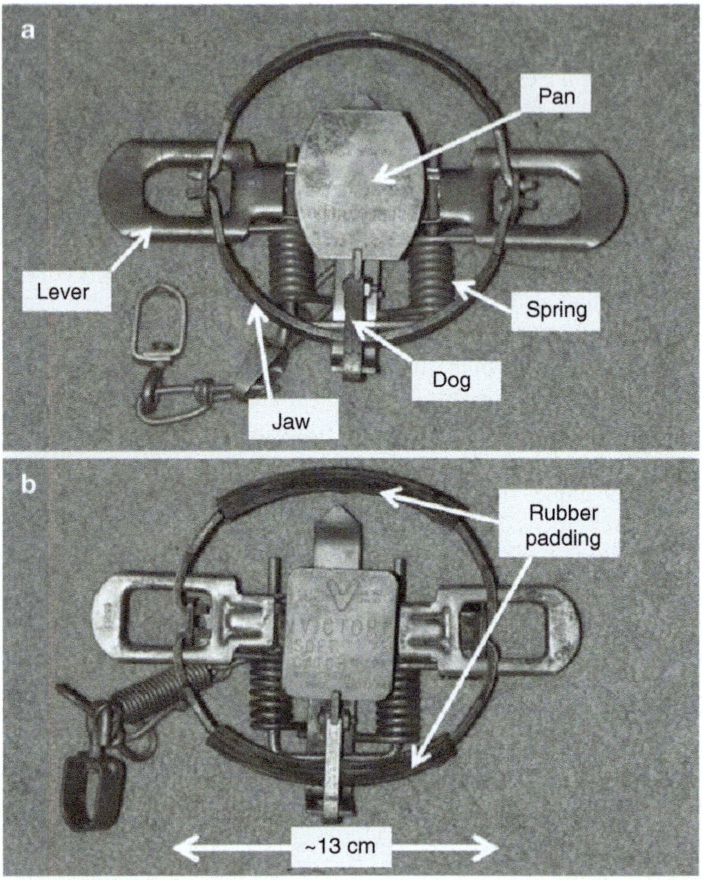

Fig. 30.1 Examples of coil spring leghold traps: (**a**) unmodified and (**b**) modified with rubber padding on inner surface of jaws ("padded jaw" or "Soft Catch™"). The traps are displayed in the "set" or "open" position. Primary components of a leghold trap are depicted on the image of the unmodified leghold trap (see AFWA 2006 for a review of the function of the trap components)

when sprung. The jaws of the trap and triggering components (the pan and dog) are comparable for all types of leghold traps (Fig. 30.1). Leghold traps are now manufactured as two types: coil spring traps (two coil springs each cause a lever to move upward, closing the trap's jaws) and longspring traps (depending on the style, either one or two longsprings close the jaws of the trap). When referring to a leghold trap, the type of trap (i.e., coil spring or longspring) is preceded with a number—usually from one to four—with smaller numbers indicating traps with smaller jaw spreads (i.e., distance between the inner sides of the jaws when the trap is set) (e.g., a No. 2 coil spring or No. 11 longspring). (Note: A No. 11 longspring trap and No. 1 longspring trap have the same jaw spread, with the No. 11 denoting the trap as having two longsprings and the No. 1 indicating the trap as having a single longspring, a convention applied to denote the use of one or two springs for all sizes of longspring traps.)

30.3.2 Modified Leghold Traps

These traps are configured and function identically to the leghold trap (see Proulx 1999), but the jaws are modified in a manner intended to increase efficiency (i.e., minimize the rate at which a captured animal pulls free of the trap) and minimize injury to the trapped appendage. Modifications to the jaws include the following: (1) laminated—an additional strip of metal is welded to the top and/or bottom of each jaw; (2) double jaws—each outer jaw (traditional jaw) is paired with a smaller, inner jaw; (3) offset jaws—the striking surface of the jaws is not in contact when closed [i.e., there is a space (offset) of 3–6 mm between the jaws of a closed trap]; and (4) padded jaws—rubber padding is inserted between the jaws (Fig. 30.1b).

30.3.3 Cage Traps

Traps constructed of wire-mesh framing with one or two doors. These traps are available in various dimensions, with the dimensions of a trap used dependent on the species intended to be trapped. Animals are captured in this trap by entering through doors and then stepping on a trigger, which causes the door(s) to close. These traps are analogous in design to *box traps*.

30.3.4 Snares

Snares are lengths of stranded steel cable configured into a loop that captures an animal by tightening over its neck, body, or limb. Tightening of the loop around the animal is accomplished either passively (i.e., non-powered snare—the loop is tightened by the movement of the animal) or actively (i.e., powered snare—tightening of the loop is initiated by a spring-powered device activated by contact with the animal). Snares used with the intent of restraining an animal by the neck should have "stops" designed to prevent excessive tightening of the cable to reduce the chance of asphyxiating captured individuals.

30.3.5 Suitcase-Type Traps

These are large traps originally designed for American beavers (*Castor canadensis*). The Hancock Live Trap (Fig. 30.2) and the Bailey Beaver Live Trap are specific types of traps within this category that have been evaluated for use in capturing river otters. Both traps have large movable metal frames covered in chain-link material that close around an animal [i.e., an animal is captured within, not between, the trap jaws—the Hancock trap has a single movable (closing) jaw, whereas both jaws of

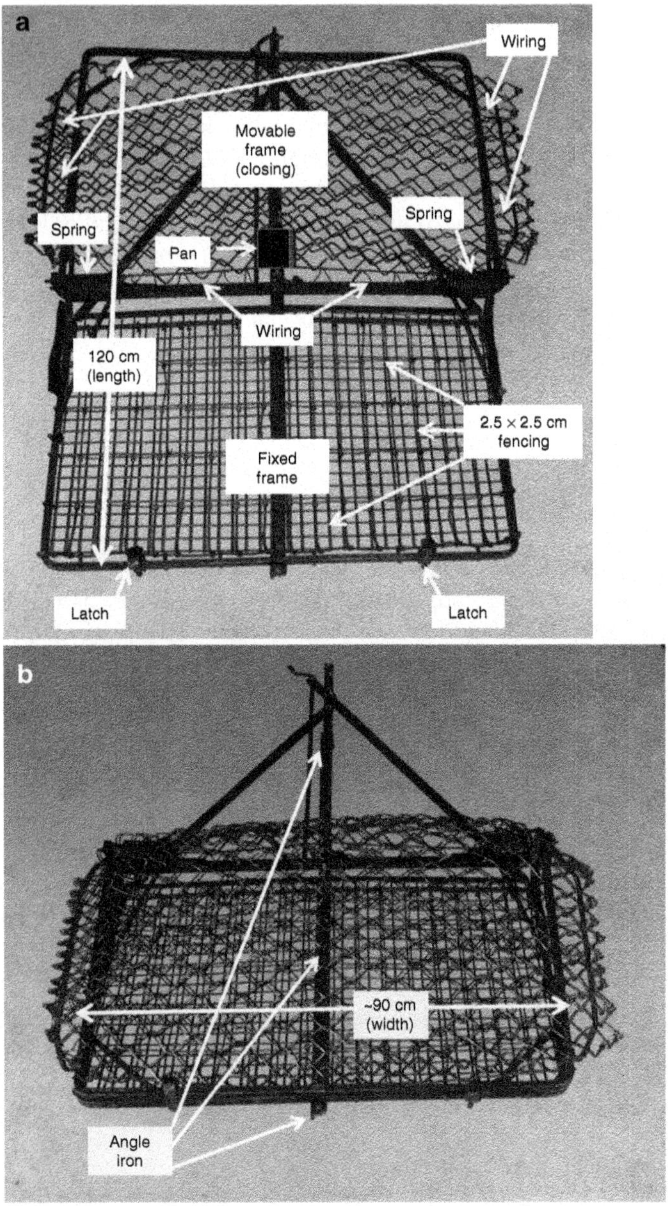

Fig. 30.2 A Hancock trap as modified by Serfass (1984) to lay flat in the "set" or "open" position (**a**) and the trap in the closed position (**b**). The trap is held flat in the open position by affixing a length of angle iron along the back of the trap [note: The movable frame (closing side) of an unmodified Hancock trap is at an angle of about 130° to the fixed frame (non-closing side) when the trap is in the set position]. To minimize chances of river otters escaping, Melquist and Hornocker (1979) recommended (1) adding springs on the inner side of "latches," which are intended to prevent a captured animal from forcing open the movable side of the opening (the springs better ensure that latches remain over the frame of the movable sides of a closed trap) and (2) using wire to close gaps along the margins of the trap frame. A further modification to prevent escape or injury of a captured river otter involves covering the 5 × 10-cm wire grid on the fixed side of the trap frame with vinyl coated 2.5 × 2.5-cm welded wire fencing (Serfass 1984)

the Bailey trap are movable and close simultaneously]. The Bailey trap has been shown to be ineffective in capturing river otters (Northcott and Slade 1976).

30.4 Killing Traps

30.4.1 Rotating-Jaw Traps

Also commonly referred to as bodygrip, bodygripping, or Conibear™-type traps, these traps have two rotating jaws powered by one or two springs (Fig. 30.3). As with leghold traps, numbering associated with these traps is a reference to the size (inner distance between jaws) of the trap, with a smaller number indicating less distance between the jaws (e.g., 110 Conibear, 220 Conibear, and 330 Conibear represent traps of progressively increasing distance between the jaws). Animals entering an open trap are intended to be killed when the jaws forcefully close and crush a vital region of the body—for the most humane death as possible, the preferred areas intended to be struck by the jaws are the neck or upper chest.

30.4.2 Killing Snares

Killing snares are configured in the same manner as snares used for restraint, and the loop likewise becomes tightened around an animal either passively or actively. However, snares designed to kill are intended to capture an animal around the neck

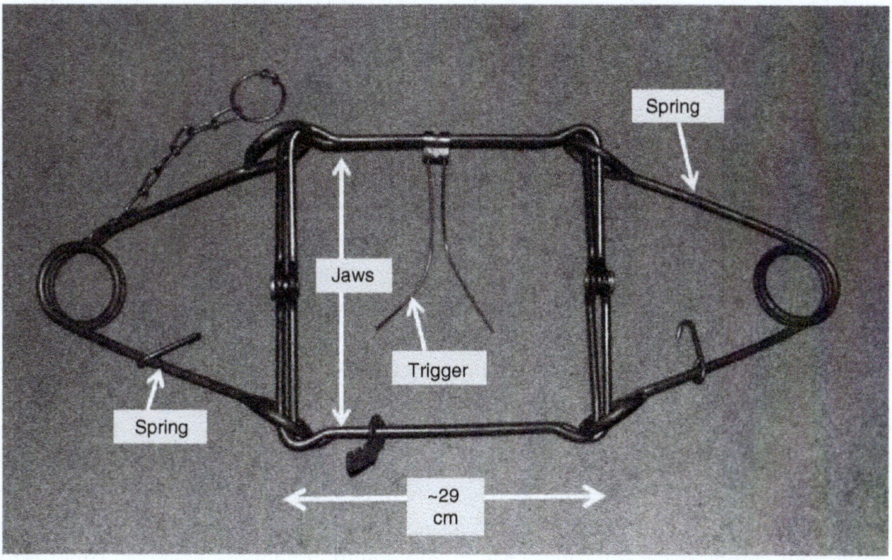

Fig. 30.3 A 330 rotating-jaw trap (also called bodygrip, bodygripping, or Conibear™-type traps) in the closed (not set) position. This type of trap is intended to quickly kill a captured animal and is frequently used by trappers to capture river otters for fur (Responsive Management 2015a)

and do not have stops to restrict tightening of the loop. Thus, the loop continues to tighten as the animal struggles until it is asphyxiated. Powered killing snares also kill by asphyxiation, but the snare tightens more quickly, ideally causing a quicker death. (Note: Stops can be used to limit closure of the loop to a circumference that minimizes capture or harm to smaller, nontarget species.)

30.4.3 Drowning (or Submersion) Traps/Sets

Leghold traps and snares can be set in a manner to drown animals captured in or near the water (AFWA 2006). Traps are either set underwater (at a depth that prevents the captured animal from reaching the surface) or along the shoreline (attached to a cable that leads the trapped animal into water deep enough to keep it from reaching the surface).

Note: Proulx (1999) and Iossa et al. (2007) provide extensive and detailed reviews of animal welfare for restraining and killing traps used to capture mammals. In BMPs for trapping in the USA (AFWA 2006), the AFWA explains trap components and trapping setting techniques for capturing furbearing animals. The information in these documents provides an important basis for developing insight necessary to inform discussion pertaining to animal welfare issues related to trapping wild mammals.

30.5 Animal Welfare and Trapping

Establishment of animal welfare standards for trapping has developed through the use of standardized scores for injuries sustained by individuals captured in restraining traps. These scores are based on the ISO trauma scale (ISO 10990-4 1999), which is categorized into four levels for each injury sustained:

Mild trauma (scores range from 2 to 10 points—injuries such as claw loss and abrasions)

Moderate trauma (scores range from 25 to 30 points—injuries such as loss of single digit and eye laceration)

Moderately severe trauma (a score of 50 points—injuries such as loss of two digits and a simple fracture below the carpus or tarsus)

Severe trauma (a score of 100 points—an injury such as loss of three or more digits to resulting in death)

A composite score of individual's injuries is used to assess if a trap meets appropriate welfare standards (see Iossa et al. 2007, for an extensive review of animal welfare standards based on scoring of trap-caused injuries following the ISO trauma scale). To achieve AIHTS, killing traps are expected to cause death in ≤5 min for 70% of trapped individuals for the species being evaluated. However, Proulx et al. (2012) and Proulx and Rodtka (2015) argue that ≤3 min should be applied as the minimum standard for time until death (irreversible unconsciousness). Time until death of animals caught in drowning sets is dependent on the onset of hypoxia, which typically will be a prolonged period (i.e., potentially much longer than the ≤5 min standard established for death of animals captured in killing traps) for the semiaquatic mammals typically targeted by trappers using this method of trapping (Gilbert and Gofton 1982; Iossa et al. 2007).

Animal welfare issues associated with fur trapping—especially the use of leghold traps—have been the primary motivation for the development of trapping standards. Nonetheless, projects that involve the live-trapping of wild animals for research and conservation purposes often involve the same types of traps used by fur trappers and likewise deserve scrutiny to understand, and mitigate, the effects on the animals during capture and handling. Outcomes of research and conservation projects likely will be enhanced when traps and trapping procedures are efficient and cause minimal injury to captured individuals (e.g., less time and expense associated with capturing an appropriate number of animals to fulfill project objectives and in rehabilitating injured animals). Hence, in addition to what should be direct concern based on animal welfare, project investigators are also motivated by practical issues related to ensuring the well-being of live-trapped animals in relation to intended research or conservation outcomes. In contrast, the intent of fur trappers is to kill trapped animals for the pelt or other products derived from the carcass—i.e., although there may be a humanitarian concern to reduce suffering to the trapped animal, there is no practical motivation for a fur trapper to be concerned about injuries incurred to an animal during trapping unless the injuries somehow impact the value of the fur or other products. In fact, in the absence of regulation, practical issues would dictate that fur trappers adopt the most efficient trapping methods—those yielding the highest capture rates at the least expense—in lieu of animal welfare concerns. This dichotomy in practical issues between live-trapping for research/conservation and trapping for fur serves to emphasize an important reason, in addition to the fact that live-trapping for research/conservation purposes is conducted much less frequently than trapping for fur, that establishing animal welfare standards for trapping has been focused on fur trapping.

30.6 Live-Trapping Otters for Research and Conservation

Although killing traps may be used to lethally collect specimens for research purposes, the focus of this section is directed toward the use of restraining traps to livetrap otters (animal welfare issues related to using kill traps are discussed in the ensuing Sect. 30.9 “Trapping River Otters for Fur”). A variety of restraining traps and associated trap-setting procedures have been assessed for use in live-trapping river otters, and sometimes these methods have subsequently been applied and refined to livetrap other otter species [e.g., for reintroducing the Eurasian otter into Spain (Fernandez-Moran et al. 2002) and reintroducing the Eurasian otter into the Netherlands (Koelewijn et al. 2009)]. Animal welfare concerns for live-trapping wild animals should necessarily apply not only to the traps used but to how traps are set, how procedures are used to restrain animals for removal from traps, and the immediate post-trapping handling of animals (hereafter this collective is referred to as the "trapping system"). Various leghold traps and Hancock™ traps have primarily been used for trapping river otters for research/conservation purposes, with results of the applications and outcomes (e.g., trap-setting procedures, review of injuries, and capture rates) published in various formats. In contrast, there are no peer-reviewed assessments of injury rates for other traps that could potentially be considered for use in live-trapping river otters (e.g., cage traps and snares), although cage traps have been used in studies requiring the live-capture Cape clawless otters (*Aonyx capensis*) (Van der Zee 1982; Arden-Clarke 1986) and spotted-necked otters (*Hydrictis maculicollis*) (Perrin and Carranza 1999). The following review focuses on published cases of various traps used to live-trap river otters, with respect to injuries and the trapping systems employed, but also includes mention of trap types that may theoretically be used but which have not been frequently used or evaluated for use with river otters or other otter species. This discussion of traps may have similar merits and/or liabilities for otters species other than river otters.

30.6.1 Leghold Traps

Serfass et al. (1996) compared injuries caused to teeth, feet, and legs of river otters captured using No. 1.5 coil spring traps with padded jaws (hereafter padded trap; Fig. 30.1b) with one factory spring replaced with a No. 2 spring (captured in Pennsylvania by authors and Maryland by personnel of the Maryland Department of Natural Resources; $n = 38$), No. 11 longspring traps (captured in Louisiana by a supplier licensed to capture and sell river otters; $n = 17$), and various unidentified types of leghold traps (captured in Michigan, New Hampshire, and New York by private trappers; $n = 29$) for a river otter reintroduction project. Trap-setting techniques were similar for No. 1.5 coil spring traps with padded jaws and No. 11 traps {traps were set and anchored in the water [anchor (i.e., the trap attachment)]}. Traps were attached with a segment of chain typically 1.5 m in length, enabling river otters to swim while captured (see Serfass et al. 1996, for details and precautions associated with this trap-setting technique to avoid drowning captured animals). In contrast, trap-setting procedures followed by private trappers are poorly reported, but traps were presumed to be primarily set and attached on the shoreline (i.e., not in the water as Serfass et al. 1996). Few severe injuries to limbs occurred among river otters captured in padded traps [1 (4%) had an injury requiring an amputation (a single digit) in comparison to amputations in 12 (71%; ≥1 digit) and 9 (37.5%; ≥1digit ($n = 7$), a foot, and a leg) river otters caught in No. 11 traps and by private trappers using unspecified traps/trap-setting techniques, respectively]. River otters caught in padded traps and No. 11 traps sustained fewer, and less severe, dental injuries than those obtained from private trappers. Regardless of trap type, injuries (to appendages and the teeth) sustained by juvenile river otters were much less than for adults (Serfass et al. 1996).

A study in coastal Alaska used No. 11 double-jaw longspring traps set on land (anchored with trap chains ≤70 cm in length) to live-capture 30 river otters (Blundell et al. 1999). This project used a trauma scale developed by Olsen et al. (1996) and Jotham and Phillips (1994) to score injuries to the teeth and appendages [scores for an individual could range from 0 (no injuries) to 100 (death)] but did not provide details of specific injuries contributing to scoring or the number of individuals acquiring injuries to the teeth and/or appendages. Traps were monitored a minimum of two to three times daily—a transmitter was attached to traps, and this was activated when traps were sprung. The scoring system and number of daily trap checks present a challenge for meaningful comparison with Serfass et al. (1996), who used different metrics to quantify injuries, and traps were checked once daily. More frequent trap checks may reduce frequency and extent of injuries by minimizing time an animal is restrained by a trap. Five (17%) of the river otters captured in No. 11 double-jaw traps by Blundell et al. (1999) attained serious injuries to appendages, whereas only one (3%) of those caught in padded traps by Serfass et al. (1996) would have been scored as having a serious injury. Injuries to the teeth considered serious were low in Blundell et al. (1999) and also likely to be low for Serfass et al. (1996), but actual comparison is not possible because of the different scoring systems followed by the respective projects. Melquist and Hornocker (1979) captured nine river otters in leghold traps [five captures in No. 2 coil spring traps and four captures in No. 3 jump traps (no longer manufactured to our knowledge)]. Injuries to river otters caught in No. 2 coil spring traps were described as minor (no details provided), but escape rates were reportedly high. Two of the river otters (both juveniles) caught in No. 3 jump traps sustained broken hind limbs (the bones broken were not reported).

30.6.2 Hancock Trap

The Hancock trap was originally designed for live-trapping American beavers. Northcott and Slade (1976) and Melquist and Hornocker (1979) described important modifications necessary for the trap to be suitable for river otters (i.e., to prevent escape). Two further modifications were made by Serfass (1984): the first enabled the trap to lay flat for concealment when set in shallow water (as manufactured the movable side of the trap is at an angle to the fixed side), and the second involved covering the fixed side of the trap (comprised wires spanning opposing sides of the trap frame to form a rigid 5 × 10-cm grid) with vinyl coated 2.5 × 2.5-cm welded wire fencing (Fig. 30.2a, b). When constrained, river otters often vigorously attempt to escape by scratching or biting to breach any perceived weak areas in a cage, cage-type trap, or other confinement, potentially causing injury to forepaws and teeth. The spacing of wires on the fixed side of the trap created a grid comprised of openings likely large enough to become the focus of escape efforts by river otters (the head of most river otters will fit through a 5 × 10-cm opening), which was overcome by the second modification. Also, when set flat in shallow water [made possible by the first modification suggested by Serfass (1984)], the fixed side of the trap is not exposed to a captured animal, and although exposed, the chain-link on the (closing) movable side of the trap compresses and is thus less likely to cause teeth damage if bitten (Fig. 30.2b). The chain-link of the movable side of a Hancock trap [the top of the trap when closed as configured by Serfass (1984)] can expand upward to about 30 cm from the bottom of the trap. Care must be taken to monitor changes in water levels to ensure that the top of the trap remains above the surface (i.e., to avoid drowning a trapped animal).

Melquist and Hornocker (1979) tested a variety of traps and considered a properly modified Hancock trap the most favorable for use with river otters—there was no mention of occurrence of injuries (or lack thereof) among 21 captures, which included 2 adult-sized river otters captured simultaneously. In Blundell et al.'s (1999) comparison of Hancock traps and No. 11 double-jaw leghold traps for capturing river otters ($n = 11$ for Hancock traps, and $n = 30$ for leghold traps), serious injuries to the teeth occurred much more frequently in Hancock traps, but serious injuries to appendages were higher for leghold trap (no injuries to appendages occurred in river otters caught in Hancock traps versus about 17% in those caught in leghold traps). In contrast, Serfass (1984) indicated no injuries to six river otters captured in Hancock traps modified as described by Melquist and Hornocker (1979) and Serfass (1984). Dental injuries reported by Blundell et al. (1999) may have occurred because modifications were not made to the fixed side of the Hancock trap.

In comparison to leghold traps, Hancock traps have received limited use and evaluation for live-trapping river otters, possibly fostered by the somewhat negative evaluation by Blundell et al. (1999). The much larger size, higher cost, and limited availability of the Hancock trap (in comparison to leghold traps) also present various practical limitations to its use. Another practical concern relates to the potential for larger animals (including people and pets) to accidentally trigger and be injured by being caught between the frames of the hard-closing trap. Likewise, there is potential for otters to be caught between the frames of this trap, especially if >1 otter visits the trap site. Regardless, the virtues of the Hancock trap for live-capturing river otters [e.g., no injuries when modified as reported by Serfass (1984) and good capture efficiency reported by Melquist and Hornocker (1979) and Blundell et al. (1999)] merit its further evaluation, particularly as an alternative for live-trapping river otters or other otter species in areas where use of leghold traps is limited or prohibited.

30.6.3 Other Traps

Various types of leghold traps and the Hancock trap are the only traps used with any regularity for live-capturing river otters. Other traps that have potential for use with river otters have either had limited or no evaluation. In addition to leghold and Hancock traps, Melquist and Hornocker (1979) also conducted brief evaluations of a powered foot snare and several cage-type traps (one from a trap manufacturer and three constructed specifically for the project: culvert, barrel, and floating traps) but reported little meaningful information on capture or injury rates. Cape clawless otters and spotted-necked otters have been successfully captured in what were described as "standard carnivore traps" (800 × 800 × 1400 mm cage traps with a single door) [Van der Zee (1982) and Arden-Clarke (1986)—capture of Cape class otters; Perrin and Carranza (1999)—capture of spotted-necked otters]. No information was provided on injuries or lack thereof to the captured animals. To our knowledge, body/neck snares have not been evaluated with live-capturing river otters. Severe injuries caused to wolves (*Canis lupus*) and coyotes (*Canis latrans*) caught in neck/body snares suggest that extreme caution should be used in developing protocols for evaluating the suitability of snares or any other untested traps to livetrap otters. Concerns for snaring these species have been raised by Proulx and Rodtka (2015) and Proulx et al. (2015), and general concerns for animals captured in snares were raised by Rochlitz (2010). Cage-type traps have been successfully used to capture a variety of carnivore species with minimal or no injury and deserve further research attention to determine if otters can be captured efficiently and relatively unharmed using this type of trap.

30.7 Restraint of Captured Otters for Release from Traps

Development and refinement of protocols for efficiently reducing stress and injury to captured animals being released from traps are sometimes overlooked as a component of the trapping system. Restraining an animal for release from a trap is accomplished either by physical or chemical restraint (delivery of a drug, i.e., a chemical immobilant) to enable handling of an animal. Physical restraint is any approach that confines the movement of an animal—a trap represents a physical restraint, but the term is most often applied to devices used to further restrict the movement of an animal restrained in a trap. Physical restraint should facilitate either the direct release of a trapped animal or delivery of a chemical restraint to immobilize the animal for release from the trap and to enable subsequent evaluations (e.g., physical examination, ear tagging, or transport to a captive facility). Methods to physically restrain river otters while captured in live traps will be the focus of the ensuing discussion.

30.7.1 Leghold Traps

Techniques for physically restraining river otters captured in leghold traps necessarily vary by trap-setting technique. Shirley et al. (1983) and Serfass et al. (1996) describe the use of long-handled nets to restrain river otters captured in leghold traps attached to chains (typically 0.6–1.25 m in length, but potentially longer) anchored in the water. River otters had limited access to the shoreline but were able to swim within the radius of trap chains and the captured animals were netted while in the water. The use of nets for physical restraint necessitates evaluation as to whether the structure of netting will cause the trap restraining an animal to become entangled in

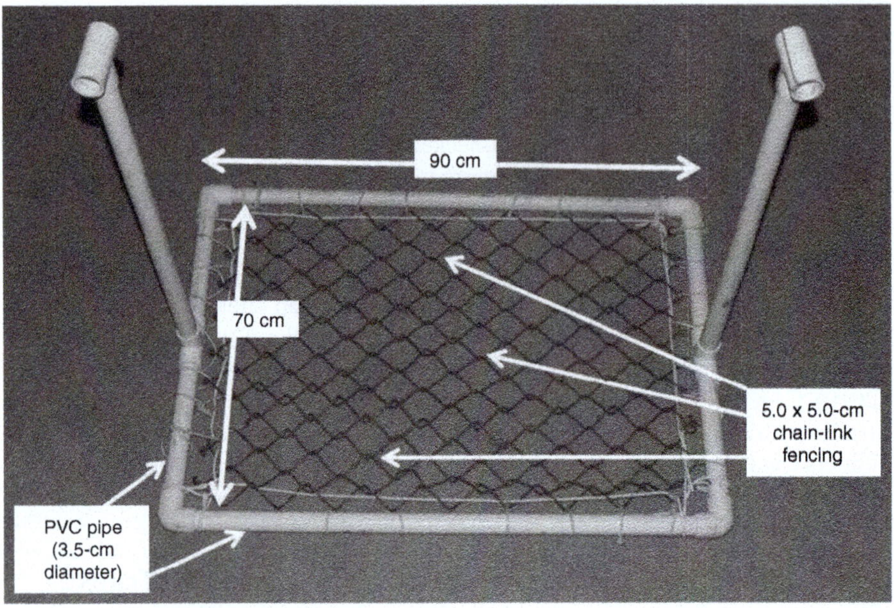

Fig. 30.4 Hold-down device used to physically restrain river otters captured in leghold traps. Trapped river otters initially are restrained in nets the netted river otter is further restrained with the hold-down device to better enable delivery of chemical restraint (see Serfass 1984; Serfass et al. 1996). The hold-down device is constructed of polyvinyl chloride (PVC) pipe (wood and metal framing also have been used) surrounding vinyl coated chain-link fencing. Handles of hold-down device detach for transport

the net. Such entanglement may result in injury and additional stress to a captured animal. The likelihood of a trapped animal becoming entangled in a net will vary based on construction of nets (e.g., fibers used, the thickness of those fibers, and mesh size—an assessment is easily accomplished by placing leghold traps inside various netting to determine if entanglement occurs). Serfass et al. (1996) describes a process for bringing the netted animal to the shoreline and application of a second form of physical restraint (use of a hold-down device; Fig. 30.4), for quick, efficient, and safe (for the animal and investigator) delivery of a chemical immobilant.

The use of capture poles (e.g., Ketch-All™ poles, San Luis Obispo, California 93401, USA) is common for restraining animals captured in leghold traps but has limited application for river otters—the circumference of a river otter's neck tends to be larger than that of the head (particularly in adults) and, unless excessively tightened, the noose of the capture pole generally will slip off the head. In lieu of physical restraint, Blundell et al. (1999) successfully delivered darts with chemical restraint through a blow gun, and Fernandez-Moran et al. (2002) also used such an approach for delivering chemical immobilants to Eurasian otters captured in No. 1.5 padded traps. Remote delivery of chemical immobilants reduces stress and potential injury that could be contributed by physical restraint, but consideration should be given for the possibility for an animal becoming free of the trap following delivery but before being restrained by the chemical immobilant. Such scenarios were not reported by either Blundell et al. (1999) or Fernandez-Moran et al. (2002), but should be considered, and would be of particular concern for otters, which if escaping the trap would likely enter the water and potentially drown after the drug takes effect.

30.7.2 Hancock Live Traps

Chemical immobilants can easily be delivered to animals captured in Hancock traps by injecting with a hand syringe (hand injection) through the chain-link mesh on the movable side of trap (Serfass 1984; Blundell et al. 1999). Movement of a trapped animal can be further restricted to better facilitate injection by compressing the chain-link comprising the movable side of the trap (i.e., the investigator will stand on the chain-link on opposing sides to the animal in a manner that confines but does not exert excessive downforce). Serfass (1984) set Hancock traps exclusively in shallow water and recommends that traps be pulled from the water prior to delivering chemical restraint to the captured animal to prevent it from ingesting water during induction.

30.8 Concluding Comments: Live Traps

Meaningful comparisons of outcomes of the relatively few reports of live-trapping river otters are a challenge. There seldom have been direct comparisons of traps where associated trapping systems have been controlled, including periods between trap checking. For example, the live-trapping study conducted by Blundell et al. (1999) occurred in an area (coastal Alaska) that enabled use of transmitters to remotely determine if traps were sprung, which facilitated monitoring each trap site at least two or three times a day. In contrast, Serfass et al. (1996) conducted their live-trapping study in northeastern Pennsylvania where trapping sites were widely distributed across the landscape, which logistically limited checking traps sites to once every 24 h. In such cases disparities in trap-check frequency may have influenced outcomes as much or more than the trap and trapping system applied. For example, longer times between the checking of traps could correlate positively with more injuries. Regional difference in environmental conditions and associated difference in trapping conditions could likewise compromise meaningful comparisons of traps and trapping systems.

Because of the large number of wild river otters captured for reintroduction projects in the USA (>4000; Bricker et al. 2016), there may be an impression that techniques for live-trapping the species are well established. However, the majority of the animals used for reintroduction projects were captured in southern Louisiana through arrangements with an individual licensed to trap and sell river otters. Hence, there were no assessments of mortality rates, injuries that prevented sale of otters for reintroduction, or, with the exceptions of Serfass et al. (1996), assessments of injuries sustained by animals that were reintroduced. Hancock traps have been used infrequently, even though a few assessments of this trap indicated its potential for use in live-trapping river otters. Clearly more rigorous studies are needed for assessing both practical and animal welfare issues for traps and trapping systems most appropriate for use in live-trapping river otters.

Outcomes of live-trapping studies conducted in the USA and the few studies conducted elsewhere (e.g., Fernandez-Moran et al. 2002; Koelewijn et al. 2009) can serve as a basis for assessing best methods to livetrap other species of otters. However, researchers should understand that physical and behavioral differences of other otter species could affect responses to being trapped and be open to investigating potentially new and more innovative approaches for live-trapping otters. Researchers investigating other species of otters also should be certain that live-trapping studies are designed in a manner that enables meaningful comparisons of the traps and trapping systems being evaluated.

The development of noninvasive techniques for otters [e.g., camera trapping (Stevens and Serfass 2008) and extraction of DNA from feces (Fike et al. 2004; Beheler et al. 2005) and hair (Depue and Ben-David 2010)] has limited the need for more invasive field techniques that may cause physical harm and stress to animals, such as live-trapping. Regardless, the use of radiotelemetry remains an important part of many studies of wild animals and provides insight about animal behaviors and movement patterns not always assessable by noninvasive techniques. Conducting radiotelemetry studies is inherently dependent on capturing and handling individual animals to attach transmitters, which argues for the continued use of live-trapping of wild animals (including otters) for some field investigations. In the case of otters, live-trapping is in need of further refinement (for species that previously have been livetrapped) and development through appropriately designed studies for species that have not been the focus of studies involving live-trapping.

30.9 Trapping River Otters for Fur

Killing otters to obtain their pelts for the fur trade is an international venture undertaken legally and illegally, depending on species and geopolitical jurisdiction. Illegal methods of killing otters will vary based on what is most expedient for perpetrators. Illegally killing of otters in some parts of the world is considered to be severely impacting populations of some species [e.g., populations of otter species inhabiting southeastern Asia are believed to be declining because of intense demand for their pelts in China (Foster-Turley and Santiapillai 1990; Gomez et al. 2016)]; but few details are available regarding the extent of the illegal trade or approaches used to kill otters. Regardless, individuals involved in the illegal killing of otters (or any wildlife) are not going to adhere to any prescribed standards of animal welfare.

Legal killing of otters presumes some standards are in place to limit depletion of populations [e.g., regulations for periods when killing can occur (closed seasons) and number of individuals that can be killed (quotas)] and to limit pain and suffering. Of the world's 13 species of otters, all are listed as Convention on International Trade in

Endangered Species of Wild Fauna and Flora (CITES) Appendix I or II because of respective concerns of endangered or threatened conservation statues, except for the river otter, which is listed under Appendix II as a "look-alike species" (*A designation for a species legally part of international trade that is of similar appearance to one or more species not legally traded. Hence, the designation serves as a precaution against inclusion of specimens or parts of a protected species from being illegally exported by being posed as those of a similar species that is legally traded.*). However, trade of Appendix II (non-look-alike) species is permissible if conditions are met demonstrating that there will be no detriment to the survival of the species in the wild.

Among the world's otter species, the river otter is the only species possessing a population status considered suitable for meeting conditions that will enable sustainable killing of individuals for the pelt trade throughout large portions of its range. As an otter species legally trapped throughout much of its range for pelts that are frequently traded internationally, the river otter is thus of predominant concern regarding the humaneness of techniques and equipment used to capture and kill individuals. Prior to European settlement, the river otter occupied aquatic habitat throughout the Continental USA and Canada (Hall 1981). By the early to mid-1900s, the species had experienced substantial population declines, or complete extirpations in some areas. These declines occurred throughout large portions of the river otter's historic range in the USA but to a lesser extent in Canada. These losses resulted from the combined detrimental effects of overkilling by trappers, disturbances to riparian habitats (e.g., deforestation), and water pollution (Bricker et al. 2016). The combination of more restrictive trapping regulations including prohibition of trapping river otters in some USA states, successful reintroduction projects in 22 states, and improvements in the conditions of riparian and aquatic habitats contributed to the recovery of river otter populations in many areas of NA (Bricker et al. 2016). Legal trapping of river otters has expanded as populations have recovered. About 171,000 river otters in the USA and about 83,000 river otters in Canada were trapped for their pelts between 2006 and 2012. River otters are a primary target species for about 9% of trappers in the USA (Responsive Management 2015a).

Trappers use a variety of devices to capture river otters. Trapping devices are selected for various reasons, including practical (e.g., cost of traps and associated equipment), social (e.g., personal preference, influence of peers, and tradition), habitat conditions, regulations imposed by a particular jurisdiction within a country, and international agreements, including the AIHTS in Canada (Fur Institute of Canada 2015) and BMPs in the USA (AFWA 2014). Growing public concern over animal welfare issues have raised specific attention to the ethics and humaneness of trapping wildlife for fur, and this has come alongside a realization that minimizing injury to a trapped animal should also be a consideration when selecting a trapping device. AIHTS and BMPs focus on physical injuries in assessing animal welfare issues regarding trapping. Iossa et al. (2007) make a compelling argument that stress and various other physiological indices should be used in such assessments. Rothschild et al. (2008) and Taylor et al. (2016) assessed stress (glucocorticoid) levels, and Kimber and Kollias (2005) evaluated biochemistry values of blood in river otters following their live-capture and placement into captivity as part of reintroduction projects. These studies demonstrated no long-term adverse stress responses and also concluded that blood values were not a good indicator of the level of physical injury. No such studies have been undertaken for river otters as part of the AIHTS and BMP trap certification processes in relation to fur trapping.

30.9.1 Restraining Traps

Leghold traps, which are the most common type of restraining trap used by trappers to capture river otters, have received extensive review through the process of developing BMPs and are thus the focus of this discussion on restraining traps. An adequate critique of leghold traps in relation to animal welfare issues requires including an assessment of various trapping systems that may be employed. For example, methods used to attach [anchor] traps at trap sites should be included in critiques. Other often overlooked factors for such critiques include trapper willingness to implement recommendations (especially when formal regulations are not in place to mandate use of a particular trap and trapping system, as with BMPs), variation in regulations for legal trap types and trapping systems imposed by wildlife management authorities (for the USA, wildlife management for most species, including river otters, is at the state-wildlife-agency level), the capabilities and effort put forth by the various management authorities to enforce regulations, and variation in response to being restrained in a trap among species and by individuals of a species.

Coil spring traps (unmodified only) with jaw spreads ≥5 in. (13 cm) and long-spring traps (either unmodified or modified to have double jaws) with jaw spreads of ≥3 7/8 × 3 7/16 in. (10 × 9 cm) meet BMP criteria for river otter (AFWA 2014). However, AFWA (2014) also states "Many currently-used trap models meet specifications." Details about testing of approved traps are not provided or description of the criteria used to establish the suitability of "many currently used trap models." Likewise, no reasons are provided for not specifically listing certain types of traps as acceptable (e.g., modified coil spring traps). These omissions may be related to a trap not yet having been tested, the trap having been tested and failed humane requirements, or having been tested and failed other BMP criteria (e.g., efficiency—a trap is not judged to be efficient if <60% of individuals for the target species remain captured after activating the trap).

30.9.1.1 Physical Injury

Other than published reports of river otters captured for research and conservation purposes (see Sect. 30.6), we were unable to find published descriptions of injuries sustained by river otters captured in leghold traps. Review of the published studies of river otters captured in leghold traps as part of conservation and/or research projects indicated considerable variation in injuries caused among various leghold traps (see Sect. 30.6.1). This contrasts with portrayals in BMP recommendations for leghold traps as being suitable for river otters. In fact, virtually all of the styles and sizes of leghold traps considered efficient in trapping river otters prior to development of BMPs are now approved as meeting BMP criteria. BMP evaluations to determine a trap as suitable appear to be based on controlling other factors related to trapping (e.g., how a trap is set and the time required to check traps). Review of the published reports on live-trapping river otters suggests that such factors (in addition to the type of trap used) are likely to influence injuries to a trapped animal. Such variations appear to be discounted in assessments for determining BMPs, where participating trappers are monitored to ensure compliance with prescribed trapping procedures. There is no evidence, for example, that the trapping procedures followed by trappers participating in BMP evaluations will become expectations (i.e., in the form of regulations) for fur trapping. Objective evaluation to determine if BMPs will be useful in enhancing welfare standards for animals caught in leghold traps is virtually impossible from published information related to the development of BMPs for river otters or other furbearers.

30.9.2 Killing Captured Animals

Methods for killing an animal captured by trappers using restraining traps are often overlooked in humane assessments of trapping. Generally, trappers are recommended to shoot the trapped animals between the eyes with a .22 caliber gun (International Association of Fish and Wildlife Agencies [IAFWA] 2005). However, trapper's magazines often recommend drowning, suffocation (standing or kneeling on the animal's chest), or hitting on the head with clubs as a way to minimize damage to the fur (i.e., avoid the blood that would get on the pelt if the animal is shot) (Fox and Papouchis 2004). The IAFWA (2005) also recommends these methods as humane forms of killing trapped animals.

30.9.3 Killing Traps: Bodygrip Traps

The published BMPs for otters list any bodygrip trap within sizes designated as 220, 280, and 330 as acceptable for use with river otters. Traps of this type are considered to meet humane standards if 70% of the animals are dead within 5 min after being captured (Iossa et al. 2007; Proulx et al. 2012; Proulx and Rodtka 2015). Such standards omit discussion of humane considerations for the 30% of animals potentially not dead after 5 min or the suffering that occurs to those that do meet the 5 min standard. Testing to assess these standards has in some cases taken place in captive settings where anesthetized animals are positioned between the jaws of a set trap and then the trap is sprung. Such an approach does not necessarily represent conditions seen in natural settings, where the trap is less likely to close on the preferred part of the body (to expedite the time until death). We were unable to find published details of testing outcomes for assessments of bodygrip traps for river otters.

30.9.4 Drowning Traps/Sets

Trappers commonly use "drowning traps/sets" when capturing semiaquatic furbearers, such as river otters. River otters reportedly have the capacity to remain underwater for up to 8 min (Smithsonian n.d.), exceeding the acceptable time established for death using bodygrip traps to meet humane requirements. However, BMPs make no mention of any evaluations conducted to assess animal welfare standards for this type of trapping of river otters, but the BMP does state that performance standards are comparable to killing devices for other aquatic furbearers (AFWA 2014). In fact, this type of trapping system is recommended for river otters, with the only BMP standard being that the trapping system must not allow the animal to reach the surface after being submerged.

30.9.5 Killing Snares

Trappers legally use snares to capture river otters in some USA states and Canadian provinces. However, there are no published evaluations of the humaneness of capturing river otters in snares nor are these devices considered in AIHTS or BMP evaluations of trap performance criteria. Proulx et al. (2015) reviewed issues pertaining to the use of snares to kill canids [gray wolves (*Canis lupus*), coyotes (*C. latrans*), and red foxes (*Vulpes vulpes*)] in Canada, concluding that death to the animals was prolonged or some animals remained alive (i.e., did not meet humane standards for death applied to other killing traps), injuries were sometimes severe (e.g., deep lacerations where the snare tightened around the neck), and killing snares are nonselective—often capturing a variety of nontarget animals. From these outcomes, Proulx et al. (2015) recommended that use of killing snares be disallowed unless modifications can be achieved that improve the humaneness of this trapping system. In contrast, use of snares is being promoted in the USA (e.g., Vantassel et al. 2010). Given a well-developed musculature in the neck, river otters, like canids, are unlikely to be killed quickly or at all when caught in a snare. Snares, incorporated into drowning sets, would eventually cause death by asphyxiation. In the absence of contrary evidence, the evaluation of killing snares by Proulx et al. (2015) for canids establishes an important basis for regarding this trapping system as likely to be inhumane (by any standards) for capturing river otters.

30.9.6 Unintended Captures

River otters are sometimes caught accidentally by trappers intending to catch other semiaquatic furbearers or those that frequent riparian habitats. Responsive Management (2015a) conducted an extensive survey of trapping in the USA, which included assessment of species captured, types of traps used for a particular species, and furbearing species captured unintentionally (i.e., not the primary target of the trapper). Unintentional capture of river otters was reported by 29.5% of trappers targeting American beavers. Large bodygrip traps, various leghold traps, and snares are used for beavers, with the No. 330 bodygrip trap predominating (about 78% of beaver trappers reported using that trap). Traps and trap sets used for beavers are in some ways comparable to what would be expected for use with river otters and, thus, represent similar issues pertaining to a humane death—time to death caused by closure and/or drowning in bodygrip traps, time until drowning in drowning sets, and potential injuries from snares. River otters also were reported to be unintentionally caught by trappers primarily pursuing American mink (*Neovison vison*), muskrat (*Ondatra zibethicus*), and raccoon (*Procyon lotor*), but less frequently than by beaver trappers (<6% for each of these species). However, trappers trapping American mink and muskrat in leghold traps often may not anchor the trap sufficiently (either by using stakes or weight) to retain a trapped river otter at the capture site (i.e., the river otter escapes with the trap attached to its leg), contributing to both humane concerns and potential for underrepresenting the extent of unintentional captures. Also, many trappers included in the Responsive Management (2015a) survey undoubtedly were not trapping in areas occupied by river otters. Expected rates of unintentional captures would thus be higher if not diluted by inclusion of trappers trapping in areas unoccupied by river otters. Realistic insight on expectations for the extent of unintentional captures is needed and could be gained by focusing only on the subset of trappers trapping in areas occupied by river otters.

30.10 Concluding Thoughts: Trapping for Fur

Trapping river otters for pelts appears to be "maintainable" (i.e., local populations appear to be able to withstand the numeric impacts) at the landscape-level scale in NA—although local, trapping-induced extirpations likely occur in marginal habitats and reintroduction projects may have been unnecessary in some areas of the USA if trapping had not limited expansion of natural populations. We note, for example, that there has been rapid post-release expansion of reintroduced populations, which initially were legally protected from trapping [see Bricker et al. 2016) for a detailed review of trapping and reintroductions of river otters], whereas native populations remained stationary or expanded slowly where trapping was permitted. Regardless, debate over trapping river otters is largely based on opposing values pertaining to what is appropriate and "ethical use of wildlife" and specific animal welfare concerns pertaining to the capture of animals in traps. However, those involved in supporting trapping in NA comprise a large, integrated wildlife management system that includes governmental wildlife agencies (and associated wildlife professionals), nongovernmental organizations representing these agencies [e.g., AFWA (http://www.fishwildlife.org/)], some university wildlife researchers, manufacturers of hunting and trapping-related equipment, and supporting political entities—a set of interactions referred to by Gill (2004) as an "Iron Triangle," whereby those not within the "Iron Triangle" have a limited voice in wildlife policy decision-making. These relationships constitute a "conservation-industrial complex," which collectively offers considerable financial, political, and organizational resources to promote a value system based on sustainably killing wild animals.

The so-called North American Model of Wildlife Conservation [NAM; first articulated by Geist et al. (2001)] demonstrates the promotional capabilities of the wildlife management system in NA. The NAM is comprised of seven primary elements (Geist et al. 2001; Organ et al. 2012), each repeatedly depicted by various media in a manner that supports and justifies consumptive use of wildlife, managed by public, state-level conservation agencies, as the "cornerstone" of wildlife conservation in NA. Two of the primary elements of NAM: wildlife products should not be commercialized (i.e., sold as part of a market-based system) and the Public Trust Doctrine (PTD) are particularly relevant to discussions of trapping and the management of furbearing animals in the USA. Trapping for fur is a large, international, commercial enterprise of which trade in furbearers captured in the USA is a prominent part, an obvious contradiction to the primary element of NAM opposing commercialization of wildlife. The PTD is based on the concept that certain natural resources, including wildlife, cannot be owned by individuals but are instead to be conserved by the government in a manner that benefits current and future generations of citizens. An implicit assumption of the PTD is that the values and interests of all citizens be considered in approaches used to conserve and manage PTD-based natural resources (Treves et al. 2015). However, the values and interests of those engaged in hunting and trapping have been disproportionately favored in wildlife management decision-making at the state-agency level.

Over about the last 15 years, NAM has been widely portrayed as both a historical account of how wildlife were conserved in NA in the past and a prescriptive model for how wildlife should be conserved in the future (Peterson and Nelson 2016). Without question progenitors of NAM clearly endorse recreational, regulated killing of wildlife (the focus is on hunting, but trapping also has been established within the framework) of certain species of wildlife (i.e., those defined as game species, which includes "furbearing" animals such as the river otter) as the fundamental aspect of wildlife conservation. The repetitiveness by which NAM has been portrayed in numerous and varied forums (e.g., Mahoney 2004; Prukop and Regan 2005; Geist 2006; Mahoney et al. 2008; Organ et al. 2010, 2012) has aspects suggesting a marketing effort to promote fundamental concepts of NAM to both conservation professionals and the general public, an approach seemingly designed to homogenize acceptance of consumptive use as fundamental to properly managing wildlife. Foundations for such marketing efforts are anchored in social-science surveys conducted by private organizations that conduct public opinion surveys for state wildlife agencies about hunting and trapping and include investigations providing outcomes such as "How to Talk to the Public About Hunting: Research-Based Communication Strategies" (Responsive Management 2015b).

As with the seemingly overarching purpose of NAM, furbearer trapping also has been promoted to gain acceptance among wildlife professionals and the public. Muth et al. (2006) provided evidence that the majority of conservation professionals supported outlawing the use of the leghold trap and expressed concern that new recruits into the wildlife profession with "…non-traditional wildlife management backgrounds, such as women, ethnic minorities, non-hunters and non-trappers, and urban residents may possess a different value system regarding consumptive use of wildlife than their older counterparts." One mechanism that evolved concurrently with NAM is "Conservation Leaders for Tomorrow"—a program designed to instill NAM's principles by instructing both nonhunting/trapping university students (enrolled in wildlife-related degree programs) and natural resource professionals about the virtues of hunting and trapping in conservation (Conservation Leaders for Tomorrow 2015).

Likewise, seminars at various conferences sponsored by AFWA and The Wildlife Society (TWS) promote the importance of fur trapping in modern wildlife management to students interested in careers in wildlife conservation as well as practicing wildlife professionals [e.g., Trapping Matters Workshop 2016; AFWA Trapping Matters Workshop 2015) and an IAFWA-produced video (see IAFWA 2015)]. The AFWA provides "quick tips" for supporters of trapping on how best to communicate the role and benefits of regulated trapping in wildlife management. These "quick tips" encourage discussions to promote trapping by focusing on the following themes (AFWA 2015):

1. Regulated trapping does not cause wildlife to become threatened or endangered.
2. Trapping is managed through scientifically based regulations enforced by conservation officers.
3. State wildlife agencies continue to refine approaches to trapping methods that include issues pertaining to animal welfare [e.g., Best Management Practices (BMPs)].
4. Regulated trapping provides many benefits to the public (e.g., reducing wildlife damage to crops and minimizing threats to human health and safety).
5. Trapped animals are used for clothing and food.

These themes are mimicked with more elaboration in various publications authored by individuals actively engaged in promoting support for trapping and BMPs—e.g., "Trapping and furbearer management in North American wildlife conservation" appearing in various editions as a standalone publication of the Northeast (USA) Furbearer Technical Committee (Organ et al. 2015) and under the same title but different text as part of a special issue of the *International Journal of Environmental Studies* featuring NAM (White et al. 2015). Recommendations of strategies to gain public acceptance of specific aspects of trapping occur unabashedly in scientific publications of TWS (e.g., use of snares: *"In states where cable-traps are currently prohibited, a drastic regulatory change would likely result in immediate protest from anti-trapping organizations. For example, focusing on regulatory liberalization of snaring in water where beavers are causing damage would likely be more successful than an immediate regulatory change that allowed all forms of cable-trapping."*; Vantassel et al. (2010)). These and other examples raise ethical questions about public employees (many of whom are involved in the articulation of NAM and BMPs) promoting personal values to the public being represented, the role of science versus personal values in formulating wildlife management policy, and, most importantly for this discussion, whether BMPs are focused on improving the welfare of trapped animals or as an opportunity to promote trapping, both in the USA and internationally.

Science is referred to as the basis for developing and implementing furbearer management policy in the USA. However, the process of developing BMPs and promoting the process of fur trapping also includes considerable emphasis on the economic and cultural values of trapping furbearers to some local communities (e.g., Organ et al. 2015; White et al. 2015); topics having practical and emotional relevance but little to do with science in addressing concerns about animal welfare. Traps recommended under the BMP for river otters include virtually all of those used prior to BMPs, and no traps are recommended as inappropriate for the species. Although the BMP for river otters has been recently updated and available on the AFWA web site, no specific details of trap testing outcomes are provided on the site or are readily available for critique. Review and interpretation of outcomes used to establish BMPs are thus seemingly conducted primarily by those involved with the BMP initiative, implying that the public should accept unquestioningly the process and outcomes (a "good faith" approach) associated with selecting traps that adequately meet humane expectations for the public's furbearers. Organ et al. (2014) seemingly support the PTD (as applied in NAM) as being in congruence with this "good faith" management scheme by citing the following statement from Scott (1999): *"Additionally, if a trustee has special skills or expertise (e.g., wildlife professional), they have a duty to use these heightened capacities to enhance the conservation of resources under their management in the interests of trust beneficiaries."* Such a statement seemingly implies that wildlife professionals employed by state wildlife agencies will act in an unbiased manner and objectively represent the interests of all stakeholders in decision-making related to trapping wildlife for fur, a process that is not in evidence when considering promotional efforts to gain public acceptance of fur trapping nor by the system of wildlife conservation championed by proponents of NAM. Treves et al. (2015) effectively identify and review concerns pertinent to the application of public trust responsibilities by state wildlife agencies—specifically pertaining to the conservation of predators. Preeminent among these concerns is the narrow and preferential focus on consumptive use of wildlife embedded in the version of PTD portrayed by proponents of NAM (Batcheller et al. 2010). In contrast, Sax (1970) interpreted proper application of PTD as incorporating interests from a broad constituent base, advocating preserving public, environmental assets for future generations and defending society from undemocratic allocations of environmental assets (modified from Treves et al. 2015). Treves et al. (2015) define undemocratic allocation in part as those that "… reflect tyranny of minority or majority,…," a situation indicative of the wildlife conservation system advocated by NAM whereby consumptive users (who represent a fraction of the overall population in NA) have the predominate voice in decision-making pertaining to wildlife policy. Although humane issues have not received specific attention in discussions of PTD, application of PTD in the narrow sense promoted by Batcheller et al. (2010) and Organ et al. (2012) may nonetheless diminish attention and action in addressing humane concerns pertaining to trapping (or other consumptive uses of wildlife), especially if such concerns collide with entrenched values systems and interests associated with the NA system of wildlife conservation.

The number of states allowing legal trapping of river otters has expanded in recent years (Bricker et al. 2016). Prior to initiation of trapping seasons, strikingly similar negative media portrayals of river otters occurred in several states (Serfass et al. 2014), characteristically beginning with praise for implementation of progressive wildlife conservation policies by state wildlife agencies (i.e., implementing successful river otter reintroduction projects) and ending by proposing that a trapping season may be necessary to alleviate conflict associated with rapidly growing numbers of river otters. Conflict was portrayed as river otters predating on fish in private ponds, and being harmful to gamefish populations, thus causing complaints by anglers (Serfass et al. 2014). However, the extent of these conflicts was seldom quantified by state wildlife agencies or exaggerated in states portrayed as having public resentment toward river otters (Bricker et al. 2016). These negative portrayals appeared to have the intent of lessening public opposition for proposed plans to initiate river otter trapping seasons. State wildlife agencies appear to have allied with some media in the negative messaging. Fostering an acrimonious situation to achieve a wildlife management outcome (i.e., a trapping season on river otters) to benefit a particular constituency (i.e., trappers) would breach PTD obligations of state wildlife agencies to conserve wildlife in a manner that considers the interest of all citizens, not to manipulate public opinion through a marketing effort to achieve a management outcome. Further, labelling an animal as a pest or problem lessens public concern for its welfare (Rochlitz 2010). The marketing approaches seemingly being followed to promote support for fur trapping in general and river otters specifically cast doubt on the objectivity of decision-makers involved in the development of BMPs in placing animal welfare at a level equivalent to traditional wildlife management practices in the USA.

Trapping wild animals for fur is a contentious issue in the USA and elsewhere and will not be accepted by most animal welfare groups, regardless of approaches used to enhance the humaneness of a trap or trapping system. Nonetheless, opposition to fur trapping (especially when using leghold traps) from animal welfare groups in Europe and NA provided the primary impetus for developing universally standardized approaches in defining animal welfare standards for trapping animals with restraining and killing traps (i.e., standards established by the ISO). Establishment of ISO criteria provides a basis for evaluating the humanness of restraining and killing traps used for both research and fur trapping. Although this appeared to be a positive step in recognizing the need to address welfare concerns for trapped animals, the process of trap testing to define traps meeting ISO standards and, more importantly, the actual humaneness of the traps and associated trapping systems and the evaluative process are in need of further scrutiny, particularly in the USA.

References

AFWA Trapping Matters Workshop (2015) Southeast Association of Fish and Wildlife Agencies-pdf. http://www.seafwa.org/Documents%20and%20Settings/46/Site%20Documents/Conferences/2015/2015%20Conference/2015%20Conference/TrappingMattersWorkshop_SEAFWA.pdf. Accessed 9 April 2017

Agreement on International Humane Trapping Standards (AIHTS) (2012) Joint management committee meeting http://www.fishwildlife.org/files/AIHTS_JMC2012report.pdf. Accessed 8 May 2016

Animal Legal and Historical Center (2010) Detailed discussions of fur animals and fur productions. Michigan State University College of Law. https://www.animallaw.info/article/detailed-discussion-fur-animals-and-fur-production. Accessed 1 July 2016

Arden-Clarke CHG (1986) Population density, home range size and spatial organization of the cape clawless otter, *Aonyx capensis*, in a marine habitat. J Zool Soc Lond 209:201–211

Association of Fish and Wildlife Agencies (AFWA) (2006) Best management practices for trapping in the United States. http://www.fishwildlife.org/files/Introduction_BMPs.pdf. Accessed 8 July 2016

Association of Fish and Wildlife Agencies (AFWA) (2014) Best management practices for trapping river otter in the United States. http://www.fishwildlife.org/files/Otter_BMP_2014_F.pdf Accessed 8 July 2016

Association of Fish and Wildlife Agencies (AFWA) (2015) Key messages on trapping. http://wwwfishwildlifeorg/files/AFWA-Furbearer_KeyMessages_2015pdf. Accessed 8 July 2016

Batcheller G R, Bambery MC, Bies L, Decker T., Dyke S, Guynn D, McEnroe M, O'Brien M, Organ J F, Riley S J, Roehm, G (2010) The public trust doctrine: implications for wildlife management and conservation in the United States and Canada. Technical Review 10–01.The Wildlife Society, the Association of Fish and Wildlife Agencies, the Western association of fish and wildlife agencies, and the Wildlife Management Institute, Bethesda, MD, USA

Beheler AS, Fike JA, Dharmarajan G, Rhodes OE Jr, Serfass TL (2005) Ten new polymorphic microsatellite loci for North American river otters (*Lontra canadensis*) and their utility in related mustleids. Mol Ecol Notes 5:602–604

Blundell GM, Kern JW, Bowyer RT, Duffy LK (1999) Capturing river otters: a comparison of Hancock and leg-hold traps. Wildl Soc Bull 27:184–192

Bricker EA, Serfass TL, Ammer FK, Hanley ZL, Pearce KJ, Stevens SS (2016) Conservation status of the North American otter in the United States and Canada: assessing management practices and public perceptions of the species. In: Linh San ED, Somers MJ (eds) Small carnivores: evolution, ecology, behaviour and conservation. Wiley-Blackwell, Oxford

Conservation Leaders for Tomorrow (CLfT) (2015) The history of CLfT. http://clftorg/history. Accessed 9 July 2016

Depue JE, Ben-David M (2010) Hair sampling techniques for river otters. J Wildl Manag 71:671–674

Fernandez-Moran J, Saavedra D, Mantexa-Vilanova X (2002) Reintroduction of the Eurasian otter (*Lutra lutra*) in northeastern Spain: trapping, handling, and medical management. J Zoo Wildl Med 33:222–227

Fike JA, Serfass TL, Beheler AS, Rhodes OE Jr (2004) Genotyping error rates associated with alternative sources of DNA for the Northern American River Otter IUCN Otter Spec Group Bull 21A

Foster-Turley P, Santiapillai C (1990) Action plan for Asian otters. In: Foster-Turley P, Macdonald SM, Mason CF (eds) Otters: an action plan for their conservation. IUCN, Gland, Switzerland

Fox CH (2004) Trapping in North America: a historical overview. In: Fox CH, Papouchis CM (eds) Cull of the wild. A contemporary analysis of wildlife trapping in the United States. Bang Publishing, Brainerd, Minnesota, pp 1–22

Fox CH, Papouchis CM (2004) Refuting the myths. In: Fox CH, Papouchis CM (eds) Cull of the wild. A contemporary analysis of wildlife trapping in the United States. Bang Publishing, Brainerd, Minnesota, pp 23–29

Fur Institute of Canada (2015) AIHTS—humane trapping standards and animal welfare. http://furca/fur-trapping/humane-trapping-standards-and-animal-welfare/. Accessed 8 May 2016

Geist V (2006) The North American model of wildlife conservation: a means of creating wealth and protecting public health while generating biodiversity. In: Gaining ground: in pursuit of ecological sustainability. International Fund for Animal Welfare; University of Limerick, Limerick, Ireland; Guelph, Canada, pp 285–293

Geist V, Mahoney SP, Organ JF (2001) Why hunting has defined the North American model of wildlife conservation. Trans N Am Wildl Nat Resour Conf 66:175–185

Gilbert FF, Gofton N (1982) Terminal dives in mink, muskrat, and beaver. Physiol Behav 8:835–840

Gill RB (2004) Challenges of change: natural resource management professionals engage their future. In: Manfredo MJ, Vaske JJ, Bruyere BL, Field DR, Brown PJ (eds) Society and natural resources: a summary of knowledge. Modern Litho, Jefferson City, MO, pp 35–46

Gomez L, Leupen BTC, Theng M, Fernandez K, Savage M (2016) Illegal otter trade: an analysis of seizures in selected Asian countries (1980–2015). TRAFFIC, Petaling Jaya, Selangor, Malaysia

Hall ER (1981) The mammals of North America, 2nd edn. Wiley, New York

Harrop SR (2000) The international regulation of animal welfare and conservation issues through standards dealing with the trapping of wildlife animals. J Environ Law 12:333–360

International Association of Fish and Wildlife Agencies (IAFWA) (2005) Trapper education manual: a guide for trappers in the United States. http://www.fishwildlife.org/files/Trapping_Student_Manual_2005.pdf. Accessed 21 June 2016

International Association of Fish and Wildlife Agencies (IAFWA) (2015) Trapping matters [video]. https://www.youtube.com/watch?v=2R61brsbSy8. Accessed 9 April 2017

Iossa G, Soulsbury CD, Harris S (2007) Mammal trapping: a review of animal welfare standards of killing and restraining traps. Anim Welf 16:335–352

ISO 10090-4 (1999) International standards for animal (mammal traps): methods for testing restraining traps. https://www.iso.org/obp/ui/#iso:std:iso:10990:-5:en. Accessed 21 June 2016

Jotham N, Phillips RL (1994) Developing international trap standards: a progress report. Proc Vertebr Pest Conf 16:308–310

Kimber K, Kollias GV (2005) Evaluation of injury severity and hematologic and plasma biochemistry values for recently captured North American river otters (*Lontra canadensis*). J Zoo Wildl Med 36:371–384

Koelewijn HP, Perez-Haro M, Jansman HAH, Boerwinkel MC, Bovenschen J, Lammertsma DR, Niewold FJJ, Kuiters AT (2009) The reintroduction of the Eurasian otter (*Lutra lutra*) into the Netherlands: hidden life revealed by noninvasive genetic monitoring. Conserv Genet 11:601–614

Kruuk H (2006) Otters: ecology, behavior, and conservation. Oxford University Press, New York

Mahoney SP (2004) The seven sisters: pillars of the North American wildlife conservation model. Bugle 21:141–144

Mahoney SP, Geist V, Organ J, Regan R, Batcheller GR, Sparrowe RD, McDonald JE, Bambery C, Dart J, Kennamer JE, Keck R, Hobbs D, Fielder D, De Gayner G, Frampton J (2008) The North American model of wildlife conservation: enduring achievement and legacy. In: J Nobile, MD Duda (eds). Strengthening America's hunting heritage and wildlife conservation in the 21st century: challenges and opportunities, pp 7–24. White house conference on North American wildlife policy. Sporting conservation council. Retrieved from http://www.fws.gov/whhcc/doc/2009-29110-16517_Report_2009-10-13-17-19-00.pdf. Accessed 1 Dec 2016

Melquist WE, Hornocker MG (1979) Methods and techniques for studying and censoring river otter populations. Univ Idaho for Wildlife and Range Esp Stri Tech Rep 8:17

Muth RM, Zwick RR, Mather ME, Organ JF, Daigle JJ, Jonker SA (2006) Unnecessary source of pain and suffering or necessary management tool: attitudes of conservation professionals toward outlawing leghold traps. Wildl Soc Bull 34:706–715
Northcott TH, Slade D (1976) A live-trapping technique for river otters. J Wildl Manag 40:163–164
Ó Néill L, De Jongh AD, Ozolinš J, De Jong T, Rochford J (2007) Minimizing leg-hold trapping trauma for otters using mobile phone technology. J Wildl Manag 71:2776–2780
Olsen GH, Linscombe RG, Wright VL, Holmes RA (1996) Reducing injuries to terrestrial furbearers by using padded foothold traps. Wildl Soc Bull 16:303–307
Organ JF, Decker, T, Langlois S, Mirick PG (2015) Trapping and Furbearer Management in North American Wildlife Conservation. http://www.fishwildlife.org/files/Trap-Fur-Mgmt_final.pdf. Accessed 9 April 2017
Organ JF, Decker DJ, Stevens SS, Lama TM, Doyle-Capitman C (2014) Public trust principles and trust administration functions in the North American model of wildlife conservation: contributions of research. Hum Dimens Wildl 16:407–416
Organ JF, Geist V, Mahoney SP, Williams S, Krausman PR, Batcheller GR, Decker TA, Carmichael R, Nanjappa P, Regan R, Medellin RA, Cantu R, McCabe RE, Craven S, Vecellio GM, Decker DJ (2012) The North American model of wildlife conservation. Technical review 12-04. The Wildlife Society, Bethesda, MD
Organ JR, Mahoney S, Geist V (2010) Born in the hands of hunters: the North American model of wildlife conservation. Wildl Prof 4:22–27
Perrin MR, Carranza ID (1999) Capture, immobilization and measurements of the spotted-necked otter in Natal Drakensberg, South Africa. S Afr J Wildl Res 29:52–52
Peterson NM, Nelson MP (2016) Why the North American model of wildlife conservation is problematic for modern wildlife management. Hum Dimens Wildl. doi:10.1080/10871209.2016.1234009
Princen SBM (2004) Exporting regulatory standards: the cases of trapping and data protection. In: Knodt MD, Princen SBM (eds) Understanding the EU's external relations. Routledge, London, pp 142–157
Proulx G (1999) Review of current mammal trap technology in North America. In: Proulx G (ed) Mammal trapping. Alpha Wildlife Publications, Alberta, Canada
Proulx G, Cattet MRL, Powell RA (2012) Humane and efficient capture and handling methods for carnivores. In: Boitani L, Powell RA (eds) Carnivore ecology and conservation: a handbook of techniques. Oxford University Press, London, pp 70–129
Proulx G, Rodtka D, Barrett MW, Cattet M, Dekker D, Moffatt E, Powell RA (2015) Humaneness and selectivity of killing neck snares used to capture canids in Canada: a review. Can Wildl Biol Manag 4:55–65
Proulx G, Rodtka D (2015) Predator bounties in western Canada cause animal suffering and compromise wildlife conservation efforts. Animals 5:1034–1046
Prukop J, Regan RJ (2005) In my opinion: the value of the North American model of wildlife conservation—an International Association of Fish and Wildlife Agencies position. Wildl Soc Bull 33:374–377
Responsive Management (2015a) Trap use, furbearers trapped, and trapper characteristics in the United States in 2015. Harrisonburg, Virginia
Responsive Management (2015b) How to talk to the public about hunting: research based communication strategies. Harrisonburg, Virginia
Rochlitz, I (2010) The impact of snares on animal welfare. Cambridge University Animal Welfare Information Service, Cambridge, United Kingdom. http://onekind.org/uploads/publications/onekind_report_on_snaring_chapter_1.pdf. Accessed 1 Dec 2016
Rothschild DM, Serfass TL, Seddon WL, Hegde L, Fritz RS (2008) Using fecal glucocorticoids to assess stress levels in captive river otters. J Wildl Manag 72:138–142
Sax JL (1970) The public trust doctrine in natural resource law: effective judicial intervention. Michigan Law Review 68:471–566
Scott A (1999) Trust law, sustainability, and responsible action. Ecol Econ 31:139–154

Serfass TL (1984) Ecology and feeding relationships of river otters in north eastern Pennsylvania. MS thesis, East Stroudsburg University, East Stroudsburg, Pennsylvania, USA

Serfass TL, Brooks RP, Swimley TJ, Rymon LM, Hayden AH, Serfass TL, Brooks RP, Swimley TJ, Rymon LM, Hayden AH (1996) Considerations for capturing, handling, and translocating river otters. Wildl Soc Bull 24:25–31

Serfass TL, Bohrman JA, Stevens SS, Bruskotter JT (2014) Otters and anglers can share the stream! The role of social science in dissuading negative messaging about reintroduced predators. Hum Dimens Wildl 19:532–544

Shirley MG, Linscombe RG, Sevin LR (1983) A live trapping and handling technique for river otter. Proc Annu Conf Southeast Assoc Fish Wildl Agencies 37:182–189

Smithsonian (n.d.) North American river otter (*Lontra canadensis*). http://naturalhistory.si.edu/mna/image_info.cfm?species_id=142. Accessed 9 Aug 2016

Stevens SS, Serfass TL (2008) Visitation patterns and behaviors or neartic river otters (*Lontra canadensis*) at latrines. Northeast Nat 15:1–12

Trapping Matters Workshop (2016) Oregon Chapter of The Wildlife Society homepage. https://ortws.wordpress.com/2016/05/31/trapping-matters-workshop/. Accessed 11 April 2017

Taylor R, Serfass T, Wong J (2016) Stress levels in captured river otters (*Lontra canadensis*) decreased after transportation to reintroduction sites. J Zoo Wildl Med 47:1057–1060

Treves A, Chaprom G, López-Bao JV, Shoemaker C, Goeckner AR, Bruskotter JT (2015) Predators and the public trust. Biol Rev. doi:10.1111/brv.12227

United States Department of Commerce (1997) European Union Humane trapping standards agreement. http://tcc.export.gov/Trade_Agreements/All_Trade_Agreements/exp_002820.asp. Accessed 21 June 2016

White BH, Decker T, O'Brien MJ, Organ JF, Roberts NM (2015) Trapping and furbearer management in North American wildlife conservation. J Environ Stud 72:756–769

Van Der Zee D (1982) Density of cape clawless otters *Aonyx capensis* (Shinz, 1821) in the Tsitsikamma coastal National Park. S Afr J Wildl Res 12:8–13

Vantassel S, Hiller TL, Powell KDJ, Hygnstrom SE (2010) Using advancements in cable-trapping to overcome barriers to furbearer management in the United States. J Wildl Manag 74:934–939

Chapter 31
Otters in Captivity

Jan Reed-Smith and Shawn Larson

Abstract In 2015 there were a minimum of 1621 otters of eight species (Asian small-clawed otter *Aonyx cinereus*, Sea otter *Enhydra lutris*, Spotted-necked otter *Hydrictis maculicollis*, N.A. river otter *Lontra canadensis*, Marine otter *Lontra felina*, Neotropical otter *Lontra longicaudis,* Eurasian otter *Lutra lutra,* Smooth-coated otter *Lutrogale perspicillata*, and Giant otter *Pteronura brasiliensis*) held in major zoos and aquariums. Husbandry, or the care and management of otters in captivity, has improved gradually over the last two decades. Fifty years ago recommendations were made that outlined the need for large, complex land areas and other features key to ensuring high levels of otter welfare, which many ex situ facilities, such as zoos, aquariums, and rehabilitation facilities, have now adopted. Increased welfare of captive otters is due partially from improved environmental conditions such as better habitat design, and partially from improved understanding of otter nutritional and health needs. This chapter discusses otter species kept in captivity and focuses on their care and well-being. This includes husbandry improvements over the last few decades, as well as some of the continuing concerns regarding otter welfare in zoos, aquariums, and rehabilitation situations. We address the benefit of multi-institutional research into welfare issues and the potential benefit of improving habitat designs allowing otters a greater choice and control over their environment.

31.1 Introduction

In 2015 there were a minimum of 1621 otters of eight species (Asian small-clawed otter *Aonyx cinereus*, sea otter *Enhydra lutris*, spotted-necked otter *Hydrictis maculicollis*, North American (NA) river otter *Lontra canadensis*, marine otter *Lontra*

J. Reed-Smith (✉)
Otters in Zoos and Aquariums Task Force, IUCN/SSC Otter Specialist Group, Lake Odessa, 48849 Michigan, USA
e-mail: jrsotter@iserv.net

S. Larson
Seattle Aquarium, Seattle, Washington, USA

A. Butterworth (ed.), *Marine Mammal Welfare*, Animal Welfare 17,
DOI 10.1007/978-3-319-46994-2_31

felina, neotropical otter *Lontra longicaudis*, Eurasian otter *Lutra lutra*, smooth-coated otter *Lutrogale perspicillata*, and giant otter *Pteronura brasiliensis*) (ISIS 2015) held in major zoos and aquariums. The number of animals actually held in captivity is in reality much higher as the International Species Information System (ISIS) database reflects only those animals reported by member institutions. For example, the Asian (Oriental) small-clawed otter, the most commonly held species in captivity, has 1567 animals listed in the 2015 international studbook (Duncan 2016 Pers Comm).

The second most commonly held species is the North American river otter with 305 individuals held, and the least common is the marine otter with one captive individual (ISIS 2015). There are currently at least 78 sea otters in captivity in the United States, Canada, Europe, and Japan (Casson 2015; Ishihara 2016, Pers Comm.). These numbers are likely to be underestimates as reported facility holdings represent many of the largest zoos and aquariums and do not take into account otters held at nonmember institutions, small facilities, wildlife parks, or rehabilitation facilities.

31.2 Zoos and Aquariums

The creation of studbooks (genealogical registries of animal's parentage and birth location) has benefited the ex situ management of otters greatly. These records have allowed zoos and aquariums to manage the gene pool to limit inbreeding and the tracking of breeding success, as well as other life history parameters such as longevity. There are several examples of this successful management: In 2000 roughly 39% of the NA river otters in the ISIS database were captive born, and by 2015 50% were captive born ($n = 259$) (Hamilton 2016, Pers Comm). The median life expectancy for this species in captivity is 11.7 years, and the longevity record is 21.64 years (Hamilton 2016). In contrast the life span for this species in the wild is approximately 8–9 years. Robin (1987) reported that in 1985, 50% of Eurasian otters were captive born and in 2015 approximately 77% were captive born (Rey 2016, Pers Comm.) indicating a shift in the origin of captive animals with time; median life expectancy in captivity is 10 years and the longevity record is 19 years (Capber 2007). In 2015, 96% of the Asian small-clawed otters were reported to be captive born, whereas the 1961–1970 studbook data reported only 26% as captive born; life expectancy in the wild is unknown, while the median life span for this species in the US captive population is 14 years for males and 11.2 years for females (Duncan 2016, Pers Comm.).

Sea otters were first held in captivity in 1955 at the Woodland Park Zoo in Seattle, Washington (Brennan and Houck 1996). Because of their success in housing sea otters, these animals have become increasingly popular exhibit animals over the decades. The numbers of sea otters held in captivity grew steadily in the last half of the twentieth century in the United States, with one event dramatically increasing their numbers. In 1989, as a result of the Exxon Valdese Oil Spill in Prince William

Sound, Alaska, 37 of the hundreds of rescued northern sea otters were deemed non-releasable and placed in facilities around the world (Gruber and Hogan 1990; VanBlaricom et al. 2015). Since then the number of captive sea otters has remained steady in the United States at approximately 40 animals with approximately equal numbers of northern (*E.l. kenyoni*) and southern sea otters (*E.l. nereis*). More recently, in the twenty-first century, there has been a shift with increasing numbers of wild southern and decreasing numbers of northern sea otters being placed into captivity due to an increase in the numbers of stranded and non-releasable southern sea otter pups (Johnson and Mayer 2015).

Sea otters in captivity exhibit a wide range of activities associated with a full life history, such as performing a range of daily behaviors, reaching expected longevity, and breeding successfully. This is thought to be due, in part, to the fact that the captive environments may mimic the small home ranges and tight family social groups found in wild sea otters (Larson and Bodkin 2015). Sea otters are highly social and should not be kept alone, ideally being kept in female-only groups or groups of females with one male. Some facilities keep male-only groups, and this mix often stimulates aggression, which must be managed by providing access to large pools with areas for individuals to escape from each other or other methods such as contraceptive drugs to reduce testosterone and aggression between mature males. The longevity record for this species is 27 years, while life span in the wild for males is 10–15 years and for females is 15–20 years (Casson 2015). Sea otters have been successfully bred in captivity since the late 1970s, with the first captive sea otter pups raised from conception to adulthood (2 years of age) in 1979. Eleven pups have since been born and ten have survived to adulthood (Fig. 31.1). In recent years North American facilities housing sea otters, in conjunction with the US Fish and Wildlife Service, have agreed to a moratorium on breeding of captive sea otters to preserve space required for rescued non-releasable sea otters from threatened populations listed under the Endangered Species Act (ESA) (VanBlaricom et al. 2015).

Fig. 31.1 Sea otter (*Enhydra lutris*) and pup. *Image credit*: P. McMahon, Seattle Aquarium

Fig. 31.2 Eurasian otter (*Lutra lutra*) enclosure at Otter-Zentrum, Hankensbüttel, Germany; an example of natural exhibits. *Image credit*: J. Reed-Smith

Husbandry, or the care and management of otters, has improved gradually over the last two decades. Early recommendations by Duplaix-Hall (1972, 1975) and Crandall (1964, 1974) first pointed out the need for large, complex land areas and other features key to ensuring high levels of otter welfare (Fig. 31.2). Later publications expanded and reemphasized these features (Reuther 1991; Reed-Smith 1994, 2012; Melissen 2000). These publications, increased experience with various otter species, and a growing use of a wide variety of enrichment (Coe 2009) and training methods leading to voluntary participation by otters in routine health examinations (Morabito and Dunn 2007, 2008; Scherrens 2014) have led to overall improvement of otter care and welfare.

However, there are still some areas of concern regarding overall welfare of otters in ex-situ facilities, which include:

- Too many facilities are housing otters in small barren- or concrete-based enclosures (Fig. 31.3) and/or inappropriately devoting too much space to water for the majority of otter species (the sea otter is the exception).
- Exposure to loud noises or inappropriate light cycles (Morgan and Tromborg 2007) and to large crowds when otters have no access to places of refuge. Inadequate denning sites for pregnant females continue to be an issue for some species and in some situations. For a good overview of zoo animal welfare and

Fig. 31.3 An otter enclosure which has only concrete as a surface. *Image credit*: J. Reed-Smith

the possible ramifications of captive environments that should be considered consult (Morgan and Tromborg 2007).

- Latrine (areas where otters deposit feces and urine) management, scent, and access to variable substrates are important for otter welfare. Otters are scent oriented and latrines play an important signaling role (except for the sea otter which is entirely marine and does not scent mark), such that the daily cleaning and disinfection of these areas can disrupt this communication tool (Morgan and Tromborg 2007). The ability to role/rub on various surfaces is behaviorally important to the maintenance of a healthy coat. Otters also like to dig and root around, and the health of their feet can be compromised by continually damp or wet surfaces (except for the sea otter). It is important to consider whether otters that are kept primarily on concrete surfaces are able to perform many of these activities. If not, then this is probably not appropriate and alternative substrates should be provided.
- Appropriate exhibit size. Too often otters are viewed as small animals and thus kept in small spaces. Instead their comparatively large home ranges in the wild should be considered, and sufficient space must be provided.
- Too often otters are confined for long periods in small, unstimulating holding spaces. Many facilities have turned to innovative enrichment techniques to deal with this, but exploring options that allow otters access 24 h a day to both the

Fig. 31.4 (**a** and **b**) Pueblo Zoo offers *L. canadensis* "on (a) and off exhibit (b)" yards allowing for 24-h access, temporary separation, or mixing of group composition. *Image credit*: J. Reed-Smith

holding and enclosure areas should be considered and further tested (Figs. 31.4a,b and 31.6a,b).

- There are reported perceptions by field biologists that Asian small-clawed otter cubs held in nonregulated situations (wildlife facilities and as pets) outside the United States sometimes look stunted, which has led to speculation that unregulated inbreeding, poor nutrition, and/or chronic stress may be occurring in these populations (Morgan and Tromborg 2007).
- The keeping of Asian small-clawed otters as pets should be discouraged. This is a practice of growing concern particularly in parts of Asia.
- Inadequate medical care at poorly regulated facilities.

Zoos and aquariums have improved the health care of otters consistently over the years. Partially this is due to improved environmental conditions from better exhibit design and partially to improved understanding of otter nutritional needs and health concerns. This is an area where veterinarians and nutritionists have been able to contribute greatly to the welfare of otters. Improvements in this area also have been aided by the trend to adapt operant conditioning (training) techniques to solicit cooperation from otters in health-care procedures, eliminating the need to forcibly capture and restrain or anesthetize the animal. These training practices have proven to be effective, as well as potentially enriching for the otters, by stimulating their inquisitive natures (Morabito and Dunn 2007, 2008; Scherrens 2014). As a result of reproductive physiology work conducted in zoos, data has been collected on verifying pregnancy and gestational length allowing facilities to better anticipate births and prepare for them (Larson et al. 2003; Da Silva and Larson 2005; Bateman et al. 2009). This research has also led to better understanding of other aspects of reproductive physiology and the use and impact of contraception (Larson et al. 2003, 2012; Bateman and Swanson 2007, 2013, 2014; Bateman et al. 2009, 2011).

The ability to engage in a wide range of behaviors is generally the first focus of captive animal welfare evaluation and often includes aspects of behavior and sociability. As mentioned above the tendency of many facilities to use too much concrete or other hard surfacing in exhibits continues to be a problem. This limits the animals' ability to engage in exploratory, foraging, digging, grooming, social, and play behavior. The lack of complexity in many exhibits inhibits these behaviors, and this may, or may not, lead to stereotypical behavioral patterns. Many institutions have enrichment programs in place to stimulate a wide range of behaviors and to address stereotypies that may, or may not, be due to stress. This is well intended, sometimes successful (Foster-Turley and Markowitz 1982; Nelson 2010), but too often becomes routine and of limited use (Morabito and Bashaw 2012). Further studies and documentation of the success of these programs are required. To date, the majority of enrichment studies have focused on changes in how, or how often, otters are fed (Foster-Turley and Markowitz 1982; Ross 2002; Hawke et al. 2004; Hasenjager 2011). Results as to its efficacy are mixed due to limited sample size, species-specific applications, and differing methodologies. At least in one study, Morabito and Bashaw (2012) suggested that pairing an auditory cue with feedings could reduce anticipatory behaviors associated with meal times. Further study into ways to enrich otter feeding through altered methods/timing of food delivery and allowance of more choice and engagement from the otters is required (Foster-Turley and Markowitz 1982; Gothard 2007; Morabito and Bashaw 2012).

A new direction in habitat design is currently being promoted (Coe 2005, 2012, 2014, 2016 Pers Comm; CLR Design 2016), which suggests a move toward allowing animals greater choice and control over their environment through concepts such as "rotation exhibits" (Coe 2014), "raceway networks" (Coe 2014), and what Coe (2005) calls the "unzoo." If these concepts are applied to otters in the future, particularly allowing them to move between exhibits of compatible species, overall welfare may improve via extension of their ability to choose where they want to be, an increase in their daily ranging abilities, an increase in their living-space complexity, the provision of opportunities for group members to join and separate, and by stimulating bored otters through enhanced foraging (Gothard 2007) (Figs. 31.4a,b, 31.5, and 31.6a,b). These housing options should be further explored. Other ongoing issues regarding the expression of natural behaviors, which are being actively debated, include the housing of otters in inappropriate social groupings and the inability of individuals to escape stressful situations.

Fig. 31.5 Wild spotted-necked otter or speckle-throated otter (*Hydrictis maculicollis*) in vegetated habitat selected for resting by many otter species. *Image credit*: J. Reed-Smith

Fig. 31.6 (**a**) Asian small-clawed otter (*Aonyx cinereus*) transferring between their exhibit and that of macaques via a tunnel at Rockhampton Zoo, Australia. *Image credit*: K. Hickey (**b**) Otters exiting a macaque enclosure at Rockhampton Zoo, Australia. *Image credit*: K. Hickey

31.3 Rehabilitation Programs

The successful rehabilitation of otter cubs/pups requires an extended time period (typically 12 months or more), adequate facilities, limited exposure to humans, no exposure to dogs, the ability to pair singletons with other orphans, and access to a safe place where a "soft" release (with supplemental feeding for the initial release period) can be supported (Green 1991; Yoxon 2003, 2013; McTurk and Spelman 2005; Haire 2009, 2011). This approach of soft release is often successful (Haire 2009; Yoxon 2003, 2013; Thibodeaux 2016, Pers Comm.) but unfortunately is not utilized by the majority of rehabilitators. While it is known that cubs are sometimes released prematurely (too young) leading to problems, there is almost nothing reported about the fate of these animals. The process of rehabilitating injured sea otters is well documented and in general is thought to be successful, specifically when surrogate mothers (adult captive females otters) are used to raise the orphaned pups (VanBlaricom et al. 2015; Johnson and Mayer 2015).

31.4 Conclusions: Assessing the Welfare of Otters in Captivity

The value of assessing animal welfare is widely accepted; however all too often inadequate steps are taken to assess welfare objectively. Measurement tools can assist staff in understanding whether their efforts to stimulate natural behaviors are successful. As an example Morabito and Bradshaw (2012) evaluated the influence of increasing the number of feedings for NA river otters, and they suggested that this approach may be more successful at reducing feeding-related stereotypies by pairing food delivery with a specific cue. This would eliminate the power of other cues such as keeper approach stimulating inappropriate anticipatory behavior.

More empirical research is required across institutions to assess current otter conditions. A practical first step to assessing otter welfare in zoos and aquariums could be to begin with "activity budgets" and "space usage." There has been much written on the difficulty of defining welfare (Maple and Purdue 2013) as well as the difficulties in interpreting results via measures of physiological and behavioral changes (Mason and Mendl 1993). Dawkins (2004) argues for simplifying our questions when measuring animal welfare to two basic questions asked concurrently: (1) "Are the animals healthy?" and (2) "Do they have what they want?". Addressing these questions in an objective, standardized, and multi-institutional manner would provide tools for facilities to move from acceptable or good otter welfare to great otter welfare. Creating factual, documented activity budgets for individual otters is a first step. A good second step would be to assess enclosure usage patterns for each animal by mapping where, and how long, each otter is spending its time. These two tools together would allow managers to evaluate enrichment programs, to understand the impact of exhibit design, and to make improvements based on the actions and reactions of their specific captive otters.

References

HL, Swanson WF (2007) Comparative assessment of semen processing and cryopreservation methods for genome resource banking of North American river otter spermatozoa. Proceeding of the 10th international otter colloquium. Hwacheon, South Korea, 10–16 Oct. p 47

Bateman HL, Bond JB, Campbell M, Barrie M, Riggs G, Snyder B, Swanson WF (2009) Characterization of basal seminal traits and reproductive endocrine profiles in North American River otters and Asian small-clawed otters. Zoo Biol 28:107–126

Bateman HL, Pelican K, Swanson WF (2011) Assessing the efficacy of a GnRH agonist implant (Deslorelin) to reduce reproductive and stress hormone levels and behavioural aggression in single-sex groups of captive Asian small-clawed otters. Abstract. Hystrix, Italian J Mammal supp. p 88

Bateman HL, Swanson WF (2013) Reliability of fecal progesterone analysis for pregnancy diagnosis and parturition timing in North American river and Asian small-clawed otters. Proceeding of wild musteloid conference. Published by Wildlife Conservation Research Unit, Department of Zoology, University of Oxford, 8–21 March 2013

Bateman HL, Swanson WF (2014) Re-establishing reproductive cyclicity in a female Asian small-clawed otter following Suprelorin-induced contraception: a case study. Reprod Fertil Dev 27:172

Brennan EJ, Houck J (1996) Sea otters in captivity: the need for coordinated management as a conservation strategy. Endanger Species Update 13:61–67

Capber F (2007) Veterinary care of Eurasian otters (*Lutra lutra*) at the otter breeding center Hunawihr (France). IUCN Otter Species Group Bull 24(1):47–62

Casson CJ (2015) Sea otter studbook report. In: Larson SE (ed) Sea otter conservation workshop IX proceedings. Seattle Aquarium, Seattle, Washington. 27–29 March 2015. pp 17–18

CLR Design (2016) Evolutionary thinking in habitats. Personal communication. http://www.clrdesign.com/. Accessed 9 Dec 2016

Coe J (2005) The Unzoo alternative. ARAZPA/SEAZA joint conference, 5 April 2005, Melbourne, Australia. https://www.researchgate.net/publication/263470731_The_Unzoo_Alternative. Accessed 9 Dec 2016

Coe J (2009) Collaborative enrichment. 9th international conference on environmental enrichment, Torquay, UK, 31 May–05 June 2009. Jon Coe Design, Healesville, Australia. http://www.joncoedesign.com/pub/PDFs/CollaborativeEnrichment2009.pdf. Accessed 9 Dec 2016

Coe J (2012) Design and architectures: Third generation conservation, post-immersion and beyond. Future of zoos symposium, 10–11 February 2012, Canislus College, Buffalo, NY. http://www.zoolex.org/publication/coe/design+architecture2012.pdf. Accessed 9 Dec 2016

Coe J (2014) Next generation rotation exhibits: raceway networks and space to explore. Power point presentation, Zoo and aquarium annual conference/New Zealand branch conference, 25–28 March 2014, Auckland Zoo, New Zealand. http://www.joncoedesign.com/pub/Talks/RotationZAA2014/index.htm. Accessed 9 Dec 2016

Crandall LS (1964 & 1974) The management of wild mammals in captivity: otters. Univ. Chicago Press, Chicago, London, pp 337–348

Dawkins MS (2004) Using behaviour to assess animal welfare. Anim Welf 13:S3–S7

Da Silva IM, Larson S (2005) Predicting reproduction in captive sea otters (*Enhydra lutris*). Zoo Biol 24:73–81

Duplaix-Hall N (1972) Notes on maintaining river otters in captivity. Int Zoo Yearbook 12:178–181

Duplaix-Hall N (1975) River otters in captivity: a review. In: Martin RD (ed) Breeding endangered species in captivity. Academic, London, pp 315–327

Foster-Turley P, Markowitz H (1982) A captive behavior enrichment study with Asian small-clawed river otters (*Aonyx cinerea*). Zoo Biol 1:29–43

Gothard N (2007) What is the proximate cause of begging behaviour in a group of Asian short-clawed otters? IUCN Otter Species Group Bull 24(1):14–35

Green RE (1991) Otters: a case study in rehabilitation. Wild aid guide to Scotland. Skye Environmental Centre Ltd, Isle of Skye

Gruber JA, Hogan ME (1990) Transfer and placement of non-releasable sea otters in aquariums outside Alaska. In: Bayha K, Kormendy J (eds) Sea otter symposium: proceedings of a symposium to evaluate the response effort on behalf of sea otters after the T/V Exxon valdez oil spill into Prince William sound. Anchorage, Alaska, 17–19 April 1990. US fish and wildlife service, biological report 90 (12):428–431. US Department of the Interior Fish and Wildlife Service, Washington, DC

Haire M (2009) Study of post-release data in the North American River otter (*Lutra canadensis*). Wildl Rehab Bull 27(1):3–17

Haire, M (2011) Successful hand-rearing and rehabilitation of north american river otter (*Lontra canadensis*), sections 1–3. http://www.otterspecialistgroup.org/Library/TaskForces/OCT.html. Accessed 9 Dec 2016

Hasenjager MJ (2011) Effects of increased feeding frequency on captive North American river otter (*Lontra canadensis*) Behavior. MSc thesis, Michigan State University, p 86

Hawke L, Lauer P, Bartholomeusz D, Steen Z (2004) Effect of increased food dispersal and random feeding time/place on stereotyped behaviors in otters at Adelaide Zoo. http://torontozoo.com/meet_animals/enrichment/otter_enrichment.htm. Accessed 9 Dec 2016

ISIS (2015) International species information system. http://www2.isis.org/Pages/Home.aspx. Accessed 9 Dec 2016

Johnson A, Mayer K (2015) The value of rescuing, treating, and releasing live-Stranded Sea otters. In: Larson SE, Bodkin JL, Van Blaricom GR (eds) Sea otter conservation. Academic, New York, pp 235–255

Larson S, Belting T, Rifenbury K, Fisher G, Boutelle SM (2012) Preliminary findings of fecal gonadal hormone concentrations in six captive sea otters (*Enhyra lutris*) after Deslorelin implantation. Zoo Biol 00:1–14. doi:10.1002/zoo.21032

Larson SL, Bodkin JL (2015) The conservation of sea otters: a prelude. In: Larson SE, Bodkin JL, VanBlaricom GR (eds) Sea otter conservation. Elsevier Scientific Publications, USA, pp 2–20

Larson S, Casson CJ, Wasser S (2003) Noninvasive reproductive steroid hormone estimates from fecal samples of captive female sea otters (*Enhydra lutris*). Gen Comp Endocrinol 134:18–25
Maple TL, Perdue BM (2013) Zoo animal welfare. Springer, New York
Mason G, Mendl M (1993) Why is there no simple way of measuring animal welfare? Anim Welf 2(4):301–319
McTurk D, Spelman L (2005) Hand-rearing and rehabilitation of orphaned wild Giant otters, *Pteronura brasiliensis*, on the Rupununi River, Guyana, South America. Zoo Biol 24:153–167
Melissen A (2000) Eurasian otter (*Lutra lutra*) husbandry guidelines. Otterpark Aqualutra, Netherlands
Morabito PM, Dunn M (2007) Injection training 1.1 North American river otters (*Lontra canadensis*) using a PVC chute. Anim Behav Manag Alliance Wellspring 8(4):12–13
Morabito PM, Dunn M (2008) Injection training 1.1 North American river otters using a PVC chute. Animal Keepers' Forum 35:106–108
Morabito PM, Bashaw MJ (2012) A survey of abnormal repetitive behaviors in North American River otters housed in zoos. J Appl Anim Welf Sci 15(3):208–221
Morgan KN, Tromborg CT (2007) Sources of stress in captivity. Appl Anim Behav Sci 102:262–302
Nelson KO (2010) The Effects of Environmental Enrichment on a Stereotyped Swimming Pattern for a North American River Otter (*Lontra canadensis*). River Otter J XIX(1):8–11. River Otter Alliance. http://www.otternet.com/ROA/directory.htm. Accessed 9 Dec 2016
Reed-Smith J (1994) North American river otter (*Lontra canadensis*) husbandry notebook. John Ball Zoo, Michigan
Reed-Smith J (2012) North American river otter (*Lontra canadensis*) husbandry notebook, 4th edn. otterspecialistgroup.org http://www.otterspecialistgroup.org/Library/TaskForces/OCT/NARO_4th_Edition_Section_1_Chapters_1_to_6.pdf. Accessed 9 Dec 2016
Reuther C (1991) Otters in captivity—a review with special reference to *Lutra lutra*. In: Reuther C, Röchert R (eds) Proceedings V international otter colloquium, Hankensbüttel, Germany . Habitat (Aktion Fischotterschutz) 6:269–307
Robin K (1987) European river otter studbook. IUCN Otter Specialist Group Bull 2:42–44
Ross R (2002) The effect of a simple feeding enrichment strategy on the behavior of two Asian small-clawed otters (*Aonyx cinerea*). Aquat Mamm 28(2):113–120
Scherrens A (2014) Inspiring passion for the natural world, Birmingham Zoo AZA otter keeper workshop 2014 presentation, Denver, Colorado, 15–19 March 2014
VanBlaricom GR, Belting TF, Triggs LH (2015) Sea otter sin captivity: applications and implications of husbandry development, public display, scientific research and management, and rescue and rehabilitation for sea otter conservation. In: Larson SE, Bodkin JL, VanBlaricom GR (eds) Sea Otter Conservation. Elsevier Scientific, USA, pp 198–234
Yoxon G (2003) Caring for wild otters (2003) In: Conroy J, Gutleb A, Ruiz Olmo J, Yoxon G (eds) proceedings of the European otter conference 'Return of the otter in Europe—where and how? 30 June–05 July 2003. J International Otter Survival Fund 2
Yoxon G (2013) Exploring the depths: Eurasian otter rescue and rehabilitation. In: Parsons ECM, Bauer A, McCafferty D, Simmonds MP, Wright AJ (eds) An introduction to marine mammal conservation and biology. Jones & Bartlett, Burlington, MA

Chapter 32
Pathology of Marine Mammals: What It Can Tell Us About Environment and Welfare

Antonio Fernández, Yara Bernaldo de Quirós, Simona Sacchini, and Eva Sierra

Abstract The study of animal welfare continues to struggle with two persistent, interrelated problems: how to define animal welfare, and how to determine which measures should be used to evaluate it. One potential indicator of an animal's welfare is the presence or absence of stress, or anything that seriously threatens homeostasis. The actual or perceived threat to an organism, the stressor, and the response will determine when stress becomes distress. Pathologically speaking, this occurs when tissue damage (lesions) and disease appears as a result of a severe (acute) or prolonged (chronic) stress response. Veterinary Pathology is a diagnostic tool, which looks for and identifies lesions involved in disease/s as well as determines cause/s of death. In this respect Veterinary Pathology could be a diagnostic tool of cetaceans' (whales, dolphins and porpoises) welfare. In recent years, attention has been focused on how human activities may affect cetaceans, particularly through use of improving methods and techniques to identify and classify lesions, and to understand mechanisms and causes, in order to associate stressors with distress. In this chapter we discuss three test cases (beaked whales mass stranding and antisubmarine mid-frequency active sonar, active stranding and capture myopathy, fingerprints in the brain of dolphins). We show what animal pathology can do to contribute to animal welfare assessment in stranded cetaceans, ranging from death to improved population welfare.

32.1 Introduction

Moberg and Mench (2000) stated that the study of animal welfare continues to struggle with two persistent, interrelated problems: how to define animal welfare and how to determine which measures should be used to evaluate welfare. One potential indicator of an animal's welfare is the presence or absence of stress.

A. Fernández (✉) • Y.B. de Quirós • S. Sacchini • E. Sierra
Veterinary Histology and Pathology, Institute of Animal Health, University of Las Palmas de Gran Canaria, Canary Islands, Spain
e-mail: antonio.fernandez@ulpgc.es

A. Butterworth (ed.), *Marine Mammal Welfare*, Animal Welfare 17,
DOI 10.1007/978-3-319-46994-2_32

The term stress represents the effects of anything that seriously threatens homeostasis (Selye 1956). The actual or perceived threat to an organism is referred to as the stressor, and the response to the stressor is called the stress response. Therefore, we could differentiate between a nonthreatening stress response (often referred to as good stress) and a biological state where the stress response has a deleterious effect on the individual's welfare (bad stress). Simply stated, our challenge is to determine when stress becomes distress (Moberg and Mench 2000) or, pathologically speaking, when tissue damage (lesions) and disease appear as a result of a severe (acute) or prolonged (chronic) stress response (Selye 1956). During distress, this impairment of function places the animal in a state that renders it vulnerable to a number of pathologies as an indicator of a threat to animal welfare (Moberg and Mench 2000).

The European College of Veterinary Pathologists (ECVP) defines veterinary pathology as a discipline-oriented specialization which uses the morphological recognition and functional interpretation of pathological conditions in animals, through analysis of the pathomorphological substrate, to attempt to define specific pathological processes, including aetiology and pathogenesis. Basically, veterinary pathology is a diagnostic tool, which looks for and identifies lesions involved in disease(s) as well as determines cause(s) of death. If we consider cause(s) as stressors and lesions as indicators of distress, why may we not consider also veterinary pathology as a diagnostic tool of animal welfare and, in our particular case, in cetaceans (whales, dolphins and porpoises)?

In the Canary Islands waters, more than 30 different cetacean species have been identified. Of these, at least 26 have been found stranded on the coasts of the Canary Islands. There are historical references to cetacean stranding in the Canary Islands, but a more systematic study and scientific analysis has begun to take place only in the last two decades. A specific unit of veterinary pathology based at the Institute of Animal Health (University Las Palmas) has been carrying out studies and research on stranded cetaceans in the Canary Islands, as well as in different parts of world. Necropsies and ancillary multidisciplinary laboratory analyses are carried out on each animal, attempting to determine whether either natural or human-related activities have been involved in stranding or the cause of death. Even though the information is obtained from individual animals, a large amount of data has been gained regarding species, pathologies and aetiologies, which has contributed to a better understanding of the health status of these species.

The veterinary and forensic pathological methods applied to each stranded cetacean are only a piece of a multidisciplinary comparative bio-eco-health discipline (including welfare). Cetaceans can be considered one of the best worldwide environmental health bioindicators of the oceans, and applying "holistic pathological analysis" can contribute scientifically to the conservation of these marine mammals. We are convinced that these cetacean studies may represent good examples of the concept "one health, one environment" which include human and animal welfare, as a parameter to be evaluated as human interactions with cetaceans may become a welfare conflict of interests.

The Canary Islands, as is true for many coastal areas in the world, are densely populated, and the anthropogenic impact on the marine environment includes the effects of maritime traffic, transport of cargo and people, the fishing industry and tourists observing cetaceans (whale watching). Marine life, including cetaceans, is affected by chemical pollution (Garcia-Alvarez et al. 2014) caused by dumping waste in the ocean (urban, industrial and agricultural) and acoustic pollution caused by maritime traffic, hydrocarbon prospecting and extraction and civil and military use of sonar (Fernández et al. 2005). In order to try to assess the impact of these potentially detrimental activities on cetaceans, systematic pathological studies have been carried out.

Briefly, between October 1999 and September 2005, these analyses allowed us to classify 128 out of 233 stranded cetaceans (54.9%) into either anthropogenic or non-anthropogenic pathological categories (Arbelo et al. 2013). The most important anthropogenic causes of death included interactions with fishing gear (including by-catch) (Fig. 32.1a), foreign body pathology (Fig. 32.1b,c), atypical strandings of beaked whales associated with military manoeuvres and sonar (Fernandez et al.

Fig. 32.1 (**a**, **b**) Striped dolphin (*Stenella coeruleoalba*) stranded on Tenerife (Canary Islands, Spain) in 2001. The first stomach was distended and filled with more than 50 plastic bags. (**c**) Oesophageal perforation by a hook in an Atlantic-spotted dolphin (*Stenella frontalis*) stranded on Gran Canaria (Canary Islands, Spain) in 2001. (**d**) A sperm whale (*Physeter macrocephalus*) died in the Canary Islands in 1996 after being struck by a high-speed ferry (*Image credit: IUSA, ULPGC*)

2004, Fernández et al. 2005) and collisions with vessels (particularly in sperm whales) (Fig. 32.1d) (Sierra et al. 2014). On the other hand, natural (non-anthropogenic) causes of death included loss of nutritional status (starvation) and a range of infectious and non-infectious diseases. The conclusions of this work indicated that anthropogenic causes were linked to 33% of the 128 stranded cetaceans investigated.

In a follow-up study, between 2006 and 2012, using improved forensic field methodologies and laboratory technologies, it was possible to classify 236/320 (73.75%) of the cetaceans studied into either the anthropogenic or the non-anthropogenic pathological categories (Díaz-Delgado 2015). Direct human activity was responsible for approximately 18% of cetacean deaths in this study, while "natural" pathologies would account for approximately 82%. The decreasing of anthropogenic causes was attributed partially to policy measures, which were taken to ban military acoustic activities in the Canary Islands.

In recent years, we have been focussing our attention on how human activities may affect cetaceans, particularly through the use of improving methods and techniques to identify and classify lesions, and to understand mechanisms and causes, in order to associate stressors with distress. At the same time, we, as veterinary pathologists, are actually interested in showing a wide, sometimes nonscientific, audience how dealing scientifically with the death of a cetacean can help to preserve the life of the animal, either individually (individual welfare) or collectively (through understanding of issues affecting populations and hence conservation).

In this way, we believe that collection of important information from these animals, which are found dead in the wild, can lead from bad news to good news. We present here three examples of what we can do through the diagnosis of lesions (representing distress) and increasing our understanding of cause(s) (representing stressor(s)) in stranded cetaceans.

Firstly, we start with the investigation of a mass stranding of beaked whales (bad news) which results in the official ban of military sonar around the Canary Islands (good news).

Secondly, we describe live stranded cetaceans and their management during stranding, and their rehabilitation, as an example with a range of outcomes from good news – rehabilitation, to bad news—death due to capture myopathy.

Thirdly, we describe our search for *fingerprints of distress* in different nuclei of the cetacean central nervous system (CNS), which may provide information on acute and/or chronic diseases and/or stressors.

32.2 Example 1: Sound as a Stressor (Beaked Whales (BW) Mass Stranding and Antisubmarine Mid-frequency Active Sonar (MFAS))

Marine mammals use sound for communication and foraging. Thus, they may respond to anthropogenic sound exposure as a stressor. There are several reports of marine mammals showing behavioural, acoustic and physiological responses to anthropogenic sound (Nowacek et al. 2007). After the development of military anti-submarine mid-frequency active sonar in the 1950s, and coincidental in time and geographically with military exercises where this kind of sonar was used, beaked whales, especially Cuvier's beaked whale (*Ziphius cavirostris*), started to strand in large numbers (D'Amico et al. 2009).

Beaked whales were for long an almost unknown species. The little knowledge of these species came mostly from stranded whales. However, something changed in the late 1990s and early twenty-first century; beaked whales were stranded in large numbers along several km of the same coastline and coincidental in time and space with military manoeuvres. The first stranding of this type (an atypical mass stranding) that caught public attention occurred in Greece in May 1996, when 12 Cuvier's beaked whales were stranded in temporal and spatial association with North Atlantic Treaty Organization (NATO) exercises (Frantzis 1998).

The animals were stranded alive within a 36 h period and spread along 38 km of coast. This spread in time and location is different from typical mass strandings, where animals arrive together and simultaneously at the same spot of the beach (Frantzis 1998). During the military exercises, low- and mid-frequency military sonar was utilized. Four years later, in March 2000, 14 beaked whales and 3 additional cetaceans of other species were stranded within a 36 h period and spread along 240 km of coast, coincidental in time and space with the US naval exercises in Bahamas, where up to five ships used mid-frequency sonar (Evans and England 2001). In that same year, in May, three Cuvier's whales were stranded between 10 and 14 May on two different islands of the Madeira Archipelago, Portugal. This stranding event was coincidental in time and space with NATO naval exercises, although NATO did not provide information about sonar use during this exercise (Cox et al. 2006).

The strandings of these rare species, in temporal and spatial association with military exercises with deployment of mid-frequency military sonar, became a scientific and public concern. The occurrence of these strandings was studied retrospectively. Indeed, there have been previous mass strandings of beaked whales; however, Cuvier's beaked whales, the most commonly involved species in atypical mass strandings, had not mass stranded prior to the 1950s (D'Amico et al. 2009). High-power, mid-frequency active sonar was employed in 1954 for the first time, and the first atypical mass stranding registered in the worldwide database on cetacean strandings of the National Museum of Natural History of the Smithsonian Institution involving Cuvier's took place in Genoa, Italy, in 1963 (Cox et al. 2006).

From the pathological point of view, no systematic, full-scale necropsies were able to be carried out on the beaked whales stranded in 1996 in Greece, or in 2000 in Bahamas and Madeira; thus, only few pathological findings were reported (Cox et al. 2006). At that point, acoustic trauma was hypothesized as the cause of the stranding (Evans and England 2001). However, this hypothesis was about to take an unexpected U-turn.

At 3 am in 24 September 2002, military mid-frequency sonar started to be used around 40 km from the coast of Fuerteventura, in the Canary Islands, Spain, as part of NEOTAPON 02, a military exercise. At 7 am, the first report of beaked whales beaching along the coast of Fuerteventura was received. Warships were visible from

Fig. 32.2 (**a, b**) Cuvier's beaked whales (*Ziphius cavirostris*) beaching on Fuerteventura (Canary Islands, Spain) in 2002 during "NEOTAPON 02" military exercises. (*Image credits: Cabildo de Fuerteventura*) (**c**) Subcutaneous tissue (Almería 2006). Gas bubbles in subcutaneous veins (fresh tissue). (**d**) Heart (Almeria 2011). Gas bubbles in pericardial veins (fresh tissue) (*Image credits: IUSA, ULPGC*)

the beach where some of the beaked whales were stranded, and response efforts were concentrated (Fig. 32.2). At 10 am, military sonar was stopped from being used, and 14 beaked whales were found stranded or beached during the three following days and spread along the coast of the islands of Fuerteventura and Lanzarote. The event overwhelmed the logistics of the local marine mammal-stranding response. Since acoustic trauma was the only previously postulated hypothesis for this kind of strandings, the pathological examination of the heads of the animals which had died was prioritized.

Postmortem examination and internal organ sampling were performed at the stranding site, while the heads of these animals were preserved at 4 °C and shipped to the necropsy room of the veterinary school in Gran Canaria for a detailed pathological exam within 24 hours. Similar to the animals stranded in Bahamas and Madeira, the beaked whales stranded in the Canary Islands presented with haemorrhages around the acoustic jaw fat, the ears and the brain. However, the pathological examination of these animals also revealed new findings: gas bubble-associated lesions and fat embolism consistent with a severe decompression-like sickness (Fernandez et al. 2005, Jepson et al. 2003). This was based on the following hypothesis: in vivo bubble formation (a likely cause of distress) associated with sonar exposure (stressor) that may have been caused by nitrogen supersaturation of the tissues following a modified or altered diving behaviour in response to sonar exposure (the stressor). This new pathologic entity in cetaceans, now also reported in by-catch turtles, would particularly affect deep, long duration, repetitive-diving species like beaked whales.

In 11–15 July 2004, international naval exercises called MEDSHARK/Majestic Eagle '04 took place 100 nautical miles northeast to the Canary Islands. Several ships played antisubmarine mid-frequency sonar during this military exercise. Five days later and in a 6-day period (21–26 July), four beaked whales were found floating or stranded in the Canary Islands. The beaked whales were found in a state of decomposition, suggesting that the animals had died several days before and had died at sea. The sea currents during and after the naval exercise would have driven the carcasses from the naval exercise to the Canary Islands. The decomposition state of the beaked whales made it impossible to establish the cause definitively as in vivo gas embolism; however, systemic fat embolism was found in all the animals examined. Since these animals most likely died at sea, fat emboli were not caused by a traumatic beaching (Fernández et al. 2012).

The results of the necropsies of the stranding on the Canary Islands in 2002 were published (Jepson et al. 2003). In 2004, the scientific committee of the International Whaling Commission concluded that there was "compelling evidence implicating military sonars as having a direct impact on whales, in particular on beaked whales" (IWC 2004). These scientific results provoked a European Parliament nonbinding resolution in 2004, where the European Parliament urged its member states "to immediately restrict the use of high-intensity active naval sonars in waters falling under their jurisdiction" following the precautionary principle. The Spanish Government established a moratorium of high-intensity active naval sonar within 50 nautical miles east to the Canary Islands.

Decompression sickness (DCS) is a syndrome described in human divers. Although the pathophysiological mechanism of decompression sickness is not fully understood, it is accepted that the pivotal mechanism is the formation of gas bubbles by gas-phase separation in tissues supersaturated with gas (Vann et al. 2011). However, decompression sickness had never been described before in marine mammals. How was it possible that marine mammals would suffer from decompression sickness if they had developed anatomical and physiological adaptations through evolution to protect them against decompression sickness? Were those gas bubbles really coming from nitrogen-supersaturated tissues, or were they a by-product of putrefaction? Were those bubbles dissection artefacts? Could the stranding process cause the fat emboli?

The authors proposed several mechanisms that could lead to decompression sickness (a very probable cause of distress) in marine mammals, such as an alteration of the diving behaviour of beaked whales in response to sonar (stressor) causing nitrogen supersaturation (stressor within the tissues) (Fernandez et al. 2005). The interpretation of these new findings became a scientific controversy, and beaked whales became a focus of research, regarding the anatomy, physiology, diving behaviour, acoustic responses and pathology. The development of new technology such as noninvasive time-depth recorders and digital acoustic recording tags has been critical to recent increased understanding of diving behaviour, responses to acoustic stimuli and the responses of these wild marine mammals to a sound stressor (Johnson and Tyack 2003).

Over the last decade, we have learnt that beaked whales, more specifically Cuvier's whales, are the most extreme deep divers of this species group, with records for the longest (137.5 min) and deepest dives (2992 m) (Schorr et al. 2014). They present a very stereotypical diving behaviour: deep dives of 1400 m depth and 67 min long on average, followed by a series of shallow dives (Schorr et al. 2014). They are more sensitive to mid-frequency sonar than other species. They show strong behavioural reactions to mid-frequency sonar playback (the stressor) at low-received levels (89–127 dB re 1 μPa). Temporary auditory damage is considered to occur with sound pressure levels higher than 180 dB. When exposed to these sonar levels, the animals stopped echolocating and performed longer dives with slower ascent rates when swimming away from the sound source than routine foraging dives. However, observation of whale behaviour to distant sonar exercises did not cause the same reaction in the beaked whales (DeRuiter et al. 2013).

Mathematical models have been used to predict end-dive nitrogen tissue tensions for beaked whales. The end-dive nitrogen tissue tensions calculated by the model for the beaked whales would cause a significant proportion of decompression sickness in terrestrial mammals (Hooker et al. 2009). Repetitive shallow dives as a behavioural response to mid-frequency sonar (stressor) have been identified as a risk for decompression sickness (a cause of distress) (Zimmer and Tyack 2007). Recent theoretical models did predict natural high-tissue end-dive nitrogen levels for beaked whales (Kvadsheim et al. 2012), although the behavioural reaction of beaked whales to mid-frequency sonar did not imply a significant increase in those levels. However, these authors conclude that a combination of behavioural and physiological responses to sonar cannot be ruled out as a cause of altered tissue and dive nitrogen tension and, in turn, decompression sickness risk (Kvadsheim et al. 2012).

From the pathological perspective, new atypical mass strandings over the last decade have provided new clues that have reinforced the decompression sickness hypothesis. On 25–26 January 2006, mid-frequency antisubmarine sonar was used within a NATO Task Force Group (SNMG2) exercise in the Cartagena area (southeast coast of Spain). On 26–27 January 2006, four Cuvier's beaked whales were stranded along Almería's coast (southeast Spain). Two of the animals were observed stranded alive and displaying abnormal behaviour, and these animals died soon after being found. Complete pathological studies were performed on the four animals. Three of them were in a fresh condition at the time of the necropsy, and at this time, special precautions were taken to look grossly for gas bubbles. For the first time, disseminated gas embolism was confirmed grossly in these fresh tissues (Fig. 32.2c,d). Histopathological findings were similar to previous atypical beaked whale strandings, with widely disseminated haemorrhages in lipid-rich tissues and extensive gas and fat emboli.

In between 28 February and 10 March 2011, an international naval exercise named "NOBLE MARINER 11" took place in the Mediterranean Sea and Atlantic Ocean off the coast of Spain. One Cuvier's beaked whale was stranded alive on 03 March. Refloating attempts failed and the animal was euthanized. Necropsy was conducted within 24 h of death, demonstrating acute gas and fat embolic lesions and haemorrhages in lipid-rich tissues (Fig. 32.2c,d). For the first time, gas analyses were performed in a beaked whale stranded in temporal and spatial association with military exercises. Nitrogen was the main compound in the gas bubbles found in the coronary and renal veins as well as right ventricle and right atrium (Bernaldo de Quirós 2011). These results were consistent with nitrogen gas derived (gas off) from supersaturated tissues as in decompression sickness (Bernaldo de Quirós et al. 2013).

Since 2004, after the Spanish government imposed a moratorium on mid-frequency military sonar, there have been no mass strandings of beaked whales in the Canary Islands (Fernández et al. 2013). In this specific case, pathological findings promoted political action and resulted in a direct impact on the welfare and conservation of beaked whales.

32.3 Example 2: Animal Welfare (Active Stranding and Capture Myopathy)

Cetaceans strand for many reasons, and stranding can be a passive or an active process. In passive strandings, cetaceans are found dead on the shore, and it is usually presumed that they died offshore, although in a minority of cases, a detailed necropsy can indicate that the cetacean found dead had probably initially stranded alive. However, active strandings occur when cetaceans are still alive when they strand on the beach/coast, and live strandings can be further associated with rescue efforts, including animal handling, medical and nursing treatments on the shore, transportation to a rehabilitation centre, one or more attempts at refloating back to sea or subsequent death (Fig. 32.3a).

While describing the histopathology of the alarm reaction in small odontocetes, Cowan and Curry (2008) defined the term stress as *demand for adaptation,* in which adaptation is considered the condition of adjustment of physiological systems to maintain homeostasis. In this sense, McEwen (2000) defined homeostasis as the set of physiological and anatomical systems concerned with maintaining the essentials of the internal milieu, the essential internal workings of the animal body, such as thermoregulation, blood gases, acid base, fluid levels, metabolite levels and blood pressure.

Some of the proposed generic threats to homeostasis include environmental extremes, extreme physical exertion and depletion of essential resources. All these situations/stressors are present at the time an animal strands, wherein these stressors elicit behavioural and physiological changes for adaptation in the live-stranded cetaceans. Stranded cetaceans certainly are likely to be highly stressed and exhibit the general and well-documented mammalian physiological responses to stress,

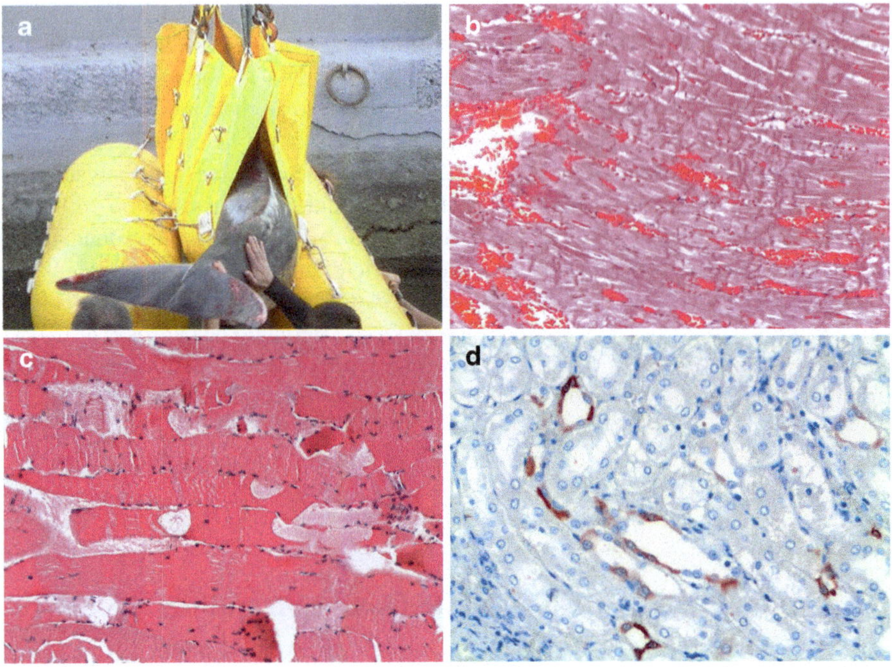

Fig. 32.3 (**a**) An adult male pygmy sperm whale (*Kogia breviceps*) stranded alive in Gran Canaria on February 2007. After an initial assessment, the decision was made to transport the animal to a rehabilitation facility, although the animal died during transportation. (**b**) The cardiac histopathological lesions from the pygmy sperm whale showed prominent contraction band necrosis, a characteristic muscle lesion associated with elevated concentrations of catecholamines, haematoxylin and eosin. 10x. (**c**) Acute monophasic rhabdomyolysis, diffuse, severe. Skeletal muscle from an adult male striped dolphin (*Stenella coeruleoalba*) stranded alive in La Gomera on June 2001. Segmental and hyalinized myofibres are hypercontracted and the cytoplasm is disrupted. Haematoxylin and eosin, 10x. (**d**) Myoglobinuric nephrosis. Kidney from an adult male spinner dolphin (*Stenella longirostris*) stranded alive in Gran Canaria on August 2004. Anti-myoglobin antibody labels the tubular cell cytoplasm. Dilated distal tubules showed flattened epithelial cells. Immunohistochemistry anti-myoglobin, 20x (*Image credit: Instituto Universitario de Sanidad Animal, Universidad de Las Palmas de Gran Canaria*)

including adrenocortical responses (release of stress hormones) having effects on thyroid hormone balance (Tsur et al. 1997) (see also 32.4.2). The stress or adaptive response is mediated by activation of the stress system, including the hypothalamo-pituitary-adrenocortical axis (HPA axis) and the sympathetic nervous system. Blood tests in captured cetaceans have shown activation of the HPA axis and the consequent production and release of glucocorticoids, mineralocorticoids and catecholamines (Fair et al. 2014).

These adaptations integrate the "stress response" and encompass a range of responses from sensory detection of the threat to the multi-subsystem, self-organizing physiological changes that facilitate fight or flight (alarm reaction) or freezing. This stage of alarm reaction is followed by the phase of resistance, in which the homeostasis can be restored to protect the organism from the most deleterious consequences of stress. These physiological adaptations have some benefits; however, an extreme or prolonged stress response can potentially result in some significant changes, compromising the survival chance of the affected animals (distress). Though the stress response is shared by most mammals, including cetaceans, it can differ markedly between species and individuals according to differences in physiology, hormonal status and previous experiences. This is especially true in cetaceans, an order comprising a biological diversity of whales and dolphins with different structural and physiological characteristics. In addition, it has been proposed that these species might be more vulnerable to the effects of stress in comparison with other species, due to their intelligence and emotional/social abilities (Curry 1999).

During the active stranding process, a stress-related capture myopathy (CM) syndrome can occur in individuals, potentially resulting in high mortality, as has been reported in other wild mammals and birds (Spraker 1993, Turnbull and Cowan 1998). It has been presumed that, in cetaceans, the muscle damage is a direct consequence of (and not the cause of) stranding and reflects extreme exertion, trauma, crush injury and stress (Colgrove 1978; Gales 1992; Herraeaz et al. 2007). Muscular activity is simultaneously activated in the fear response (Spraker 1993). Capture myopathy is defined as a non-infectious, metabolic muscle disease of wild mammals and birds associated with the stress of capture, restraint and transportation. Four clinical syndromes of CM have been described in wild animals: (1) capture shock syndrome, (2) ataxic myoglobinuric syndrome, (3) delayed peracute syndrome and (4) ruptured muscle syndrome (Spraker 1993).

Stress, excessive muscular activity, trauma and prolonged muscle compression during the stranding, restraint and the transportation are considered the most probable causes of the muscular damage seen in stranded animals. Pathological findings in CM consist of rhabdomyolysis affecting both cardiac and skeletal muscle and areas of necrosis distributed across different organs including the brain, lung, liver, intestine, pancreas and lymph nodes. Rhabdomyolysis is a pathology caused by muscle injury resulting in acute renal failure. This syndrome is caused by accumulation of muscle breakdown products (mainly myoglobin, Mb) in the bloodstream, and it is associated with oxidative stress with primary role of mitochondria (outside their cellular environments, Mb is known to cause oxidative damage). In the ataxic myoglobinuric syndrome, additional renal lesions are characterized by moderate to severe tubular necrosis, with intratubular protein casts. The urinary bladder usually contains small amounts of brownish urine, containing Mb. Thus, skeletal muscle necrosis and acute renal failure are common sequelae of CM. Concerning the etiopathogenesis of muscular lesions in CM, it has traditionally been related to cellular hypoxia and lactic acidosis being associated with exhaustion of the energy store reserves useful to escape (Spraker 1993), although stress plays an important role in myonecrosis development. Capture and restraint of a wild animal (injured/ill or healthy) is extremely stressful and commonly implies devastating consequences, mainly by ischaemia and reperfusion mechanisms. As stated above, live-stranded cetaceans are highly stressed during the stranding, and as with the prototypic mammalian response to stress, elevated blood cortisol and aldosterone (stress hormones) concentrations have been observed in cetaceans subjected to capture, handling and restraint (Fair and Becker 2000, St. Aubin and Geraci 1990). While live stranding in cetaceans represents an extreme and multifactorial condition, a syndrome of stress myopathy has been proposed as a cause of death associated with the long-time frame of protracted rescues (Geraci and Lounsbury 2005, Herraez et al. 2007, Simpson and Cornell 1983). The presence of acute degenerative skeletal muscle and myocardial and renal lesions with myoglobinuria, as previously described in 25 of 51 live-stranded cetaceans following human capture/rescue interactions, indicates that many live-stranded cetaceans experience CM similar to that of terrestrial wildlife (Herraez et al. 2013, Herraez et al. 2007).

Biochemical markers in animals with CM include elevated serum muscle enzyme activities, variable degrees of myopathy and, in ataxic myoglobinuric syndrome, elevated serum urea associated with a myoglobinuria. Clinical signs of CM vary depending on the species and individuals and can include dyspnoea, weakness, hyperthermia, muscle tremors and/or rigidity. A wide spectrum of pathological findings is present in CM, including areas of necrosis in the brain, lung, liver, intestine, pancreas and lymph nodes. However, the hallmark lesions of CM are acute to subacute muscle degeneration affecting both skeletal and cardiac muscle and tubular nephrosis, often associated with tubular myoglobin (Fig. 32.3b,c,d) (Spraker 1993). The time elapsed from the stressful episode to the time of death determines to a degree the temporality, i.e. which of the possible histopathological findings are seen.

There is no treatment for CM; thus, the driver should be to prevent CM from occurring or to minimize its effects. Management of the stranded animal should include awareness of the situation, and the risks of CM, and efforts to reduce the striated muscle damage due to the release of catecholamines and sustained high blood cortisol and aldosterone levels, as well as the effects of extreme temperatures and exertion. The animal's reaction during handling could be different between species and individuals, since it will be governed by a complex interaction of genetic factors and previous experiences (Van Bressem et al. 1999). Neither age nor sex predisposition has been completely elucidated, although mature males with large muscle mass are more susceptible to CM. In this sense, species- and individual-specific therapy should be conducted by veterinarians.

In summary, active stranding refers to events in which a cetacean comes ashore alive, representing a stressful, extreme and multifactorial condition which is usually followed by human capture/rescue interactions and transportation for rehabilitation or in some cases euthanasia or death. Live-stranded cetaceans are usually debilitated when rescued, their condition deteriorates over the period of capture, and they most commonly die after a period of captivity (Herraez et al. 2013).

Responses to stress can be beneficial and constructive, but if the organism is unable to successfully manage these responses or the allostatic system (controlled deviation of homeostatic system) becomes activated too frequently and for prolonged long periods of time, it can also be destructive and elicit undesirable effects. In consequence, the pathophysiology of CM could have a role in mortality following live-stranding events, and CM can compromise subsequent rehabilitation, the effectiveness of therapy and care of the recovering animals as well as their welfare status.

32.4 Example 3: Fingerprints in the Brain of Dolphins (How Can We See Them?)

Many of the classical measures used to evaluate neuropathological changes do not tell us if meaningful cellular changes have occurred, and thus, the link between animal neuropathological alterations, welfare and stress is still mostly unclear. Our knowledge of the location of certain brain nuclei enables us to understand that these are activated in situations of acute, chronic and/or severe stress. The feasibility of postmortem analysis of archival brain specimens to look for markers of previous activity may answer basic questions in comparative neurobiology and neuropathology, in a way that has not previously been possible.

32.4.1 Cetacean's Neuroanatomy: Big Brains for a Big Dilemma

There is a significant lack of information concerning cetacean neuroanatomy, due to the limited access to fresh brain samples of such unconventional mammals. In the course of evolution, cetaceans have undergone numerous modifications with respect to their ancestral land status. The brain size of toothed whales with respect to body size allocates them between that of apes and humans in relation to brain volume and is significantly larger than those of any nonhuman primate (Marino et al. 2007). Thus, the overall size of the brain is large—especially the size of the neocortex, the auditory centres, the cerebellum and the sensory component of the trigeminal nerve. The olfactory bulbs, as well as all the olfactory structures, are almost always regressed or absent in toothed whales (Fig. 32.4a,b). In contrast, the auditory processing areas reflect the necessity of effective underwater hearing for echolocation (sonar), navigation and communication (Oelschlager 2008). In toothed whales, the acoustic brain centres are large: the medial geniculate body is about 7 times larger, the inferior colliculus 12 times and the lateral lemniscus 250

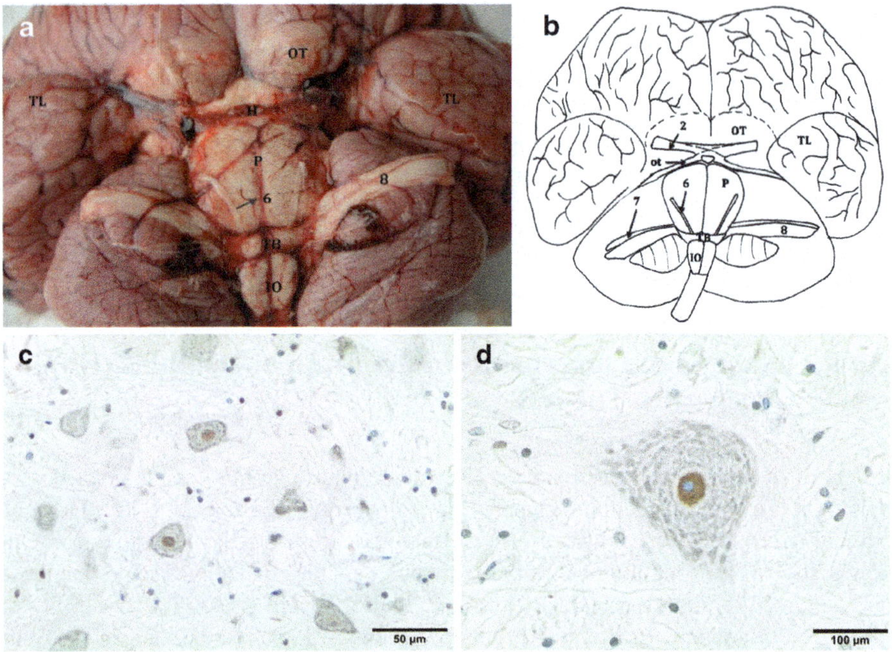

Fig. 32.4 (**a**, **b**) Ventral view of the brain of a common dolphin (*Delphinus delphis*). H, hypothalamus; IO, inferior olive; ot, optic tract; OT, olfactory tubercle; P, pons; TB, trapezoid body; TL, temporal lobe; 2, optic nerve; 6, abducens nerve; 7, facial nerve; 8, vestibulocochlear nerve. Intranuclear c-fos labelling in multipolar small spherical neurons (**c**) and in giant neurons (**d**) of the ventral cochlear nucleus (VCN) of the striped dolphin (*Stenella coeruleoalba*). 3,3-Diaminobenzidine (DAB) counterstained with thionine (*Image credit: Instituto Universitario de Sanidad Animal, Universidad de Las Palmas de Gran Canaria*)

times larger in absolute terms than the equivalent structures in the human brain (Bullock and Gurevich 1979).

32.4.2 Cetacean's Neuroanatomy of Stress: The Hypothalamus and the Locus Coeruleus

The stress or adaptive response is mediated by activation of the stress system, including the HPA axis and the sympathetic nervous system, which results in increased secretion of glucocorticoids (cortisol) and catecholamines (epinephrine and norepinephrine), respectively (Han et al. 2013). The stress regulatory control of the HPA axis is essential to health and survival. The activation of the HPA axis begins at the level of the hypothalamus, a tiny area of the ventral brain. The hypothalamus controls the autonomic nervous system via a set of neurons that directly innervate both the parasympathetic and sympathetic preganglionic neurons, as well as various cell groups in the brainstem that control autonomic reflexes, maintaining the body's homeostasis (Saper and Lowell 2014). Despite its small size, this region of the diencephalon contains the comprehensive system for basic life support. The hypothalamus is also a key structure of the brain in the regulation of the release of hormones from the pituitary gland. The hypothalamus, particularly the paraventricular nucleus (PVN), is activated during an emergency reaction, probably through inputs from the amygdala and other parts of the central nervous system involved in processing fearful or threatening stimuli (Nicol 2000). In reaction to stressful stimuli, the hypothalamus acts releasing the corticotropin-releasing factor (CRF) or corticotropin-releasing hormone (CRH), produced primarily by the PVN of the paraventricular nucleus, as well as in the magnocellular population of the PVN and supraoptic nucleus (SON). CRF is released into the hypophyseal portal veins and to the anterior pituitary, which subsequently releases adrenocorticotropic hormone (ACTH). ACTH then stimulates secretion of the glucocorticoids, mainly cortisol and corticosterone, from the zona fasciculata of the adrenal glands. The stress response has classically been characterized by two temporal waves of stress mediator actions. The first rapid one includes noradrenaline and CRF and promotes vigilance, alertness, appraisal of the situation and the choice of an optimal strategy to face the challenge. The CRF is released within seconds of recognition of the emergency with an effect on several brain regions; therefore, it acts as a neuromodulator in the brainstem and neocortical regions of stress responses. CRF is also expressed in neuronal populations in the amygdala, hippocampus and the *locus coeruleus*. The second wave, attributed to corticosteroids, mediates the alterations of gene expression and cell function (Joels and Baram 2009). In the hypothalamus, vasopressin (ADH) interacts with CRF, promoting ACTH release from the pituitary in response to stress.

In the toothed whales, the PVN (Sacchini 2015) remains in the preoptic region where it reaches its maximum size and then regresses in the anterior or suprachiasmatic region. PVN is made up of two cell populations: a magnocellular population composed of large neurons and a parvocellular population uniformly composed of small neurons (Sacchini 2015).

The toothed whales' supraoptic nucleus (SON) extends from the preoptic area to the hypothalamic tuberal area (Sacchini 2015). SON bounds all the optical structures (the proximal portion of the optic nerve, optic chiasm and distal optic tract) with its neurons and neuronal processes. The SON is divided into a retrochiasmal region (SONr) laterally and a principal region (SONp) medially and dorsally. The SON is made up of a magnocellular population of large neurons. ADH immunoreactivity is strongly visible in the magnocellular population, both in the PVN and the SON. The parvocellular neurons of the PVN express a mild cytoplasmic positivity to CRF (Sacchini 2015).

The *locus coeruleus* (LC) is a densely packed cluster of norepinephrine producer neurons, located in the upper part of the rostral rhombencephalon, near the floor of the fourth ventricle. It is the largest catecholaminergic nucleus of the brain, and it supplies norepinephrine to the entire central nervous system. LC is involved in attention, behavioural activation and arousal. Again, the population of neurons of the LC innervates a wide swath of distal targets, making possible the kind of global and coordinated effects on information processing that an emotional state like fear requires (Adolphs 2013).

The toothed whales' LC extends from the caudal level of the motor nucleus of the trochlear nerve to the rostral level of the motor nucleus on the trigeminal nerve. The toothed whales' LC presents five subdivisions: A6d, restricted to the caudal periaqueductal grey; A6v, external to the periaqueductal grey; A7, located in the pontine tegmentum; A5, confined to the caudal ventral tegmental portion, near to the superior and inferior olivary complex and caudal to the motor trigeminal nucleus; and A4, latero-dorsally, close to the superior cerebellar peduncle. Neurons are very large and polygonal in shape and exhibit cytoplasmic neuromelanin. They are also strongly immunopositive for thyroxine hydroxylase (TH). Manger et al. examined the LC in a bottlenose dolphin specimen (Manger et al. 2003) and stated that the A4 subdivision was absent. Nevertheless, toothed whales seem to present the A4 division, making the LC projections wider, in order to innervate a more extensive part of the brain, contributing to keep it in a constant mood of alertness (Sacchini 2015).

32.4.3 Cetacean's Neuroanatomy of Fear: The Amygdaloid Complex

Recent studies in rodents have shown that there are highly specific brain circuits for fear, whereas findings from human neuroimaging seem to make the opposite claim (Adolphs 2013). The amygdala or amygdaloid complex is a noticeable bilateral structure in the medial temporal lobe and is composed of at least 13 different nuclei and cortical areas, subdivided into the deep nuclei or basolateral amygdala (BLA), the superficial nuclei, the anterior amygdaloid area (AAA), the amygdalohippocampal area (AHA), the intercalated nuclei (I) and the central nucleus of the amygdala (CeA).

The amygdaloid complex is part of the neural circuitry that is critical for emotion. The amygdala initiates the fight or flight response through inputs into the hypothalamus and to brainstem control centres of the sympathetic nervous system. Beyond its role in emotional reactivity, it is important in emotional learning, whereby cues acquire significance through association with rewarding or aversive events. Finally, the amygdala regulates additional cognitive processes, such as memory or attention (Gallagher and Chiba 1996). The BLA receives most of the sensory inputs that specify fear associations. The CeA is widely considered the main output regulator for mediating fear responses. The CeA mediates the behavioural and physiological reactions associated with fear, anxiety and the HPA responses by modulating brain CRF activity. CRF in the CeA during acute stress enhances memory consolidation (Joels and Baram 2009).

The flexible modulation of different downstream fear components by the CeA depends on an intricate inhibitory control balance internal to the amygdala (Adolphs

2013). An interconnected sheath of GABAergic neurons, the intercalated neurons (I), is found interposed between the BLA and CeA, providing an important source of inhibition (Janak and Tye 2015). Whereas some of CeA neurons can inhibit cholinergic targets mediating cortical arousal, they can at the same time promote freezing through projections to the periaqueductal grey (Gozzi et al. 2010). In the amygdala, the excitatory actions of vasopressin might contribute to the behavioural stress response. Vasopressin might also modulate emotional memory and anxiety (Joels and Baram 2009).

Previous published work on toothed whales' amygdaloid complex (Breathnach and Goldby 1954, Jansen 1953, Morgane 1972) was based on archaic classification, which differs considerably from currently used nomenclature. In the toothed whales' amygdala, there are 12 nuclei divided into three main groups (Sacchini 2015). The basolateral complex or deep nuclei are made up of the lateral nucleus (NL); the basal nucleus (NB) with its magnocellular, intermediate and parvocellular division; the basal accessory nucleus (NBA) with its magnocellular and parvocellular division; and the paralaminar nucleus (PL). The superficial or cortical nuclei are composed of the anterior cortical nucleus (CoA), the posterior cortical nucleus (Cop), the periamygdaloid cortex (PAC) and the medial nucleus (M). Finally, the other amygdaloid nuclei are composed of the CeA, the AAA, the AHA and the I. Neither the nucleus of the lateral olfactory tract (NLOT) nor the bed nucleus of the accessory olfactory tract (BAOT) is present in the toothed whales' amygdala (Sacchini 2015), possibly due to the involution of the olfactory structures. The CeA extends mainly dorsal to the NL and NB. The CeA contains small neurons, spherical, poligonal or spindle in shape. The neuropil and somas of the CeA are heterogenically positive to Calbindin D-28k. The subdivision, volume and location of CeA and, in general, of the amygdaloid complex of toothed whales are very similar to that described in other species such as primates (Sacchini 2015).

32.4.4 Fingerprints in the Cetacean Brain: Is It Possible to Get a Real Identikit?

In 1955, a front-page article in The New York Times reported that "the brain that worked out the theory of relativity and made possible the development of nuclear fission" had been removed for scientific study. The pathologist Thomas Harvey carried out the autopsy and took Albert Einstein's brain out. Harvey sent samples of the coveted organ to a number of different researchers, including Harvard University's Marian Diamond, who found that the genius had an average number of normal-sized brain cells. As Einstein stated "Not everything that counts can be counted and not everything that can be counted counts", only some decades later were most of Einstein's brain peculiarities identified (Colombo et al. 2006, Falk et al. 2012, Hines 1998). The neuronal/glial ratio for the Einstein brain was significantly smaller than the mean for the control population (Diamond et al. 1985). Furthermore, Einstein's astrocytic processes showed larger size and higher numbers of interlaminar terminal masses (Colombo et al. 2006). Recent studies have shown that glial cells may play a critical role in regulating synaptic function and plasticity (Filosa et al. 2009), perhaps explaining the extraordinary abilities of the scientist.

If intra-individual brain structure variability can reflect changes in behavioural performance and capacity (perhaps measurable and so a potential "good news" indicator) or can reflect behavioural changes associated with neurodegenerative and other brain-related disorders (a potential "bad news" indicator), are we also able to draw a specific fingerprint-like pattern of the behaviour of neuronal and non-neuronal components of the brain (reflecting good or bad news)?

Stress may result in some significant neurobiological changes that place animal and human well-being at risk (resulting in distress). It is not something new that chronic exposure to stress hormones, from the prenatal time to the ageing period, can have deleterious effects on brain structures involved in cognition and mental health (Lupien et al. 2016). Exposure to chronic stress in adolescence is sufficient to induce lasting changes in neuroendocrine drive and behaviour, potentially altering the developmental trajectory of stress circuits (Wulsin et al. 2016). Distress can jeopardize the brain and behaviour across the lifespan in animals and humans.

Some molecular mechanisms are generally used by the neurons with the purpose of protecting them from ischemic injuries or toxic substance, for example. Nevertheless, if the noxious agent is not removed, the cells begin an irreversible process of degeneration. Cellular recovery from acute injury involves the expression of early stress genes such as c-jun, c-fos and stress proteins, like heat shock proteins (HSPs). Finally, chronic/severe stress increases the vulnerability of neurons, accompanied by a detectable increase in pro-apoptotic and/or apoptotic neurons (Fig. 32.5).

Multiple studies state that a variety of different stimuli have been shown to induce the expression of the proto-oncogene c-fos in the brain, and thus, c-fos immunostaining is now used to map brain metabolism under different physiological and non-physiological conditions (Herrera and Robertson 1996). Transient c-fos expression in the central nervous system was first observed after seizure activity and following noxious stimulation in the spinal cord (Herrera and Robertson 1996). C-fos gene expression is related to various signal cascades involved in biochemical processes such as neuronal plasticity, cell growth and mitosis (Barros et al. 2015). Circulating adrenal steroid hormones in the brain may result in ever-changing patterns of gene expression, too (McEwen et al. 2016). A dramatic induction of c-fos is also detected in response to stressful stimuli in different rat brain areas. Few brain areas show increased expression of c-jun, but these regions show induction of c-fos (Cullinan et al. 1995).

Protein degradation is a dynamic and complex process in which ubiquitin has a key regulatory role (Tai and Schuman 2008). Ubiquitin, an essential protein in nonlysosomal proteolytic system, is expressed after metabolic stress to the cell (Ide et al. 1999). The expression of brain heat shock proteins has been used for a

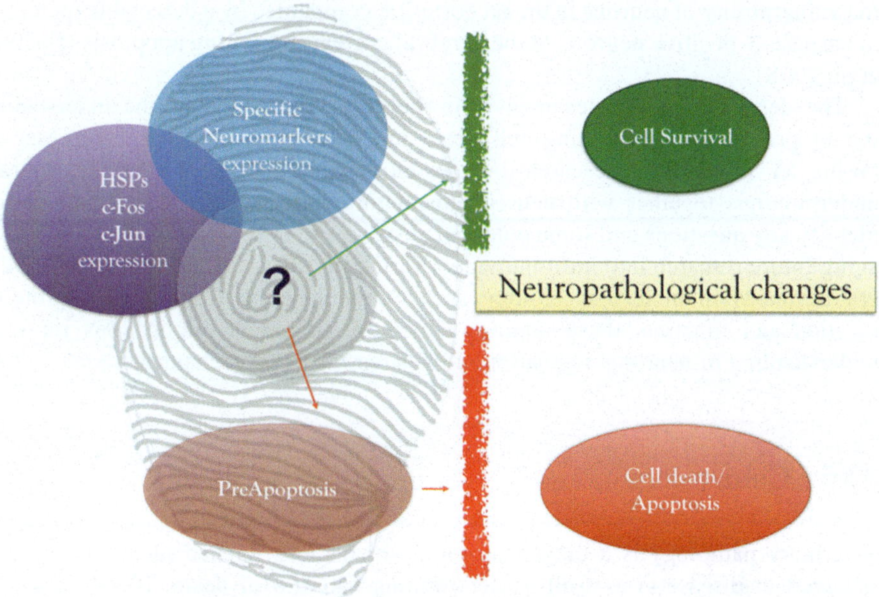

Fig. 32.5 Schematic representation of fingerprint-like cellular identikit, of "normal" or distress suffering neurons. The "?" denotes potential measurable neuropathological changes (*Image credit: Instituto Universitario de Sanidad Animal, Universidad de Las Palmas de Gran Canaria*)

long time for its potential to protect the brain from ischemic injury (Giffard et al. 2004) and deleterious stresses (Samson et al. 2007). Levels of plasma corticosterone and brain norepinephrine significantly increase, as well as the expression of brain heat shock proteins (HSP70) after different durations of noise stress exposure.

Time-dependent stress response to noise exposure is a complex mechanism involving highly interconnected systems such as HPA axis and heat shock proteins and may have serious implications in vital organs, particularly in the brain (Samson et al. 2007). There is an important role of HSP70 which increases after chronic noise exposure, together with a decrease of brain norepinephrine and corticosterone (Samson et al. 2007). Noise seems to increase expression of HSP70, due to the cellular injury, probably caused by the elevated levels of free radicals (Samson et al. 2007). Heat shock system plays a vital role in the protective mechanism of the brain, and the HSP70 is considered as a critical determinant of the brain's stress tolerance (Calabrese et al. 2006). Previous studies in mice have demonstrated that noise exposure, a great stressor, leads to dramatic pathophysiological changes within the central auditory pathway (dorsal and ventral cochlear nuclei and the central nucleus of the inferior colliculus) in addition to cochlear damage, in a time-dependent manner (Coordes et al. 2012). Apoptosis-related pathophysiological changes within the central auditory pathway could be associated with the harmful effects of stress (distress). Moreover, chronic stress increases the vulnerability of neurons in the rat cortex, accompanied by a detectable increase in caspase-3-positive neurons in the cerebral cortex, suggesting apoptosis (Bachis et al. 2008).

The identification of genes involved in apoptosis is a key goal for the understanding of programmes regulating cell death in development and tissue regression (Wang et al. 2015). Research on postmortem brains may integrate neural underpinnings together with neurochemical fingerprints (Fig. 32.5), in order to identify key questions and some potential answers and to set challenges for encouraging future research into intra-individual variability. Through the support of modern advances in molecular biology and immunolabeling for assessing protein location and function, these findings may have important implications for our understanding of neuronal regulation of the stress/distress mechanisms.

32.5 Conclusions

Veterinary pathology is a diagnostic tool, which looks for and identifies lesions involved in disease(s) as well as determining cause(s) of death. If we consider cause(s) as stressors and lesions as indicators of distress, why may we not consider also veterinary pathology as a diagnostic tool of animal welfare and, in our particular case, in cetaceans (whales, dolphins and porpoises)?

Response to stress can be beneficial and constructive, but it can also be destructive and elicit undesirable effects. In consequence, the pathophysiology of CM could have a role in mortality following live-stranding events and compromise subsequent rehabilitation, the effectiveness of therapy and care of the recovering animals, as well as their welfare status. Research on the postmortem brain may integrate neural underpinnings together with neurochemical fingerprints, in order to identify key questions and some potential answers and to set challenges for encouraging future research into intra-individual variability. The moratorium on mid-frequency military sonar is a good example on how pathological findings have driven political action and resulted in a direct impact on the welfare and conservation of cetaceans.

References

Adolphs R (2013) The biology of fear. Curr Biol 23:R79–R93

Arbelo M, Monteros AE, Herraez P, Andrada M, Sierra E, Rodriguez F et al (2013) Pathology and causes of death of stranded cetaceans in the canary islands (1999-2005). Dis Aquat Org 103:87–99

Bachis A, Cruz MI, Nosheny RL, Mocchetti I (2008) Chronic unpredictable stress promotes neuronal apoptosis in the cerebral cortex. Neurosci Lett 442:104–108

Barros VN, Mundim M, Galindo LT, Bittencourt S, Porcionatto M, Mello LE (2015) The pattern of c-fos expression and its refractory period in the brain of rats and monkeys. Front Cell Neurosci 9:72

Bernaldo de Quirós Y (2011) Methodology and analysis of gas embolism: Experimental models and stranded cetaceans. University of Las Palmas de Gran Canaria, Las Palmas de Gran Canaria, Doctoral thesis

Bernaldo de Quirós Y, González-Díaz O, Møllerløkken A, Brubakk AO, Hjelde A, Saavedra P et al (2013) Differentiation at autopsy between in vivo gas embolism and putrefaction using gas composition analysis. Int J Legal Med 127:437–445

Breathnach AS, Goldby F (1954) The amygdaloid nuclei, hippocampus and other parts of the rhinencephalon in the porpoise (Phocaena phocaena). J Anat 88:267–291

Bullock TH, Gurevich VS (1979) Soviet literature on the nervous system and psychobiology of cetacea. In: John RS, Ronald JB (eds) International review of neurobiology, Volume, vol 21. Academic Press, pp 47–127

Calabrese V, Colombrita C, Sultana R, Scapagnini G, Calvani M, Butterfield et al. DA (2006) Redox modulation of heat shock protein expression by acetylcarnitine in aging brain: Relationship to antioxidant status and mitochondrial function. Antioxid Redox Signal 8: 404–416

Colgrove GS (1978) Suspected transportation-associated myopathy in a dolphin. J Am Vet Med Assoc 173:1121–1123

Colombo JA, Reisin HD, Miguel-Hidalgo JJ, Rajkowska G (2006) Cerebral cortex astroglia and the brain of a genius: A propos of A. Einstein's. Brain Res Rev 52:257–263

Coordes A, Groschel M, Ernst A, Basta D (2012) Apoptotic cascades in the central auditory pathway after noise exposure. J Neurotrauma 29:1249–1254

Cowan DF, Curry BE (2008) Histopathology of the alarm reaction in small odontocetes. J Comp Pathol 139:24–33

Cox TM, Ragen TJ, Read AJ, Vos E, Baird RW, Balcomb et al. K (2006) Understanding the impacts of anthropogenic sound on beaked whales. J Cetacean Res Manag 7: 117–187

Cullinan WE, Herman JP, Battaglia DF, Akil H, Watson SJ (1995) Pattern and time course of immediate early gene expression in rat brain following acute stress. Neuroscience 64:477–505

Curry B (1999) Stress in mammals: The potential influence of fishery-induced stress on dolphins in the eastern tropical pacific ocean. In: COMMERCE USDO (ed) NOAA Technical Memorandum NMFS-SWFSC-260. National Marine Fisheries Service, NOAA, La Jolla, CA

D'Amico A, Gisiner R, Ketten D, Hammock JA, Jonhson C, Tyack PL et al (2009) Beaked whale strandings and naval exercises. Aquat Mammals 35:435–444

DeRuiter SL, Southall BL, Calambokidis J, Zimmer WMX, Sadykova D, Falcone et al. EA (2013) First direct measurements of behavioural responses by cuvier's beaked whales to mid-frequency active sonar. Biol Lett 9: 20130223–20130223

Diamond MC, Scheibel AB, Murphy GM Jr, Harvey T (1985) On the brain of a scientist: Albert Einstein. Exp Neurol 88:198–204

Díaz-Delgado J (2015) Patología y causas de la muerte de los cetáceos varados en las islas canarias (2006–2012). Doctoral thesis, University of las Palmas de Gran Canaria, Las Palmas de Gran Canaria.

Evans DL, England GR (2001) Joint interim report: Bahamas marine mammal stranding event of 15–16 march 2000. US Department of Commerce and US Navy, Washington DC

Fair P, Becker P (2000) Review of stress in marine mammals. J Aquat Ecosyst Stress Recover 7:335–354

Fair PA, Schaefer AM, Romano TA, Bossart GD, Lamb SV, Reif JS (2014) Stress response of wild bottlenose dolphins (*Tursiops truncatus*) during capture-release health assessment studies. Gen Comp Endocrinol 206:203–212

Falk D, Lepore FE, Noe A (2012) The cerebral cortex of albert einstein: A description and preliminary analysis of unpublished photographs. Brain. Oxford Journals http://brain. oxfordjournals. org/content/early/2012/11/14/brain. aws 295 Accessed 11 12 16

Fernandez A, Arbelo M, Deaville R, Patterson IAP, Castro P, Baker et al. JR (2004) Pathology: Whales, sonar and decompression sickness. Nature 428: 438

Fernández A, Arbelo M, Martin V (2013) Whales: No mass stranding since sonar ban. Nature 487:317

Fernandez A, Edwards JF, Rodriguez F, de los Monteros AE, Herraez P, Castro et al. P (2005) "Gas and fat embolic syndrome" involving a mass stranding of beaked whales (family Ziphiidae) exposed to anthropogenic sonar signals. Vet Pathol 42: 446–457

Fernández A, Edwards JF, Rodriguez F, Espinosa de Los Monteros A, Herraez P, Castro P et al (2005) "Gas and fat embolic syndrome" involving a mass stranding of beaked whales (family Ziphiidae) exposed to anthropogenic sonar signals. Vet Pathol 42:446–457

Fernández A, Sierra E, Martin V, Méndez M, Sacchini S, Bernaldo de Quirós Y et al (2012) Last "atypical" beaked whales mass stranding in the canary islands (July, 2004). Journal of Marine Science Research and Develpment:2

Filosa A, Paixao S, Honsek SD, Carmona MA, Becker L, Feddersen et al. B (2009) Neuron-glia communication via epha4/ephrin-a3 modulates ltp through glial glutamate transport. Nat Neurosci 12: 1285–1292

Frantzis A (1998) Does acoustic testing strand whales? Nature 392:29–29

Gales NJ (1992) Mass stranding of striped dolphin, *Stenella coeruleoalba*, at Augusta, western Australia: Notes on clinical pathology and general observations. J Wildl Dis 28:651–655

Gallagher M, Chiba AA (1996) The amygdala and emotion. Curr Opin Neurobiol 6:221–227

Garcia-Alvarez N, Boada LD, Fernandez A, Zumbado M, Arbelo M, Sierra E et al (2014) Assessment of the levels of polycyclic aromatic hydrocarbons and organochlorine contaminants in bottlenose dolphins (*Tursiops truncatus*) from the eastern Atlantic ocean. Mar Environ Res 100:48–56

Geraci JR, Lounsbury VJ (2005) Cetaceans – single strandings. In: Marine Mammals Ashore: A Field Guide for Strandings, Second edn. National Aquarium in Baltimore, Baltimore, MD, USA, pp 75–112

Giffard RG, Xu L, Zhao H, Carrico W, Ouyang Y, Qiao et al. Y (2004) Chaperones, protein aggregation, and brain protection from hypoxic/ischemic injury. J Exp Biol 207: 3213–3220

Gozzi A, Jain A, Giovanelli A, Bertollini C, Crestan V, Schwarz et al. AJ (2010) A neural switch for active and passive fear. Neuron 67: 656–666

Han RI, Clark CH, Black A, French A, Culshaw GJ, Kempson et al. SA (2013) Morphological changes to endothelial and interstitial cells and to the extra-cellular matrix in canine myxomatous mitral valve disease (endocardiosis). Vet J 197: 388–394

Herraez P, Espinosa de Los Monteros A, Fernandez A, Edwards JF, Sacchini S, Sierra E (2013) Capture myopathy in live-stranded cetaceans. Vet J 196:181–188

Herraez P, SierraE AM, Jaber JR, de Los Monteros AE, Fernandez A (2007) Rhabdomyolysis and myoglobinuric nephrosis (capture myopathy) in a striped dolphin. J Wildl Dis 43:770–774

Herrera DG, Robertson HA (1996) Activation of c-fos in the brain. Prog Neurobiol 50:83–107

Hines T (1998) Further on Einstein's brain. Exp Neurol 150:343–344

Hooker SK, Baird RW, Fahlman A (2009) Could beaked whales get the bends? Effect of diving behaviour and physiology on modelled gas exchange for three species: *Ziphius cavirostris*, *Mesoplodon densirostris* and *Hyperoodon ampullatus*. Respir Physiol Neurobiol 167:235–246

Ide T, Takada K, Qiu JH, Saito N, Kawahara N, Asai et al. A (1999) Ubiquitin stress response in postischemic hippocampal neurons under nontolerant and tolerant conditions. J Cereb Blood Flow Metab 19: 750–756

IWC (2004) Report of the scientific committee. In: The international whaling commission's 56th annual meeting. Sorrento, Italy, pp 37–39

Janak PH, Tye KM (2015) From circuits to behaviour in the amygdala. Nature 517:284–292

Jansen JKS (1953) A note on the amygdaloid complex in the fin whale (*Balaenoptera physalus* (L.)). In: J. Jansen (ed). I kommisjon hos J. Dybwad, Oslo. http://aquaticcommons.org/5151/1/The_Fin_Whale-1971.pdf Accessed 11 Dec 2016

Jepson PD, Arbelo M, Deaville R, Patterson IAP, Castro P, Baker et al. JR (2003) Gas-bubble lesions in stranded cetaceans - was sonar responsible for a spate of whale deaths after an Atlantic military exercise? Nature 425: 575–576

Joels M, Baram TZ (2009) The neuro-symphony of stress. Nat Rev Neurol 10:459–466

Johnson MP, Tyack PL (2003) A digital acoustic recording tag for measuring the response of wild marine mammals to sound. IEEE J Ocean Eng 28:3–12

Kvadsheim P, Miller PJO, Tyack PL, Sivle LD, Lam FPA, Fahlman A (2012) Estimated tissue and blood n2 levels and risk of decompression sickness in deep-, intermediate-, and shallow-diving toothed whales during exposure to naval sonar. Front in Physiol 3:125

Lupien SJ, Ouellet-Morin I, Herba CM, Juster R, McEwen BS (2016) From vulnerability to neurotoxicity: A developmental approach to the effects of stress on the brain and behavior. In: Spengler D, Binder E (eds) Epigenetics and neuroendocrinology: Clinical focus on psychiatry. Springer International Publishing, Cham, pp 3–48

Manger PR, Ridgway SH, Siegel JM (2003) The locus coeruleus complex of the bottlenose dolphin (*Tursiops truncatus*) as revealed by tyrosine hydroxylase immunohistochemistry. J Sleep Res 12:149–155

Marino L, Connor RC, Fordyce RE, Herman LM, Hof PR, Lefebvre et al. L (2007) Cetaceans have complex brains for complex cognition. PLoS Biol 5: e139

McEwen BS (2000) Allostasis and allostatic load: Implications for neuropsychopharmacology. Neuropsychopharmacology 22:108–124

McEwen BS, Nasca C, Gray JD (2016) Stress effects on neuronal structure: Hippocampus, amygdala, and prefrontal cortex. Neuropsychopharmacology 41:3–23

Moberg GP, Mench JA (2000) The biology of animal stress: Basic principles and implications for animal welfare. CABI Publishing, New York, NW, USA

Morgane PJ (1972) Comparative anatomy of the cetacean nervous system. p 117–244. Editors B Cozzi, S Huggenberger, H Oelschlager, Academic Press, Elsevier Publishing

Nicol C (2000) The biology of animal stress: Basic principles and implications for animal welfare. Appl Anim Behav Sci 72:375–378

Nowacek DP, Thorne LH, Johnston DW, Tyack PL (2007) Responses of cetaceans to anthropogenic noise. Mammal Rev 37:81–115

Oelschlager HH (2008) The dolphin brain, a challenge for synthetic neurobiology. Brain Res Bull 75:450–459

Sacchini S (2015) Macroscopic and microscopic, histochemical and immunohistochemical characterization of the central nucleus of the amygdala, supraoptic and paraventricular nuclei of the hypothalamus, and the locus coeruleus of the brain of toothed whales. PhD thesis, pending of publication.

Samson J, Sheeladevi R, Ravindran R, Senthilvelan M (2007) Stress response in rat brain after different durations of noise exposure. Neurosci Res 57:143–147

Saper CB, Lowell BB (2014) The hypothalamus. Curr Biol 24:R1111–R1116

Schorr GS, Falcone EA, Moretti DJ, Andrews RD (2014) First long-term behavioral records from Cuvier's beaked whales (*Ziphius cavirostris*) reveal record-breaking dives. PLoS One 9:e92633

Selye H (1956) The stress of life. McGraw-Hill, New York

Sierra E, Fernandez A, Espinosa de Los Monteros A Arbelo M, Diaz-Delgado J, Andrada M et al. (2014) Histopathological muscle findings may be essential for a definitive diagnosis of suspected sharp trauma associated with ship strikes in stranded cetaceans PLoS One 9 e88780

Simpson JG, Cornell LH (1983) Diseases associated with stranding and captivity. In: Pathobiology of marine mammal diseases. CRC Press, Boca Raton, FL, USA

Spraker TR (1993) Stress and capture myopathy in artiodactylids. In: Zoo and wild animal medicine, current therapy 3, vol 3. W.B. Saunders, Philadelphia, PA, USA, pp 481–488

St. Aubin DJ, Geraci JR (1990) Adrenal responsiveness to stimulation by adrenocorticotropic hormone (acth) in captive beluga whales (*Delphinapterus leucas*). In: Smith, t.G., st. Aubin, d.J., and geraci, j.R., editors. Advances in Research on the beluga whale (*Delphinapterus leucas*). Can Bull Fish Aquat Sci 224:149–157

Tai H-C, Schuman EM (2008) Ubiquitin, the proteasome and protein degradation in neuronal function and dysfunction. Nat Rev Neurol 9:826–838

Tsur I, Yakobson B, Elad D, Moffett D, Kennedy S (1997) Morbillivirus infection in a bottlenose dolphin from the Mediterranean sea. Eur J Vet Pathol 2:83–85

Turnbull BS, Cowan DF (1998) Myocardial contraction band necrosis in stranded cetaceans. J Comp Pathol 118:317–327

Van Bressem MF, Van Waerebeek K, Raga JA (1999) A review of virus infections of cataceans and the potential impact of morbilliviruses, poxviruses and papillomaviruses on host population dynamics. Dis Aquat Org 38:53–65

Vann RD, Butler FK, Mitchell SJ, Moon RE (2011) Decompression illness. Lancet 377:153–164

Wang Y, Wang W, Li Z, Hao S, Wang B (2015) A novel perspective on neuron study: Damaging and promoting effects in different neurons induced by mechanical stress. Biomech Model Mechanobiol. doi:10.1007/s10237-015-0743-4

Wulsin AC, Wick-Carlson D, Packard BA, Morano R, Herman JP (2016) Adolescent chronic stress causes hypothalamo–pituitary–adrenocortical hypo-responsiveness and depression-like behavior in adult female rats. Psychoneuroendocrinology 65:109–117

Zimmer WMX, Tyack PL (2007) Repetitive shallow dives pose decompression risk in deep-diving beaked whales. Mar Mamm Sci 23:888–925

Chapter 33
Epilogue

Andy Butterworth

The image credit as Eva Butterworth 2017

A. Butterworth
Department of Clinical Veterinary Science, University of Bristol Veterinary School, Langford, N Somerset, UK
e-mail: andy.butterworth@bris.ac.uk

A. Butterworth (ed.), *Marine Mammal Welfare*, Animal Welfare 17,
DOI 10.1007/978-3-319-46994-2_33

Day 1. It's getting darker, the water is cooling at the surface, and I can feel it cold against my back as I surface. I've been feeding for about half an hour, and the small fish I am gathering into my mouth are quite easy to collect in the still water tonight. The other humpback whale, which is feeding next to me, is quietly resting on the surface at the moment, and I can hear his blows intermittently when I also surface. I swim and then allow the water to flow into my mouth and to expand my jaws and open the ridged throat grooves and folds which allow me to scoop the fish and small food creatures into my mouth and then to concentrate them as they are held against the baleen plates in my mouth. I move slowly to the left and right as I move forward, sensing the density of the fish like wavering shadow clouds in the water. After taking several scoops, I rest for a short time and then start swimming actively to move toward another density of fish, which I can sense and see in the water. I feel a slight pulling sensation on the right side of my jaw and then a more distinct sense of something tight across the tissues of my mouth as it closes. This is not a sensation I have felt before, and quite suddenly, I am aware of a strong intense pulling force on the side of my mouth, so strong that it actually pulls me to the right side. Something has got hold of my jaw and is slowing me down, pulling me to the right, and I can now feel a burning cutting sensation in the tissues along the edge of my jaw. I'm not prepared for this sudden attack—I did not sense any killer whales or other large animals close to me before—all I was aware of was open water and the schools of fish which moved in clouds ahead and around me. I react in the way I always do when I am surprised by something potentially dangerous—I flip my tail and dive down—the water here is not very deep and soon I am close to the seabed. The tension on the side of my mouth has not reduced; in fact, it is so strong that it is now pulling me to the side as I try to swim, and the more I push with my powerful tail fluke and long pectoral fins, I feel the cutting, burning sensation at the side of my mouth. Now I start to become frightened; I am not used to much being dangerous for me—of course the big human floating things concern me because they are noisy and large and fast and hard, and the killer whales can try to nip my fins and flukes, but this is something new, something I don't know anything about. Whatever it is, it's silent, and small (I can't sense any swimming movements from the attacker) and very, very strong, and now, extremely painful. The thing is cutting deep into my mouth, and I can feel it bending and damaging my baleen plates, cutting into the soft tissues of my mouth, and also I can taste blood (my blood) in the water. Also, worryingly, I cannot fully close my mouth, and I can feel the water flowing in, and out, as I swim. I decide to surface fast in the hope that the attacker will be thrown off, and as I breach at the surface, I feel myself freed a little from the intense pulling tension. Maybe I've lost it! But no, soon I can sense another feeling—not only is the thing attached at my mouth, but I can feel strands of it, whatever it is, against my sides. I don't like this at all; I'm scared; I spin myself around in the water—this takes a few moments as I weigh about 35 tonnes (I'm only medium sized) and I'm 12 m long, not fully grown! I make three spinning turns, hoping to throw the thing off. Oh no! I can feel that one of my pectoral fins is now held against my side. This is very, very scary—how can I turn, steer, and propel myself? I start to panic and rush to the surface to make a series of short snorty breaths and blows—my huge heart is beating fast, and I propel myself along the surface as fast as I can, powering with my tail, but hampered by the fact that I can't completely close my mouth. My pectoral fin is not working at all, 4 m in length, and it is held tight against me, and despite my huge muscles, I am unable to pull it away from my side. Two of the whales from my pod have been trying to keep up with me, clearly concerned by my strange behavior and are sending acoustic signals to me—what's wrong? What's wrong? I'm so frightened; all I can do is swim and spin and make shallow dives and short snorty breaths—and repeatedly send out an alarm signal to the other whales around me.

Day 30. I've been living with the thing for a whole moon cycle now. I've lost quite a bit of weight—whereas before I was fat and sleek with a very thick layer of blubber, I am now finding it very difficult to feed because my jaw hurts so much, the baleen has been damaged inside, I can't completely close my mouth (which I need to be able to do, to dive properly, and to swim fast, and to allow my baleen filter to work well), and the thing has wrapped itself around my body a number of times (I can't say how many) and has trapped my right fin against my side—so I can only limp through the sea. I trail a long tail of "the thing"—it's heavy with pieces of litter and weed, and it slows me down so much that my most common companion, and the rest of the pod—usually three others with whom I usually move through the sea—have had to leave me. I am very alone, but other humpback pods and other species of whales sometimes come close to me and send me acoustic messages. But they can all see that I am sick and unable to feed and move easily. I feel very tired most of the time, the lack of food, the fear, the cold (my blubber is not as thick as it should be), and the constant pain (the thing has cut right into the tissues in my mouth and also deep into my side along the line from my mouth to the root of my pectoral fin). The cut is so deep that I can see blood in the water sometimes, and the pain is very, very severe—I can feel the thing against my insides and feel a grating feeling as it moves against my bones around my fin. The swelling and infection around the thing are sapping my strength and endurance.

Day 90. The thing has been with me for three moon cycles now. I'm very tired, I fin slowly—and sometimes I am lucky enough to sense a school of fish or a cloud of small crustacea and to summon enough energy to open my sore and damaged mouth and to envelop them—but my mouth does not work well, the thing has formed a tight band across the middle of the inside of my jaw, and even the mass of soft fish and krill bodies hurt me when I close my mouth against them. I have lost a lot of weight, and the cold water is beginning to really affect me, so I have turned north, back toward warmer water—and I have reached a place now where I can just about tolerate the temperature—but the feeding is bad, and I am very alone. I feel very sick; the infection around the thing is causing me to feel emptied of energy, confused, and sometimes disoriented. I have tried scraping myself along a pebble seabed to try and relieve myself of the thing, but now it is so deeply sunk into my skin, blubber, muscle, and the bone of my jaw and pectoral fin that I know it cannot be released—unless it chooses to let go of me itself. I used to send out distress signals when other whales came within hearing distance—but I have stopped now, as this simply caused whales to come to see what was happening, but they could do nothing. I know that I am dying, but I fight on with the small hope that the thing will break off, let go, or decide to leave me.

Day 115. I am lying silent in shallow water. I have no strength left. I can feel seabirds landing on me to peck at the open wound caused by the thing. I have not been able to feed for 20 days and nights, and my blubber is very thin and my skin is damaged and sore in places where the trailing part of the thing has touched and cut into me again and again. I feel my life sliding away; at least the constant pain, the deep burning constant pain of the thing will be finished.

Bibliography (Papers Which Influenced the Text of the Epilogue)

Andersen MS, Forney KA, Cole TVN, Eagle T, Angliss R, Long K, Barre L, Van Atta L, Borggaard D, Rowles T, Norberg B, Whaley J, Engleby L (2008) Differentiating serious and non-serious injury of marine mammals: Report of the serious injury technical workshop, 10–13 September 2007, Seattle, Washington. U.S. Dep. Commer., NOAA Tech. Memo. NMFS-OPR-39. 94 p

Brown SL, Reid D, Rogan E (2015) Spatial and temporal assessment of potential risk to cetaceans from static fishing gears. Mar Pol 51:267–280

Carretta JV, Barlow J, Enriquez L (2008) Acoustic pingers eliminate beaked whale bycatch in a gill net fishery. Mar Mam Sci 24:956–961

Cassoff RM, Moore KM, McLellan WA, Barco SG, Rotstein DS, Moore MJ (2011) Lethal entanglement in baleen whales. Dis Aquat Org 96:175–185; Cole T, Hartley D, Garron M (2006) Mortality and serious injury determinations for baleen whale stocks along the eastern seaboard of the United States, 2000–2004. Reference document 06-04. Northeast Fisheries Science Center, Gloucester, MA

Coupland RM (1992) The red cross classification of war wounds: the E.X.C.F.V.M. Scoring system. World J Surg 16(5):910–917

Dau BK, Gilardi KV, Gulland FM, Higgins A, Holcomb JB, Leger JS, Ziccardi MH (2009) Fishing gear-related injury in California marine wildlife. J Wildl Dis 45(2):355–362

Hunt KE, Rolland RM, Kraus SD, Wasser SK (2006) Analysis of fecal glucocorticoids in the North Atlantic right whale (*Eubalaena glacialis*). Gen Compare Endocrine 148:260–272

Jepson PD, Barbieri M, Barco SG, Bernaldo de Quiros Y, Bogomolni A, Danil K, Rowles T (2013) Peracute underwater entrapment of pinnipeds and cetaceans. Dis Aquat Org 103:229–264

Kirkwood JK, Bennett PM, Jepson PD, Kuiken T, Simpson VR, Baker JR (1997) Entanglement in fishing gear and other causes of death in cetaceans stranded on the coasts of England and Wales. Vet Rec 141:94–98

Knowlton AR, Kraus SD (2001) Mortality and serious injury of northern right whales (*Eubalaena glacialis*) in the western North Atlantic Ocean. J Cetacean Res Manag 2:193–208

Knowlton AR, Robbins J, Landry S, McKenna HA, Kraus SD, Werner TB (2015) Effects of fishing rope strength on the severity of large whale entanglements. Conserv Biol. doi:10.1111/cobi.12590

Lambertsen RH, Rasmussen KJ, Lancaster WC, Hintz RJ (2005) Functional morphology of the mouth of the bowhead whale and its implications for conservation. J Mammal 86:342–352

Leaper R, Papastavrou V, Sadler L (2006) Consideration of factors affecting time to death for whales following entanglement in fishing gear. Report, University of Bristol. Published by the Whale and Dolphin Conservation Society, Brookfield House, St Paul St, Chippenham SN15 1LJ, UK (2008)

Lien J (1994) Entrapment of large cetaceans in passive inshore fishing gear in Newfoundland and Labrador (1979-1990). Rep Int Whal Comm 15:149–157

MARPOL (2016) International convention for the prevention of pollution from ships (MARPOL) International Maritime Organization. http://www.imo.org/en/About/Conventions/ListOfConventions/Pages/International-Convention-for-the-Prevention-of-Pollution-from-Ships-(MARPOL).aspx. Retrieved 27 June 2016

Moore M (2013) Welfare of whales bycaught in fishing gear or struck by vessels. Anim Welf 22:117–121

Moore MJ, van der Hoop JM (2012) The painful side of trap and fixed net fisheries: chronic entanglement of large whales. J Mar Biol. doi:10.1155/2012/230653

Moore MJ, van der Hoop J, Barco SG, Costidis AM, Gulland FM, Jepson JD, Moore KT, Raverty S, McLellan WA (2013) Criteria and case definitions for serious injury and death of pinnipeds and cetaceans caused by anthropogenic trauma. Dis Aquat Org 103:229–264

NOAA/NMFS (2007) Serious injury technical workshop, Seattle, 2007 Severity indexes page 7. http://www.nmfs.noaa.gov/pr/pdfs/interactions/serious_injury_techmemo2008.pdf. Accessed 2 April 2016

Pearson E, Mellish S, Sanders B, Litchfield C (2014) Marine wildlife entanglement: assessing knowledge, attitudes, and relevant behaviour in the Australian community. Mar Pollut Bull 89:136–148

Read AJ (2008) The looming crisis: interactions between marine mammals and fisheries. J Mammal 89:541–548

Robbins J, Knowlton AR, Landry S (2015) Apparent survival of North Atlantic right whales after entanglement in fishing gear. Biol Conserv 191:421–427

Sandoe P, Simonsen HB (1992) Assessing animal welfare: where does science end and philosophy begin? Anim Welf 1:257–267

Soulsbury CD, Iossa G, Harris S (2008) The animal welfare implications of cetacean deaths in fisheries. A University of Bristol report to the Whale and Dolphin Conservation Society (WDC)

Wilcox C, Mallos NJ, Leonard GH, Rodriguez A, Hardesty BD (2016) Using expert elicitation to estimate the impacts of plastic pollution on marine wildlife. Mar Policy 65(2016):107–114

World Society for the Protection of Animals (WSPA) (2012) In: Butterworth A, Clegg I, Bass C (eds) Marine debris: a global picture of the impact on animal welfare and of animal-focused solutions. WSPA International, London. http://oceansandplastics.info/wp-content/uploads/2015/07/REP_WSPA_2012.pdf. Retrieved 27 June 2016

Index

A. Butterworth (ed.), *Marine Mammal Welfare*, Animal Welfare 17,
DOI 10.1007/978-3-319-46994-2

Q

R

The manufacturer's authorised representative in the EU is Springer Nature Customer Service Centre GmbH, Europaplatz 3, 69115 Heidelberg, Germany. If you have any concerns regarding our products, please contact ProductSafety@springernature.com

Printed and bound by CPI Group (UK) Ltd, Croydon, CR0 4YY
01/10/2026
02225869-0007